Two-dimensional Crossing and Product Polynomial Systems

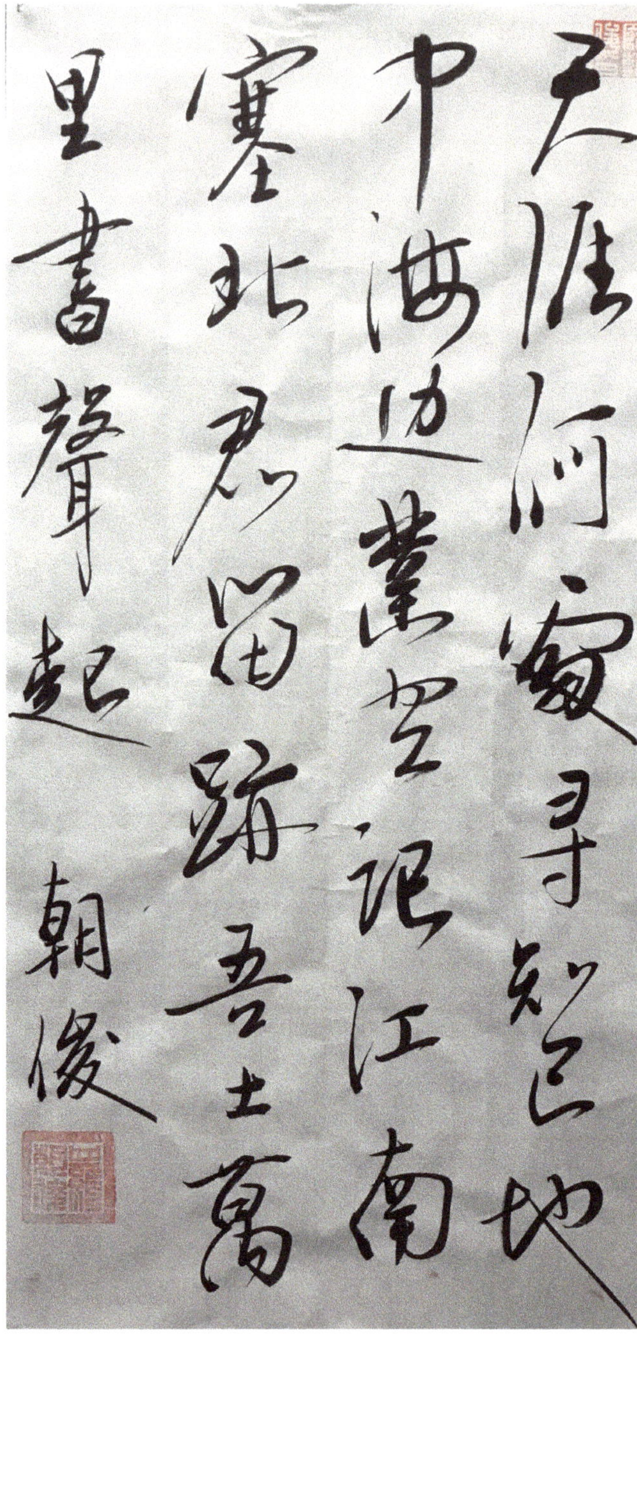

天涯何处寻知己，地中海边凿空记，江南塞北君留迹，吾土万里书声起。

——朝俊甲辰秋 送【米格尔。圣胡安】好友

Albert C. J. Luo

Two-dimensional Crossing and Product Polynomial Systems

高等教育出版社
HIGHER EDUCATION PRESS

Springer

Albert C. J. Luo
Department of Mechanical
and Mechatronics Engineering
Southern Illinois University Edwardsville
Edwardsville, IL, USA

ISBN 978-981-96-5714-8 ISBN 978-981-96-5715-5 (eBook)
https://doi.org/10.1007/978-981-96-5715-5

Jointly published with Higher Education Press Limited Company
The print edition is not for sale in the mainland of China. Customers from the mainland of China please order the print book from: Higher Education Press Limited Company.

This Springer imprint is published by the registered company Springer Nature Singapore Pte Ltd.
The registered company address is: 152 Beach Road, #21-01/04 Gateway East, Singapore 189721, Singapore

Preface

This book is a monograph about hybrid networks of singular and non-singular, 1-dimensional flows and equilibriums in crossing and product polynomial systems. The singular equilibriums and 1-dimensional flows with infinite-equilibriums in product polynomial systems are presented in the theorem. The singular equilibriums are singular saddles and centers, parabola-saddles, and double-inflection-saddles. The singular 1-dimensional flows are singular hyperbolic-flows, hyperbolic-to-hyperbolic-secant flows, inflection-source and sink flows, inflection-saddle flows. The higher-order singular 1-dimensional flows and singular equilibriums are for the appearing bifurcations of lower-order singular and non-singular 1-dimesnional flows and equilibriums. The infinite-equilibriums are the switching bifurcations for two associated networks of singular and non-singular, 1-dimensional flows and equilibriums. The infinite-equilibriums include:

- two types of inflection-parabola-saddles, positive and negative-saddles,
- CCW and CW circular saddles, hyperbolic parabola sink-to-source, and
- circular parabola-sink-to-source;
- two types of inflection-saddles, parabola-saddles, parabola-sinks and sources;
- two types of concave-parabola-saddles, hyperbolic-saddles, hyperbolic-sink-to-source;
- concave inflection-sources and sinks, inflection asymptote-saddles, inflection asymptote-sinks and sources, concave inflection-saddles.

The corresponding mathematical conditions are presented, and the theory for nonlinear dynamics of crossing and product polynomial systems is presented through a theorem. The mathematical proof is completed through the local analysis and the first integral manifolds. The illustrations of singular 1-diemsnional flows and equilibriums are completed, and the sampled networks of non-singular 1-dimesnional flows and equilibriums are presented in this book.

In this book, three chapters are included. Chapter 1 presents the nonlinear dynamics of singular flows and equilibriums with the corresponding infinite-equilibriums in crossing and product polynomial systems through a theorem. The appearing and switching bifurcations based on the singular flows and equilibriums and infinite-equilibriums are presented. Chapter 2 proved the mathematical condition in the theorem through local analysis and the first integral manifolds. In Chap. 3, the singular 1-dimensional flows and singular equilibriums with infinite-equilibriums are presented for a better understanding of singular dynamics in self and product polynomial systems. The singular saddles and centers, parabola-saddles, and inflections-saddles are presented. The singular hyperbolic-flows, hyperbolic-to-hyperbolic-secant flows, inflection-source and sink flows, inflection-saddle flows are presented. In addition, hybrid networks of non-singular 1-dimensional flows and equilibriums are presented, and such hybrid networks appear, and two associated networks are switched through the singular 1-dimensional flows and equilibriums with the infinite-equilibriums.

Finally, the author hopes the materials presented herein can provide a better understanding of the extended Hilbert sixteen problem and the corresponding dynamics.

Edwardsville, IL, USA Albert C. J. Luo

Contents

Chapter 1
Crossing and Product Polynomial Systems

In this chapter, the dynamics of crossing and product polynomial systems will be presented through a theorem. The singular equilibriums and 1-dimensional flows with infinite-equilibriums in product polynomial systems are presented in the theorem. The singular equilibriums are singular saddles and centers, parabola-saddles and inflections-saddles. The singular 1-dimensional flows are singular hyperbolic-flows, hyperbolic-to-hyperbolic-secant flows, inflection-source and sink flows, inflection-saddle flows. The corresponding first integral manifolds are obtained. The hybrid networks of singular equilibriums and 1-dimensional flows in the crossing and product polynomial systems are discussed in the theorem, and the corresponding infinite-equilibriums are given for switching bifurcations. The hybrid network of simple equilibriums and 1-dimensional flows with infinite-equilibriums in the crossing and product polynomial systems is stated in the theorem. The appearing and switching bifurcations of simple and singular 1-dimensional flows and equilibriums are stated in the theorem. Finally, the switching bifurcations of hybrid networks of simple and singular equilibriums and flows is also stated. The corresponding dynamical behaviors are presented through the following theorem.

1.1 Theorem 1.1

(i) Consider a two-dimensional, nonlinear dynamical system as

$$\begin{aligned} \dot{x}_{j_2} &= a_{j_2 j_2 0}(x_{j_1} - a_{j_2 j_1 1})^{m_2}, \\ \dot{x}_{j_1} &= a_{j_1 j_1 0}(x_{j_1} - a_{j_1 j_1 1})^{m_1}(x_{j_2} - a_{j_1 j_2 1})^{n_1}. \end{aligned} \tag{1.1}$$

(i_1) For $a_{j_2 j_1 1} \neq a_{j_1 j_1 1}$ the first integral manifold is

A. C. J. Luo, *Two-dimensional Crossing and Product Polynomial Systems*,
https://doi.org/10.1007/978-981-96-5715-5_1

$$\sum_{\alpha,\beta=0}^{\alpha+\beta=m_2} \frac{m_2!}{\alpha!\beta!}(a_{j_1j_11}-a_{j_2j_11})^{\alpha}$$
$$\times \frac{1}{\beta-m_1+1}\left[(x_{j_1}-a_{j_1j_11})^{\beta-m_1+1}-(x_{j_10}-a_{j_1j_11})^{\beta-m_1+1}\right]$$
$$=\frac{a_{j_1j_10}}{a_{j_2j_20}}\frac{1}{n_1+1}\left[(x_{j_2}-a_{j_1j_21})^{n_1+1}-(x_{j_20}-a_{j_1j_21})^{n_1+1}\right],$$
for $\beta - m_1 + 1 \neq 0$;
$$\frac{1}{\beta-m_1+1}\left[(x_{j_1}-a_{j_1j_11})^{\beta-m_1+1}-(x_{j_10}-a_{j_1j_11})^{\beta-m_1+1}\right]$$
$$=\ln\frac{|x_{j_1}-a_{j_1j_11}|}{|x_{j_10}-a_{j_1j_11}|},$$
for $\beta - m_1 + 1 = 0$. (1.2)

(i_{1a}) There is an equilibrium for 1-dimensional flow at

$$(x^*_{j_1}, x^*_{j_2}) = (a_{j_1j_11}, a_{j_1j_21}). \tag{1.3}$$

(i_{1a1}) For $m_1 = 2m_{11}+1$ and $n_1 = 2n_{11}+1$, the equilibrium of $(x^*_{j_1}, x^*_{j_2}) = (a_{j_1j_11}, a_{j_1j_21})$ in the x_{j_1}-direction has the following properties.

- *For $a_{j_1j_10} > 0$ and $a_{j_2j_20}(a_{j_1j_21}-a_{j_2j_21})^{n_2} > 0$,*

$$(a_{j_1j_11}, a_{j_1j_21}) = \underbrace{([(2m_{11}+1),\,(2n_{11}+1)]\text{DP:UP,pF})}_{[(2m_{11}+1),(2n_{11}+1)]\text{-hyperbolic flow }(+)}. \tag{1.4}$$

The flow at $(x^*_{j_1}, x^*_{j_2}) = (a_{j_1j_11}, a_{j_1j_21})$ is a ([(2m_{11}+1), (2n_{11} + 1)]DP:UP,pF)-positive hyperbolic flow.

- *For $a_{j_1j_10} < 0$ and $a_{j_2j_20}(a_{j_1j_21}-a_{j_2j_21})^{n_2} > 0$,*

$$(a_{j_1j_11}, a_{j_1j_21}) = \underbrace{([(2m_{11}+1),\,(2n_{11}+1)]\text{UP:DP,pF})}_{[(2m_{11}+1),(2n_{11}+1)]\text{-hyperbolic-secant flow }(+)}. \tag{1.5}$$

The flow at $(x^*_{j_1}, x^*_{j_2}) = (a_{j_1j_11}, a_{j_1j_21})$ is a ([(2m_{11}+1), (2n_{11} + 1)]UP:DP,pF)-positive hyperbolic-secant flow.

- *For $a_{j_1j_10} > 0$ and $a_{j_2j_20}(a_{j_1j_21}-a_{j_2j_21})^{n_2} < 0$,*

$$(a_{j_1j_11}, a_{j_1j_21}) = \underbrace{([(2m_{11}+1),\,(2n_{11}+1)]\text{UP:DP,nF})}_{[(2m_{11}+1),(2n_{11}+1)]\text{-hyperbolic-secant flow }(-)}. \tag{1.6}$$

The flow at $(x^*_{j_1}, x^*_{j_2}) = (a_{j_1j_11}, a_{j_1j_21})$ is a ([(2m_{11}+1), (2n_{11} + 1)]UP:DP,nF)-negative hyperbolic-secant flow.

- *For* $a_{j_1j_10} < 0$ *and* $a_{j_2j_20}(a_{j_1j_21} - a_{j_2j_21})^{n_2} < 0$,

$$(a_{j_1j_11}, a_{j_1j_21}) = \underbrace{([(2m_{11}{+}1), (2n_{11}+1)]\text{DP:UP,nF})}_{[(2m_{11}+1),(2n_{11}+1)]\text{-hyperbolic flow }(-)}. \tag{1.7}$$

The flow at $(x^*_{j_1}, x^*_{j_2}) = (a_{j_1j_11}, a_{j_1j_21})$ *is a* ([(2m11+1), (2n11 + 1)]DP:UP,pF)-*negative hyperbolic flow.*

(i_{1a2}) For $m_1 = 2m_{11}$ *and* $n_1 = 2n_{11} + 1$, *the equilibrium of* $(x^*_{j_1}, x^*_{j_2}) = (a_{j_1j_11}, a_{j_1j_21})$ *for the* x_{j_1}*-direction has the following properties.*

- *For* $a_{j_1j_10} > 0$ *and* $a_{j_2j_20}(a_{j_1j_21} - a_{j_2j_21})^{n_2} > 0$,

$$(a_{j_1j_11}, a_{j_1j_21}) = \underbrace{([(2m_{11}), (2n_{11}+1)]\text{UP:UP,pF})}_{[(2m_{11}),(2n_{11}+1)]\text{-hyperbolic-secant-to-hyperbolic flow }(+)}. \tag{1.8}$$

The flow at $(x^*_{j_1}, x^*_{j_2}) = (a_{j_1j_11}, a_{j_1j_21})$ *is a* ([(2m11), (2n11 + 1)]UP:UP,pF)-*positive hyperbolic-secant to hyperbolic flow.*

- *For* $a_{j_1j_10} < 0$ *and* $a_{j_2j_20}(a_{j_1j_21} - a_{j_2j_21})^{n_2} > 0$,

$$(a_{j_1j_11}, a_{j_1j_21}) = \underbrace{([(2m_{11}), (2n_{11}+1)]\text{DP:DP,pF})}_{[(2m_{11}),(2n_{11}+1)]\text{-hyperbolic-to-hyperbolic-secant flow }(+)}. \tag{1.9}$$

The flow at $(x^*_{j_1}, x^*_{j_2}) = (a_{j_1j_11}, a_{j_1j_21})$ *is a* ([(2m11), (2n11 + 1)]DP:DP,pF)-*positive hyperbolic to hyperbolic-secant flow.*

- *For* $a_{j_1j_10} > 0$ *and* $a_{j_2j_20}(a_{j_1j_21} - a_{j_2j_21})^{n_2} < 0$,

$$(a_{j_1j_11}, a_{j_1j_21}) = \underbrace{([(2m_{11}), (2n_{11}+1)]\text{DP:DP,nF})}_{[(2m_{11}),(2n_{11}+1)]\text{-hyperbolic-to-hyperbolic-secant flow }(-)}. \tag{1.10}$$

The flow at $(x^*_{j_1}, x^*_{j_2}) = (a_{j_1j_11}, a_{j_1j_21})$ *is a* ([(2m11), (2n11 + 1)]DP:DP,nF)-negative *hyperbolic-to-hyperbolic-secant flow.*

- *For* $a_{j_1j_10} < 0$ *and* $a_{j_2j_20}(a_{j_1j_21} - a_{j_2j_21})^{n_2} < 0$,

$$(a_{j_1j_11}, a_{j_1j_21}) = \underbrace{([(2m_{11}), (2n_{11}+1)]\text{UP:UP,nF})}_{[(2m_{11}),(2n_{11}+1)]\text{-hyperbolic-secant-to-hyperbolic flow }(-)}. \tag{1.11}$$

The flow at $(x^*_{j_1}, x^*_{j_2}) = (a_{j_1j_11}, a_{j_1j_21})$ *is a* ([(2m11), (2n11 + 1)]UP:UP,nF)-*negative hyperbolic-secant-to-hyperbolic flow.*

(i_{1a3}) For $m_1 = 2m_{11} + 1$ and $n_1 = 2n_{11}$, the equilibrium of $(x^*_{j_1}, x^*_{j_2}) = (a_{j_1j_1}1, a_{j_1j_2}1)$ for the x_{j_1}-direction has the following properties.

- *For $a_{j_1j_1}0 > 0$ and $a_{j_2j_2}0(a_{j_1j_2}1 - a_{j_2j_2}1)^{n_2} > 0$,*

$$(a_{j_1j_1}1, a_{j_1j_2}1) = \underbrace{([(2m_{11}{+}1), (2n_{11})]_{\mathrm{DI:II}}\mathrm{SO,pF})}_{[(2m_{11}+1),(2n_{11})]\text{-decreasing-to-increasing inflection source flow }(+)}. \quad (1.12)$$

The flow at $(x^*_{j_1}, x^*_{j_2}) = (a_{j_1j_1}1, a_{j_1j_2}1)$ is a $([(2m_{11}{+}1), (2n_{11})]_{\mathrm{DI:II}}\mathrm{SO,pF})$-positive decreasing-to-increasing inflection source flow.

- *For $a_{j_1j_1}0 < 0$ and $a_{j_2j_2}0(a_{j_1j_2}1 - a_{j_2j_2}1)^{n_2} > 0$,*

$$(a_{j_1j_1}1, a_{j_1j_2}1) = \underbrace{([(2m_{11}{+}1), (2n_{11})]_{\mathrm{II:DI}}\mathrm{SI,pF})}_{[(2m_{11}+1),(2n_{11})]\text{-increasing-to-decreasing inflection sink flow }(+)}. \quad (1.13)$$

The flow at $(x^*_{j_1}, x^*_{j_2}) = (a_{j_1j_1}1, a_{j_1j_2}1)$ is a $([(2m_{11}{+}1), (2n_{11})]_{\mathrm{II:DI}}\mathrm{SI,pF})$-positive increasing-to-decreasing inflection source flow.

- *For $a_{j_1j_1}0 > 0$ and $a_{j_2j_2}0(a_{j_1j_2}1 - a_{j_2j_2}1)^{n_2} < 0$,*

$$(a_{j_1j_1}1, a_{j_1j_2}1) = \underbrace{([(2m_{11}{+}1), (2n_{11})]_{\mathrm{II:DI}}\mathrm{SO,nF})}_{[(2m_{11}+1),(2\mathrm{n}_{11})]\text{-increasing-to-decreasing inflection source flow }(-)}. \quad (1.14)$$

The flow at $(x^*_{j_1}, x^*_{j_2}) = (a_{j_1j_1}1, a_{j_1j_2}1)$ is a $([(2m_{11}{+}1), (2n_{11})]_{\mathrm{II:DI}}\mathrm{SO,nF})$-negative increasing-to-decreasing inflection source flow.

- *For $a_{j_1j_1}0 < 0$ and $a_{j_2j_2}0(a_{j_1j_2}1 - a_{j_2j_2}1)^{n_2} < 0$,*

$$(a_{j_1j_1}1, a_{j_1j_2}1) = \underbrace{([(2m_{11}{+}1), (2n_{11})]_{\mathrm{DI:II}}\mathrm{SI,nF})}_{[(2m_{11}+1),(2n_{11})]\text{-decreasing-to-increasing inflection sink flow }(-)}. \quad (1.15)$$

The flow at $(x^*_{j_1}, x^*_{j_2}) = (a_{j_1j_1}1, a_{j_1j_2}1)$ is a $([(2m_{11}{+}1), (2n_{11})]_{\mathrm{DI:II}}\mathrm{SI,nF})$-negative decreasing-to-increasing inflection sink flow.

(i_{1a4}) For $m_1 = 2m_{11}$ and $n_1 = 2n_{11}$, the equilibrium of $(x^*_{j_1}, x^*_{j_2}) = (a_{j_1j_1}1, a_{j_1j_2}1)$ for the x_{j_1}-direction has the following properties.

- *For $a_{j_1j_1}0 > 0$ and $a_{j_2j_2}0(a_{j_1j_2}1 - a_{j_2j_2}1)^{n_2} > 0$,*

$$(a_{j_1j_1}1, a_{j_1j_2}1) = \underbrace{([(2m_{11}), (2n_{11})]_{\mathrm{II:II}}\mathrm{US,pF})}_{[(2m_{11}),(2n_{11})]\text{-increasing-inflection upper-saddle flow }(+)}. \quad (1.16)$$

The flow at $(x^*_{j_1}, x^*_{j_2}) = (a_{j_1j_1}1, a_{j_1j_2}1)$ is a $([(2m_{11}), (2n_{11})]_{\mathrm{II:II}}\mathrm{US,pF})$-positive increasing-inflection upper-saddle flow.

- *For $a_{j_1j_1}0 < 0$ and $a_{j_2j_2}0(a_{j_1j_2}1 - a_{j_2j_2}1)^{n_2} > 0$,*

$$(a_{j_1j_11}, a_{j_1j_21}) = \underbrace{([(2m_{11}), (2n_{11})]_{\mathrm{DI:DI}}\mathrm{LS,pF})}_{[(2m_{11}),(2n_{11})]\text{-decreasing-inflection lower-saddle flow }(+)}. \quad (1.17)$$

The flow at $(x^*_{j_1}, x^*_{j_2}) = (a_{j_1j_11}, a_{j_1j_21})$ *is a* $([(2m_1), (2n_1)]_{\mathrm{DI:DI}}\mathrm{LS,pF})$-*positive decreasing-inflection lower-saddle flow.*

- *For* $a_{j_1j_10} > 0$ *and* $a_{j_2j_20}(a_{j_1j_21} - a_{j_2j_21})^{n_2} < 0$,

$$(a_{j_1j_11}, a_{j_1j_21}) = \underbrace{([(2m_{11}), (2n_{11})]_{\mathrm{DI:DI}}\mathrm{US,nF})}_{[(2m_{11}),(2n_{11})]\text{-decreasing-inflection upper-saddle flow }(-)}. \quad (1.18)$$

The flow at $(x^*_{j_1}, x^*_{j_2}) = (a_{j_1j_11}, a_{j_1j_21})$ *is a* $([(2m_{11}), (2n_{11})]_{\mathrm{DI:DI}}\mathrm{US,nF})$-*negative decreasing-inflection upper-saddle flow.*

- *For* $a_{j_1j_10} < 0$ *and* $a_{j_2j_20}(a_{j_1j_21} - a_{j_2j_21})^{n_2} < 0$,

$$(a_{j_1j_11}, a_{j_1j_21}) = \underbrace{([(2m_{11}), (2n_{11})]_{\mathrm{II:II}}\mathrm{LS,nF})}_{[(2m_{11}),(2n_{11})]\text{-increasing-inflection lower-saddle flow }(-)}. \quad (1.19)$$

The flow at $(x^*_{j_1}, x^*_{j_2}) = (a_{j_1j_11}, a_{j_1j_21})$ *is a* $([(2m_{11}), (2n_{11})]_{\mathrm{II:II}}\mathrm{LS,nF})$-*negative increasing-inflection lower-saddle flow.*

(i_{1b}) There is an equilibrium for crossing-univariate equilibrium at

$$(x^*_{j_2}, x^*_{j_1}) = (a_{j_1j_21}, a_{j_2j_11}). \quad (1.20)$$

(i_{1b1}) For $n_1 = 2n_{11} + 1$ *and* $m_2 = 2m_{21} + 1$, *the equilibrium of* $(x^*_{j_2}, x^*_{j_1}) = (a_{j_1j_21}, a_{j_2j_11})$ *has the following properties.*

- *For* $a_{j_1j_10}(a_{j_2j_11} - a_{j_1j_11})^{m_1} > 0$ *and* $a_{j_2j_20} > 0$,

$$(a_{j_1j_21}, a_{j_2j_11}) = \underbrace{((2n_{11}+1)^{\mathrm{th}}\mathrm{UP}_+, (2m_{21}+1)^{\mathrm{th}}\mathrm{UP}_+)}_{((2n_{12}+1),(2m_{21}+1))\text{-positive saddle}}. \quad (1.21)$$

The flow at $(x^*_{j_2}, x^*_{j_1}) = (a_{j_1j_21}, a_{j_2j_11})$ *is a* $((2n_{11}+1)^{\mathrm{th}}\mathrm{UP}_+, (2m_{21}+1)^{\mathrm{th}}\mathrm{UP}_+)$-*positive saddle.*

- *For* $a_{j_1j_10}(a_{j_2j_11} - a_{j_1j_11})^{m_1} < 0$ *and* $a_{j_2j_20} > 0$,

$$(a_{j_1j_21}, a_{j_2j_11}) = \underbrace{((2n_{11}+1)^{\mathrm{th}}\mathrm{DP}_+, (2m_{21}+1)^{\mathrm{th}}\mathrm{DP}_-)}_{((2n_{12}+1),(2m_{21}+1))\text{-CCW center}}. \quad (1.22)$$

The flow at $(x^*_{j_2}, x^*_{j_1}) = (a_{j_1j_21}, a_{j_2j_11})$ *is a* $((2n_{11}+1)^{\mathrm{th}}\mathrm{DP}_+, (2m_{21}+1)^{\mathrm{th}}\mathrm{DP}_-)$-*counter-clockwise center.*

- *For* $a_{j_1j_10}(a_{j_2j_11}-a_{j_1j_11})^{m_1}>0$ *and* $a_{j_2j_20}<0$,

$$(a_{j_1j_21},a_{j_2j_11})=\underbrace{((2n_{11}+1)^{\text{th}}\text{DP}_{-},(2m_{21}+1)^{\text{th}}\text{DP}_{+})}_{((2n_{11}+1),(2m_{21}+1))\text{-CW center}}. \tag{1.23}$$

The flow at $(x_{j_2}^*,x_{j_1}^*)=(a_{j_1j_21},a_{j_2j_11})$ *is a* $((2n_{11}+1)^{\text{th}}\text{DP}_{-},(2m_{21}+1)^{\text{th}}\text{DP}_{+})$-*clockwise center.*

- *For* $a_{j_1j_10}(a_{j_2j_11}-a_{j_1j_11})^{m_1}<0$ *and* $a_{j_2j_20}<0$,

$$(a_{j_1j_21},a_{j_2j_11})=\underbrace{((2n_{11}+1)^{\text{th}}\text{UP}_{-},(2m_{21}+1)^{\text{th}}\text{UP}_{-})}_{((2n_{11}+1),(2m_{21}+1))\text{-negative saddle}}. \tag{1.24}$$

The flow at $(x_{j_2}^*,x_{j_1}^*)=(a_{j_1j_21},a_{j_2j_11})$ *is a* $((2n_{11}+1)^{\text{th}}\text{UP}_{-},(2m_{21}+1)^{\text{th}}\text{UP}_{-})$-*negative saddle.*

(i_{1b2}*) For* $n_1=2n_{11}$ *and* $m_2=2m_{21}+1$, *the equilibrium of* $(x_{j_2}^*,x_{j_1}^*)=(a_{j_1j_21},a_{j_2j_11})$ *has the following properties.*

- *For* $a_{j_1j_10}(a_{j_2j_11}-a_{j_1j_11})^{m_1}>0$ *and* $a_{j_2j_20}>0$,

$$(a_{j_1j_21},a_{j_2j_11})=\underbrace{((2n_{11})^{\text{th}}\text{US},(2m_{21}+1)^{\text{th}}\text{UP})}_{((2n_{11}),(2m_{21}+1))\text{-up-parabola upper-saddle}}. \tag{1.25}$$

The flow at $(x_{j_2}^*,x_{j_1}^*)=(a_{j_1j_21},a_{j_2j_11})$ *is a* $((2n_{11})^{\text{th}}\text{US},(2m_{21}+1)^{\text{th}}\text{UP})$-*up-parabola upper-saddle.*

- *For* $a_{j_1j_10}(a_{j_2j_11}-a_{j_1j_11})^{m_1}<0$ *and* $a_{j_2j_20}>0$,

$$(a_{j_1j_21},a_{j_2j_11})=\underbrace{((2n_{11})^{\text{th}}\text{LS},(2m_{21}+1)^{\text{th}}\text{DP})}_{((2n_{11}),(2m_{21}+1))\text{-down-parabola lower-saddle}}. \tag{1.26}$$

The flow at $(x_{j_2}^*,x_{j_1}^*)=(a_{j_1j_21},a_{j_2j_11})$ *is a* $((2n_{11})^{\text{th}}\text{LS},(2m_{21}+1)^{\text{th}}\text{DP})$-*down-parabola lower-saddle.*

- *For* $a_{j_1j_10}(a_{j_2j_21}-a_{j_1j_21})^{n_1}>0$ *and* $a_{j_2j_20}<0$,

$$(a_{j_1j_21},a_{j_2j_11})=\underbrace{((2n_{11})^{\text{th}}\text{US},(2m_{21}+1)^{\text{th}}\text{DP})}_{((2n_{11}),(2m_{21}+1))\text{-down-parabola upper-saddle}}. \tag{1.27}$$

The flow at $(x_{j_2}^*,x_{j_1}^*)=(a_{j_1j_21},a_{j_2j_11})$ *is a* $((2n_{11})^{\text{th}}\text{US},(2m_{21}+1)^{\text{th}}\text{DP})$-*down-parabola upper-saddle.*

- *For* $a_{j_1j_10}(a_{j_2j_11}-a_{j_1j_11})^{m_1}<0$ *and* $a_{j_2j_20}<0$,

$$(a_{j_1j_21}, a_{j_2j_11}) = \underbrace{((2n_{11})^{\text{th}}\text{LS}, (2m_{21}+1)^{\text{th}}\text{UP})}_{((2n_{11}),(2m_{21}+1))\text{-up-parabola lower-saddle}}. \tag{1.28}$$

The flow at $(x^*_{j_2}, x^*_{j_1}) = (a_{j_1j_21}, a_{j_2j_11})$ *is* $((2n_{11})^{\text{th}}\text{LS}, (2m_{21}+1)^{\text{th}}\text{UP})$*-up-parabola lower-saddle.*

(i_{1b3}) For $n_1 = 2n_{11} + 1$ *and* $m_2 = 2m_{21}$*, the equilibrium of* $(x^*_{j_2}, x^*_{j_1}) = (a_{j_1j_21}, a_{j_2j_11})$ *has the following properties.*

- *For* $a_{j_1j_10}(a_{j_2j_11} - a_{j_1j_11})^{m_1} > 0$ *and* $a_{j_2j_20} > 0$*,*

$$(a_{j_1j_21}, a_{j_2j_11}) = \underbrace{((2n_{11}+1)^{\text{th}}\text{UP}, (2m_{21})^{\text{th}}\text{US})}_{((2n_{11}+1),(2m_{21}))\text{-up-parabola upper-saddle}}. \tag{1.29}$$

The flow at $(x^*_{j_2}, x^*_{j_1}) = (a_{j_1j_21}, a_{j_2j_11})$ *is a* $((2n_{11}+1)^{\text{th}}\text{UP}, (2m_{21})^{\text{th}}\text{US})$*-up-parabola upper-saddle.*

- *For* $a_{j_1j_10}(a_{j_2j_11} - a_{j_1j_11})^{m_1} < 0$ *and* $a_{j_2j_20} > 0$*,*

$$(a_{j_1j_21}, a_{j_2j_11}) = \underbrace{((2n_{11}+1)^{\text{th}}\text{DP}, (2m_{21})^{\text{th}}\text{US})}_{((2n_{11}+1),(2m_{21}))\text{-down-parabola upper-saddle}}. \tag{1.30}$$

The flow at $(x^*_{j_2}, x^*_{j_1}) = (a_{j_1j_21}, a_{j_2j_11})$ *is a* $((2n_{11}+1)^{\text{th}}\text{DP}, (2m_{21})^{\text{th}}\text{US})$*-down-parabola upper-saddle.*

- *For* $a_{j_1j_10}(a_{j_2j_11} - a_{j_1j_11})^{m_1} > 0$ *and* $a_{j_2j_20} < 0$*,*

$$(a_{j_1j_21}, a_{j_2j_11}) = \underbrace{((2n_{11}+1)^{\text{th}}\text{DP}, (2m_{21})^{\text{th}}\text{LS})}_{((2n_{11}+1),(2m_{21}))\text{-down-parabola lower-saddle}}. \tag{1.31}$$

The flow at $(x^*_{j_2}, x^*_{j_1}) = (a_{j_1j_21}, a_{j_2j_11})$ *is a* $((2n_{11}+1)^{\text{th}}\text{DP}, (2m_{21})^{\text{th}}\text{LS})$*-down-parabola lower-saddle.*

- *For* $a_{j_1j_10}(a_{j_2j_11} - a_{j_1j_11})^{m_1} < 0$ *and* $a_{j_2j_20} < 0$*,*

$$(a_{j_1j_21}, a_{j_2j_11}) = \underbrace{((2n_{11}+1)^{\text{th}}\text{UP}, (2m_{21})^{\text{th}}\text{LS})}_{((2n_{11}+1),(2m_{21}))\text{-up-parabola lower-saddle}}. \tag{1.32}$$

The flow at $(x^*_{j_2}, x^*_{j_1}) = (a_{j_1j_21}, a_{j_2j_11})$ *is a* $((2n_{11}+1)^{\text{th}}\text{UP}, (2m_{21})^{\text{th}}\text{LS})$*-up-parabola upper-saddle.*

(i_{1b4}) For $n_1 = 2n_{11}$ *and* $m_2 = 2m_{21}$*, the equilibrium of* $(x^*_{j_2}, x^*_{j_1}) = (a_{j_1j_21}, a_{j_2j_11})$ *has the following properties.*

- *For* $a_{j_1j_10}(a_{j_2j_11} - a_{j_1j_11})^{m_1} > 0$ *and* $a_{j_2j_20} > 0$*,*

$$(a_{j_1j_21}, a_{j_2j_11}) = \underbrace{((2n_{11})^{\text{th}}\text{II}_+, (2m_{21})^{\text{th}}\text{II}_+)}_{((2n_{11}),(2m_{21}))\text{-double-inflection saddle}}. \tag{1.33}$$

The flow at $(x_{j_2}^*, x_{j_1}^*) = (a_{j_1j_2 1}, a_{j_2j_1 1})$ *is a* $((2n_{11})^{\text{th}}\text{II}_+, (2m_{21})^{\text{th}}\text{II}_+)$*-double-inflection saddle.*

- *For* $a_{j_1j_1 0}(a_{j_2j_1 1} - a_{j_1j_1 1})^{m_1} < 0$ *and* $a_{j_2j_2 0} > 0$,

$$(a_{j_1j_2 1}, a_{j_2j_1 1}) = \underbrace{((2n_{11})^{\text{th}}\text{DI}_+, (2m_{21})^{\text{th}}\text{DI}_-)}_{((2n_{11}),(2m_{21}))\text{-double-inflection saddle}}. \tag{1.34}$$

The flow at $(x_{j_2}^*, x_{j_1}^*) = (a_{j_1j_2 1}, a_{j_2j_1 1})$ *is a* $((2n_{11})^{\text{th}}\text{DI}_+, (2m_{21})^{\text{th}}\text{DI}_-)$*-double-inflection saddle.*

- *For* $a_{j_1j_1 0}(a_{j_2j_1 1} - a_{j_1j_1 1})^{m_1} > 0$ *and* $a_{j_2j_2 0} < 0$,

$$(a_{j_1j_2 1}, a_{j_2j_1 1}) = \underbrace{((2n_{11})^{\text{th}}\text{DI}_-, (2m_{21})^{\text{th}}\text{DI}_+)}_{((2n_{11}),(2m_{21}))\text{-double-inflection saddle}}. \tag{1.35}$$

The flow at $(x_{j_2}^*, x_{j_1}^*) = (a_{j_1j_2 1}, a_{j_2j_1 1})$ *is a* $((2n_{11})^{\text{th}}\text{DI}_-, (2m_{21})^{\text{th}}\text{DI}_+)$*-double-inflection saddle.*

- *For* $a_{j_1j_1 0}(a_{j_2j_1 1} - a_{j_1j_1 1})^{m_1} < 0$ *and* $a_{j_2j_2 0} < 0$,

$$(a_{j_1j_2 1}, a_{j_2j_1 1}) = \underbrace{((2n_{11})^{\text{th}}\text{II}_-, (2m_{21})^{\text{th}}\text{II}_-)}_{((2n_{11}),(2m_{21}))\text{-double-inflection saddle}}. \tag{1.36}$$

The flow at $(x_{j_2}^*, x_{j_1}^*) = (a_{j_1j_2 1}, a_{j_2j_1 1})$ *is a* $((2n_{11})^{\text{th}}\text{II}_-, (2m_{21})^{\text{th}}\text{II}_-)$*-double-inflection saddle.*

(i_2) For $a_{j_2j_1 1} = a_{j_1j_1 1}$ *the first integral manifold is*

$$\begin{aligned}
&\frac{1}{m_2 - m_1 + 1}\big[(x_{j_1} - a_{j_2j_1 1})^{m_2-m_1+1} - (x_{j_1 0} - a_{j_2j_1 1})^{m_2-m_1+1}\big] \\
&= \frac{a_{j_1j_1 0}}{a_{j_2j_2 0}} \frac{1}{n_1 + 1}\big[(x_{j_2} - a_{j_1j_2 1})^{n_1+1} - (x_{j_2 0} - a_{j_1j_2 1})^{n_1+1}\big], \\
&\text{for } m_2 - m_1 + 1 \neq 0; \\
&\frac{1}{m_2 - m_1 + 1}\big[(x_{j_1} - a_{j_1j_1 1})^{m_2-m_1+1} - (x_{j_1 0} - a_{j_1j_1 1})^{m_2-m_1+1}\big] = \ln \frac{|x_{j_1} - a_{j_1j_1 1}|}{|x_{j_1 0} - a_{j_1j_1 1}|}, \\
&\text{for } m_2 - m_1 + 1 = 0.
\end{aligned} \tag{1.37}$$

(i_{2a}) For $m_2 \geq m_1$*, the first integral manifold is*

$$\begin{aligned}
&\frac{1}{m_2 - m_1 + 1}\big[(x_{j_1} - a_{j_2j_1 1})^{m_2-m_1+1} - (x_{j_1 0} - a_{j_2j_1 1})^{m_2-m_1+1}\big] \\
&= \frac{a_{j_1j_1 0}}{a_{j_2j_2 0}} \frac{1}{n_1 + 1}\big[(x_{j_2} - a_{j_1j_2 1})^{n_1+1} - (x_{j_2 0} - a_{j_1j_2 1})^{n_1+1}\big].
\end{aligned} \tag{1.38}$$

(i_{2a1}) There are 16 types of infinite-equilibriums at $x^*_{j_1} = a_{j_1j_1 1} = a_{j_2j_1 1}$.

(i_{2a1a}) For $m_1 = 2m_{11}+1$ *and* $m_2 = 2m_{21}+1$, *the infinite-equilibrium of* $x^*_{j_1} = a_{j_1j_1 1} = a_{j_2j_1 1}$ *with* $(\overline{x}_{j_2} \neq a_{j_1j_2 1})$ *has the following properties.*

- *For* $a_{j_1j_1 0}(\overline{x}_{j_2} - a_{j_1j_2 1})^{n_1} > 0$ and $a_{j_2j_2 0} > 0$,

$$(a_{j_1j_1 1}, \overline{x}_{j_2}) = \underbrace{((2m_{11}+1)^{\text{th}}\text{SO}, [2(m_{21}-m_{11})]^{\text{th}}\text{II})}_{((2m_{11}+1),[2(m_{21}-m_{11})])\text{-increasing inflection source}}. \tag{1.39}$$

The infinite-equilibrium is a $((2m_{11}+1)^{\text{th}}\text{SO}, [2(m_{21}-m_{11})]^{\text{th}}\text{II})$*-increasing-inflection source at* $x^*_{j_1} = a_{j_1j_1 1} = a_{j_2j_1 1}$.

- *For* $a_{j_1j_1 0}(\overline{x}_{j_2} - a_{j_1j_2 1})^{n_1} < 0$ *and* $a_{j_2j_2 0} > 0$,

$$(a_{j_1j_1 1}, \overline{x}_{j_2}) = \underbrace{((2m_{11}+1)^{\text{th}}\text{SI}, [2(m_{21}-m_{11})]^{\text{th}}\text{DI})}_{((2m_{11}+1),[2(m_{21}-m_{11})])\text{-decreasing inflection sink}}. \tag{1.40}$$

The infinite-equilibrium is a $((2m_{11}+1)^{\text{th}}\text{SI}, [2(m_{21}-m_{11})]^{\text{th}}\text{DI})$*-decreasing-inflection sink at* $x^*_{j_1} = a_{j_1j_1 1} = a_{j_2j_1 1}$.

- *For* $a_{j_1j_1 0}(\overline{x}_{j_2} - a_{j_1j_2 1})^{n_1} > 0$ *and* $a_{j_2j_2 0} < 0$,

$$(a_{j_1j_1 1}, \overline{x}_{j_2}) = \underbrace{((2m_{11}+1)^{\text{th}}\text{SO}, [2(m_{21}-m_{11})]^{\text{th}}\text{DI})}_{((2m_{11}+1),[2(m_{21}-m_{11})])\text{-decreasing inflection source}}. \tag{1.41}$$

The infinite-equilibrium is a $((2m_{11}+1)^{\text{th}}\text{SO}, [2(m_{21}-m_{11})]^{\text{th}}\text{DI})$*-decreasing-inflection source at* $x^*_{j_1} = a_{j_1j_1 1} = a_{j_2j_1 1}$.

- *For* $a_{j_1j_1 0}(\overline{x}_{j_2} - a_{j_1j_2 1})^{n_1} < 0$ *and* $a_{j_2j_2 0} < 0$,

$$(a_{j_1j_1 1}, \overline{x}_{j_2}) = \underbrace{((2m_{11}+1)^{\text{th}}\text{SI}, [2(m_{21}-m_{11})]^{\text{th}}\text{II})}_{((2m_{11}+1),[2(m_{21}-m_{11})])\text{-increasing inflection sink}}. \tag{1.42}$$

The infinite-equilibrium is a $((2m_{11}+1)^{\text{th}}\text{SI}, [2(m_{21}-m_{11})]^{\text{th}}\text{II})$*-increasing-inflection source at* $x^*_{j_1} = a_{j_1j_1 1} = a_{j_2j_1 1}$.

(i_{2a1b}) For $m_1 = 2m_{11}$ *and* $m_2 = 2m_{21}+1$, *the infinite-equilibrium of* $x^*_{j_1} = a_{j_1j_1 1} = a_{j_2j_1 1}$ *with* $(\overline{x}_{j_2} \neq a_{j_1j_2 1})$ *has the following properties.*

- *For* $a_{j_1j_1 0}(\overline{x}_{j_2} - a_{j_1j_2 1})^{n_1} > 0$ *and* $a_{j_2j_2 0} > 0$,

$$(a_{j_1j_1 1}, \overline{x}_{j_2}) = \underbrace{((2m_{11})^{\text{th}}\text{US}, [2(m_{21}-m_{11})+1]^{\text{th}}\text{UP})}_{((2m_{11}),[2(m_{21}-m_{11})+])\text{-up-parabola upper-saddle}}. \tag{1.43}$$

The infinite-equilibrium is a $((2m_{11})^{\text{th}}\text{US}, [2(m_{21}-m_{11})+1]^{\text{th}}\text{UP})$*-up-parabola upper-saddle at* $x^*_{j_1} = a_{j_1j_1 1} = a_{j_2j_1 1}$.

- *For* $a_{j_1j_10}(\overline{x}_{j_2} - a_{j_1j_21})^{n_1} < 0$ *and* $a_{j_2j_20} > 0$,

$$(a_{j_1j_11}, \overline{x}_{j_2}) = \underbrace{((2m_{11})^{\text{th}}\text{LS}, [2(m_{21} - m_{11}) + 1]^{\text{th}}\text{DP})}_{((2m_{11}),[2(m_{21}-m_{11})+])\text{-down-parabola lower-saddle}}. \tag{1.44}$$

The infinite-equilibrium is a $((2m_{11})^{\text{th}}\text{LS}, [2(m_{21} - m_{11}) + 1]^{\text{th}}\text{DP})$*-down-parabola lower-saddle at* $x^*_{j_1} = a_{j_1j_11} = a_{j_2j_11}$.

- *For* $a_{j_1j_10}(\overline{x}_{j_2} - a_{j_1j_21})^{n_1} > 0$ *and* $a_{j_2j_20} < 0$,

$$(a_{j_1j_11}, \overline{x}_{j_2}) = \underbrace{((2m_{11})^{\text{th}}\text{US}, [2(m_{21} - m_{11}) + 1]^{\text{th}}\text{DP})}_{((2m_{11}),[2(m_{21}-m_{11})+])\text{-down-parabola upper-saddle}}. \tag{1.45}$$

The infinite-equilibrium is a $((2m_{11})^{\text{th}}\text{US}, [2(m_{21} - m_{11}) + 1]^{\text{th}}\text{DP})$*-down-parabola upper-saddle at* $x^*_{j_1} = a_{j_1j_11} = a_{j_2j_11}$.

- *For* $a_{j_1j_10}(\overline{x}_{j_2} - a_{j_1j_21})^{n_1} < 0$ *and* $a_{j_2j_20} < 0$,

$$(a_{j_1j_11}, \overline{x}_{j_2}) = \underbrace{((2m_{11})^{\text{th}}\text{LS}, [2(m_{21} - m_{11}) + 1]^{\text{th}}\text{UP})}_{((2m_{11}),[2(m_{21}-m_{11})+])\text{-up-parabola lower-saddle}}. \tag{1.46}$$

The infinite-equilibrium is a $((2m_{11})^{\text{th}}\text{LS}, [2(m_{21}-m_{11})+1]^{\text{th}}\text{UP})$*-up-parabola lower-saddle at* $x^*_{j_1} = a_{j_1j_11} = a_{j_2j_11}$.

(i_{2a1c}) For $m_1 = 2m_{11} + 1$ *and* $m_2 = 2m_{21}$, *the infinite-equilibrium of* $x^*_{j_1} = a_{j_1j_11} = a_{j_2j_11}$ *with* $(\overline{x}_{j_2} \neq a_{j_1j_21})$ *has the following properties.*

- *For* $a_{j_1j_10}(\overline{x}_{j_2} - a_{j_1j_21})^{n_1} > 0$ *and* $a_{j_2j_20} > 0$,

$$(a_{j_1j_11}, \overline{x}_{j_2}) = \underbrace{((2m_{11} + 1)^{\text{th}}\text{SO}, [2(m_{21} - m_{11}) - 1]^{\text{th}}\text{UP})}_{((2m_{11}+1),[2(m_{21}-m_{11})-1])\text{-up-parabola source}}. \tag{1.47}$$

The infinite-equilibrium is a $((2m_{11} + 1)^{\text{th}}\text{SO}, [2(m_{21} - m_{11}) - 1]^{\text{th}}\text{UP})$*-up-parabola source at* $x^*_{j_1} = a_{j_1j_11} = a_{j_2j_11}$.

- *For* $a_{j_1j_10}(\overline{x}_{j_2} - a_{j_1j_21})^{n_1} < 0$ *and* $a_{j_2j_20} > 0$,

$$(a_{j_1j_11}, \overline{x}_{j_2}) = \underbrace{((2m_{11} + 1)^{\text{th}}\text{SI}, [2(m_{21} - m_{11}) - 1]^{\text{th}}\text{DP})}_{((2m_{11}),[2(m_{21}-m_{11})-1])\text{-down-parabola sink}}. \tag{1.48}$$

The infinite-equilibrium is a $((2m_{11} + 1)^{\text{th}}\text{SI}, [2(m_{21} - m_{11}) - 1]^{\text{th}}\text{DP})$*-down-parabola sink at* $x^*_{j_1} = a_{j_1j_11} = a_{j_2j_11}$.

- *For* $a_{j_1j_10}(\overline{x}_{j_2} - a_{j_1j_21})^{n_1} > 0$ *and* $a_{j_2j_20} < 0$,

$$(a_{j_1j_11}, \overline{x}_{j_2}) = \underbrace{((2m_{11}+1)^{\text{th}}\text{SO}, [2(m_{21}-m_{11})-1]^{\text{th}}\text{DP})}_{((2m_{11}),[2(m_{21}-m_{11})-1])\text{-down-parabola source}}. \tag{1.49}$$

The infinite-equilibrium is a $((2m_{11}+1)^{\text{th}}\text{SO}, [2(m_{21}-m_{11})-1]^{\text{th}}\text{DP})$*-down-parabola source at* $x^*_{j_1} = a_{j_1j_11} = a_{j_2j_11}$.

- *For* $a_{j_1j_10}(\overline{x}_{j_2} - a_{j_1j_21})^{n_1} < 0$ *and* $a_{j_2j_20} < 0$,

$$(a_{j_1j_11}, \overline{x}_{j_2}) = \underbrace{((2m_{11}+1)^{\text{th}}\text{SI}, [2(m_{21}-m_{11})-1]^{\text{th}}\text{UP})}_{((2m_{11}+1),[2(m_{21}-m_{11})-1])\text{-up-parabola sink}}. \tag{1.50}$$

The infinite-equilibrium is a $((2m_{11}+1)^{\text{th}}\text{SI}, [2(m_{21}-m_{11})-1]^{\text{th}}\text{UP})$*-up-parabola sink at* $x^*_{j_1} = a_{j_1j_11} = a_{j_2j_11}$.

(i_{2ald}) *For* $m_1 = 2m_{11}$ *and* $m_2 = 2m_{21}$*, the infinite-equilibrium of* $x^*_{j_1} = a_{j_1j_11} = a_{j_2j_11}$ *with* $(\overline{x}_{j_2} \neq a_{j_1j_21})$ *has the following properties.*

- *For* $a_{j_1j_10}(\overline{x}_{j_2} - a_{j_1j_21})^{n_1} > 0$ *and* $a_{j_2j_20} > 0$,

$$(a_{j_1j_11}, \overline{x}_{j_2}) = \underbrace{((2m_{11})^{\text{th}}\text{US}, [2(m_{21}-m_{11})]^{\text{th}}\text{II})}_{((2m_{11}),[2(m_{21}-m_{11})])\text{-increasing-inflection upper-saddle}}. \tag{1.51}$$

The infinite-equilibrium is a $((2m_{11})^{\text{th}}\text{US}, [2(m_{21}-m_{11})]^{\text{th}}\text{II})$*-increasing-inflection upper-saddle at* $x^*_{j_1} = a_{j_1j_11} = a_{j_2j_11}$.

- *For* $a_{j_1j_10}(\overline{x}_{j_2} - a_{j_1j_21})^{n_1} < 0$ *and* $a_{j_2j_20} > 0$,

$$(a_{j_1j_11}, \overline{x}_{j_2}) = \underbrace{((2m_{11})^{\text{th}}\text{LS}, [2(m_{21}-m_{11})]^{\text{th}}\text{DI})}_{((2m_{11}),[2(m_{21}-m_{11})])\text{-decreasing-inflection lower-saddle}}. \tag{1.52}$$

The infinite-equilibrium is a $((2m_{11})^{\text{th}}\text{LS}, [2(m_{21}-m_{11})]^{\text{th}}\text{DI})$*-decreasing-inflection lower-saddle at* $x^*_{j_1} = a_{j_1j_11} = a_{j_2j_11}$.

- *For* $a_{j_1j_10}(\overline{x}_{j_2} - a_{j_1j_21})^{n_1} > 0$ *and* $a_{j_2j_20} < 0$,

$$(a_{j_1j_11}, \overline{x}_{j_2}) = \underbrace{((2m_{11})^{\text{th}}\text{US}, [2(m_{21}-m_{11})]^{\text{th}}\text{DI})}_{((2m_{11}),[2(m_{21}-m_{11})])\text{-decreasing-inflection upper-saddle}}. \tag{1.53}$$

The infinite-equilibrium is a $((2m_{11})^{\text{th}}\text{US}, [2(m_{21}-m_{11})]^{\text{th}}\text{DI})$*-decreasing-inflection upper-saddle at* $x^*_{j_1} = a_{j_1j_11} = a_{j_2j_11}$.

- *For* $a_{j_1j_10}(\overline{x}_{j_2} - a_{j_1j_21})^{n_1} < 0$ *and* $a_{j_2j_20} < 0$,

$$(a_{j_1j_11}, \overline{x}_{j_2}) = \underbrace{((2m_{11})^{\text{th}}\text{LS}, [2(m_{21}-m_{11})]^{\text{th}}\text{II})}_{((2m_{11}),[2(m_{21}-m_{11})])\text{-increasing-inflection lower-saddle}}. \tag{1.54}$$

The infinite-equilibrium is a $((2m_{11})^{\text{th}}\text{LS}, [2(m_{21}-m_{11})]^{\text{th}}\text{II})$*-increasing-inflection lower-saddle at* $x_{j_1}^* = a_{j_1j_1 1} = a_{j_2j_1 1}$.

(i_{2a2}) *There are 32 types of equilibrium* $(x_{j_2}^*, x_{j_1}^*) = (a_{j_1j_2 1}.a_{j_2j_1 1})$ *on the infinite-equilibrium of* $x_{j_1}^* = a_{j_1j_1 1} = a_{j_2j_1 1}$.

(i_{2a2a}) *For* $m_1 = 2m_{11}+1$, $n_1 = 2n_{11}+1$ *and* $m_2 = 2m_{21}+1$, *the equilibrium of* $(x_{j_2}^*, x_{j_1}^*) = (a_{j_1j_2 1}, a_{j_2j_1 1})$ *on the infinite-equilibrium of* $x_{j_1}^* = a_{j_1j_1 1} = a_{j_2j_1 1}$ *has the following properties.*

- *For* $a_{j_1j_1 0} > 0$ *and* $a_{j_2j_2 0} > 0$,

$$(a_{j_1j_2 1}, a_{j_2j_1 1}) = \underbrace{([(2m_{11}+1),(2n_{11}+1)]^{\text{th}}\text{UP}_{-}\text{:UP}_{+},[2(m_{21}-m_{11})]^{\text{th}}{}_{\text{DI:II}}\text{US})}_{([(2m_{11}+1),(2n_{11}+1)],[2(m_{21}-m_{11})])\text{-inflection up-parabola upper-saddle}}. \quad (1.55)$$

The equilibrium is a $([(2m_{11}+1),(2n_{11}+1)]^{\text{th}}\text{UP}_{-}\text{:UP}_{+},[2(m_{21}-m_{11})]^{\text{th}}{}_{\text{DI:II}}\text{US})$*-inflection up-parabola upper-saddle at* $(x_{j_2}^*, x_{j_1}^*) = (a_{j_1j_2 1}, a_{j_2j_1 1})$.

- *For* $a_{j_1j_1 0} < 0$ *and* $a_{j_2j_2 0} > 0$,

$$(a_{j_1j_2 1}, a_{j_2j_1 1}) = \underbrace{([(2m_{11}+1),(2n_{11}+1)]^{\text{th}}\text{DP}_{-}\text{:DP}_{+},[2(m_{21}-m_{11})]^{\text{th}}{}_{\text{II:DI}}\text{LS})}_{([(2m_{11}+1),(2n_{11}+1)],[2(m_{21}-m_{11})])\text{-inflection down-parabola lower-saddle}}. \quad (1.56)$$

The equilibrium is a $([(2m_{11}+1),(2n_{11}+1)]^{\text{th}}\text{DP}_{-}\text{:DP}_{+},[2(m_{21}-m_{11})]^{\text{th}}{}_{\text{II:DI}}\text{LS})$*-inflection down-parabola lower-saddle at* $(x_{j_2}^*, x_{j_1}^*) = (a_{j_1j_2 1}, a_{j_2j_1 1})$.

- *For* $a_{j_1j_1 0} > 0$ *and* $a_{j_2j_2 0} < 0$,

$$(a_{j_1j_2 1}, a_{j_2j_1 1}) = \underbrace{([(2m_{11}+1),(2n_{11}+1)]^{\text{th}}\text{DP}_{+}\text{:DP}_{-},[2(m_{21}-m_{11})]^{\text{th}}{}_{\text{II:DI}}\text{US})}_{([(2m_{11}+1),(2n_{11}+1)],[2(m_{21}-m_{11})])\text{-inflection down-parabola upper-saddle}}. \quad (1.57)$$

The equilibrium is a $([(2m_{11}+1),(2n_{11}+1)]^{\text{th}}\text{DP}_{+}\text{:DP}_{-},[2(m_{21}-m_{11})]^{\text{th}}{}_{\text{II:DI}}\text{US})$*-inflection down-parabola upper-saddle at* $(x_{j_2}^*, x_{j_1}^*) = (a_{j_1j_2 1}, a_{j_2j_1 1})$.

- *For* $a_{j_1j_1 0} < 0$ *and* $a_{j_2j_2 0} < 0$,

$$(a_{j_1j_2 1}, a_{j_2j_1 1}) = \underbrace{([(2m_{11}+1),(2n_{11}+1)]^{\text{th}}\text{UP}_{+}\text{:UP}_{-},[2(m_{21}-m_{11})]^{\text{th}}{}_{\text{DI:II}}\text{LS})}_{([(2m_{11}+1),(2n_{11}+1)],[2(m_{21}-m_{11})])\text{-inflection up-parabola lower-saddle}}. \quad (1.58)$$

The equilibrium is a $([(2m_{11}+1),(2n_{11}+1)]^{\text{th}}\text{UP}_{+}\text{:UP}_{-},[2(m_{21}-m_{11})]^{\text{th}}{}_{\text{DI:II}}\text{LS})$*-inflection up-parabola lower-saddle at* $(x_{j_2}^*, x_{j_1}^*) = (a_{j_1j_21}, a_{j_2j_11})$.

(i_{2a2b}) *For* $m_1 = 2m_{11}$, $n_1 = 2n_{11}+1$ *and* $m_2 = 2m_{21}+1$, *the equilibrium of* $(x_{j_2}^*, x_{j_1}^*) = (a_{j_1j_21}, a_{j_2j_11})$ *on the infinite-equilibrium of* $x_{j_1}^* = a_{j_1j_11} = a_{j_2j_11}$ *has the following properties.*

- *For* $a_{j_1j_10} > 0$ *and* $a_{j_2j_20} > 0$,

$$(a_{j_1j_21}, a_{j_2j_11}) = \underbrace{([(2m_{11}),(2n_{11}+1)]^{\text{th}}\text{DP}_{-}\text{:UP}_{+},[2(m_{21}-m_{11})+1]^{\text{th}}{}_{\text{DP}}\text{LS:}_{\text{UP}}\text{US})}_{([(2m_{11}),(2n_{11}+1)],[2(m_{21}-m_{11})+1])\text{-positive hyperbolic parabola lower-to-upper saddle}}. \tag{1.59}$$

The equilibrium is a $([(2m_{11}),(2n_{11}+1)]^{\text{th}}\text{DP}_{-}\text{:UP}_{+},[2(m_{21}-m_{11})+1]^{\text{th}}{}_{\text{DP}}\text{LS:}_{\text{UP}}\text{US})$*-positive hyperbolic parabola lower-to-upper saddle at* $(x_{j_2}^*, x_{j_1}^*) = (a_{j_1j_21}, a_{j_2j_11})$

.

- *For* $a_{j_1j_10} < 0$ *and* $a_{j_2j_20} > 0$,

$$(a_{j_1j_21}, a_{j_2j_11}) = \underbrace{([(2m_{11}),(2n_{11}+1)]^{\text{th}}\text{UP}_{-}\text{:DP}_{+},[2(m_{21}-m_{11})+1]^{\text{th}}{}_{\text{UP}}\text{US:}_{\text{DP}}\text{LS})}_{([(2m_{11}),(2n_{11}+1)],[2(m_{21}-m_{11})+1])\text{-CCW circular parabola upper-to-lower saddle}}. \tag{1.60}$$

The equilibrium is a $([(2m_{11}),(2n_{11}+1)]^{\text{th}}\text{UP}_{-}\text{:DP}_{+},[2(m_{21}-m_{11})+1]^{\text{th}}{}_{\text{UP}}\text{US:}_{\text{DP}}\text{LS})$-CCW *circular parabola upper-to-lower saddle at* $(x_{j_2}^*, x_{j_1}^*) = (a_{j_1j_21}, a_{j_2j_11})$

.

- *For* $a_{j_1j_10} > 0$ *and* $a_{j_2j_20} < 0$,

$$(a_{j_1j_21}, a_{j_2j_11}) = \underbrace{([(2m_{11}),(2n_{11}+1)]^{\text{th}}\text{UP}_{+}\text{:DP}_{-},[2(m_{21}-m_{11})+1]^{\text{th}}{}_{\text{UP}}\text{LS:}_{\text{DP}}\text{US})}_{([(2m_{11}),(2n_{11}+1)],[2(m_{21}-m_{11})+1])\text{-CW circular parabola lower-to-upper saddle}}. \tag{1.61}$$

The equilibrium is a $([(2m_{11}),(2n_{11}+1)]^{\text{th}}\text{UP}_{+}\text{:DP}_{-},[2(m_{21}-m_{11})+1]^{\text{th}}{}_{\text{UP}}\text{LS:}_{\text{DP}}\text{US})$-CW *circular parabola lower-to-upper saddle at* $(x_{j_2}^*, x_{j_1}^*) = (a_{j_1j_21}, a_{j_2j_11})$.

- *For* $a_{j_1j_10} < 0$ *and* $a_{j_2j_20} < 0$,

$$(a_{j_1j_21}, a_{j_2j_11}) = \underbrace{([(2m_{11}),(2n_{11}+1)]^{\text{th}}\text{DP}_{+}\text{:UP}_{-},[2(m_{21}-m_{11})+1]^{\text{th}}{}_{\text{DP}}\text{US:}_{\text{UP}}\text{LS})}_{([(2m_{11}),(2n_{11}+1)],[2(m_{21}-m_{11})+1])\text{-negative hyperbolic parabola upper-to-lower saddle}}. \tag{1.62}$$

The equilibrium is a $([(2m_{11}),(2n_{11}+1)]^{\text{th}}\text{DP}_{-}\text{:UP}_{+},[2(m_{21}-m_{11})+1]^{\text{th}}{}_{\text{DP}}\text{US:}{}_{\text{UP}}\text{LS})$*-negative hyperbolic parabola upper-to-lower saddle* at $(x^*_{j_2}, x^*_{j_1}) = (a_{j_1j_21}, a_{j_2j_11})$.

(i_{2a2c}*) For* $m_1 = 2m_{11}+1$, $n_1 = 2n_{11}+1$ *and* $m_2 = 2m_{21}$, *the equilibrium of* $(x^*_{j_2}, x^*_{j_1}) = (a_{j_1j_21}, a_{j_2j_11})$ *on the infinite-equilibrium of* $x^*_{j_1} = a_{j_1j_11} = a_{j_2j_11}$ *has the following properties.*

- *For* $a_{j_1j_10} > 0$ *and* $a_{j_2j_20} > 0$,

$$(a_{j_1j_21}, a_{j_2j_11}) = \underbrace{([(2m_{11}+1),(2n_{11}+1)]^{\text{th}}\text{DP}_{+}\text{:UP}_{+},[2(m_{21}-m_{11})-1]^{\text{th}}{}_{\text{DP}}\text{SI:}{}_{\text{UP}}\text{SO})}_{([(2m_{11}+1),(2n_{11}+1)],[2(m_{21}-m_{11})-1])\text{-hyperbolic parabola sink-to-source}}. \tag{1.63}$$

The equilibrium is a $([(2m_{11}+1),(2n_{11}+1)]^{\text{th}}\text{DP}_{+}\text{:UP}_{+},[2(m_{21}-m_{11})-1]^{\text{th}}{}_{\text{DP}}\text{SI:}{}_{\text{UP}}\text{SO})$*-hyperbolic parabola sink-to-source at* $(x^*_{j_2}, x^*_{j_1}) = (a_{j_1j_21}, a_{j_2j_11})$.

- *For* $a_{j_1j_10} < 0$ *and* $a_{j_2j_20} > 0$,

$$(a_{j_1j_21}, a_{j_2j_11}) = \underbrace{([(2m_{11}+1),(2n_{11}+1)]^{\text{th}}\text{UP}_{+}\text{:DP}_{+},[2(m_{21}-m_{11})-1]^{\text{th}}{}_{\text{UP}}\text{SO:}{}_{\text{DP}}\text{SI})}_{([(2m_{11}+1),(2n_{11}+1)],[2(m_{21}-m_{11})-1])\text{-circular parabola source-to-sink}}. \tag{1.64}$$

The equilibrium is a $([(2m_{11}+1),(2n_{11}+1)]^{\text{th}}\text{UP}_{+}\text{:DP}_{+},[2(m_{21}-m_{11})-1]^{\text{th}}{}_{\text{UP}}\text{SO:}{}_{\text{DP}}\text{SI})$*-circular parabola source-to-sink at* $(x^*_{j_2}, x^*_{j_1}) = (a_{j_1j_21}, a_{j_2j_11})$.

- *For* $a_{j_1j_10} > 0$ *and* $a_{j_2j_20} < 0$,

$$(a_{j_1j_21}, a_{j_2j_11}) = \underbrace{([(2m_{11}+1),(2n_{11}+1)]^{\text{th}}\text{UP}_{-}\text{:DP}_{-},[2(m_{21}-m_{11})-1]^{\text{th}}{}_{\text{UP}}\text{SI:}{}_{\text{DP}}\text{SO})}_{([(2m_{11}+1),(2n_{11}+1)],[2(m_{21}-m_{11})-1])\text{-circular parabola sink-to-source}}. \tag{1.65}$$

The equilibrium is a $([(2m_{11}+1),(2n_{11}+1)]^{\text{th}}\text{UP}_{-}\text{:DP}_{-},[2(m_{21}-m_{11})-1]^{\text{th}}{}_{\text{UP}}\text{SI:}{}_{\text{DP}}\text{SO})$*-circular parabola sink-to-source at* $(x^*_{j_2}, x^*_{j_1}) = (a_{j_1j_21}, a_{j_2j_11})$.

- *For* $a_{j_1j_10} < 0$ *and* $a_{j_2j_20} < 0$,

$$(a_{j_1j_21}, a_{j_2j_11}) = \underbrace{([(2m_{11}+1),(2n_{11}+1)]^{\text{th}}\text{DP}_{-}\text{:UP}_{-},[2(m_{21}-m_{11})-1]^{\text{th}}{}_{\text{DP}}\text{SO:}{}_{\text{UP}}\text{SI})}_{([(2m_{11}+1),(2n_{11}+1)],[2(m_{21}-m_{11})-1])\text{-hyperbolic parabola source-to-sink}}. \tag{1.66}$$

The equilibrium is a $([(2m_{11}+1),(2n_{11}+1)]^{\text{th}}\text{DP}_-\text{:UP}_-,[2(m_{21}-m_{11})-1]^{\text{th}}{}_{\text{DP}}\text{SO:}_{\text{UP}}\text{SI})$*-hyperbolic parabola source-to-sink at* $(x^*_{j_2}, x^*_{j_1}) = (a_{j_1j_2 1}, a_{j_2j_1 1})$.

(i_{2a2d}) *For* $m_1 = 2m_{11}$, $n_1 = 2n_{11}+1$ *and* $m_2 = 2m_{21}$, *the equilibrium of* $(x^*_{j_2}, x^*_{j_1}) = (a_{j_1j_2 1}, a_{j_2j_1 1})$ *on the infinite-equilibrium of* $x^*_{j_1} = a_{j_1j_1 1} = a_{j_2j_1 1}$ *has the following properties.*

- *For* $a_{j_1j_1 0} > 0$ *and* $a_{j_2j_2 0} > 0$,

$$(a_{j_1j_2 1}, a_{j_2j_1 1}) = \underbrace{([(2m_{11}),(2n_{11}+1)]^{\text{th}}\text{UP}_+\text{:UP}_+,[2(m_{21}-m_{11})]^{\text{th}}{}_{\text{DI:II}}\text{US})}_{([(2m_{11}),(2n_{11}+1)],[2(m_{21}-m_{11})])\text{-inflection up-parabola upper-saddle}}. \tag{1.67}$$

The equilibrium is a $([(2m_{11}),(2n_{11}+1)]^{\text{th}}\text{UP}_+\text{:UP}_+,[2(m_{21}-m_{11})]^{\text{th}}{}_{\text{DI:II}}\text{US})$*-inflection up-parabola upper-saddle at* $(x^*_{j_2}, x^*_{j_1}) = (a_{j_1j_2 1}, a_{j_2j_1 1})$.

- *For* $a_{j_1j_1 0} < 0$ *and* $a_{j_2j_2 0} > 0$,

$$(a_{j_1j_2 1}, a_{j_2j_1 1}) = \underbrace{([(2m_{11}),(2n_{11}+1)]^{\text{th}}\text{DP}_+\text{:DP}_+,[2(m_{21}-m_{11})]^{\text{th}}{}_{\text{II:DI}}\text{US})}_{([(2m_{11}),(2n_{11}+1)],[2(m_{21}-m_{11})])\text{-inflection down-parabola upper-saddle}}. \tag{1.68}$$

The equilibrium is a $([(2m_{11}),(2n_{11}+1)]^{\text{th}}\text{DP}_+\text{:DP}_+,[2(m_{21}-m_{11})]^{\text{th}}{}_{\text{DI:II}}\text{US})$*-inflection down-parabola upper-saddle at* $(x^*_{j_2}, x^*_{j_1}) = (a_{j_1j_2 1}, a_{j_2j_1 1})$.

- *For* $a_{j_1j_1 0} > 0$ *and* $a_{j_2j_2 0} < 0$,

$$(a_{j_1j_2 1}, a_{j_2j_1 1}) = \underbrace{([(2m_{11}),(2n_{11}+1)]^{\text{th}}\text{DP}_-\text{:DP}_-,[2(m_{21}-m_{11})]^{\text{th}}{}_{\text{II:DI}}\text{LS})}_{([(2m_{11}),(2n_{11}+1)],[2(m_{21}-m_{11})])\text{-inflection down-parabola lower-saddle}}. \tag{1.69}$$

The equilibrium is a $([(2m_{11}),(2n_{11}+1)]^{\text{th}}\text{DP}_-\text{:DP}_-,[2(m_{21}-m_{11})]^{\text{th}}{}_{\text{DI:II}}\text{LS})$*-inflection down-parabola lower-saddle at* $(x^*_{j_2}, x^*_{j_1}) = (a_{j_1j_2 1}, a_{j_2j_1 1})$.

- *For* $a_{j_1j_1 0} < 0$ *and* $a_{j_2j_2 0} < 0$,

$$(a_{j_1j_2 1}, a_{j_2j_1 1}) = \underbrace{([(2m_{11}),(2n_{11}+1)]^{\text{th}}\text{UP}_-\text{:UP}_-,[2(m_{21}-m_{11})]^{\text{th}}{}_{\text{DI:II}}\text{LS})}_{([(2m_{11}),(2n_{11}+1)],[2(m_{21}-m_{11})])\text{-inflection up-parabola lower-saddle}}. \tag{1.70}$$

The equilibrium is a $([(2m_{11}),(2n_{11}+1)]^{\text{th}}\text{UP}_+\text{:UP}_+,[2(m_{21}-m_{11})]^{\text{th}}{}_{\text{DI:II}}\text{US})$*-inflection up-parabola lower-saddle at* $(x^*_{j_2}, x^*_{j_1}) = (a_{j_1j_2 1}, a_{j_2j_1 1})$.

(i_{2a2e}) *For* $m_1 = 2m_{11} + 1$, $n_1 = 2n_{11}$ *and* $m_2 = 2m_{21} + 1$, *the equilibrium of* $(x^*_{j_2}, x^*_{j_1}) = (a_{j_1j_21}, a_{j_2j_11})$ *on the infinite-equilibrium of* $x^*_{j_1} = a_{j_1j_11} = a_{j_2j_11}$ *has the following properties.*

- *For* $a_{j_1j_10} > 0$ *and* $a_{j_2j_20} > 0$,

$$(a_{j_1j_21}, a_{j_2j_11}) = \underbrace{([(2m_{11}+1),(2n_{11})]^{\text{th}}\text{II}_{-}\text{:II}_{+},[2(m_{21} - m_{11})]^{\text{th}}{}_{\text{II:II}}\text{SO})}_{([(2m_{11}+1),(2n_{11})],[2(m_{21}-m_{11})])\text{-increasing-inflection source}}. \tag{1.71}$$

The equilibrium is a $([(2m_{11}+1),(2n_{11})]^{\text{th}}\text{II}_{-}\text{:II}_{+},[2(m_{21} - m_{11})]^{\text{th}}{}_{\text{II:II}}SO)$-*increasing-inflection source at* $(x^*_{j_2}, x^*_{j_1}) = (a_{j_1j_21}, a_{j_2j_11})$.

- *For* $a_{j_1j_10} < 0$ *and* $a_{j_2j_20} > 0$,

$$(a_{j_1j_21}, a_{j_2j_11}) = \underbrace{([(2m_{11}+1),(2n_{11})]^{\text{th}}\text{DI}_{-}\text{:DI}_{+},[2(m_{21} - m_{11})]^{\text{th}}{}_{\text{DI:DI}}\text{SI})}_{([(2m_{11}+1),(2n_{11})],[2(m_{21}-m_{11})])\text{-decreasing-inflection sink}}. \tag{1.72}$$

The equilibrium is a $([(2m_{11}+1),(2n_{11})]^{\text{th}}\text{DI}_{-}\text{:DI}_{+},[2(m_{21} - m_{11})]^{\text{th}}{}_{\text{DI:DI}}\text{SI})$-*decreasing-inflection sink at* $(x^*_{j_2}, x^*_{j_1}) = (a_{j_1j_21}, a_{j_2j_11})$.

- *For* $a_{j_1j_10} > 0$ *and* $a_{j_2j_20} < 0$,

$$(a_{j_1j_21}, a_{j_2j_11}) = \underbrace{([(2m_{11}+1),(2n_{11})]^{\text{th}}\text{DI}_{+}\text{:DI}_{-},[2(m_{21} - m_{11})]^{\text{th}}{}_{\text{DI:DI}}\text{SO})}_{([(2m_{11}+1),(2n_{11})],[2(m_{21}-m_{11})])\text{-decreasing-inflection source}}. \tag{1.73}$$

The equilibrium is a $([(2m_{11}+1),(2n_{11})]^{\text{th}}\text{DI}_{+}\text{:DI}_{-},[2(m_{21} - m_{11})]^{\text{th}}{}_{\text{DI:DI}}\text{SO})$-*decreasing-inflection source at* $(x^*_{j_2}, x^*_{j_1}) = (a_{j_1j_21}, a_{j_2j_11})$.

- *For* $a_{j_1j_10} < 0$ *and* $a_{j_2j_20} < 0$,

$$(a_{j_1j_21}, a_{j_2j_11}) = \underbrace{([(2m_{11}+1),(2n_{11})]^{\text{th}}\text{II}_{+}\text{:II}_{-},[2(m_{21} - m_{11})]^{\text{th}}{}_{\text{II:II}}\text{SI})}_{([(2m_{11}+1),(2n_{11})],[2(m_{21}-m_{11})])\text{-increasing-inflection sink}}. \tag{1.74}$$

The equilibrium is a $([(2m_{11}+1),(2n_{11})]^{\text{th}}\text{II}_{+}\text{:II}_{-},[2(m_{21} - m_{11})]^{\text{th}}{}_{\text{II:II}}\text{SI})$-*increasing-inflection sink at* $(x^*_{j_2}, x^*_{j_1}) = (a_{j_1j_21}, a_{j_2j_11})$.

(i_{2a2f}) *For* $m_1 = 2m_{11}$, $n_1 = 2n_{11}$ *and* $m_2 = 2m_{21} + 1$, *the equilibrium of* $(x^*_{j_2}, x^*_{j_1}) = (a_{j_1j_21}, a_{j_2j_11})$ *on the infinite-equilibrium of* $x^*_{j_1} = a_{j_1j_11} = a_{j_2j_11}$ *has the following properties.*

- *For* $a_{j_1j_10} > 0$ *and* $a_{j_2j_20} > 0$,

$$(a_{j_1j_21}, a_{j_2j_11}) = \underbrace{([(2m_{11}),(2n_{11})]^{\text{th}}\text{DI}_{-}\text{:II}_{+},[2(m_{21}-m_{11})+1]^{\text{th}}{}_{\text{UP:UP}}\text{US})}_{([(2m_{11}),(2n_{11})],[2(m_{21}-m_{11})+1])\text{-double up-parabola upper-saddle}}. \tag{1.75}$$

The equilibrium is a $([(2m_{11}),(2n_{11})]^{\text{th}}\text{DI}_{-}\text{:DI}_{+},[2(m_{21}-m_{11})+1]^{\text{th}}{}_{\text{UP:UP}}\text{US})$*-double up-parabola upper-saddle at* $(x_{j_2}^*, x_{j_1}^*) = (a_{j_1j_21}, a_{j_2j_11})$.

- *For* $a_{j_1j_10} < 0$ *and* $a_{j_2j_20} > 0$,

$$(a_{j_1j_21}, a_{j_2j_11}) = \underbrace{([(2m_{11}),(2n_{11})]^{\text{th}}\text{II}_{-}\text{:DI}_{+},[2(m_{21}-m_{11})+1]^{\text{th}}{}_{\text{DP:DP}}\text{LS})}_{([(2m_{11}),(2n_{11})],[2(m_{21}-m_{11})+1])\text{-double down-parabola lower-saddle}}. \tag{1.76}$$

The equilibrium is a $([(2m_{11}),(2n_{11})]^{\text{th}}\text{II}_{-}\text{:DI}_{+},[2(m_{21}-m_{11})+1]^{\text{th}}{}_{\text{DP:DP}}\text{LS})$*-double down-parabola lower-saddle at* $(x_{j_2}^*, x_{j_1}^*) = (a_{j_1j_21}, a_{j_2j_11})$.

- *For* $a_{j_1j_10} > 0$ *and* $a_{j_2j_20} < 0$,

$$(a_{j_1j_21}, a_{j_2j_11}) = \underbrace{([(2m_{11}),(2n_{11})]^{\text{th}}\text{II}_{+}\text{:DI}_{-},[2(m_{21}-m_{11})+1]^{\text{th}}{}_{\text{DP:DP}}\text{US})}_{([(2m_{11}),(2n_{11})],[2(m_{21}-m_{11})+1])\text{-double down-parabola upper-saddle}}. \tag{1.77}$$

The equilibrium is a $([(2m_{11}),(2n_{11})]^{\text{th}}\text{II}_{+}\text{:DI}_{-},[2(m_{21}-m_{11})+1]^{\text{th}}{}_{\text{DP:DP}}\text{US})$*-double down-parabola upper-saddle at* $(x_{j_2}^*, x_{j_1}^*) = (a_{j_1j_21}, a_{j_2j_11})$.

- *For* $a_{j_1j_10} < 0$ *and* $a_{j_2j_20} < 0$,

$$(a_{j_1j_21}, a_{j_2j_11}) = \underbrace{([(2m_{11}),(2n_{11})]^{\text{th}}\text{DI}_{+}\text{:II}_{-},[2(m_{21}-m_{11})+1]^{\text{th}}{}_{\text{UP:UP}}\text{LS})}_{([(2m_{11}),(2n_{11})],[2(m_{21}-m_{11})+1])\text{-double up-parabola lower-saddle}}. \tag{1.78}$$

The equilibrium is a $([(2m_{11}),(2n_{11})]^{\text{th}}\text{DI}_{+}\text{:II}_{-},[2(m_{21}-m_{11})+1]^{\text{th}}{}_{\text{UP:UP}}\text{LS})$*-double up-parabola lower-saddle at* $(x_{j_2}^*, x_{j_1}^*) = (a_{j_1j_21}, a_{j_2j_11})$.

(i_{2a2g}) *For* $m_1 = 2m_{11}+1$, $n_1 = 2n_{11}$ *and* $m_2 = 2m_{21}$, *the equilibrium of* $(x_{j_2}^*, x_{j_1}^*) = (a_{j_1j_21}, a_{j_2j_11})$ *on the infinite-equilibrium of* $x_{j_1}^* = a_{j_1j_11} = a_{j_2j_11}$ *has the following properties.*

- *For* $a_{j_1j_10} > 0$ *and* $a_{j_2j_20} > 0$,

$$(a_{j_1j_21}, a_{j_2j_11}) = \underbrace{([(2m_{11}+1),(2n_{11})]^{\text{th}}\text{DI}_{+}\text{:II}_{+},[2(m_{21}-m_{11})-1]^{\text{th}}{}_{\text{UP:UP}}\text{SO})}_{([(2m_{11}+1),(2n_{11})],[2(m_{21}-m_{11})-1])\text{-double up-parabola source}}. \tag{1.79}$$

The equilibrium is a $([(2m_{11}+1),(2n_{11})]^{\text{th}}\text{DI}_{+}\text{:II}_{+},[2(m_{21}-m_{11})-1]^{\text{th}}{}_{\text{UP:UP}}\text{SO})$*-double up-parabola source at* $(x_{j_2}^*, x_{j_1}^*) = (a_{j_1j_21}, a_{j_2j_11})$.

- *For* $a_{j_1j_10} < 0$ *and* $a_{j_2j_20} > 0$,

$$(a_{j_1j_21}, a_{j_2j_11}) = \underbrace{([(2m_{11}+1),(2n_{11})]^{\text{th}}\text{II}_+\text{:DI}_+,[2(m_{21}-m_{11})-1]^{\text{th}}{}_{\text{DP:DP}}\text{SI})}_{([(2m_{11}+1),(2n_{11})],[2(m_{21}-m_{11})-1])\text{-double down-parabola sink}}. \tag{1.80}$$

The equilibrium is a $([(2m_{11}+1),(2n_{11})]^{\text{th}}\text{II}_+\text{:DI}_+,[2(m_{21}-m_{11})-1]^{\text{th}}{}_{\text{DP:DP}}\text{SI})$*-double down-parabola sink at* $(x^*_{j_2}, x^*_{j_1}) = (a_{j_1j_21}, a_{j_2j_11})$.

- *For* $a_{j_1j_10} > 0$ *and* $a_{j_2j_20} < 0$,

$$(a_{j_1j_21}, a_{j_2j_11}) = \underbrace{([(2m_{11}+1),(2n_{11})]^{\text{th}}\text{II}_-\text{:DI}_-,[2(m_{21}-m_{11})-1]^{\text{th}}{}_{\text{DP:DP}}\text{SO})}_{([(2m_{11}+1),(2n_{11})],[2(m_{21}-m_{11})+1])\text{-double down-parabola source}}. \tag{1.81}$$

The equilibrium is a $([(2m_{11}+1),(2n_{11})]^{\text{th}}\text{II}_-\text{:DI}_-,[2(m_{21}-m_{11})-1]^{\text{th}}{}_{\text{DP:DP}}\text{SO})$*-double down-parabola source at* $(x^*_{j_2}, x^*_{j_1}) = (a_{j_1j_21}, a_{j_2j_11})$.

- *For* $a_{j_1j_10} < 0$ *and* $a_{j_2j_20} < 0$,

$$(a_{j_1j_21}, a_{j_2j_11}) = \underbrace{([(2m_{11}+1),(2n_{11})]^{\text{th}}\text{DI}_-\text{:II}_-,[2(m_{21}-m_{11})-1]^{\text{th}}{}_{\text{UP:UP}}\text{SI})}_{([(2m_{11}+1),(2n_{11})],[2(m_{21}-m_{11})+1])\text{-double up-parabola sink}}. \tag{1.82}$$

The equilibrium is a $([(2m_{11}+1),(2n_{11})]^{\text{th}}\text{DI}_-\text{:II}_-,[2(m_{21}-m_{11})-1]^{\text{th}}{}_{\text{UP:UP}}\text{SI})$*-double up-parabola sink at* $(x^*_{j_2}, x^*_{j_1}) = (a_{j_1j_21}, a_{j_2j_11})$.

(i_{2a2h}*) For* $m_1 = 2m_{11}$, $n_1 = 2n_{11}$ *and* $m_2 = 2m_{21}$*, the equilibrium of* $(x^*_{j_2}, x^*_{j_1}) = (a_{j_1j_21}, a_{j_2j_11})$ *on the infinite-equilibrium of* $x^*_{j_1} = a_{j_1j_11} = a_{j_2j_11}$ *has the following properties.*

- *For* $a_{j_1j_10} > 0$ *and* $a_{j_2j_20} > 0$,

$$(a_{j_1j_21}, a_{j_2j_11}) = \underbrace{([(2m_{11}),(2n_{11})]^{\text{th}}\text{II}_+\text{:II}_+,[2(m_{21}-m_{11})]^{\text{th}}{}_{\text{II:II}}\text{US})}_{([(2m_{11}),(2n_{11})],[2(m_{21}-m_{11})])\text{-increasing-inflection upper-saddle}}. \tag{1.83}$$

The equilibrium is a $([(2m_{11}),(2n_{11})]^{\text{th}}\text{II}_+\text{:II}_+,[2(m_{21}-m_{11})]^{\text{th}}{}_{\text{II:II}}\text{US})$*-increasing-inflection upper-saddle at* $(x^*_{j_2}, x^*_{j_1}) = (a_{j_1j_21}, a_{j_2j_11})$.

- *For* $a_{j_1j_10} < 0$ *and* $a_{j_2j_20} > 0$,

$$(a_{j_1j_21}, a_{j_2j_11}) = \underbrace{([(2m_{11}),(2n_{11})]^{\text{th}}\text{DI}_+\text{:DI}_+,[2(m_{21}-m_{11})]^{\text{th}}{}_{\text{DI:DI}}\text{LS})}_{([(2m_{11}),(2n_{11})],[2(m_{21}-m_{11})])\text{-decreasing-inflection lower-saddle}}. \tag{1.84}$$

The equilibrium is a $([(2m_{11}),(2n_{11})]^{\text{th}}\text{DI}_{+}\text{:DI}_{+},[2(m_{21}-m_{11})]^{\text{th}}{}_{\text{DI:DI}}\text{LS})$-*decreasing-inflection lower-saddle at* $(x^*_{j_2},x^*_{j_1})=(a_{j_1j_21},a_{j_2j_11})$.

- *For* $a_{j_1j_10}>0$ *and* $a_{j_2j_20}<0$,

$$(a_{j_1j_21},a_{j_2j_11})=\underbrace{([(2m_{11}),(2n_{11})]^{\text{th}}\text{DI}_{-}\text{:DI}_{-},[2(m_{21}-m_{11})]^{\text{th}}{}_{\text{DI:DI}}\text{US})}_{([(2m_{11}),(2n_{11})],[2(m_{21}-m_{11})])\text{-decreasing-inflection upper-saddle}}. \tag{1.85}$$

The equilibrium is a $([(2m_{11}),(2n_{11})]^{\text{th}}\text{DI}_{-}\text{:DI}_{-},[2(m_{21}-m_{11})]^{\text{th}}{}_{\text{DI:DI}}\text{US})$-*decreasing-inflection upper-saddle at* $(x^*_{j_2},x^*_{j_1})=(a_{j_1j_21},a_{j_2j_11})$.

- *For* $a_{j_1j_10}<0$ *and* $a_{j_2j_20}<0$,

$$(a_{j_1j_21},a_{j_2j_11})=\underbrace{([(2m_{11}),(2n_{11})]^{\text{th}}\text{II}_{-}\text{:II}_{-},[2(m_{21}-m_{11})]^{\text{th}}{}_{\text{II:II}}\text{LS})}_{([(2m_{11}),(2n_{11})],[2(m_{21}-m_{11})])\text{-increasing-inflection lower-saddle}}. \tag{1.86}$$

The equilibrium is a $([(2m_{11}),(2n_{11})]^{\text{th}}\text{II}_{-}\text{:II}_{-},[2(m_{21}-m_{11})]^{\text{th}}{}_{\text{II:II}}\text{LS})$-*increasing-inflection lower-saddle at* $(x^*_{j_2},x^*_{j_1})=(a_{j_1j_21},a_{j_2j_11})$.

(i_{2b}) For $m_2<m_1$, *the first integral manifold is*

$$\begin{aligned}
&-\frac{1}{m_1-m_2-1}\left[\frac{1}{(x_{j_1}-a_{j_1j_11})^{m_1-m_2+1}}-\frac{1}{(x_{j_10}-a_{j_1j_11})^{m_1-m_2+1}}\right]\\
&=\frac{a_{j_1j_10}}{a_{j_2j_20}}\frac{1}{n_1+1}\left[(x_{j_2}-a_{j_1j_21})^{n_1+1}-(x_{j_20}-a_{j_1j_21})^{n_1+1}\right],\\
&\text{for } m_2-m_1+1\neq 0;\\
&\frac{1}{m_2-m_1+1}\left[(x_{j_1}-a_{j_1j_11})^{m_2-m_1+1}-(x_{j_10}-a_{j_1j_11})^{m_2-m_1+1}\right]=\ln\frac{|x_{j_1}-a_{j_1j_11}|}{|x_{j_10}-a_{j_1j_11}|},\\
&\text{for } m_2-m_1+1=0.
\end{aligned} \tag{1.87}$$

(i_{2b1}) There are 16 types of infinite-equilibriums at $x^*_{j_1}=a_{j_1j_11}=a_{j_2j_11}$.

(i_{2b1a}) For $m_1=2m_{11}+1$ *and* $m_2=2m_{21}+1$, *the infinite-equilibrium of* $x^*_{j_1}=a_{j_1j_11}=a_{j_2j_11}$ *and* $\overline{x}_{j_2}\neq a_{j_1j_21}$ *has the following properties.*

- *For* $a_{j_1j_10}(\overline{x}_{j_2}-a_{j_1j_21})^{n_1}>0$ *and* $a_{j_2j_20}>0$,

$$(a_{j_1j_11},\overline{x}_{j_2})=\underbrace{((2m_{11}+1)^{\text{th}}\text{SO},[2(m_{11}-m_{21})]^{\text{th}}\text{IC})}_{((2m_{11}+1),[2(m_{11}-m_{21})])\text{-increasing concave source}}. \tag{1.88}$$

The infinite-equilibrium is a $((2m_{11}+1)^{\text{th}}\text{SO},[2(m_{11}-m_{21})]^{\text{th}}\text{IC})$-*increasing-concave source at* $x^*_{j_1}=a_{j_1j_11}=a_{j_2j_11}$.

- *For* $a_{j_1j_10}(\overline{x}_{j_2}-a_{j_1j_21})^{n_1}<0$ *and* $a_{j_2j_20}>0$,

$$(a_{j_1j_11}, \overline{x}_{j_2}) = \underbrace{((2m_{11}+1)^{\text{th}}\text{SI},[2(m_{11}-m_{21})]^{\text{th}}\text{DC})}_{((2m_{11}+1),[2(m_{11}-m_{21})])\text{-decreasing concave sink}}. \tag{1.89}$$

The infinite-equilibrium is a $((2m_{11}+1)^{\text{th}}\text{SI},[2(m_{11}-m_{21})]^{\text{th}}\text{DC})$*-decreasing-concave sink at* $x_{j_1}^* = a_{j_1j_11} = a_{j_2j_11}$.

- *For* $a_{j_1j_10}(\overline{x}_{j_2} - a_{j_1j_21})^{n_1} > 0$ *and* $a_{j_2j_20} < 0$,

$$(a_{j_1j_11}, \overline{x}_{j_2}) = \underbrace{((2m_{11}+1)^{\text{th}}\text{SO},[2(m_{11}-m_{21})]^{\text{th}}\text{DC})}_{((2m_{11}+1),[2(m_{11}-m_{21})])\text{-decreasing concave source}}. \tag{1.90}$$

The infinite-equilibrium is a $((2m_{11}+1)^{\text{th}}\text{SO},[2(m_{11}-m_{21})]^{\text{th}}\text{DC})$*-decreasing-concave source at* $x_{j_1}^* = a_{j_1j_11} = a_{j_2j_11}$.

- *For* $a_{j_1j_10}(\overline{x}_{j_2} - a_{j_1j_21})^{n_1} < 0$ *and* $a_{j_2j_20} < 0$,

$$(a_{j_1j_11}, \overline{x}_{j_2}) = \underbrace{((2m_{11}+1)^{\text{th}}\text{SI},[2(m_{11}-m_{21})]^{\text{th}}\text{IC})}_{((2m_{11}+1),[2(m_{11}-m_{21})])\text{increasing concave sink}}. \tag{1.91}$$

The infinite-equilibrium is a $((2m_{11}+1)^{\text{th}}\text{SI},[2(m_{11}-m_{21})]^{\text{th}}\text{IC})$*-increasing-concave sink at* $x_{j_1}^* = a_{j_1j_11} = a_{j_2j_11}$.

(i_{2b1b}) *For* $m_1 = 2m_{11}$ *and* $m_2 = 2m_{21}+1$, *the infinite-equilibrium of* $x_{j_1}^* = a_{j_1j_11} = a_{j_2j_11}$ *with* $\overline{x}_{j_2} \neq a_{j_1j_21}$ *has the following properties.*

- *For* $a_{j_1j_10}(\overline{x}_{j_2} - a_{j_1j_21})^{n_1} > 0$ *and* $a_{j_2j_20} > 0$,

$$(a_{j_1j_11}, \overline{x}_{j_2}) = \underbrace{((2m_{11})^{\text{th}}\text{US},[2(m_{11}-m_{21})-1]^{\text{th}}\text{DA})}_{((2m_{11}),[2(m_{11}-m_{21})-1])\text{-down-asymptote upper-saddle}}. \tag{1.92}$$

The infinite-equilibrium is a $((2m_{11})^{\text{th}}\text{US},[2(m_{11}-m_{21})-1]^{\text{th}}\text{DA})$*-down-asymptote upper-saddle at* $x_{j_1}^* = a_{j_1j_11} = a_{j_2j_11}$.

- *For* $a_{j_1j_10}(\overline{x}_{j_2} - a_{j_1j_21})^{n_1} < 0$ *and* $a_{j_2j_20} > 0$,

$$(a_{j_1j_11}, \overline{x}_{j_2}) = \underbrace{((2m_{11})^{\text{th}}\text{LS},[2(m_{11}-m_{21})-1]^{\text{th}}\text{UA})}_{((2m_{11}),[2(m_{11}-m_{21})-1])\text{-up-asymptote lower-saddle}}. \tag{1.93}$$

The infinite-equilibrium is a $((2m_{11})^{\text{th}}\text{LS},[2(m_{11}-m_{21})-1]^{\text{th}}\text{UA})$*-up-asymptote lower-saddle at* $x_{j_1}^* = a_{j_1j_11} = a_{j_2j_11}$.

- *For* $a_{j_1j_10}(\overline{x}_{j_2} - a_{j_1j_21})^{n_1} > 0$ *and* $a_{j_2j_20} < 0$,

$$(a_{j_1j_11}, \overline{x}_{j_2}) = \underbrace{((2m_{11})^{\text{th}}\text{US},[2(m_{11}-m_{21})-1]^{\text{th}}\text{UA})}_{((2m_{11}),[2(m_{11}-m_{21})-1])\text{-up-asymptote upper-saddle}}. \tag{1.94}$$

The infinite-equilibrium is a $((2m_{11})^{\text{th}}\text{US},[2(m_{11}-m_{21})-1]^{\text{th}}\text{UA})$*-up-asymptote upper-saddle at* $x_{j_1}^* = a_{j_1j_11} = a_{j_2j_11}$.

- *For* $a_{j_1j_10}(\overline{x}_{j_2} - a_{j_1j_21})^{n_1} < 0$ *and* $a_{j_2j_20} < 0$,

$$(a_{j_1j_11}, \overline{x}_{j_2}) = \underbrace{((2m_{11})^{\text{th}}\text{LS},[2(m_{11}-m_{21})-1]^{\text{th}}\text{DA})}_{((2m_{11}),[2(m_{11}-m_{21})-1])\text{-down-asymptote lower-saddle}}. \quad (1.95)$$

The infinite-equilibrium is a $((2m_{11})^{\text{th}}\text{LS},[2(m_{11}-m_{21})-1]^{\text{th}}\text{DA})$*-down-asymptote lower-saddle at* $x_{j_1}^* = a_{j_1j_11} = a_{j_2j_11}$.

(i_{2b1c}*) For* $m_1 = 2m_{11}+1$ *and* $m_2 = 2m_{21}$*, the infinite-equilibrium of* $x_{j_1}^* = a_{j_1j_11} = a_{j_2j_11}$ *and* $\overline{x}_{j_2} \neq a_{j_1j_21}$ *has the following properties.*

- *For* $a_{j_1j_10}(\overline{x}_{j_2} - a_{j_1j_21})^{n_1} > 0$ *and* $a_{j_2j_20} > 0$,

$$(a_{j_1j_11}, \overline{x}_{j_2}) = \underbrace{((2m_{11}+1)^{\text{th}}\text{SO},[2(m_{11}-m_{21})+1]^{\text{th}}\text{DA})}_{((2m_{11}+1),[2(m_{11}-m_{21})+1])\text{-down-asymptote source}}. \quad (1.96)$$

The infinite-equilibrium is a $((2m_{11}+1)^{\text{th}}\text{SO},[2(m_{11}-m_{21})+1]^{\text{th}}\text{DA})$*-down-asymptote source at* $x_{j_1}^* = a_{j_1j_11} = a_{j_2j_11}$.

- *For* $a_{j_1j_10}(\overline{x}_{j_2} - a_{j_1j_21})^{n_1} < 0$ *and* $a_{j_2j_20} > 0$,

$$(a_{j_1j_11}, \overline{x}_{j_2}) = \underbrace{((2m_{11}+1)^{\text{th}}\text{SI},[2(m_{11}-m_{21})+1]^{\text{th}}\text{UA})}_{((2m_{11}+1),[2(m_{11}-m_{21})+1])\text{-up-asymptote sink}}. \quad (1.97)$$

The infinite-equilibrium is a $((2m_{11}+1)^{\text{th}}\text{SI},[2(m_{11}-m_{21})+1]^{\text{th}}\text{UA})$*-up-asymptote sink at* $x_{j_1}^* = a_{j_1j_11} = a_{j_2j_11}$.

- *For* $a_{j_1j_10}(\overline{x}_{j_2} - a_{j_1j_21})^{n_1} > 0$ *and* $a_{j_2j_20} < 0$,

$$(a_{j_1j_11}, \overline{x}_{j_2}) = \underbrace{((2m_{11}+1)^{\text{th}}\text{SO},[2(m_{11}-m_{21})+1]^{\text{th}}\text{UA})}_{((2m_{11}+1),[2(m_{11}-m_{21})+1])\text{-up-asymptote source}}. \quad (1.98)$$

The infinite-equilibrium is a $((2m_{11}+1)^{\text{th}}\text{SO},[2(m_{11}-m_{21})+1]^{\text{th}}\text{UA})$*-up-asymptote source at* $x_{j_1}^* = a_{j_1j_11} = a_{j_2j_11}$.

- *For* $a_{j_1j_10}(\overline{x}_{j_2} - a_{j_1j_21})^{n_1} < 0$ *and* $a_{j_2j_20} < 0$,

$$(a_{j_1j_11}, \overline{x}_{j_2}) = \underbrace{((2m_{11}+1)^{\text{th}}\text{SI},[2(m_{11}-m_{21})+1]^{\text{th}}\text{DA})}_{((2m_{11}+1),[2(m_{11}-m_{21})+1])\text{-down-asymptote sink}}. \quad (1.99)$$

The infinite-equilibrium is a $((2m_{11}+1)^{\text{th}}\text{SI},[2(m_{11}-m_{21})+1]^{\text{th}}\text{DA})$*-down-asymptote lower-saddle at* $x_{j_1}^{*}=a_{j_1j_11}=a_{j_2j_11}$.

(i_{2b1d}) *For* $m_1=2m_{11}$ *and* $m_2=2m_{21}$*, the infinite-equilibrium of* $x_{j_1}^{*}=a_{j_1j_11}=a_{j_2j_11}$ *with* $\overline{x}_{j_1}\neq a_{j_1j_11}$ *has the following properties.*

- *For* $a_{j_1j_10}(\overline{x}_{j_2}-a_{j_1j_21})^{n_1}>0$ *and* $a_{j_2j_20}>0$,

$$(a_{j_1j_11},\overline{x}_{j_2})=\underbrace{((2m_{11})^{\text{th}}\text{US},[2(m_{11}-m_{21})]^{\text{th}}\text{IC})}_{((2m_{11}),[2(m_{11}-m_{21})])\text{-increasing-concave upper-saddle}}. \tag{1.100}$$

The infinite-equilibrium is a $((2m_{11})^{\text{th}}\text{US},[2(m_{11}-m_{21})]^{\text{th}}\text{IC})$*-increasing-concave upper-saddle at* $x_{j_1}^{*}=a_{j_1j_11}=a_{j_2j_11}$.

- *For* $a_{j_1j_10}(\overline{x}_{j_2}-a_{j_1j_21})^{n_1}<0$ *and* $a_{j_2j_20}>0$,

$$(a_{j_1j_11},\overline{x}_{j_2})=\underbrace{((2m_{11})^{\text{th}}\text{LS},[2(m_{11}-m_{21})]^{\text{th}}\text{DC})}_{((2m_{11}),[2(m_{11}-m_{21})])\text{-decreasing-concave lower-saddle}}. \tag{1.101}$$

The infinite-equilibrium is a $((2m_{11})^{\text{th}}\text{LS},[2(m_{11}-m_{21})]^{\text{th}}\text{DC})$*-decreasing-concave lower-saddle at* $x_{j_1}^{*}=a_{j_1j_11}=a_{j_2j_11}$.

- *For* $a_{j_1j_10}(\overline{x}_{j_2}-a_{j_1j_21})^{n_1}>0$ *and* $a_{j_2j_20}<0$,

$$(a_{j_1j_11},\overline{x}_{j_2})=\underbrace{((2m_{11})^{\text{th}}\text{US},[2(m_{11}-m_{21})]^{\text{th}}\text{DC})}_{((2m_{11}),[2(m_{11}-m_{21})])\text{-decreasing-concave upper-saddle}}. \tag{1.102}$$

The infinite-equilibrium is a $((2m_{11})^{\text{th}}\text{US},[2(m_{11}-m_{21})]^{\text{th}}\text{DC})$*-decreasing-concave upper-saddle at* $x_{j_1}^{*}=a_{j_1j_11}=a_{j_2j_11}$.

- *For* $a_{j_1j_10}(\overline{x}_{j_2}-a_{j_1j_21})^{n_1}<0$ *and* $a_{j_2j_20}<0$,

$$(a_{j_1j_11},\overline{x}_{j_2})=\underbrace{((2m_{11})^{\text{th}}\text{LS},[2(m_{11}-m_{21})]^{\text{th}}\text{IC})}_{((2m_{11}),[2(m_{11}-m_{21})])\text{-increasing-concave lower-saddle}}. \tag{1.103}$$

The infinite-equilibrium is a $((2m_{11})^{\text{th}}\text{LS},[2(m_{11}-m_{21})]^{\text{th}}\text{IC})$*-increasing-concave lower-saddle at* $x_{j_1}^{*}=a_{j_1j_11}=a_{j_2j_11}$.

(i_{2b2}) *There are 32 types of equilibrium* $(x_{j_2}^{*},x_{j_1}^{*})=(a_{j_1j_21},a_{j_2j_11})$ *on the infinite-equilibrium of* $x_{j_1}^{*}=a_{j_1j_11}=a_{j_2j_11}$.

(i_{2b2a}) *For* $m_1=2m_{11}+1$, $n_1=2n_{11}+1$*, and* $m_2=2m_{21}+1$*, the equilibriums of* $(x_{j_2}^{*},x_{j_1}^{*})=(a_{j_1j_21},a_{j_2j_11})$ *on the equilibrium of* $x_{j_1}^{*}=a_{j_1j_11}=a_{j_2j_11}$ *have the following properties.*

- *For* $a_{j_1j_10}>0$ *and* $a_{j_2j_20}>0$,

$$\begin{aligned}&(a_{j_1j_2 1}, a_{j_2j_1 1})\\&=\underbrace{([(2m_{11}+1),(2n_{11}+1)]^{\text{th}}\text{UP}_{-}\text{:UP}_{+},[2(m_{11}-m_{21})]^{\text{th}}{}_{\text{DC:IC}}\text{US})}_{([(2m_{11}+1),(2n_{11}+1)],[2(m_{11}-m_{21})])\text{-concave up-parabola upper-saddle}}.\end{aligned} \tag{1.104}$$

The equilibrium is a $([(2m_{11}+1),(2n_{11}+1)]^{\text{th}}\text{UP}_{-}\text{:UP}_{+},[2(m_{11}-m_{21})]^{\text{th}}{}_{\text{DC:IC}}\text{US})$*-concave up-parabola upper-saddle at* $(x_{j_2}^*, x_{j_1}^*) = (a_{j_1j_2 1}, a_{j_2j_1 1})$.

- *For* $a_{j_1j_1 0} < 0$ *and* $a_{j_2j_2 0} > 0$,

$$\begin{aligned}&(a_{j_1j_2 1}, a_{j_2j_1 1})\\&=\underbrace{([(2m_{11}+1),(2n_{11}+1)]^{\text{th}}\text{DP}_{-}\text{:DP}_{+},[2(m_{11}-m_{21})]^{\text{th}}{}_{\text{IC:DC}}\text{LS})}_{([(2m_{11}+1),(2n_{11}+1)],[2(m_{11}-m_{21})])\text{-concave down-parabola lower-saddle}}.\end{aligned} \tag{1.105}$$

The equilibrium is a $([(2m_{11}+1),(2n_{11}+1)]^{\text{th}}\text{DP}_{-}\text{:DP}_{+},[2(m_{11}-m_{21})]^{\text{th}}{}_{\text{IC:DC}}\text{LS})$*-concave down-parabola lower-saddle at* $(x_{j_2}^*, x_{j_1}^*) = (a_{j_1j_2 1}, a_{j_2j_1 1})$.

- *For* $a_{j_1j_1 0} > 0$ *and* $a_{j_2j_2 0} < 0$,

$$\begin{aligned}&(a_{j_1j_2 1}, a_{j_2j_1 1})\\&=\underbrace{([(2m_{11}+1),(2n_{11}+1)]^{\text{th}}\text{DP}_{+}\text{:DP}_{+},[2(m_{11}-m_{21})]^{\text{th}}{}_{\text{IC:DC}}\text{US})}_{([(2m_{11}+1),(2n_{11}+1)],[2(m_{11}-m_{21})])\text{-concave down-parabola upper-saddle}}.\end{aligned} \tag{1.106}$$

The equilibrium is a $([(2m_{11}+1),(2n_{11}+1)]^{\text{th}}\text{DP}_{+}\text{:DP}_{+},[2(m_{11}-m_{21})]^{\text{th}}{}_{\text{IC:DC}}\text{US})$*-concave down-parabola upper-saddle at* $(x_{j_2}^*, x_{j_1}^*) = (a_{j_1j_2 1}, a_{j_2j_1 1})$.

- *For* $a_{j_1j_1 0} < 0$ *and* $a_{j_2j_2 0} < 0$,

$$\begin{aligned}&(a_{j_1j_2 1}, a_{j_2j_1 1})\\&=\underbrace{([(2m_{11}+1),(2n_{11}+1)]^{\text{th}}\text{UP}_{+}\text{:UP}_{-},[2(m_{11}-m_{21})]^{\text{th}}{}_{\text{DC:IC}}\text{LS})}_{([(2m_{11}+1),(2n_{11}+1)],[2(m_{11}-m_{21})])\text{-concave up-parabola lower-saddle}}.\end{aligned} \tag{1.107}$$

The equilibrium is a $([(2m_{11}+1),(2n_{11}+1)]^{\text{th}}\text{UP}_{+}\text{:UP}_{-},[2(m_{11}-m_{21})]^{\text{th}}{}_{\text{DC:IC}}\text{LS})$*-concave up-parabola lower-saddle at* $(x_{j_2}^*, x_{j_1}^*)=(a_{j_1j_2 1}, a_{j_2j_1 1})$.

(i_{2b2b}) For $m_1 = 2m_{11}, n_1 = 2n_{11} + 1$ *and* $m_2 = 2m_{21} + 1$, *the equilibriums of* $(x_{j_2}^*, x_{j_1}^*) = (a_{j_1j_2 1}, a_{j_2j_1 1})$ *on the infinite-equilibrium of* $x_{j_1}^* = a_{j_1j_1 1} = a_{j_2j_1 1}$ *have the following properties.*

- *For* $a_{j_1j_10} > 0$ *and* $a_{j_2j_20} > 0$,

$$(a_{j_1j_11}, a_{j_2j_21}) \\ = \underbrace{([(2m_{11}),(2n_{11}+1)]^{\text{th}}\text{DP}_-:\text{UP}_+,[2(m_{11}-m_{21})-1]^{\text{th}}{}_{\text{UA}}\text{LS:}_{\text{DA}}\text{US})}_{([(2m_{11}),(2n_{11}+1)],[2(m_{11}-m_{21})-1])\text{-hyperbolic asymptote lower-to-upper saddle}}. \tag{1.108}$$

The equilibrium is a $([(2m_{11}),(2n_{11}+1)]^{\text{th}}\text{DP}_-:\text{UP}_+,[2(m_{11}-m_{21})-1]^{\text{th}}{}_{\text{UA}}\text{LS:}_{\text{DA}}\text{US})$*-hyperbolic asymptote lower-to-upper saddle at* $(x^*_{j_2}, x^*_{j_1}) = (a_{j_1j_21}, a_{j_2j_11})$.

- *For* $a_{j_1j_10} < 0$ *and* $a_{j_2j_20} > 0$,

$$(a_{j_1j_11}, a_{j_2j_21}) \\ = \underbrace{([(2m_{11}),(2n_{11}+1)]^{\text{th}}\text{UP}_-:\text{DP}_+,[2(m_{11}-m_{21})-1]^{\text{th}}{}_{\text{DA}}\text{US:}_{\text{UA}}\text{LS})}_{([(2m_{11}),(2n_{11}+1)],[2(m_{11}-m_{21})-1])\text{-hyperbolic-secant asymptote upper-to-lower saddle}}. \tag{1.109}$$

The equilibrium is a $([(2m_{11}),(2n_{11}+1)]^{\text{th}}\text{UP}_-:\text{DP}_+,[2(m_{11}-m_{21})-1]^{\text{th}}{}_{\text{DA}}\text{US:}_{\text{UA}}\text{LS})$*-hyperbolic-secant asymptote upper-to-lower saddle at* $(x^*_{j_2}, x^*_{j_1}) = (a_{j_1j_21}, a_{j_2j_11})$.

- *For* $a_{j_1j_10} > 0$ *and* $a_{j_2j_20} < 0$,

$$(a_{j_1j_11}, a_{j_2j_21}) \\ = \underbrace{([(2m_{11}),(2n_{11}+1)]^{\text{th}}\text{UP}_+:\text{DP}_-,[2(m_{11}-m_{21})-1]^{\text{th}}{}_{\text{DA}}\text{LS:}_{\text{UA}}\text{US})}_{([(2m_{11}),(2n_{11}+1)],[2(m_{11}-m_{21})-1])\text{-hyperbolic-secant asymptote lower-to-upper saddle}}. \tag{1.110}$$

The equilibrium is a $([(2m_{11}),(2n_{11}+1)]^{\text{th}}\text{UP}_+:\text{DP}_-,[2(m_{11}-m_{21})-1]^{\text{th}}{}_{\text{DA}}\text{LS:}_{\text{UA}}\text{US})$*-hyperbolic-secant asymptote lower-to-upper saddle at* $(x^*_{j_2}, x^*_{j_1}) = (a_{j_1j_21}, a_{j_2j_11})$.

- *For* $a_{j_1j_10} < 0$ *and* $a_{j_2j_20} < 0$,

$$(a_{j_1j_11}, a_{j_2j_21}) \\ = \underbrace{([(2m_{11}),(2n_{11}+1)]^{\text{th}}\text{DP}_+:\text{UP}_-,[2(m_{11}-m_{21})-1]^{\text{th}}{}_{\text{UA}}\text{US:}_{\text{DA}}\text{LS})}_{([(2m_{11}),(2n_{11}+1)],[2(m_{11}-m_{21})-1])\text{-hyperbolic asymptote upper-to-lower saddle}}. \tag{1.111}$$

The equilibrium is a $([(2m_{11}),(2n_{11}+1)]^{\text{th}}\text{DP}_+:\text{UP}_-,[2(m_{11}-m_{21})-1]^{\text{th}}{}_{\text{UA}}\text{US:}_{\text{DA}}\text{LS})$*-hyperbolic asymptote upper-to-lower saddle at* $(x^*_{j_2}, x^*_{j_1}) = (a_{j_1j_21}, a_{j_2j_11})$.

(i_{2b2c}) For $m_1 = 2m_{11} + 1$, $n_1 = 2n_{11} + 1$, and $m_2 = 2m_{21}$, the equilibriums of $(x^*_{j_2}, x^*_{j_1}) = (a_{j_1j_2 1}, a_{j_2j_1 1})$ on the infinite-equilibrium of $x^*_{j_1} = a_{j_1j_1 1} = a_{j_2j_1 1}$ have the following properties.

- *For $a_{j_1j_1 0} > 0$ and $a_{j_2j_2 0} > 0$,*

$$(a_{j_1j_1 1}, a_{j_2j_2 1}) = \underbrace{([(2m_{11}+1),(2n_{11}+1)]^{\text{th}}\text{DP}_+\text{:UP}_+,[2(m_{11}-m_{21})+1]^{\text{th}}{}_{\text{UA}}\text{SI:}{}_{\text{DA}}\text{SO})}_{([(2m_{11}+1),(2n_{11}+1)],[2(m_{11}-m_{21})+1])\text{-hyperbolic asymptote sink-to-source}}. \tag{1.112}$$

The equilibrium is a $([(2m_{11}+1),(2n_{11}+1)]^{\text{th}}\text{DP}_+\text{:UP}_+,[2(m_{11}-m_{21})+1]^{\text{th}}{}_{\text{UA}}\text{SI:}{}_{\text{DA}}\text{SO})$*-hyperbolic-asymptote sink-to-source at $(x^*_{j_2}, x^*_{j_1}) = (a_{j_1j_2 1}, a_{j_2j_1 1})$.*

- *For $a_{j_1j_1 0} < 0$ and $a_{j_2j_2 0} > 0$,*

$$(a_{j_1j_1 1}, a_{j_2j_2 1}) = \underbrace{([(2m_{11}+1),(2n_{11}+1)]^{\text{th}}\text{UP}_+\text{:DP}_+,[2(m_{11}-m_{21})+1]^{\text{th}}{}_{\text{DA}}\text{SO:}{}_{\text{UA}}\text{SI})}_{([(2m_{11}+1),(2n_{11}+1)],[2(m_{11}-m_{21})+1])\text{-hyperbolic-secant asymptote source-to-sink}}. \tag{1.113}$$

The equilibrium is a $([(2m_{11}+1),(2n_{11}+1)]^{\text{th}}\text{UP}_+\text{:DP}_+,[2(m_{11}-m_{21})+1]^{\text{th}}{}_{\text{DA}}\text{SO:}{}_{\text{UA}}\text{SI})$*-hyperbolic-secant asymptote source-to-sink at $(x^*_{j_2}, x^*_{j_1}) = (a_{j_1j_2 1}, a_{j_2j_1 1})$.*

- *For $a_{j_1j_1 0} > 0$ and $a_{j_2j_2 0} < 0$,*

$$(a_{j_1j_1 1}, a_{j_2j_2 1}) = \underbrace{([(2m_{11}+1),(2n_{11}+1)]^{\text{th}}\text{UP}_-\text{:DP}_-,[2(m_{11}-m_{21})+1]^{\text{th}}{}_{\text{DA}}\text{SI:}{}_{\text{UA}}\text{SO})}_{([(2m_{11}+1),(2n_{11}+1)],[2(m_{11}-m_{21})+1])\text{-hyperbolic-secant asymptote sink-to-source}}. \tag{1.114}$$

The equilibrium is a $([(2m_{11}+1),(2n_{11}+1)]^{\text{th}}\text{UP}_-\text{:DP}_-,[2(m_{11}-m_{21})+1]^{\text{th}}{}_{\text{DA}}\text{SI:}{}_{\text{UA}}\text{SO})$*-hyperbolic-secant asymptote sink-to-source at $(x^*_{j_2}, x^*_{j_1}) = (a_{j_1j_2 1}, a_{j_2j_1 1})$.*

- *For $a_{j_1j_1 0} < 0$ and $a_{j_2j_2 0} < 0$,*

$$(a_{j_1j_1 1}, a_{j_2j_2 1}) = \underbrace{([(2m_{11}+1),(2n_{11}+1)]^{\text{th}}\text{DP}_-\text{:UP}_-,[2(m_{11}-m_{21})+1]^{\text{th}}{}_{\text{UA}}\text{SO:}{}_{\text{DA}}\text{SI})}_{([(2m_{11}+1),(2n_{11}+1)],[2(m_{11}-m_{21})+1])\text{-hyperbolic asymptote source-to-sink}}. \tag{1.115}$$

The equilibrium is a $([(2m_{11}+1),(2n_{11}+1)]^{\text{th}}\text{DP}_-\text{:UP}_-[2(m_{11}-m_{21})+1]^{\text{th}}{}_{\text{UA}}\text{SO:}{}_{\text{DA}}\text{SI})$ *hyperbolic-asymptote lower-saddle at $(x^*_{j_2}, x^*_{j_1}) = (a_{j_1j_2 1}, a_{j_2j_1 1})$.*

(i_{2b2d}) For $m_1 = 2m_{11}$, $n_1 = 2n_{11}+1$ and $m_2 = 2m_{21}$, the equilibriums of $(x^*_{j_2}, x^*_{j_1}) = (a_{j_1j_21}, a_{j_2j_11})$ on the infinite-equilibrium of $x^*_{j_2} = a_{j_2j_21} = a_{j_1j_21}$ have the following properties.

- *For $a_{j_1j_10} > 0$ and $a_{j_2j_20} > 0$,*

$$(a_{j_1j_21}, a_{j_2j_11}) = \underbrace{([(2m_{11}),(2n_{11}+1)]^{\text{th}}\text{UP}_+\text{:UP}_+,[2(m_{11}-m_{21})]^{\text{th}}{}_{\text{DC:IC}}\text{US})}_{([(2m_{11}),(2n_{11}+1)],[2(m_{11}-m_{21})])\text{-positve concave up-parabola upper-saddle}}. \tag{1.116}$$

The equilibrium is a $([(2m_{11}),(2n_{11}+1)]^{\text{th}}\text{UP}_+\text{:UP}_+,[2(m_{11}-m_{21})]^{\text{th}}{}_{\text{DC:IC}}\text{US})$-positive-concave up-parabola upper-saddle at $(x^*_{j_2}, x^*_{j_1}) = (a_{j_1j_21}, a_{j_2j_11})$.

- *For $a_{j_1j_10} < 0$ and $a_{j_2j_20} > 0$,*

$$(a_{j_1j_21}, a_{j_2j_11}) = \underbrace{([(2m_{11}),(2n_{11}+1)]^{\text{th}}\text{DP}_+\text{:DP}_+,[2(m_{11}-m_{21})]^{\text{th}}{}_{\text{IC:DC}}\text{US})}_{([(2m_{11}),(2n_{11}+1)],[2(m_{11}-m_{21})])\text{-positve concave down-parabola upper-saddle}}. \tag{1.117}$$

The equilibrium is a $([(2m_{11}),(2n_{11}+1)]^{\text{th}}\text{DP}_+\text{:DP}_+,[2(m_{11}-m_{21})]^{\text{th}}{}_{\text{IC:DC}}\text{US})$-positive-concave down-parabola upper-saddle at $(x^*_{j_2}, x^*_{j_1}) = (a_{j_1j_21}, a_{j_2j_11})$.

- *For $a_{j_1j_10} > 0$ and $a_{j_2j_20} < 0$,*

$$(a_{j_1j_21}, a_{j_2j_11}) = \underbrace{([(2m_{11}),(2n_{11}+1)]^{\text{th}}\text{DP}_-\text{:DP}_-,[2(m_{11}-m_{21})]^{\text{th}}{}_{\text{IC:DC}}\text{LS})}_{([(2m_{11}),(2n_{11}+1)],[2(m_{11}-m_{21})])\text{-negative concave down-parabola lower-saddle}}. \tag{1.118}$$

The equilibrium is a $([(2m_{11}),(2n_{11}+1)]^{\text{th}}\text{DP}_-\text{:DP}_-,[2(m_{11}-m_{21})]^{\text{th}}{}_{\text{IC:DC}}\text{LS})$-negative-concave down-parabola lower-saddle at $(x^*_{j_2}, x^*_{j_1}) = (a_{j_1j_21}, a_{j_2j_11})$.

- *For $a_{j_1j_10} < 0$ and $a_{j_2j_20} < 0$,*

$$(a_{j_1j_21}, a_{j_2j_11}) = \underbrace{([(2m_{11}),(2n_{11}+1)]^{\text{th}}\text{UP}_-\text{:UP}_-,[2(m_{11}-m_{21})]^{\text{th}}{}_{\text{DC:IC}}\text{LS})}_{([(2m_{11}),(2n_{11}+1)],[2(m_{11}-m_{21})])\text{-negative concave up-parabola lower-saddle}}. \tag{1.119}$$

The equilibrium is a $([(2m_{11}),(2n_{11}+1)]^{\text{th}}\text{UP}_-\text{:UP}_-,[2(m_{11}-m_{21})]^{\text{th}}{}_{\text{DC:IC}}\text{LS})$-negative-concave up-parabola lower-saddle at $(x^*_{j_2}, x^*_{j_1}) = (a_{j_1j_21}, a_{j_2j_11})$.

(i_{2b2e}) For $m_1 = 2m_{11}+1$, $n_1 = 2n_{11}$ and $m_2 = 2m_{21}+1$, the equilibrium of $(x_{j_2}^*, x_{j_1}^*) = (a_{j_1j_21}, a_{j_2j_11})$ on the infinite-equilibrium of $x_{j_1}^* = a_{j_1j_11} = a_{j_2j_11}$ have the following properties.

- *For $a_{j_1j_10} > 0$ and $a_{j_2j_20} > 0$,*

$$(a_{j_1j_21}, a_{j_2j_11}) = \underbrace{([(2m_{11}+1),(2n_{11})]^{\text{th}}\text{II}_{-}\text{:II}_{+},[2(m_{11}-m_{21})]^{\text{th}}{}_{\text{IC:IC}}\text{SO})}_{([(2m_{11}+1),(2n_{11})],[2(m_{11}-m_{21})])\text{-increasing-concave inflection-source}}. \quad (1.120)$$

The equilibrium is a $([(2m_{11}+1),(2n_{11})]^{\text{th}}\text{II}_{-}\text{:II}_{+},[2(m_{11}-m_{21})]^{\text{th}}{}_{\text{IC:IC}}\text{SO})$-increasing-concave inflection-source at $(x_{j_2}^*, x_{j_1}^*) = (a_{j_1j_21}, a_{j_2j_11})$.

- *For $a_{j_1j_10} < 0$ and $a_{j_2j_20} > 0$,*

$$(a_{j_1j_21}, a_{j_2j_11}) = \underbrace{([(2m_{11}+1),(2n_{11})]^{\text{th}}\text{DI}_{-}\text{:DI}_{+},[2(m_{11}-m_{21})]^{\text{th}}{}_{\text{DC:DC}}\text{SI})}_{([(2m_{11}+1),(2n_{11})],[2(m_{11}-m_{21})])\text{-decreasing-concave inflection-sink}}. \quad (1.121)$$

The equilibrium is a $([(2m_{11}+1),(2n_{11})]^{\text{th}}\text{DI}_{-}\text{:DI}_{+},[2(m_{11}-m_{21})]^{\text{th}}{}_{\text{DC:DC}}\text{SI})$-decreasing-concave inflection-sink at $(x_{j_2}^*, x_{j_1}^*) = (a_{j_1j_21}, a_{j_2j_11})$.

- *For $a_{j_1j_10} > 0$ and $a_{j_2j_20} < 0$,*

$$(a_{j_1j_21}, a_{j_2j_11}) = \underbrace{([(2m_{11}+1),(2n_{11})]^{\text{th}}\text{DI}_{+}\text{:DI}_{-},[2(m_{11}-m_{21})]^{\text{th}}{}_{\text{DC:DC}}\text{SO})}_{([(2m_{11}+1),(2n_{11})],[2(m_{11}-m_{21})])\text{-decreasing-concave inflection-source}}. \quad (1.122)$$

The equilibrium is a $([(2m_{11}+1),(2n_{11})]^{\text{th}}\text{DI}_{+}\text{:DI}_{-},[2(m_{11}-m_{21})]^{\text{th}}{}_{\text{DC:DC}}\text{SO})$-decreasing concave inflection-source at $(x_{j_2}^*, x_{j_1}^*) = (a_{j_1j_21}, a_{j_2j_11})$.

- *For $a_{j_1j_10} < 0$ and $a_{j_2j_20} < 0$,*

$$(a_{j_1j_21}, a_{j_2j_11}) = \underbrace{([(2m_{11}+1),(2n_{11})]^{\text{th}}\text{II}_{+}\text{:II}_{-},[2(m_{11}-m_{21})]^{\text{th}}{}_{\text{IC:IC}}\text{SI})}_{([(2m_{11}+1),(2n_{11})],[2(m_{11}-m_{21})])\text{-increasing-concave inflection-sink}}. \quad (1.123)$$

The equilibrium is a $([(2m_{11}+1),(2n_{11})]^{\text{th}}\text{II}_{+}\text{:II}_{-},[2(m_{11}-m_{21})]^{\text{th}}{}_{\text{IC:IC}}\text{SI})$-increasing-concave inflection-sink at $(x_{j_2}^*, x_{j_1}^*) = (a_{j_1j_21}, a_{j_2j_11})$.

(i_{2b2f}) For $m_1 = 2m_{11}, n_1 = 2n_{11}$, *and* $m_2 = 2m_{21}+1$, *the equilibrium of* $(x^*_{j_2}, x^*_{j_1}) = (a_{j_1j_2 1}, a_{j_2j_1 1})$ *on the infinite-equilibrium of* $x^*_{j_1} = a_{j_1j_1 1} = a_{j_2j_1 1}$ *have the following properties.*

- *For* $a_{j_1j_1 0} > 0$ *and* $a_{j_2j_2 0} > 0$,

$$(a_{j_1j_2 1}, a_{j_2j_1 1}) = \underbrace{([(2m_{11}),(2n_{11})]^{\text{th}}\text{DI}_{-}\text{:II}_{+},[2(m_{11}-m_{21})-1]^{\text{th}}{}_{\text{DA:DA}}\text{US})}_{([(2m_{11}),(2n_{11})],[2(m_{11}-m_{21})-1])\text{-inflection down-asymptote upper-saddle}}. \quad (1.124)$$

The equilibrium is a $([(2m_{11}),(2n_{11})]^{\text{th}}\text{DI}_{-}\text{:II}_{+},[2(m_{11}-m_{21})-1]^{\text{th}}{}_{\text{DA:DA}}\text{US})$-*inflection down-asymptote upper-saddle at* $(x^*_{j_1}, x^*_{j_2}) = (a_{j_1j_1 1}, a_{j_1j_2 1})$.

- *For* $a_{j_1j_1 0} < 0$ *and* $a_{j_2j_2 0} > 0$,

$$(a_{j_1j_2 1}, a_{j_2j_1 1}) = \underbrace{([(2m_{11}),(2n_{11})]^{\text{th}}\text{II}_{-}\text{:DI}_{+},[2(m_{11}-m_{21})-1]^{\text{th}}{}_{\text{UA:UA}}\text{LS})}_{([(2m_{11}),(2n_{11})],[2(m_{11}-m_{21})-1])\text{-inflection up-asymptote lower-saddle}}. \quad (1.125)$$

The equilibrium is a $([(2m_{11}),(2n_{11})]^{\text{th}}\text{II}_{-}\text{:DI}_{+},[2(m_{11}-m_{21})-1]^{\text{th}}{}_{\text{DA:DA}}\text{LS})$-*inflection up-asymptote lower-saddle at* $(x^*_{j_2}, x^*_{j_1}) = (a_{j_1j_2 1}, a_{j_2j_1 1})$.

- *For* $a_{j_1j_1 0} > 0$ *and* $a_{j_2j_2 0} < 0$,

$$(a_{j_1j_2 1}, a_{j_2j_1 1}) = \underbrace{([(2m_{11}),(2n_{11})]^{\text{th}}\text{II}_{+}\text{:DI}_{-},[2(m_{11}-m_{21})-1]^{\text{th}}{}_{\text{UA:UA}}\text{US})}_{([(2m_{11}),(2n_{11})],[2(m_{11}-m_{21})-1])\text{-inflection up-asymptote upper-saddle}}. \quad (1.126)$$

The equilibrium is a $([(2m_{11}),(2n_{11})]^{\text{th}}\text{II}_{+}\text{:DI}_{-},[2(m_{11}-m_{21})-1]^{\text{th}}{}_{\text{UA:UA}}\text{US})$-*inflection up-asymptote upper-saddle at* $(x^*_{j_2}, x^*_{j_1}) = (a_{j_1j_2 1}, a_{j_2j_1 1})$.

- *For* $a_{j_1j_1 0} < 0$ *and* $a_{j_2j_2 0} < 0$,

$$(a_{j_1j_2 1}, a_{j_2j_1 1}) = \underbrace{([(2m_{11}),(2n_{11})]^{\text{th}}\text{DI}_{+}\text{:II}_{-},[2(m_{11}-m_{21})-1]^{\text{th}}{}_{\text{DA:DA}}\text{LS})}_{([(2m_{11}),(2n_{11})],[2(m_{11}-m_{21})-1])\text{-inflection down-asymptote lower-saddle}}. \quad (1.127)$$

The equilibrium is a $([(2m_{11}),(2n_{11})]^{\text{th}}\text{DI}_{+}\text{:II}_{-},[2(m_{11}-m_{21})-1]^{\text{th}}{}_{\text{DA:DA}}\text{LS})$-*inflection down-asymptote lower-saddle at* $(x^*_{j_2}, x^*_{j_1}) = (a_{j_1j_2 1}, a_{j_2j_1 1})$.

(i_{2b2g}) For $m_1 = 2m_{11} + 1$, $n_1 = 2n_{11}$ and $m_2 = 2m_{21}$, the equilibrium $(x^*_{j_2}, x^*_{j_1}) = (a_{j_1j_21}, a_{j_2j_11})$ on the infinite-equilibrium of $x^*_{j_1} = a_{j_1j_11} = a_{j_2j_11}$ have the following properties.

- *For $a_{j_1j_10} > 0$ and $a_{j_2j_20} > 0$,*

$$(a_{j_1j_21}, a_{j_2j_11}) = \underbrace{([(2m_{11}+1),(2n_{11})]^{\text{th}}\text{DI}_+\text{:II}_+,[2(m_{11}-m_{21})+1]^{\text{th}}{}_{\text{DA:DA}}\text{SO})}_{([(2m_{11}+1),(2n_{11})],[2(m_{11}-m_{21})+1])\text{-inflection down-asymptote source}}. \tag{1.128}$$

The equilibrium is a $([(2m_{11}+1),(2n_{11})]^{\text{th}}\text{DI}_+\text{:II}_+,[2(m_{11}-m_{21})+1]^{\text{th}}{}_{\text{DA:DA}}\text{SO})$*-inflection down-asymptote source at* $(x^*_{j_2}, x^*_{j_1}) = (a_{j_1j_21}, a_{j_2j_11})$.

- *For $a_{j_1j_10} < 0$ and $a_{j_2j_20} > 0$,*

$$(a_{j_1j_21}, a_{j_2j_11}) = \underbrace{([(2m_{11}+1),(2n_{11})]^{\text{th}}\text{II}_+\text{:DI}_+,[2(m_{11}-m_{21})+1]^{\text{th}}{}_{\text{UA:UA}}\text{SI})}_{([(2m_{11}+1),(2n_{11})],[2(m_{11}-m_{21})+1])\text{-inflection up-asymptote sink}}. \tag{1.129}$$

The equilibrium is a $([(2m_{11}+1),(2n_{11})]^{\text{th}}\text{II}_+\text{:DI}_+,[2(m_{11}-m_{21})+1]^{\text{th}}{}_{\text{UA:UA}}\text{SI})$*-inflection up-asymptote sink at* $(x^*_{j_2}, x^*_{j_1}) = (a_{j_1j_21}, a_{j_2j_11})$.

- *For $a_{j_1j_10} > 0$ and $a_{j_2j_20} < 0$,*

$$(a_{j_1j_21}, a_{j_2j_11}) = \underbrace{([(2m_{11}+1),(2n_{11})]^{\text{th}}\text{II}_-\text{:DI}_-,[2(m_{11}-m_{21})+1]^{\text{th}}{}_{\text{UA:UA}}\text{SO})}_{([(2m_{11}+1),(2n_{11})],[2(m_{11}-m_{21})+1])\text{-inflection up-asymptote source}}. \tag{1.130}$$

The equilibrium is a $([(2m_{11}+1),(2n_{11})]^{\text{th}}\text{II}_-\text{:DI}_-,[2(m_{11}-m_{21})+1]^{\text{th}}{}_{\text{UA:UA}}\text{SO})$*-inflection up-asymptote source at* $(x^*_{j_2}, x^*_{j_1}) = (a_{j_1j_21}, a_{j_2j_11})$.

- *For $a_{j_1j_10} < 0$ and $a_{j_2j_20} < 0$,*

$$(a_{j_1j_21}, a_{j_2j_11}) = \underbrace{([(2m_{11}+1),(2n_{11})]^{\text{th}}\text{DI}_-\text{:II}_-,[2(m_{11}-m_{21})+1]^{\text{th}}{}_{\text{DA:DA}}\text{SI})}_{([(2m_{11}+1),(2n_{11})],[2(m_{11}-m_{21})+1])\text{-inflection down-asymptote sink}}. \tag{1.131}$$

The equilibrium is a $([(2m_{11}+1),(2n_{11})]^{\text{th}}\text{DI}_-\text{:II}_-,[2(m_{11}-m_{21})+1]^{\text{th}}{}_{\text{DA:DA}}\text{SI})$*-inflection down-asymptote sink at* $(x^*_{j_2}, x^*_{j_1}) = (a_{j_1j_21}, a_{j_2j_11})$.

(i_{2b2h}) For $m_1 = 2m_{11}$, $n_l = 2n_{11}$ and $m_2 = 2m_{21}$, the equilibrium of $(x^*_{j_2}, x^*_{j_1}) = (a_{j_1j_21}, a_{j_2j_11})$ on the infinite-equilibrium of $x^*_{j_1} = a_{j_1j_11} = a_{j_2j_11}$ have the following properties.

- *For $a_{j_1j_10} > 0$ and $a_{j_2j_20} > 0$,*

$$(a_{j_1j_21}, a_{j_2j_11}) = \underbrace{([(2m_{11}),(2n_{11})]^{\text{th}}\text{II}_{+}\text{:II}_{+},[2(m_{11}-m_{21})]^{\text{th}}{}_{\text{IC:IC}}\text{US})}_{([(2m_{11}),(2n_{11})],[2(m_{11}-m_{21})])\text{-increasing-concave inflection-upper-saddle}}. \tag{1.132}$$

The equilibrium is a $([(2m_{11}),(2n_{11})]^{\text{th}}\text{II}_{+}\text{:II}_{+},[2(m_{11}-m_{21})]^{\text{th}}{}_{\text{IC:IC}}\text{US})$-increasing-concave inflection-upper-saddle at $(x^*_{j_2}, x^*_{j_1}) = (a_{j_1j_21}, a_{j_2j_11})$.

- *For $a_{j_1j_10} < 0$ and $a_{j_2j_20} > 0$,*

$$(a_{j_1j_21}, a_{j_2j_11}) = \underbrace{([(2m_{11}),(2n_{11})]^{\text{th}}\text{DI}_{+}\text{:DI}_{+},[2(m_{11}-m_{21})]^{\text{th}}{}_{\text{DC:DC}}\text{LS})}_{([(2m_{11}),(2n_{11})],[2(m_{11}-m_{21})])\text{-decreasing-concave inflection-lower-saddle}}. \tag{1.133}$$

The equilibrium is a $([(2m_{11}),(2n_{11})]^{\text{th}}\text{DI}_{+}\text{:DI}_{+},[2(m_{11}-m_{21})]^{\text{th}}{}_{\text{DC:DC}}\text{LS})$-decreasing-concave inflection-lower-saddle at $(x^*_{j_2}, x^*_{j_1}) = (a_{j_1j_21}, a_{j_2j_11})$.

- *For $a_{j_1j_10} > 0$ and $a_{j_2j_20} < 0$,*

$$(a_{j_1j_21}, a_{j_2j_11}) = \underbrace{([(2m_{11}),(2n_{11})]^{\text{th}}\text{DI}_{-}\text{:DI}_{-},[2(m_{11}-m_{21})]^{\text{th}}{}_{\text{DC:DC}}\text{US})}_{([(2m_{11}),(2n_{11})],[2(m_{11}-m_{21})])\text{-decreasing-concave inflection-upper-saddle}}. \tag{1.134}$$

The equilibrium is a $([(2m_{11}),(2n_{11})]^{\text{th}}\text{DI}_{-}\text{:DI}_{-},[2(m_{11}-m_{21})]^{\text{th}}{}_{\text{DC:DC}}\text{US})$-decreasing-concave inflection-lower-saddle at $(x^*_{j_2}, x^*_{j_1}) = (a_{j_1j_21}, a_{j_2j_11})$.

- *For $a_{j_1j_10} < 0$ and $a_{j_2j_20} < 0$,*

$$(a_{j_1j_21}, a_{j_2j_11}) = \underbrace{([(2m_{11}),(2n_{11})]^{\text{th}}\text{II}_{-}\text{:II}_{-},[2(m_{11}-m_{21})]^{\text{th}}{}_{\text{IC:IC}}\text{LS})}_{([(2m_{11}),(2n_{11})],[2(m_{11}-m_{21})])\text{-increasing-concave inflection-lower-saddle}}. \tag{1.135}$$

The equilibrium is a $([(2m_{11}),(2n_{11})]^{\text{th}}\text{II}_{-}\text{:II}_{-},[2(m_{11}-m_{21})]^{\text{th}}{}_{\text{IC:IC}}\text{LS})$-increasing-concave inflection-lower-saddle at $(x^*_{j_2}, x^*_{j_1}) = (a_{j_1j_21}, a_{j_2j_11})$.

(ii) Consider

$$
\begin{aligned}
\dot{x}_{j_2} &= a_{j_2 j_2 0} \prod_{v=1}^{p_2} (x_{j_1} - a_{j_2 j_1 v})^{m_{j_2 v}} \\
\dot{x}_{j_1} &= a_{j_1 j_1 0} \prod_{r=1}^{p_1} (x_{j_1} - a_{j_1 j_1 r})^{m_{j_1 r}} \prod_{u=1}^{q_1} (x_{j_2} - a_{j_1 j_2 u})^{n_{j_1 u}},
\end{aligned} \tag{1.136}
$$

where

$$
\sum_{r=1}^{p_1} m_{j_1 r} = m_1, \sum_{u=1}^{q_1} n_{j_2 u} = n_1; \sum_{v=1}^{p_2} m_{j_2 v} = m_2. \tag{1.137}
$$

(ii$_1$) For $a_{j_1 j_1 r} \neq a_{j_2 j_1 v}$ $(r = 1, 2, \cdots, p_1; v = 1, 2, \cdots, p_2)$, the first integral manifold is

$$
\begin{aligned}
&\sum_{\nu_0=m_1}^{m_2} b_{j_2 j_1 \nu_0} \sum_{\substack{\mu_1 \in [0, \nu_0 - m_1], \\ \nu_1 \in [m_1 - m_{j_1 r_1}, \beta_0 - m_{j_1 r_1}]}}^{\mu_1+\nu_1=\nu_0-m_{j_1 s_1}} \ \sum_{\substack{\mu_2 \in [0, \nu_1 - m_1] \\ \nu_2 \in [m_1 - \Sigma_{i=1}^2 m_{j_1 r_i}, \nu_1 - \Sigma_{i=1}^2 m_{j_1 r_i}]}}^{\mu_2+\nu_2=\nu_1-\Sigma_{i=1}^2 m_{j_1 r_i}} \cdots \\
&\quad \sum_{\substack{\mu_{p_1-1} \in [0, \nu_{p_1-2} - m_1] \\ \nu_{p_1-1} \in [m_1 - \Sigma_{i=1}^{p_1-1} m_{j_1 r_i}, \nu_{p_1-2} - \Sigma_{i=1}^{p_1-1} m_{j_1 r_i}]}}^{\mu_{p_1-1}+\nu_{p_1-1}=\nu_{p_1-2}-\Sigma_{i=1}^{p_1-1} m_{j_1 r_i}} \ \prod_{j=1}^{p_1-1} \frac{(\nu_{j-1} - \Sigma_{i=1}^{j} m_{j_1 r_i})!}{\mu_j! \nu_j!} (a_{j_1 j_1 r_{j+1}} - a_{j_1 j_1 r_j})^{\mu_j} \\
&\quad \times \frac{1}{\nu_{p_1-1} - m_{j_1 r_{p_1}} + 1} \left[(x_{j_1} - a_{j_1 j_1 r_\rho})^{\nu_{p_1-1} - m_{j_1 r_{p_1}} + 1} - (x_{j_1 0} - a_{j_1 j_1 r_\rho})^{\nu_{p_1-1} - m_{j_1 r_{p_1}} + 1} \right] \\
&\quad - \sum_{\nu_0=m_1}^{m_2} b_{j_2 j_1 \nu_0} \sum_{k=2}^{p_1-1} \Bigg(\sum_{\substack{\mu_1 \in [0, \nu_0 - m_1], \\ \nu_1 \in [m_1 - m_{j_1 r_1}, \beta_0 - m_{j_1 r_1}]}}^{\mu_1+\nu_1=\nu_0-m_{j_1 s_1}} \ \sum_{\substack{\mu_2 \in [0, \nu_1 - m_1] \\ \nu_2 \in [m_1 - \Sigma_{i=1}^2 m_{j_1 r_i}, \nu_1 - \Sigma_{i=1}^2 m_{j_1 r_i}]}}^{\mu_2+\nu_2=\nu_1-\Sigma_{i=1}^2 m_{j_1 r_i}} \cdots \\
&\quad \times \Bigg\{ \sum_{\substack{\mu_{k-1} \in (\nu_{k-2} - m_1 + m_{j_1 r_k}, \nu_{k-2} - \Sigma_{i=1}^{k-1} m_{j_1 r_i}] \\ \nu_{k-1} \in [0, m_1 - \Sigma_{i=1}^{k} m_{j_1 r_i})}}^{\mu_{k-1}+\nu_{k-1}=\nu_{k-2}-\Sigma_{i=1}^{k-1} m_{j_1 r_i}} \ \prod_{j=1}^{k-1} \frac{(\nu_{j-1} - \Sigma_{i=1}^{j} m_{j_1 r_i})!}{\mu_j! \nu_j!} (a_{j_1 j_1 r_{j+1}} - a_{j_1 j_1 r_j})^{\mu_j} \\
&\quad \times \sum_{\rho=k}^{p_1} \sum_{l_\rho=1}^{\bar{m}_{j_1 r_\rho}} \frac{A_{l_\rho \bar{m}_{j_1 r_\rho}}^{(r_k)}}{l_\rho - 1} \left[\frac{1}{(x_{j_1} - a_{j_1 j_1 r_\rho})^{l_\rho - 1}} - \frac{1}{(x_{j_1 0} - a_{j_1 j_1 r_\rho})^{l_\rho - 1}} \right] \\
&\quad + \sum_{\substack{\mu_{k-1} \in (\nu_{k-2} - m_1, \nu_{k-2} - m_1 + m_{j_1 r_k}] \\ \nu_{k-1} \in [m_1 - \Sigma_{i=1}^{k} m_{j_1 r_i}, m_1 - \Sigma_{i=1}^{k-1} m_{j_1 r_i})}}^{\mu_{k-1}+\nu_{k-1}=\nu_{k-2}-\Sigma_{i=1}^{k-1} m_{j_1 r_i}} \ \prod_{j=1}^{k-1} \frac{(\nu_{j-1} - \Sigma_{i=1}^{j} m_{j_1 r_i})!}{\mu_j! \nu_j!} (a_{j_2 j_2 s_{j+1}} - a_{j_2 j_2 s_j})^{\alpha_j}
\end{aligned}
$$

$$
\times \sum_{\rho=k+1}^{p_1} \sum_{l_\rho=1}^{\bar{m}_{j_1 r_\rho}} \frac{B^{(r_k)}_{l_\rho \bar{m}_{j_1 r_\rho}}}{l_\rho - 1} \left[\frac{1}{(x_{j_1} - a_{j_1 j_1 r_\rho})^{l_\rho - 1}} - \frac{1}{(x_{j_1 0} - a_{j_1 j_1 r_\rho})^{l_\rho - 1}} \right] \Bigg\} \Bigg)
$$

$$
- \Bigg\{ \sum_{\nu_0=0}^{m_{j_1 r_1}-1} b_{j_2 j_1 \nu_0} \sum_{\rho=1}^{p_1} \sum_{l_\rho=1}^{\bar{m}_{j_1 r_\rho}} \frac{A^{(r_1)}_{l_\rho \bar{m}_{j_1 r_\rho}}}{l_\rho - 1} \left[\frac{1}{(x_{j_1} - a_{j_1 j_1 r_\rho})^{l_\rho - 1}} - \frac{1}{(x_{j_1 0} - a_{j_1 j_1 r_\rho})^{l_\rho - 1}} \right]
$$

$$
+ \sum_{\nu_0=m_{j_1 r_1}}^{m_1-1} b_{j_2 j_1 \nu_0} \sum_{\rho=2}^{p_1} \sum_{l_\rho=1}^{\bar{m}_{j_1 r_\rho}} \frac{B^{(r_1)}_{l_\rho \bar{m}_{j_1 r_\rho}}}{l_\rho - 1} \left[\frac{1}{(x_{j_1} - a_{j_1 j_1 r_\rho})^{l_\rho - 1}} - \frac{1}{(x_{j_1 0} - a_{j_1 j_1 r_\rho})^{l_\rho - 1}} \right] \Bigg\}
$$

$$
= \frac{a_{j_1 j_1 0}}{a_{j_2 j_2 0}} \Bigg\{ \frac{1}{n_1+1} [(x_{j_2} - a_{j_1 j_2 u_1})^{n_1+1} - (x_{j_2 0} - a_{j_1 j_2 u_1})^{n_1+1}]
$$

$$
+ \sum_{\alpha=1}^{n_1 - n_{r_1}} \frac{1}{n_1 - \alpha + 1} b_{j_1 j_2 \alpha} [(x_{j_1} - a_{j_1 j_2 u_1})^{n_1 - \alpha + 1} - (x_{j_1 0} - a_{j_1 j_2 u_1})^{n_1 - \alpha + 1}] \Bigg\}, \tag{1.138}
$$

where

$$
\begin{aligned}
& A^{(r_k)}_{l_\rho \bar{m}_{j_2 s_\rho}} = \frac{1}{(\bar{m}_{j_1 r_\rho} - l_\rho)!} \frac{d^{\bar{m}_{j_1 r_\rho} - l_\rho}}{dx_{j_1}^{\bar{m}_{j_1 r_\rho} - l_\rho}} \frac{1}{\prod_{\sigma=k, \sigma \neq \rho}^{p_1} (x_{j_1} - a_{j_1 j_1 r_\sigma})^{m_{j_1 r_\sigma}}} \Bigg|_{x^*_{j_1} = a_{j_1 j_1 r_\rho}} \\
& (l_\rho = 1, 2, \ldots, \bar{m}_{j_2 s r_\rho}, \rho = k, k+1, \ldots, p_1), \\
& A^{(r_k)}_{\bar{m}_{j_1 r_\rho} \bar{m}_{j_1 r_\rho}} = \frac{1}{\prod_{\sigma=k, \sigma \neq \rho}^{p_1} (a_{j_1 j_1 r_\rho} - a_{j_1 j_1 r_\sigma})^{m_{j_1 r_\sigma}}}, \\
& B^{(r_k)}_{l_\rho \bar{m}_{j_1 r_\rho}} = \frac{1}{(\bar{m}_{j_1 r_\rho} - l_\rho)!} \frac{d^{\bar{m}_{j_1 r_\rho} - l_\rho}}{dx_{j_1}^{\bar{m}_{j_1 r_\rho} - l_\rho}} \frac{(x_{j_1} - a_{j_1 j_1 r_k})^{\nu_{k-1} - \sum_{i=1}^{k} m_{j_1 r_i}}}{\prod_{\sigma=k+1, \sigma \neq \rho}^{p_1} (x_{j_1} - a_{j_1 j_1 r_\sigma})^{m_{j_1 r_\sigma}}} \Bigg|_{x^*_{j_1} = a_{j_1 j_1 r_\rho}} \\
& (l_\rho = 1, 2, \ldots, \bar{m}_{j_1 r_\rho}, \rho = k+1, \ldots, p_2), \\
& B^{(r_k)}_{\bar{m}_{j_1 r_\rho} \bar{m}_{j_1 r_\rho}} = \frac{(a_{j_1 j_1 r_\rho} - a_{j_1 j_1 r_k})^{\nu_{k-1} - \sum_{i=1}^{k} m_{j_1 r_i}}}{\prod_{\sigma=k+1, \sigma \neq \rho}^{p_1} (a_{j_1 j_1 r_\rho} - a_{j_1 j_1 r_\sigma})^{m_{j_1 r_\sigma}}}, \\
& \{\bar{m}_{j_1 r_\rho} | \rho = k, k+1, \ldots, p_1\} = \Big\{ \sum\nolimits_{i=1}^{k} m_{j_1 r_i} - \nu_{k-1}, m_{j_1 r_{k+1}}, \ldots, m_{j_1 r_{p_1}} \Big\}, \\
& \text{for } k \in \{1, 2, \ldots, p_1 - 1\};
\end{aligned} \tag{1.139}
$$

and

$$
\begin{aligned}
& b_{j_1 j_2 \alpha} = \sum_{l_{q_1}, t_{q_1}=0}^{l_{q_1} + t_{q_1} = n_{j_1 u_{q_1}}} \cdots \sum_{l_2, t_2=0}^{l_2 + t_2 = n_{j_1 u_2}} [\prod_{k=2}^{q_1} \frac{n_{j_1 u_k}!}{l_k! t_k!} (a_{j_1 j_2 u_1} - a_{j_1 j_2 u_k})^{l_k}] \delta^{n_1 - \alpha}_{t_2 + \cdots + t_{q_1}}; \\
& b_{j_1 j_2 0} = 1.
\end{aligned}
$$

$$
\begin{aligned}
b_{j_2j_1v_0} &= \sum_{t_1,l_1=0}^{t_1+l_1=m_{j_2v_1}} \cdots \sum_{t_{p_2},l_{p_2}=0}^{t_{p_2}+l_{p_2}=m_{j_1v_{p_2}}} \prod_{k=1}^{p_2} \frac{m_{j_2v_k}!}{t_k!l_k!}(a_{j_1j_1r_1}-a_{j_2j_1k})^{t_k}\delta_{v_0}^{\Sigma_{k=1}^{p_2}l_k},\\
&\frac{1}{v_{p_1-1}-m_{j_1p_1}+1}\left[(x_{j_1}-a_{j_1j_1r_{p_1}})^{v_{p_1-1}-m_{j_1p_1}+1}-(x_{j_10}-a_{j_1j_1r_{p_1}})^{v_{p_1-1}-m_{j_1p_1}+1}\right]\\
&= \ln\frac{|x_{j_1}-a_{j_1j_1r_{p_1}}|}{|x_{j_10}-a_{j_1j_1r_{p_1}}|},\ \text{for } v_{p_1-1}-m_{j_1p_1}+1=0;\\
&\frac{1}{l_k-1}\left[\frac{1}{(x_{j_1}-a_{j_1j_1r_k})^{l_k-1}}-\frac{1}{(x_{j_10}-a_{j_1j_1r_k})^{l_k-1}}\right] = -\ln\frac{|x_{j_1}-a_{j_1j_1r_k}|}{|x_{j_10}-a_{j_1j_1r_k}|},\\
&\text{for } l_k=1\ (k=1,2,\cdots,p_1-1).
\end{aligned}
\tag{1.140}
$$

The network matrices of singular equilibriums and 1-dimensional flows are defined as

$$
\begin{aligned}
&\bigcup_{\eta=j_1}^{j_2}\bigcup_{u=1}^{q_1}\bigcup_{l=1}^{p_1+p_2}(c_{\eta j_1l},a_{j_1j_2u}) = \bigcup_{r=1}^{p_1}\bigcup_{u=1}^{q_1}\underbrace{(a_{j_1j_1r},a_{j_1j_2u})}_{([m_{j_1r},n_{j_1u}]\text{-XX,xF})} \oplus \bigcup_{u=1}^{q_1}\bigcup_{v=1}^{p_2}\underbrace{(a_{j_1j_2u},a_{j_2j_1v})}_{(n_{j_1u},m_{j_2v})\text{-XX}},\\
&\left\{c_{\eta j_1l}|\eta\in\{j_1,j_2\},l=1,2,\cdots,p_1+p_2\right\}\\
&= \text{sort}\left\{a_{j_1j_1r},a_{j_2j_1v}|r=1,2,\cdots p_1;v=1,2,\cdots,p_2\right\},\\
&c_{\eta_1j_1l_1}<c_{\eta_2j_1l_2},\ l_1<l_2,\ \eta_1,\eta_2\in\{j_1,j_2\};
\end{aligned}
\tag{1.141}
$$

where

$$
\begin{aligned}
&\bigcup_{r=1}^{p_1}\bigcup_{u=1}^{q_1}\underbrace{(a_{j_1j_1r},a_{j_1j_2u})}_{([m_{j_1r},n_{j_1u}]\text{-XX,xF})}\\
&= \begin{bmatrix}
\underbrace{(a_{j_1j_11},a_{j_1j_2q_1})}_{([m_{j_11},n_{j_1q_1}]\text{XX,xF})} & \cdots & \underbrace{(a_{j_1j_1(p_1-1)},a_{j_1j_2q_1})}_{([m_{1(p_1-1)},n_{j_1q_1}]\text{XX,xF})} & \underbrace{(a_{j_1j_1p_1},a_{j_1j_2q_1})}_{([m_{j_1p_1},n_{j_1q_1}]\text{XX,xF})}\\
\underbrace{(a_{j_1j_11},a_{j_1j_2(q_1-1)})}_{([m_{j_11},n_{j_1(q_1-1)}]\text{XX,xF})} & \cdots & \underbrace{(a_{j_1j_1(p_1-1)},a_{j_1j_2(q_1-1})}_{([m_{j_1(p_1-1)},n_{j_1(q_1-1)}]\text{XX,xF})} & \underbrace{(a_{j_1j_1p_1},a_{j_1j_2(q_1-1)})}_{([m_{j_1p_1},n_{j_1(q_1-1)}]\text{XX,xF})}\\
\vdots & \vdots & \vdots & \vdots\\
\underbrace{(a_{j_1j_11},a_{j_1j_21})}_{([m_{j_11},n_{j_11}]\text{XX,xF})} & \cdots & \underbrace{(a_{j_1j_1(p_1-1)},a_{j_1j_21})}_{([m_{j_1(p_1-1)},n_{j_11}]\text{XX,xF})} & \underbrace{(a_{j_1j_1p_1},a_{j_1j_21})}_{([m_{j_1p_1},n_{j_11}]\text{XX,xF})}
\end{bmatrix}_{q_1\times p_1},
\end{aligned}
\tag{1.142}
$$

and

$$
\bigcup_{u=1}^{q_1}\bigcup_{v=1}^{p_2}\underbrace{(a_{j_1j_2u}, a_{j_2j_1v})}_{(n_{j_1u},m_{j_2v})\text{-XX}}
= \begin{bmatrix}
\underbrace{(a_{j_1j_2q_1}, a_{j_2j_1 1})}_{(n_{j_1q_1},m_{j_2(p_2-1)})\text{-YY}} & \cdots & \underbrace{(a_{j_1j_2q_1}, a_{j_2j_1(p_2-1)})}_{(n_{j_1q_1},m_{j_2(p_2-1)})\text{-YY}} & \underbrace{(a_{j_1j_2q_1}, a_{j_2j_1p_2})}_{(n_{j_1q_1},m_{j_2p_2})\text{-YY}} \\
\underbrace{(a_{j_1j_2(q_1-1)}, a_{j_2j_1 1})}_{(n_{j_1q_1},m_{j_2(p_2-1)})\text{-YY}} & \cdots & \underbrace{(a_{j_1j_2(q_1-1)}, a_{j_2j_1(p_2-1)})}_{(n_{j_1q_1},m_{j_2(p_2-1)})\text{-YY}} & \underbrace{(a_{j_1j_2(q_1-1)}, a_{j_2j_1p_2})}_{(n_{j_2(q_1-1)},m_{j_2p_2})\text{-YY}} \\
\vdots & \vdots & \vdots & \vdots \\
\underbrace{(a_{j_1j_2 1}, a_{j_2j_1 1})}_{(n_{j_2 1},m_{j_2 1})\text{-YY}} & \cdots & \underbrace{(a_{j_1j_2 1}, a_{j_2j_1(p_2-1)})}_{(n_{j_1 1},m_{j_2p_2})\text{-YY}} & \underbrace{(a_{j_1j_2 1}, a_{j_2j_1p_2})}_{(n_{j_1 1},m_{j_2p_2})\text{-YY}}
\end{bmatrix}_{q_1\times p_2}. \tag{1.143}
$$

(ii_{1a}) There is an equilibrium for 1-dimensional flow as

$$(x^*_{j_1}, x^*_{j_2}) = (a_{j_1j_1r_1}, a_{j_1j_2u_1}). \tag{1.144}$$

(ii_{1a1}) For $m_{j_1r_1} = 2m_{j_1r_1 1} + 1$ and $n_{j_1u_1} = 2n_{j_1u_1 1} + 1$, the equilibrium of $(x^*_{j_1}, x^*_{j_2}) = (a_{j_1j_1r_1}, a_{j_1j_2u_1})$ has the following properties.

- *For $a_{j_1j_1 0}\prod_{r_2=1,r_2\neq r_1}^{p_1}(a_{j_1j_1r_1} - a_{j_1j_1r_2})^{m_{j_1r_2}}\prod_{u_2=1,u_2\neq u_1}^{q_1}(a_{j_1j_2u_1} - a_{j_1j_2u_2})^{n_{j_1u_2}} > 0$ and $a_{j_2j_2 0}\prod_{v=1}^{p_2}(a_{j_1j_1r_1} - a_{j_2j_1v})^{m_{j_2v}} > 0$,*

$$(a_{j_1j_1r_1}, a_{j_1j_2u_1}) = \underbrace{([(2m_{j_1r_1 1}+1), (2n_{j_1u_1 1} + 1)]\text{DP:UP,pF})}_{[(2m_{j_1r_1 1}+1),(2n_{j_1u_1 1}+1)]\text{-hyperbolic flow } (+)}. \tag{1.145}$$

  The flow at $(x^*_{j_1}, x^*_{j_2})=(a_{j_1j_1r_1}, a_{j_1j_2u_1})$ is a $([(2m_{j_1r_1 1}+1), (2n_{j_1u_1 1}+1)]$DP:UP, pF) -positive hyperbolic flow .

- *For $a_{j_1j_1 0}\prod_{r_2=1,r_2\neq r_1}^{p_1}(a_{j_1j_1r_1} - a_{j_1j_1r_2})^{m_{j_1r_2}}\prod_{u_2=1,u_2\neq u_1}^{q_1}(a_{j_1j_2u_1} - a_{j_1j_2u_2})^{n_{j_1u_2}} < 0$ and $a_{j_2j_2 0}\prod_{v=1}^{p_2}(a_{j_1j_1r_1} - a_{j_2j_1v})^{m_{j_2v}} > 0$,*

$$(a_{j_1j_1r_1}, a_{j_1j_2u_1}) = \underbrace{([(2m_{j_1r_1 1}+1), (2n_{j_1u_1 1} + 1)]\text{UP:DP,pF})}_{[(2m_{j_1r_1 1}+1),(2n_{j_1u_1 1} + 1)]\text{-hyperbolic-secant flow } (+)}. \tag{1.146}$$

  The flow at $(x^*_{j_1}, x^*_{j_2})=(a_{j_1j_1r_1}, a_{j_1j_2u_1})$ is a $([(2m_{j_1r_1 1}+1), (2n_{j_1u_1 1} + 1)]$UP:DP,pF)-positive hyperbolic-secant flow.

- *For $a_{j_1j_1 0}\prod_{r_2=1,r_2\neq r_1}^{p_1}(a_{j_1j_1r_1} - a_{j_1j_1r_2})^{m_{j_1r_2}}\prod_{u_2=1,u_2\neq u_1}^{q_1}(a_{j_1j_2u_1} - a_{j_1j_2u_2})^{n_{j_1u_2}} > 0$ and $a_{j_2j_2 0}\prod_{v=1}^{p_2}(a_{j_1j_1r_1} - a_{j_2j_1v})^{m_{j_2v}} < 0$,*

$$(a_{j_1j_1r_1}, a_{j_1j_2u_1}) = \underbrace{([(2m_{j_1r_1 1}+1), (2n_{j_1u_1 1} + 1)]\text{UP:DP,nF})}_{[(2m_{j_1r_1 1}+1),(2n_{j_1u_1 1} + 1)]\text{-hyperbolic-secant flow } (-)}. \tag{1.147}$$

The flow at $(x_{j_1}^*, x_{j_2}^*)=(a_{j_1j_1r_1}, a_{j_1j_2u_1})$ *is a* ([$(2m_{j_1r_1}1+1)$, $(2n_{j_1u_1}1+1)$]UP:DP,nF)-*negative hyperbolic-secant flow.*

- *For* $a_{j_1j_10}\prod_{r_2=1,r_2\neq r_1}^{p_1}(a_{j_1j_1r_1}-a_{j_1j_1r_2})^{m_{j_1r_2}}\prod_{u_2=1,u_2\neq u_1}^{q_1}(a_{j_1j_2u_1}-a_{j_1j_2u_2})^{n_{j_1u_2}}<0$ *and* $a_{j_2j_20}\prod_{v=1}^{p_2}(a_{j_1j_1r_1}-a_{j_2j_1v})^{m_{j_2v}}<0$,

$$(a_{j_1j_1r_1}, a_{j_1j_2u_1}) = \underbrace{([(2m_{j_1r_1}1+1),\ (2n_{j_1u_1}1+1)]\text{DP:UP,nF})}_{[(2m_{j_1r_1}1+1),(2n_{j_1u_1}1+1)]\text{-hyperbolic flow }(-)}. \tag{1.148}$$

The flow at $(x_{j_1}^*, x_{j_2}^*)=(a_{j_1j_1r_1}, a_{j_1j_2u_1})$ *is a* ([$(2m_{j_1r_1}1+1)$, $(2n_{j_1u_1}1+1)$]DP:UP,pF)-*negative hyperbolic flow.*

(ii_{1a2}) *For* $m_{j_1r_1}=2m_{j_1r_1}1$ *and* $n_{j_1u_1}=2n_{j_1u_1}1+1$, *the equilibrium of* $(x_{j_1}^*, x_{j_2}^*)=(a_{j_1j_1r_1}, a_{j_1j_2u_1})$ *for the* x_{j_1}*-direction has the following properties.*

- *For* $a_{j_1j_10}\prod_{r_2=1,r_2\neq r_1}^{p_1}(a_{j_1j_1r_1}-a_{j_1j_1r_2})^{m_{j_1r_2}}\prod_{u_2=1,u_2\neq u_1}^{q_1}(a_{j_1j_2u_1}-a_{j_1j_2u_2})^{n_{j_1u_2}}>0$ *and* $a_{j_2j_20}\prod_{v=1}^{p_2}(a_{j_1j_1r_1}-a_{j_2j_1v})^{m_{j_2v}}>0$,

$$(a_{j_1j_1r_1}, a_{j_1j_2u_1}) = \underbrace{([(2m_{j_1r_1}1),\ (2n_{j_1u_1}1+1)]\text{UP:UP,pF})}_{[(2m_{j_1r_1}1),(2n_{j_1u_1}1+1)]\text{-hyperbolic-secant-to-hyperbolic flow }(+)}. \tag{1.149}$$

The flow at $(x_{j_1}^*, x_{j_2}^*)=(a_{j_1j_1r_1}, a_{j_1j_2u_1})$ *is a* ([$(2m_{j_1r_1}1)$, $(2n_{j_1u_1}1+1)$]UP:UP,pF)-*positive hyperbolic-secant to hyperbolic flow.*

- *For* $a_{j_1j_10}\prod_{r_2=1,r_2\neq r_1}^{p_1}(a_{j_1j_1r_1}-a_{j_1j_1r_2})^{m_{j_1r_2}}\prod_{u_2=1,u_2\neq u_1}^{q_1}(a_{j_1j_2u_1}-a_{j_1j_2u_2})^{n_{j_1u_2}}<0$ *and* $a_{j_2j_20}\prod_{v=1}^{p_2}(a_{j_1j_1r_1}-a_{j_2j_1v})^{m_{j_2v}}>0$,

$$(a_{j_1j_1r_1}, a_{j_1j_2u_1}) = \underbrace{([(2m_{j_1r_1}1),\ (2n_{j_1u_1}1+1)]\text{DP:DP,pF})}_{[(2m_{j_1r_1}1),(2n_{j_1u_1}1+1)]\text{-hyperbolic-to-hyperbolic-secant flow }(+)}. \tag{1.150}$$

The flow at $(x_{j_1}^*, x_{j_2}^*)=(a_{j_1j_1r_1}, a_{j_1j_2u_1})$ *is a* ([$(2m_{j_1r_1}1)$, $(2n_{j_1u_1}1+1)$]DP:DP,pF)-*positive hyperbolic to hyperbolic-secant flow.*

- *For* $a_{j_1j_10}\prod_{r_2=1,r_2\neq r_1}^{p_1}(a_{j_1j_1r_1}-a_{j_1j_1r_2})^{m_{j_1r_2}}\prod_{u_2=1,u_2\neq u_1}^{q_1}(a_{j_1j_2u_1}-a_{j_1j_2u_2})^{n_{j_1u_2}}>0$ *and* $a_{j_2j_20}\prod_{v=1}^{p_2}(a_{j_1j_1r_1}-a_{j_2j_1v})^{m_{j_2v}}<0$,

$$(a_{j_1j_1r_1}, a_{j_1j_2u_1}) = \underbrace{([(2m_{j_1r_1}1),\ (2n_{j_1u_1}1+1)]\text{DP:DP,nF})}_{[(2m_{j_1r_1}1),(2n_{j_1u_1}1+1)]\text{-hyperbolic-to-hyperbolic-secant flow }(-)}. \tag{1.151}$$

The flow at $(x_{j_1}^*, x_{j_2}^*)=(a_{j_1j_1r_1}, a_{j_1j_2u_1})$ *is a* ([$(2m_{j_1r_1}1)$, $(2n_{j_1u_1}1+1)$]DP:DP,nF)-*negative hyperbolic-to-hyperbolic-secant flow.*

- *For* $a_{j_1j_10}\prod_{r_2=1,r_2\neq r_1}^{p_1}(a_{j_1j_1r_1}-a_{j_1j_1r_2})^{m_{j_1r_2}}\prod_{u_2=1,u_2\neq u_1}^{q_1}(a_{j_1j_2u_1}-a_{j_1j_2u_2})^{n_{j_1u_2}}<0$ *and* $a_{j_2j_20}\prod_{v=1}^{p_2}(a_{j_1j_1r_1}-a_{j_2j_1v})^{m_{j_2v}}<0$,

$$(a_{j_1j_1r_1}, a_{j_1j_2u_1}) = \underbrace{([(2m_{j_1r_1}1),\ (2n_{j_1u_1}1+1)]\text{UP:UP,nF})}_{[(2m_{j_1r_1}1),(2n_{j_1u_1}1+1)]\text{-hyperbolic-secant-to-hyperbolic flow }(-)}. \tag{1.152}$$

The flow at $(x_{j_1}^*, x_{j_2}^*) = (a_{j_1j_1r_1}, a_{j_1j_2u_1})$ *is a* ([(2$m_{j_1r_11}$), (2$n_{j_1u_11}$ + 1)]UP:UP,nF)-*negative hyperbolic-secant-to-hyperbolic flow.*

(ii_{1a3}) *For* $m_{j_1r_1} = 2m_{j_1r_11} + 1$ *and* $n_{j_1u_1} = 2n_{j_1u_11}$, *the equilibrium of* $(x_{j_1}^*, x_{j_2}^*) = (a_{j_1j_1r_1}, a_{j_1j_2u_1})$ *for the* x_{j_1}*-direction has the following properties.*

- *For* $a_{j_1j_10}\prod_{r_2=1,r_2\neq r_1}^{p_1}(a_{j_1j_1r_1} - a_{j_1j_1r_2})^{m_{j_1r_2}}\prod_{u_2=1,u_2\neq u_1}^{q_1}(a_{j_1j_2u_1} - a_{j_1j_2u_2})^{n_{j_1u_2}} > 0$ *and* $a_{j_2j_20}\prod_{v=1}^{p_2}(a_{j_1j_1r_1} - a_{j_2j_1v})^{m_{j_2v}} > 0$,

$$(a_{j_1j_1r_1}, a_{j_1j_2u_1}) = \underbrace{([(2m_{j_1r_11}+1), (2n_{j_1u_11})]_{\text{DI:II}}\text{SO,pF})}_{[(2m_{j_1r_11}+1),(2n_{j_1u_11})]\text{-decreasing-to-increasing inflection source flow }(+)}. \tag{1.153}$$

The flow at $(x_{j_1}^*, x_{j_2}^*) = (a_{j_1j_1r_1}, a_{j_1j_2u_1})$ *is a* $([(2m_{j_1r_11}+1), (2n_{j_1u_11})]_{\text{DI:II}}\text{SO,pF})$-*positive decreasing-to-increasing inflection source flow.*

- *For* $a_{j_1j_10}\prod_{r_2=1,r_2\neq r_1}^{p_1}(a_{j_1j_1r_1} - a_{j_1j_1r_2})^{m_{j_1r_2}}\prod_{u_2=1,u_2\neq u_1}^{q_1}(a_{j_1j_2u_1} - a_{j_1j_2u_2})^{n_{j_1u_2}} < 0$ *and* $a_{j_2j_20}\prod_{v=1}^{p_2}(a_{j_1j_1r_1} - a_{j_2j_1v})^{m_{j_2v}} > 0$,

$$(a_{j_1j_1r_1}, a_{j_1j_2u_1}) = \underbrace{([(2m_{j_1r_11}+1), (2n_{j_1u_11})]_{\text{II:DI}}\text{SI,pF})}_{[(2m_{j_1r_11}+1),(2n_{j_1u_11})]\text{-increasing-to-decreasing inflection sink flow }(+)}. \tag{1.154}$$

The flow at $(x_{j_1}^*, x_{j_2}^*) = (a_{j_1j_1r_1}, a_{j_1j_2u_1})$ *is a* $([(2m_{j_1r_11}+1), (2n_{j_1u_11})]_{\text{II:DI}}\text{SI,pF})$-*positive increasing-to-decreasing inflection source flow.*

- *For* $a_{j_1j_10}\prod_{r_2=1,r_2\neq r_1}^{p_1}(a_{j_1j_1r_1} - a_{j_1j_1r_2})^{m_{j_1r_2}}\prod_{u_2=1,u_2\neq u_1}^{q_1}(a_{j_1j_2u_1} - a_{j_1j_2u_2})^{n_{j_1u_2}} > 0$ *and* $a_{j_2j_20}\prod_{v=1}^{p_2}(a_{j_1j_1r_1} - a_{j_2j_1v})^{m_{j_2v}} < 0$,

$$(a_{j_1j_1r_1}, a_{j_1j_2u_1}) = \underbrace{([(2m_{j_1r_11}+1), (2n_{j_1u_11})]_{\text{II:DI}}\text{SO,nF})}_{[(2m_{j_1r_11}+1),(2n_{j_1u_11})]\text{-increasing-to-decreasing inflection source flow }(-)}. \tag{1.155}$$

The flow at $(x_{j_1}^*, x_{j_2}^*) = (a_{j_1j_1r_1}, a_{j_1j_2u_1})$ *is a* $([(2m_{j_1r_11}+1), (2n_{j_1u_11})]_{\text{II:DI}}\text{SO,nF})$-*negative increasing-to-decreasing inflection source flow.*

- *For* $a_{j_1j_10}\prod_{r_2=1,r_2\neq r_1}^{p_1}(a_{j_1j_1r_1} - a_{j_1j_1r_2})^{m_{j_1r_2}}\prod_{u_2=1,u_2\neq u_1}^{q_1}(a_{j_1j_2u_1} - a_{j_1j_2u_2})^{n_{j_1u_2}} < 0$ *and* $a_{j_2j_20}\prod_{v=1}^{p_2}(a_{j_1j_1r_1} - a_{j_2j_1v})^{m_{j_2v}} < 0$,

$$(a_{j_1j_1r_1}, a_{j_1j_2u_1}) = \underbrace{([(2m_{j_1r_11}+1), (2n_{j_1u_11})]_{\text{DI:II}}\text{SI,nF})}_{[(2m_{j_1r_11}+1),(2n_{j_1u_11})]\text{-decreasing-to-increasing inflection sink flow }(-)}. \tag{1.156}$$

The flow at $(x_{j_1}^*, x_{j_2}^*) = (a_{j_1j_1r_1}, a_{j_1j_2u_1})$ *is a* $([(2m_{j_1r_11}+1), (2n_{j_1u_11})]_{\text{DI:II}}\text{SI,nF})$-*negative decreasing-to-increasing inflection sink flow.*

(ii_{1a4}) For $m_{j_1r_1} = 2m_{j_1r_1}1$ and $n_{j_1u_1} = 2n_{j_1u_1}1$, the equilibrium of $(x^*_{j_1}, x^*_{j_2}) = (a_{j_1j_1r_1}, a_{j_1j_2u_1})$ for the x_{j_1}-direction has the following properties.

- *For $a_{j_1j_10}\prod_{r_2=1,r_2\neq r_1}^{p_1}(a_{j_1j_1r_1} - a_{j_1j_1r_2})^{m_{j_1r_2}}\prod_{u_2=1,u_2\neq u_1}^{q_1}(a_{j_1j_2u_1} - a_{j_1j_2u_2})^{n_{j_1u_2}} > 0$ and $a_{j_2j_20}\prod_{v=1}^{p_2}(a_{j_1j_1r_1} - a_{j_2j_1v})^{m_{j_2v}} > 0$,*

$$(a_{j_1j_1r_1}, a_{j_1j_2u_1}) = \underbrace{([(2m_{j_1r_1}1), (2n_{j_1u_1}1)]_{\text{II:II}}\text{US,pF})}_{[(2m_{j_1r_1}1),(2n_{j_1u_1}1)]\text{-increasing-inflection upper-saddle flow }(+)}. \tag{1.157}$$

  The flow at $(x^*_{j_1}, x^*_{j_2}) = (a_{j_1j_1r_1}, a_{j_1j_2u_1})$ is a $([(2m_{j_1r_1}1), (2n_{j_1u_1}1)]_{\text{II:II}}\text{US,pF})$-positive increasing-inflection upper-saddle flow.

- *For $a_{j_1j_10}\prod_{r_2=1,r_2\neq r_1}^{p_1}(a_{j_1j_1r_1} - a_{j_1j_1r_2})^{m_{j_1r_2}}\prod_{u_2=1,u_2\neq u_1}^{q_1}(a_{j_1j_2u_1} - a_{j_1j_2u_2})^{n_{j_1u_2}} < 0$ and $a_{j_2j_20}\prod_{v=1}^{p_2}(a_{j_1j_1r_1} - a_{j_2j_1v})^{m_{j_2v}} > 0$,*

$$(a_{j_1j_1r_1}, a_{j_1j_2u_1}) = \underbrace{([(2m_{j_1r_1}1), (2n_{j_1u_1}1)]_{\text{DI:DI}}\text{LS,pF})}_{[(2m_{j_1r_1}1),(2n_{j_1u_1}1)]\text{-decreasing-inflection lower-saddle flow }(+)}. \tag{1.158}$$

  The flow at $(x^*_{j_1}, x^*_{j_2}) = (a_{j_1j_1r_1}, a_{j_1j_2u_1})$ is a $([(2m_{j_1r_1}1), (2n_{j_1u_1}1)]_{\text{DI:DI}}\text{LS,pF})$-positive decreasing-inflection lower-saddle flow.

- *For $a_{j_1j_10}\prod_{r_2=1,r_2\neq r_1}^{p_1}(a_{j_1j_1r_1} - a_{j_1j_1r_2})^{m_{j_1r_2}}\prod_{u_2=1,u_2\neq u_1}^{q_1}(a_{j_1j_2u_1} - a_{j_1j_2u_2})^{n_{j_1u_2}} > 0$ and $a_{j_2j_20}\prod_{v=1}^{p_2}(a_{j_1j_1r_1} - a_{j_2j_1v})^{m_{j_2v}} < 0$,*

$$(a_{j_1j_1r_1}, a_{j_1j_2u_1}) = \underbrace{([(2m_{j_1r_1}1), (2n_{j_1u_1}1)]_{\text{DI:DI}}\text{US,nF})}_{[(2m_{j_1r_1}1),(2n_{j_1u_1}1)]\text{-decreasing-inflection upper-saddle flow }(-)}. \tag{1.159}$$

  The flow at $(x^*_{j_1}, x^*_{j_2}) = (a_{j_1j_1r_1}, a_{j_1j_2u_1})$ is a $([(2m_{j_1r_1}1), (2n_{j_1u_1}1)]_{\text{DI:DI}}\text{US,nF})$-negative decreasing-inflection upper-saddle flow.

- *For $a_{j_1j_10}\prod_{r_2=1,r_2\neq r_1}^{p_1}(a_{j_1j_1r_1} - a_{j_1j_1r_2})^{m_{j_1r_2}}\prod_{u_2=1,u_2\neq u_1}^{q_1}(a_{j_1j_2u_1} - a_{j_1j_2u_2})^{n_{j_1u_2}} < 0$ and $a_{j_2j_20}\prod_{v=1}^{p_2}(a_{j_1j_1r_1} - a_{j_2j_1v})^{m_{j_2v}} < 0$,*

$$(a_{j_1j_1r_1}, a_{j_1j_2u_1}) = \underbrace{([(2m_{j_1r_1}1), (2n_{j_1u_1}1)]_{\text{II:II}}\text{LS,nF})}_{[(2m_{j_1r_1}1),(2n_{j_1u_1}1)]\text{-increasing-inflection lower-saddle flow }(-)}. \tag{1.160}$$

  The flow at $(x^*_{j_1}, x^*_{j_2}) = (a_{j_1j_1r_1}, a_{j_1j_2u_1})$ is a $([(2m_{j_1r_1}1), (2n_{j_1u_1}1)]_{\text{II:II}}\text{LS,nF})$-negative increasing-inflection lower-saddle flow.

(ii_{1b}) There is an equilibrium for self-univariate equilibrium as

$$(x^*_{j_2}, x^*_{j_1}) = (a_{j_1j_2u_1}, a_{j_2j_1v_1}). \tag{1.161}$$

(ii_{1b1}) For $n_{j_1u_1} = 2n_{j_1u_1}1 + 1$ and $m_{j_2v_1} = 2m_{j_2v_1}1 + 1$, the equilibrium of $(x^*_{j_2}, x^*_{j_1}) = (a_{j_1j_2u_1}, a_{j_2j_1v_1})$ has the following properties.

- *For* $a_{j_1j_10}\prod_{r=1}^{p_1}(a_{j_2j_1v_1}-a_{j_1j_1r})^{m_{j_1r}}\prod_{u_2=1,u_2\neq u_1}^{q_1}(a_{j_1j_2u_1}-a_{j_1j_2u_2})^{n_{j_1u_2}}>0$ *and* $a_{j_2j_20}\prod_{v_2=1,v_2\neq v_1}^{p_2}(a_{j_2j_1v_1}-a_{j_2j_1v_2})^{m_{j_2v_2}}>0$,

$$(a_{j_1j_2u_1},a_{j_2j_1v_1})=\underbrace{((2n_{j_1u_1}1+1)^{\text{th}}\text{UP}_+,(2m_{j_2v_1}1+1)^{\text{th}}\text{UP}_+)}_{((2n_{j_1u_1}1+1),(2m_{j_2v_1}1+1))\text{-positive saddle}}. \tag{1.162}$$

The flow at $(x_{j_2}^*,x_{j_1}^*)=(a_{j_1j_2u_1},a_{j_2j_1v_1})$ *is a* $((2n_{j_1u_1}1+1)^{\text{th}}\text{UP}_+,(2m_{j_2v_1}1+1)^{\text{th}}\text{UP}_+)$*-positive saddle.*

- *For* $a_{j_1j_10}\prod_{r=1}^{p_1}(a_{j_2j_1v_1}-a_{j_1j_1r})^{m_{j_1r}}\prod_{u_2=1,u_2\neq u_1}^{q_1}(a_{j_1j_2u_1}-a_{j_1j_2u_2})^{n_{j_1u_2}}<0$ *and* $a_{j_2j_20}\prod_{v_2=1,v_2\neq v_1}^{p_2}(a_{j_2j_1v_1}-a_{j_2j_1v_2})^{m_{j_2v_2}}>0$,

$$(a_{j_1j_2u_1},a_{j_2j_1v_1})=\underbrace{((2n_{j_1u_1}1+1)^{\text{th}}\text{DP}_+,(2m_{j_2v_1}1+1)^{\text{th}}\text{DP}_-)}_{((2n_{j_1u_1}1+1),(2m_{j_2v_1}1+1))\text{-CCW center}}. \tag{1.163}$$

The flow at $(x_{j_2}^*,x_{j_1}^*)=(a_{j_1j_2u_1},a_{j_2j_1v_1})$ *is a* $((2n_{j_1u_1}1+1)^{\text{th}}\text{DP}_+,(2m_{j_2v_1}1+1)^{\text{th}}\text{DP}_-)$*-counter-clockwise center.*

- *For* $a_{j_1j_10}\prod_{r=1}^{p_1}(a_{j_2j_1v_1}-a_{j_1j_1r})^{m_{j_1r}}\prod_{u_2=1,u_2\neq u_1}^{q_1}(a_{j_1j_2u_1}-a_{j_1j_2u_2})^{n_{j_1u_2}}>0$ *and* $a_{j_2j_20}\prod_{v_2=1,v_2\neq v_1}^{p_2}(a_{j_2j_1v_1}-a_{j_2j_1v_2})^{m_{j_2v_2}}<0$,

$$(a_{j_1j_2u_1},a_{j_2j_1v_1})=\underbrace{((2n_{j_1u_1}1+1)^{\text{th}}\text{DP}_-,(2m_{j_2v_1}1+1)^{\text{th}}\text{DP}_+)}_{((2n_{j_1u_1}1+1),(2m_{j_2v_1}1+1))\text{-CW center}}. \tag{1.164}$$

The flow at $(x_{j_2}^*,x_{j_1}^*)=(a_{j_1j_2u_1},a_{j_2j_1v_1})$ *is a* $((2n_{j_1u_1}1+1)^{\text{th}}\text{DP}_-,(2m_{j_2v_1}1+1)^{\text{th}}\text{DP}_+)$*-clockwise center.*

- *For* $a_{j_1j_10}\prod_{r=1}^{p_1}(a_{j_2j_1v_1}-a_{j_1j_1r})^{m_{j_1r}}\prod_{u_2=1,u_2\neq u_1}^{q_1}(a_{j_1j_2u_1}-a_{j_1j_2u_2})^{n_{j_1u_2}}<0$ *and* $a_{j_2j_20}\prod_{v_2=1,v_2\neq v_1}^{p_2}(a_{j_2j_1v_1}-a_{j_2j_1v_2})^{m_{j_2v_2}}<0$,

$$(a_{j_1j_2u_1},a_{j_2j_1v_1})=\underbrace{((2n_{j_1u_1}1+1)^{\text{th}}\text{UP}_-,(2m_{j_2v_1}1+1)^{\text{th}}\text{UP}_-)}_{((2n_{j_1u_1}1+1),(2m_{j_2v_1}1+1))\text{-negative saddle}}. \tag{1.165}$$

The flow at $(x_{j_2}^*,x_{j_1}^*)=(a_{j_1j_2u_1},a_{j_2j_1v_1})$ *is a* $((2n_{j_1u_1}1+1)^{\text{th}}\text{UP}_-,(2m_{j_2v_1}1+1)^{\text{th}}\text{UP}_-)$*-negative saddle.*

(ii_{1b2}) For $n_{j_1u_1}=2n_{j_1u_1}1$ *and* $m_{j_2v_1}=2m_{j_2v_1}1+1$*, the equilibrium of* $(x_{j_2}^*,x_{j_1}^*)=(a_{j_1j_2u_1},a_{j_2j_1v_1})$ *has the following properties.*

- *For* $a_{j_1j_10}\prod_{r=1}^{p_1}(a_{j_2j_1v_1}-a_{j_1j_1r})^{m_{j_1r}}\prod_{u_2=1,u_2\neq u_1}^{q_1}(a_{j_1j_2u_1}-a_{j_1j_2u_2})^{n_{j_1u_2}}>0$ *and* $a_{j_2j_20}\prod_{v_2=1,v_2\neq v_1}^{p_2}(a_{j_2j_1v_1}-a_{j_2j_1v_2})^{m_{j_2v_2}}>0$,

$$(a_{j_1j_2u_1},a_{j_2j_1v_1})=\underbrace{((2n_{j_1u_1}1)^{\text{th}}\text{US},(2m_{j_2v_1}1+1)^{\text{th}}\text{UP})}_{((2n_{j_1u_1}1),(2m_{j_2v_1}1+1))\text{-up-parabola upper-saddle}}. \tag{1.166}$$

The flow at $(x_{j_2}^*, x_{j_1}^*) = (a_{j_1 j_2 u_1}, a_{j_2 j_1 v_1})$ *is a* $((2n_{j_1 u_1 1})^{\text{th}}\text{US}, (2m_{j_2 v_1 1}+1)^{\text{th}}\text{UP})$-*up-parabola upper-saddle.*

- *For* $a_{j_1 j_1 0}\prod_{r=1}^{p_1}(a_{j_2 j_1 v_1}-a_{j_1 j_1 r})^{m_{j_1 r}}\prod_{u_2=1,u_2\neq u_1}^{q_1}(a_{j_1 j_2 u_1}-a_{j_1 j_2 u_2})^{n_{j_1 u_2}}<0$ *and* $a_{j_2 j_2 0}\prod_{v_2=1,v_2\neq v_1}^{p_2}(a_{j_2 j_1 v_1}-a_{j_2 j_1 v_2})^{m_{j_2 v_2}}>0$,

$$(a_{j_1 j_2 u_1}, a_{j_2 j_1 v_1}) = \underbrace{((2n_{j_1 u_1 1})^{\text{th}}\text{LS}, (2m_{j_2 v_1 1}+1)^{\text{th}}\text{DP})}_{((2n_{j_1 u_1 1}),(2m_{j_2 v_1 1}+1))\text{-down-parabola lower-saddle}}. \quad (1.167)$$

The flow at $(x_{j_2}^*, x_{j_1}^*) = (a_{j_1 j_2 u_1}, a_{j_2 j_1 v_1})$ *is a* $((2n_{j_1 u_1 1})^{\text{th}}\text{LS}, (2m_{j_2 v_1 1}+1)^{\text{th}}\text{DP})$-*down-parabola lower-saddle.*

- *For* $a_{j_1 j_1 0}\prod_{r=1}^{p_1}(a_{j_2 j_1 v_1}-a_{j_1 j_1 r})^{m_{j_1 r}}\prod_{u_2=1,u_2\neq u_1}^{q_1}(a_{j_1 j_2 u_1}-a_{j_1 j_2 u_2})^{n_{j_1 u_2}}>0$ *and* $a_{j_2 j_2 0}\prod_{v_2=1,v_2\neq v_1}^{p_2}(a_{j_2 j_1 v_1}-a_{j_2 j_1 v_2})^{m_{j_2 v_2}}<0$,

$$(a_{j_1 j_2 u_1}, a_{j_2 j_1 v_1}) = \underbrace{((2n_{j_1 u_1 1})^{\text{th}}\text{US}, (2m_{j_2 v_1 1}+1)^{\text{th}}\text{DP})}_{((2n_{j_1 u_1 1}),(2m_{j_2 v_1 1}+1))\text{-down-parabola upper-saddle}}. \quad (1.168)$$

The flow at $(x_{j_2}^*, x_{j_1}^*) = (a_{j_1 j_2 u_1}, a_{j_2 j_1 v_1})$ *is a* $((2n_{j_1 u_1 1})^{\text{th}}\text{US}, (2m_{j_2 v_1 1}+1)^{\text{th}}\text{DP})$-*down-parabola upper-saddle.*

- *For* $a_{j_1 j_1 0}\prod_{r=1}^{p_1}(a_{j_2 j_1 v_1}-a_{j_1 j_1 r})^{m_{j_1 r}}\prod_{u_2=1,u_2\neq u_1}^{q_1}(a_{j_1 j_2 u_1}-a_{j_1 j_2 u_2})^{n_{j_1 u_2}}<0$ *and* $a_{j_2 j_2 0}\prod_{v_2=1,v_2\neq v_1}^{p_2}(a_{j_2 j_1 v_1}-a_{j_2 j_1 v_2})^{m_{j_2 v_2}}<0$,

$$(a_{j_1 j_2 u_1}, a_{j_2 j_1 v_1}) = \underbrace{((2n_{j_1 u_1 1})^{\text{th}}\text{LS}, (2m_{j_2 v_1 1}+1)^{\text{th}}\text{UP})}_{((2n_{j_1 u_1 1}),(2m_{j_2 v_1 1}+1))\text{-up-parabola lower-saddle}}. \quad (1.169)$$

The flow at $(x_{j_2}^*, x_{j_1}^*) = (a_{j_1 j_2 u_1}, a_{j_2 j_1 v_1})$ *is a* $((2n_{j_1 u_1 1})^{\text{th}}\text{LS}, (2m_{j_2 v_1 1}+1)^{\text{th}}\text{UP})$-*up-parabola lower-saddle.*

(ii$_{1b3}$*) For* $n_{j_1 u_1} = 2n_{j_1 u_1 1}+1$ *and* $m_{j_2 v_1} = 2m_{j_2 v_1 1}$*, the equilibrium of* $(x_{j_2}^*, x_{j_1}^*) = (a_{j_1 j_2 u_1}, a_{j_2 j_1 v_1})$ *has the following properties.*

- *For* $a_{j_1 j_1 0}\prod_{r=1}^{p_1}(a_{j_2 j_1 v_1}-a_{j_1 j_1 r})^{m_{j_1 r}}\prod_{u_2=1,u_2\neq u_1}^{q_1}(a_{j_1 j_2 u_1}-a_{j_1 j_2 u_2})^{n_{j_1 u_2}}>0$ *and* $a_{j_2 j_2 0}\prod_{v_2=1,v_2\neq v_1}^{p_2}(a_{j_2 j_1 v_1}-a_{j_2 j_1 v_2})^{m_{j_2 v_2}}>0$,

$$(a_{j_1 j_2 u_1}, a_{j_2 j_1 v_1}) = \underbrace{((2n_{j_1 u_1 1}+1)^{\text{th}}\text{UP}, (2m_{j_2 v_1 1})^{\text{th}}\text{US})}_{((2n_{j_1 u_1 1}+1),(2m_{j_2 v_1 1}))\text{-up-parabola upper-saddle}}. \quad (1.170)$$

The flow at $(x_{j_2}^*, x_{j_1}^*) = (a_{j_1 j_2 u_1}, a_{j_2 j_1 v_1})$ *is a* $((2n_{j_1 u_1 1}+1)^{\text{th}}\text{UP}, (2m_{j_2 v_1 1})^{\text{th}}\text{US})$-*up-parabola upper-saddle.*

- *For* $a_{j_1 j_1 0}\prod_{r=1}^{p_1}(a_{j_2 j_1 v_1}-a_{j_1 j_1 r})^{m_{j_1 r}}\prod_{u_2=1,u_2\neq u_1}^{q_1}(a_{j_1 j_2 u_1}-a_{j_1 j_2 u_2})^{n_{j_1 u_2}}<0$ *and* $a_{j_2 j_2 0}\prod_{v_2=1,v_2\neq v_1}^{p_2}(a_{j_2 j_1 v_1}-a_{j_2 j_1 v_2})^{m_{j_2 v_2}}>0$,

$$(a_{j_1j_2u_1}, a_{j_2j_1v_1}) = \underbrace{((2n_{j_1u_1}1 + 1)^{\text{th}}\text{DP}, (2m_{j_2v_1}1)^{\text{th}}\text{US})}_{((2n_{j_1u_1}1+1),(2m_{j_2v_1}1))\text{-down-parabola upper-saddle}}. \tag{1.171}$$

The flow at $(x_{j_2}^*, x_{j_1}^*) = (a_{j_1j_2u_1}, a_{j_2j_1v_1})$ *is a* $((2n_{j_1u_1}1+1)^{\text{th}}\text{DP}, (2m_{j_2v_1}1)^{\text{th}}\text{US})$-*down-parabola upper-saddle.*

- *For* $a_{j_1j_10}\prod_{r=1}^{p_1}(a_{j_2j_1v_1} - a_{j_1j_1r})^{m_{j_1r}}\prod_{u_2=1,u_2\neq u_1}^{q_1}(a_{j_1j_2u_1} - a_{j_1j_2u_2})^{n_{j_1u_2}} > 0$ *and* $a_{j_2j_20}\prod_{v_2=1,v_2\neq v_1}^{p_2}(a_{j_2j_1v_1} - a_{j_2j_1v_2})^{m_{j_2v_2}} < 0$,

$$(a_{j_1j_2u_1}, a_{j_2j_1v_1}) = \underbrace{((2n_{j_1u_1}1 + 1)^{\text{th}}\text{DP}, (2m_{j_2v_1}1)^{\text{th}}\text{LS})}_{((2n_{j_1u_1}1+1),(2m_{j_2v_1}1))\text{-down-parabola lower-saddle}}. \tag{1.172}$$

The flow at $(x_{j_2}^*, x_{j_1}^*) = (a_{j_1j_2u_1}, a_{j_2j_1v_1})$ *is a* $((2n_{j_1u_1}1+1)^{\text{th}}\text{DP}, (2m_{j_2v_1}1)^{\text{th}}\text{LS})$-*down-parabola lower-saddle.*

- *For* $a_{j_1j_10}\prod_{r=1}^{p_1}(a_{j_2j_1v_1} - a_{j_1j_1r})^{m_{j_1r}}\prod_{u_2=1,u_2\neq u_1}^{q_1}(a_{j_1j_2u_1} - a_{j_1j_2u_2})^{n_{j_1u_2}} < 0$ *and* $a_{j_2j_20}\prod_{v_2=1,v_2\neq v_1}^{p_2}(a_{j_2j_1v_1} - a_{j_2j_1v_2})^{m_{j_2v_2}} < 0$,

$$(a_{j_1j_2u_1}, a_{j_2j_1v_1}) = \underbrace{((2n_{j_1u_1}1 + 1)^{\text{th}}\text{UP}, (2m_{j_2v_1}1)^{\text{th}}\text{LS})}_{((2n_{j_1u_1}1+1),(2m_{j_2v_1}1))\text{-up-parabola lower-saddle}}. \tag{1.173}$$

The flow at $(x_{j_2}^*, x_{j_1}^*) = (a_{j_1j_2u_1}, a_{j_2j_1v_1})$ *is a* $((2n_{j_1u_1}1+1)^{\text{th}}\text{UP}, (2m_{j_2v_1}1)^{\text{th}}\text{LS})$-*up-parabola lower-saddle.*

(ii_{1b4}) *For* $n_{j_1u_1} = 2n_{j_1u_1}1$ *and* $m_{j_2v_1} = 2m_{j_2v_1}1$, *the equilibrium of* $(x_{j_2}^*, x_{j_1}^*) = (a_{j_1j_2u_1}, a_{j_2j_1v_1})$ *has the following properties.*

- *For* $a_{j_1j_10}\prod_{r=1}^{p_1}(a_{j_2j_1v_1} - a_{j_1j_1r})^{m_{j_1r}}\prod_{u_2=1,u_2\neq u_1}^{q_1}(a_{j_1j_2u_1} - a_{j_1j_2u_2})^{n_{j_1u_2}} > 0$ *and* $a_{j_2j_20}\prod_{v_2=1,v_2\neq v_1}^{p_2}(a_{j_2j_1v_1} - a_{j_2j_1v_2})^{m_{j_2v_2}} > 0$,

$$(a_{j_1j_2u_1}, a_{j_2j_1v_1}) = \underbrace{((2n_{j_1u_1}1)^{\text{th}}\text{II}_+, (2m_{j_2v_1}1)^{\text{th}}\text{II}_+)}_{((2n_{j_1u_1}1),(2m_{j_2v_1}1))\text{-double inflection-saddle}}. \tag{1.174}$$

The flow at $(x_{j_2}^*, x_{j_1}^*) = (a_{j_1j_2u_1}, a_{j_2j_1v_1})$ *is a* $((2n_{j_1u_1}1)^{\text{th}}\text{II}_+, (2m_{j_2v_1}1)^{\text{th}}\text{II}_+)$-*double inflection-saddle.*

- *For* $a_{j_1j_10}\prod_{r=1}^{p_1}(a_{j_2j_1v_1} - a_{j_1j_1r})^{m_{j_1r}}\prod_{u_2=1,u_2\neq u_1}^{q_1}(a_{j_1j_2u_1} - a_{j_1j_2u_2})^{n_{j_1u_2}} < 0$ *and* $a_{j_2j_20}\prod_{v_2=1,v_2\neq v_1}^{p_2}(a_{j_2j_1v_1} - a_{j_2j_1v_2})^{m_{j_2v_2}} > 0$,

$$(a_{j_1j_2u_1}, a_{j_2j_1v_1}) = \underbrace{((2n_{j_1u_1}1)^{\text{th}}\text{DI}_+, (2m_{j_2v_1}1)^{\text{th}}\text{DI}_-)}_{((2n_{j_1u_1}1),(2m_{j_2v_1}1))\text{-double inflection-saddle}}. \tag{1.175}$$

The flow at $(x_{j_2}^*, x_{j_1}^*) = (a_{j_1j_2u_1}, a_{j_2j_1v_1})$ *is a* $((2n_{j_1u_1}1)^{\text{th}}\text{DI}_+, (2m_{j_2v_1}1)^{\text{th}}\text{DI}_-)$-*double inflection-saddle.*

- *For* $a_{j_1j_10}\prod_{r=1}^{p_1}(a_{j_2j_1v_1}-a_{j_1j_1r})^{m_{j_1r}}\prod_{u_2=1,u_2\neq u_1}^{q_1}(a_{j_1j_2u_1}-a_{j_1j_2u_2})^{n_{j_1u_2}}>0$ *and* $a_{j_2j_20}\prod_{v_2=1,v_2\neq v_1}^{p_2}(a_{j_2j_1v_1}-a_{j_2j_1v_2})^{m_{j_2v_2}}<0$,

$$(a_{j_1j_2u_1},a_{j_2j_1v_1})=\underbrace{((2n_{j_1u_1}1)^{\text{th}}\text{DI}_-,(2m_{j_2v_1}1)^{\text{th}}\text{DI}_+)}_{((2n_{j_1u_1}1),(2m_{j_2v_1}1))\text{-double inflection-saddle}}. \tag{1.176}$$

The flow at $(x_{j_2}^*,x_{j_1}^*)=(a_{j_1j_2u_1},a_{j_2j_1v_1})$ *is a* $((2n_{j_1u_1}1)^{\text{th}}\text{DI}_-,(2m_{j_2v_1}1)^{\text{th}}\text{DI}_+)$*-double-inflection-saddle.*

- *For* $a_{j_1j_10}\prod_{r=1}^{p_1}(a_{j_2j_1v_1}-a_{j_1j_1r})^{m_{j_1r}}\prod_{u_2=1,u_2\neq u_1}^{q_1}(a_{j_1j_2u_1}-a_{j_1j_2u_2})^{n_{j_1u_2}}<0$ *and* $a_{j_2j_20}\prod_{v_2=1,v_2\neq v_1}^{p_2}(a_{j_2j_1v_1}-a_{j_2j_1v_2})^{m_{j_2v_2}}<0$,

$$(a_{j_1j_2u_1},a_{j_2j_1v_1})=\underbrace{((2n_{j_1u_1}1)^{\text{th}}\text{II}_-,(2m_{j_2v_1}1)^{\text{th}}\text{II}_-)}_{((2n_{j_1u_1}1),(2m_{j_2v_1}1))\text{-double inflection-saddle}}. \tag{1.177}$$

The flow at $(x_{j_2}^*,x_{j_1}^*)=(a_{j_1j_2u_1},a_{j_2j_1v_1})$ *is a* $((2n_{j_1u_1}1)^{\text{th}}\text{II}_-,(2m_{j_2v_1}1)^{\text{th}}\text{II}_-)$*-double inflection-saddle.*

(ii$_2$) For $a_{j_1j_1r_1}=a_{j_2j_1v_1}$ $(r_1\in\{1,2,\cdots,p_1\};\ v_1\in\{1,2,\cdots,p_2\})$

$$\begin{aligned}\dot{x}_{j_2}&=a_{j_2j_20}(x_{j_1}-a_{j_2j_1v_1})^{m_{j_1v_1}}\prod_{v_1=1,v_2\neq v_1}^{p_2}(x_{j_1}-a_{j_2j_1v_2})^{m_{j_1v_2}},\\ \dot{x}_{j_1}&=a_{j_1j_10}(x_{j_1}-a_{j_1j_1r_1})^{m_{j_1r_1}}\prod_{r_2=1,r_2\neq r_1}^{p_1}(x_{j_1}-a_{j_1j_1r_2})^{m_{j_1r_2}}\prod_{u=1}^{q_1}(x_{j_2}-a_{j_1j_2u})^{n_{j_1u}},\end{aligned} \tag{1.178}$$

The first integral manifold is

$$\begin{aligned}&\sum_{v_0=m_1}^{m_2}b_{j_2j_1v_0}\sum_{\substack{\mu_1\in[0,v_0-m_1],\\ v_1\in[m_1-m_{j_1r_1},\beta_0-m_{j_1r_1}]}}^{\mu_1+v_1=v_0-m_{j_1s_1}}\ \sum_{\substack{\mu_2\in[0,v_1-m_1]\\ v_2\in[m_1-\Sigma_{i=1}^2m_{j_1r_i},v_1-\Sigma_{i=1}^2m_{j_1r_i}]}}^{\mu_2+v_2=v_1-\Sigma_{i=1}^2m_{j_1r_i}}\cdots\\ &\sum_{\substack{\mu_{p_2-1}\in[0,v_{q_2-2}-m_1]\\ v_{p_2-1}\in[m_1-\Sigma_{i=1}^{p_2-1}m_{j_1r_i},v_{p-2}-\Sigma_{i=1}^{q_2-1}m_{j_1r_i}]}}^{\mu_{p_2-1}+v_{p_2-1}=v_{p_2-2}-\Sigma_{i=1}^{p_2-1}m_{j_1r_i}}\ \prod_{j=1}^{p_2-1}\frac{(v_{j-1}-\Sigma_{i=1}^{j}m_{j_1r_i})!}{\mu_j!v_j!}(a_{j_1j_1r_{j+1}}-a_{j_1j_1r_j})^{\mu_j}\\ &\times\frac{1}{v_{p_2-1}-m_{j_1r_{p_2}}+1}\left[(x_{j_1}-a_{j_1j_1r_\rho})^{v_{p_2-1}-m_{j_1r_{p_2}}+1}-(x_{j_10}-a_{j_1j_1r_\rho})^{v_{p_2-1}-m_{j_1r_{p_2}}+1}\right]\\ &-\sum_{\beta_0=p_2}^{p_1}b_{j_2j_1v_0}\sum_{k=2}^{p_2-1}\Bigg(\sum_{\substack{\alpha_1\in[0,\beta_0-n_2),\\ \beta_1\in(n_2-n_{j_2s_1},\beta_0-n_{j_2s_1}]}}^{\alpha_1+\beta_1=\beta_0-n_{j_2s_1}}\ \sum_{\substack{\alpha_2\in[0,\beta_1-n_2]\\ \beta_2\in(n_2-\Sigma_{i=1}^2n_{j_2s_i},\beta_1-\Sigma_{i=1}^2n_{j_2s_i})}}^{\alpha_2+\beta_2=\beta_1-\Sigma_{i=1}^2n_{j_2s_i}}\cdots\end{aligned}$$

$$\times\left\{\sum_{\substack{\mu_{k-1}\in(\nu_{k-2}-m_1+m_{j_1r_k},\nu_{k-2}-\Sigma_{i=1}^{k-1}m_{j_1r_i}]\\ \nu_{k-1}\in[0,m_1-\Sigma_{i=1}^{k}m_{j_1r_i})}}^{\mu_{k-1}+\nu_{k-1}\in\nu_{k-2}-\Sigma_{i=1}^{k-1}m_{j_1r_i}}\prod_{j=1}^{k-1}\frac{(\nu_{j-1}-\Sigma_{i=1}^{j}m_{j_1r_i})!}{\mu_j!\nu_j!}(a_{j_1j_1r_{j+1}}-a_{j_1j_1r_j})^{\mu_j}\right.$$

$$\times\sum_{\rho=k}^{p_2}\sum_{l_\rho\in 1}^{m_{j_1r_\rho}}\frac{A_{l_\rho m_{j_1r_\rho}}^{(r_k)}}{l_\rho-1}\left[\frac{1}{(x_{j_1}-a_{j_1j_1r_\rho})^{l_\rho-1}}-\frac{1}{(x_{j_20}-a_{j_2j_2s_\rho})^{l_\rho-1}}\right]$$

$$+\sum_{\substack{\mu_{k-1}=(\nu_{k-2}-m_1,\nu_{k-2}-m_1+m_{j_1r_k}]\\ \nu_{k-1}\in[m_1-\Sigma_{i=1}^{k}m_{j_1r_i},m_1-\Sigma_{i=1}^{k-1}m_{j_1r_i})}}^{\mu_{k-1}+\nu_{k-1}=\nu_{k-2}-\Sigma_{i=1}^{k-1}m_{j_1r_i}}\prod_{j=1}^{k-1}\frac{(\nu_{j-1}-\Sigma_{i=1}^{j}m_{j_1r_i})!}{\mu_j!\nu_j!}(a_{j_2j_2s_{j+1}}-a_{j_2j_2s_j})^{\alpha_j}$$

$$\left.\left.\times\sum_{\rho=k+1}^{p_2}\sum_{l_\rho=1}^{\bar{m}_{j_1r_\rho}}\frac{B_{l_\rho\bar{m}_{j_1r_\rho}}^{(r_k)}}{l_\rho-1}\left[\frac{1}{(x_{j_1}-a_{j_1j_1r_\rho})^{l_\rho-1}}-\frac{1}{(x_{j_10}-a_{j_1j_1r_\rho})^{l_\rho-1}}\right]\right\}\right)+\Pi(\nu_1)$$

$$=\frac{a_{j_1j_10}}{a_{j_2j_20}}\left\{\frac{1}{n_1+1}[(x_{j_2}-a_{j_1j_2u_1})^{n_1+1}-(x_{j_20}-a_{j_1j_2u_1})^{n_1+1}]\right.$$

$$\left.+\sum_{\alpha=1}^{n_1-n_{r_1}}\frac{1}{n_1-\alpha+1}b_{j_1j_2\alpha}[(x_{j_1}-a_{j_1j_2u_1})^{n_1-\alpha+1}-(x_{j_10}-a_{j_1j_2u_1})^{n_1-\alpha+1}]\right\},\tag{1.179}$$

where

$$\Pi(\nu_1)=-\left(\sum_{\beta_0=0}^{m_{j_1r_1}-1}b_{j_2j_1\nu_0}\sum_{\rho=1}^{p_1}\sum_{l_\rho=1}^{\overline{m}_{j_1r_\rho}}\frac{A_{l_\rho\overline{m}_{j_1r_\rho}}^{(r_1)}}{l_\rho-1}\left[\frac{1}{(x_{j_1}-a_{j_1j_1r_\rho})^{l_\rho-1}}-\frac{1}{(x_{j_10}-a_{j_1j_1r_\rho})^{l_\rho-1}}\right]\right.$$

$$\left.+\sum_{\beta_0=m_{j_1r_1}}^{m_1-1}b_{j_1j_2\nu_0}\sum_{\rho=2}^{p_1}\sum_{l_\rho=1}^{\overline{m}_{j_1r_\rho}}\frac{B_{l_\rho\overline{m}_{j_1r_\rho}}^{(r_1)}}{l_\rho-1}\left[\frac{1}{(x_{j_1}-a_{j_1j_1r_\rho})^{l_\rho-1}}-\frac{1}{(x_{j_10}-a_{j_1j_1r_\rho})^{l_\rho-1}}\right]\right)$$

for $m_{j_2\nu_1}<m_{j_1r_1}$;

$$\Pi(\nu_1)=-\sum_{\nu_0=m_{j_1r_1}}^{m_1-1}b_{j_2j_1\nu_0}\sum_{\rho=2}^{p_1}\sum_{l_\rho=1}^{\overline{m}_{j_1r_\rho}}\frac{B_{l_\rho\overline{m}_{j_1r_\rho}}^{(r_1)}}{l_\rho-1}\left[\frac{1}{(x_{j_1}-a_{j_1j_1r_\rho})^{l_\rho-1}}-\frac{1}{(x_{j_10}-a_{j_1j_1r_\rho})^{l_\rho-1}}\right],$$

for $m_{j_2\nu_1}=m_{j_1r_1}$;

$$\Pi(\nu_1)=-\sum_{\nu_0=m_{j_2\nu_1}}^{m_1-1}b_{j_2j_1\nu_0}\sum_{\rho=2}^{p_1}\sum_{l_\rho=1}^{\overline{m}_{j_1r_\rho}}\frac{B_{l_\rho\overline{m}_{j_1r_\rho}}^{(r_1)}}{l_\rho-1}\left[\frac{1}{(x_{j_1}-a_{j_1j_1r_\rho})^{l_\rho-1}}-\frac{1}{(x_{j_10}-a_{j_1j_1r_\rho})^{l_\rho-1}}\right],$$

for $m_{j_2\nu_1}>m_{j_1r_1}$; (1.180)

and

$$b_{j_2j_1\nu_0}=\sum_{t_2,l_2=0}^{t_2+l_2=m_{j_2\nu_2}}\cdots\sum_{t_{p_2},l_{p_2}=0}^{t_{p_2}+l_{p_2}=m_{j_2\nu_{p_2}}}\prod_{k=2}^{p_2}\frac{m_{j_2\nu_k}!}{t_k!l_k!}(a_{j_1j_1r_1}-a_{j_2j_1\nu_k})^{t_k}\delta_{\beta_0}^{\Sigma_{k=2}^{p_2}l_k+m_{j_2\nu_1}},$$

$$\frac{1}{\beta_{p_1-1} - m_{j_1p_1} + 1}\left[(x_{j_1} - a_{j_1j_1r_{p_1}})^{\beta_{p_1-1}-m_{j_1p_1}+1} - (x_{j_10} - a_{j_1j_1r_{p_1}})^{\beta_{p_1-1}-m_{j_1p_1}+1}\right]$$
$$= \ln\frac{|x_{j_1} - a_{j_1j_1r_{p_1}}|}{|x_{j_10} - a_{j_1j_1r_{p_1}}|}, \quad \text{for } \beta_{p_1-1} - m_{j_1p_1} + 1 = 0;$$
$$\frac{1}{l_k - 1}\left[\frac{1}{(x_{j_1} - a_{j_1j_1r_k})^{l_k-1}} - \frac{1}{(x_{j_10} - a_{j_1j_1r_k})^{l_k-1}}\right] = -\ln\frac{|x_{j_1} - a_{j_1j_1r_k}|}{|x_{j_10} - a_{j_1j_1r_k}|},$$
$$\text{for } l_k = 1\ (k = 1, 2, \ldots, p_1 - 1). \tag{1.181}$$

Define a network matrix of hybrid singular 1-dimensional flows and singular equilibriums plus a singular infinite-equilibrium as

$$\begin{aligned}\text{NM} = &\underbrace{(a_{j_1j_2u_1}, a_{j_2j_1v_1})}_{([m_{j_1r_1}, n_{j_1u_1}]^{\text{th}}\text{YY},[(m_{j_2v_1} - m_{j_1r_1})]^{\text{th}}\text{XX})}\\ &\oplus \bigcup_{u_2=1,u_2\neq u_1}^{q_1} \underbrace{(a_{j_1j_2u_2}, a_{j_2j_1v_1})}_{([m_{j_1r_1}, n_{j_1u_2}]^{\text{th}}\text{YY},[(m_{j_2v_1} - m_{j_1r_1})]^{\text{th}}\text{XX})}\\ &\oplus \bigcup_{u=1}^{q_1}\bigcup_{r_2=1,r_2\neq r_1}^{p_1} \underbrace{(a_{j_1j_1r_2}, a_{j_1j_2u})}_{([m_{j_1r_2}, n_{j_1u}]\text{XX},\text{xx})}\\ &\oplus \bigcup_{u=1}^{q_1}\bigcup_{v_1=1,v_2\neq v_1}^{p_2} \underbrace{(a_{j_1j_2u}, a_{j_2j_1v_2})}_{(n_{j_1u}, m_{j_2v_2})^{\text{th}}\text{-UU}}.\end{aligned} \tag{1.182}$$

(ii_{2a}) The infinite-equilibriums and the switching equilibriums for $m_{j_1r_1} \leq m_{j_2v_1}$ are presented.

(ii_{2a1}) For $m_{j_1r_1} \leq m_{j_2v_1}$, 16 types of infinite-equilibriums exist at $x^*_{j_1} = a_{j_1j_1r_1} = a_{j_2j_1v_1}$.

(ii_{2a1a}) For $m_{j_1r_1} = 2m_{j_1r_11} + 1$ and $m_{j_2v_1} = 2m_{j_1v_11} + 1$, the infinite-equilibrium of $x^*_{j_1} = a_{j_2j_1v_1} = a_{j_1j_1r_1}$ with $\overline{x}_{j_2} \neq a_{j_1j_2u} (u = 1, 2, \cdots, q_1)$ has the following properties.

- *For $a_{j_1j_10}\prod_{r_2=1,r_2\neq r_1}^{p_1}(a_{j_1j_1r_1} - a_{j_1j_1r_2})^{m_{j_1r_2}}\prod_{u=1}^{q_1}(\overline{x}_{j_2} - a_{j_1j_2u})^{n_{j_1u}} > 0$ and $a_{j_2j_20}\prod_{v_2=1,v_2\neq v_1}^{p_2}(a_{j_2j_1v_1} - a_{j_2j_1v_2})^{m_{j_2v_2}} > 0$,*

$$(a_{j_1j_1r_1}, \overline{x}_{j_2}) = \underbrace{((2m_{j_1r_11} + 1)^{\text{th}}\text{SO},[2(m_{j_2v_11} - m_{j_1r_11})]^{\text{th}}\text{II})}_{((2m_{j_1r_11}+1),[2(m_{j_2v_11} - m_{j_1r_11})])\text{-increasing-inflection source}}. \tag{1.183}$$

The infinite-equilibrium is a $((2m_{j_1r_11} + 1)^{\text{th}}\text{SO},[2(m_{j_2v_11} - m_{j_1r_11})]^{\text{th}}\text{II})$-increasing-inflection source at $x^*_{j_1} = a_{j_2j_1v_1} = a_{j_1j_1r_1}$.

- *For $a_{j_1j_10}\prod_{r_2=1,r_2\neq r_1}^{p_1}(a_{j_1j_1r_1} - a_{j_1j_1r_2})^{m_{j_1r_2}}\prod_{u=1}^{q_1}(\overline{x}_{j_2} - a_{j_1j_2u})^{n_{j_1u}} < 0$ and $a_{j_2j_20}\prod_{v_2=1,v_2\neq v_1}^{p_2}(a_{j_2j_1v_1} - a_{j_2j_1v_2})^{m_{j_2v_2}} > 0$,*

$$(a_{j_1j_1r_1}, \overline{x}_{j_2}) = \underbrace{((2m_{j_1r_1}1 + 1)^{\text{th}}\text{SI},[2(m_{j_2v_1}1 - m_{j_1r_1}1)]^{\text{th}}\text{DI})}_{((2m_{j_1r_1}1+1),[2(m_{j_2v_1}1-m_{j_1r_1}1)])\text{-decreasing-inflection sink}}. \tag{1.184}$$

The infinite-equilibrium is a $((2m_{j_1r_1}1 + 1)^{\text{th}}\text{SI},[2(m_{j_2v_1}1 - m_{j_1r_1}1)]^{\text{th}}\text{DI})$*-decreasing-inflection sink at* $x^*_{j_1} = a_{j_2j_1v_1} = a_{j_1j_1r_1}$.

- *For* $a_{j_1j_10}\prod_{r_2=1,r_2\neq r_1}^{p_1}(a_{j_1j_1r_1} - a_{j_1j_1r_2})^{m_{j_1r_2}}\prod_{u=1}^{q_1}(\overline{x}_{j_2} - a_{j_1j_2u})^{n_{j_1u}} > 0$ *and* $a_{j_2j_20}\prod_{v_2=1,v_2\neq v_1}^{p_2}(a_{j_2j_1v_1} - a_{j_2j_1v_2})^{m_{j_2v_2}} < 0$,

$$(a_{j_1j_1r_1}, \overline{x}_{j_2}) = \underbrace{((2m_{j_1r_1}1 + 1)^{\text{th}}\text{SO},[2(m_{j_2v_1}1 - m_{j_1r_1}1)]^{\text{th}}\text{DI})}_{((2m_{j_1r_1}1+1),[2(m_{j_2v_1}1-m_{j_1r_1}1)])\text{-decreasing-inflection source}}. \tag{1.185}$$

The infinite-equilibrium is a $((2m_{j_1r_1}1 + 1)^{\text{th}}\text{SO},[2(m_{j_2v_1}1 - m_{j_1r_1}1)]^{\text{th}}\text{DI})$*-decreasing-inflection sink at* $x^*_{j_1} = a_{j_2j_1v_1} = a_{j_1j_1r_1}$.

- *For* $a_{j_1j_10}\prod_{r_2=1,r_2\neq r_1}^{p_1}(a_{j_1j_1r_1} - a_{j_1j_1r_2})^{m_{j_1r_2}}\prod_{u=1}^{q_1}(\overline{x}_{j_2} - a_{j_1j_2u})^{n_{j_1u}} < 0$ *and* $a_{j_2j_20}\prod_{v_2=1,v_2\neq v_1}^{p_2}(a_{j_2j_1v_1} - a_{j_2j_1v_2})^{m_{j_2v_2}} < 0$,

$$(a_{j_1j_1r_1}, \overline{x}_{j_2}) = \underbrace{((2m_{j_1r_1}1 + 1)^{\text{th}}\text{SI},[2(m_{j_2v_1}1 - m_{j_1r_1}1)]^{\text{th}}\text{II})}_{((2m_{j_1r_1}1+1),[2(m_{j_2v_1}1-m_{j_1r_1}1)])\text{-increasing-inflection sink}}. \tag{1.186}$$

The infinite-equilibrium is a $((2m_{j_1r_1}1 + 1)^{\text{th}}\text{SI},[2(m_{j_2v_1}1 - m_{j_1r_1}1)]^{\text{th}}\text{II})$*-increasing-inflection sink at* $x^*_{j_1} = a_{j_2j_1v_1} = a_{j_1j_1r_1}$.

(ii_{2a1b}) *For* $m_{j_1r_1} = 2m_{j_1r_1}1$ *and* $m_{j_2v_1} = 2m_{j_1v_1}1 + 1$, *the infinite-equilibrium of* $x^*_{j_1} = a_{j_2j_1v_1} = a_{j_1j_1r_1}$ *with* $\overline{x}_{j_2} \neq a_{j_1j_2u}(u = 1, 2, \ldots, q_1)$ *has the following properties.*

- *For* $a_{j_1j_10}\prod_{r_2=1,r_2\neq r_1}^{p_1}(a_{j_1j_1r_1} - a_{j_1j_1r_2})^{m_{j_1r_2}}\prod_{u=1}^{q_1}(\overline{x}_{j_2} - a_{j_1j_2u})^{n_{j_1u}} > 0$ *and* $a_{j_2j_20}\prod_{v_2=1,v_2\neq v_1}^{p_2}(a_{j_2j_1v_1} - a_{j_2j_1v_2})^{m_{j_2v_2}} > 0$,

$$(a_{j_1j_1r_1}, \overline{x}_{j_2}) = \underbrace{((2m_{j_1r_1}1)^{\text{th}}\text{US},[2(m_{j_2v_1}1 - m_{j_1r_1}1) + 1]^{\text{th}}\text{UP})}_{((2m_{j_1r_1}1),[2(m_{j_2v_1}1-m_{j_1r_1}1)+1])\text{-up-parabola upper-saddle}}. \tag{1.187}$$

The infinite-equilibrium is a $((2m_{j_1r_1}1)^{\text{th}}\text{US},[2(m_{j_2v_1}1 - m_{j_1r_1}1) + 1]^{\text{th}}\text{UP})$*-up-parabola upper-saddle at* $x^*_{j_1} = a_{j_2j_1v_1} = a_{j_1j_1r_1}$.

- *For* $a_{j_1j_10}\prod_{r_2=1,r_2\neq r_1}^{p_1}(a_{j_1j_1r_1} - a_{j_1j_1r_2})^{m_{j_1r_2}}\prod_{u=1}^{q_1}(\overline{x}_{j_2} - a_{j_1j_2u})^{n_{j_1u}} < 0$ *and* $a_{j_2j_20}\prod_{v_2=1,v_2\neq v_1}^{p_2}(a_{j_2j_1v_1} - a_{j_2j_1v_2})^{m_{j_2v_2}} > 0$,

$$(a_{j_1j_1r_1}, \overline{x}_{j_2}) = \underbrace{((2m_{j_1r_1}1)^{\text{th}}\text{LS},[2(m_{j_2v_1}1 - m_{j_1r_1}1) + 1]^{\text{th}}\text{DP})}_{((2m_{j_1r_1}1),[2(m_{j_2v_1}1-m_{j_1r_1}1)+1])\text{-down-parabola lower-saddle}}. \tag{1.188}$$

The infinite-equilibrium is a $((2m_{j_1r_1}1)^{\text{th}}\text{LS},[2(m_{j_2v_1}1 - m_{j_1r_1}1) + 1]^{\text{th}}\text{DP})$*-down-parabola lower-saddle at* $x^*_{j_1} = a_{j_2j_1v_1} = a_{j_1j_1r_1}$.

- *For* $a_{j_1j_10}\prod_{r_2=1,r_2\neq r_1}^{p_1}(a_{j_1j_1r_1}-a_{j_1j_1r_2})^{m_{j_1r_2}}\prod_{u=1}^{q_1}(\overline{x}_{j_2}-a_{j_1j_2u})^{n_{j_1u}}>0$ *and* $a_{j_2j_20}\prod_{v_2=1,v_2\neq v_1}^{p_2}(a_{j_2j_1v_1}-a_{j_2j_1v_2})^{m_{j_2v_2}}<0$,

$$(a_{j_1j_1r_1},\overline{x}_{j_2})=\underbrace{((2m_{j_1r_1}1)^{\text{th}}\text{US},[2(m_{j_2v_1}1-m_{j_1r_1}1)+1]^{\text{th}}\text{DP})}_{((2m_{j_1r_1}1),[2(m_{j_2v_1}1-m_{j_1r_1}1)+1])\text{-down-parabola upper-saddle}}. \tag{1.189}$$

The infinite-equilibrium is a $((2m_{j_1r_1}1)^{\text{th}}\text{US},[2(m_{j_2v_1}1-m_{j_1r_1}1)+1]^{\text{th}}\text{DP})$*-down-parabola upper-saddle at* $x_{j_1}^*=a_{j_2j_1v_1}=a_{j_1j_1r_1}$.

- *For* $a_{j_1j_10}\prod_{r_2=1,r_2\neq r_1}^{p_1}(a_{j_1j_1r_1}-a_{j_1j_1r_2})^{m_{j_1r_2}}\prod_{u=1}^{q_1}(\overline{x}_{j_2}-a_{j_1j_2u})^{n_{j_1u}}<0$ *and* $a_{j_2j_20}\prod_{v_2=1,v_2\neq v_1}^{p_2}(a_{j_2j_1v_1}-a_{j_2j_1v_2})^{m_{j_2v_2}}<0$,

$$(a_{j_1j_1r_1},\overline{x}_{j_2})=\underbrace{((2m_{j_1r_1}1)^{\text{th}}\text{LS},[2(m_{j_2v_1}1-m_{j_1r_1}1)+1]^{\text{th}}\text{UP})}_{((2m_{j_1r_1}1),[2(m_{j_2v_1}1-m_{j_1r_1}1)+1])\text{-up-parabola lower-saddle}}. \tag{1.190}$$

The infinite-equilibrium is a $((2m_{j_1r_1}1)^{\text{th}}\text{LS},[2(m_{j_2v_1}1-m_{j_1r_1}1)+1]^{\text{th}}\text{UP})$*-up-parabola lower-saddle at* $x_{j_1}^*=a_{j_2j_1v_1}=a_{j_1j_1r_1}$.

(ii$_{2a1c}$) For $m_{j_1r_1}=2m_{j_1r_1}1+1$ *and* $m_{j_2v_1}=2m_{j_1v_1}1$*, the infinite-equilibrium of* $x_{j_1}^*=a_{j_2j_1v_1}=a_{j_1j_1r_1}$ *with* $\overline{x}_{j_2}\neq a_{j_1j_2u}(u=1,2,\ldots,q_1)$ *has the following properties.*

- *For* $a_{j_1j_10}\prod_{r_2=1,r_2\neq r_1}^{p_1}(a_{j_1j_1r_1}-a_{j_1j_1r_2})^{m_{j_1r_2}}\prod_{u=1}^{q_1}(\overline{x}_{j_2}-a_{j_1j_2u})^{n_{j_1u}}>0$ *and* $a_{j_2j_20}\prod_{v_2=1,v_2\neq v_1}^{p_2}(a_{j_2j_1v_1}-a_{j_2j_1v_2})^{m_{j_2v_2}}>0$,

$$(a_{j_1j_1r_1},\overline{x}_{j_2})=\underbrace{((2m_{j_1r_1}1+1)^{\text{th}}\text{SO},[2(m_{j_2v_1}1-m_{j_1r_1}1)-1]^{\text{th}}\text{UP})}_{((2m_{j_1r_1}1+1),[2(m_{j_2v_1}1-m_{j_1r_1}1)-1])\text{-up-parabola source}}. \tag{1.191}$$

The infinite-equilibrium is a $((2m_{j_1r_1}1+1)^{\text{th}}\text{SO},[2(m_{j_2v_1}1-m_{j_1r_1}1)-1]^{\text{th}}\text{UP})$*-up-parabola source at* $x_{j_1}^*=a_{j_2j_1v_1}=a_{j_1j_1r_1}$.

- *For* $a_{j_1j_10}\prod_{r_2=1,r_2\neq r_1}^{p_1}(a_{j_1j_1r_1}-a_{j_1j_1r_2})^{m_{j_1r_2}}\prod_{u=1}^{q_1}(\overline{x}_{j_2}-a_{j_1j_2u})^{n_{j_1u}}<0$ *and* $a_{j_2j_20}\prod_{v_2=1,v_2\neq v_1}^{p_2}(a_{j_2j_1v_1}-a_{j_2j_1v_2})^{m_{j_2v_2}}>0$,

$$(a_{j_1j_1r_1},\overline{x}_{j_2})=\underbrace{((2m_{j_1r_1}1+1)^{\text{th}}\text{SI},[2(m_{j_2v_1}1-m_{j_1r_1}1)-1]^{\text{th}}\text{DP})}_{((2m_{j_1r_1}1+1),[2(m_{j_2v_1}1-m_{j_1r_1}1)-1])\text{-down-parabola sink}}. \tag{1.192}$$

The infinite-equilibrium is a $((2m_{j_1r_1}1+1)^{\text{th}}\text{SI},[2(m_{j_2v_1}1-m_{j_1r_1}1)-1]^{\text{th}}\text{DP})$*-down-parabola sink at* $x_{j_1}^*=a_{j_2j_1v_1}=a_{j_1j_1r_1}$.

- *For* $a_{j_1j_10}\prod_{r_2=1,r_2\neq r_1}^{p_1}(a_{j_1j_1r_1}-a_{j_1j_1r_2})^{m_{j_1r_2}}\prod_{u=1}^{q_1}(\overline{x}_{j_2}-a_{j_1j_2u})^{n_{j_1u}}>0$ *and* $a_{j_2j_20}\prod_{v_2=1,v_2\neq v_1}^{p_2}(a_{j_2j_1v_1}-a_{j_2j_1v_2})^{m_{j_2v_2}}<0$,

$$(a_{j_1j_1r_1},\overline{x}_{j_2})=\underbrace{((2m_{j_1r_1}1+1)^{\text{th}}\text{SO},[2(m_{j_2v_1}1-m_{j_1r_1}1)-1]^{\text{th}}\text{DP})}_{((2m_{j_1r_1}1+1),[2(m_{j_2v_1}1-m_{j_1r_1}1)-1])\text{-down-parabola source}}. \tag{1.193}$$

The infinite-equilibrium is a $((2m_{j_1r_1}1{+}1)^{\text{th}}\text{SO},[2(m_{j_2v_1}1{-}m_{j_1r_1}1){-}1]^{\text{th}}\text{DP})$-*down-parabola source at* $x_{j_1}^{*}{=}a_{j_2j_1v_1}{=}a_{j_1j_1r_1}$.

- *For* $a_{j_1j_10}\prod_{r_2=1,r_2\neq r_1}^{p_1}(a_{j_1j_1r_1}-a_{j_1j_1r_2})^{m_{j_1r_2}}\prod_{u=1}^{q_1}(\bar{x}_{j_2}-a_{j_1j_2u})^{n_{j_1u}}<0$ *and* $a_{j_2j_20}\prod_{v_2=1,v_2\neq v_1}^{p_2}(a_{j_2j_1v_1}-a_{j_2j_1v_2})^{m_{j_2v_2}}<0$,

$$(a_{j_1j_1r_1},\bar{x}_{j_2})=\underbrace{((2m_{j_1r_1}1+1)^{\text{th}}\text{SI},[2(m_{j_2v_1}1-m_{j_1r_1}1)-1]^{\text{th}}\text{UP})}_{((2m_{j_1r_1}1+1),[2(m_{j_2v_1}1-m_{j_1r_1}1)-1])\text{-up-parabola sink}}. \quad (1.194)$$

The infinite-equilibrium is a $((2m_{j_1r_1}1{+}1)^{\text{th}}\text{SI},[2(m_{j_2v_1}1{-}m_{j_1r_1}1){-}1]^{\text{th}}\text{UP})$-*up-parabola sink at* $x_{j_1}^{*}{=}a_{j_2j_1v_1}{=}a_{j_1j_1r_1}$.

(ii$_{2a1d}$) For $m_{j_1r_1}=2m_{j_1r_1}1$ *and* $m_{j_2v_1}=2m_{j_1v_1}1$, *the infinite-equilibrium of* $x_{j_1}^{*}=a_{j_2j_1v_1}=a_{j_1j_1r_1}$ *with* $\bar{x}_{j_2}\neq a_{j_1j_2u}(u=1,2,\ldots,q_1)$ *has the following properties.*

- *For* $a_{j_1j_10}\prod_{r_2=1,r_2\neq r_1}^{p_1}(a_{j_1j_1r_1}-a_{j_1j_1r_2})^{m_{j_1r_2}}\prod_{u=1}^{q_1}(\bar{x}_{j_2}-a_{j_1j_2u})^{n_{j_1u}}>0$ *and* $a_{j_2j_20}\prod_{v_2=1,v_2\neq v_1}^{p_2}(a_{j_2j_1v_1}-a_{j_2j_1v_2})^{m_{j_2v_2}}>0$,

$$(a_{j_1j_1r_1},\bar{x}_{j_2})=\underbrace{((2m_{j_1r_1}1)^{\text{th}}\text{US},[2(m_{j_2v_1}1-m_{j_1r_1}1)]^{\text{th}}\text{II})}_{((2m_{j_1r_1}1),[2(m_{j_2v_1}1-m_{j_1r_1}1)])\text{-increasing-inflection upper-saddle}}. \quad (1.195)$$

The infinite-equilibrium is a $((2m_{j_1r_1}1)^{\text{th}}\text{US},[2(m_{j_2v_1}1-m_{j_1r_1}1)]^{\text{th}}\text{II})$-*increasing-inflection upper-saddle at* $x_{j_1}^{*}=a_{j_2j_1v_1}=a_{j_1j_1r_1}$.

- *For* $a_{j_1j_10}\prod_{r_2=1,r_2\neq r_1}^{p_1}(a_{j_1j_1r_1}-a_{j_1j_1r_2})^{m_{j_1r_2}}\prod_{u=1}^{q_1}(\bar{x}_{j_2}-a_{j_1j_2u})^{n_{j_1u}}<0$ *and* $a_{j_2j_20}\prod_{v_2=1,v_2\neq v_1}^{p_2}(a_{j_2j_1v_1}-a_{j_2j_1v_2})^{m_{j_2v_2}}>0$,

$$(a_{j_1j_1r_1},\bar{x}_{j_2})=\underbrace{((2m_{j_1r_1}1)^{\text{th}}\text{LS},[2(m_{j_2v_1}1-m_{j_1r_1}1)]^{\text{th}}\text{DI})}_{((2m_{j_1r_1}1),[2(m_{j_2v_1}1-m_{j_1r_1}1)])\text{-decreasing-inflection lower-saddle}}. \quad (1.196)$$

The infinite-equilibrium is a $((2m_{j_1r_1}1)^{\text{th}}\text{LS},[2(m_{j_2v_1}1-m_{j_1r_1}1)]^{\text{th}}\text{DI})$-*decreasing-inflection lower-saddle at* $x_{j_1}^{*}=a_{j_2j_1v_1}=a_{j_1j_1r_1}$.

- *For* $a_{j_1j_10}\prod_{r_2=1,r_2\neq r_1}^{p_1}(a_{j_1j_1r_1}-a_{j_1j_1r_2})^{m_{j_1r_2}}\prod_{u=1}^{q_1}(\bar{x}_{j_2}-a_{j_1j_2u})^{n_{j_1u}}>0$ *and* $a_{j_2j_20}\prod_{v_2=1,v_2\neq v_1}^{p_2}(a_{j_2j_1v_1}-a_{j_2j_1v_2})^{m_{j_2v_2}}<0$,

$$(a_{j_1j_1r_1},\bar{x}_{j_2})=\underbrace{((2m_{j_1r_1}1)^{\text{th}}\text{US},[2(m_{j_2v_1}1-m_{j_1r_1}1)]^{\text{th}}\text{DI})}_{((2m_{j_1r_1}1),[2(m_{j_2v_1}1-m_{j_1r_1}1)])\text{-decreasing-inflection upper-saddle}}. \quad (1.197)$$

The infinite-equilibrium is a $((2m_{j_1r_1}1)^{\text{th}}\text{US},[2(m_{j_2v_1}1-m_{j_1r_1}1)]^{\text{th}}\text{DI})$-*decreasing-inflection upper-saddle at* $x_{j_1}^{*}=a_{j_2j_1v_1}=a_{j_1j_1r_1}$.

- *For* $a_{j_1j_10}\prod_{r_2=1,r_2\neq r_1}^{p_1}(a_{j_1j_1r_1}-a_{j_1j_1r_2})^{m_{j_1r_2}}\prod_{u=1}^{q_1}(\bar{x}_{j_2}-a_{j_1j_2u})^{n_{j_1u}}<0$ *and* $a_{j_2j_20}\prod_{v_2=1,v_2\neq v_1}^{p_2}(a_{j_2j_1v_1}-a_{j_2j_1v_2})^{m_{j_2v_2}}<0$,

$$(a_{j_1j_1r_1},\bar{x}_{j_2}) = \underbrace{((2m_{j_1r_1}1)^{\text{th}}\text{LS},[2(m_{j_2v_1}1 - m_{j_1r_1}1)]^{\text{th}}\text{II})}_{((2m_{j_1r_1}1),[2(m_{j_2v_1}1-m_{j_1r_1}1)])\text{-increasing-inflection lower-saddle}}. \tag{1.198}$$

The infinite-equilibrium is a $((2m_{j_1r_1}1)^{\text{th}}\text{SI},[2(m_{j_2v_1}1-m_{j_1r_1}1)]^{\text{th}}\text{II})$*-increasing-inflection lower-saddle at* $x_{j_1}^* = a_{j_2j_1v_1} = a_{j_1j_1r_1}$.

(ii_{2a2}) There are 32 types of equilibrium $(x_{j_1}^*, x_{j_2}^*) = (a_{j_1j_1r_1}, a_{j_1j_2u_1})$ *on the infinite-equilibrium of* $x_{j_1}^* = a_{j_2j_1v_1} = a_{j_1j_1r_1}$.

(ii_{2a2a}) For $m_{j_1r_1} = 2m_{j_1r_1}1 + 1$, $n_{j_1u_1} = 2n_{j_1u_1}1 + 1$ *and* $m_{j_2v_1} = 2m_{j_1v_1}1 + 1$, *the equilibrium of* $(x_{j_2}^*, x_{j_1}^*) = (a_{j_1j_2u_1}, a_{j_2j_1v_1})$ *on the infinite-equilibrium* $x_{j_1}^* = a_{j_2j_1v_1} = a_{j_1j_1r_1}$ *has the following properties.*

- *For* $a_{j_1j_10}\prod_{r_2=1,r_2\neq r_1}^{p_1}(a_{j_1j_1r_1} - a_{j_1j_1r_2})^{m_{j_1r_2}}\prod_{u_2=1,u_2\neq u_1}^{q_1}(a_{j_1j_2u_1} - a_{j_1j_2u_2})^{n_{j_1u_2}} > 0$ *and* $a_{j_2j_20}\prod_{v_2=1,v_2\neq v_1}^{p_2}(a_{j_2j_1v_1} - a_{j_2j_1v_2})^{m_{j_2v_2}} > 0$,

$$(a_{j_1j_2u_1}, a_{j_2j_1v_1}) = \underbrace{([(2m_{j_1r_1}1+1),(2n_{j_1u_1}1+1)]^{\text{th}}\text{UP}_{-}\text{:UP}_{+},[2(m_{j_2v_1}1 - m_{j_1r_1}1)]^{\text{th}}{}_{\text{DI:II}}\text{US})}_{([(2m_{j_1r_1}1+1),(2n_{j_1u_1}1+1)],[2(m_{j_2v_1}1-m_{j_1r_1}1)])\text{-inflection up-parabola upper-saddle}}. \tag{1.199}$$

The equilibrium is a $([(2m_{j_1r_1}1+1),(2n_{j_1u_1}1+1)]^{\text{th}}\text{UP}_{-}\text{:UP}_{+},[2(m_{j_2v_1}1-m_{j_1r_1}1)]^{\text{th}}{}_{\text{DI:II}}\text{US})$*-inflection up-parabola upper-saddle at* $(x_{j_2}^*, x_{j_1}^*) = (a_{j_1j_2u_1}, a_{j_2j_1v_1})$.

- *For* $a_{j_1j_10}\prod_{r_2=1,r_2\neq r_1}^{p_1}(a_{j_1j_1r_1} - a_{j_1j_1r_2})^{m_{j_1r_2}}\prod_{u_2=1,u_2\neq u_1}^{q_1}(a_{j_1j_2u_1} - a_{j_1j_2u_2})^{n_{j_1u_2}} < 0$ *and* $a_{j_2j_20}\prod_{v_2=1,v_2\neq v_1}^{p_2}(a_{j_2j_1v_1} - a_{j_2j_1v_2})^{m_{j_2v_2}} > 0$,

$$(a_{j_1j_2u_1}, a_{j_2j_1v_1}) = \underbrace{([(2m_{j_1r_1}1+1),(2n_{j_1u_1}1+1)]^{\text{th}}\text{DP}_{-}\text{:DP}_{+},[2(m_{j_2v_1}1 - m_{j_1r_1}1)]^{\text{th}}{}_{\text{II:DI}}\text{LS})}_{([(2m_{j_1r_1}1+1),(2n_{j_1u_1}1+1)],[2(m_{j_2v_1}1-m_{j_1r_1}1)])\text{-inflection down-parabola lower-saddle}}. \tag{1.200}$$

The equilibrium is a $([(2m_{j_1r_1}1+1),(2n_{j_1u_1}1+1)]^{\text{th}}\text{DP}_{-}\text{:DP}_{+},[2(m_{j_2v_1}1-m_{j_1r_1}1)]^{\text{th}}{}_{\text{II:DI}}\text{LS})$*-inflection down-parabola lower-saddle at* $(x_{j_2}^*, x_{j_1}^*) = (a_{j_1j_2u_1}, a_{j_2j_1v_1})$.

- *For* $a_{j_1j_10}\prod_{r_2=1,r_2\neq r_1}^{p_1}(a_{j_1j_1r_1} - a_{j_1j_1r_2})^{m_{j_1r_2}}\prod_{u_2=1,u_2\neq u_1}^{q_1}(a_{j_1j_2u_1} - a_{j_1j_2u_2})^{n_{j_1u_2}} > 0$ *and* $a_{j_2j_20}\prod_{v_2=1,v_2\neq v_1}^{p_2}(a_{j_2j_1v_1} - a_{j_2j_1v_2})^{m_{j_2v_2}} < 0$,

$$(a_{j_1j_2u_1}, a_{j_2j_1v_1}) = \underbrace{([(2m_{j_1r_1}1+1),(2n_{j_1u_1}1+1)]^{\text{th}}\text{DP}_{+}\text{:DP}_{-},[2(m_{j_2v_1}1 - m_{j_1r_1}1)]^{\text{th}}{}_{\text{II:DI}}\text{US})}_{([(2m_{j_1r_1}1+1),(2n_{j_1u_1}1+1)],[2(m_{j_2v_1}1-m_{j_1r_1}1)])\text{-inflection down-parabola upper-saddle}}. \tag{1.201}$$

The equilibrium is a $([(2m_{j_1r_11}+1),\,(2n_{j_1u_11}+1)]^{\text{th}}\text{DP}_+\text{:DP}_-,[2(m_{j_2v_11}-m_{j_1r_11})]^{\text{th}}{}_{\text{II:DI}}\text{US})$*-inflection down-parabola upper-saddle at* $(x^*_{j_2},x^*_{j_1})=(a_{j_1j_2u_1},a_{j_2j_1v_1})$.

- *For* $a_{j_1j_10}\prod_{r_2=1,r_2\neq r_1}^{p_1}(a_{j_1j_1r_1}-a_{j_1j_1r_2})^{m_{j_1r_2}}\prod_{u_2=1,u_2\neq u_1}^{q_1}(a_{j_1j_2u_1}-a_{j_1j_2u_2})^{n_{j_1u_2}}<0$ *and* $a_{j_2j_20}\prod_{v_2=1,v_2\neq v_1}^{p_2}(a_{j_2j_1v_1}-a_{j_2j_1v_2})^{m_{j_2v_2}}<0$,

$$
\begin{aligned}
&(a_{j_1j_2u_1},a_{j_2j_1v_1})\\
&\quad=\underbrace{([(2m_{j_1r_11}+1),\,(2n_{j_1u_11}+1)]^{\text{th}}\text{UP}_-\text{:UP}_-,[2(m_{j_2v_11}-m_{j_1r_11})]^{\text{th}}{}_{\text{DI:II}}\text{LS})}_{([(2m_{j_1r_11}+1),(2n_{j_1u_11}+1)],[2(m_{j_2v_11}-m_{j_1r_11})])\text{-inlfection up-parabola lower-saddle}}.
\end{aligned}
\tag{1.202}
$$

The equilibrium is a $([(2m_{j_1r_11}+1),\,(2n_{j_1u_11}+1)]^{\text{th}}\text{UP}_-\text{:UP}_-,[2(m_{j_2v_11}-m_{j_1r_11})]^{\text{th}}{}_{\text{DI:II}}\text{LS})$*-inflection up-parabola lower-saddle at* $(x^*_{j_2},x^*_{j_1})=(a_{j_1j_2u_1},a_{j_2j_1v_1})$.

(ii$_{2a2b}$) For $m_{j_1r_1}=2m_{j_1r_11}$, $n_{j_1u_1}=2n_{j_1u_11}+1$ *and* $m_{j_2v_1}=2m_{j_1v_11}+1$, *the equilibrium of* $(x^*_{j_2},x^*_{j_1})=(a_{j_1j_2u_1},a_{j_2j_1v_1})$ *on the infinite-equilibrium* $x^*_{j_1}=a_{j_2j_1v_1}=a_{j_1j_1r_1}$ *has the following properties.*

- *For* $a_{j_1j_10}\prod_{r_2=1,r_2\neq r_1}^{p_1}(a_{j_1j_1r_1}-a_{j_1j_1r_2})^{m_{j_1r_2}}\prod_{u_2=1,u_2\neq u_1}^{q_1}(a_{j_1j_2u_1}-a_{j_1j_2u_2})^{n_{j_1u_2}}>0$ *and* $a_{j_2j_20}\prod_{v_2=1,v_2\neq v_1}^{p_2}(a_{j_2j_1v_1}-a_{j_2j_1v_2})^{m_{j_2v_2}}>0$,

$$
\begin{aligned}
&(a_{j_1j_2u_1},a_{j_2j_1v_1})\\
&=\underbrace{([(2m_{j_1r_11}),\,(2n_{j_1u_11}+1)]^{\text{th}}\text{DP}_-\text{:UP}_+,[2(m_{j_2v_11}-m_{j_1r_11})+1]^{\text{th}}{}_{\text{DP}}\text{LS:}{}_{\text{UP}}\text{US})}_{([2m_{j_1r_11}+1),(2n_{j_1u_11}+1)],[2(m_{j_2v_11}-m_{j_1r_11})+1])\text{-positive hyperbolic lower-to-upper saddle}}.
\end{aligned}
\tag{1.203}
$$

The equilibrium is a $[(2m_{j_1r_11}),\,(2n_{j_1u_11}+1)]^{\text{th}}\text{DP}_-\text{:UP}_+,[2(m_{j_2v_11}-m_{j_1r_11})+1]^{\text{th}}{}_{\text{DP}}\text{LS:}{}_{\text{UP}}\text{US})$*-positive hyperbolic lower-to-upper saddle at* $(x^*_{j_2},x^*_{j_1})=(a_{j_1j_2u_1},a_{j_2j_1v_1})$.

- *For* $a_{j_1j_10}\prod_{r_2=1,r_2\neq r_1}^{p_1}(a_{j_1j_1r_1}-a_{j_1j_1r_2})^{m_{j_1r_2}}\prod_{u_2=1,u_2\neq u_1}^{q_1}(a_{j_1j_2u_1}-a_{j_1j_2u_2})^{n_{j_1u_2}}<0$ *and* $a_{j_2j_20}\prod_{v_2=1,v_2\neq v_1}^{p_2}(a_{j_2j_1v_1}-a_{j_2j_1v_2})^{m_{j_2v_2}}>0$,

$$
\begin{aligned}
&(a_{j_1j_2u_1},a_{j_2j_1v_1})\\
&=\underbrace{([(2m_{j_1r_11}),\,(2n_{j_1u_11}+1)]^{\text{th}}\text{UP}_-\text{:DP}_+,[2(m_{j_2v_11}-m_{j_1r_11})+1]^{\text{th}}{}_{\text{UP}}\text{US:}{}_{\text{DP}}\text{LS})}_{([2m_{j_1r_11}+1),(2n_{j_1u_11}+1)],[2(m_{j_2v_11}-m_{j_1r_11})+1])\text{-CCW circular upper-to-lower saddle}}.
\end{aligned}
\tag{1.204}
$$

The equilibrium is a $[(2m_{j_1r_11}),\,(2n_{j_1u_11}+1)]^{\text{th}}\text{UP}_-\text{:DP}_+,[2(m_{j_2v_11}-m_{j_1r_11})+1]^{\text{th}}{}_{\text{UP}}\text{US:}{}_{\text{DP}}\text{LS})$*-CCW circular upper-to-lower saddle at* $(x^*_{j_2},x^*_{j_1})=(a_{j_1j_2u_1},a_{j_2j_1v_1})$.

- *For* $a_{j_1j_10}\prod_{r_2=1,r_2\neq r_1}^{p_1}(a_{j_1j_1r_1}-a_{j_1j_1r_2})^{m_{j_1r_2}}\prod_{u_2=1,u_2\neq u_1}^{q_1}(a_{j_1j_2u_1}-a_{j_1j_2u_2})^{n_{j_1u_2}}>0$ *and* $a_{j_2j_20}\prod_{v_2=1,v_2\neq v_1}^{p_2}(a_{j_2j_1v_1}-a_{j_2j_1v_2})^{m_{j_2v_2}}<0$,

$$(a_{j_1j_2u_1},a_{j_2j_1v_1}) = \underbrace{([(2m_{j_1r_1}1),(2n_{j_1u_1}1+1)]^{\text{th}}\text{UP}_{+}\text{:DP}_{-},[2(m_{j_2v_1}1-m_{j_1r_1}1)+1]^{\text{th}}{}_{\text{UP}}\text{LS:}_{\text{DP}}\text{US})}_{([2m_{j_1r_1}1+1),(2n_{j_1u_1}1+1)],[2(m_{j_2v_1}1-m_{j_1r_1}1)-1])\text{-CW circular lower-to-upper saddle}}. \tag{1.205}$$

The equilibrium is a $[(2m_{j_1r_1}1),(2n_{j_1u_1}1+1)]^{\text{th}}\text{UP}_{-}\text{:DP}_{+},[2(m_{j_2v_1}1-m_{j_1r_1}1)+1]^{\text{th}}{}_{\text{UP}}\text{LS:}_{\text{DP}}\text{US})$*-CW circular lower-to-upper saddle at* $(x_{j_2}^*,x_{j_1}^*)=(a_{j_1j_2u_1},a_{j_2j_1v_1})$.

- *For* $a_{j_1j_10}\prod_{r_2=1,r_2\neq r_1}^{p_1}(a_{j_1j_1r_1}-a_{j_1j_1r_2})^{m_{j_1r_2}}\prod_{u_2=1,u_2\neq u_1}^{q_1}(a_{j_1j_2u_1}-a_{j_1j_2u_2})^{n_{j_1u_2}}<0$ *and* $a_{j_2j_20}\prod_{v_2=1,v_2\neq v_1}^{p_2}(a_{j_2j_1v_1}-a_{j_2j_1v_2})^{m_{j_2v_2}}<0$,

$$(a_{j_1j_2u_1},a_{j_2j_1v_1}) = \underbrace{([(2m_{j_1r_1}1),(2n_{j_1u_1}1+1)]^{\text{th}}\text{DP}_{+}\text{:UP}_{-},[2(m_{j_2v_1}1-m_{j_1r_1}1)+1]^{\text{th}}{}_{\text{DP}}\text{US:}_{\text{UP}}\text{LS})}_{([2m_{j_1r_1}1+1),(2n_{j_1u_1}1+1)],[2(m_{j_2v_1}1-m_{j_1r_1}1)+1])\text{-negative hyperbolic upper-to-lower saddle}}. \tag{1.206}$$

The equilibrium is a $[(2m_{j_1r_1}1),(2n_{j_1u_1}1+1)]^{\text{th}}\text{DP}_{-}\text{:UP}_{+},[2(m_{j_2v_1}1-m_{j_1r_1}1)+1]^{\text{th}}{}_{\text{DP}}\text{LS:}_{\text{UP}}\text{US})$*-negative hyperbolic upper-to-lower saddle at* $(x_{j_2}^*,x_{j_1}^*)=(a_{j_1j_2u_1},a_{j_2j_1v_1})$.

(ii_{2a2c}) *For* $m_{j_1r_1}=2m_{j_1r_1}1+1$, $n_{j_1u_1}=2n_{j_1u_1}1+1$ *and* $m_{j_2v_1}=2m_{j_1v_1}1$, *the equilibrium of* $(x_{j_1}^*,x_{j_2}^*)=(a_{j_1j_1r_1},a_{j_1j_2u_1})$ *on the infinite-equilibrium* $x_{j_1}^*=a_{j_2j_1v_1}=a_{j_1j_1r_1}$ *has the following properties.*

- *For* $a_{j_1j_10}\prod_{r_2=1,r_2\neq r_1}^{p_1}(a_{j_1j_1r_1}-a_{j_1j_1r_2})^{m_{j_1r_2}}\prod_{u_2=1,u_2\neq u_1}^{q_1}(a_{j_1j_2u_1}-a_{j_1j_2u_2})^{n_{j_1u_2}}>0$ *and* $a_{j_2j_20}\prod_{v_2=1,v_2\neq v_1}^{p_2}(a_{j_2j_1v_1}-a_{j_2j_1v_2})^{m_{j_2v_2}}>0$,

$$(a_{j_1j_2u_1},a_{j_2j_1v_1}) = \underbrace{([(2m_{j_1r_1}1+1),(2n_{j_1u_1}1+1)]^{\text{th}}\text{DP}_{+}\text{:UP}_{+},[2(m_{j_2v_1}1-m_{j_1r_1}1)-1]^{\text{th}}{}_{\text{DP}}\text{SI:}_{\text{UP}}\text{SO})}_{([2m_{j_1r_1}1+1),(2n_{j_1u_1}1+1)],[2(m_{j_2v_1}1-m_{j_1r_1}1)-1])\text{-hyperbolic parabola sink-to-source}}. \tag{1.207}$$

The equilibrium is a $[(2m_{j_1r_1}1),(2n_{j_1u_1}1+1)]^{\text{th}}\text{DP}_{-}\text{:UP}_{+},[2(m_{j_2v_1}1-m_{j_1r_1}1)-1]^{\text{th}}{}_{\text{DP}}\text{SI:}_{\text{UP}}\text{SO})$ *-hyperbolic parabola sink-to-source at* $(x_{j_2}^*,x_{j_1}^*)=(a_{j_1j_2u_1},a_{j_2j_1v_1})$.

- *For* $a_{j_1j_10}\prod_{r_2=1,r_2\neq r_1}^{p_1}(a_{j_1j_1r_1}-a_{j_1j_1r_2})^{m_{j_1r_2}}\prod_{u_2=1,u_2\neq u_1}^{q_1}(a_{j_1j_2u_1}-a_{j_1j_2u_2})^{n_{j_1u_2}}<0$ *and* $a_{j_2j_20}\prod_{v_2=1,v_2\neq v_1}^{p_2}(a_{j_2j_1v_1}-a_{j_2j_1v_2})^{m_{j_2v_2}}>0$,

$$
\begin{aligned}
&(a_{j_1j_2u_1}, a_{j_2j_1v_1})\\
&=\underbrace{([(2m_{j_1r_1}1+1),\,(2n_{j_1u_1}1+1)]^{\text{th}}\text{UP}_{+}\text{:DP}_{+},[2(m_{j_2v_1}1-m_{j_1r_1}1)-1]^{\text{th}}{}_{\text{UP}}\text{SO:}_{\text{DP}}\text{SI})}_{([2m_{j_1r_1}1+1),(2n_{j_1u_1}1+1)],[2(m_{j_2v_1}1-m_{j_1r_1}1)-1])\text{-circular parabola source-to-sink}}.
\end{aligned}
\tag{1.208}
$$

The equilibrium is a $[(2m_{j_1r_1}1),\,(2n_{j_1u_1}1+1)]^{\text{th}}\text{DP}_{-}\text{:UP}_{+},[2(m_{j_2v_1}1-m_{j_1r_1}1)-1]^{\text{th}}$ ${}_{\text{UP}}\text{SO:}_{\text{UP}}\text{SI})$ *-circular parabola source-to-sink at* $(x^*_{j_2}, x^*_{j_1}) = (a_{j_1j_2u_1}, a_{j_2j_1v_1})$.

- *For* $a_{j_1j_10}\prod_{r_2=1,r_2\neq r_1}^{p_1}(a_{j_1j_1r_1}-a_{j_1j_1r_2})^{m_{j_1r_2}}\prod_{u_2=1,u_2\neq u_1}^{q_1}(a_{j_1j_2u_1}-a_{j_1j_2u_2})^{n_{j_1u_2}}>0$ *and* $a_{j_2j_20}\prod_{v_2=1,v_2\neq v_1}^{p_2}(a_{j_2j_1v_1}-a_{j_2j_1v_2})^{m_{j_2v_2}}<0$,

$$
\begin{aligned}
&(a_{j_1j_2u_1}, a_{j_2j_1v_1})\\
&=\underbrace{([(2m_{j_1r_1}1+1),\,(2n_{j_1u_1}1+1)]^{\text{th}}\text{UP}_{-}\text{:DP}_{-},[2(m_{j_2v_1}1-m_{j_1r_1}1)-1]^{\text{th}}{}_{\text{UP}}\text{SI:}_{\text{DP}}\text{SO})}_{([2m_{j_1r_1}1+1),(2n_{j_1u_1}1+1)],[2(m_{j_2v_1}1-m_{j_1r_1}1)-1])\text{-circular parabola sink-to-source}}.
\end{aligned}
\tag{1.209}
$$

The equilibrium is a $[(2m_{j_1r_1}1),\,(2n_{j_1u_1}1+1)]^{\text{th}}\text{DP}_{-}\text{:UP}_{+},[2(m_{j_2v_1}1-m_{j_1r_1}1)-1]^{\text{th}}$ ${}_{\text{UP}}\text{SI:}_{\text{UP}}\text{SO})$ *-circular parabola sink-to-source at* $(x^*_{j_2}, x^*_{j_1}) = (a_{j_1j_2u_1}, a_{j_2j_1v_1})$.

- *For* $a_{j_1j_10}\prod_{r_2=1,r_2\neq r_1}^{p_1}(a_{j_1j_1r_1}-a_{j_1j_1r_2})^{m_{j_1r_2}}\prod_{u_2=1,u_2\neq u_1}^{q_1}(a_{j_1j_2u_1}-a_{j_1j_2u_2})^{n_{j_1u_2}}<0$ *and* $a_{j_2j_20}\prod_{v_2=1,v_2\neq v_1}^{p_2}(a_{j_2j_1v_1}-a_{j_2j_1v_2})^{m_{j_2v_2}}<0$,

$$
\begin{aligned}
&(a_{j_1j_2u_1}, a_{j_2j_1v_1})\\
&=\underbrace{([(2m_{j_1r_1}1+1),\,(2n_{j_1u_1}1+1)]^{\text{th}}\text{DP}_{-}\text{:UP}_{-},[2(m_{j_2v_1}1-m_{j_1r_1}1)-1]^{\text{th}}{}_{\text{DP}}\text{SO:}_{\text{UP}}\text{SI})}_{([2m_{j_1r_1}1+1),(2n_{j_1u_1}1+1)],[2(m_{j_2v_1}1-m_{j_1r_1}1)-1])\text{-hyperbolic parabola source-to-sink}}.
\end{aligned}
\tag{1.210}
$$

The equilibrium is a $[(2m_{j_1r_1}1),\,(2n_{j_1u_1}1+1)]^{\text{th}}\text{DP}_{-}\text{:UP}_{-},[2(m_{j_2v_1}1-m_{j_1r_1}1)-1]^{\text{th}}$ ${}_{\text{DP}}\text{SO:}_{\text{UP}}\text{SI})$ *-hyperbolic parabola source-to-sink at* $(x^*_{j_2}, x^*_{j_1}) = (a_{j_1j_2u_1}, a_{j_2j_1v_1})$.

(ii$_{2a2d}$*) For* $m_{j_1r_1}=2m_{j_1r_1}1$, $n_{j_1u_1}=2n_{j_1u_1}1+1$ *and* $m_{j_2v_1}=2m_{j_1v_1}1$, *the equilibrium of* $(x^*_{j_2}, x^*_{j_1}) = (a_{j_1j_2u_1}, a_{j_2j_1v_1})$ *on the infinite-equilibrium* $x^*_{j_1}=a_{j_2j_1v_1}=a_{j_1j_1r_1}$ *has the following properties.*

- *For* $a_{j_1j_10}\prod_{r_2=1,r_2\neq r_1}^{p_1}(a_{j_1j_1r_1}-a_{j_1j_1r_2})^{m_{j_1r_2}}\prod_{u_2=1,u_2\neq u_1}^{q_1}(a_{j_1j_2u_1}-a_{j_1j_2u_2})^{n_{j_1u_2}}>0$ *and* $a_{j_2j_20}\prod_{v_2=1,v_2\neq v_1}^{p_2}(a_{j_2j_1v_1}-a_{j_2j_1v_2})^{m_{j_2v_2}}>0$,

$$
\begin{aligned}
&(a_{j_1j_2u_1}, a_{j_2j_1v_1})\\
&=\underbrace{([(2m_{j_1r_1}1),\,(2n_{j_1u_1}1+1)]^{\text{th}}\text{UP}_{+}\text{:UP}_{+},[2(m_{j_2v_1}1-m_{j_1r_1}1)]^{\text{th}}{}_{\text{DI:II}}\text{US})}_{([(2m_{j_1r_1}1),(2n_{j_1u_1}1+1)],[2(m_{j_2v_1}1-m_{j_1r_1}1)]\text{)-inflection up-parabola upper-saddle}}.
\end{aligned}
\tag{1.211}
$$

The equilibrium is a $([(2m_{j_1r_1}1),\,(2n_{j_1u_1}1+1)]^{\text{th}}\text{UP}_{+}\text{:UP}_{+},[2(m_{j_2v_1}1-m_{j_1r_1}1)]^{\text{th}}$ ${}_{\text{DI:II}}\text{US})$*-inflection up-parabola upper-saddle at* $(x^*_{j_2}, x^*_{j_1}) = (a_{j_1j_2u_1}, a_{j_2j_1v_1})$.

- *For* $a_{j_1j_10}\prod_{r_2=1,r_2\neq r_1}^{p_1}(a_{j_1j_1r_1}-a_{j_1j_1r_2})^{m_{j_1r_2}}\prod_{u_2=1,u_2\neq u_1}^{q_1}(a_{j_1j_2u_1}-a_{j_1j_2u_2})^{n_{j_1u_2}}<0$ *and* $a_{j_2j_20}\prod_{v_2=1,v_2\neq v_1}^{p_2}(a_{j_2j_1v_1}-a_{j_2j_1v_2})^{m_{j_2v_2}}>0$,

$$(a_{j_1j_2u_1},a_{j_2j_1v_1})=\underbrace{([(2m_{j_1r_1}1),(2n_{j_1u_1}1+1)]^{\text{th}}\text{DP}_{+}\text{:DP}_{+},[2(m_{j_2v_1}1-m_{j_1r_1}1)]^{\text{th}}{}_{\text{II:DI}}\text{US})}_{([(2m_{j_1r_1}1),(2n_{j_1u_1}1+1)],[2(m_{j_2v_1}1-m_{j_1r_1}1)])\text{-inflection up-parabola upper-saddle}}. \tag{1.212}$$

The equilibrium is a $([(2m_{j_1r_1}1),(2n_{j_1u_1}1+1)]^{\text{th}}\text{DP}_{+}\text{:DP}_{+},[2(m_{j_2v_1}1-m_{j_1r_1}1)]^{\text{th}}{}_{\text{II:DI}}\text{US})$*-inflection down-parabola upper-saddle at* $(x_{j_2}^*,x_{j_1}^*)=(a_{j_1j_2u_1},a_{j_2j_1v_1})$.

- *For* $a_{j_1j_10}\prod_{r_2=1,r_2\neq r_1}^{p_1}(a_{j_1j_1r_1}-a_{j_1j_1r_2})^{m_{j_1r_2}}\prod_{u_2=1,u_2\neq u_1}^{q_1}(a_{j_1j_2u_1}-a_{j_1j_2u_2})^{n_{j_1u_2}}>0$ *and* $a_{j_2j_20}\prod_{v_2=1,v_2\neq v_1}^{p_2}(a_{j_2j_1v_1}-a_{j_2j_1v_2})^{m_{j_2v_2}}<0$,

$$(a_{j_1j_2u_1},a_{j_2j_1v_1})=\underbrace{([(2m_{j_1r_1}1),(2n_{j_1u_1}1+1)]^{\text{th}}\text{DP}_{-}\text{:DP}_{-},[2(m_{j_2v_1}1-m_{j_1r_1}1)]^{\text{th}}{}_{\text{II:DI}}\text{LS})}_{([(2m_{j_1r_1}1),(2n_{j_1u_1}1+1)],[2(m_{j_2v_1}1-m_{j_1r_1}1)])\text{-inflection down-parabola lower-saddle}}. \tag{1.213}$$

The equilibrium is a $([(2m_{j_1r_1}1),(2n_{j_1u_1}1+1)]^{\text{th}}\text{DP}_{-}\text{:DP}_{-},[2(m_{j_2v_1}1-m_{j_1r_1}1)]^{\text{th}}{}_{\text{II:DI}}\text{LS})$*-inflection down-parabola lower-saddle at* $(x_{j_2}^*,x_{j_1}^*)=(a_{j_1j_2u_1},a_{j_2j_1v_1})$.

- *For* $a_{j_1j_10}\prod_{r_2=1,r_2\neq r_1}^{p_1}(a_{j_1j_1r_1}-a_{j_1j_1r_2})^{m_{j_1r_2}}\prod_{u_2=1,u_2\neq u_1}^{q_1}(a_{j_1j_2u_1}-a_{j_1j_2u_2})^{n_{j_1u_2}}<0$ *and* $a_{j_2j_20}\prod_{v_2=1,v_2\neq v_1}^{p_2}(a_{j_2j_1v_1}-a_{j_2j_1v_2})^{m_{j_2v_2}}<0$,

$$(a_{j_1j_2u_1},a_{j_2j_1v_1})=\underbrace{([(2m_{j_1r_1}1),(2n_{j_1u_1}1+1)]^{\text{th}}\text{UP}_{-}\text{:UP}_{-},[2(m_{j_2v_1}1-m_{j_1r_1}1)]^{\text{th}}{}_{\text{DI:II}}\text{LS})}_{([(2m_{j_1r_1}1),(2n_{j_1u_1}1+1)],[2(m_{j_2v_1}1-m_{j_1r_1}1)])\text{-inflection up-parabola lower-saddle}}. \tag{1.214}$$

The equilibrium is a $([(2m_{j_1r_1}1),(2n_{j_1u_1}1+1)]^{\text{th}}\text{UP}_{+}\text{:UP}_{+},[2(m_{j_2v_1}1-m_{j_1r_1}1)]^{\text{th}}{}_{\text{DI:II}}\text{LS})$*-inflection up-parabola lower-saddle at* $(x_{j_2}^*,x_{j_1}^*)=(a_{j_1j_2u_1},a_{j_2j_1v_1})$.

(ii_{2a2e}*) For* $m_{j_1r_1}=2m_{j_1r_1}1+1$, $n_{j_1u_1}=2n_{j_1u_1}1$ *and* $m_{j_2v_1}=2m_{j_1v_1}1+1$*, the equilibrium of* $(x_{j_2}^*,x_{j_1}^*)=(a_{j_1j_2u_1},a_{j_2j_1v_1})$ *on the infinite-equilibrium* $x_{j_1}^*=a_{j_2j_1v_1}=a_{j_1j_1r_1}$ *has the following properties.*

- *For* $a_{j_1j_10}\prod_{r_2=1,r_2\neq r_1}^{p_1}(a_{j_1j_1r_1}-a_{j_1j_1r_2})^{m_{j_1r_2}}\prod_{u_2=1,u_2\neq u_1}^{q_1}(a_{j_1j_2u_1}-a_{j_1j_2u_2})^{n_{j_1u_2}}>0$ *and* $a_{j_2j_20}\prod_{v_2=1,v_2\neq v_1}^{p_2}(a_{j_2j_1v_1}-a_{j_2j_1v_2})^{m_{j_2v_2}}>0$,

$$(a_{j_1j_2u_1},a_{j_2j_1v_1})=\underbrace{([(2m_{j_1r_1}1+1),(2n_{j_1u_1}1)]^{\text{th}}\text{II}_{-}\text{:II}_{+},[2(m_{j_2v_1}1-m_{j_1r_1}1)]^{\text{th}}{}_{\text{II:II}}\text{SO})}_{([(2m_{j_1r_1}1+1),(2n_{j_1u_1}1)],[2(m_{j_2v_1}1-m_{j_1r_1}1)])\text{-increasing-inflection source}}. \tag{1.215}$$

The equilibrium is a $([(2m_{j_1r_11}+1),(2n_{j_1u_11})]^{\text{th}}\text{II}_-{:}\text{II}_+,[2(m_{j_2v_11}-m_{j_1r_11})]^{\text{th}}{}_{\text{II:II}}\text{SO})$*-increasing-inflection source at* $(x^*_{j2},x^*_{j_1})=(a_{j_1j_2u_1},a_{j_2j_1v_1})$.

- *For* $a_{j_1j_10}\prod_{r_2=1,r_2\neq r_1}^{p_1}(a_{j_1j_1r_1}-a_{j_1j_1r_2})^{m_{j_1r_2}}\prod_{u_2=1,u_2\neq u_1}^{q_1}(a_{j_1j_2u_1}-a_{j_1j_2u_2})^{n_{j_1u_2}}<0$ *and* $a_{j_2j_20}\prod_{v_2=1,v_2\neq v_1}^{p_2}(a_{j_2j_1v_1}-a_{j_2j_1v_2})^{m_{j_2v_2}}>0$,

$$(a_{j_1j_2u_1},a_{j_2j_1v_1})=\underbrace{([(2m_{j_1r_11}+1),(2n_{j_1u_11})]^{\text{th}}\text{DI}_-{:}\text{DI}_+,[2(m_{j_2v_11}-m_{j_1r_11})]^{\text{th}}{}_{\text{DI:DI}}\text{SI})}_{([(2m_{j_1r_1}1+1),(2n_{j_1u_1}1)],[2(m_{j_2v_1}1-m_{j_1r_1}1)])\text{-decreasing-inflection sink}}. \tag{1.216}$$

The equilibrium is a $([(2m_{j_1r_11}+1),(2n_{j_1u_11})]^{\text{th}}\text{DI}_-{:}\text{DI}_+,[2(m_{j_2v_11}-m_{j_1r_11})]^{\text{th}}{}_{\text{DI:DI}}\text{SI})$*-decreasing-inflection sink at* $(x^*_{j_2},x^*_{j_1})=(a_{j_1j_2u_1},a_{j_2j_1v_1})$.

- *For* $a_{j_1j_10}\prod_{r_2=1,r_2\neq r_1}^{p_1}(a_{j_1j_1r_1}-a_{j_1j_1r_2})^{m_{j_1r_2}}\prod_{u_2=1,u_2\neq u_1}^{q_1}(a_{j_1j_2u_1}-a_{j_1j_2u_2})^{n_{j_1u_2}}>0$ *and* $a_{j_2j_20}\prod_{v_2=1,v_2\neq v_1}^{p_2}(a_{j_2j_1v_1}-a_{j_2j_1v_2})^{m_{j_2v_2}}<0$,

$$(a_{j_1j_2u_1},a_{j_2j_1v_1})=\underbrace{([(2m_{j_1r_11}+1),(2n_{j_1u_11})]^{\text{th}}\text{DI}_+{:}\text{DI}_-,[2(m_{j_2v_11}-m_{j_1r_11})]^{\text{th}}{}_{\text{DI:DI}}\text{SO})}_{([(2m_{j_1r_1}1+1),(2n_{j_1u_1}1)],[2(m_{j_2v_1}1-m_{j_1r_1}1)])\text{-decreasing-inflection sink}}. \tag{1.217}$$

The equilibrium is a $([(2m_{j_1r_11}+1),(2n_{j_1u_11})]^{\text{th}}\text{DI}_+{:}\text{DI}_1,[2(m_{j_2v_11}-m_{j_1r_11})]^{\text{th}}{}_{\text{DI:DI}}\text{SO})$*-decreasing-inflection source at* $(x^*_{j_2},x^*_{j_1})=(a_{j_1j_2u_1},a_{j_2j_1v_1})$.

- *For* $a_{j_1j_10}\prod_{r_2=1,r_2\neq r_1}^{p_1}(a_{j_1j_1r_1}-a_{j_1j_1r_2})^{m_{j_1r_2}}\prod_{u_2=1,u_2\neq u_1}^{q_1}(a_{j_1j_2u_1}-a_{j_1j_2u_2})^{n_{j_1u_2}}<0$ *and* $a_{j_2j_20}\prod_{v_2=1,v_2\neq v_1}^{p_2}(a_{j_2j_1v_1}-a_{j_2j_1v_2})^{m_{j_2v_2}}<0$,

$$(a_{j_1j_2u_1},a_{j_2j_1v_1})=\underbrace{([(2m_{j_1r_11}+1),(2n_{j_1u_11})]^{\text{th}}\text{II}_+{:}\text{II}_-,[2(m_{j_2v_11}-m_{j_1r_11})]^{\text{th}}{}_{\text{II:II}}\text{SI})}_{([(2m_{j_1r_1}1+1),(2n_{j_1u_1}1)],[2(m_{j_2v_1}1-m_{j_1r_1}1)])\text{-increasing-inflection sink}}. \tag{1.218}$$

The equilibrium is a $([(2m_{j_1r_11}+1),(2n_{j_1u_11})]^{\text{th}}\text{II}_+{:}\text{II}_-,[2(m_{j_2v_11}-m_{j_1r_11})]^{\text{th}}{}_{\text{II:II}}\text{SI})$*-increasing-inflection sink at* $(x^*_{j_2},x^*_{j_1})=(a_{j_1j_2u_1},a_{j_2j_1v_1})$.

(ii_{2a2f}) For $m_{j_1r_1}=2m_{j_1r_11}$, $n_{j_1u_1}=2n_{j_1u_11}$ *and* $m_{j_2v_1}=2m_{j_1v_11}+1$, *the equilibrium of* $(x^*_{j_2},x^*_{j_1})=(a_{j_1j_2u_1},a_{j_2j_1v_1})$ *on the infinite-equilibrium* $x^*_{j_1}=a_{j_2j_1v_1}=a_{j_1j_1r_1}$ *has the following properties.*

- *For* $a_{j_1j_10}\prod_{r_2=1,r_2\neq r_1}^{p_1}(a_{j_1j_1r_1}-a_{j_1j_1r_2})^{m_{j_1r_2}}\prod_{u_2=1,u_2\neq u_1}^{q_1}(a_{j_1j_2u_1}-a_{j_1j_2u_2})^{n_{j_1u_2}}>0$ *and* $a_{j_2j_20}\prod_{v_2=1,v_2\neq v_1}^{p_2}(a_{j_2j_1v_1}-a_{j_2j_1v_2})^{m_{j_2v_2}}>0$,

$$
\begin{aligned}
&(a_{j_1j_2u_1}, a_{j_2j_1v_1})\\
&= \underbrace{([(2m_{j_1r_1}1), (2n_{j_1u_1}1)]^{\text{th}}\text{DI}_{-}\text{:II}_{+},[2(m_{j_2v_1}1 - m_{j_1r_1}1) + 1]^{\text{th}}{}_{\text{UP:UP}}\text{US})}_{([(2m_{j_1r_1}1),(2n_{j_1u_1}1)],[2(m_{j_2v_1}1-m_{j_1r_1}1)+1])\text{-up-parabola upper-saddle}}.
\end{aligned}
\tag{1.219}
$$

The equilibrium is a $([(2m_{j_1r_1}1), (2n_{j_1u_1}1)]^{\text{th}}\text{DI}_{-}\text{:II}_{+},[2(m_{j_2v_1}1 - m_{j_1r_1}1) + 1]^{\text{th}}{}_{\text{UP:UP}}\text{US})$*-up-parabola upper-saddle at* $(x_{j_2}^*, x_{j_1}^*) = (a_{j_1j_2u_1}, a_{j_2j_1v_1})$.

- *For* $a_{j_1j_10}\prod_{r_2=1,r_2\neq r_1}^{p_1}(a_{j_1j_1r_1} - a_{j_1j_1r_2})^{m_{j_1r_2}}\prod_{u_2=1,u_2\neq u_1}^{q_1}(a_{j_1j_2u_1} - a_{j_1j_2u_2})^{n_{j_1u_2}} < 0$ *and* $a_{j_2j_20}\prod_{v_2=1,v_2\neq v_1}^{p_2}(a_{j_2j_1v_1} - a_{j_2j_1v_2})^{m_{j_2v_2}} > 0$,

$$
\begin{aligned}
&(a_{j_1j_2u_1}, a_{j_2j_1v_1})\\
&= \underbrace{([(2m_{j_1r_1}1), (2n_{j_1u_1}1)]^{\text{th}}\text{II}_{-}\text{:DI}_{+},[2(m_{j_2v_1}1 - m_{j_1r_1}1) + 1]^{\text{th}}{}_{\text{DP:DP}}\text{LS})}_{([(2m_{j_1r_1}1),(2n_{j_1u_1}1)],[2(m_{j_2v_1}1-m_{j_1r_1}1)+1])\text{-down-parabola lower-saddle}}.
\end{aligned}
\tag{1.220}
$$

The equilibrium is a $([(2m_{j_1r_1}1), (2n_{j_1u_1}1)]^{\text{th}}\text{DI}_{-}\text{:II}_{+},[2(m_{j_2v_1}1 - m_{j_1r_1}1) + 1]^{\text{th}}{}_{\text{DP:DP}}\text{LS})$*-down-parabola lower-saddle at* $(x_{j_2}^*, x_{j_1}^*) = (a_{j_1j_2u_1}, a_{j_2j_1v_1})$.

- *For* $a_{j_1j_10}\prod_{r_2=1,r_2\neq r_1}^{p_1}(a_{j_1j_1r_1} - a_{j_1j_1r_2})^{m_{j_1r_2}}\prod_{u_2=1,u_2\neq u_1}^{q_1}(a_{j_1j_2u_1} - a_{j_1j_2u_2})^{n_{j_1u_2}} > 0$ *and* $a_{j_2j_20}\prod_{v_2=1,v_2\neq v_1}^{p_2}(a_{j_2j_1v_1} - a_{j_2j_1v_2})^{m_{j_2v_2}} < 0$,

$$
\begin{aligned}
&(a_{j_1j_2u_1}, a_{j_2j_1v_1})\\
&= \underbrace{([(2m_{j_1r_1}1), (2n_{j_1u_1}1)]^{\text{th}}\text{II}_{+}\text{:DI}_{-},[2(m_{j_2v_1}1 - m_{j_1r_1}1) + 1]^{\text{th}}{}_{\text{DP:DP}}\text{US})}_{([(2m_{j_1r_1}1),(2n_{j_1u_1}1)],[2(m_{j_2v_1}1-m_{j_1r_1}1)+1])\text{-down-parabola upper-saddle}}.
\end{aligned}
\tag{1.221}
$$

The equilibrium is a $([(2m_{j_1r_1}1), (2n_{j_1u_1}1)]^{\text{th}}\text{DI}_{-}\text{:II}_{+},[2(m_{j_2v_1}1 - m_{j_1r_1}1) + 1]^{\text{th}}{}_{\text{DP:DP}}\text{US})$*-down-parabola upper-saddle at* $(x_{j_2}^*, x_{j_1}^*) = (a_{j_1j_2u_1}, a_{j_2j_1v_1})$.

- *For* $a_{j_1j_10}\prod_{r_2=1,r_2\neq r_1}^{p_1}(a_{j_1j_1r_1} - a_{j_1j_1r_2})^{m_{j_1r_2}}\prod_{u_2=1,u_2\neq u_1}^{q_1}(a_{j_1j_2u_1} - a_{j_1j_2u_2})^{n_{j_1u_2}} < 0$ *and* $a_{j_2j_20}\prod_{v_2=1,v_2\neq v_1}^{p_2}(a_{j_2j_1v_1} - a_{j_2j_1v_2})^{m_{j_2v_2}} < 0$,

$$
\begin{aligned}
&(a_{j_1j_2u_1}, a_{j_2j_1v_1})\\
&= \underbrace{([(2m_{j_1r_1}1), (2n_{j_1u_1}1)]^{\text{th}}\text{DI}_{+}\text{:II}_{-},[2(m_{j_2v_1}1 - m_{j_1r_1}1) + 1]^{\text{th}}{}_{\text{UP:UP}}\text{LS})}_{([(2m_{j_1r_1}1),(2n_{j_1u_1}1)],[2(m_{j_2v_1}1-m_{j_1r_1}1)+1])\text{-up-parabola lower-saddle}}.
\end{aligned}
\tag{1.222}
$$

The equilibrium is a $([(2m_{j_1r_1}1), (2n_{j_1u_1}1)]^{\text{th}}\text{DI}_{-}\text{:II}_{+},[2(m_{j_2v_1}1 - m_{j_1r_1}1) + 1]^{\text{th}}{}_{\text{UP:UP}}\text{US})$*-up-parabola lower-saddle at* $(x_{j_2}^*, x_{j_1}^*) = (a_{j_1j_2u_1}, a_{j_2j_1v_1})$.

(ii_{2a2g}) For $m_{j_1r_1} = 2m_{j_1r_1 1} + 1$, $n_{j_1u_1} = 2n_{j_1u_1 1}$ and $m_{j_2v_1} = 2m_{j_1v_1 1}$, the equilibrium of $(x_{j_2}^*, x_{j_1}^*) = (a_{j_1j_2u_1}, a_{j_2j_1v_1})$ on the infinite-equilibrium $x_{j_1}^* = a_{j_2j_1v_1} = a_{j_1j_1r_1}$ has the following properties.

- *For* $a_{j_1j_10}\prod_{r_2=1,r_2\neq r_1}^{p_1}(a_{j_1j_1r_1}-a_{j_1j_1r_2})^{m_{j_1r_2}}\prod_{u_2=1,u_2\neq u_1}^{q_1}(a_{j_1j_2u_1}-a_{j_1j_2u_2})^{n_{j_1u_2}} > 0$ *and* $a_{j_2j_20}\prod_{v_2=1,v_2\neq v_1}^{p_2}(a_{j_2j_1v_1}-a_{j_2j_1v_2})^{m_{j_2v_2}} > 0$,

$$(a_{j_1j_2u_1}, a_{j_2j_1v_1}) = \underbrace{([(2m_{j_1r_1 1}+1),(2n_{j_1u_1 1})]^{\text{th}}\text{DI}_+{:}\text{II}_+,[2(m_{j_2v_1 1}-m_{j_1r_1 1})-1]^{\text{th}}{}_{\text{UP:UP}}\text{SO})}_{([(2m_{j_1r_1 1}+1),(2n_{j_1u_1 1})],[2(m_{j_2v_1 1}-m_{j_1r_1 1})-1])\text{-up-parabola source}}. \tag{1.223}$$

The equilibrium is a $([(2m_{j_1r_1 1}+1),(2n_{j_1u_1 1})]^{\text{th}}\text{DI}_+{:}\text{II}_+,[2(m_{j_2v_1 1}-m_{j_1r_1 1})-1]^{\text{th}}{}_{\text{UP:UP}}\text{SO})$*-up-parabola source at* $(x_{j_2}^*, x_{j_1}^*) = (a_{j_1j_2u_1}, a_{j_2j_1v_1})$.

- *For* $a_{j_1j_10}\prod_{r_2=1,r_2\neq r_1}^{p_1}(a_{j_1j_1r_1}-a_{j_1j_1r_2})^{m_{j_1r_2}}\prod_{u_2=1,u_2\neq u_1}^{q_1}(a_{j_1j_2u_1}-a_{j_1j_2u_2})^{n_{j_1u_2}} < 0$ *and* $a_{j_2j_20}\prod_{v_2=1,v_2\neq v_1}^{p_2}(a_{j_2j_1v_1}-a_{j_2j_1v_2})^{m_{j_2v_2}} > 0$,

$$(a_{j_1j_2u_1}, a_{j_2j_1v_1}) = \underbrace{([(2m_{j_1r_1 1}+1),(2n_{j_1u_1 1})]^{\text{th}}\text{II}_+{:}\text{DI}_+,[2(m_{j_2v_1 1}-m_{j_1r_1 1})-1]^{\text{th}}{}_{\text{DP:DP}}\text{SI})}_{([(2m_{j_1r_1 1}+1),(2n_{j_1u_1 1})],[2(m_{j_2v_1 1}-m_{j_1r_1 1})-1])\text{-down-parabola sink}}. \tag{1.224}$$

The equilibrium is a $([(2m_{j_1r_1 1}+1),(2n_{j_1u_1 1})]^{\text{th}}\text{II}_+{:}\text{DI}_+,[2(m_{j_2v_1 1}-m_{j_1r_1 1})-1]^{\text{th}}{}_{\text{DP:DP}}\text{SI})$*-down-parabola sink at* $(x_{j_2}^*, x_{j_1}^*) = (a_{j_1j_2u_1}, a_{j_2j_1v_1})$.

- *For* $a_{j_1j_10}\prod_{r_2=1,r_2\neq r_1}^{p_1}(a_{j_1j_1r_1}-a_{j_1j_1r_2})^{m_{j_1r_2}}\prod_{u_2=1,u_2\neq u_1}^{q_1}(a_{j_1j_2u_1}-a_{j_1j_2u_2})^{n_{j_1u_2}} > 0$ *and* $a_{j_2j_20}\prod_{v_2=1,v_2\neq v_1}^{p_2}(a_{j_2j_1v_1}-a_{j_2j_1v_2})^{m_{j_2v_2}} < 0$,

$$(a_{j_1j_2u_1}, a_{j_2j_1v_1}) = \underbrace{([(2m_{j_1r_1 1}+1),(2n_{j_1u_1 1})]^{\text{th}}\text{II}_-{:}\text{DI}_-,[2(m_{j_2v_1 1}-m_{j_1r_1 1})-1]^{\text{th}}{}_{\text{DP:DP}}\text{SO})}_{([(2m_{j_1r_1 1}+1),(2n_{j_1u_1 1})],[2(m_{j_2v_1 1}-m_{j_1r_1 1})-1])\text{-down-parabola source}}. \tag{1.225}$$

The equilibrium is a $([(2m_{j_1r_1 1}+1),(2n_{j_1u_1 1})]^{\text{th}}\text{II}_-{:}\text{DI}_-,[2(m_{j_2v_1 1}-m_{j_1r_1 1})-1]^{\text{th}}{}_{\text{DP:DP}}\text{SO})$*-down-parabola source at* $(x_{j_2}^*, x_{j_1}^*) = (a_{j_1j_2u_1}, a_{j_2j_1v_1})$.

- *For* $a_{j_1j_10}\prod_{r_2=1,r_2\neq r_1}^{p_1}(a_{j_1j_1r_1}-a_{j_1j_1r_2})^{m_{j_1r_2}}\prod_{u_2=1,u_2\neq u_1}^{q_1}(a_{j_1j_2u_1}-a_{j_1j_2u_2})^{n_{j_1u_2}} < 0$ *and* $a_{j_2j_20}\prod_{v_2=1,v_2\neq v_1}^{p_2}(a_{j_2j_1v_1}-a_{j_2j_1v_2})^{m_{j_2v_2}} < 0$,

$$(a_{j_1j_2u_1}, a_{j_2j_1v_1}) = \underbrace{([(2m_{j_1r_1 1}+1),(2n_{j_1u_1 1})]^{\text{th}}\text{DI}_-{:}\text{II}_-,[2(m_{j_2v_1 1}-m_{j_1r_1 1})-1]^{\text{th}}{}_{\text{UP:UP}}\text{SI})}_{([(2m_{j_1r_1 1}+1),(2n_{j_1u_1 1})],[2(m_{j_2v_1 1}-m_{j_1r_1 1})-1])\text{-up-parabola sink}}. \tag{1.226}$$

The equilibrium is a $([(2m_{j_1r_1}1+1),(2n_{j_1u_1}1)]^{\text{th}}\text{DI}_{-}\text{:II}_{-},[2(m_{j_2v_1}1-m_{j_1r_1}1)-1]^{\text{th}}{}_{\text{UP:UP}}\text{SI})$*-up-parabola source at* $(x^*_{j_2},x^*_{j_1})=(a_{j_1j_2u_1},a_{j_2j_1v_1})$.

(ii$_{2a2h}$*) For* $m_{j_1r_1}=2m_{j_1r_1}1$, $n_{j_1u_1}=2n_{j_1u_1}1$ *and* $m_{j_2v_1}=2m_{j_1v_1}1$*, the equilibrium of* $(x^*_{j_1},x^*_{j_2})=(a_{j_1j_1r_1},a_{j_1j_2u_1})$ *on the infinite-equilibrium* $x^*_{j_1}=a_{j_2j_1v_1}=a_{j_1j_1r_1}$ *has the following properties.*

- *For* $a_{j_1j_1 0}\prod_{r_2=1,r_2\neq r_1}^{p_1}(a_{j_1j_1r_1}-a_{j_1j_1r_2})^{m_{j_1r_2}}\prod_{u_2=1,u_2\neq u_1}^{q_1}(a_{j_1j_2u_1}-a_{j_1j_2u_2})^{n_{j_1u_2}}>0$ *and* $a_{j_2j_2 0}\prod_{v_2=1,v_2\neq v_1}^{p_2}(a_{j_2j_1v_1}-a_{j_2j_1v_2})^{m_{j_2v_2}}>0$,

$$(a_{j_1j_2u_1},a_{j_2j_1v_1})=\underbrace{([(2m_{j_1r_1}1),(2n_{j_1u_1}1)]^{\text{th}}\text{II}_{+}\text{:II}_{+},[2(m_{j_2v_1}1-m_{j_1r_1}1)]^{\text{th}}{}_{\text{II:II}}\text{US})}_{([(2m_{j_1r_1}1),(2n_{j_1u_1}1)],[2(m_{j_2v_1}1-m_{j_1r_1}1)])\text{-increasing-inflection upper-saddle}}. \quad (1.227)$$

The equilibrium is a $([(2m_{j_1r_1}1),(2n_{j_1u_1}1)]^{\text{th}}\text{II}_{+}\text{:II}_{+},[2(m_{j_2v_1}1-m_{j_1r_1}1)]^{\text{th}}{}_{\text{II:II}}\text{US})$*-increasing-inflection upper-saddle at* $(x^*_{j_2},x^*_{j_1})=(a_{j_1j_2u_1},a_{j_2j_1v_1})$.

- *For* $a_{j_1j_1 0}\prod_{r_2=1,r_2\neq r_1}^{p_1}(a_{j_1j_1r_1}-a_{j_1j_1r_2})^{m_{j_1r_2}}\prod_{u_2=1,u_2\neq u_1}^{q_1}(a_{j_1j_2u_1}-a_{j_1j_2u_2})^{n_{j_1u_2}}<0$ *and* $a_{j_2j_2 0}\prod_{v_2=1,v_2\neq v_1}^{p_2}(a_{j_2j_1v_1}-a_{j_2j_1v_2})^{m_{j_2v_2}}>0$,

$$(a_{j_1j_2u_1},a_{j_2j_1v_1})=\underbrace{([(2m_{j_1r_1}1),(2n_{j_1u_1}1)]^{\text{th}}\text{DI}_{+}\text{:DI}_{+},[2(m_{j_2v_1}1-m_{j_1r_1}1)]^{\text{th}}{}_{\text{DI:DI}}\text{LS})}_{([(2m_{j_1r_1}1),(2n_{j_1u_1}1)],[2(m_{j_2v_1}1-m_{j_1r_1}1)])\text{-decreasing-inflection lower-saddle}}. \quad (1.228)$$

The equilibrium is a $([(2m_{j_1r_1}1),(2n_{j_1u_1}1)]^{\text{th}}\text{DI}_{+}\text{:DI}_{+},[2(m_{j_2v_1}1-m_{j_1r_1}1)]^{\text{th}}{}_{\text{DI:DI}}\text{LS})$*-decreasing-inflection lower-saddle at* $(x^*_{j_2},x^*_{j_1})=(a_{j_1j_2u_1},a_{j_2j_1v_1})$.

- *For* $a_{j_1j_1 0}\prod_{r_2=1,r_2\neq r_1}^{p_1}(a_{j_1j_1r_1}-a_{j_1j_1r_2})^{m_{j_1r_2}}\prod_{u_2=1,u_2\neq u_1}^{q_1}(a_{j_1j_2u_1}-a_{j_1j_2u_2})^{n_{j_1u_2}}>0$ *and* $a_{j_2j_2 0}\prod_{v_2=1,v_2\neq v_1}^{p_2}(a_{j_2j_1v_1}-a_{j_2j_1v_2})^{m_{j_2v_2}}<0$,

$$(a_{j_1j_2u_1},a_{j_2j_1v_1})=\underbrace{([(2m_{j_1r_1}1),(2n_{j_1u_1}1)]^{\text{th}}\text{DI}_{-}\text{:DI}_{-},[2(m_{j_2v_1}1-m_{j_1r_1}1)]^{\text{th}}{}_{\text{DI:DI}}\text{US})}_{([(2m_{j_1r_1}1),(2n_{j_1u_1}1)],[2(m_{j_2v_1}1-m_{j_1r_1}1)])\text{-decreasing-inflection upper-saddle}}. \quad (1.229)$$

The equilibrium is a $([(2m_{j_1r_1}1),(2n_{j_1u_1}1)]^{\text{th}}\text{DI}_{-}\text{:DI}_{-},[2(m_{j_2v_1}1-m_{j_1r_1}1)]^{\text{th}}{}_{\text{DI:DI}}\text{US})$*-decreasing-inflection upper-saddle at* $(x^*_{j_2},x^*_{j_1})=(a_{j_1j_2u_1},a_{j_2j_1v_1})$.

- *For* $a_{j_1j_1 0}\prod_{r_2=1,r_2\neq r_1}^{p_1}(a_{j_1j_1r_1}-a_{j_1j_1r_2})^{m_{j_1r_2}}\prod_{u_2=1,u_2\neq u_1}^{q_1}(a_{j_1j_2u_1}-a_{j_1j_2u_2})^{n_{j_1u_2}}<0$ *and* $a_{j_2j_2 0}\prod_{v_2=1,v_2\neq v_1}^{p_2}(a_{j_2j_1v_1}-a_{j_2j_1v_2})^{m_{j_2v_2}}<0$,

$$(a_{j_1j_2u_1},a_{j_2j_1v_1})=\underbrace{([(2m_{j_1r_1}1),(2n_{j_1u_1}1)]^{\text{th}}\text{II}_{-}\text{:II}_{-},[2(m_{j_2v_1}1-m_{j_1r_1}1)]^{\text{th}}{}_{\text{II:II}}\text{LS})}_{([(2m_{j_1r_1}1),(2n_{j_1u_1}1)],[2(m_{j_2v_1}1-m_{j_1r_1}1)])\text{-increasing-inflection lower-saddle}}. \quad (1.230)$$

The equilibrium is a $([(2m_{j_1r_11}),(2n_{j_1u_11})]^{\text{th}}\text{II}_{+}{:}\text{II}_{+},[2(m_{j_2v_11}-m_{j_1r_11})]^{\text{th}}{}_{\text{II:II}}\text{LS})$*-increasing-inflection lower-saddle at* $(x^*_{j_2},x^*_{j_1})=(a_{j_1j_2u_1},a_{j_2j_1v_1})$.

(ii$_{2b}$) *The infinite-equilibrium and switching equilibriums for* $m_{j_1r_1}>m_{j_2v_1}$ are presented.

(ii$_{2b1}$) *For* $m_{j_1r_1}>m_{j_2v_1}$, *16 types of infinite-equilibriums exist at* $x^*_{j_1}=a_{j_1j_1r_1}=a_{j_2j_1v_1}$.

(ii$_{2b1a}$) *For* $m_{j_1r_1}=2m_{j_1r_11}+1$ *and* $m_{j_2v_1}=2m_{j_1v_11}+1$, *the infinite-equilibrium of* $x^*_{j_1}=a_{j_2j_1v_1}=a_{j_1j_1r_1}$ *with* $\overline{x}_{j_2}\neq a_{j_1j_2u},a_{j_2j_2s}$ $(u=1,2,\ldots,q_1;s=1,2,\ldots,q_2)$ *has the following properties.*

- *For* $a_{j_1j_10}\prod_{r_2=1,r_2\neq r_1}^{p_1}(a_{j_1j_1r_1}-a_{j_1j_1r_2})^{m_{j_1r_2}}\prod_{u=1}^{q_1}(\overline{x}_{j_2}-a_{j_1j_2u})^{n_{j_1u}}>0$ *and* $a_{j_2j_20}\prod_{v_2=1,v_2\neq v_1}^{p_2}(a_{j_2j_1v_1}-a_{j_2j_1v_2})^{m_{j_2v_2}}>0$,

$$(a_{j_1j_1r_1},\overline{x}_{j_2})=\underbrace{((2m_{j_1r_11}+1)^{\text{th}}\text{SO},[2(m_{j_1r_11}-m_{j_2v_11})]^{\text{th}}\text{IC})}_{((2m_{j_1r_11}+1),[2(m_{j_1r_11}-m_{j_2v_11})])\text{-increasing-concave source}}.\tag{1.231}$$

The infinite-equilibrium is a $((2m_{j_1r_11}+1)^{\text{th}}\text{SO},[2(m_{j_1r_11}-m_{j_2v_11})]^{\text{th}}\text{IC})$*-increasing-concave source at* $x^*_{j_1}=a_{j_2j_1v_1}=a_{j_1j_1r_1}$.

- *For* $a_{j_1j_10}\prod_{r_2=1,r_2\neq r_1}^{p_1}(a_{j_1j_1r_1}-a_{j_1j_1r_2})^{m_{j_1r_2}}\prod_{u=1}^{q_1}(\overline{x}_{j_2}-a_{j_1j_2u})^{n_{j_1u}}<0$ *and* $a_{j_2j_20}\prod_{v_2=1,v_2\neq v_1}^{p_2}(a_{j_2j_1v_1}-a_{j_2j_1v_2})^{m_{j_2v_2}}>0$,

$$(a_{j_1j_1r_1},\overline{x}_{j_2})=\underbrace{((2m_{j_1r_11}+1)^{\text{th}}\text{SI},[2(m_{j_1r_11}-m_{j_2v_11})]^{\text{th}}\text{DC})}_{((2m_{j_1r_11}+1),[2(m_{j_1r_11}-m_{j_2v_11})])\text{-decreasing-concave sink}}.\tag{1.232}$$

The infinite-equilibrium is a $((2m_{j_1r_11}+1)^{\text{th}}\text{SI},[2(m_{j_1r_11}-m_{j_2v_11})]^{\text{th}}\text{DC})$*-decreasing-concave sink at* $x^*_{j_1}=a_{j_2j_1v_1}=a_{j_1j_1r_1}$.

- *For* $a_{j_1j_10}\prod_{r_2=1,r_2\neq r_1}^{p_1}(a_{j_1j_1r_1}-a_{j_1j_1r_2})^{m_{j_1r_2}}\prod_{u=1}^{q_1}(\overline{x}_{j_2}-a_{j_1j_2u})^{n_{j_1u}}>0$ *and* $a_{j_2j_20}\prod_{v_2=1,v_2\neq v_1}^{p_2}(a_{j_2j_1v_1}-a_{j_2j_1v_2})^{m_{j_2v_2}}<0$,

$$(a_{j_1j_1r_1},\overline{x}_{j_2})=\underbrace{((2m_{j_1r_11}+1)^{\text{th}}\text{SO},[2(m_{j_1r_11}-m_{j_2v_11})]^{\text{th}}\text{DC})}_{((2m_{j_1r_11}+1),[2(m_{j_1r_11}-m_{j_2v_11})])\text{-decreasing-concave source}}.\tag{1.233}$$

The infinite-equilibrium is a $((2m_{j_1r_11}+1)^{\text{th}}\text{SO},[2(m_{j_1r_11}-m_{j_2v_11})]^{\text{th}}\text{DC})$*-decreasing-concave sink at* $x^*_{j_1}=a_{j_2j_1v_1}=a_{j_1j_1r_1}$.

- *For* $a_{j_1j_10}\prod_{r_2=1,r_2\neq r_1}^{p_1}(a_{j_1j_1r_1}-a_{j_1j_1r_2})^{m_{j_1r_2}}\prod_{u=1}^{q_1}(\overline{x}_{j_2}-a_{j_1j_2u})^{n_{j_1u}}<0$ *and* $a_{j_2j_20}\prod_{v_2=1,v_2\neq v_1}^{p_2}(a_{j_2j_1v_1}-a_{j_2j_1v_2})^{m_{j_2v_2}}<0$,

$$(a_{j_1j_1r_1},\overline{x}_{j_2})=\underbrace{((2m_{j_1r_11}+1)^{\text{th}}\text{SI},[2(m_{j_1r_11}-m_{j_2v_11})]^{\text{th}}\text{IC})}_{((2m_{j_1r_11}+1),[2(m_{j_1r_11}-m_{j_2v_11})])\text{-increasing-concave sink}}.\tag{1.234}$$

The infinite-equilibrium is a $((2m_{j_1r_1}1+1)^{\text{th}}\text{SI},[2(m_{j_1r_1}1-m_{j_2v_1}1)]^{\text{th}}\text{IC})$*-increasing-concave sink at* $x_{j_1}^*=a_{j_2j_1v_1}=a_{j_1j_1r_1}$.

(ii_{2b1b}) *For* $m_{j_1r_1}=2m_{j_1r_1}1$ *and* $m_{j_2v_1}=2m_{j_1v_1}1+1$*, the infinite-equilibrium of* $x_{j_1}^*=a_{j_2j_1v_1}=a_{j_1j_1r_1}$ *with* $\overline{x}_{j_2}\neq a_{j_1j_2u},a_{j_2j_2s}$ $(u=1,2,\cdots,q_1;s=1,2,\cdots,q_2)$ *has the following properties.*

- *For* $a_{j_1j_10}\prod_{r_2=1,r_2\neq r_1}^{p_1}(a_{j_1j_1r_1}-a_{j_1j_1r_2})^{m_{j_1r_2}}\prod_{u=1}^{q_1}(\overline{x}_{j_2}-a_{j_1j_2u})^{n_{j_1u}}>0$ *and* $a_{j_2j_20}\prod_{v_2=1,v_2\neq v_1}^{p_2}(a_{j_2j_1v_1}-a_{j_2j_1v_2})^{m_{j_2v_2}}>0$,

$$(a_{j_1j_1r_1},\overline{x}_{j_2})=\underbrace{((2m_{j_1r_1}1)^{\text{th}}\text{US},[2(m_{j_1r_1}1-m_{j_2v_1}1)-1]^{\text{th}}\text{DA})}_{((2m_{j_1r_1}1),[2(m_{j_1r_1}1-m_{j_2v_1}1)-1])\text{-down-asymptote upper-saddle}}. \tag{1.235}$$

The infinite-equilibrium is a $((2m_{j_1r_1}1)^{\text{th}}\text{US},[2(m_{j_1r_1}1-m_{j_2v_1}1)-1]^{\text{th}}\text{DA})$*-down-asymptote upper-saddle at* $x_{j_1}^*=a_{j_2j_1v_1}=a_{j_1j_1r_1}$.

- *For* $a_{j_1j_10}\prod_{r_2=1,r_2\neq r_1}^{p_1}(a_{j_1j_1r_1}-a_{j_1j_1r_2})^{m_{j_1r_2}}\prod_{u=1}^{q_1}(\overline{x}_{j_2}-a_{j_1j_2u})^{n_{j_1u}}<0$ *and* $a_{j_2j_20}\prod_{v_2=1,v_2\neq v_1}^{p_2}(a_{j_2j_1v_1}-a_{j_2j_1v_2})^{m_{j_2v_2}}>0$,

$$(a_{j_1j_1r_1},\overline{x}_{j_2})=\underbrace{((2m_{j_1r_1}1)^{\text{th}}\text{LS},[2(m_{j_1r_1}1-m_{j_2v_1}1)-1]^{\text{th}}\text{UA})}_{((2m_{j_1r_1}1),[2(m_{j_1r_1}1-m_{j_2v_1}1)-1])\text{-up-asymptote lower-saddle}}. \tag{1.236}$$

The infinite-equilibrium is a $((2m_{j_1r_1}1)^{\text{th}}\text{LS},[2(m_{j_1r_1}1-m_{j_2v_1}1)-1]^{\text{th}}\text{UA})$*-up-asymptote lower-saddle at* $x_{j_1}^*=a_{j_2j_1v_1}=a_{j_1j_1r_1}$.

- *For* $a_{j_1j_10}\prod_{r_2=1,r_2\neq r_1}^{p_1}(a_{j_1j_1r_1}-a_{j_1j_1r_2})^{m_{j_1r_2}}\prod_{u=1}^{q_1}(\overline{x}_{j_2}-a_{j_1j_2u})^{n_{j_1u}}>0$ *and* $a_{j_2j_20}\prod_{v_2=1,v_2\neq v_1}^{p_2}(a_{j_2j_1v_1}-a_{j_2j_1v_2})^{m_{j_2v_2}}<0$,

$$(a_{j_1j_1r_1},\overline{x}_{j_2})=\underbrace{((2m_{j_1r_1}1)^{\text{th}}\text{US},[2(m_{j_1r_1}1-m_{j_2v_1}1)-1]^{\text{th}}\text{UA})}_{((2m_{j_1r_1}1),[2(m_{j_1r_1}1-m_{j_2v_1}1)-1])\text{-up-asymptote upper-saddle}}. \tag{1.237}$$

The infinite-equilibrium is a $((2m_{j_1r_1}1)^{\text{th}}\text{US},[2(m_{j_1r_1}1-m_{j_2v_1}1)-1]^{\text{th}}\text{UA})$*-up-asymptote upper-saddle at* $x_{j_1}^*=a_{j_2j_1v_1}=a_{j_1j_1r_1}$.

- *For* $a_{j_1j_10}\prod_{r_2=1,r_2\neq r_1}^{p_1}(a_{j_1j_1r_1}-a_{j_1j_1r_2})^{m_{j_1r_2}}\prod_{u=1}^{q_1}(\overline{x}_{j_2}-a_{j_1j_2u})^{n_{j_1u}}<0$ *and* $a_{j_2j_20}\prod_{v_2=1,v_2\neq v_1}^{p_2}(a_{j_2j_1v_1}-a_{j_2j_1v_2})^{m_{j_2v_2}}<0$,

$$(a_{j_1j_1r_1},\overline{x}_{j_2})=\underbrace{((2m_{j_1r_1}1)^{\text{th}}\text{LS},[2(m_{j_1r_1}1-m_{j_2v_1}1)-1]^{\text{th}}\text{DA})}_{((2m_{j_1r_1}1),[2(m_{j_1r_1}1-m_{j_2v_1}1)-1])\text{-down-asymptote lower-saddle}}. \tag{1.238}$$

The infinite-equilibrium is a $((2m_{j_1r_1}1)^{\text{th}}\text{LS},[2(m_{j_1r_1}1-m_{j_2v_1}1)-1]^{\text{th}}\text{DA})$*-down-asymptote lower-saddle at* $x_{j_1}^*=a_{j_2j_1v_1}=a_{j_1j_1r_1}$.

($\text{ii}_{2\text{b}1\text{c}}$) *For* $m_{j_1r_1}=2m_{j_1r_1}1+1$ *and* $m_{j_2v_1}=2m_{j_1v_1}1$*, the infinite-equilibrium of* $x_{j_1}^*=a_{j_2j_1v_1}=a_{j_1j_1r_1}$ *with* $\overline{x}_{j_2}\neq a_{j_1j_2u},a_{j_2j_2s}$ $(u=1,2,\cdots,q_1;s=1,2,\cdots,q_2)$ *has the following properties.*

- *For* $a_{j_1 j_1 0}\prod_{r_2=1,r_2\neq r_1}^{p_1}(a_{j_1 j_1 r_1}-a_{j_1 j_1 r_2})^{m_{j_1 r_2}}\prod_{u=1}^{q_1}(\overline{x}_{j_2}-a_{j_1 j_2 u})^{n_{j_1 u}}>0$ *and* $a_{j_2 j_2 0}\prod_{v_2=1,v_2\neq v_1}^{p_2}(a_{j_2 j_1 v_1}-a_{j_2 j_1 v_2})^{m_{j_2 v_2}}>0$,

$$(a_{j_1 j_1 r_1},\overline{x}_{j_2})=\underbrace{((2m_{j_1 r_1 1}+1)^{\text{th}}\text{SO},[2(m_{j_1 r_1 1}-m_{j_2 v_1 1})+1]^{\text{th}}\text{DA})}_{((2m_{j_1 r_1 1}+1),[2(m_{j_1 r_1 1}-m_{j_2 v_1 1})+1])\text{-down-asymptote source}}. \tag{1.239}$$

The infinite-equilibrium is a $((2m_{j_1 r_1 1}+1)^{\text{th}}\text{SO},[2(m_{j_1 r_1 1}-m_{j_2 v_1 1})+1]^{\text{th}}\text{DA})$*-up-parabola source at* $x_{j_1}^{*}=a_{j_2 j_1 v_1}=a_{j_1 j_1 r_1}$.

- *For* $a_{j_1 j_1 0}\prod_{r_2=1,r_2\neq r_1}^{p_1}(a_{j_1 j_1 r_1}-a_{j_1 j_1 r_2})^{m_{j_1 r_2}}\prod_{u=1}^{q_1}(\overline{x}_{j_2}-a_{j_1 j_2 u})^{n_{j_1 u}}<0$ *and* $a_{j_2 j_2 0}\prod_{v_2=1,v_2\neq v_1}^{p_2}(a_{j_2 j_1 v_1}-a_{j_2 j_1 v_2})^{m_{j_2 v_2}}>0$,

$$(a_{j_1 j_1 r_1},\overline{x}_{j_2})=\underbrace{((2m_{j_1 r_1 1}+1)^{\text{th}}\text{SI},[2(m_{j_1 r_1 1}-m_{j_2 v_1 1})+1]^{\text{th}}\text{UA})}_{((2m_{j_1 r_1 1}+1),[2(m_{j_1 r_1 1}-m_{j_2 v_1 1})+1])\text{-up-asymptote sink}}. \tag{1.240}$$

The infinite-equilibrium is a $((2m_{j_1 r_1 1}+1)^{\text{th}}\text{SI},[2(m_{j_1 r_1 1}-m_{j_2 v_1 1})+1]^{\text{th}}\text{UA})$*-up-asymptote sink at* $x_{j_1}^{*}=a_{j_2 j_1 v_1}=a_{j_1 j_1 r_1}$.

- *For* $a_{j_1 j_1 0}\prod_{r_2=1,r_2\neq r_1}^{p_1}(a_{j_1 j_1 r_1}-a_{j_1 j_1 r_2})^{m_{j_1 r_2}}\prod_{u=1}^{q_1}(\overline{x}_{j_2}-a_{j_1 j_2 u})^{n_{j_1 u}}>0$ *and* $a_{j_2 j_2 0}\prod_{v_2=1,v_2\neq v_1}^{p_2}(a_{j_2 j_1 v_1}-a_{j_2 j_1 v_2})^{m_{j_2 v_2}}<0$,

$$(a_{j_1 j_1 r_1},\overline{x}_{j_2})=\underbrace{((2m_{j_1 r_1 1}+1)^{\text{th}}\text{SO},[2(m_{j_1 r_1 1}-m_{j_2 v_1 1})+1]^{\text{th}}\text{UA})}_{((2m_{j_1 r_1 1}+1),[2(m_{j_1 r_1 1}-m_{j_2 v_1 1})+1])\text{-up-asymptote source}}. \tag{1.241}$$

The infinite-equilibrium is a $((2m_{j_1 r_1 1}+1)^{\text{th}}\text{SO},[2(m_{j_1 r_1 1}-m_{j_2 v_1 1})+1]^{\text{th}}\text{UA})$*-up-asymptote source at* $x_{j_1}^{*}=a_{j_2 j_1 v_1}=a_{j_1 j_1 r_1}$.

- *For* $a_{j_1 j_1 0}\prod_{r_2=1,r_2\neq r_1}^{p_1}(a_{j_1 j_1 r_1}-a_{j_1 j_1 r_2})^{m_{j_1 r_2}}\prod_{u=1}^{q_1}(\overline{x}_{j_2}-a_{j_1 j_2 u})^{n_{j_1 u}}<0$ *and* $a_{j_2 j_2 0}\prod_{v_2=1,v_2\neq v_1}^{p_2}(a_{j_2 j_1 v_1}-a_{j_2 j_1 v_2})^{m_{j_2 v_2}}<0$,

$$(a_{j_1 j_1 r_1},\overline{x}_{j_2})=\underbrace{((2m_{j_1 r_1 1}+1)^{\text{th}}\text{SI},[2(m_{j_1 r_1 1}-m_{j_2 v_1 1})+1]^{\text{th}}\text{DA})}_{((2m_{j_1 r_1 1}+1),[2(m_{j_1 r_1 1}-m_{j_2 v_1 1})+1])\text{-down-asymptote sink}}. \tag{1.242}$$

The infinite-equilibrium is a $((2m_{j_1 r_1 1}+1)^{\text{th}}\text{SI},[2(m_{j_1 r_1 1}-m_{j_2 v_1 1})+1]^{\text{th}}\text{DA})$*-up-parabola sink at* $x_{j_1}^{*}=a_{j_2 j_1 v_1}=a_{j_1 j_1 r_1}$.

($\text{ii}_{2\text{b1d}}$) *For* $m_{j_1 r_1}=2m_{j_1 r_1 1}$ *and* $m_{j_2 v_1}=2m_{j_1 v_1 1}$, *the infinite-equilibrium of* $x_{j_1}^{*}=a_{j_2 j_1 v_1}=a_{j_1 j_1 r_1}$ *with* $\overline{x}_{j_2}\neq a_{j_1 j_2 u},a_{j_2 j_2 s}$ $(u=1,2,\cdots,q_1;s=1,2,\cdots,q_2)$ *has the following properties.*

- *For* $a_{j_1 j_1 0}\prod_{r_2=1,r_2\neq r_1}^{p_1}(a_{j_1 j_1 r_1}-a_{j_1 j_1 r_2})^{m_{j_1 r_2}}\prod_{u=1}^{q_1}(\overline{x}_{j_2}-a_{j_1 j_2 u})^{n_{j_1 u}}>0$ *and* $a_{j_2 j_2 0}\prod_{v_2=1,v_2\neq v_1}^{p_2}(a_{j_2 j_1 v_1}-a_{j_2 j_1 v_2})^{m_{j_2 v_2}}>0$,

$$(a_{j_1j_1r_1}, \overline{x}_{j_2}) = \underbrace{((2m_{j_1r_1}1)^{\text{th}}\text{US},[2(m_{j_1r_1}1 - m_{j_2v_1}1)]^{\text{th}}\text{IC})}_{((2m_{j_1r_1}1),[2(m_{j_1r_1}1-m_{j_2v_1}1)])\text{-increasing-concave upper-saddle}}. \quad (1.243)$$

The infinite-equilibrium is a $((2m_{j_1r_1}1)^{\text{th}}\text{US},[2(m_{j_1r_1}1 - m_{j_2v_1}1)]^{\text{th}}\text{IC})$*-increasing-concave upper-saddle at* $x_{j_1}^* = a_{j_2j_1v_1} = a_{j_1j_1r_1}$.

- *For* $a_{j_1j_10}\prod_{r_2=1,r_2\neq r_1}^{p_1}(a_{j_1j_1r_1} - a_{j_1j_1r_2})^{m_{j_1r_2}}\prod_{u=1}^{q_1}(\overline{x}_{j_2} - a_{j_1j_2u})^{n_{j_1u}} < 0$ *and* $a_{j_2j_20}\prod_{v_2=1,v_2\neq v_1}^{p_2}(a_{j_2j_1v_1} - a_{j_2j_1v_2})^{m_{j_2v_2}} > 0$,

$$(a_{j_1j_1r_1}, \overline{x}_{j_2}) = \underbrace{((2m_{j_1r_1}1)^{\text{th}}\text{LS},[2(m_{j_1r_1}1 - m_{j_2v_1}1)]^{\text{th}}\text{DC})}_{((2m_{j_1r_1}1),[2(m_{j_1r_1}1-m_{j_2v_1}1)])-\textit{de}\text{creasing-concave lower-saddle}}. \quad (1.244)$$

The infinite-equilibrium is a $((2m_{j_1r_1}1)^{\text{th}}\text{LS},[2(m_{j_1r_1}1 - m_{j_2v_1}1)]^{\text{th}}\text{DC})$*-decreasing-concave lower-saddle at* $x_{j_1}^* = a_{j_2j_1v_1} = a_{j_1j_1r_1}$.

- *For* $a_{j_1j_10}\prod_{r_2=1,r_2\neq r_1}^{p_1}(a_{j_1j_1r_1} - a_{j_1j_1r_2})^{m_{j_1r_2}}\prod_{u=1}^{q_1}(\overline{x}_{j_2} - a_{j_1j_2u})^{n_{j_1u}} > 0$ *and* $a_{j_2j_20}\prod_{v_2=1,v_2\neq v_1}^{p_2}(a_{j_2j_1v_1} - a_{j_2j_1v_2})^{m_{j_2v_2}} < 0$,

$$(a_{j_1j_1r_1}, \overline{x}_{j_2}) = \underbrace{((2m_{j_1r_1}1)^{\text{th}}\text{US},[2(m_{j_1r_1}1 - m_{j_2v_1}1)]^{\text{th}}\text{DC})}_{((2m_{j_1r_1}1),[2(m_{j_1r_1}1-m_{j_2v_1}1)])-\textit{de}\text{creasing-concave upper-saddle}}. \quad (1.245)$$

The infinite-equilibrium is a $((2m_{j_1r_1}1)^{\text{th}}\text{US},[2(m_{j_1r_1}1 - m_{j_2v_1}1)]^{\text{th}}\text{DC})$*-decreasing-concave upper-saddle at* $x_{j_1}^* = a_{j_2j_1v_1} = a_{j_1j_1r_1}$.

- *For* $a_{j_1j_10}\prod_{r_2=1,r_2\neq r_1}^{p_1}(a_{j_1j_1r_1} - a_{j_1j_1r_2})^{m_{j_1r_2}}\prod_{u=1}^{q_1}(\overline{x}_{j_2} - a_{j_1j_2u})^{n_{j_1u}} < 0$ *and* $a_{j_2j_20}\prod_{v_2=1,v_2\neq v_1}^{p_2}(a_{j_2j_1v_1} - a_{j_2j_1v_2})^{m_{j_2v_2}} < 0$,

$$(a_{j_1j_1r_1}, \overline{x}_{j_2}) = \underbrace{((2m_{j_1r_1}1)^{\text{th}}\text{LS},[2(m_{j_1r_1}1 - m_{j_2v_1}1)]^{\text{th}}\text{IC})}_{((2m_{j_1r_1}1),[2(m_{j_1r_1}1-m_{j_2v_1}1)])\text{-increasing-concave lower-saddle}}. \quad (1.246)$$

The infinite-equilibrium is a $((2m_{j_1r_1}1)^{\text{th}}\text{LS},[2(m_{j_1r_1}1 - m_{j_2v_1}1)]^{\text{th}}\text{IC})$*-increasing-concave lower-saddle at* $x_{j_1}^* = a_{j_2j_1v_1} = a_{j_1j_1r_1}$.

(ii$_{2b2}$*) There are 32 types of equilibrium* $(x_{j_2}^*, x_{j_1}^*) = (a_{j_1j_2u_1}, a_{j_2j_1v_1})$ *on the infinite-equilibrium of* $x_{j_1}^* = a_{j_2j_1v_1} = a_{j_1j_1r_1}$.

(ii$_{2b2a}$*) For* $m_{j_1r_1} = 2m_{j_1r_1}1 + 1, n_{j_1u_1} = 2n_{j_1u_1}1 + 1$ *and* $m_{j_2v_1} = 2m_{j_1v_1}1 + 1$*, the* equilibrium $(x_{j_2}^*, x_{j_1}^*) = (a_{j_1j_2u_1}, a_{j_2j_1v_1})$ *on the infinite-equilibrium of* $x_{j_1}^* = a_{j_2j_1v_1} = a_{j_1j_1r_1}$ *has the following properties.*

- *For* $a_{j_1j_10}\prod_{r_2=1,r_2\neq r_1}^{p_1}(a_{j_1j_1r_1} - a_{j_1j_1r_2})^{m_{j_1r_2}}\prod_{u_2=1,u_2\neq u_1}^{q_1}(a_{j_1j_2u_1} - a_{j_1j_2u_2})^{n_{j_1u_2}} > 0$ *and* $a_{j_2j_20}\prod_{v_2=1,v_2\neq v_1}^{p_2}(a_{j_2j_1v_1} - a_{j_2j_1v_2})^{m_{j_2v_2}} > 0$,

$$
\begin{aligned}
&(a_{j_1j_2u_1}, a_{j_2j_1v_1}) \\
&= \underbrace{([(2m_{j_1r_1}1+1), (2n_{j_1u_1}1+1)]^{\text{th}}\text{UP}_-{:}\text{UP}_+,[2(m_{j_1r_1}1-m_{j_2v_1}1)]^{\text{th}}{}_{\text{DC:IC}}\text{US})}_{([(2m_{j_1r_1}1+1),(2n_{j_1u_1}1+1)],[2(m_{j_1r_1}1-m_{j_2v_1}1)])\text{-concave up-parabola upper-saddle}}.
\end{aligned}
\tag{1.247}
$$

The equilibrium is a $([(2m_{j_1r_1}1+1), (2n_{j_1u_1}1+1)]^{\text{th}}\text{UP}_-{:}\text{UP}_+,[2(m_{j_1r_1}1-m_{j_2v_1}1)]^{\text{th}}{}_{\text{DC:IC}}\text{US})$ *-concave up-parabola upper-saddle at* $(x_{j_2}^*, x_{j_1}^*) = (a_{j_1j_2u_1}, a_{j_2j_1v_1})$.

- *For* $a_{j_1j_10}\prod_{r_2=1,r_2\neq r_1}^{p_1}(a_{j_1j_1r_1}-a_{j_1j_1r_2})^{m_{j_1r_2}}\prod_{u_2=1,u_2\neq u_1}^{q_1}(a_{j_1j_2u_1}-a_{j_1j_2u_2})^{n_{j_1u_2}}<0$ *and* $a_{j_2j_20}\prod_{v_2=1,v_2\neq v_1}^{p_2}(a_{j_2j_1v_1}-a_{j_2j_1v_2})^{m_{j_2v_2}}>0$,

$$
\begin{aligned}
&(a_{j_1j_2u_1}, a_{j_2j_1v_1}) \\
&= \underbrace{([(2m_{j_1r_1}1+1), (2n_{j_1u_1}1+1)]^{\text{th}}\text{DP}_-{:}\text{DP}_+,[2(m_{j_1r_1}1-m_{j_2v_1}1)]^{\text{th}}{}_{\text{IC:DC}}\text{LS})}_{([(2m_{j_1r_1}1+1),(2n_{j_1u_1}1+1)],[2(m_{j_1r_1}1-m_{j_2v_1}1)])\text{-concave down-parabola lower-saddle}}.
\end{aligned}
\tag{1.248}
$$

The equilibrium is a $([(2m_{j_1r_1}1+1), (2n_{j_1u_1}1+1)]^{\text{th}}\text{UP}_-{:}\text{UP}_+,[2(m_{j_2v_1}1-m_{j_1r_1}1)]^{\text{th}}{}_{\text{IC:DC}}\text{LS}$ *-concave down-parabola lower-saddle at* $(x_{j_2}^*, x_{j_1}^*) = (a_{j_1j_2u_1}, a_{j_2j_1v_1})$.

- *For* $a_{j_1j_10}\prod_{r_2=1,r_2\neq r_1}^{p_1}(a_{j_1j_1r_1}-a_{j_1j_1r_2})^{m_{j_1r_2}}\prod_{u_2=1,u_2\neq u_1}^{q_1}(a_{j_1j_2u_1}-a_{j_1j_2u_2})^{n_{j_1u_2}}>0$ *and* $a_{j_2j_20}\prod_{v_2=1,v_2\neq v_1}^{p_2}(a_{j_2j_1v_1}-a_{j_2j_1v_2})^{m_{j_2v_2}}<0$,

$$
\begin{aligned}
&(a_{j_1j_2u_1}, a_{j_2j_1v_1}) \\
&= \underbrace{([(2m_{j_1r_1}1+1), (2n_{j_1u_1}1+1)]^{\text{th}}\text{DP}_+{:}\text{DP}_-,[2(m_{j_1r_1}1-m_{j_2v_1}1)]^{\text{th}}{}_{\text{IC:DC}}\text{US})}_{([(2m_{j_1r_1}1+1),(2n_{j_1u_1}1+1)],[2(m_{j_1r_1}1-m_{j_2v_1}1)])\text{-concave down-parabola upper-saddle}}.
\end{aligned}
\tag{1.249}
$$

The equilibrium is a $([(2m_{j_1r_1}1+1), (2n_{j_1u_1}1+1)]^{\text{th}}\text{DP}_+{:}\text{DP}_-,[2(m_{j_2v_1}1-m_{j_1r_1}1)]^{\text{th}}{}_{\text{IC:DC}}\text{US})$ *-concave down-parabola upper-saddle at* $(x_{j_2}^*, x_{j_1}^*) = (a_{j_1j_2u_1}, a_{j_2j_1v_1})$.

- *For* $a_{j_1j_10}\prod_{r_2=1,r_2\neq r_1}^{p_1}(a_{j_1j_1r_1}-a_{j_1j_1r_2})^{m_{j_1r_2}}\prod_{u_2=1,u_2\neq u_1}^{q_1}(a_{j_1j_2u_1}-a_{j_1j_2u_2})^{n_{j_1u_2}}<0$ *and* $a_{j_2j_20}\prod_{v_2=1,v_2\neq v_1}^{p_2}(a_{j_2j_1v_1}-a_{j_2j_1v_2})^{m_{j_2v_2}}<0$,

$$
\begin{aligned}
&(a_{j_1j_2u_1}, a_{j_2j_1v_1}) \\
&= \underbrace{([(2m_{j_1r_1}1+1), (2n_{j_1u_1}1+1)]^{\text{th}}\text{UP}_+{:}\text{UP}_-,[2(m_{j_1r_1}1-m_{j_2v_1}1)]^{\text{th}}{}_{\text{DC:IC}}\text{LS})}_{([(2m_{j_1r_1}1+1),(2n_{j_1u_1}1+1)],[2(m_{j_1r_1}1-m_{j_2v_1}1)])\text{-concave up-paraobla upper-saddle}}.
\end{aligned}
\tag{1.250}
$$

The equilibrium is a $([(2m_{j_1r_1}1+1), (2n_{j_1u_1}1+1)]^{\text{th}}\text{UP}_+{:}\text{UP}_-,[2(m_{j_2v_1}1-m_{j_1r_1}1)]^{\text{th}}{}_{\text{DC:IC}}\text{LS})$ *-concave up-parabola lower-saddle at* $(x_{j_2}^*, x_{j_1}^*) = (a_{j_1j_2u_1}, a_{j_2j_1v_1})$.

(ii$_{2b2b}$) For $m_{j_1r_1}=2m_{j_1r_1}1, n_{j_1u_1}=2n_{j_1u_1}1+1$ *and* $m_{j_2v_1}=2m_{j_1v_1}1+1$, *the* equilibrium $(x_{j_2}^*, x_{j_1}^*) = (a_{j_1j_2u_1}, a_{j_2j_1v_1})$ *on the infinite-equilibrium of* $x_{j_1}^* = a_{j_2j_1v_1} = a_{j_1j_1r_1}$ *has the following properties.*

- *For* $a_{j_1j_10}\prod_{r_2=1,r_2\neq r_1}^{p_1}(a_{j_1j_1r_1}-a_{j_1j_1r_2})^{m_{j_1r_2}}\prod_{u_2=1,u_2\neq u_1}^{q_1}(a_{j_1j_2u_1}-a_{j_1j_2u_2})^{n_{j_1u_2}}>0$ *and* $a_{j_2j_20}\prod_{v_2=1,v_2\neq v_1}^{p_2}(a_{j_2j_1v_1}-a_{j_2j_1v_2})^{m_{j_2v_2}}>0$,

$$(a_{j_1j_2u_1},a_{j_2j_1v_1}) = \underbrace{([(2m_{j_1r_1}1),(2n_{j_1u_1}1+1)]^{\text{th}}\text{DP}_{-}\text{:UP}_{+},[2(m_{j_1r_1}1-m_{j_2v_1}1)-1]^{\text{th}}{}_{\text{UA}}\text{LS:}_{\text{DA}}\text{US})}_{([(2m_{j_1r_1}1),(2n_{j_1u_1}1+1)],[2(m_{j_1r_1}1-m_{j_2v_1}1)]-1)-\textit{hyperbolic}\text{ asymptote lower-to-upper saddle}}. \tag{1.251}$$

The equilibrium is a $([(2m_{j_1r_1}1),(2n_{j_1u_1}1+1)]^{\text{th}}\text{DP}_{-}\text{:UP}_{+},[2(m_{j_1r_1}1-m_{j_2v_1}1)-1]^{\text{th}}{}_{\text{UA}}\text{LS:}_{\text{DA}}\text{US})$*-hyperbolic asymptote lower-to-upper saddle at* $(x_{j_2}^*,x_{j_1}^*)=(a_{j_1j_2u_1},a_{j_2j_1v_1})$.

- *For* $a_{j_1j_10}\prod_{r_2=1,r_2\neq r_1}^{p_1}(a_{j_1j_1r_1}-a_{j_1j_1r_2})^{m_{j_1r_2}}\prod_{u_2=1,u_2\neq u_1}^{q_1}(a_{j_1j_2u_1}-a_{j_1j_2u_2})^{n_{j_1u_2}}<0$ *and* $a_{j_2j_20}\prod_{v_2=1,v_2\neq v_1}^{p_2}(a_{j_2j_1v_1}-a_{j_2j_1v_2})^{m_{j_2v_2}}>0$,

$$(a_{j_1j_2u_1},a_{j_2j_1v_1}) = \underbrace{([(2m_{j_1r_1}1),(2n_{j_1u_1}1+1)]^{\text{th}}\text{UP}_{-}\text{:DP}_{+},[2(m_{j_1r_1}1-m_{j_2v_1}1)-1]^{\text{th}}{}_{\text{DA}}\text{US:}_{\text{UA}}\text{LS})}_{([(2m_{j_1r_1}1),(2n_{j_1u_1}1+1)],[2(m_{j_1r_1}1-m_{j_2v_1}1)]-1)\text{-hyperbolic-secant asymptote upper-to-lower saddle}}. \tag{1.252}$$

The equilibrium is a $([(2m_{j_1r_1}1),(2n_{j_1u_1}1+1)]^{\text{th}}\text{UP}_{-}\text{:DP}_{+},[2(m_{j_1r_1}1-m_{j_2v_1}1)-1]^{\text{th}}{}_{\text{DA}}\text{US:}_{\text{UA}}\text{LS})$*-hyperbolic-secant asymptote upper-to-lower saddle at* $(x_{j_2}^*,x_{j_1}^*)=(a_{j_1j_2u_1},a_{j_2j_1v_1})$.

- *For* $a_{j_1j_10}\prod_{r_2=1,r_2\neq r_1}^{p_1}(a_{j_1j_1r_1}-a_{j_1j_1r_2})^{m_{j_1r_2}}\prod_{u_2=1,u_2\neq u_1}^{q_1}(a_{j_1j_2u_1}-a_{j_1j_2u_2})^{n_{j_1u_2}}>0$ *and* $a_{j_2j_20}\prod_{v_2=1,v_2\neq v_1}^{p_2}(a_{j_2j_1v_1}-a_{j_2j_1v_2})^{m_{j_2v_2}}<0$,

$$(a_{j_1j_2u_1},a_{j_2j_1v_1}) = \underbrace{([(2m_{j_1r_1}1),(2n_{j_1u_1}1+1)]^{\text{th}}\text{UP}_{+}\text{:DP}_{-},[2(m_{j_1r_1}1-m_{j_2v_1}1)-1]^{\text{th}}{}_{\text{DA}}\text{LS:}_{\text{UA}}\text{US})}_{([(2m_{j_1r_1}1),(2n_{j_1u_1}1+1)],[2(m_{j_1r_1}1-m_{j_2v_1}1)]-1)\text{-hyperbolic-secant asymptote lower-to-upper saddle}}. \tag{1.253}$$

The equilibrium is a $([(2m_{j_1r_1}1),(2n_{j_1u_1}1+1)]^{\text{th}}\text{UP}_{+}\text{:DP}_{-},[2(m_{j_1r_1}1-m_{j_2v_1}1)-1]^{\text{th}}{}_{\text{DA}}\text{LS:}_{\text{UA}}\text{US})$*-hyperbolic-secant asymptote upper-to-lower saddle at* $(x_{j_2}^*,x_{j_1}^*)=(a_{j_1j_2u_1},a_{j_2j_1v_1})$.

- *For* $a_{j_1j_10}\prod_{r_2=1,r_2\neq r_1}^{p_1}(a_{j_1j_1r_1}-a_{j_1j_1r_2})^{m_{j_1r_2}}\prod_{u_2=1,u_2\neq u_1}^{q_1}(a_{j_1j_2u_1}-a_{j_1j_2u_2})^{n_{j_1u_2}}<0$ *and* $a_{j_2j_20}\prod_{v_2=1,v_2\neq v_1}^{p_2}(a_{j_2j_1v_1}-a_{j_2j_1v_2})^{m_{j_2v_2}}<0$,

$$(a_{j_1j_2u_1}, a_{j_2j_1v_1}) = \underbrace{([(2m_{j_1r_1}1), (2n_{j_1u_1}1+1)]^{\text{th}}\text{DP}_+\text{:UP}_-,[2(m_{j_1r_1}1-m_{j_2v_1}1)-1]^{\text{th}}{}_{\text{UA}}\text{US:}{}_{\text{DA}}\text{LS})}_{([(2m_{j_1r_1}1),(2n_{j_1u_1}1+1)],[2(m_{j_1r_1}1-m_{j_2v_1}1)]-1)\text{-hyperbolic asymptote upper-to-lower saddle}}. \tag{1.254}$$

The equilibrium is a ($[(2m_{j_1r_1}1), (2n_{j_1u_1}1+1)]^{\text{th}}\text{DP}_+\text{:UP}_-,[2(m_{j_1r_1}1-m_{j_2v_1}1)-1]^{\text{th}}{}_{\text{UA}}\text{US:}{}_{\text{DA}}\text{LS}$)*-hyperbolic-secant asymptote-upper-to-lower saddle at* $(x_{j_2}^*, x_{j_1}^*) = (a_{j_1j_2u_1}, a_{j_2j_1v_1})$.

(ii$_{2b2c}$) For $m_{j_1r_1} = 2m_{j_1r_1}1+1, n_{j_1u_1} = 2n_{j_1u_1}1+1$ *and* $m_{j_2v_1} = 2m_{j_1v_1}1$, *the equilibrium* $(x_{j_2}^*, x_{j_1}^*) = (a_{j_1j_2u_1}, a_{j_2j_1v_1})$ *on the infinite-equilibrium of* $x_{j_1}^* = a_{j_2j_1v_1} = a_{j_1j_1r_1}$ *has the following properties.*

- *For* $a_{j_1j_10}\prod_{r_2=1,r_2\neq r_1}^{p_1}(a_{j_1j_1r_1}-a_{j_1j_1r_2})^{m_{j_1r_2}}\prod_{u_2=1,u_2\neq u_1}^{q_1}(a_{j_1j_2u_1}-a_{j_1j_2u_2})^{n_{j_1u_2}} > 0$ *and* $a_{j_2j_20}\prod_{v_2=1,v_2\neq v_1}^{p_2}(a_{j_2j_1v_1}-a_{j_2j_1v_2})^{m_{j_2v_2}} > 0$,

$$(a_{j_1j_2u_1}, a_{j_2j_1v_1}) = \underbrace{([(2m_{j_1r_1}1+1), (2n_{j_1u_1}1+1)]^{\text{th}}\text{DP}_+\text{:UP}_+,[2(m_{j_1r_1}1-m_{j_2v_1}1)+1]^{\text{th}}{}_{\text{UA}}\text{SI:}{}_{\text{DA}}\text{SO})}_{([(2m_{j_1r_1}1+1),(2n_{j_1u_1}1+1)],[2(m_{j_1r_1}1-m_{j_2v_1}1)]+1)\text{-hyperbolic asymptote sink-to-source}}. \tag{1.255}$$

The equilibrium is a ($[(2m_{j_1r_1}1+1), (2n_{j_1u_1}1+1)]^{\text{th}}\text{DP}_+\text{:UP}_+,[2(m_{j_1r_1}1-m_{j_2v_1}1)+1]^{\text{th}}{}_{\text{UA}}\text{SI:}{}_{\text{DA}}\text{SO}$)*-hyperbolic asymptote sink-to-source at* $(x_{j_2}^*, x_{j_1}^*) = (a_{j_1j_2u_1}, a_{j_2j_1v_1})$.

- *For* $a_{j_1j_10}\prod_{r_2=1,r_2\neq r_1}^{p_1}(a_{j_1j_1r_1}-a_{j_1j_1r_2})^{m_{j_1r_2}}\prod_{u_2=1,u_2\neq u_1}^{q_1}(a_{j_1j_2u_1}-a_{j_1j_2u_2})^{n_{j_1u_2}} < 0$ *and* $a_{j_2j_20}\prod_{v_2=1,v_2\neq v_1}^{p_2}(a_{j_2j_1v_1}-a_{j_2j_1v_2})^{m_{j_2v_2}} > 0$,

$$(a_{j_1j_2u_1}, a_{j_2j_1v_1}) = \underbrace{([(2m_{j_1r_1}1+1), (2n_{j_1u_1}1+1)]^{\text{th}}\text{UP}_+\text{:DP}_+,[2(m_{j_1r_1}1-m_{j_2v_1}1)+1]^{\text{th}}{}_{\text{DA}}\text{SO:}{}_{\text{DA}}\text{SI})}_{([(2m_{j_1r_1}1+1),(2n_{j_1u_1}1+1)],[2(m_{j_1r_1}1-m_{j_2v_1}1)]+11)\text{-hyperbolic-secant asymptote source-to-sink}}. \tag{1.256}$$

The equilibrium is a ($[(2m_{j_1r_1}1+1), (2n_{j_1u_1}1+1)]^{\text{th}}\text{UP}_+\text{:DP}_+,[2(m_{j_1r_1}1-m_{j_2v_1}1)+1]^{\text{th}}{}_{\text{DA}}\text{SO:}{}_{\text{UA}}\text{SI}$)*-hyperbolic-secant asymptote source-to-sink at* $(x_{j_2}^*, x_{j_1}^*) = (a_{j_1j_2u_1}, a_{j_2j_1v_1})$.

- *For* $a_{j_1j_10}\prod_{r_2=1,r_2\neq r_1}^{p_1}(a_{j_1j_1r_1}-a_{j_1j_1r_2})^{m_{j_1r_2}}\prod_{u_2=1,u_2\neq u_1}^{q_1}(a_{j_1j_2u_1}-a_{j_1j_2u_2})^{n_{j_1u_2}} > 0$ *and* $a_{j_2j_20}\prod_{v_2=1,v_2\neq v_1}^{p_2}(a_{j_2j_1v_1}-a_{j_2j_1v_2})^{m_{j_2v_2}} < 0$,

$$(a_{j_1j_2u_1}, a_{j_2j_1v_1}) = \underbrace{([(2m_{j_1r_1}1+1), (2n_{j_1u_1}1+1)]^{\text{th}}\text{UP}_-\text{:DP}_-,[2(m_{j_1r_1}1-m_{j_2v_1}1)+1]^{\text{th}}{}_{\text{DA}}\text{SI:}{}_{\text{DA}}\text{SO})}_{([(2m_{j_1r_1}1+1),(2n_{j_1u_1}1+1)],[2(m_{j_1r_1}1-m_{j_2v_1}1)]+1)\text{-hyperbolic-secant asymptote sink-to-source}}. \tag{1.257}$$

The equilibrium is a ($[(2m_{j_1r_1 1}+1),(2n_{j_1u_1 1}+1)]^{\text{th}}\text{UP}_-{:}\text{DP}_-,[2(m_{j_1r_1 1}-m_{j_2v_1 1})+1]^{\text{th}}\ {}_{\text{DA}}\text{SI}_{\text{UA}}\text{SO}$)*-hyperbolic-secant asymptote sink-to-source at* $(x_{j_2}^*,x_{j_1}^*)=(a_{j_1j_2u_1},a_{j_2j_1v_1})$.

- *For* $a_{j_1j_10}\prod_{r_2=1,r_2\neq r_1}^{p_1}(a_{j_1j_1r_1}-a_{j_1j_1r_2})^{m_{j_1r_2}}\prod_{u_2=1,u_2\neq u_1}^{q_1}(a_{j_1j_2u_1}-a_{j_1j_2u_2})^{n_{j_1u_2}}<0$ *and* $a_{j_2j_20}\prod_{v_2=1,v_2\neq v_1}^{p_2}(a_{j_2j_1v_1}-a_{j_2j_1v_2})^{m_{j_2v_2}}<0$,

$$(a_{j_1j_2u_1},a_{j_2j_1v_1})=\underbrace{([(2m_{j_1r_1 1}+1),(2n_{j_1u_1 1}+1)]^{\text{th}}\text{DP}_+{:}\text{UP}_+,[2(m_{j_1r_1 1}-m_{j_2v_1 1})+1]^{\text{th}}{}_{\text{UA}}\text{SO}{:}{}_{\text{DA}}\text{SI})}_{([(2m_{j_1r_1}1+1),(2n_{j_1u_1}1+1)],[2(m_{j_1r_1}1-m_{j_2v_1}1)]+1)\text{-hyperbolic asymptote source-to-sink}}. \tag{1.258}$$

The equilibrium is a ($[(2m_{j_1r_1 1}+1),(2n_{j_1u_1 1}+1)]^{\text{th}}\text{DP}_-{:}\text{UP}_-,[2(m_{j_1r_1 1}-m_{j_2v_1 1})+1]^{\text{th}}\ {}_{\text{UA}}\text{SI}{:}{}_{\text{DA}}\text{SO}$)*-hyperbolic asymptote sink-to-source at* $(x_{j_2}^*,x_{j_1}^*)=(a_{j_1j_2u_1},a_{j_2j_1v_1})$.

(ii$_{2b2d}$) *For* $m_{j_1r_1}=2m_{j_1r_1 1},n_{j_1u_1}=2n_{j_1u_1 1}+1$ *and* $m_{j_2v_1}=2m_{j_1v_1 1}$, *the* equilibrium $(x_{j_2}^*,x_{j_1}^*)=(a_{j_1j_2u_1},a_{j_2j_1v_1})$ *on the infinite-equilibrium of* $x_{j_1}^*=a_{j_2j_1v_1}=a_{j_1j_1r_1}$ *has the following properties.*

- *For* $a_{j_1j_10}\prod_{r_2=1,r_2\neq r_1}^{p_1}(a_{j_1j_1r_1}-a_{j_1j_1r_2})^{m_{j_1r_2}}\prod_{u_2=1,u_2\neq u_1}^{q_1}(a_{j_1j_2u_1}-a_{j_1j_2u_2})^{n_{j_1u_2}}>0$ *and* $a_{j_2j_20}\prod_{v_2=1,v_2\neq v_1}^{p_2}(a_{j_2j_1v_1}-a_{j_2j_1v_2})^{m_{j_2v_2}}>0$,

$$(a_{j_1j_2u_1},a_{j_2j_1v_1})=\underbrace{([(2m_{j_1r_1 1}),(2n_{j_1u_1 1}+1)]^{\text{th}}\text{UP}_+{:}\text{UP}_+,[2(m_{j_1r_1 1}-m_{j_2v_1 1})]^{\text{th}}{}_{\text{DC:IC}}\text{US})}_{([(2m_{j_1r_1}1),(2n_{j_1u_1}1+1)],[2(m_{j_1r_1}1-m_{j_2v_1}1)])\text{-concave up-parabola upper-saddle}}. \tag{1.259}$$

The equilibrium is a ($[(2m_{j_1r_1 1}),(2n_{j_1u_1 1}+1)]^{\text{th}}\text{UP}_+{:}\text{UP}_+,[2(m_{j_1r_1 1}-m_{j_2v_1 1})]^{\text{th}}\ {}_{\text{DC:IC}}\text{US}$)*-concave up-parabola upper-saddle at* $(x_{j_2}^*,x_{j_1}^*)=(a_{j_1j_2u_1},a_{j_2j_1v_1})$.

- *For* $a_{j_1j_10}\prod_{r_2=1,r_2\neq r_1}^{p_1}(a_{j_1j_1r_1}-a_{j_1j_1r_2})^{m_{j_1r_2}}\prod_{u_2=1,u_2\neq u_1}^{q_1}(a_{j_1j_2u_1}-a_{j_1j_2u_2})^{n_{j_1u_2}}<0$ *and* $a_{j_2j_20}\prod_{v_2=1,v_2\neq v_1}^{p_2}(a_{j_1j_1r_1}-a_{j_2j_1v_2})^{m_{j_2v_2}}>0$,

$$(a_{j_1j_2u_1},a_{j_2j_1v_1})=\underbrace{([(2m_{j_1r_1 1}),(2n_{j_1u_1 1}+1)]^{\text{th}}\text{DP}_+{:}\text{DP}_+,[2(m_{j_1r_1 1}-m_{j_2v_1 1})]^{\text{th}}{}_{\text{IC:DC}}\text{US})}_{([(2m_{j_1r_1}1),(2n_{j_1u_1}1+1)],[2(m_{j_1r_1}1-m_{j_2v_1}1)])\text{-concave down-parabola upper-saddle}}. \tag{1.260}$$

The equilibrium is a ($[(2m_{j_1r_1 1}),(2n_{j_1u_1 1}+1)]^{\text{th}}\text{DP}_+{:}\text{DP}_+,[2(m_{j_1r_1 1}-m_{j_2v_1 1})]^{\text{th}}\ {}_{\text{IC:DC}}\text{US}$)*-concave down-parabola upper-saddle at* $(x_{j_2}^*,x_{j_1}^*)=(a_{j_1j_2u_1},a_{j_2j_1v_1})$.

- *For* $a_{j_1j_10}\prod_{r_2=1,r_2\neq r_1}^{p_1}(a_{j_1j_1r_1}-a_{j_1j_1r_2})^{m_{j_1r_2}}\prod_{u_2=1,u_2\neq u_1}^{q_1}(a_{j_1j_2u_1}-a_{j_1j_2u_2})^{n_{j_1u_2}}>0$ *and* $a_{j_2j_20}\prod_{v_2=1,v_2\neq v_1}^{p_2}(a_{j_2j_1v_1}-a_{j_2j_1v_2})^{m_{j_2v_2}}<0$,

$$\begin{aligned}&(a_{j_1j_2u_1}, a_{j_2j_1v_1})\\&=\underbrace{([(2m_{j_1r_1}1), (2n_{j_1u_1}1+1)]^{\text{th}}\text{DP}_{-}\text{:DP}_{-},[2(m_{j_1r_1}1-m_{j_2v_1}1)]^{\text{th}}{}_{\text{IC:DC}}\text{LS})}_{([(2m_{j_1r_1}1),(2n_{j_1u_1}1+1)],[2(m_{j_1r_1}1-m_{j_2v_1}1)])\text{-concave down-parabola lower-saddle}}.\end{aligned} \tag{1.261}$$

The equilibrium is a $([(2m_{j_1r_1}1), (2n_{j_1u_1}1+1)]^{\text{th}}\text{DP}_{-}\text{:DP}_{-},[2(m_{j_1r_1}1-m_{j_2v_1}1)]^{\text{th}}{}_{\text{IC:DC}}\text{LS})$*-concave down-parabola lower-saddle at* $(x^*_{j_2}, x^*_{j_1}) = (a_{j_1j_2u_1}, a_{j_2j_1v_1})$.

- *For* $a_{j_1j_10}\prod_{r_2=1,r_2\neq r_1}^{p_1}(a_{j_1j_1r_1}-a_{j_1j_1r_2})^{m_{j_1r_2}}\prod_{u_2=1,u_2\neq u_1}^{q_1}(a_{j_1j_2u_1}-a_{j_1j_2u_2})^{n_{j_1u_2}}<0$ *and* $a_{j_2j_20}\prod_{v_2=1,v_2\neq v_1}^{p_2}(a_{j_2j_1v_1}-a_{j_2j_1v_2})^{m_{j_2v_2}}<0$,

$$\begin{aligned}&(a_{j_1j_2u_1}, a_{j_2j_1v_1})\\&=\underbrace{([(2m_{j_1r_1}1), (2n_{j_1u_1}1+1)]^{\text{th}}\text{UP}_{-}\text{:UP}_{-},[2(m_{j_1r_1}1-m_{j_2v_1}1)]^{\text{th}}{}_{\text{DC:IC}}\text{LS})}_{([(2m_{j_1r_1}1),(2n_{j_1u_1}1+1)],[2(m_{j_1r_1}1-m_{j_2v_1}1)])\text{-concave up-parabola lower-saddle}}.\end{aligned} \tag{1.262}$$

The equilibrium is a $([(2m_{j_1r_1}1), (2n_{j_1u_1}1+1)]^{\text{th}}\text{UP}_{-}\text{:UP}_{-},[2(m_{j_1r_1}1-m_{j_2v_1}1)]^{\text{th}}{}_{\text{DC:IC}}\text{LS})$ *concave up-parabola lower-saddle at* $(x^*_{j_2}, x^*_{j_1}) = (a_{j_1j_2u_1}, a_{j_2j_1v_1})$.

(ii$_{2b2e}$) For $m_{j_1r_1}=2m_{j_1r_1}1+1, n_{j_1u_1}=2n_{j_1u_1}1$ *and* $m_{j_2v_1}=2m_{j_1v_1}1+1$, *the* equilibrium $(x^*_{j_2}, x^*_{j_1}) = (a_{j_1j_2u_1}, a_{j_2j_1v_1})$ *on the infinite-equilibrium of* $x^*_{j_1}=a_{j_2j_1v_1}=a_{j_1j_1r_1}$ *has the following properties.*

- *For* $a_{j_1j_10}\prod_{r_2=1,r_2\neq r_1}^{p_1}(a_{j_1j_1r_1}-a_{j_1j_1r_2})^{m_{j_1r_2}}\prod_{u_2=1,u_2\neq u_1}^{q_1}(a_{j_1j_2u_1}-a_{j_1j_2u_2})^{n_{j_1u_2}}>0$ *and* $a_{j_2j_20}\prod_{v_2=1,v_2\neq v_1}^{p_2}(a_{j_2j_1v_1}-a_{j_2j_1v_2})^{m_{j_2v_2}}>0$,

$$\begin{aligned}&(a_{j_1j_2u_1}, a_{j_2j_1v_1})\\&=\underbrace{([(2m_{j_1r_1}1+1), (2n_{j_1u_1}1)]^{\text{th}}\text{II}_{-}\text{:II}_{+},[2(m_{j_1r_1}1-m_{j_2v_1}1)]^{\text{th}}{}_{\text{IC:IC}}\text{SO})}_{([(2m_{j_1r_1}1+1),(2n_{j_1u_1}1)],[2(m_{j_1r_1}1-m_{j_2v_1}1)])\text{-increasing-concave inflection-source}}.\end{aligned} \tag{1.263}$$

The equilibrium is a $([(2m_{j_1r_1}1+1), (2n_{j_1u_1}1)]^{\text{th}}\text{II}_{-}\text{:II}_{+},[2(m_{j_1r_1}1-m_{j_2v_1}1)]^{\text{th}}{}_{\text{IC:IC}}\text{SO})$*-increasing-concave inflection-source at* $(x^*_{j_2}, x^*_{j_1}) = (a_{j_1j_2u_1}, a_{j_2j_1v_1})$.

- *For* $a_{j_1j_10}\prod_{r_2=1,r_2\neq r_1}^{p_1}(a_{j_1j_1r_1}-a_{j_1j_1r_2})^{m_{j_1r_2}}\prod_{u_2=1,u_2\neq u_1}^{q_1}(a_{j_1j_2u_1}-a_{j_1j_2u_2})^{n_{j_1u_2}}<0$ *and* $a_{j_2j_20}\prod_{v_2=1,v_2\neq v_1}^{p_2}(a_{j_2j_1v_1}-a_{j_2j_1v_2})^{m_{j_2v_2}}>0$,

$$\begin{aligned}&(a_{j_1j_2u_1}, a_{j_2j_1v_1})\\&=\underbrace{([(2m_{j_1r_1}1+1), (2n_{j_1u_1}1)]^{\text{th}}\text{DI}_{-}\text{:DI}_{+},[2(m_{j_1r_1}1-m_{j_2v_1}1)]^{\text{th}}{}_{\text{DC:DC}}\text{SI})}_{([(2m_{j_1r_1}1+1),(2n_{j_1u_1}1)],[2(m_{j_1r_1}1-m_{j_2v_1}1)])\text{-decreasing-concave inflection-sink}}.\end{aligned} \tag{1.264}$$

The equilibrium is a $([(2m_{j_1r_1}1+1),(2n_{j_1u_1}1)]^{\text{th}}\text{DI}_{-}{:}\text{DI}_{+},[2(m_{j_1r_1}1-m_{j_2v_1}1)]^{\text{th}}{}_{\text{DC:DC}}\text{SI})$*-decreasing-concave inflection-sink at* $(x_{j_2}^*,x_{j_1}^*)=(a_{j_1j_2u_1},a_{j_2j_1v_1})$.

- *For* $a_{j_1j_10}\prod_{r_2=1,r_2\neq r_1}^{p_1}(a_{j_1j_1r_1}-a_{j_1j_1r_2})^{m_{j_1r_2}}\prod_{u_2=1,u_2\neq u_1}^{q_1}(a_{j_1j_2u_1}-a_{j_1j_2u_2})^{n_{j_1u_2}}>0$ *and* $a_{j_2j_20}\prod_{v_2=1,v_2\neq v_1}^{p_2}(a_{j_2j_1v_1}-a_{j_2j_1v_2})^{m_{j_2v_2}}<0$,

$$
\begin{aligned}
&(a_{j_1j_2u_1},a_{j_2j_1v_1})\\
&=\underbrace{([(2m_{j_1r_1}1+1),(2n_{j_1u_1}1)]^{\text{th}}\text{DI}_{+}{:}\text{DI}_{-},[2(m_{j_1r_1}1-m_{j_2v_1}1)]^{\text{th}}{}_{\text{DC:DC}}\text{SO})}_{([(2m_{j_1r_1}1+1),(2n_{j_1u_1}1)],[2(m_{j_1r_1}1-m_{j_2v_1}1)])\text{-decreasing-concave inflection-source}}.
\end{aligned}
\tag{1.265}
$$

The equilibrium is a $([(2m_{j_1r_1}1+1),(2n_{j_1u_1}1)]^{\text{th}}\text{DI}_{+}{:}\text{DI}_{-},[2(m_{j_1r_1}1-m_{j_2v_1}1)]^{\text{th}}{}_{\text{DC:DC}}\text{SO})$*-decreasing-concave inflection-source at* $(x_{j_2}^*,x_{j_1}^*)=(a_{j_1j_2u_1},a_{j_2j_1v_1})$.

- *For* $a_{j_1j_10}\prod_{r_2=1,r_2\neq r_1}^{p_1}(a_{j_1j_1r_1}-a_{j_1j_1r_2})^{m_{j_1r_2}}\prod_{u_2=1,u_2\neq u_1}^{q_1}(a_{j_1j_2u_1}-a_{j_1j_2u_2})^{n_{j_1u_2}}<0$ *and* $a_{j_2j_20}\prod_{v_2=1,v_2\neq v_1}^{p_2}(a_{j_2j_1v_1}-a_{j_2j_1v_2})^{m_{j_2v_2}}<0$,

$$
\begin{aligned}
&(a_{j_1j_2u_1},a_{j_2j_1v_1})\\
&=\underbrace{([(2m_{j_1r_1}1+1),(2n_{j_1u_1}1)]^{\text{th}}\text{II}_{+}{:}\text{II}_{-},[2(m_{j_1r_1}1-m_{j_2v_1}1)]^{\text{th}}{}_{\text{IC:IC}}\text{SI})}_{([(2m_{j_1r_1}1+1),(2n_{j_1u_1}1)],[2(m_{j_2v_1}1-m_{j_1r_1}1)])\text{-increasing-concave inflection-sink}}.
\end{aligned}
\tag{1.266}
$$

The equilibrium is a $([(2m_{j_1r_1}1+1),(2n_{j_1u_1}1)]^{\text{th}}\text{II}_{+}{:}\text{II}_{-},[2(m_{j_1r_1}1-m_{j_2v_1}1)]^{\text{th}}{}_{\text{IC:IC}}\text{SI})$*-increasing-concave inflection-sink at* $(x_{j_2}^*,x_{j_1}^*)=(a_{j_1j_2u_1},a_{j_2j_1v_1})$.

(ii_{2b2f}*) For* $m_{j_1r_1}=2m_{j_1r_1}1$,$n_{j_1u_1}=2n_{j_1u_1}1$ *and* $m_{j_2v_1}=2m_{j_1v_1}1+1$*, the equilibrium* $(x_{j_2}^*,x_{j_1}^*)=(a_{j_1j_2u_1},a_{j_2j_1v_1})$ *on the infinite-equilibrium of* $x_{j_1}^*=a_{j_2j_1v_1}=a_{j_1j_1r_1}$ *has the following properties.*

- *For* $a_{j_1j_10}\prod_{r_2=1,r_2\neq r_1}^{p_1}(a_{j_1j_1r_1}-a_{j_1j_1r_2})^{m_{j_1r_2}}\prod_{u_2=1,u_2\neq u_1}^{q_1}(a_{j_1j_2u_1}-a_{j_1j_2u_2})^{n_{j_1u_2}}>0$ *and* $a_{j_2j_20}\prod_{v_2=1,v_2\neq v_1}^{p_2}(a_{j_2j_1v_1}-a_{j_2j_1v_2})^{m_{j_2v_2}}>0$,

$$
\begin{aligned}
&(a_{j_1j_2u_1},a_{j_2j_1v_1})\\
&=\underbrace{([(2m_{j_1r_1}1),(2n_{j_1u_1}1)]^{\text{th}}\text{DI}_{-}{:}\text{II}_{+},[2(m_{j_1r_1}1-m_{j_2v_1}1)-1]^{\text{th}}{}_{\text{DA:DA}}\text{US})}_{([(2m_{j_1r_1}1),(2n_{j_1u_1}1)],[2(m_{j_1r_1}1-m_{j_2v_1}1)-1])\text{-inflection down-asymptote upper-saddle}}.
\end{aligned}
\tag{1.267}
$$

The equilibrium is a $([(2m_{j_1r_1}1),(2n_{j_1u_1}1)]^{\text{th}}\text{DI}_{-}{:}\text{II}_{+},[2(m_{j_1r_1}1-m_{j_2v_1}1)-1]^{\text{th}}{}_{\text{DA:DA}}\text{US})$*-inflection down-asymptote upper-saddle at* $(x_{j_2}^*,x_{j_1}^*)=(a_{j_1j_2u_1},a_{j_2j_1v_1})$.

- *For* $a_{j_1j_10}\prod_{r_2=1,r_2\neq r_1}^{p_1}(a_{j_1j_1r_1}-a_{j_1j_1r_2})^{m_{j_1r_2}}\prod_{u_2=1,u_2\neq u_1}^{q_1}(a_{j_1j_2u_1}-a_{j_1j_2u_2})^{n_{j_1u_2}}<0$ *and* $a_{j_2j_20}\prod_{v_2=1,v_2\neq v_1}^{p_2}(a_{j_2j_1v_1}-a_{j_2j_1v_2})^{m_{j_2v_2}}>0$,

$$(a_{j_1j_2u_1}, a_{j_2j_1v_1}) = \underbrace{([(2m_{j_1r_1}1), (2n_{j_1u_1}1)]^{\text{th}}\text{II}_-\text{:DI}_+, [2(m_{j_1r_1}1 - m_{j_2v_1}1) - 1]^{\text{th}}{}_{\text{UA:UA}}\text{LS})}_{([(2m_{j_1r_1}1),(2n_{j_1u_1}1)],[2(m_{j_1r_1}1-m_{j_2v_1}1)-1])\text{-inflection up-asymptote lower-saddle}}. \tag{1.268}$$

The equilibrium is a $([(2m_{j_1r_1}1), (2n_{j_1u_1}1)]^{\text{th}}\text{II}_-\text{:DI}_+, [2(m_{j_1r_1}1 - m_{j_2v_1}1) - 1]^{\text{th}}{}_{\text{UA:UA}}\text{LS})$*-inflection up-asymptote lower-saddle at* $(x^*_{j_2}, x^*_{j_1}) = (a_{j_1j_2u_1}, a_{j_2j_1v_1})$.

- *For* $a_{j_1j_10}\prod_{r_2=1,r_2\neq r_1}^{p_1}(a_{j_1j_1r_1} - a_{j_1j_1r_2})^{m_{j_1r_2}}\prod_{u_2=1,u_2\neq u_1}^{q_1}(a_{j_1j_2u_1} - a_{j_1j_2u_2})^{n_{j_1u_2}} > 0$ *and* $a_{j_2j_20}\prod_{v_2=1,v_2\neq v_1}^{p_2}(a_{j_2j_1v_1} - a_{j_2j_1v_2})^{m_{j_2v_2}} < 0$,

$$(a_{j_1j_2u_1}, a_{j_2j_1v_1}) = \underbrace{([(2m_{j_1r_1}1), (2n_{j_1u_1}1)]^{\text{th}}\text{II}_+\text{:DI}_-, [2(m_{j_1r_1}1 - m_{j_2v_1}1) - 1]^{\text{th}}{}_{\text{UA:UA}}\text{US})}_{([(2m_{j_1r_1}1),(2n_{j_1u_1}1)],[2(m_{j_1r_1}1-m_{j_2v_1}1)-1])\text{-inflection up-asymptote upper-saddle}}. \tag{1.269}$$

The equilibrium is a $([(2m_{j_1r_1}1), (2n_{j_1u_1}1)]^{\text{th}}\text{II}_+\text{:DI}_-, [2(m_{j_1r_1}1 - m_{j_2v_1}1) - 1]^{\text{th}}{}_{\text{UA:UA}}\text{US})$*-inflection up-asymptote upper-saddle at* $(x^*_{j_2}, x^*_{j_1}) = (a_{j_1j_2u_1}, a_{j_2j_1v_1})$.

- *For* $a_{j_1j_10}\prod_{r_2=1,r_2\neq r_1}^{p_1}(a_{j_1j_1r_1} - a_{j_1j_1r_2})^{m_{j_1r_2}}\prod_{u_2=1,u_2\neq u_1}^{q_1}(a_{j_1j_2u_1} - a_{j_1j_2u_2})^{n_{j_1u_2}} < 0$ *and* $a_{j_2j_20}\prod_{v_2=1,v_2\neq v_1}^{p_2}(a_{j_2j_1v_1} - a_{j_2j_1v_2})^{m_{j_2v_2}} < 0$,

$$(a_{j_1j_2u_1}, a_{j_2j_1v_1}) = \underbrace{([(2m_{j_1r_1}1), (2n_{j_1u_1}1)]^{\text{th}}\text{DI}_+\text{:II}_-, [2(m_{j_1r_1}1 - m_{j_2v_1}1) - 1]^{\text{th}}{}_{\text{DA:DA}}\text{LS})}_{([(2m_{j_1r_1}1),(2n_{j_1u_1}1)],[2(m_{j_1r_1}1-m_{j_2v_1}1)-1])\text{-inflection down-asymptote lower-saddle}}. \tag{1.270}$$

The equilibrium is a $([(2m_{j_1r_1}1), (2n_{j_1u_1}1)]^{\text{th}}\text{DI}_+\text{:II}_-, [2(m_{j_1r_1}1 - m_{j_2v_1}1) - 1]^{\text{th}}{}_{\text{DA:DA}}\text{LS})$*-inflection down-asymptote lower-saddle at* $(x^*_{j_2}, x^*_{j_1}) = (a_{j_1j_2u_1}, a_{j_2j_1v_1})$.

(ii_{2b2g}) *For* $m_{j_1r_1} = 2m_{j_1r_1}1 + 1, n_{j_1u_1} = 2n_{j_1u_1}1$ *and* $m_{j_2v_1} = 2m_{j_1v_1}1$, *the* equilibrium $(x^*_{j_2}, x^*_{j_1}) = (a_{j_1j_2u_1}, a_{j_2j_1v_1})$ *on the infinite-equilibrium of* $x^*_{j_1} = a_{j_2j_1v_1} = a_{j_1j_1r_1}$ *has the following properties.*

- *For* $a_{j_1j_10}\prod_{r_2=1,r_2\neq r_1}^{p_1}(a_{j_1j_1r_1} - a_{j_1j_1r_2})^{m_{j_1r_2}}\prod_{u_2=1,u_2\neq u_1}^{q_1}(a_{j_1j_2u_1} - a_{j_1j_2u_2})^{n_{j_1u_2}} > 0$ *and* $a_{j_2j_20}\prod_{v_2=1,v_2\neq v_1}^{p_2}(a_{j_2j_1v_1} - a_{j_2j_1v_2})^{m_{j_2v_2}} > 0$,

$$(a_{j_1j_2u_1}, a_{j_2j_1v_1}) = \underbrace{([(2m_{j_1r_1}1 + 1), (2n_{j_1u_1}1)]^{\text{th}}\text{DI}_+\text{:II}_+, [2(m_{j_1r_1}1 - m_{j_2v_1}1) + 1]^{\text{th}}{}_{\text{DA:DA}}\text{SO})}_{([(2m_{j_1r_1}1+1),(2n_{j_1u_1}1)],[2(m_{j_1r_1}1-m_{j_2v_1}1)]+1)\text{-inflection down-asymptote source}}. \tag{1.271}$$

The equilibrium is a $([(2m_{j_1r_1}1+1),(2n_{j_1u_1}1)]^{\text{th}}\text{DI}_{+}\text{:II}_{+},[2(m_{j_1r_1}1-m_{j_2v_1}1)+1]^{\text{th}}{}_{\text{DA:DA}}\text{SO})$*-inflection down-asymptote source at* $(x_{j_2}^*,x_{j_1}^*)=(a_{j_1j_2u_1},a_{j_2j_1v_1})$.

- *For* $a_{j_1j_1 0}\prod_{r_2=1,r_2\neq r_1}^{p_1}(a_{j_1j_1r_1}-a_{j_1j_1r_2})^{m_{j_1r_2}}\prod_{u_2=1,u_2\neq u_1}^{q_1}(a_{j_1j_2u_1}-a_{j_1j_2u_2})^{n_{j_1u_2}}<0$ *and* $a_{j_2j_2 0}\prod_{v_2=1,v_2\neq v_1}^{p_2}(a_{j_2j_1v_1}-a_{j_2j_1v_2})^{m_{j_2v_2}}>0$,

$$\begin{aligned}&(a_{j_1j_2u_1},a_{j_2j_1v_1})\\&=\underbrace{([(2m_{j_1r_1}1+1),(2n_{j_1u_1}1)]^{\text{th}}\text{II}_{+}\text{:DI}_{+},[2(m_{j_1r_1}1-m_{j_2v_1}1)+1]^{\text{th}}{}_{\text{UA:UA}}\text{SI})}_{([(2m_{j_1r_1}1+1),(2n_{j_1u_1}1)],[2(m_{j_1r_1}1-m_{j_2v_1}1)]+1)\text{-inflection up-asymptote sink}}.\end{aligned}\tag{1.272}$$

The equilibrium is a $([(2m_{j_1r_1}1+1),(2n_{j_1u_1}1)]^{\text{th}}\text{II}_{+}\text{:DI}_{+},[2(m_{j_1r_1}1-m_{j_2v_1}1)+1]^{\text{th}}{}_{\text{UA:UA}}\text{SI})$*-inflection up-asymptote sink at* $(x_{j_1}^*,x_{j_2}^*)=(a_{j_1j_1r_1},a_{j_2j_2s_1})$.

- *For* $a_{j_1j_1 0}\prod_{r_2=1,r_2\neq r_1}^{p_1}(a_{j_1j_1r_1}-a_{j_1j_1r_2})^{m_{j_1r_2}}\prod_{u_2=1,u_2\neq u_1}^{q_1}(a_{j_1j_2u_1}-a_{j_1j_2u_2})^{n_{j_1u_2}}>0$ *and* $a_{j_2j_2 0}\prod_{v_2=1,v_2\neq v_1}^{p_2}(a_{j_2j_1v_1}-a_{j_2j_1v_2})^{m_{j_2v_2}}<0$,

$$\begin{aligned}&(a_{j_1j_2u_1},a_{j_2j_1v_1})\\&=\underbrace{([(2m_{j_1r_1}1+1),(2n_{j_1u_1}1)]^{\text{th}}\text{II}_{-}\text{:DI}_{-},[2(m_{j_1r_1}1-m_{j_2v_1}1)+1]^{\text{th}}{}_{\text{UA:UA}}\text{SO})}_{([(2m_{j_1r_1}1+1),(2n_{j_1u_1}1)],[2(m_{j_1r_1}1-m_{j_2v_1}1)]+1)\text{-inflection up-asymptote source}}.\end{aligned}\tag{1.273}$$

The equilibrium is a $([(2m_{j_1r_1}1+1),(2n_{j_1u_1}1)]^{\text{th}}\text{II}_{-}\text{:DI}_{-},[2(m_{j_1r_1}1-m_{j_2v_1}1)+1]^{\text{th}}{}_{\text{UA:UA}}\text{SO})$*-inflection up-asymptote source at* $(x_{j_2}^*,x_{j_1}^*)=(a_{j_1j_2u_1},a_{j_2j_1v_1})$.

- *For* $a_{j_1j_1 0}\prod_{r_2=1,r_2\neq r_1}^{p_1}(a_{j_1j_1r_1}-a_{j_1j_1r_2})^{m_{j_1r_2}}\prod_{u_2=1,u_2\neq u_1}^{q_1}(a_{j_1j_2u_1}-a_{j_1j_2u_2})^{n_{j_1u_2}}<0$ *and* $a_{j_2j_2 0}\prod_{v_2=1,v_2\neq v_1}^{p_2}(a_{j_2j_1v_1}-a_{j_2j_1v_2})^{m_{j_2v_2}}<0$,

$$\begin{aligned}&(a_{j_1j_2u_1},a_{j_2j_1v_1})\\&=\underbrace{([(2m_{j_1r_1}1+1),(2n_{j_1u_1}1)]^{\text{th}}\text{DI}_{-}\text{:II}_{-},[2(m_{j_1r_1}1-m_{j_2v_1}1)+1]^{\text{th}}{}_{\text{DA:DA}}\text{SI})}_{([(2m_{j_1r_1}1+1),(2n_{j_1u_1}1)],[2(m_{j_1r_1}1-m_{j_2v_1}1)]+1)\text{-inflection down-asymptote sink}}.\end{aligned}\tag{1.274}$$

The equilibrium is a $([(2m_{j_1r_1}1+1),(2n_{j_1u_1}1)]^{\text{th}}\text{DI}_{-}\text{:II}_{-},[2(m_{j_1r_1}1-m_{j_2v_1}1)+1]^{\text{th}}{}_{\text{DA:DA}}\text{SI})$*-inflection down-asymptote sink at* $(x_{j_2}^*,x_{j_1}^*)=(a_{j_1j_2u_1},a_{j_2j_1v_1})$.

(ii$_{2b2h}$*) For* $m_{j_1r_1}=2m_{j_1r_1}1, n_{j_1u_1}=2n_{j_1u_1}1$ *and* $m_{j_2v_1}=2m_{j_1v_1}1$*, the equilibrium of* $(x_{j_1}^*,x_{j_2}^*)=(a_{j_1j_1r_1},a_{j_1j_2u_1})$ *on the infinite-equilibrium of* $x_{j_1}^*=a_{j_2j_1v_1}=a_{j_1j_1r_1}$ *has the following properties.*

- *For* $a_{j_1j_1 0}\prod_{r_2=1,r_2\neq r_1}^{p_1}(a_{j_1j_1r_1}-a_{j_1j_1r_2})^{m_{j_1r_2}}\prod_{u_2=1,u_2\neq u_1}^{q_1}(a_{j_1j_2u_1}-a_{j_1j_2u_2})^{n_{j_1u_2}}>0$ *and* $a_{j_2j_2 0}\prod_{v_2=1,v_2\neq v_1}^{p_2}(a_{j_2j_1v_1}-a_{j_2j_1v_2})^{m_{j_2v_2}}>0$,

$$(a_{j_1j_2u_1}, a_{j_2j_1v_1}) = \underbrace{([(2m_{j_1r_1}1), (2n_{j_1u_1}1)]^{\text{th}}\text{II}_+{:}\text{II}_+,[2(m_{j_1r_1}1 - m_{j_2v_1}1)]^{\text{th}}{}_{\text{IC:IC}}\text{US})}_{([(2m_{j_1r_1}1),(2n_{j_1u_1}1)],[2(m_{j_1r_1}1-m_{j_2v_1}1)])\text{-increasing-concave inflection upper-saddle}}. \quad (1.275)$$

The equilibrium is a $([(2m_{j_1r_1}1), (2n_{j_1u_1}1)]^{\text{th}}\text{II}_+{:}\text{II}_+,[2(m_{j_1r_1}1 - m_{j_2v_1}1)]^{\text{th}}{}_{\text{IC:IC}}\text{US})$*-increasing-concave inflection upper-saddle at* $(x^*_{j_2}, x^*_{j_1}) = (a_{j_1j_2u_1}, a_{j_2j_1v_1})$.

- *For* $a_{j_1j_10}\prod_{r_2=1,r_2\neq r_1}^{p_1}(a_{j_1j_1r_1} - a_{j_1j_1r_2})^{m_{j_1r_2}}\prod_{u_2=1,u_2\neq u_1}^{q_1}(a_{j_1j_2u_1} - a_{j_1j_2u_2})^{n_{j_1u_2}} < 0$ *and* $a_{j_2j_20}\prod_{v_2=1,v_2\neq v_1}^{p_2}(a_{j_2j_1v_1} - a_{j_2j_1v_2})^{m_{j_2v_2}} > 0$,

$$(a_{j_1j_2u_1}, a_{j_2j_1v_1}) = \underbrace{([(2m_{j_1r_1}1), (2n_{j_1u_1}1)]^{\text{th}}\text{DI}_+{:}\text{DI}_+,[2(m_{j_1r_1}1 - m_{j_2v_1}1)]^{\text{th}}{}_{\text{DC:DC}}\text{LS})}_{([(2m_{j_1r_1}1),(2n_{j_1u_1}1)],[2(m_{j_1r_1}1-m_{j_2v_1}1)])\text{-decreasing-concave inflection lower-saddle}}. \quad (1.276)$$

The equilibrium is a $([(2m_{j_1r_1}1), (2n_{j_1u_1}1)]^{\text{th}}\text{DI}_+{:}\text{DI}_+,[2(m_{j_1r_1}1 - m_{j_2v_1}1)]^{\text{th}}{}_{\text{DC:DC}}\text{LS})$*-decreasing-concave inflection lower-saddle at* $(x^*_{j_2}, x^*_{j_1}) = (a_{j_1j_2u_1}, a_{j_2j_1v_1})$.

- *For* $a_{j_1j_10}\prod_{r_2=1,r_2\neq r_1}^{p_1}(a_{j_1j_1r_1} - a_{j_1j_1r_2})^{m_{j_1r_2}}\prod_{u_2=1,u_2\neq u_1}^{q_1}(a_{j_1j_2u_1} - a_{j_1j_2u_2})^{n_{j_1u_2}} > 0$ *and* $a_{j_2j_20}\prod_{v_2=1,v_2\neq v_1}^{p_2}(a_{j_2j_1v_1} - a_{j_2j_1v_2})^{m_{j_2v_2}} < 0$,

$$(a_{j_1j_2u_1}, a_{j_2j_1v_1}) = \underbrace{([(2m_{j_1r_1}1), (2n_{j_1u_1}1)]^{\text{th}}\text{DI}_+{:}\text{DI}_+,[2(m_{j_1r_1}1 - m_{j_2v_1}1)]^{\text{th}}{}_{\text{DC:DC}}\text{US})}_{([(2m_{j_1r_1}1),(2n_{j_1u_1}1)],[2(m_{j_1r_1}1-m_{j_2v_1}1)])\text{-decreasing-concave inflection upper-saddle}}. \quad (1.277)$$

The equilibrium is a $([(2m_{j_1r_1}1), (2n_{j_1u_1}1)]^{\text{th}}\text{DI}_+{:}\text{DI}_+,[2(m_{j_1r_1}1 - m_{j_2v_1}1)]^{\text{th}}{}_{\text{DC:DC}}\text{US})$*-decreasing-concave inflection upper-saddle at* $(x^*_{j_2}, x^*_{j_1}) = (a_{j_1j_2u_1}, a_{j_2j_1v_1})$.

- *For* $a_{j_1j_10}\prod_{r_2=1,r_2\neq r_1}^{p_1}(a_{j_1j_1r_1} - a_{j_1j_1r_2})^{m_{j_1r_2}}\prod_{u_2=1,u_2\neq u_1}^{q_1}(a_{j_1j_2u_1} - a_{j_1j_2u_2})^{n_{j_1u_2}} < 0$ *and* $a_{j_2j_20}\prod_{v_2=1,v_2\neq v_1}^{p_2}(a_{j_2j_1v_1} - a_{j_2j_1v_2})^{m_{j_2v_2}} < 0$,

$$(a_{j_1j_2u_1}, a_{j_2j_1v_1}) = \underbrace{([(2m_{j_1r_1}1), (2n_{j_1u_1}1)]^{\text{th}}\text{II}_-{:}\text{II}_-,[2(m_{j_1r_1}1 - m_{j_2v_1}1)]^{\text{th}}{}_{\text{IC:IC}}\text{LS})}_{([(2m_{j_1r_1}1),(2n_{j_1u_1}1)],[2(m_{j_1r_1}1-m_{j_2v_1}1)])\text{-increasing-concave inflection lower-saddle}}. \quad (1.278)$$

The equilibrium is a $([(2m_{j_1r_1}1), (2n_{j_1u_1}1)]^{\text{th}}\text{II}_-{:}\text{II}_-,[2(m_{j_1r_1}1 - m_{j_2v_1}1)]^{\text{th}}{}_{\text{IC:IC}}\text{LS})$*-increasing-concave inflection lower-saddle at* $(x^*_{j_2}, x^*_{j_1}) = (a_{j_1j_2u_1}, a_{j_2j_1v_1})$.

(iii) Consider

$$\dot{x}_{j_2} = a_{j_2j_20}\prod_{v=1}^{m_2}(x_{j_1} - a_{j_1j_2v}),$$
$$\dot{x}_{j_1} = a_{j_1j_10}\prod_{r=1}^{m_1}(x_{j_1} - a_{j_1j_1r})\prod_{u=1}^{n_1}(x_{j_2} - a_{j_1j_2u}). \tag{1.279}$$

The first integral manifold is

$$\begin{aligned}
&\sum_{\nu_0=m_1}^{m_2} b_{j_2j_1\nu_0} \sum_{\substack{\mu_1+\nu_1=\nu_0-1 \\ \mu_1\in[0,\beta_0-m_1], \\ \nu_1\in[m_1-1,\nu_0-1]}} \sum_{\substack{\mu_2+\nu_2=\nu_1-2 \\ \mu_2\in[0,\nu_1-m_1] \\ \nu_2\in[m_1-2,\nu_1-2]}} \cdots \sum_{\substack{\mu_{m_1-1}+\nu_{m_1-1}=\nu_{m_1-2}-m_1+1 \\ \mu_{m_1-1}\in[0,\nu_{m_1-2}-m_1] \\ \nu_{m_1-1}\in[1,\nu_{m_1-2}-m_1+1]}} \prod_{j=1}^{m_1-1} \frac{(\nu_{j-1}-j)!}{\mu_j!\nu_j!} \\
&\quad \times (a_{j_1j_1r_{j+1}} - a_{j_1j_1r_j})^{\mu_j} \frac{1}{\nu_{m_1-1}}\left[(x_{j_1} - a_{j_1j_1m_1})^{\nu_{m_1-1}} - (x_{j_10} - a_{j_1j_1m_1})^{\nu_{m_1-1}}\right] \\
&\quad - \sum_{\nu_0=m_1}^{m_2} b_{j_2j_1\nu_0} \sum_{k=2}^{m_1-1}\Bigg[\sum_{\substack{\mu_1+\nu_1=\nu_0-1 \\ \mu_1\in[0,\nu_0-m_1), \\ \nu_1\in(m_1-1,\nu_0-1]}} \sum_{\substack{\alpha_2+\nu_2=\nu_1-2 \\ \mu_2\in[0,\nu_1-m_1] \\ \mu_2\in(m_1-2,\nu_1-2)}} \cdots \\
&\quad \times \Bigg(\sum_{\substack{\mu_{k-1}+\nu_{k-1}=\nu_{k-2}-k+1 \\ \mu_{k-1}\in(\nu_{k-2}-m_1+1,\nu_{k-2}-k+1] \\ \nu_{k-1}\in[0,m_1-k)}} \prod_{j=1}^{k-1} \frac{(\nu_{j-1}-j)!}{\mu_j!\nu_j!}(a_{j_1j_1r_{j+1}} - a_{j_1j_1r_j})^{\mu_j} \\
&\quad \times \sum_{\rho=k}^{m_1} A_{\rho 1}^{(r_k)} \ln\frac{|x_{j_1} - a_{j_1j_1r_\rho}|}{|x_{j_10} - a_{j_1j_1r_\rho}|} \\
&\quad + \sum_{\substack{\mu_{k-1}+\nu_{k-1}=\nu_{k-2}-k+1 \\ \mu_{k-1}\in(\nu_{k-2}-m_1,\nu_{k-2}-m_1+1] \\ \nu_{k-1}\in[m_1-k,m_1-k+1)}} \prod_{j=1}^{k-1} \frac{(\nu_{j-1}-j)!}{\mu_j!\nu_j!}(a_{j_1j_1r_{j+1}} - a_{j_1j_1r_j})^{\mu_j} \\
&\quad \times \sum_{\rho=k+1}^{m_1} B_{\rho 1}^{(r_k)} \ln\frac{|x_{j_1} - a_{j_1j_1r_\rho}|}{|x_{j_10} - a_{j_1j_1r_\rho}|}\Bigg)\Bigg] + b_{j_2j_10}\sum_{\rho=1}^{m_1} A_{\rho 1}^{(r_1)} \ln\frac{|x_{j_1} - a_{j_1j_1r_\rho}|}{|x_{j_10} - a_{j_1j_1r_\rho}|} \\
&\quad + \sum_{\nu_0=1}^{m_1-1} b_{j_2j_1\nu_0} \sum_{\rho=2}^{m_1} B_{\rho 1}^{(r_1)} \ln\frac{|x_{j_1} - a_{j_1j_1r_\rho}|}{|x_{j_10} - a_{j_1j_1r_\rho}|} \\
&= \frac{a_{j_1j_10}}{a_{j_2j_20}}\Bigg\{\frac{1}{n_1+1}\left[(x_{j_2} - a_{j_1j_2u_1})^{n_1+1} - (x_{j_20} - a_{j_1j_2u_1})^{n_1+1}\right] \\
&\quad + \sum_{\alpha=1}^{n_1-1} \frac{1}{n_1-\alpha+1} b_{j_1j_2\alpha}\left[(x_{j_2} - a_{j_1j_2u_1})^{n_1-\alpha+1} - (x_{j_20} - a_{j_1j_2u_1})^{n_1-\alpha+1}\right]\Bigg\},
\end{aligned} \tag{1.280}$$

where

$$b_{j_1j_2\alpha} = \sum_{l_{n_1},t_{n_1}=0}^{l_{n_1}+t_{n_1}=1} \cdots \sum_{l_2,t_2=0}^{l_2+t_2=1} [\prod_{k=2}^{n_1} (a_{j_1j_2u_1} - a_{j_1j_2u_k})^{l_k}]\delta_{t_2+\cdots+t_{n_1}}^{n_1-\alpha};$$

$$b_{j_1j_20} = 1.$$

$$b_{j_2j_1v_0} = \sum_{t_1,l_1=0}^{t_1+l_1=1} \cdots \sum_{t_{m_2},l_{m_2}=0}^{t_{m_2}+l_{m_2}=1} \prod_{k=1}^{m_2} (a_{j_1j_1r_1} - a_{j_2j_1v_k})^{t_k}\delta_{v_0}^{\Sigma_{r=1}^{m_2} l_r};$$

$$\lim_{v_{m_1-1}\to 0} \frac{1}{v_{m_1-1}}\left[(x_{j_1} - a_{j_1j_1r_{m_1}})^{v_{m_1-1}} - (x_{j_10} - a_{j_1j_1r_{m_1}})^{v_{m_1-1}}\right] = \ln\frac{|x_{j_1} - a_{j_1j_1r_{m_1}}|}{|x_{j_10} - a_{j_1j_1r_{m_1}}|};$$

$$A_{\rho 1}^{(r_k)} = \frac{1}{\prod_{\sigma=k,\sigma\neq\rho}^{m_1} (a_{j_1j_1r_\rho} - a_{j_1j_1r_\sigma})}, \text{ for } \rho \in \{k, k+1, \ldots, m_1\},$$

$$B_{\rho 1}^{(r_k)} = \frac{(a_{j_1j_1s_r} - a_{j_1j_1r_k})^{\beta_{k-1}-k}}{\prod_{\sigma=k+1,\sigma\neq\rho}^{m_2} (a_{j_1j_1r_\rho} - a_{j_1j_1r_\sigma})}, \text{ for } \rho \in \{k+1, k+2, \ldots, m_1\};$$

$$k \in \{1, 2, \ldots, m_1 - 1\}. \tag{1.281}$$

(iii₁) For $a_{j_1j_2u} \neq a_{j_2j_2s}(s = 1, 2, \ldots, n_2;\ u = 1, 2, \ldots, n_1)$*, the simple 1-dimensional flows and simple equilibriums exist.*

(iii₁ₐ) The equilibrium of $(x_{j_1}^*, x_{j_2}^*) = (a_{j_1j_1r_1}, a_{j_1j_2u_1})(r_1 = 1, 2, \ldots, m_1;\ u_1 = 1, 2, \ldots, n_1)$ *for the* x_{j_1}*-direction has the following properties.*

- *For* $a_{j_1j_10}\prod_{r_2=1,r_2\neq r_1}^{m_1}(a_{j_1j_1r_1} - a_{j_1j_1r_2})\prod_{u_2=1,u_2\neq u_1}^{n_1}(a_{j_1j_2u_2} - a_{j_1j_2u_2}) > 0$ *and* $a_{j_2j_20}\prod_{v=1}^{m_1}(a_{j_1j_1r_1} - a_{j_2j_1v}) > 0$,

$$(a_{j_1j_1r_1}, a_{j_1j_2u_1}) = \underbrace{(\text{DP:UP,pF})}_{\text{hyperbolic flow }(+)}. \tag{1.282}$$

- *For* $a_{j_1j_10}\prod_{r_2=1,r_2\neq r_1}^{m_1}(a_{j_1j_1r_1} - a_{j_1j_1r_2})\prod_{u_2=1,u_2\neq u_1}^{n_1}(a_{j_1j_2u_2} - a_{j_1j_2u_2}) < 0$ *and* $a_{j_2j_20}\prod_{v=1}^{m_1}(a_{j_1j_1r_1} - a_{j_2j_1v}) > 0$,

$$(a_{j_1j_1r_1}, a_{j_1j_2u_1}) = \underbrace{(\text{UP:DP,pF})}_{\text{hyperbolic-secant flow }(+)}. \tag{1.283}$$

- *For* $a_{j_1j_10}\prod_{r_2=1,r_2\neq r_1}^{m_1}(a_{j_1j_1r_1} - a_{j_1j_1r_2})\prod_{u_2=1,u_2\neq u_1}^{n_1}(a_{j_1j_2u_2} - a_{j_1j_2u_2}) > 0$ *and* $a_{j_2j_20}\prod_{v=1}^{m_1}(a_{j_1j_1r_1} - a_{j_2j_1v}) < 0$,

$$(a_{j_1j_1r_1}, a_{j_1j_2u_1}) = \underbrace{\text{(UP:DP,nF)}}_{\text{hyperbolic-secant flow }(-)} . \tag{1.284}$$

- *For* $a_{j_1j_10}\prod_{r_2=1,r_2\neq r_1}^{m_1}(a_{j_1j_1r_1}-a_{j_1j_1r_2})\prod_{u_2=1,u_2\neq u_1}^{n_1}(a_{j_1j_2u_2}-a_{j_1j_2u_2}) < 0$ *and* $a_{j_2j_20}\prod_{v=1}^{m_1}(a_{j_1j_1r_1}-a_{j_2j_1v}) < 0,$

$$(a_{j_1j_1r_1}, a_{j_1j_2u_1}) = \underbrace{\text{(DP:UP,nF)}}_{\text{hyperbolic flow }(-)} . \tag{1.285}$$

(iii$_{1b}$*) The equilibrium of* $(x_{j_2}^*, x_{j_1}^*) = (a_{j_2j_1u_1}, a_{j_2j_1v_1})(u_1 = 1, 2, \ldots, n_1; v_1 = 1, 2, \ldots, m_2)$ *has the following properties.*

- *For* $a_{j_1j_10}\prod_{r=1}^{m_1}(a_{j_2j_1v_1}-a_{j_1j_1r})\prod_{u_2=1,u_2\neq u_1}^{n_1}(a_{j_1j_2u_1}-a_{j_1j_2u_2}) > 0$ *and* $a_{j_2j_20}\prod_{v_2=1,v_2\neq v_1}^{m_2}(a_{j_2j_1v_1}-a_{j_2j_1v_2}) > 0,$

$$(a_{j_2j_1u_1}, a_{j_2j_1v_1}) = \underbrace{(\text{UP}_+,\text{UP}_+)}_{\text{positve saddle}} . \tag{1.286}$$

- *For* $a_{j_1j_10}\prod_{r=1}^{m_1}(a_{j_2j_1v_1}-a_{j_1j_1r})\prod_{u_2=1,u_2\neq u_1}^{n_1}(a_{j_1j_2u_1}-a_{j_1j_2u_2}) < 0$ *and* $a_{j_2j_20}\prod_{v_2=1,v_2\neq v_1}^{m_2}(a_{j_2j_1v_1}-a_{j_2j_1v_2}) > 0,$

$$(a_{j_2j_1u_1}, a_{j_2j_1v_1}) = \underbrace{(\text{DP}_+,\text{DP}_-)}_{\text{CCW center}} . \tag{1.287}$$

- *For* $a_{j_1j_10}\prod_{r=1}^{m_1}(a_{j_2j_1v_1}-a_{j_1j_1r})\prod_{u_2=1,u_2\neq u_1}^{n_1}(a_{j_1j_2u_1}-a_{j_1j_2u_2}) > 0$ *and* $a_{j_2j_20}\prod_{v_2=1,v_2\neq v_1}^{m_2}(a_{j_2j_1v_1}-a_{j_2j_1v_2}) < 0,$

$$(a_{j_2j_1u_1}, a_{j_2j_1v_1}) = \underbrace{(\text{DP}_-,\text{DP}_+)}_{\text{CW center}} . \tag{1.288}$$

- *For* $a_{j_1j_10}\prod_{r=1}^{m_1}(a_{j_2j_1v_1}-a_{j_1j_1r})\prod_{u_2=1,u_2\neq u_1}^{n_1}(a_{j_1j_2u_1}-a_{j_1j_2u_2}) < 0$ *and*

$a_{j_2j_20}\prod_{v_2=1,v_2\neq v_1}^{m_2}(a_{j_2j_1v_1}-a_{j_2j_1v_2}) < 0,$

$$(a_{j_2j_1u_1}, a_{j_2j_1v_1}) = \underbrace{(\text{UP}_-,\text{UP}_-)}_{\text{negative saddle}} . \tag{1.289}$$

Define

$$\bigcup_{l=1}^{m_1+m_2}\bigcup_{u=1}^{n_1}(c_{j_1j_1l}, a_{j_1j_2u})$$

$$
\begin{aligned}
&= \bigcup_{r=1}^{m_1}\bigcup_{u=1}^{n_1}\underbrace{(a_{j_1j_1r},a_{j_1j_2u})}_{(XX,xF)} \oplus \bigcup_{v=1}^{m_2}\bigcup_{u=1}^{n_1}\underbrace{(a_{j_1j_2u},a_{j_2j_1v})}_{YY},\\
&\{c_{j_1j_1l}|l=1,2,\cdots,m_1+m_2\}\\
&= \text{sort}\{a_{j_1j_1r},a_{j_2j_1v}|r=1,2,\cdots,m_1;v=1,2,\cdots,m_2\},\\
&\text{for } c_{j_1j_1l_1}<c_{j_1j_1l_2} \text{ for } l_1<l_2;
\end{aligned}
\tag{1.290}
$$

where

$$
\begin{aligned}
&\bigcup_{r=1}^{m_1}\bigcup_{u=1}^{n_1}\underbrace{(a_{j_1j_1r},a_{j_1j_2u})}_{(\text{xF,XX})}\\
&= \begin{bmatrix}
(a_{j_1j_11},a_{j_1j_2n_1}) & \cdots & (a_{j_1j_1(m_1-1)},a_{j_1j_2n_1}) & (a_{j_1j_1m_1},a_{j_1j_2n_1})\\
(a_{j_1j_11},a_{j_1j_2(n_1-1)}) & \cdots & (a_{j_1j_1(m_1-1)},a_{j_1j_2(n_1-1)}) & (a_{j_1j_1m_1},a_{j_1j_2(n_1-1)})\\
\vdots & \vdots & \vdots & \vdots\\
(a_{j_1j_11},a_{j_1j_21}) & \cdots & (a_{j_1j_1(m_1-1)},a_{j_1j_21}) & (a_{j_1j_1m_1},a_{j_1j_21})
\end{bmatrix}_{n_1\times m_1}.
\end{aligned}
\tag{1.291}
$$

and

$$
\begin{aligned}
&\bigcup_{v=1}^{m_2}\bigcup_{u=1}^{n_1}\underbrace{(a_{j_1j_2u},a_{j_2j_1v})}_{\text{YY}}\\
&= \begin{bmatrix}
(a_{j_1j_2n_1},a_{j_2j_11}) & \cdots & (a_{j_1j_2(n_1-1)},a_{j_2j_1m_2}) & (a_{j_1j_2n_1},a_{j_2j_1m_2})\\
(a_{j_1j_2(n_1-1)},a_{j_2j_11}) & \cdots & (a_{j_1j_2(n_1-1)},a_{j_2j_1(m_2-1)}) & (a_{j_1j_2(n_1-1)},a_{j_2j_1m_2})\\
\vdots & \vdots & \vdots & \vdots\\
(a_{j_1j_21},a_{j_2j_11}) & \cdots & (a_{j_1j_21},a_{j_2j_1(m_2-1)}) & (a_{j_1j_21},a_{j_2j_1m_2})
\end{bmatrix}_{n_1\times m_2}.
\end{aligned}
\tag{1.292}
$$

(iii$_2$) For $a_{j_1j_1r_1}=a_{j_2j_1v_1}$, the dynamical system becomes

$$
\begin{aligned}
\dot{x}_{j_1} &= a_{j_1j_10}(x_{j_1}-a_{j_1j_1r_1})\prod_{r_2=1,r_2\neq r_1}^{m_1}(x_{j_1}-a_{j_1j_1r_2})\prod_{u=1,}^{n_1}(x_{j_2}-a_{j_1j_2u}),\\
\dot{x}_{j_2} &= a_{j_2j_20}(x_{j_1}-a_{j_2j_1v_1})\prod_{v_2=1,v_2\neq v_1}^{m_2}(x_{j_1}-a_{j_2j_1v_2}).
\end{aligned}
\tag{1.293}
$$

The first integral manifolds can be determined from Eq. (1.280) through the equivalent dynamical system as

$$\begin{aligned}
\dot{x}_{j_1} &= a_{j_1j_10}\prod_{r_2=1,r_2\neq r_1}^{m_1}(x_{j_1}-a_{j_1j_1r_2})\prod_{u=1,}^{n_1}(x_{j_2}-a_{j_1j_2u}),\\
\dot{x}_{j_2} &= a_{j_2j_20}\prod_{v_2=1,v_2\neq v_1}^{m_2}(x_{j_1}-a_{j_2j_1v_2}).\\
&m_1\Rightarrow m_1-1,\, r_{j+1}\Rightarrow r_j\ (j=1,2,\ldots,m_1-1);\\
&m_2\Rightarrow m_2-1,\, v_{i+1}\Rightarrow v_i\ (i=1,2,\ldots,m_2-1).
\end{aligned}\tag{1.294}$$

Define a network matrix as

$$\begin{aligned}
NM &= \underbrace{(a_{j_1j_2u_1},a_{j_2j_1v_1})}_{\text{(XX,YY)}}\oplus\bigcup_{u_2=1,u_2\neq u_1}^{n_1}\underbrace{(a_{j_1j_2u_1},a_{j_2j_1v_1})}_{\text{(XX,YY)}}\\
&\oplus\bigcup_{u=1}^{n_1}\bigcup_{r_2=1,r_2\neq r_1}^{m_1}\underbrace{(a_{j_1j_1r_2},a_{j_1j_2u})}_{\text{(XX,xF)}}\\
&\oplus\bigcup_{u=1}^{n_1}\bigcup_{v_2=1,v_2\neq v_1}^{m_2}\underbrace{(a_{j_1j_2u},a_{j_2j_1v_2})}_{\text{ZZ}}.
\end{aligned}\tag{1.295}$$

(iii_{2a}) The infinite-equilibrium of $x^*_{j_1}=a_{j_2j_1v_1}=a_{j_1j_1r_1}$ *with* $\overline{x}_{j_2}\neq a_{j_1j_2u_1}$ *has the following properties.*

- *For* $a_{j_1j_10}\prod_{r_2=1,r_2\neq r_1}^{m_1}(a_{j_1j_1r_1}-a_{j_1j_1r_2})\prod_{u=1}^{n_1}(\overline{x}_{j_2}-a_{j_1j_2u})>0$ *and* $a_{j_2j_20}\prod_{v_2=1,v_2\neq v_1}^{m_2}(a_{j_2j_1v_1}-a_{j_2j_1v_2})>0$,

$$(a_{j_1j_1r_1},\overline{x}_{j_2})=\underbrace{\text{(SO,II)}}_{\text{increasing-inflection source}}.\tag{1.296}$$

The infinite-equilibrium is a (SO,II)*-increasing-inflection source at* $x^*_{j_1}=a_{j_2j_1v_1}=a_{j_1j_1r_1}$.

- *For* $a_{j_1j_10}\prod_{r_2=1,r_2\neq r_1}^{m_1}(a_{j_1j_1r_1}-a_{j_1j_1r_2})\prod_{u=1}^{n_1}(\overline{x}_{j_2}-a_{j_1j_2u})<0$ *and* $a_{j_2j_20}\prod_{v_2=1,v_2\neq v_1}^{m_2}(a_{j_2j_1v_1}-a_{j_2j_1v_2})>0$,

$$(a_{j_1j_1r_1},\overline{x}_{j_2})=\underbrace{\text{(SI,DI)}}_{\text{decreasing-inflection sink}}.\tag{1.297}$$

The infinite-equilibrium is a (SI,DI)*-decreasing-inflection sink at* $x^*_{j_1}=a_{j_2j_1v_1}=a_{j_1j_1r_1}$.

- *For* $a_{j_1j_10}\prod_{r_2=1,r_2\neq r_1}^{m_1}(a_{j_1j_1r_1}-a_{j_1j_1r_2})\prod_{u=1}^{n_1}(\overline{x}_{j_2}-a_{j_1j_2u})>0$ *and* $a_{j_2j_20}\prod_{v_2=1,v_2\neq v_1}^{m_2}(a_{j_2j_1v_1}-a_{j_2j_1v_2})<0$,

$$(a_{j_1j_1r_1}, \overline{x}_{j_2}) = \underbrace{\text{(SO,DI)}}_{\text{decreasing-inflection source}} . \tag{1.298}$$

The infinite-equilibrium is a (SO,DI)*-decreasing-inflection sink at* $x^*_{j_1} = a_{j_2j_1v_1} = a_{j_1j_1r_1}$.

- *For* $a_{j_1j_10}\prod_{r_2=1,r_2\neq r_1}^{m_1}(a_{j_1j_1r_1}-a_{j_1j_1r_2})\prod_{u=1}^{n_1}(\overline{x}_{j_2}-a_{j_1j_2u}) < 0$ *and* $a_{j_2j_20}\prod_{v_2=1,v_2\neq v_1}^{m_2}(a_{j_2j_1v_1}-a_{j_2j_1v_2}) < 0$,

$$(a_{j_1j_1r_1}, \overline{x}_{j_2}) = \underbrace{\text{(SI,II)}}_{\text{increasing-inflection sink}} . \tag{1.299}$$

The infinite-equilibrium is a (SI,II)*-increasing-inflection sink at* $x^*_{j_1} = a_{j_2j_1v_1} = a_{j_1j_1r_1}$.

(iii_{2b}) *The equilibrium of* $(x^*_{j_2}, x^*_{j_1}) = (a_{j_1j_2u_1}, a_{j_2j_1v_1})$ *on the infinite-equilibrium of* $x^*_{j_1} = a_{j_1j_1r_1} = a_{j_2j_1v_1}$ *has the following properties.*

- *For* $a_{j_1j_10}\prod_{r_2=1,r_2\neq r_1}^{m_1}(a_{j_1j_1r_1}-a_{j_1j_1r_2})\prod_{u_2=1,u_2\neq u_1}^{n_2}(a_{j_1j_2u_1}-a_{j_1j_2u_2}) > 0$ *and* $a_{j_2j_20}\prod_{v_2=1,v_2\neq v_1}^{m_2}(a_{j_2j_1v_1}-a_{j_2j_1v_2}) > 0$,

$$(a_{j_1j_2u_1}, a_{j_2j_1v_1}) = \underbrace{(\text{UP}_- : \text{UP}_{+,\text{DI:II}}\text{US})}_{\text{inflection up-parabola upper-saddle}} . \tag{1.300}$$

The equilibrium is a $(\text{UP}_- : \text{UP}_{+,\text{DI:II}}\text{US})$*-inflection up-parabola upper-saddle at* $(x^*_{j_2}, x^*_{j_1}) = (a_{j_1j_2u_1}, a_{j_2j_1v_1})$.

- *For* $a_{j_1j_10}\prod_{r_2=1,r_2\neq r_1}^{m_1}(a_{j_1j_1r_1}-a_{j_1j_1r_2})\prod_{u_2=1,u_2\neq u_1}^{n_2}(a_{j_1j_2u_1}-a_{j_1j_2u_2}) < 0$ *and* $a_{j_2j_20}\prod_{v_2=1,v_2\neq v_1}^{m_2}(a_{j_2j_1v_1}-a_{j_2j_1v_2}) > 0$,

$$(a_{j_1j_2u_1}, a_{j_2j_1v_1}) = \underbrace{(\text{DP}_- : \text{DP}_{+,\text{II:DI}}\text{LS})}_{\text{inflection down-parabola lower-saddle}} . \tag{1.301}$$

The equilibrium is a $(\text{DP}_- : \text{DP}_{+,\text{II:DI}}\text{LS})$*-inflection down-parabola lower-saddle at* $(x^*_{j_2}, x^*_{j_1}) = (a_{j_1j_2u_1}, a_{j_2j_1v_1})$.

- *For* $a_{j_1j_10}\prod_{r_2=1,r_2\neq r_1}^{m_1}(a_{j_1j_1r_1}-a_{j_1j_1r_2})\prod_{u_2=1,u_2\neq u_1}^{n_2}(a_{j_1j_2u_1}-a_{j_1j_2u_2}) > 0$ *and* $a_{j_2j_20}\prod_{v_2=1,v_2\neq v_1}^{m_2}(a_{j_2j_1v_1}-a_{j_2j_1v_2}) < 0$,

$$(a_{j_1j_2u_1}, a_{j_2j_1v_1}) = \underbrace{(\text{DP}_+ : \text{DP}_{-,\text{II:DI}}\text{US})}_{\text{inflection down-parabola upper-saddle}} . \tag{1.302}$$

The equilibrium is a $(\text{DP}_+ : \text{DP}_{-,\text{II:DI}}\text{US})$*-inflection down-parabola upper-saddle at* $(x^*_{j_2}, x^*_{j_1}) = (a_{j_1j_2u_1}, a_{j_2j_1v_1})$.

- *For* $a_{j_1j_10}\prod_{r_2=1,r_2\neq r_1}^{m_1}(a_{j_1j_1r_1}-a_{j_1j_1r_2})\prod_{u_2=1,u_2\neq u_1}^{n_2}(a_{j_1j_2u_1}-a_{j_1j_2u_2})<0$ *and* $a_{j_2j_20}\prod_{v_2=1,v_2\neq v_1}^{m_2}(a_{j_2j_1v_1}-a_{j_2j_1v_2})<0$,

$$(a_{j_1j_2u_1},a_{j_2j_1v_1})=\underbrace{(\mathrm{UP}_+:\mathrm{UP}_{-},_{\mathrm{DI:II}}\mathrm{LS})}_{\text{inflection up-parabola lower-saddle}}\ . \tag{1.303}$$

The equilibrium is a $(\mathrm{UP}_+:\mathrm{UP}_{-},_{\mathrm{DI:II}}\mathrm{LS})$*-inflection up-parabola lower-saddle at* $(x_{j_2}^*,x_{j_1}^*)=(a_{j_1j_2u_1},a_{j_2j_1v_1})$.

(iv) The 1-dimensional flow at $(x_{j_1}^*,x_{j_2}^*)=(a_{j_1j_1 1},a_{j_1j_2 1})$ *with* $a_{j_2j_2 1}\neq a_{j_1j_2 1}$ *has four classes of 1-dimensional flow bifurcations with* $\dot{x}_{j_2}^*=a_{j_2j_20}(a_{j_1j_11}-a_{j_2j_11})^{m_2}$.

(iv$_1$) For $m_1=2m_{11}+1$ *and* $n_1=2n_{11}+1$, *the equilibrium of* $(x_{j_1}^*,x_{j_2}^*)=(a_{j_1j_11},a_{j_1j_21})$ *for the* x_{j_1}*-direction has the following bifurcation properties with*

$$\begin{aligned}
&2m_{11}+1-\sum_{r_1=1}^{p_1}m_{j_1r_1}^{(1)}=2\sum_{\rho_1=1}^{l_1}m_{\rho_1},\\
&2n_{11}+1-\sum_{u_1=1}^{q_1}n_{j_1u_1}^{(1)}=2\sum_{\sigma_1=1}^{l_2}n_{\sigma_1};\\
&\sum_{r_2=1}^{\bar{p}_1}m_{j_1r_2}^{(2)}=2m_{11}+1,\ \sum_{u_2=1}^{\bar{q}_1}n_{j_1u_2}^{(2)}=2n_{11}+1;\\
&\sum_{r_i=1}^{p_1}m_{j_1r_i}^{(i)}=2m_{11}+1,\ \sum_{u_i=1}^{q_1}n_{j_1u_i}^{(2)}=2n_{11}+1,\ \text{ for } i=1,2;
\end{aligned}\tag{1.304}$$

l_1,l_2*-quadratic polynomials are without real roots, and the* ρ_1^{th} *and* σ_1^{th}*-quadratic polynomials are with powers* m_{ρ_1} *and* n_{σ_1}, *respectively.*

(iv$_{1a}$) For $a_{j_1j_10}>0$ *and* $a_{j_2j_20}(a_{j_1j_11}-a_{j_2j_11})^{m_2}>0$, *there is a* ([(2$m_{11}$+1), (2$n_{11}$+1)] DP:UP,pF)*-positive hyperbolic flow as*

$$(a_{j_1j_11},a_{j_1j_21})=\underbrace{([(2m_{11}{+}1),(2n_{11}+1)]\mathrm{DP{:}UP,pF})}_{[(2m_{11}+1),(2n_{11}+1)]\text{-hyperbolic flow }(+)}. \tag{1.305}$$

There are three following ([(2m_{11}+1), (2n_{11} + 1)]DP:UP,pF)*-positive hyperbolic flow appearing and switching bifurcations.*

(iv$_{1a1}$) The ([(2m_{11}+1), (2n_{11} + 1)]DP:UP,pF)*-positive hyperbolic flow appearing bifurcation is from a* (DP:UP, pF)*-positive hyperbolic-flow to a* (2m_{11}+1) × (2n_{11}+1)*-hyperbolic and hyperbolic-secant flow network as*

$$\underbrace{(a_{j_1j_11},a_{j_1j_21})}_{\text{hyperbolic flow }(+)}\rightleftharpoons\underbrace{([(2m_{11}{+}1),(2n_{11}+1)]\mathrm{DP{:}UP,pF})}_{[(2m_{11}+1),(2n_{11}+1)]\text{-hyperbolic flow }(+)}$$

$$\rightleftharpoons \bigcup_{r=1}^{2m_{11}+1} \bigcup_{u=1}^{2n_{11}+1} \underbrace{(a_{j_1 j_1 r}, a_{j_1 j_2 u})}_{\text{YY}}, \tag{1.306}$$

where, without any equilibriums,

$$\bigcup_{r=1}^{2m_{11}+1} \bigcup_{u=1}^{2n_{11}+1} \underbrace{(a_{j_1 j_1 r}, a_{j_1 j_2 u})}_{\text{YY}}$$
$$= \begin{bmatrix} \underbrace{(\text{DP:UP,pF})}_{\text{hyperboli flow }(+)} & \cdots & \underbrace{(\text{UP:DP,pF})}_{\text{hyperbolic-secant flow }(+)} & \underbrace{(\text{DP:UP,pF})}_{\text{hyperbolic flow }(+)} \\ \underbrace{(\text{UP:DP,pF})}_{\text{hyperbolic-secant flow }(+)} & \cdots & \underbrace{(\text{DP:UP,pF})}_{\text{hyperbolic flow }(+)} & \underbrace{(\text{UP:DP,pF})}_{\text{hyperbolic-secant flow }(+)} \\ \vdots & \vdots & \vdots & \vdots \\ \underbrace{(\text{DP:UP,pF})}_{\text{hyperbolic flow }(+)} & \cdots & \underbrace{(\text{UP:DP,pF})}_{\text{hyperbolic-secant flow }(+)} & \underbrace{(\text{DP:UP,pF})}_{\text{hyperbolic flow }(+)} \end{bmatrix}_{(2n_{11}+1)\times(2m_{11}+1)} . \tag{1.307}$$

(iv_{1a2}) The $([(2m_{11}+1), (2n_{11}+1)]$DP:UP,pF)*-positive hyperbolic flow appearing bifurcation is from a* $q_1 \times p_1$ *to* $\overline{q}_1 \times \overline{p}_1$ *1-dimensional flow networks as*

$$\bigcup_{r_1=1}^{p_1} \bigcup_{u_1=1}^{q_1} \underbrace{(a_{j_1 j_1 r_1}, a_{j_1 j_2 u_1})}_{([m_{j_1 r_1}, n_{j_1 u_1}]\text{-XX,pF})} \rightleftharpoons \underbrace{([(2m_{11}+1), (2n_{11}+1)]\text{DP:UP,pF})}_{[(2m_{11}+1),(2n_{11}+1)]\text{-hyperbolic flow }(+)}$$
$$\rightleftharpoons \bigcup_{r_2=1}^{\overline{p}_1} \bigcup_{u_2=1}^{\overline{q}_1} \underbrace{(a_{j_1 j_1 r_2}, a_{j_1 j_2 u_2})}_{([m_{j_1 r_2}, n_{j_1 u_2}]\text{-YY,pF})} . \tag{1.308}$$

(iv_{1a3}) The $([(2m_{11}+1), (2n_{11}+1)]$DP:UP,pF)*-positive hyperbolic flow switching bifurcation is for the switching of two networks of* $q_1 \times p_1$ *1-dimensional flows as*

$$\bigcup_{r_1=1}^{p_1} \bigcup_{u_1=1}^{q_1} \underbrace{(a_{j_1 j_1 r_1}, a_{j_1 j_2 u_1})}_{([m_{j_1 r_1}, n_{j_1 u_1}]\text{-XX,pF})} \rightleftharpoons \underbrace{([(2m_{11}+1), (2n_{11}+1)]\text{DP:UP,pF})}_{[(2m_{11}+1),(2n_{11}+1)]\text{-hyperbolic flow }(+)}$$
$$\rightleftharpoons \bigcup_{r_2=1}^{p_1} \bigcup_{u_2=1}^{q_1} \underbrace{(a_{j_1 j_1 r_2}, a_{j_1 j_2 u_2})}_{([m_{j_1 r_2}, n_{j_1 u_2}]\text{-YY,pF})} . \tag{1.309}$$

(iv_{1b}) For $a_{j_1j_10} < 0$ and $a_{j_2j_20}(a_{j_1j_11} - a_{j_2j_11})^{m_2} > 0$, there is a ([(2$m_{11}$+1), (2$n_{11}$+1)] UP:DP,pF)*-positive hyperbolic-secant flow as*

$$(a_{j_1j_11}, a_{j_1j_21}) = \underbrace{([(2m_{11}+1), (2n_{11}+1)]\text{UP:DP,pF})}_{[(2m_{11}+1),(2n_{11}+1)]\text{-hyperbolic-secant flow }(+)}. \tag{1.310}$$

There are three following ([(2m_{11}+1), (2n_{11} + 1)]UP:DP,pF)*-positive hyperbolic-secant flow appearing and switching bifurcations.*

(iv_{1b1}) The ([(2m_{11}+1), (2n_{11}+1)]UP : DP, pF)*-positive hyperbolic-secant flow appearing bifurcation is from a (*UP:DP, pF*)-positive hyperbolic-secant flow to a* (2m_1+1) × (2n_1+1)*-hyperbolic and hyperbolic-secant flow network as*

$$\begin{aligned}\underbrace{(a_{j_1j_11}, a_{j_1j_21})}_{\text{hyperbolic-secant flow }(+)} &\rightleftharpoons \underbrace{([(2m_{11}+1), (2n_{11}+1)]\text{UP:DP,pF})}_{[(2m_{11}+1),(2n_{11}+1)]\text{-hyperbolic-secant flow }(+)}\\ &\rightleftharpoons \bigcup_{r=1}^{2m_{11}+1}\bigcup_{u=1}^{2n_{11}+1}\underbrace{(a_{j_1j_1r}, a_{j_1j_2u})}_{\text{YY}},\end{aligned} \tag{1.311}$$

where, without any equilibriums,

$$\bigcup_{r=1}^{2m_{11}+1}\bigcup_{u=1}^{2n_{11}+1}\underbrace{(a_{j_1j_1r}, a_{j_1j_2u})}_{\text{YY}} = \begin{bmatrix} \underbrace{(\text{UP:DP,pF})}_{\text{hyperbolic-secant flow }(+)} & \cdots & \underbrace{(\text{DP:UP,pF})}_{\text{hyperbolic flow }(+)} & \underbrace{(\text{UP:DP,pF})}_{\text{hyperbolic-secant flow }(+)}\\ \underbrace{(\text{DP:UP,pF})}_{\text{hyperbolic flow }(+)} & \cdots & \underbrace{(\text{UP:DP,pF})}_{\text{hyperbolic-secant flow }(+)} & \underbrace{(\text{DP:UP,pF})}_{\text{hyperbolic flow }(+)}\\ \vdots & \vdots & \vdots & \vdots\\ \underbrace{(\text{UP:DP,pF})}_{\text{hyperbolic-secant flow }(+)} & \cdots & \underbrace{(\text{DP:UP,pF})}_{\text{hyperbolic flow }(+)} & \underbrace{(\text{UP:DP,pF})}_{\text{hyperbolic-secant flow }(+)} \end{bmatrix}_{(2n_{11}+1)\times(2m_{11}+1)}. \tag{1.312}$$

(iv_{1b2}) The ([(2m_{11}+1), (2n_{11} + 1)]UP:DP,pF)*-positive hyperbolic-secant flow appearing bifurcation is from a $q_1 \times p_1$ to $\overline{q}_1 \times \overline{p}_1$ 1-dimensional flow networks as*

$$\bigcup_{r_1=1}^{p_1}\bigcup_{u_1=1}^{q_1}\underbrace{(a_{j_1j_1r_1}, a_{j_1j_2u_1})}_{([m_{j_1r_1}, n_{j_1u_1}]\text{-XX,pF})} \rightleftharpoons \underbrace{([(2m_{11}+1), (2n_{11}+1)]\text{UP:DP,pF})}_{[(2m_{11}+1),(2n_{11}+1)]\text{-hyperbolic-secant flow }(+)}$$

$$\rightleftharpoons \bigcup_{r_2=1}^{\overline{p}_1} \bigcup_{u_2=1}^{\overline{q}_1} \underbrace{(a_{j_1 j_1 r_2}, a_{j_1 j_2 u_2})}_{([m_{j_1 r_2}, n_{j_1 u_2}]\text{-YY,pF})} . \tag{1.313}$$

(iv_{Ib3}) The ([$(2m_{11}+1)$, $(2n_{11}+1)$]UP:DP,pF)*-positive hyperbolic-secant flow switching, bifurcation is for the switching of two networks of $q_1 \times p_1$ 1-dimensional flows as*

$$\begin{aligned} \bigcup_{r_1=1}^{p_1} \bigcup_{u_1=1}^{q_1} \underbrace{(a_{j_1 j_1 r_1}, a_{j_1 j_2 u_1})}_{([m_{j_1 r_1}, n_{j_1 u_1}]\text{-XX,pF})} &\rightleftharpoons \underbrace{([(2m_{11}+1), (2n_{11}+1)]\text{UP:DP,pF})}_{[(2m_{11}+1),(2n_{11}+1)]\text{-hyperbolic flow }(+)} \\ &\rightleftharpoons \bigcup_{r_2=1}^{p_1} \bigcup_{u_2=1}^{q_1} \underbrace{(a_{j_1 j_1 r_2}, a_{j_1 j_2 u_2})}_{([m_{j_1 r_2}, n_{j_1 u_2}]\text{-YY,pF})} . \end{aligned} \tag{1.314}$$

(iv_{Ic}) For $a_{j_1 j_1 0} > 0$ and $a_{j_2 j_2 0}(a_{j_1 j_1 1} - a_{j_2 j_1 1})^{m_2} < 0$, there is a ([$(2m_{11}+1)$, $(2n_{11}+1)$] UP:DP,nF)*-negative hyperbolic-secant flow as*

$$(a_{j_1 j_1 1}, a_{j_1 j_2 1}) = \underbrace{([(2m_{11}+1), (2n_{11}+1)]\text{UP:DP,nF})}_{[(2m_{11}+1),(2n_{11}+1)]\text{-hyperbolic-secant flow }(-)} . \tag{1.315}$$

There are three following ([$(2m_{11}+1)$, $(2n_{11}+1)$]UP:DP,nF)*-negative hyperbolic-secant flow appearing and switching bifurcations.*

(iv_{Ic1}) The ([$(2m_{11}+1)$, $(2n_{11}+1)$]UP:DP,nF)*-negative hyperbolic-secant flow appearing bifurcation is from a (*UP:DP, nF*)-negative hyperbolic-secant flow to a* $(2m_1+1) \times (2n_1+1)$*-hyperbolic and hyperbolic-secant flow network as*

$$\begin{aligned} \underbrace{(a_{j_1 j_1 1}, a_{j_1 j_2 1})}_{\text{hyperbolic-secant flow }(-)} &\rightleftharpoons \underbrace{([(2m_{11}+1), (2n_{11}+1)]\text{UP:DP,nF})}_{[(2m_{11}+1),(2n_{11}+1)]\text{-hyperbolic-secant flow }(-)} \\ &\rightleftharpoons \bigcup_{r=1}^{2m_{11}+1} \bigcup_{u=1}^{2n_{11}+1} \underbrace{(a_{j_1 j_1 r}, a_{j_1 j_2 u})}_{\text{YY}}, \end{aligned} \tag{1.316}$$

where, without any equilibriums,

$$\bigcup_{r=1}^{2m_{11}+1} \bigcup_{u=1}^{2n_{11}+1} \underbrace{(a_{j_1 j_1 r}, a_{j_1 j_2 u})}_{\text{YY}}$$

$$
= \begin{bmatrix} \underbrace{(\text{UP:DP,nF})}_{\text{hyperbolic-secant flow }(-)} & \cdots & \underbrace{(\text{DP:UP,nF})}_{\text{hyperbolic flow }(-)} & \underbrace{(\text{UP:DP,nF})}_{\text{hyperbolic-secant flow }(-)} \\ \underbrace{(\text{DP:UP,nF})}_{\text{hyperbolic flow }(-)} & \cdots & \underbrace{(\text{UP:DP,nF})}_{\text{hyperbolic-secant flow }(-)} & \underbrace{(\text{DP:UP,nF})}_{\text{hyperbolic flow }(-)} \\ \vdots & \vdots & \vdots & \vdots \\ \underbrace{(\text{UP:DP,nF})}_{\text{hyperbolic-secant flow }(-)} & \cdots & \underbrace{(\text{DP:UP,nF})}_{\text{hyperbolic flow }(-)} & \underbrace{(\text{UP:DP,nF})}_{\text{hyperbolic-secant flow }(-)} \end{bmatrix}_{(2n_{11}+1)\times(2m_{11}+1)} . \tag{1.317}
$$

(iv_{1c2}) The ([$(2m_{11}+1)$, $(2n_{11}+1)$]UP:DP,nF)*-negative hyperbolic-secant flow appearing bifurcation is from a $q_1 \times p_1$ to $\overline{q}_1 \times \overline{p}_1$ 1-dimensional flow networks as*

$$
\bigcup_{r_1=1}^{p_1}\bigcup_{u_1=1}^{q_1} \underbrace{(a_{j_1j_1r_1}, a_{j_1j_2u_1})}_{([m_{j_1r_1}, n_{j_1u_1}]\text{-XX,nF})} \rightleftharpoons \underbrace{([(2m_{11}+1), (2n_{11}+1)]\text{UP:DP,nF})}_{[(2m_{11}+1),(2n_{11}+1)]\text{-hyperbolic-secant flow }(-)}
$$
$$
\rightleftharpoons \bigcup_{r_2=1}^{\overline{p}_1}\bigcup_{u_2=1}^{\overline{q}_1} \underbrace{(a_{j_1j_1r_2}, a_{j_1j_2u_2})}_{([m_{j_1r_2}, n_{j_1u_2}]\text{-YY,nF})} . \tag{1.318}
$$

(iv_{1c3}) The ([$(2m_{11}+1)$, $(2n_{11}+1)$]UP:DP,nF)*-negative hyperbolic-secant flow switching bifurcation is for the switching of two networks of $q_1 \times p_1$ 1-dimensional flows as*

$$
\bigcup_{r_1=1}^{p_1}\bigcup_{u_1=1}^{q_1} \underbrace{(a_{j_1j_1r_1}, a_{j_1j_2u_1})}_{([m_{j_1r_1}, n_{j_1u_1}]\text{-XX,nF})} \rightleftharpoons \underbrace{([(2m_{11}+1), (2n_{11}+1)]\text{UP:DP,nF})}_{[(2m_{11}+1),(2n_{11}+1)]\text{-hyperbolic flow }(-)}
$$
$$
\rightleftharpoons \bigcup_{r_2=1}^{p_1}\bigcup_{u_2=1}^{q_1} \underbrace{(a_{j_1j_1r_2}, a_{j_1j_2u_2})}_{([m_{j_1r_2}, n_{j_1u_2}]\text{-YY,nF})} . \tag{1.319}
$$

(iv_{1d}) For $a_{j_1j_10} < 0$ and $a_{j_2j_20}(a_{j_1j_11} - a_{j_2j_11})^{m_2} < 0$, there is a ([$(2m_{11}+1)$, $(2n_{11}+1)$] DP:UP,nF)*-negative hyperbolic flow as*

$$
(a_{j_1j_11}, a_{j_1j_21}) = \underbrace{([(2m_{11}+1), (2n_{11}+1)]\text{DP:UP,nF})}_{[(2m_{11}+1),(2n_{11}+1)]\text{-hyperbolic flow }(-)} . \tag{1.320}
$$

There are three following ([(2m_{11}+1), (2n_{11} + 1)]DP:UP,nF)*-negative hyperbolic flow appearing and switching bifurcations.*

(iv$_{1d1}$) The ([(2m_{11}+1), (2n_{11} + 1)]DP:UP,nF)*-negative hyperbolic flow appearing bifurcation is from a (DP:UP, nF)-negative hyperbolic-secant flow to a* (2m_1+1) × (2n_1+1)*-hyperbolic and hyperbolic-secant flow network as*

$$
\begin{aligned}
\underbrace{(a_{j_1j_11}, a_{j_1j_21})}_{\text{hyperbolic-secant flow }(-)} &\rightleftharpoons \underbrace{([(2m_{11}+1), (2n_{11}+1)]\text{DP:UP,nF})}_{[(2m_{11}+1),(2n_{11}+1)]\text{-hyperbolic flow }(-)} \\
&\rightleftharpoons \bigcup_{r=1}^{2m_{11}+1} \bigcup_{u=1}^{2n_{11}+1} \underbrace{(a_{j_1j_1r}, a_{j_1j_2u})}_{\text{YY}},
\end{aligned}
\tag{1.321}
$$

where, without any equilibriums,

$$
\begin{aligned}
&\bigcup_{r=1}^{2m_{11}+1} \bigcup_{u=1}^{2n_{11}+1} \underbrace{(a_{j_1j_1r}, a_{j_1j_2u})}_{\text{YY}} \\
&= \begin{bmatrix}
\underbrace{(\text{UP:DP,nF})}_{\text{hyperbolic-secant flow }(-)} & \cdots & \underbrace{(\text{DP:UP,nF})}_{\text{hyperbolic flow }(-)} & \underbrace{(\text{UP:DP,nF})}_{\text{hyperbolic-secant flow }(-)} \\
\underbrace{(\text{DP:UP,nF})}_{\text{hyperbolic flow }(-)} & \cdots & \underbrace{(\text{UP:DP,nF})}_{\text{hyperbolic-secant flow }(-)} & \underbrace{(\text{DP:UP,nF})}_{\text{hyperbolic flow }(-)} \\
\vdots & \vdots & \vdots & \vdots \\
\underbrace{(\text{UP:DP,nF})}_{\text{hyperbolic-secant flow }(-)} & \cdots & \underbrace{(\text{DP:UP,nF})}_{\text{hyperbolic flow }(-)} & \underbrace{(\text{UP:DP,nF})}_{\text{hyperbolic-secant flow }(-)}
\end{bmatrix}_{(2n_{11}+1)\times(2m_{11}+1)}.
\end{aligned}
\tag{1.322}
$$

(iv$_{1d2}$) The ([(2m_{11}+1), (2n_{11} + 1)]DP:UP,nF)*-negative hyperbolic flow appearing bifurcation is from a $q_1 \times p_1$ to $\overline{q}_1 \times \overline{p}_1$ 1-dimensional flow networks as*

$$
\begin{aligned}
\bigcup_{r_1=1}^{p_1} \bigcup_{u_1=1}^{q_1} \underbrace{(a_{j_1j_1r_1}, a_{j_1j_2u_1})}_{([m_{j_1r_1}, n_{j_1u_1}]\text{-XX,nF})} &\rightleftharpoons \underbrace{([(2m_{11}+1), (2n_{11}+1)]\text{DP:UP,nF})}_{[(2m_{11}+1),(2n_{11}+1)]\text{-hyperbolic flow }(-)} \\
&\rightleftharpoons \bigcup_{r_2=1}^{\overline{p}_1} \bigcup_{u_2=1}^{\overline{q}_1} \underbrace{(a_{j_1j_1r_2}, a_{j_1j_2u_2})}_{([m_{j_1r_2}, n_{j_1u_2}]\text{-YY,nF})}.
\end{aligned}
\tag{1.323}
$$

(iv$_{1d3}$) The ([(2m_{11}+1), (2n_{11} + 1)]DP:UP,nF)*-negative hyperbolic flow switching bifurcation is for the switching of two networks of $q_1 \times p_1$ 1-dimensional flows as*

$$\bigcup_{r_1=1}^{p_1}\bigcup_{u_1=1}^{q_1}\underbrace{(a_{j_1j_1r_1},a_{j_1j_2u_1})}_{([m_{j_1r_1},n_{j_1u_1}]\text{-XX,nF})}\rightleftharpoons\underbrace{([(2m_{11}+1),(2n_{11}+1)]\text{DP:UP,nF})}_{[(2m_{11}+1),(2n_{11}+1)]\text{-hyperbolic flow }(-)}$$
$$\rightleftharpoons\bigcup_{r_2=1}^{p_1}\bigcup_{u_2=1}^{q_1}\underbrace{(a_{j_1j_1r_2},a_{j_1j_2u_2})}_{([m_{j_1r_2},n_{j_1u_2}]\text{-YY,nF})}. \tag{1.324}$$

(iv$_2$) For $m_1=2m_{11}$ and $n_1=2n_{11}+1$, the equilibrium of $(x_{j_1}^*,x_{j_2}^*)=(a_{j_1j_1 1},a_{j_1j_2 1})$ for the x_{j_1}-direction has the following bifurcation properties with

$$2m_{11}-\sum_{r_1=1}^{p_1}m_{j_1r_1}^{(1)}=2\sum_{\rho_1=1}^{l_1}m_{\rho_1},\ 2n_{11}+1-\sum_{u_1=1}^{q_1}n_{j_1u_1}^{(1)}=2\sum_{\sigma_1=1}^{l_2}n_{\sigma_1};$$
$$\sum_{r_2=1}^{\bar{p}_1}m_{j_1r_2}^{(2)}=2m_{11},\ \sum_{u_2=1}^{\bar{q}_1}n_{j_1u_2}^{(2)}=2n_{11}+1;$$
$$\sum_{r_i=1}^{p_1}m_{j_1r_i}^{(i)}=2m_{11},\ \sum_{u_i=1}^{q_1}n_{j_1u_i}^{(2)}=2n_{11}+1,\ \text{for } i=1,2; \tag{1.325}$$

l_1,l_2-quadratic polynomials are without real roots, and the ρ_1^{th} and σ_1^{th}-quadratic polynomials are with powers m_{ρ_1} and n_{σ_1}, respectively.

(iv$_{2a}$) For $a_{j_1j_1 0}>0$ and $a_{j_2j_2 0}(a_{j_1j_1 1}-a_{j_2j_1 1})^{m_2}>0$, there is a ([(2$m_{11}$), (2$n_{11}$ + 1)]UP:UP,pF)*-positive hyperbolic-secant-to-hyperbolic flow as*

$$(a_{j_1j_1 1},a_{j_1j_2 1})=\underbrace{([(2m_{11}),(2n_{11}+1)]\text{UP:UP,pF})}_{[(2m_{11}),(2n_{11}+1)]\text{-hyperbolic-secant-to-hyperbolic flow }(+)}. \tag{1.326}$$

There are three following ([(2m_{11}), (2n_{11} + 1)]UP:UP,pF)*-positive hyperbolic-secant-to-hyperbolic flow appearing and switching bifurcations.*

(iv$_{2a1}$) The ([(2m_{11}), (2n_{11} + 1)]UP:UP,pF)*-positive hyperbolic-secant-to-hyperbolic flow appearing bifurcation is from a* (UP, pF)*-positive up-parabola flow to a* $(2m_1)\times(2n_1+1)$*-hyperbolic and hyperbolic-secant flow network as*

$$\underbrace{(a_{j_1j_2 1},\dot{x}_{j_2})}_{\text{up-parabola flow }(+)}\rightleftharpoons\underbrace{([(2m_{11}),(2n_{11}+1)]\text{UP:UP,pF})}_{[(2m_{11}),(2n_{11}+1)]\text{-hyperbolic-secant-to-hyperbolic flow }(+)}$$
$$\rightleftharpoons\bigcup_{r=1}^{2m_{11}}\bigcup_{u=1}^{2n_{11}+1}\underbrace{(a_{j_1j_1r},a_{j_1j_2u})}_{\text{YY}}, \tag{1.327}$$

where, without any equilibriums,

$$\bigcup_{r=1}^{2m_{11}}\bigcup_{u=1}^{2n_{11}+1}\underbrace{(a_{j_1j_1r},a_{j_1j_2u})}_{\text{YY}}=\begin{bmatrix}\underbrace{(\text{UP:DP,pF})}_{\text{hyperbolic-secant flow }(+)} & \cdots & \underbrace{(\text{UP:DP,pF})}_{\text{hyperbolic-secant flow }(+)} & \underbrace{(\text{DP:UP,pF})}_{\text{hyperbolic flow }(+)}\\ \underbrace{(\text{DP:UP,pF})}_{\text{hyperbolic flow }(+)} & \cdots & \underbrace{(\text{DP:UP,pF})}_{\text{hyperbolic flow }(+)} & \underbrace{(\text{UP:UP,pF})}_{\text{hyperbolic-secant flow }(+)}\\ \vdots & \vdots & \vdots & \vdots\\ \underbrace{(\text{UP:DP,pF})}_{\text{hyperbolic-secant flow }(+)} & \cdots & \underbrace{(\text{UP:DP,pF})}_{\text{hyperbolic-secant flow }(+)} & \underbrace{(\text{DP:UP,pF})}_{\text{hyperbolic flow }(+)}\end{bmatrix}_{(2n_{11}+1)\times(2m_{11})}. \tag{1.328}$$

(iv_{2a2}) The ([($2m_{11}$), ($2n_{11}+1$)]UP:UP,pF)*-positive hyperbolic-secant-to-hyperbolic flow appearing bifurcation is from a* $q_1\times p_1$ *to* $\overline{q}_1\times\overline{p}_1$ *1-dimensional flow networks as*

$$\bigcup_{r_1=1}^{p_1}\bigcup_{u_1=1}^{q_1}\underbrace{(a_{j_1j_1r_1},a_{j_1j_2u_1})}_{([m_{j_1r_1},n_{j_1u_1}]\text{-XX,pF})}\rightleftharpoons\underbrace{([(2m_{11}),(2n_{11}+1)]\text{UP:UP,pF})}_{[(2m_{11}),(2n_{11}+1)]\text{-hyperbolic-secant-to-hyperbolic flow }(+)}\rightleftharpoons\bigcup_{r_2=1}^{\overline{p}_1}\bigcup_{u_2=1}^{\overline{q}_1}\underbrace{(a_{j_1j_1r_2},a_{j_1j_2u_2})}_{([m_{j_1r_2},n_{j_1u_2}]\text{-YY,pF})}. \tag{1.329}$$

(iv_{2a3}) The ([($2m_{11}$), ($2n_{11}+1$)]UP:UP,pF)*-positive hyperbolic-secant-to-hyperbolic flow switching bifurcation is for the switching of two networks of* $q_1\times p_1$ *1-dimensional flows as*

$$\bigcup_{r_1=1}^{p_1}\bigcup_{u_1=1}^{q_1}\underbrace{(a_{j_1j_1r_1},a_{j_1j_2u_1})}_{([m_{j_1r_1},n_{j_1u_1}]\text{-XX,pF})}\rightleftharpoons\underbrace{([(2m_{11}),(2n_{11}+1)]\text{UP:UP,pF})}_{[(2m_{11}),(2n_{11}+1)]\text{-hyperbolic-secant-to-hyperbolic flow }(+)}\rightleftharpoons\bigcup_{r_2=1}^{p_1}\bigcup_{u_2=1}^{q_1}\underbrace{(a_{j_1j_1r_2},a_{j_1j_2u_2})}_{([m_{j_1r_2},n_{j_1u_2}]\text{-YY,pF})}. \tag{1.330}$$

(iv_{2b}) For $a_{j_1j_10}<0$ *and* $a_{j_2j_20}(a_{j_1j_11}-a_{j_2j_11})^{m_2}>0$*, there is a* ([($2m_{11}$), ($2n_{11}+1$)]DP:DP,pF)*-positive hyperbolic-to-hyperbolic-secant flow as*

$$(a_{j_1j_11},a_{j_1j_21})=\underbrace{([(2m_{11}),(2n_{11}+1)]\text{DP:DP,pF})}_{[(2m_{11}),(2n_{11}+1)]\text{-hyerpbolic-to-hyperbolic-secant flow }(+)}. \tag{1.331}$$

There are three following ([(2m_{11}), (2n_{11} + 1)]DP:DP,pF)*-positive hyperbolic-to-hyperbolic-secant flow appearing and switching bifurcations.*

(iv$_{2b1}$) The ([(2m_{11}), (2n_{11} + 1)]DP:DP,pF)*-positive hyperbolic-to-hyperbolic-secant flow appearing bifurcation is from a (DP, pF)-positive down-parabola flow to a* (2m_1) × (2n_1+1)*-hyperbolic and hyperbolic-secant flow network as*

$$\underbrace{(a_{j_1j_2 1}, \dot{x}_{j_2})}_{\text{down-parabola flow }(+)} \rightleftharpoons \underbrace{([(2m_{11}),(2n_{11}+1)]\text{DP:DP,pF})}_{[(2m_{11}),(2n_{11}+1)]\text{-hyerpbolic-to-hyperbolic-secant flow }(+)} \rightleftharpoons \bigcup_{r=1}^{2m_{11}} \bigcup_{u=1}^{2n_{11}+1} \underbrace{(a_{j_1j_1r}, a_{j_1j_2u})}_{\text{YY}}, \quad (1.332)$$

where, without any equilibriums,

$$\bigcup_{r=1}^{2m_{11}} \bigcup_{u=1}^{2n_{11}+1} \underbrace{(a_{j_1j_1r}, a_{j_1j_2u})}_{\text{YY}} = \begin{bmatrix} \underbrace{(\text{DP:UP,pF})}_{\text{hyperbolic flow }(+)} & \cdots & \underbrace{(\text{DP:UP,pF})}_{\text{hyperbolic flow }(+)} & \underbrace{(\text{UP:DP,pF})}_{\text{hyperbolic-secant flow }(+)} \\ \underbrace{(\text{UP:DP,pF})}_{\text{hyperbolic-secant flow }(+)} & \cdots & \underbrace{(\text{UP:DP,pF})}_{\text{hyperbolic-secant flow }(+)} & \underbrace{(\text{DP:UP,pF})}_{\text{hyperbolic flow }(+)} \\ \vdots & \vdots & \vdots & \vdots \\ \underbrace{(\text{DP:UP,pF})}_{\text{hyperbolic flow }(+)} & \cdots & \underbrace{(\text{DP:UP,pF})}_{\text{hyperbolic flow }(+)} & \underbrace{(\text{UP:DP,pF})}_{\text{hyperbolic-secant flow }(+)} \end{bmatrix}_{(2n_{11}+1)\times(2m_{11})}. \quad (1.333)$$

(iv$_{2b2}$) The ([(2m_{11}), (2n_{11} + 1)]DP:DP,pF)*-positive hyperbolic-to-hyperbolic-secant flow appearing bifurcation is from a* $q_1 \times p_1$ *to* $\overline{q}_1 \times \overline{p}_1$ *1-dimensional flow networks as*

$$\bigcup_{r_1=1}^{p_1} \bigcup_{u_1=1}^{q_1} \underbrace{(a_{j_1j_1r_1}, a_{j_1j_2u_1})}_{([m_{j_1r_1}, n_{j_1u_1}]\text{-XX,pF})} \rightleftharpoons \underbrace{([(2m_{11}),(2n_{11}+1)]\text{DP:DP,pF})}_{[(2m_{11}),(2n_{11}+1)]\text{-hyerpbolic-to-hyperbolic-secant flow }(+)} \rightleftharpoons \bigcup_{r_2=1}^{\overline{p}_1} \bigcup_{u_2=1}^{\overline{q}_1} \underbrace{(a_{j_1j_1r_2}, a_{j_1j_2u_2})}_{([m_{j_1r_2}, n_{j_1u_2}]\text{-YY,pF})}. \quad (1.334)$$

(iv_{2b3}) The $([(2m_{11}), (2n_{11}+1)]$DP:DP,pF)*-positive hyperbolic-secant-to-hyperbolic-secant flow switching bifurcation is for the switching of two networks of* $q_1 \times p_1$ *1-dimensional flows as*

$$\bigcup_{r_1=1}^{p_1}\bigcup_{u_1=1}^{q_1}\underbrace{(a_{j_1j_1r_1}, a_{j_1j_2u_1})}_{([m_{j_1r_1}, n_{j_1u_1}]\text{-XX,pF})} \rightleftharpoons \underbrace{([(2m_{11}), (2n_{11}+1)]\text{DP:DP,pF})}_{[(2m_{11}),(2n_{11}+1)]\text{-hyerpbolic-to-hyperbolic-secant flow }(+)}$$
$$\rightleftharpoons \bigcup_{r_2=1}^{p_1}\bigcup_{u_2=1}^{q_1}\underbrace{(a_{j_1j_1r_2}, a_{j_1j_2u_2})}_{([m_{j_1r_2}, n_{j_1u_2}]\text{-YY,pF})}. \tag{1.335}$$

(iv_{2c}) For $a_{j_1j_10} > 0$ *and* $a_{j_2j_20}(a_{j_1j_11} - a_{j_2j_11})^{m_2} < 0$*, there is a* $([(2m_{11}), (2n_{11}+1)]$DP:DP,nF)*-negative hyperbolic-to-hyperbolic-secant flow as*

$$(a_{j_1j_11}, a_{j_1j_21}) = \underbrace{([(2m_{11}), (2n_{11}+1)]\text{DP:DP,nF})}_{[(2m_{11}),(2n_{11}+1)]\text{-hyperbolic-to-hyperbolic-secant flow }(-)}. \tag{1.336}$$

There are three following $([(2m_{11}), (2n_{11}+1)]$DP:DP,nF)*-negative hyperbolic-to-hyperbolic-secant flow appearing and switching bifurcations.*

(iv_{2c1}) The $([(2m_{11}), (2n_{11}+1)]$DP:DP,nF)*-negative hyperbolic-to-hyperbolic-secant flow appearing bifurcation is from a (DP, nF)-negative down-parabola flow to a* $(2m_1) \times (2n_1+1)$*-hyperbolic and hyperbolic-secant flow network as*

$$\underbrace{(a_{j_1j_21}, \dot{x}_{j_2})}_{\text{down-parabola flow }(-)} \rightleftharpoons \underbrace{([(2m_{11}), (2n_{11}+1)]\text{DP:DP,nF})}_{[(2m_{11}),(2n_{11}+1)]\text{-hyperbolic-to-hyperbolic-secant flow }(-)}$$
$$\rightleftharpoons \bigcup_{r=1}^{2m_{11}}\bigcup_{u=1}^{2n_{11}+1}\underbrace{(a_{j_1j_1r}, a_{j_1j_2u})}_{\text{YY}}, \tag{1.337}$$

where, without any equilibriums,

$$\bigcup_{r=1}^{2m_{11}}\bigcup_{u=1}^{2n_{11}+1}\underbrace{(a_{j_1j_1r}, a_{j_1j_2u})}_{\text{YY}}$$

$$= \begin{bmatrix} \underbrace{\text{(DP:UP,nF)}}_{\text{hyperbolic flow }(-)} & \cdots & \underbrace{\text{(DP:UP,nF)}}_{\text{hyperbolic flow }(-)} & \underbrace{\text{(UP:UP,nF)}}_{\text{hyperbolic-secant flow }(-)} \\ \underbrace{\text{(UP:UP,nF)}}_{\text{hyperbolic-secant flow }(-)} & \cdots & \underbrace{\text{(UP:UP,nF)}}_{\text{hyperbolic-secant flow }(-)} & \underbrace{\text{(DP:UP,nF)}}_{\text{hyperbolic flow }(-)} \\ \vdots & \vdots & \vdots & \vdots \\ \underbrace{\text{(DP:UP,nF)}}_{\text{hyperbolic flow }(-)} & \cdots & \underbrace{\text{(DP:UP,nF)}}_{\text{hyperbolic flow }(-)} & \underbrace{\text{(UP:UP,nF)}}_{\text{hyperbolic-secant flow }(-)} \end{bmatrix}_{(2n_{11}+1)\times(2m_{11})} . \tag{1.338}$$

(iv_{2c2}) The $([(2m_{11}),(2n_{11}+1)]\text{DP:DP,nF})$*-negative hyperbolic-to-hyperbolic-secant flow appearing bifurcation is from a* $q_1 \times p_1$ *to* $\overline{q}_1 \times \overline{p}_1$ *1-dimensional flow networks as*

$$\begin{aligned} \bigcup_{r_1=1}^{p_1}\bigcup_{u_1=1}^{q_1} \underbrace{(a_{j_1j_1r_1}, a_{j_1j_2u_1})}_{([m_{j_1r_1}, n_{j_1u_1}]\text{-XX,nF})} &\rightleftharpoons \underbrace{([(2m_{11}),(2n_{11}+1)]\text{DP:DP,nF})}_{[(2m_{11}),(2n_{11}+1)]\text{-hyperbolic-to-hyperbolic-secant flow }(-)} \\ &\rightleftharpoons \bigcup_{r_2=1}^{\overline{p}_1}\bigcup_{u_2=1}^{\overline{q}_1} \underbrace{(a_{j_1j_1r_2}, a_{j_1j_2u_2})}_{([m_{j_1r_2}, n_{j_1u_2}]\text{-YY,nF})} . \end{aligned} \tag{1.339}$$

(iv_{2c3}) The $([(2m_{11}),(2n_{11}+1)]\text{DP:DP,nF})$*-negative hyperbolic-to-hyperbolic-secant flow switching bifurcation is for the switching of two networks of* $q_1 \times p_1$ *1-dimensional flows as*

$$\begin{aligned} \bigcup_{r_1=1}^{p_1}\bigcup_{u_1=1}^{q_1} \underbrace{(a_{j_1j_1r_1}, a_{j_1j_2u_1})}_{([m_{j_1r_1}, n_{j_1u_1}]\text{-XX,nF})} &\rightleftharpoons \underbrace{([(2m_{11}),(2n_{11}+1)]\text{DP:DP,nF})}_{[(2m_{11}),(2n_{11}+1)]\text{-hyperbolic-to-hyperbolic-secant flow }(-)} \\ &\rightleftharpoons \bigcup_{r_2=1}^{p_1}\bigcup_{u_2=1}^{q_1} \underbrace{(a_{j_1j_1r_2}, a_{j_1j_2u_2})}_{([m_{j_1r_2}, n_{j_1u_2}]\text{-YY,nF})} . \end{aligned} \tag{1.340}$$

(iv_{2d}) For $a_{j_1j_10} < 0$ *and* $a_{j_2j_20}(a_{j_1j_11} - a_{j_2j_11})^{m_2} < 0$*, there is a* $([(2m_{11}),(2n_{11}+1)]\text{UP:UP,nF})$*-negative hyperbolic-secant-to-hyperbolic flow as*

$$(a_{j_1j_11}, a_{j_1j_21}) = \underbrace{([(2m_{11}),(2n_{11}+1)]\text{UP:UP,nF})}_{[(2m_{11}),(2n_{11}+1)]\text{-hyperbolic-secant-to-hyperbolic flow }(-)} . \tag{1.341}$$

There are three following $([(2m_{11}),(2n_{11}+1)]\text{UP:UP,nF})$*-negative hyperbolic-secant-to-hyperbolic flow appearing and switching bifurcations.*

(iv_{2d1}) The ([(2m_{11}), (2n_{11} + 1)]UP:UP,nF)*-negative hyperbolic-secant-to-hyperbolic flow appearing bifurcation is from a (UP, nF)-negative up-parabola flow to a* $(2m_1) \times (2n_1+1)$*-hyperbolic and hyperbolic-secant flow network as*

$$\underbrace{(a_{j_1j_2 1}, \dot{x}_{j_2})}_{\text{up-parabola flow }(-)} \rightleftharpoons \underbrace{([(2m_{11}), (2n_{11}+1)]\text{UP:UP,nF})}_{[(2m_{11}),(2n_{11}+1)]\text{-hyperbolic-secant-to-hyperbolic flow }(-)}$$
$$\rightleftharpoons \bigcup_{r=1}^{2m_{11}} \bigcup_{u=1}^{2n_{11}+1} \underbrace{(a_{j_1j_1r}, a_{j_1j_2u})}_{\text{YY}}, \tag{1.342}$$

where, without any equilibriums,

$$\bigcup_{r=1}^{2m_{11}} \bigcup_{u=1}^{2n_{11}+1} \underbrace{(a_{j_1j_1r}, a_{j_1j_2u})}_{\text{YY}}$$
$$= \begin{bmatrix} \underbrace{(\text{UP:UP,nF})}_{\text{hyperbolic-secant flow }(-)} & \cdots & \underbrace{(\text{DP:UP,nF})}_{\text{hyperbolic flow }(-)} & \underbrace{(\text{UP:UP,nF})}_{\text{hyperbolic-secant flow }(-)} \\ \underbrace{(\text{DP:UP,nF})}_{\text{hyperbolic flow }(-)} & \cdots & \underbrace{(\text{UP:UP,nF})}_{\text{hyperbolic-secant flow }(-)} & \underbrace{(\text{DP:UP,nF})}_{\text{hyperbolic flow }(-)} \\ \vdots & \vdots & \vdots & \vdots \\ \underbrace{(\text{UP:UP,nF})}_{\text{hyperbolic-secant flow }(-)} & \cdots & \underbrace{(\text{DP:UP,nF})}_{\text{hyperbolic flow }(-)} & \underbrace{(\text{UP:UP,nF})}_{\text{hyperbolic-secant flow }(-)} \end{bmatrix}_{(2n_{11}+1)\times(2m_{11})} . \tag{1.343}$$

(iv_{2d2}) The ([(2m_{11}), (2n_{11} + 1)]UP:UP,nF)*-negative hyperbolic-secant-to-hyperbolic flow appearing bifurcation is from a* $q_1 \times p_1$ *to* $\overline{q}_1 \times \overline{p}_1$ *1-dimensional flow networks as*

$$\bigcup_{r_1=1}^{p_1} \bigcup_{u_1=1}^{q_1} \underbrace{(a_{j_1j_1r_1}, a_{j_1j_2u_1})}_{([m_{j_1r_1}, n_{j_1u_1}]\text{-XX,nF})} \rightleftharpoons \underbrace{([(2m_{11}), (2n_{11}+1)]\text{UP:UP,nF})}_{[(2m_{11}),(2n_{11}+1)]\text{-hyperbolic-secant-to-hyperbolic flow }(-)}$$
$$\rightleftharpoons \bigcup_{r_2=1}^{\overline{p}_1} \bigcup_{u_2=1}^{\overline{q}_1} \underbrace{(a_{j_1j_1r_2}, a_{j_1j_2u_2})}_{([m_{j_1r_2}, n_{j_1u_2}]\text{-YY,nF})} . \tag{1.344}$$

(iv_{2d3}) The ([(2m_{11}), (2n_{11} + 1)]UP:UP,nF)*-negative hyperbolic-secant-to-hyperbolic flow switching bifurcation is for the switching of two networks of* $q_1 \times p_1$ *1-dimensional flows as*

$$\bigcup_{r_1=1}^{p_1}\bigcup_{u_1=1}^{q_1}\underbrace{(a_{j_1j_1r_1},a_{j_1j_2u_1})}_{([m_{j_1r_1},n_{j_1u_1}]\text{-XX,nF})} \rightleftharpoons \underbrace{([(2m_{11}),(2n_{11}+1)]\text{UP:UP,nF})}_{[(2m_{11}),(2n_{11}+1)]\text{-hyperbolic-secant-to-hyperbolic flow }(-)}$$
$$\rightleftharpoons \bigcup_{r_2=1}^{p_1}\bigcup_{u_2=1}^{q_1}\underbrace{(a_{j_1j_1r_2},a_{j_1j_2u_2})}_{([m_{j_1r_2},n_{j_1u_2}]\text{-YY,nF})}. \tag{1.345}$$

(iv$_3$) For $m_1 = 2m_{11}+1$ and $n_1 = 2n_{11}$, the equilibrium of $(x^*_{j_1}, x^*_{j_2}) = (a_{j_1j_1 1}, a_{j_1j_2 1})$ for the x_{j_1}-direction has the following bifurcation properties with

$$2m_{11}+1-\sum_{r_1=1}^{p_1} m^{(1)}_{j_1r_1} = 2\sum_{\rho_1=1}^{l_1} m_{\rho_1},$$
$$2n_{11}-\sum_{u_1=1}^{q_1} n^{(1)}_{j_1u_1} = 2\sum_{\sigma_1=1}^{l_2} n_{\sigma_1};$$
$$\sum_{r_2=1}^{\bar{p}_1} m^{(2)}_{j_1r_2} = 2m_{11}+1,\ \sum_{u_2=1}^{\bar{q}_1} n^{(2)}_{j_1u_2} = 2n_{11};$$
$$\sum_{r_i=1}^{p_1} m^{(i)}_{j_1r_i} = 2m_{11}+1,\ \sum_{u_i=1}^{q_1} n^{(2)}_{j_1u_i} = 2n_{11},\ \text{for } i=1,2; \tag{1.346}$$

l_1, l_2-quadratic polynomials are without real roots, and the ρ_1^{th} and σ_1^{th}-quadratic polynomials are with powers m_{ρ_1} and n_{σ_1}, respectively.

(iv$_{3a}$) For $a_{j_1j_1 0} > 0$ and $a_{j_2j_2 0}(a_{j_1j_1 1} - a_{j_2j_1 1})^{m_2} > 0$, there is a $([(2m_{11}+1),(2n_{11})]$ $_{\text{DI:II}}$SO,pF)*-positive decreasing-to-increasing inflection source flow as*

$$(a_{j_1j_1 1},a_{j_1j_2 1}) = \underbrace{([(2m_{11}+1),(2n_{11})]_{\text{DI:II}}\text{SO,pF})}_{[(2m_{11}+1),(2n_{11})]\text{-decreasing-to-increasing infelction source flow }(+)}. \tag{1.347}$$

There are three following $([(2m_{11}+1),(2n_{11})]_{\text{DI:II}}$SO,pF)*-positive decreasing-to-increasing inflection source flow appearing and switching bifurcations.*

(iv$_{3a1}$) The $([(2m_{11}+1),(2n_{11})]_{\text{DI:II}}$SO,pF)*-positive decreasing-to-increasing inflection source flow appearing bifurcation is from an* (SO, pF)*-positive hyperbolic-flow to a $(2m_1+1)\times(2n_1)$-hyperbolic and hyperbolic-secant flow network as*

$$\underbrace{(a_{j_1j_1 1},\dot{x}_{j_2})}_{\text{source flow }(+)} \rightleftharpoons \underbrace{([(2m_{11}+1),(2n_{11})]_{\text{DI:II}}\text{SO,pF})}_{[(2m_{11}+1),(2n_{11})]\text{-decreasing-to-increasing infelction source flow }(+)}$$
$$\rightleftharpoons \bigcup_{r=1}^{2m_{11}+1}\bigcup_{u=1}^{2n_{11}}\underbrace{(a_{j_1j_1r},a_{j_1j_2u})}_{\text{YY}}, \tag{1.348}$$

where, without any equilibriums,

$$\bigcup_{r=1}^{2m_{11}+1}\bigcup_{u=1}^{2n_{11}}\underbrace{(a_{j_1j_1r},a_{j_1j_2u})}_{\text{YY}}
=\begin{bmatrix}
\underbrace{(\text{DP:UP,pF})}_{\text{hyperbolic flow }(+)} & \cdots & \underbrace{(\text{UP:DP,pF})}_{\text{hyperbolic-secant flow }(+)} & \underbrace{(\text{DP:UP,pF})}_{\text{hyperbolic flow }(+)}\\
\underbrace{(\text{UP:DP,pF})}_{\text{hyperbolic-secant flow }(+)} & \cdots & \underbrace{(\text{DP:UP,pF})}_{\text{hyperbolic flow }(+)} & \underbrace{(\text{UP:DP,pF})}_{\text{hyperbolic-secant flow }(+)}\\
\vdots & \vdots & \vdots & \vdots\\
\underbrace{(\text{UP:DP,pF})}_{\text{hyperbolic-secant flow }(+)} & \cdots & \underbrace{(\text{DP:UP,pF})}_{\text{hyperbolic flow }(+)} & \underbrace{(\text{UP:DP,pF})}_{\text{hyperbolic-secant flow }(+)}
\end{bmatrix}_{(2n_{11})\times(2m_{11}+1)}. \tag{1.349}$$

(iv$_{3a2}$) The ([(2m_{11}+1), (2n_{11})]$_{\text{DI:II}}$SO,pF)*-positive decreasing-to-increasing inflection source flow appearing bifurcation is from a* $q_1\times p_1$ *to* $\overline{q}_1\times\overline{p}_1$ *1-dimensional flow networks as*

$$\begin{aligned}\bigcup_{r_1=1}^{p_1}\bigcup_{u_1=1}^{q_1}\underbrace{(a_{j_1j_1r_1},a_{j_1j_2u_1})}_{([m^{(1)}_{j_1r_1},n^{(1)}_{j_1u_1}]\text{-YY,pF})} &\rightleftharpoons \underbrace{([(2m_{11}+1),(2n_{11})]_{\text{DI:II}}\text{SO,pF})}_{[(2m_{11}+1),(2n_{11})]\text{-decreasing-to-increasing infelction source flow }(+)}\\
&\rightleftharpoons\bigcup_{r_2=1}^{\overline{p}_1}\bigcup_{u_2=1}^{\overline{q}_1}\underbrace{(a_{j_1j_1r_2},a_{j_1j_2u_2})}_{([m^{(1)}_{j_1r_2},n^{(1)}_{j_1u_2}]\text{-YY,pF})}.\end{aligned} \tag{1.350}$$

(iv$_{3a3}$) The ([(2m_{11}+1), (2n_{11})]$_{\text{DI:II}}$SO,pF)*-positive decreasing-to-increasing inflection source flow switching bifurcation is for the switching of two networks of* $q_1\times p_1$ *1-dimensional flows as*

$$\begin{aligned}\bigcup_{r_1=1}^{p_1}\bigcup_{u_1=1}^{q_1}\underbrace{(a_{j_1j_1r_1},a_{j_1j_2u_1})}_{([m_{j_1r_1},n_{j_1u_1}]\text{-XX,pF})} &\rightleftharpoons \underbrace{([(2m_{11}+1),(2n_{11})]_{\text{DI:II}}\text{SO,pF})}_{[(2m_{11}+1),(2n_{11})]\text{-decreasing-to-increasing infelction source flow }(+)}\\
&\rightleftharpoons\bigcup_{r_2=1}^{p_1}\bigcup_{u_2=1}^{q_1}\underbrace{(a_{j_1j_1r_2},a_{j_1j_2u_2})}_{([m_{j_1r_2},n_{j_1u_2}]\text{-YY,pF})}.\end{aligned} \tag{1.351}$$

(iv$_{3b}$) For $a_{j_1j_10}<0$ *and* $a_{j_2j_20}(a_{j_1j_11}-a_{j_2j_11})^{m_2}>0$, *there is a* ([(2$m_{11}$+1), (2$n_{11}$)]$_{\text{II:DI}}$SI, pF)*-positive increasing-to-decreasing inflection sink flow as*

$$(a_{j_1j_11}, a_{j_1j_21}) = \underbrace{([(2m_{11}+1), (2n_{11})]_{\text{II:DI}}\text{SI,pF})}_{[(2m_{11}+1),(2n_{11})]\text{-increasing-to-dereasing inflection sink flow }(+)}. \tag{1.352}$$

There are three following $([(2m_{11}+1), (2n_{11})]_{\text{II:DI}}\text{SI, pF})$*-positive increasing-to-decreasing inflection sink flow appearing and switching bifurcations.*

(iv$_{3b1}$) The $([(2m_{11}+1), (2n_{11})]_{\text{II:DI}}\text{SI,pF})$*-positive increasing-to-decreasing inflection sink flow appearing bifurcation is from an (SI, pF)-positive sink flow to a* $(2m_1+1)\times(2n_1)$*-hyperbolic and hyperbolic-secant flow network as*

$$\begin{aligned}\underbrace{(a_{j_1j_11}, \dot{x}_{j_2})}_{\text{sink flow }(+)} &\rightleftharpoons \underbrace{([(2m_{11}+1), (2n_{11})]_{\text{II:DI}}\text{SI,pF})}_{[(2m_{11}+1),(2n_{11})]\text{-increasing-to-dereasing inflection sink flow }(+)}\\ &\rightleftharpoons \bigcup_{r=1}^{2m_{11}+1}\bigcup_{u=1}^{2n_{11}}\underbrace{(a_{j_1j_1r}, a_{j_1j_2u})}_{\text{YY}},\end{aligned} \tag{1.353}$$

where, without any equilibriums,

$$\bigcup_{r=1}^{2m_{11}+1}\bigcup_{u=1}^{2n_{11}}\underbrace{(a_{j_1j_1r}, a_{j_1j_2u})}_{\text{YY}} = \begin{bmatrix} \underbrace{(\text{UP:DP,pF})}_{\text{hyperbolic-secant flow }(+)} & \cdots & \underbrace{(\text{DP:UP,pF})}_{\text{hyperbolic flow }(+)} & \underbrace{(\text{UP:DP,pF})}_{\text{hyperbolic-secant flow }(+)} \\ \underbrace{(\text{DP:UP,pF})}_{\text{hyperbolic flow }(+)} & \cdots & \underbrace{(\text{UP:DP,pF})}_{\text{hyperbolic-secant flow }(+)} & \underbrace{(\text{DP:UP,pF})}_{\text{hyperbolic flow }(+)} \\ \vdots & \vdots & \vdots & \vdots \\ \underbrace{(\text{DP:UP,pF})}_{\text{hyperbolic flow }(+)} & \cdots & \underbrace{(\text{UP:DP,pF})}_{\text{hyperbolic-secant flow }(+)} & \underbrace{(\text{DP:UP,pF})}_{\text{hyperbolic flow }(+)} \end{bmatrix}_{(2n_{11})\times(2m_{11}+1)}. \tag{1.354}$$

(iv$_{3b2}$) The $([(2m_{11}+1), (2n_{11})]_{\text{II:DI}}\text{SI,pF})$ *is from a* $q_1 \times p_1$ *to* $\overline{q}_1 \times \overline{p}_1$ *1-dimensional flow networks as*

$$\begin{aligned}\bigcup_{r_1=1}^{p_1}\bigcup_{u_1=1}^{q_1}\underbrace{(a_{j_1j_1r_1}, a_{j_1j_2u_1})}_{([m_{j_1r_1}, n_{j_1u_1}]\text{-XX,pF})} &\rightleftharpoons \underbrace{([(2m_{11}+1), (2n_{11})]_{\text{II:DI}}\text{SI,pF})}_{[(2m_{11}+1),(2n_{11})]\text{-increasing-to-dereasing inflection sink flow }(+)}\\ &\rightleftharpoons \bigcup_{r_2=1}^{\overline{p}_1}\bigcup_{u_2=1}^{\overline{q}_1}\underbrace{(a_{j_1j_1r_2}, a_{j_1j_2u_2})}_{([m_{j_1r_2}, n_{j_1u_2}]\text{-YY,pF})}.\end{aligned} \tag{1.355}$$

(iv$_{3b3}$) The $([(2m_{11}+1), (2n_{11})]_{\text{II:DI}}\text{SI,pF})$*-positive increasing-to-decreasing inflection sink flow switching bifurcation is for the switching of two networks of* $q_1 \times p_1$ *1-dimensional flows as*

$$\bigcup_{r_1=1}^{p_1}\bigcup_{u_1=1}^{q_1}\underbrace{(a_{j_1j_1r_1},a_{j_1j_2u_1})}_{([m_{j_1r_1},n_{j_1u_1}]\text{-XX,pF})} \rightleftharpoons \underbrace{([(2m_{11}+1),(2n_{11})]_{\text{II:DI}}\text{SI,pF})}_{[(2m_{11}+1),(2n_{11})]\text{-increasing-to-dereasing inflection sink flow }(+)}$$
$$\rightleftharpoons \bigcup_{r_2=1}^{p_1}\bigcup_{u_2=1}^{q_1}\underbrace{(a_{j_1j_1r_2},a_{j_1j_2u_2})}_{([m_{j_1r_2},n_{j_1u_2}]\text{-YY,pF})}. \tag{1.356}$$

(iv$_{3c}$) For $a_{j_1j_10} > 0$ and $a_{j_2j_20}(a_{j_1j_11} - a_{j_2j_11})^{m_2} < 0$, there is a $([(2m_{11}+1),(2n_{11})]_{\text{II:DI}}\text{SO,nF})$*-negative increasing-to-decreasing inflection source flow as*

$$(a_{j_1j_11},a_{j_1j_21}) = \underbrace{([(2m_{11}+1),(2n_{11})]_{\text{II:DI}}\text{SO,nF})}_{[(2m_{11}+1),(2n_{11})]\text{-increasing-to-dereasing inflection source flow }(-)}. \tag{1.357}$$

There are three following $([(2m_{11}+1),(2n_{11})]_{\text{II:DI}}\text{SO,nF})$*-negative increasing-to-decreasing inflection source flow appearing and switching bifurcations.*

(iv$_{3c1}$) *The* $([(2m_{11}+1),(2n_{11})]_{\text{II:DI}}\text{SO,nF})$*-negative increasing-to-decreasing inflection source flow appearing bifurcation is from an* (SO, nF)*-negative sink flow to a* $(2m_1+1)\times(2n_1)$*-hyperbolic and hyperbolic-secant flow network as*

$$\underbrace{(a_{j_1j_11},\dot{x}_{j_2})}_{\text{source flow }(-)} \rightleftharpoons \underbrace{([(2m_{11}+1),(2n_{11})]_{\text{II:DI}}\text{SO,nF})}_{[(2m_{11}+1),(2n_{11})]\text{-increasing-to-dereasing inflection source flow }(-)}$$
$$\rightleftharpoons \bigcup_{r=1}^{2m_{11}+1}\bigcup_{u=1}^{2n_{11}}\underbrace{(a_{j_1j_1r},a_{j_1j_2u})}_{\text{YY}}, \tag{1.358}$$

where, without any equilibriums,

$$\bigcup_{r=1}^{2m_{11}+1}\bigcup_{u=1}^{2n_{11}}\underbrace{(a_{j_1j_1r},a_{j_1j_2u})}_{\text{YY}}$$
$$= \begin{bmatrix} \underbrace{(\text{UP:DP,nF})}_{\text{hyperbolic-secant flow }(-)} & \cdots & \underbrace{(\text{DP:UP,nF})}_{\text{hyperbolic flow }(-)} & \underbrace{(\text{UP:DP,nF})}_{\text{hyperbolic-secant flow }(-)} \\ \underbrace{(\text{DP:UP,nF})}_{\text{hyperbolic flow }(-)} & \cdots & \underbrace{(\text{UP:DP,nF})}_{\text{hyperbolic-secant flow }(-)} & \underbrace{(\text{DP:UP,nF})}_{\text{hyperbolic flow }(-)} \\ \vdots & \vdots & \vdots & \vdots \\ \underbrace{(\text{UP:DP,nF})}_{\text{hyperbolic-secant flow }(-)} & \cdots & \underbrace{(\text{DP:UP,nF})}_{\text{hyperbolic flow }(-)} & \underbrace{(\text{UP:DP,nF})}_{\text{hyperbolic-secant flow }(-)} \end{bmatrix}_{(2n_{11})\times(2m_{11}+1)}. \tag{1.359}$$

(iv_{3c2}) The $([(2m_{11}+1),(2n_{11})]_{\text{II:DI}}\text{SO,nF})$*-negative increasing-to-decreasing inflection source flow appearing bifurcation is from a* $q_1 \times p_1$ *to* $\overline{q}_1 \times \overline{p}_1$ *1-dimensional flow networks as*

$$\begin{aligned}\bigcup_{r_1=1}^{p_1}\bigcup_{u_1=1}^{q_1}\underbrace{(a_{j_1j_1r_1},a_{j_1j_2u_1})}_{([m_{j_1r_1},n_{j_1u_1}]\text{-XX,nF})} &\rightleftharpoons \underbrace{([(2m_{11}+1),(2n_{11})]_{\text{II:DI}}\text{SO,nF})}_{[(2m_{11}+1),(2n_{11})]\text{-increasing-to-dereasing inflection source flow }(-)}\\ &\rightleftharpoons \bigcup_{r_2=1}^{\overline{p}_1}\bigcup_{u_2=1}^{\overline{q}_1}\underbrace{(a_{j_1j_1r_2},a_{j_1j_2u_2})}_{([m_{j_1r_2},n_{j_1u_2}]\text{-YY,nF})}. \end{aligned} \tag{1.360}$$

(iv_{3c3}) The $([(2m_{11}+1),(2n_{11})]_{\text{II:DI}}\text{SO,nF})$*-negative increasing-to-decreasing inflection source flow switching bifurcation is for the switching of two networks of* $q_1 \times p_1$ *1-dimensional flows as*

$$\begin{aligned}\bigcup_{r_1=1}^{p_1}\bigcup_{u_1=1}^{q_1}\underbrace{(a_{j_1j_1r_1},a_{j_1j_2u_1})}_{([m_{j_1r_1},n_{j_1u_1}]\text{-XX,nF})} &\rightleftharpoons \underbrace{([(2m_{11}+1),(2n_{11})]_{\text{II:DI}}\text{SO,nF})}_{[(2m_{11}+1),(2n_{11})]\text{-increasing-to-decreasing inflection source flow }(-)}\\ &\rightleftharpoons \bigcup_{r_2=1}^{p_1}\bigcup_{u_2=1}^{q_1}\underbrace{(a_{j_1j_1r_2},a_{j_1j_2u_2})}_{([m_{j_1r_2},n_{j_1u_2}]\text{-YY,nF})}. \end{aligned} \tag{1.361}$$

(iv_{3d}) For $a_{j_1j_10}<0$ *and* $a_{j_2j_20}(a_{j_1j_11}-a_{j_2j_11})^{m_2}<0$*, there is a* $([(2m_{11}+1),(2n_{11})]_{\text{DI:II}}\text{SI,nF})$*-negative increasing-to-decreasing inflection sink flow as*

$$(a_{j_1j_11},a_{j_1j_21}) = \underbrace{([(2m_{11}+1),(2n_{11})]_{\text{DI:II}}\text{SI,nF})}_{[(2m_{11}+1),(2n_{11})]\text{-decreasing-to-increasing inflection sink flow }(-)}. \tag{1.362}$$

There are three following $([(2m_{11}+1),(2n_{11})]_{\text{DI:II}}\text{SI,nF})$*-negative increasing-to-decreasing inflection sink flow appearing and switching bifurcations.*

(iv_{3d1}) *The* $([(2m_{11}+1),(2n_{11})]_{\text{DI:II}}\text{SI,nF})$*-negative increasing-to-decreasing inflection sink flow appearing bifurcation is from an* (SI, nF)*-negative sink flow to a* $(2m_{11}+1)\times(2n_{11})$*-hyperbolic and hyperbolic-secant flow network as*

$$\begin{aligned}\underbrace{(a_{j_1j_11},a_{j_1j_21})}_{\text{hyperbolic-secant flow }(-)} &\rightleftharpoons \underbrace{([(2m_{11}+1),(2n_{11})]_{\text{DI:II}}\text{SI,nF})}_{[(2m_{11}+1),(2n_{11})]\text{-decreasing-to-increasing inflection sink flow }(-)}\\ &\rightleftharpoons \bigcup_{r=1}^{2m_{11}+1}\bigcup_{u=1}^{2n_{11}}\underbrace{(a_{j_1j_1r},a_{j_1j_2u})}_{\text{YY}}, \end{aligned} \tag{1.363}$$

where, without any equilibriums,

$$\bigcup_{r=1}^{2m_{11}+1}\bigcup_{u=1}^{2n_{11}}\underbrace{(a_{j_1j_1r},a_{j_1j_2u})}_{\text{YY}}$$
$$=\left[\begin{array}{cccc}\underbrace{\text{(UP:DP,nF)}}_{\text{hyperbolic-secant flow }(-)} & \cdots & \underbrace{\text{(DP:UP,nF)}}_{\text{hyperbolic flow }(-)} & \underbrace{\text{(UP:DP,nF)}}_{\text{hyperbolic-secant flow }(-)}\\ \underbrace{\text{(DP:UP,nF)}}_{\text{hyperbolic flow }(-)} & \cdots & \underbrace{\text{(UP:DP,nF)}}_{\text{hyperbolic-secant flow }(-)} & \underbrace{\text{(DP:UP,nF)}}_{\text{hyperbolic flow }(-)}\\ \vdots & \vdots & \vdots & \vdots\\ \underbrace{\text{(DP:UP,nF)}}_{\text{hyperbolic flow }(-)} & \cdots & \underbrace{\text{(UP:DP,nF)}}_{\text{hyperbolic-secant flow }(-)} & \underbrace{\text{(DP:UP,nF)}}_{\text{hyperbolic flow }(-)}\end{array}\right]_{(2n_{11})\times(2m_{11}+1)}. \tag{1.364}$$

(iv_{3d2}) The ($[(2m_{11}+1),(2n_{11})]_{\text{DI:II}}$SI,nF)*-negative increasing-to-decreasing inflection sink flow appearing bifurcation is from a $q_1\times p_1$ to $\overline{q}_1\times\overline{p}_1$ 1-dimensional flow networks as*

$$\begin{aligned}\bigcup_{r_1=1}^{p_1}\bigcup_{u_1=1}^{q_1}\underbrace{(a_{j_1j_1r_1},a_{j_1j_2u_1})}_{([m_{j_1r_1},n_{j_1u_1}]\text{-XX,nF})} &\rightleftharpoons \underbrace{([(2m_{11}+1),(2n_{11})]_{\text{DI:II}}\text{SI,nF})}_{[(2m_{11}+1),(2n_{11})]\text{-decreasing-to-increasing inflection sink flow }(-)}\\ &\rightleftharpoons\bigcup_{r_2=1}^{\overline{p}_1}\bigcup_{u_2=1}^{\overline{q}_1}\underbrace{(a_{j_1j_1r_2},a_{j_1j_2u_2})}_{([m_{j_1r_2},n_{j_1u_2}]\text{-YY,nF})}.\end{aligned}\tag{1.365}$$

(iv_{3d3}) The ($[(2m_{11}+1),(2n_{11})]_{\text{DI:II}}$SI,nF)*-negative increasing-to-decreasing inflection sink flow switching bifurcation is for the switching of two networks of $q_1\times p_1$ 1-dimensional flows as*

$$\begin{aligned}\bigcup_{r_1=1}^{p_1}\bigcup_{u_1=1}^{q_1}\underbrace{(a_{j_1j_1r_1},a_{j_1j_2u_1})}_{([m_{j_1r_1},n_{j_1u_1}]\text{-XX,nF})} &\rightleftharpoons \underbrace{([(2m_{11}+1),(2n_{11})]_{\text{DI:II}}\text{SI,nF})}_{[(2m_{11}+1),(2n_{11})]\text{-decreasing-to-increasing inflection sink flow }(-)}\\ &\rightleftharpoons\bigcup_{r_2=1}^{p_1}\bigcup_{u_2=1}^{q_1}\underbrace{(a_{j_1j_1r_2},a_{j_1j_2u_2})}_{([m_{j_1r_2},n_{j_1u_2}]\text{-XX,nF})}.\end{aligned}\tag{1.366}$$

(iv_4) For $m_1=2m_{11}$ and $n_1=2n_{11}$, the equilibrium of $(x^*_{j_1},x^*_{j_2})=(a_{j_1j_11},a_{j_1j_21})$ for the x_{j_1}-direction has the following bifurcation properties with

$$2m_{11}-\sum_{r_1=1}^{p_1}m^{(1)}_{j_1r_1}=2\sum_{\rho_1=1}^{l_1}m_{\rho_1},\ 2n_{11}-\sum_{u_1=1}^{q_1}n^{(1)}_{j_1u_1}=2\sum_{\sigma_1=1}^{l_2}n_{\sigma_1};$$

$$\sum_{r_2=1}^{\overline{p}_1} m_{j_1 r_2}^{(2)} = 2m_{11}, \sum_{u_2=1}^{\overline{q}_1} n_{j_1 u_2}^{(2)} = 2n_{11};$$
$$\sum_{r_i=1}^{p_1} m_{j_1 r_i}^{(i)} = 2m_{11}, \sum_{u_i=1}^{q_1} n_{j_1 u_i}^{(2)} = 2n_{11}, \text{ for } i = 1, 2; \tag{1.367}$$

l_1, l_2-quadratic polynomials are without real roots, and the ρ_1^{th} and σ_1^{th}-quadratic polynomials are with powers m_{ρ_1} and n_{σ_1}, respectively.

(iv_{4a}) For $a_{j_1 j_1 0} > 0$ and $a_{j_2 j_2 0}(a_{j_1 j_1 1} - a_{j_2 j_1 1})^{m_2} > 0$, there is a ([(2$m_{11}$), (2$n_{11}$)]$_{\text{II:II}}$US ,pF)*-positive increasing-inflection upper-saddle flow as*

$$(a_{j_1 j_1 1}, a_{j_1 j_2 1}) = \underbrace{([(2m_{11}), (2n_{11})]_{\text{II:II}}\text{US,pF})}_{[(2m_{11}),(2n_{11})]\text{-increasing infelction upper-saddle flow }(+)} . \tag{1.368}$$

There are three following ([(2m_{11}), (2n_{11})]$_{\text{II:II}}$US,pF)*-positive increasing-inflection upper-saddle flow appearing and switching bifurcations.*

(v_{4a1}) The ([(2m_{11}), (2n_{11})]$_{\text{II:II}}$US,pF)*-positive increasing-inflection upper-saddle flow appearing bifurcation* is from a (pF, pF)-flow to a (2m_1) × (2n_1)-hyperbolic and hyperbolic-secant flow network as

$$\underbrace{(\dot{x}_{j_1}, \dot{x}_{j_2})}_{\text{(pF, pF)-flow}} \rightleftharpoons \underbrace{([(2m_{11}), (2n_{11})]_{\text{II:II}}\text{US,pF})}_{[(2m_{11}),(2n_{11})]\text{-increasing infelction upper-saddle flow }(+)}$$
$$\rightleftharpoons \bigcup_{r=1}^{2m_{11}} \bigcup_{u=1}^{2n_{11}} \underbrace{(a_{j_1 j_1 r}, a_{j_1 j_2 u})}_{\text{YY}}, \tag{1.369}$$

where, without any equilibriums,

$$\bigcup_{r=1}^{2m_{11}} \bigcup_{u=1}^{2n_{11}} \underbrace{(a_{j_1 j_1 r}, a_{j_1 j_2 u})}_{\text{YY}}$$

$$= \left[\begin{array}{cccc} \underbrace{(\text{UP:DP,pF})}_{\text{hyperbolic-secant flow } (+)} & \cdots & \underbrace{(\text{UP:DP,pF})}_{\text{hyperbolic-secant flow } (+)} & \underbrace{(\text{DP:UP,pF})}_{\text{hyperbolic flow } (+)} \\ \underbrace{(\text{DP:UP,pF})}_{\text{hyperbolic flow } (+)} & \cdots & \underbrace{(\text{DP:UP,pF})}_{\text{hyperbolic flow } (+)} & \underbrace{(\text{UP:DP,pF})}_{\text{hyperbolic-secant flow } (+)} \\ \vdots & \vdots & \vdots & \vdots \\ \underbrace{(\text{DP:UP,pF})}_{\text{hyperbolic flow } (+)} & \cdots & \underbrace{(\text{DP:UP,pF})}_{\text{hyperbolic flow } (+)} & \underbrace{(\text{UP:DP,pF})}_{\text{hyperbolic-secant flow } (+)} \end{array} \right]_{(2n_{11})\times(2m_{11})} . \tag{1.370}$$

(iv$_{4a2}$) The $([(2m_{11}),(2n_{11})]_{\text{II:II}}\text{US,pF})$*-positive increasing-inflection upper-saddle flow appearing bifurcation is from a* $q_1 \times p_1$ *to* $\overline{q}_1 \times \overline{p}_1$ *1-dimensional flow networks as*

$$\begin{aligned} \bigcup_{r_1=1}^{p_1}\bigcup_{u_1=1}^{q_1} \underbrace{(a_{j_1j_1r_1}, a_{j_1j_2u_1})}_{([m_{j_1r_1}, n_{j_1u_1}]\text{-XX},pF)} &\rightleftharpoons \underbrace{([(2m_{11}),(2n_{11})]_{\text{II:II}}\text{US,pF})}_{[(2m_{11}),(2n_{11})]\text{-increasing infelction upper-saddle flow } (+)} \\ &\rightleftharpoons \bigcup_{r_2=1}^{\overline{p}_1}\bigcup_{u_2=1}^{\overline{q}_1} \underbrace{(a_{j_1j_1r_2}, a_{j_1j_2u_2})}_{([m_{j_1r_2}, n_{j_1u_2}]\text{-YY},pF)} . \end{aligned} \tag{1.371}$$

(iv$_{4a3}$) The $([(2m_{11}),(2n_{11})]_{\text{II:II}}\text{US,pF})$*-positive increasing-inflection upper-saddle flow switching bifurcation is for the switching of two networks of* $q_1 \times p_1$ *1-dimensional flows as*

$$\begin{aligned} \bigcup_{r_1=1}^{p_1}\bigcup_{u_1=1}^{q_1} \underbrace{(a_{j_1j_1r_1}, a_{j_1j_2u_1})}_{([m_{j_1r_1}, n_{j_1u_1}]\text{-XX,pF})} &\rightleftharpoons \underbrace{([(2m_{11}),(2n_{11})]_{\text{II:II}}\text{US,pF})}_{[(2m_{11}),(2n_{11})]\text{-increasing infelction upper-saddle flow } (+)} \\ &\rightleftharpoons \bigcup_{r_2=1}^{p_1}\bigcup_{u_2=1}^{q_1} \underbrace{(a_{j_1j_1r_2}, a_{j_1j_2u_2})}_{([m_{j_1r_2}, n_{j_1u_2}]\text{-YY,pF})} . \end{aligned} \tag{1.372}$$

(iv$_{4b}$) For $a_{j_1j_10} < 0$ *and* $a_{j_2j_20}(a_{j_1j_21} - a_{j_2j_21})^{n_2} > 0$*, there is a* $([(2m_{11}),(2n_{11})]_{\text{DI:DI}}\text{LS, pF})$*-positive decreasing-inflection lower-saddle flow as*

$$(a_{j_1j_11}, a_{j_1j_21}) = \underbrace{([(2m_{11}),(2n_{11})]_{\text{DI:DI}}\text{LS,pF})}_{[(2m_{11}),(2n_{11})]\text{-decreasing-infelction lower-saddle flow } (+)} . \tag{1.373}$$

There are three following $([(2m_{11}),(2n_{11})]_{\text{DI:DI}}\text{LS,pF})$*-positive decreasing-inflection lower-saddle flow appearing and switching bifurcations.*

(iv_{4b1}) The ([(2m_{11}), (2n_{11})]$_{\text{DI:DI}}$LS,pF)*-positive decreasing-inflection lower-saddle flow appearing bifurcation is from an (nF, pF)-flow to a* $(2m_1) \times (2n_1)$*-decreasing-inflection lower-saddle flow network as*

$$\underbrace{(\dot{x}_{j_1}, \dot{x}_{j_2})}_{\text{(nF,pF)-flow}} \rightleftharpoons \underbrace{([(2m_{11}), (2n_{11})]_{\text{DI:DI}}\text{LS,pF})}_{[(2m_{11}),(2n_{11})]\text{-dereasing-inflection lower-saddle flow }(+)}$$
$$\rightleftharpoons \bigcup_{r=1}^{2m_{11}} \bigcup_{u=1}^{2n_{11}} \underbrace{(a_{j_1j_1r}, a_{j_1j_2u})}_{\text{YY}}, \tag{1.374}$$

where, without any equilibriums,

$$\bigcup_{r=1}^{2m_{11}} \bigcup_{u=1}^{2n_{11}} \underbrace{(a_{j_1j_1r}, a_{j_1j_2u})}_{\text{YY}}$$
$$= \begin{bmatrix} \underbrace{(\text{DP:UP,pF})}_{\text{hyperbolic flow }(+)} & \cdots & \underbrace{(\text{DP:UP,pF})}_{\text{hyperbolic flow }(+)} & \underbrace{(\text{UP:DP,pF})}_{\text{hyperbolic-secant flow }(+)} \\ \underbrace{(\text{UP:DP,pF})}_{\text{hyperbolic-secant flow }(+)} & \cdots & \underbrace{(\text{UP:DP,pF})}_{\text{hyperbolic-secant flow }(+)} & \underbrace{(\text{DP:UP,pF})}_{\text{hyperbolic flow }(+)} \\ \vdots & \vdots & \vdots & \vdots \\ \underbrace{(\text{UP:DP,pF})}_{\text{hyperbolic-secant flow }(+)} & \cdots & \underbrace{(\text{UP:DP,pF})}_{\text{hyperbolic-secant flow }(+)} & \underbrace{(\text{DP:UP,pF})}_{\text{hyperbolic flow }(+)} \end{bmatrix}_{(2n_{11})\times(2m_{11})}. \tag{1.375}$$

(iv_{4b2}) The ([(2m_{11}), (2n_{11})]$_{\text{DI:DI}}$LS,pF)*-positive decreasing-inflection lower-saddle flow appearing bifurcation is from a* $q_1 \times p_1$ *to* $\overline{q}_1 \times \overline{p}_1$ *1-dimensional flow networks as*

$$\bigcup_{r_1=1}^{p_1} \bigcup_{u_1=1}^{q_1} \underbrace{(a_{j_1j_1r_1}, a_{j_1j_2u_1})}_{([m_{j_1r_1}, n_{j_1u_1}]\text{-XX,pF})} \rightleftharpoons \underbrace{([(2m_{11}), (2n_{11})]_{\text{DI:DI}}\text{LS,pF})}_{[(2m_{11}),(2n_{11})]\text{-dereasing-inflection lower-saddle flow }(+)}$$
$$\rightleftharpoons \bigcup_{r_2=1}^{\overline{p}_1} \bigcup_{u_2=1}^{\overline{q}_1} \underbrace{(a_{j_1j_1r_2}, a_{j_1j_2u_2})}_{([m_{j_1r_2}, n_{j_1u_2}]\text{-YY,pF}))}. \tag{1.376}$$

(iv_{4b3}) The ([(2m_{11}), (2n_{11})]$_{\text{DI:DI}}$LS,pF)*-positive decreasing-inflection lower-saddle flow switching bifurcation is for the switching of two networks of* $q_1 \times p_1$ *1-dimensional flows as*

$$\bigcup_{r_1=1}^{p_1}\bigcup_{u_1=1}^{q_1}\underbrace{(a_{j_1j_1r_1},a_{j_1j_2u_1})}_{([m_{j_1r_1},n_{j_1u_1}]\text{-XX,pF})} \rightleftharpoons \underbrace{([(2m_{11}),(2n_{11})]_{\text{DI:DI}}\text{LS,pF})}_{[(2m_{11}),(2n_{11})]\text{-dereasing-inflection lower-saddle flow }(+)}$$
$$\rightleftharpoons \bigcup_{r_2=1}^{p_1}\bigcup_{u_2=1}^{q_1}\underbrace{(a_{j_1j_1r_2},a_{j_1j_2u_2})}_{([m_{j_1r_2},n_{j_1u_2}]\text{-YY,pF})}. \tag{1.377}$$

(iv_{4c}) For $a_{j_1j_10} > 0$ and $a_{j_2j_20}(a_{j_1j_11} - a_{j_2j_11})^{m_2} < 0$, there is a ([(2$m_{11}$), (2$n_{11}$)] $_{\text{DI:DI}}$US,nF)*-negative decreasing-inflection upper-saddle flow as*

$$(a_{j_1j_11},a_{j_1j_21}) = \underbrace{([(2m_{11}),(2n_{11})]_{\text{DI:DI}}\text{US,nF})}_{[(2m_{11}),(2n_{11})]\text{-dereasing-inflection upper-saddle flow }(-)}. \tag{1.378}$$

There are three following ([(2m_{11}), (2n_{11})]$_{\text{DI:DI}}$US,nF)*-negative decreasing-inflection upper-saddle flow appearing and switching bifurcations.*

(iv_{4c1}) The ([(2m_{11}), (2n_{11})]$_{\text{DI:DI}}$US,nF)*-negative decreasing-inflection upper-saddle flow appearing bifurcation is from a (pF, nF)-flow to a* $(2m_1) \times (2n_1)$*-decreasing-inflection upper-saddle flow network as*

$$\underbrace{(\dot{x}_{j_1},\dot{x}_{j_2})}_{\text{(pF,nF)-flow}} \rightleftharpoons \underbrace{([(2m_{11}),(2n_{11})]_{\text{DI:DI}}\text{US,nF})}_{[(2m_{11}),(2n_{11})]\text{-dereasing-inflection upper-saddle flow }(-)}$$
$$\rightleftharpoons \bigcup_{r=1}^{2m_{11}+1}\bigcup_{u=1}^{2n_{11}}\underbrace{(a_{j_1j_1r},a_{j_1j_2u})}_{\text{YY}}, \tag{1.379}$$

where, without any equilibriums,

$$\bigcup_{r=1}^{2m_{11}}\bigcup_{u=1}^{2n_{11}}\underbrace{(a_{j_1j_1r},a_{j_1j_2u})}_{\text{YY}}$$
$$= \begin{bmatrix} \underbrace{(\text{DP:UP,nF})}_{\text{hyperbolic flow }(-)} & \cdots & \underbrace{(\text{DP:UP,nF})}_{\text{hyperbolic flow }(-)} & \underbrace{(\text{UP:DP,nF})}_{\text{hyperbolic-secant flow }(-)} \\ \underbrace{(\text{UP:DP,nF})}_{\text{hyperbolic-secant flow }(-)} & \cdots & \underbrace{(\text{UP:DP,nF})}_{\text{hyperbolic-secant flow }(-)} & \underbrace{(\text{DP:UP,nF})}_{\text{hyperbolic flow }(-)} \\ \vdots & \vdots & \vdots & \vdots \\ \underbrace{(\text{UP:DP,nF})}_{\text{hyperbolic-secant flow }(-)} & \cdots & \underbrace{(\text{UP:DP,nF})}_{\text{hyperbolic-secant flow }(-)} & \underbrace{(\text{DP:UP,nF})}_{\text{hyperbolic flow }(-)} \end{bmatrix}_{(2n_{11})\times(2m_{11})}. \tag{1.380}$$

(iv$_{4c2}$) The $([(2m_{11}),(2n_{11})]_{\mathrm{DI:DI}}\mathrm{US,nF})$*-negative decreasing-inflection upper-saddle flow appearing bifurcation is from a* $q_1 \times p_1$ *to* $\overline{q}_1 \times \overline{p}_1$ *1-dimensional flow networks as*

$$\bigcup_{r_1=1}^{p_1}\bigcup_{u_1=1}^{q_1}\underbrace{(a_{j_1j_1r_1},a_{j_1j_2u_1})}_{([m_{j_1r_1},n_{j_1u_1}]\text{-XX,nF})} \rightleftharpoons \underbrace{([(2m_{11}),(2n_{11})]_{\mathrm{DI:DI}}\mathrm{US,nF})}_{[(2m_{11}),(2n_{11})]\text{-dereasing-inflection upper-saddle flow }(-)}$$
$$\rightleftharpoons \bigcup_{r_2=1}^{\overline{p}_1}\bigcup_{u_2=1}^{\overline{q}_1}\underbrace{(a_{j_1j_1r_2},a_{j_1j_2u_2})}_{([m_{j_1r_2},n_{j_1u_2}]\text{-YY,nF})}. \tag{1.381}$$

(iv$_{4c3}$) The $([(2m_{11}),(2n_{11})]_{\mathrm{DI:DI}}\mathrm{US,nF})$*-negative decreasing-inflection upper-saddle flow switching bifurcation is for the switching of two networks of* $q_1 \times p_1$ *1-dimensional flows as*

$$\bigcup_{r_1=1}^{p_1}\bigcup_{u_1=1}^{q_1}\underbrace{(a_{j_1j_1r_1},a_{j_1j_2u_1})}_{([m_{j_1r_1},n_{j_1u_1}]\text{-XX,nF})} \rightleftharpoons \underbrace{([(2m_{11}),(2n_{11})]_{\mathrm{DI:DI}}\mathrm{US,nF})}_{[(2m_{11}),(2n_{11})]\text{-dereasing-inflection upper-saddle flow }(-)}$$
$$\rightleftharpoons \bigcup_{r_2=1}^{p_1}\bigcup_{u_2=1}^{q_1}\underbrace{(a_{j_1j_1r_2},a_{j_1j_2u_2})}_{([m_{j_1r_2},n_{j_1u_2}]\text{-XX,nF})}. \tag{1.382}$$

(iv$_{4d}$) For $a_{j_1j_10} < 0$ *and* $a_{j_2j_20}(a_{j_1j_11} - a_{j_2j_11})^{m_2} < 0$*, there is a* $([(2m_{11}),(2n_{11})]_{\mathrm{II:II}}\mathrm{LS,nF})$*-negative increasing inflection lower-saddle flow as*

$$(a_{j_1j_11},a_{j_1j_21}) = \underbrace{([(2m_{11}),(2n_{11})]_{\mathrm{II:II}}\mathrm{LS,nF})}_{[(2m_{11}),(2n_{11})]\text{-increasing-inflection lower-saddle flow }(-)}. \tag{1.383}$$

There are three following $([(2m_{11}),(2n_{11})]_{\mathrm{II:II}}\mathrm{LS,nF})$*-negative increasing-inflection lower-saddle flow appearing and switching bifurcations.*

(iv$_{4d1}$) The $([(2m_{11}),(2n_{11})]_{\mathrm{II:II}}\mathrm{LS,nF})$*-negative increasing-inflection lower-saddle flow appearing bifurcation is from an (nF, nF)-flow to a* $(2m_{11}) \times (2n_{11})$*-hyperbolic and hyperbolic-secant flow network as*

$$\underbrace{(\dot{x}_{j_1},\dot{x}_{j_2})}_{\text{(nF,nF)-flow}} \rightleftharpoons \underbrace{([(2m_{11}),(2n_{11})]_{\mathrm{II:II}}\mathrm{LS,nF})}_{[(2m_{11}),(2n_{11})]\text{-increasing-inflection lower-saddle flow }(-)}$$
$$\rightleftharpoons \bigcup_{r=1}^{2m_{11}}\bigcup_{u=1}^{2n_{11}}\underbrace{(a_{j_1j_1r},a_{j_1j_2u})}_{\text{YY}}, \tag{1.384}$$

where, without any equilibriums,

$$\bigcup_{r=1}^{2m_{11}}\bigcup_{u=1}^{2n_{11}}\underbrace{(a_{j_1j_1r},a_{j_1j_2u})}_{\text{YY}}$$

$$=\begin{bmatrix}\underbrace{\text{(DP:UP,nF)}}_{\text{hyperbolic flow }(-)} & \cdots & \underbrace{\text{(DP:UP,nF)}}_{\text{hyperbolic flow }(-)} & \underbrace{\text{(UP:DP,nF)}}_{\text{hyperbolic-secant flow }(-)}\\ \underbrace{\text{(UP:DP,nF)}}_{\text{hyperbolic-secant flow }(-)} & \cdots & \underbrace{\text{(UP:DP,nF)}}_{\text{hyperbolic-secant flow }(-)} & \underbrace{\text{(DP:UP,nF)}}_{\text{hyperbolic flow }(-)}\\ \vdots & \vdots & \vdots & \vdots\\ \underbrace{\text{(UP:DP,nF)}}_{\text{hyperbolic-secant flow }(-)} & \cdots & \underbrace{\text{(UP:DP,nF)}}_{\text{hyperbolic-secant flow }(-)} & \underbrace{\text{(DP:UP,nF)}}_{\text{hyperbolic flow }(-)}\end{bmatrix}_{(2n_{11})\times(2m_{11})}. \tag{1.385}$$

(iv_{4d2}) The $([(2m_{11}),(2n_{11})]_{\text{II:II}}\text{LS,nF})$*-negative increasing-inflection lower-saddle flow appearing bifurcation is from a* $q_1\times p_1$ *to* $\overline{q}_1\times\overline{p}_1$ *1-dimensional flow networks as*

$$\bigcup_{r_1=1}^{p_1}\bigcup_{u_1=1}^{q_1}\underbrace{(a_{j_1j_1r_1},a_{j_1j_2u_1})}_{([m_{j_1r_1},n_{j_1u_1}]\text{-XX,nF})}\rightleftharpoons\underbrace{([(2m_{11}),(2n_{11})]_{\text{II:II}}\text{LS,nF})}_{[(2m_{11}),(2n_{11})]\text{-increasing-inflection lower-saddle flow }(-)}$$
$$\rightleftharpoons\bigcup_{r_2=1}^{\overline{p}_1}\bigcup_{u_2=1}^{\overline{q}_1}\underbrace{(a_{j_1j_1r_2},a_{j_1j_2u_2})}_{([m_{j_1r_2},n_{j_1u_2}]\text{-YY,nF}))}. \tag{1.386}$$

(iv_{4d3}) The $([(2m_{11}),(2n_{11})]_{\text{II:II}}\text{LS,nF})$*-negative increasing-inflection lower-saddle flow switching bifurcation is for the switching of two networks of* $q_1\times p_1$ *1-dimensional flows as*

$$\bigcup_{r_1=1}^{p_1}\bigcup_{u_1=1}^{q_1}\underbrace{(a_{j_1j_1r_1},a_{j_1j_2u_1})}_{([m_{j_1r_1},n_{j_1u_1}]\text{-XX,nF})}\rightleftharpoons\underbrace{([(2m_{11}),(2n_{11})]_{\text{II:II}}\text{LS,nF})}_{[(2m_{11}),(2n_{11})]\text{-increasing-inflection lower-saddle flow }(-)}$$
$$\rightleftharpoons\bigcup_{r_2=1}^{p_1}\bigcup_{u_2=1}^{q_1}\underbrace{(a_{j_1j_1r_2},a_{j_1j_2u_2})}_{([m_{j_1r_2},n_{j_1u_2}]\text{-YY,nF})}. \tag{1.387}$$

(v) The equilibrium of $(x^*_{j_2},x^*_{j_1})=(a_{j_1j_2 1},a_{j_2j_1 1})$ *with* $a_{j_2j_2 1}\neq a_{j_1j_2 1}$ *has four classes of equilibrium bifurcations.*

(v_1) For $n_1=2n_{11}+1$ *and* $m_2=2m_{21}+1$*, the equilibrium of* $(x^*_{j_2},x^*_{j_1})=(a_{j_1j_2 1},a_{j_2j_1 1})$ *for the* x_{j_1}*-direction has the following bifurcation properties with*

$$
\begin{aligned}
&2n_{11}+1-\sum_{u_1=1}^{p_1} n_{j_1u_1}^{(1)} = 2\sum_{\rho_1=1}^{l_1} n_{\rho_1},\, 2m_{21}+1-\sum_{s_1=1}^{q_2} m_{j_2v_1}^{(1)} = 2\sum_{\sigma_1=1}^{l_2} m_{\sigma_1};\\
&\sum_{u_2=1}^{\bar{p}_1} n_{j_1u_2}^{(2)} = 2n_{11}+1,\, \sum_{s_2=1}^{\bar{q}_2} m_{j_2v_2}^{(2)} = 2m_{21}+1;\\
&\sum_{u_i=1}^{p_1} n_{j_1u_i}^{(i)} = 2n_{11}+1,\, \sum_{s_i=1}^{q_2} m_{j_2v_i}^{(2)} = 2m_{21}+1, \text{ for } i=1,2;
\end{aligned}
\tag{1.388}
$$

l_1, l_2-quadratic polynomials are without real roots, and the ρ_1^{th} and σ_1^{th}-quadratic polynomials are with powers m_{ρ_1} and n_{σ_1}, respectively.

(v_{1a}) For $a_{j_1j_10}(a_{j_2j_11}-a_{j_1j_11})^{m_1}>0$ and $a_{j_2j_20}>0$, there is a $((2n_{11}+1), (2m_{21}+1))$-positive saddle equilibrium at $(x_{j_2}^*, x_{j_1}^*) = (a_{j_1j_21}, a_{j_2j_11})$ as

$$
(a_{j_1j_21}, a_{j_2j_11}) = \underbrace{((2n_{11}+1)^{\text{th}}\text{UP}_+, (2m_{21}+1)^{\text{th}}\text{UP}_+)}_{((2n_{11}+1),(2m_{21}+1))\text{-positve saddle}}. \tag{1.389}
$$

The following three $((2n_{11}+1)^{\text{th}}\text{UP}_+, (2m_{21}+1)^{\text{th}}\text{UP}_+)$-positive saddle appearing and switching *bifurcations as follows.*

(v_{1a1}) The $((2n_{11}+1)^{\text{th}}\text{UP}_+, (2m_{21}+1)^{\text{th}}\text{UP}_+)$-positive saddle appearing bifurcation is *from a* (UP_+, UP_+)*-positive saddle to a $(2n_{11}+1)\times(2m_{21}+1)$-network of positive and negative saddles, CCW and CW centers as*

$$
\begin{aligned}
\underbrace{(a_{j_1j_21}, a_{j_2j_11})}_{(\text{UP}_+,\text{UP}_+)\text{-positive saddle}} &\rightleftharpoons \underbrace{((2n_{11}+1)^{\text{th}}\text{UP}_+, (2m_{21}+1)^{\text{th}}\text{UP}_+)}_{((2n_{11}+1),(2m_{21}+1))\text{-positve saddle}}\\
&\rightleftharpoons \bigcup_{u=1}^{2n_{11}+1}\bigcup_{v=1}^{2m_{21}+1}\underbrace{(a_{j_1j_2u}, a_{j_2j_1v})}_{\text{XX}},
\end{aligned}
\tag{1.390}
$$

where, without 1-dimensional flows,

$$
\begin{aligned}
&\bigcup_{u=1}^{2n_{11}+1}\bigcup_{v=1}^{2m_{21}+1}\underbrace{(a_{j_1j_2u}, a_{j_2j_1v})}_{\text{XX}}\\
&=\begin{bmatrix}
\underbrace{(\text{UP}_+,\text{UP}_+)}_{\text{positive saddle}} & \cdots & \underbrace{(\text{DP}_-,\text{DP}_+)}_{\text{CW center}} & \underbrace{(\text{UP}_+,\text{UP}_+)}_{\text{positive saddle}}\\
\underbrace{(\text{DP}_+,\text{DP}_-)}_{\text{CCW center}} & \cdots & \underbrace{(\text{UP}_-,\text{UP}_-)}_{\text{negative saddle}} & \underbrace{(\text{DP}_+,\text{DP}_-)}_{\text{CCW center}}\\
\vdots & \vdots & \vdots & \vdots\\
\underbrace{(\text{UP}_+,\text{UP}_+)}_{\text{positive saddle}} & \cdots & \underbrace{(\text{DP}_-,\text{DP}_+)}_{\text{CW center}} & \underbrace{(\text{UP}_+,\text{UP}_+)}_{\text{positive saddle}}
\end{bmatrix}_{(2n_{11}+1)\times(2m_{21}+1)}.
\end{aligned}
\tag{1.391}
$$

(v_{1a2}) The $((2n_{11}+1)^{\text{th}}\text{UP}_{+}, (2m_{21}+1)^{\text{th}}\text{UP}_{+})$*-positive saddle appearing bifurcation is from a* $q_1 \times p_2$ *to* $\overline{q}_1 \times \overline{p}_2$ *networks of positive and negative saddles and CCW and CW centers as*

$$\bigcup_{u_1=1}^{q_1}\bigcup_{v_1=1}^{p_2}\underbrace{(a_{j_1j_2u_1}, a_{j_2j_1v_1})}_{(n_{j_1u_1},m_{j_2v_1})\text{-XX}} \rightleftharpoons \underbrace{((2n_{11}+1)^{\text{th}}\text{UP}_{+}, (2m_{21}+1)^{\text{th}}\text{UP}_{+})}_{((2n_{11}+1),(2m_{21}+1))\text{-positive saddle}}$$
$$\rightleftharpoons \bigcup_{u_2=1}^{\overline{q}_1}\bigcup_{v_2=1}^{\overline{p}_2}\underbrace{(a_{j_1j_2u_2}, a_{j_2j_1v_2})}_{(n_{j_1u_2},m_{j_2v_2})\text{-YY}}. \tag{1.392}$$

(v_{1a3}) The $((2n_{11}+1)^{\text{th}}\text{UP}_{+}, (2m_{21}+1)^{\text{th}}\text{UP}_{+})$*-positive saddle switching bifurcation is for the switching of two* $q_1 \times p_2$ *networks of positive and negative saddles and CCW and CW centers as*

$$\bigcup_{u_1=1}^{q_1}\bigcup_{v_1=1}^{p_2}\underbrace{(a_{j_1j_2u_1}, a_{j_2j_1v_1})}_{(n_{j_1u_1},m_{j_2v_1})\text{-XX}} \rightleftharpoons \underbrace{((2n_{11}+1)^{\text{th}}\text{UP}_{+}, (2m_{21}+1)^{\text{th}}\text{UP}_{+})}_{((2n_{11}+1),(2m_{21}+1))\text{-positive saddle}}$$
$$\rightleftharpoons \bigcup_{u_2=1}^{q_1}\bigcup_{v_2=1}^{p_2}\underbrace{(a_{j_1j_2u_2}, a_{j_2j_1v_2})}_{(n_{j_1u_2},m_{j_2v_2})\text{-XX}} \tag{1.393}$$

(v_{1b}) For $a_{j_1j_10}(a_{j_2j_11} - a_{j_1j_11})^{m_1} < 0$ *and* $a_{j_2j_20} > 0$*, there is a* $((2n_{11}+1), (2m_{21}+1))$-CCW center *equilibrium at* $(x^*_{j_2}, x^*_{j_1}) = (a_{j_1j_21}, a_{j_2j_11})$ *as*

$$(a_{j_1j_21}, a_{j_2j_11}) = \underbrace{((2n_{11}+1)^{\text{th}}\text{DP}_{+}, (2m_{21}+1)^{\text{th}}\text{DP}_{-})}_{((2n_{11}+1),(2m_{21}+1))\text{-CCW center}}. \tag{1.394}$$

The following three $((2n_{11}+1)^{\text{th}}\text{DP}_{+}, (2m_{21}+1)^{\text{th}}\text{DP}_{-})$-CCW *center appearing and switching bifurcations as follows.*

(v_{1b1}) The $((2n_{11}+1)^{\text{th}}\text{DP}_{+}, (2m_{21}+1)^{\text{th}}\text{DP}_{-})$-CCW *center appearing bifurcation is from a* $(\text{DP}_{+}, \text{DP}_{-})$*-CCW center to a* $(2n_{11}+1) \times (2m_{21}+1)$ *network of positive and negative saddles and CCW and CW centers as*

$$\underbrace{(a_{j_1j_21}, a_{j_2j_11})}_{(\text{DP}_{+},\text{DP}_{-})\text{-CCW center}} \rightleftharpoons \underbrace{((2n_{11}+1)^{\text{th}}\text{DP}_{+}, (2m_{21}+1)^{\text{th}}\text{DP}_{-})}_{((2n_{11}+1),(2m_{21}+1))\text{-CCW center}}$$
$$\rightleftharpoons \bigcup_{u=1}^{2n_{11}+1}\bigcup_{v=1}^{2m_{21}+1}\underbrace{(a_{j_1j_2u}, a_{j_2j_1v})}_{\text{XX}}, \tag{1.395}$$

where, without 1-dimensional flows,

$$\bigcup_{u=1}^{2n_{11}+1}\bigcup_{v=1}^{2m_{21}+1}\underbrace{(a_{j_1j_2u},a_{j_2j_1v})}_{\text{XX}} = \begin{bmatrix} \underbrace{(\text{DP}_+,\text{DP}_-)}_{\text{CCW center}} & \cdots & \underbrace{(\text{UP}_-,\text{UP}_-)}_{\text{negative saddle}} & \underbrace{(\text{DP}_+,\text{DP}_-)}_{\text{CCW center}} \\ \underbrace{(\text{UP}_+,\text{UP}_+)}_{\text{positive saddle}} & \cdots & \underbrace{(\text{DP}_-,\text{DP}_+)}_{\text{CW center}} & \underbrace{(\text{UP}_+,\text{UP}_+)}_{\text{positive saddle}} \\ \vdots & \vdots & \vdots & \vdots \\ \underbrace{(\text{DP}_+,\text{DP}_-)}_{\text{CCW center}} & \cdots & \underbrace{(\text{UP}_-,\text{UP}_-)}_{\text{negative saddle}} & \underbrace{(\text{DP}_+,\text{DP}_-)}_{\text{CCW center}} \end{bmatrix}_{(2n_{11}+1)\times(2m_{21}+1)}. \quad (1.396)$$

(v_{1b2}) The $((2n_{11}+1)^{\text{th}}\text{DP}_+, (2m_{21}+1)^{\text{th}}\text{DP}_-)$-CCW center appearing bifurcation is from a $q_1 \times p_2$ to $\overline{q}_1 \times \overline{p}_2$ networks of positive and negative saddles and CCW and CW centers as

$$\begin{aligned}\bigcup_{u_1=1}^{q_1}\bigcup_{v_1=1}^{p_2}\underbrace{(a_{j_1j_2u_1},a_{j_2j_1v_1})}_{(n_{j_1u_1},m_{j_2v_1})\text{-XX}} &\rightleftharpoons \underbrace{((2n_{11}+1)^{\text{th}}\text{DP}_+,(2m_{21}+1)^{\text{th}}\text{DP}_-)}_{((2n_{11}+1),(2m_{21}+1))\text{-CCW center}} \\ &\rightleftharpoons \bigcup_{u_2=1}^{\overline{q}_1}\bigcup_{v_2=1}^{\overline{p}_2}\underbrace{(a_{j_1j_2u_2},a_{j_2j_1v_2})}_{(n_{j_1u_2},m_{j_2v_2})\text{-YY}}.\end{aligned} \quad (1.397)$$

(v_{1b3}) The $((2n_{11}+1)^{\text{th}}\text{DP}_+, (2m_{21}+1)^{\text{th}}\text{DP}_-)$-CCW center switching bifurcation is for the switching of two $q_1 \times p_2$ networks of positive and negative saddles and CCW and CW centers as

$$\begin{aligned}\bigcup_{u_1=1}^{q_1}\bigcup_{v_1=1}^{p_2}\underbrace{(a_{j_1j_2u_1},a_{j_2j_1v_1})}_{(n_{j_1u_1},m_{j_2v_1})\text{-XX}} &\rightleftharpoons \underbrace{((2n_{11}+1)^{\text{th}}\text{DP}_+,(2m_{21}+1)^{\text{th}}\text{DP}_-)}_{((2n_{11}+1),(2m_{21}+1))\text{-CCW center}} \\ &\rightleftharpoons \bigcup_{u_2=1}^{q_1}\bigcup_{v_2=1}^{p_2}\underbrace{(a_{j_1j_2u_2},a_{j_2j_1v_2})}_{(n_{j_1u_2},m_{j_2v_2})\text{-YY}}.\end{aligned} \quad (1.398)$$

(v_{1c}) For $a_{j_1j_10}(a_{j_2j_11}-a_{j_1j_11})^{m_1}>0$ and $a_{j_2j_20}<0$, there is a $((2n_{11}+1),(2m_{21}+1))$-CW center equilibrium at $(x^*_{j_2},x^*_{j_1})=(a_{j_1j_21},a_{j_2j_11})$ as

$$(a_{j_1j_21},a_{j_2j_11}) = \underbrace{((2n_{11}+1)^{\text{th}}\text{DP}_-,(2m_{21}+1)^{\text{th}}\text{DP}_+)}_{((2n_{11}+1),(2m_{21}+1))\text{-CW center}}. \quad (1.399)$$

The following three $((2n_{11}+1)^{\text{th}}\text{DP}_-,\ (2m_{21}+1)^{\text{th}}\text{DP}_+)$*-CW center appearing and switching bifurcations as follows.*

(v_{1c1})*The* $((2n_{11}+1)^{\text{th}}\text{DP}_-,\ (2m_{21}+1)^{\text{th}}\text{DP}_+)$*-CW center appearing bifurcation is from a* $(\text{DP}_-,\ \text{DP}_+)$-CW *center to a* $(2n_{11}+1)\times(2m_{21}+1)$ *network of positive and negative saddles and CCW and CW centers as*

$$\begin{aligned}\underbrace{(a_{j_1j_21},\ a_{j_2j_11})}_{(\text{DP}_-,\text{DP}_+)\text{-CW center}} &\rightleftharpoons \underbrace{((2n_{11}+1)^{\text{th}}\text{DP}_+,\ (2m_{21}+1)^{\text{th}}\text{DP}_-)}_{((2n_{11}+1),(2m_{21}+1))\text{-CW center}}\\ &\rightleftharpoons \bigcup_{u=1}^{2n_{11}+1}\bigcup_{v=1}^{2m_{21}+1}\underbrace{(a_{j_1j_2u},\ a_{j_2j_1v})}_{\text{XX}},\end{aligned} \tag{1.400}$$

where, without 1-dimensional flows,

$$\begin{aligned}&\bigcup_{u=1}^{2n_{11}+1}\bigcup_{v=1}^{2m_{21}+1}\underbrace{(a_{j_1j_2u},\ a_{j_2j_1v})}_{\text{XX}}\\ &=\begin{bmatrix}\underbrace{(\text{DP}_-,\text{DP}_+)}_{\text{CW center}} & \cdots & \underbrace{(\text{UP}_+,\text{UP}_+)}_{\text{positive saddle}} & \underbrace{(\text{DP}_-,\text{DP}_+)}_{\text{CW center}}\\ \underbrace{(\text{UP}_-,\text{UP}_-)}_{\text{negative saddle}} & \cdots & \underbrace{(\text{DP}_+,\text{DP}_-)}_{\text{CCW center}} & \underbrace{(\text{UP}_-,\text{UP}_-)}_{\text{negative saddle}}\\ \vdots & \vdots & \vdots & \vdots\\ \underbrace{(\text{DP}_-,\text{DP}_+)}_{\text{CW center}} & \cdots & \underbrace{(\text{UP}_+,\text{UP}_+)}_{\text{positive saddle}} & \underbrace{(\text{DP}_-,\text{DP}_+)}_{\text{CW center}}\end{bmatrix}_{(2n_{11}+1)\times(2m_{21}+1)}.\end{aligned} \tag{1.401}$$

(v_{1c2}) *The* $((2n_{11}+1)^{\text{th}}\text{DP}_-,\ (2m_{21}+1)^{\text{th}}\text{DP}_+)$*-CW center appearing bifurcation is from a* $q_1\times p_2$ *to* $\overline{q}_1\times\overline{p}_2$ *networks of positive and negative saddles and CCW and CW centers as*

$$\begin{aligned}\bigcup_{u_1=1}^{q_1}\bigcup_{v_1=1}^{p_2}\underbrace{(a_{j_1j_2u_1},\ a_{j_2j_1v_1})}_{(n_{j_1u_1},m_{j_2v_1})\text{-XX}} &\rightleftharpoons \underbrace{((2n_{11}+1)^{\text{th}}\text{DP}_-,\ (2m_{21}+1)^{\text{th}}\text{DP}_+)}_{((2n_{11}+1),(2m_{21}+1))\text{-CW center}}\\ &\rightleftharpoons \bigcup_{u_2=1}^{\overline{q}_1}\bigcup_{v_2=1}^{\overline{p}_2}\underbrace{(a_{j_1j_2u_2},\ a_{j_2j_1v_2})}_{(n_{j_1u_2},m_{j_2v_2})\text{-YY}}.\end{aligned} \tag{1.402}$$

(v_{1c3}) *The* $((2n_{11}+1)^{\text{th}}\text{DP}_-,\ (2m_{21}+1)^{\text{th}}\text{DP}_+)$*-CW center switching bifurcation is for the switching of two* $q_1\times p_2$ *networks of positive and negative saddles and CCW and CW centers as*

$$\bigcup_{u_1=1}^{q_1}\bigcup_{v_1=1}^{p_2}\underbrace{(a_{j_1j_2u_1},a_{j_2j_1v_1})}_{(n_{j_1u_1},m_{j_2v_1})\text{-XX}} \rightleftharpoons \underbrace{((2n_{11}+1)^{\text{th}}\text{DP}_-,(2m_{21}+1)^{\text{th}}\text{DP}_+)}_{((2n_{11}+1),(2m_{21}+1))\text{-CW center}}$$
$$\rightleftharpoons \bigcup_{u_2=1}^{q_1}\bigcup_{v_2=1}^{p_2}\underbrace{(a_{j_1j_2u_2},a_{j_2j_1v_2})}_{(n_{j_1u_2},m_{j_2v_2})\text{-YY}}. \quad (1.403)$$

(v_{1d}) For $a_{j_1j_10}(a_{j_2j_11}-a_{j_1j_11})^{m_1}<0$ and $a_{j_2j_20}<0$, there is a $((2n_{11}+1),(2m_{21}+1))$-negative saddle equilibrium at $(x^*_{j_2},x^*_{j_1})=(a_{j_1j_21},a_{j_2j_11})$ as

$$(a_{j_1j_21},a_{j_2j_11})=\underbrace{((2n_{11}+1)^{\text{th}}\text{UP}_-,(2m_{21}+1)^{\text{th}}\text{UP}_-)}_{((2n_{11}+1),(2m_{21}+1))\text{-negative saddle}}. \quad (1.404)$$

The following three $((2n_{11}+1)^{\text{th}}\text{UP}_-,(2m_{21}+1)^{\text{th}}\text{UP}_-)$-negative saddle appearing and switching bifurcations as follows.

(v_{1d1}) *The $((2n_{11}+1)^{\text{th}}\text{UP}_-,(2m_{21}+1)^{\text{th}}\text{UP}_-)$-negative saddle appearing bifurcation is from a $(\text{UP}_-,\text{UP}_-)$-negative saddle to a $(2n_{11}+1)\times(2m_{21}+1)$-network of positive and negative saddles, CCW and CW centers as*

$$\underbrace{(a_{j_1j_21},a_{j_2j_11})}_{(\text{UP}_-,\text{UP}_-)\text{-positive saddle}} \rightleftharpoons \underbrace{((2n_{11}+1)^{\text{th}}\text{UP}_-,(2m_{21}+1)^{\text{th}}\text{UP}_-)}_{((2n_{11}+1),(2m_{21}+1))\text{-negative saddle}}$$
$$\rightleftharpoons \bigcup_{u=1}^{2n_{11}+1}\bigcup_{v=1}^{2m_{21}+1}\underbrace{(a_{j_1j_2u},a_{j_2j_1v})}_{\text{XX}}, \quad (1.405)$$

where, without 1-dimensional flows,

$$\bigcup_{u=1}^{2n_{11}+1}\bigcup_{v=1}^{2m_{21}+1}\underbrace{(a_{j_1j_2u},a_{j_2j_1v})}_{\text{XX}}$$
$$=\begin{bmatrix}\underbrace{(\text{UP}_-,\text{UP}_-)}_{\text{negative saddle}} & \cdots & \underbrace{(\text{DP}_+,\text{DP}_-)}_{\text{CCW center}} & \underbrace{(\text{UP}_-,\text{UP}_-)}_{\text{negative saddle}}\\ \underbrace{(\text{DP}_-,\text{DP}_+)}_{\text{CW center}} & \cdots & \underbrace{(\text{UP}_+,\text{UP}_+)}_{\text{positive saddle}} & \underbrace{(\text{DP}_-,\text{DP}_+)}_{\text{CW center}}\\ \vdots & \vdots & \vdots & \vdots\\ \underbrace{(\text{UP}_-,\text{UP}_-)}_{\text{negative saddle}} & \cdots & \underbrace{(\text{DP}_+,\text{DP}_-)}_{\text{CCW center}} & \underbrace{(\text{UP}_-,\text{UP}_-)}_{\text{negative saddle}}\end{bmatrix}_{(2n_{11}+1)\times(2m_{21}+1)}. \quad (1.406)$$

(v_{1d2}) *The* $((2n_{11}+1)^{\text{th}}\text{UP}_{-},(2m_{21}+1)^{\text{th}}\text{UP}_{-})$-*negative saddle appearing bifurcation is from a* $q_1 \times p_2$ *to* $\overline{q}_1 \times \overline{p}_2$ *networks of positive and negative saddles and CCW and CW centers as*

$$\begin{aligned}\bigcup_{u_1=1}^{q_1}\bigcup_{v_1=1}^{p_2}\underbrace{(a_{j_1j_2u_1},a_{j_2j_1v_1})}_{(n_{j_1u_1},m_{j_2v_1})\text{-XX}} &\rightleftharpoons \underbrace{((2n_{11}+1)^{\text{th}}\text{UP}_{-},(2m_{21}+1)^{\text{th}}\text{UP}_{-})}_{((2n_{11}+1),(2m_{21}+1))\text{-negative saddle}}\\ &\rightleftharpoons \bigcup_{u_2=1}^{\overline{q}_1}\bigcup_{v_2=1}^{\overline{p}_2}\underbrace{(a_{j_1j_2u_2},a_{j_2j_1v_2})}_{(n_{j_1u_2},m_{j_2v_2})\text{-YY}}.\end{aligned} \tag{1.407}$$

(v_{1d3}) *The* $((2n_{11}+1)^{\text{th}}\text{UP}_{-},(2m_{21}+1)^{\text{th}}\text{UP}_{-})$-*negative saddle switching bifurcation is for the switching of two* $q_1 \times p_2$ *networks of positive and negative saddles and CCW and CW centers as*

$$\begin{aligned}\bigcup_{u_1=1}^{q_1}\bigcup_{v_1=1}^{p_2}\underbrace{(a_{j_1j_2u_1},a_{j_2j_1v_1})}_{(n_{j_1u_1},m_{j_2v_1})\text{-XX}} &\rightleftharpoons \underbrace{((2n_{11}+1)^{\text{th}}\text{UP}_{-},(2m_{21}+1)^{\text{th}}\text{UP}_{-})}_{((2n_{11}+1),(2m_{21}+1))\text{-negative saddle}}\\ &\rightleftharpoons \bigcup_{u_2=1}^{q_1}\bigcup_{v_2=1}^{p_2}\underbrace{(a_{j_1j_2u_2},a_{j_2j_1v_2})}_{(n_{j_1u_2},m_{j_2v_2})\text{-XX}}.\end{aligned} \tag{1.408}$$

(v_2) *For* $n_1 = 2n_{11}$ *and* $m_2 = 2m_{21}+1$, *the equilibrium of* $(x_{j_2}^*, x_{j_1}^*) = (a_{j_1j_2 1}, a_{j_2j_1 1})$ *has the following bifurcation properties with*

$$\begin{aligned}&2n_{11} - \sum_{r_1=1}^{q_1} n_{j_1u_1}^{(1)} = 2\sum_{\rho_1=1}^{l_1} n_{\rho_1},\\ &2m_{21}+1 - \sum_{s_1=1}^{q_2} m_{j_2v_1}^{(1)} = 2\sum_{\sigma_1=1}^{l_2} m_{\sigma_1};\\ &\sum_{r_2=1}^{\overline{p}_1} n_{j_1u_2}^{(2)} = 2n_{11}, \sum_{s_2=1}^{\overline{q}_2} m_{j_2v_2}^{(2)} = 2m_{21}+1;\\ &\sum_{r_i=1}^{p_1} n_{j_1u_i}^{(i)} = 2n_{11}, \sum_{s_i=1}^{q_2} m_{j_2v_i}^{(2)} = 2m_{21}+1, \text{ for } i=1,2;\end{aligned} \tag{1.409}$$

l_1, l_2-*quadratic polynomials are without real roots, and the* ρ_1^{th} *and* σ_1^{th}-*quadratic polynomials are with powers* m_{ρ_1} *and* n_{σ_1}, *respectively.*

(v_{2a}) *For* $a_{j_1j_10}(a_{j_2j_11} - a_{j_1j_11})^{m_1} > 0$ *and* $a_{j_2j_20} > 0$, *there is a* $((2n_{11}),(2m_{21}+1))$-*up-parabola upper-saddle equilibrium at* $(x_{j_2}^*, x_{j_1}^*) = (a_{j_1j_21}, a_{j_2j_11})$ *as*

$$(a_{j_1j_21}, a_{j_2j_11}) = \underbrace{((2n_{11})^{\text{th}}\text{US}, (2m_{21}+1)^{\text{th}}\text{UP})}_{((2n_{11}),(2m_{21}+1))\text{-up-parabola upper-saddle}}. \tag{1.410}$$

The following three $((2n_{11})^{\text{th}}\text{US}, (2m_{21}+1)^{\text{th}}\text{UP})$*-up-parabola upper-saddle appearing and switching bifurcations as follows.*

(v_{2a1}) The $((2n_{11})^{\text{th}}\text{US}, (2m_{21}+1)^{\text{th}}\text{UP})$*-up-parabola upper-saddle appearing bifurcation is from a* (pF, UP)*-positive up-parabola flow to a* $(2n_{11}) \times (2m_{21}+1)$*-network of positive and negative saddles, CCW and CW centers as*

$$\begin{aligned}\underbrace{(\dot{x}_{j_1}, a_{j_2j_11})}_{\text{(pF,UP)-flow}} &\rightleftharpoons \underbrace{((2n_{11})^{\text{th}}\text{US}, (2m_{21}+1)^{\text{th}}\text{UP})}_{((2n_{11}),(2m_{21}+1))\text{-up-parabola upper-saddle}} \\ &\rightleftharpoons \bigcup_{u=1}^{2n_{11}} \bigcup_{v=1}^{2m_{21}+1} \underbrace{(a_{j_1j_2u}, a_{j_2j_1v})}_{\text{XX}},\end{aligned} \tag{1.411}$$

where, without 1-dimensional flows,

$$\begin{aligned}&\bigcup_{u=1}^{2n_{11}} \bigcup_{v=1}^{2m_{21}+1} \underbrace{(a_{j_1j_2u}, a_{j_2j_1v})}_{\text{XX}} \\ &= \begin{bmatrix} \underbrace{(\text{UP}_+,\text{UP}_+)}_{\text{positive saddle}} & \cdots & \underbrace{(\text{DP}_-,\text{DP}_+)}_{\text{CW center}} & \underbrace{(\text{UP}_+,\text{UP}_+)}_{\text{positive saddle}} \\ \underbrace{(\text{DP}_+,\text{DP}_-)}_{\text{CCW center}} & \cdots & \underbrace{(\text{UP}_-,\text{UP}_-)}_{\text{negative saddle}} & \underbrace{(\text{DP}_+,\text{DP}_-)}_{\text{CCW center}} \\ \vdots & \vdots & \vdots & \vdots \\ \underbrace{(\text{DP}_+,\text{DP}_-)}_{\text{CCW center}} & \cdots & \underbrace{(\text{UP}_-,\text{UP}_-)}_{\text{negative saddle}} & \underbrace{(\text{DP}_+,\text{DP}_-)}_{\text{CCW center}} \end{bmatrix}_{(2n_{11})\times(2m_{21}+1)}.\end{aligned} \tag{1.412}$$

(v_{2a2}) The $((2n_{11})^{\text{th}}\text{US}, (2m_{21}+1)^{\text{th}}\text{UP})$*-up-parabola upper-saddle appearing bifurcation is from a* $q_1 \times p_2$ *to* $\overline{q}_1 \times \overline{p}_2$ *networks of positive and negative saddles and CCW and CW centers as*

$$\begin{aligned}\bigcup_{u_1=1}^{q_1} \bigcup_{v_1=1}^{p_2} \underbrace{(a_{j_1j_2u_1}, a_{j_2j_1v_1})}_{(n_{j_1u_1},m_{j_2v_1})\text{-XX}} &\rightleftharpoons \underbrace{((2n_{11})^{\text{th}}\text{US}, (2m_{21}+1)^{\text{th}}\text{UP})}_{((2n_{11}),(2m_{21}+1))\text{-up-parabola upper-saddle}} \\ &\rightleftharpoons \bigcup_{u_2=1}^{\overline{q}_1} \bigcup_{v_2=1}^{\overline{p}_2} \underbrace{(a_{j_1j_2u_2}, a_{j_2j_1v_2})}_{(n_{j_1u_2},m_{j_2v_2})\text{-YY}}.\end{aligned} \tag{1.413}$$

(v_{2a3}) *The* $((2n_{11})^{\text{th}}\text{US}, (2m_{21}+1)^{\text{th}}\text{UP})$-*up-parabola upper-saddle switching bifurcation is for the switching of two* $q_1 \times p_2$ *networks of positive and negative saddles and CCW and CW centers as*

$$\bigcup_{u_1=1}^{q_1}\bigcup_{v_1=1}^{p_2}\underbrace{(a_{j_1j_2u_1}, a_{j_2j_1v_1})}_{(n_{j_1u_1},m_{j_2v_1})\text{-XX}} \rightleftharpoons \underbrace{((2n_{11})^{\text{th}}\text{US}, (2m_{21}+1)^{\text{th}}\text{UP})}_{((2n_{11}),(2m_{21}+1))\text{-up-parabola upper-saddle}}$$
$$\rightleftharpoons \bigcup_{u_2=1}^{q_1}\bigcup_{v_2=1}^{p_2}\underbrace{(a_{j_1j_2u_2}, a_{j_2j_1v_2})}_{(n_{j_1u_2},m_{j_2v_2})\text{-XX}}. \tag{1.414}$$

(v_{2b}) *For* $a_{j_1j_10}(a_{j_2j_11}-a_{j_1j_11})^{m_1}<0$ *and* $a_{j_2j_20}>0$, *there is a* $((2n_{11}), (2m_{21}+1))$-*down-parabola lower-saddle equilibrium at* $(x_{j_2}^*, x_{j_1}^*) = (a_{j_1j_21}, a_{j_2j_11})$ *as*

$$(a_{j_1j_21}, a_{j_2j_11}) = \underbrace{((2n_{11})^{\text{th}}\text{LS}, (2m_{21}+1)^{\text{th}}\text{DP})}_{((2n_{11}),(2m_{21}+1))\text{-down-parabola lower-saddle}}. \tag{1.415}$$

The following three $((2n_{11})^{\text{th}}\text{LS}, (2m_{21}+1)^{\text{th}}\text{DP})$-*down-parabola lower-saddle appearing and switching bifurcations as follows.*

(v_{2b1}) *The* $((2n_{11})^{\text{th}}\text{LS}, (2m_{21}+1)^{\text{th}}\text{DP})$-*down-parabola lower-saddle appearing bifurcation is from an* (nF, DP)-*negative down-parabola flow to a* $(2n_{11})\times(2m_{21}+1)$ *network of positive and negative saddles and CCW and CW centers as*

$$\underbrace{(\dot{x}_{j_1}, a_{j_2j_11})}_{(\text{nF,DP})\text{-flow}} \rightleftharpoons \underbrace{((2n_{11})^{\text{th}}\text{LS}, (2m_{21}+1)^{\text{th}}\text{DP})}_{((2n_{11}),(2m_{21}+1))\text{-down-parabola lower-saddle}}$$
$$\rightleftharpoons \bigcup_{u=1}^{2n_{11}}\bigcup_{v=1}^{2m_{21}+1}\underbrace{(a_{j_1j_2u}, a_{j_2j_1v})}_{\text{XX}}, \tag{1.416}$$

where, without 1-dimensional flows,

$$\bigcup_{u=1}^{2n_{11}}\bigcup_{v=1}^{2m_{21}+1}\underbrace{(a_{j_1j_2u}, a_{j_2j_1v})}_{\text{XX}}$$
$$= \begin{bmatrix} \underbrace{(\text{DP}_+,\text{DP}_-)}_{\text{CCW center}} & \cdots & \underbrace{(\text{UP}_-,\text{UP}_-)}_{\text{negative saddle}} & \underbrace{(\text{DP}_+,\text{DP}_-)}_{\text{CCW center}} \\ \underbrace{(\text{UP}_+,\text{UP}_+)}_{\text{positive saddle}} & \cdots & \underbrace{(\text{DP}_-,\text{DP}_+)}_{\text{CW center}} & \underbrace{(\text{UP}_+,\text{UP}_+)}_{\text{positive saddle}} \\ \vdots & \vdots & \vdots & \vdots \\ \underbrace{(\text{UP}_+,\text{UP}_+)}_{\text{positive saddle}} & \cdots & \underbrace{(\text{DP}_-,\text{DP}_+)}_{\text{CW center}} & \underbrace{(\text{UP}_+,\text{UP}_+)}_{\text{positive saddle}} \end{bmatrix}_{(2n_{11})\times(2m_{21}+1)}. \tag{1.417}$$

(v_{2b2}) The $((2n_{11})^{\text{th}}\text{LS}, (2m_{21}{+}1)^{\text{th}}\text{DP})$*-down-parabola lower-saddle appearing bifurcation is from a* $q_1 \times p_2$ *to* $\overline{q}_1 \times \overline{p}_2$ *networks of positive and negative saddles and CCW and CW centers as*

$$\bigcup_{u_1=1}^{q_1}\bigcup_{v_1=1}^{p_2}\underbrace{(a_{j_1j_2u_1}, a_{j_2j_1v_1})}_{(n_{j_1u_1}, m_{j_2v_1})\text{-XX}} \rightleftharpoons \underbrace{((2n_{11})^{\text{th}}\text{LS}, (2m_{21}{+}1)^{\text{th}}\text{DP})}_{((2n_{11}),(2m_{21}+1))\text{-down-parabola lower-saddle}}$$
$$\rightleftharpoons \bigcup_{u_2=1}^{\overline{q}_1}\bigcup_{v_2=1}^{\overline{p}_2}\underbrace{(a_{j_1j_2u_2}, a_{j_2j_1v_2})}_{(n_{j_1u_2}, m_{j_2v_2})\text{-YY}}. \tag{1.418}$$

(v_{2b3}) The $((2n_{11})^{\text{th}}\text{LS}, (2m_{21}{+}1)^{\text{th}}\text{DP})$*-down-parabola lower-saddle switching bifurcation is for the switching of two* $q_1 \times p_2$ *networks of positive and negative saddles and CCW and CW centers as*

$$\bigcup_{u_1=1}^{q_1}\bigcup_{v_1=1}^{p_2}\underbrace{(a_{j_1j_2u_1}, a_{j_2j_1v_1})}_{(n_{j_1u_1}, m_{j_2v_1})\text{-XX}} \rightleftharpoons \underbrace{((2n_{11})^{\text{th}}\text{LS}, (2m_{21}{+}1)^{\text{th}}\text{DP})}_{((2n_{11}),(2m_{21}+1))\text{-down-parabola lower-saddle}}$$
$$\rightleftharpoons \bigcup_{u_2=1}^{q_1}\bigcup_{v_2=1}^{p_2}\underbrace{(a_{j_1j_2u_2}, a_{j_2j_1v_2})}_{(n_{j_1u_2}, m_{j_2v_2})\text{-YY}}. \tag{1.419}$$

(v_{2c}) For $a_{j_1j_10}(a_{j_2j_11}-a_{j_1j_11})^{m_1}>0$ *and* $a_{j_2j_20}<0$*, there is a* $((2n_{11}), (2m_{21}{+}1))$*-down-parabola upper-saddle equilibrium at* $(x^*_{j_2}, x^*_{j_1}) = (a_{j_1j_21}, a_{j_2j_11})$ *as*

$$(a_{j_1j_21}, a_{j_2j_11}) = \underbrace{((2n_{11})^{\text{th}}\text{US}, (2m_{21}{+}1)^{\text{th}}\text{DP})}_{((2n_{11}),(2m_{21}+1))\text{-down-parabola upper-saddle}}. \tag{1.420}$$

The following three $((2n_{11})^{\text{th}}\text{US}, (2m_{21}{+}1)^{\text{th}}\text{DP})$*-down-parabola upper-saddle appearing and switching bifurcations as follows.*

(v_{2c1})The $((2n_{11})^{\text{th}}\text{US}, (2m_{21}{+}1)^{\text{th}}\text{DP})$*-down-parabola upper-saddle appearing bifurcation is from a* (pF, DP)*-positive down-parabola flow to a* $(2n_{11}) \times (2m_{21}+1)$ *network of positive and negative saddles and CCW and CW centers as*

$$\underbrace{(\dot{x}_{j_1}, a_{j_2j_11})}_{(\text{pF,DP})\text{-flow}} \rightleftharpoons \underbrace{((2n_{11})^{\text{th}}\text{US}, (2m_{21}{+}1)^{\text{th}}\text{DP})}_{((2n_{11}),(2m_{21}+1))\text{-down-parabola upper-saddle}}$$
$$\rightleftharpoons \bigcup_{u=1}^{2n_{11}}\bigcup_{v=1}^{2m_{21}+1}\underbrace{(a_{j_1j_2u}, a_{j_2j_1v})}_{\text{XX}}, \tag{1.421}$$

where, without 1-dimensional flows,

$$\bigcup_{u=1}^{2n_{11}}\bigcup_{v=1}^{2m_{21}+1}\underbrace{(a_{j_1j_2u},a_{j_2j_1v})}_{\text{XX}} = \begin{bmatrix} \underbrace{(\text{DP}_-,\text{DP}_+)}_{\text{CW center}} & \cdots & \underbrace{(\text{UP}_+,\text{UP}_+)}_{\text{positive saddle}} & \underbrace{(\text{DP}_-,\text{DP}_+)}_{\text{CW center}} \\ \underbrace{(\text{UP}_-,\text{UP}_-)}_{\text{negative saddle}} & \cdots & \underbrace{(\text{DP}_+,\text{DP}_-)}_{\text{CCW center}} & \underbrace{(\text{UP}_-,\text{UP}_-)}_{\text{negative saddle}} \\ \vdots & \vdots & \vdots & \vdots \\ \underbrace{(\text{UP}_-,\text{UP}_-)}_{\text{negative saddle}} & \cdots & \underbrace{(\text{DP}_+,\text{DP}_-)}_{\text{CCW center}} & \underbrace{(\text{UP}_-,\text{UP}_-)}_{\text{negative saddle}} \end{bmatrix}_{(2n_{11})\times(2m_{21}+1)}. \tag{1.422}$$

(v_{2c2}) The $((2n_{11})^{\text{th}}\text{US}, (2m_{21}+1)^{\text{th}}\text{DP})$*-down-parabola upper-saddle appearing bifurcation is from a* $q_1 \times p_2$ *to* $\overline{q}_1 \times \overline{p}_2$ *networks of positive and negative saddles and CCW and CW centers as*

$$\begin{aligned}\bigcup_{u_1=1}^{q_1}\bigcup_{v_1=1}^{p_2}\underbrace{(a_{j_1j_2u_1},a_{j_2j_1v_1})}_{(n_{j_1u_1},m_{j_2v_1})\text{-XX}} &\rightleftharpoons \underbrace{((2n_{11})^{\text{th}}\text{US}, (2m_{21}+1)^{\text{th}}\text{DP})}_{((2n_{11}),(2m_{21}+1))\text{-down-parabola upper-saddle}} \\ &\rightleftharpoons \bigcup_{u_2=1}^{\overline{q}_1}\bigcup_{v_2=1}^{\overline{p}_2}\underbrace{(a_{j_1j_2u_2},a_{j_2j_1v_2})}_{(n_{j_1u_2},m_{j_2v_2})\text{-YY}}.\end{aligned} \tag{1.423}$$

(v_{2c3}) The $((2n_{11})^{\text{th}}\text{US}, (2m_{21}+1)^{\text{th}}\text{DP})$*-down-parabola upper-saddle switching bifurcation is for the switching of two* $q_1 \times p_2$ *networks of positive and negative saddles and CCW and CW centers as*

$$\begin{aligned}\bigcup_{u_1=1}^{q_1}\bigcup_{v_1=1}^{p_2}\underbrace{(a_{j_1j_2u_1},a_{j_2j_1v_1})}_{(n_{j_1u_1},m_{j_2v_1})\text{-XX}} &\rightleftharpoons \underbrace{((2n_{11})^{\text{th}}\text{US}, (2m_{21}+1)^{\text{th}}\text{DP})}_{((2n_{11}),(2m_{21}+1))\text{-down-parabola upper-saddle}} \\ &\rightleftharpoons \bigcup_{u_2=1}^{q_1}\bigcup_{v_2=1}^{p_2}\underbrace{(a_{j_1j_2u_2},a_{j_2j_1v_2})}_{(n_{j_1u_2},m_{j_2v_2})\text{-YY}}.\end{aligned} \tag{1.424}$$

(v_{2d}) For $a_{j_1j_10}(a_{j_2j_11}-a_{j_1j_11})^{m_1}<0$ *and* $a_{j_2j_20}<0$*, there is a* $((2n_{11}), (2m_{21}+1))$*-up-parabola lower-saddle equilibrium at* $(x^*_{j_2}, x^*_{j_1}) = (a_{j_1j_21}, a_{j_2j_11})$ *as*

$$(a_{j_1j_21},a_{j_2j_11}) = \underbrace{((2n_{11})^{\text{th}}\text{LS}, (2m_{21}+1)^{\text{th}}\text{UP})}_{((2n_{11}),(2m_{21}+1))\text{-up-parabola lower-saddle}}. \tag{1.425}$$

The following three $((2n_{11})^{\text{th}}\text{LS}, (2m_{21}+1)^{\text{th}}\text{UP})$*-up-parabola lower-saddle appearing and switching bifurcations as follows.*

(v_{2d1}) The $((2n_{11})^{\text{th}}\text{LS}, (2m_{21}+1)^{\text{th}}\text{UP})$*-up-parabola lower-saddle appearing bifurcation is from an* (nF, UP)*-negative up-parabola flow to a* $(2n_{11})\times(2m_{21}+1)$*-network of positive and negative saddles, CCW and CW centers as*

$$\begin{aligned}\underbrace{(\dot{x}_{j_1}, a_{j_2j_1 1})}_{\text{(nF,UP)-flow}} &\rightleftharpoons \underbrace{((2n_{11})^{\text{th}}\text{LS}, (2m_{21}+1)^{\text{th}}\text{UP})}_{((2n_{11}),(2m_{21}+1))\text{-up-parabola lower-saddle}}\\ &\rightleftharpoons \bigcup_{u=1}^{2n_{11}}\bigcup_{v=1}^{2m_{21}+1}\underbrace{(a_{j_1j_2u}, a_{j_2j_1v})}_{\text{XX}},\end{aligned} \tag{1.426}$$

where, without 1-dimensional flows,

$$\bigcup_{u=1}^{2n_{11}}\bigcup_{v=1}^{2m_{21}+1}\underbrace{(a_{j_1j_2u}, a_{j_2j_1v})}_{\text{XX}}$$
$$= \begin{bmatrix} \underbrace{(\text{UP}_-,\text{UP}_-)}_{\text{negative saddle}} & \cdots & \underbrace{(\text{DP}_+,\text{DP}_-)}_{\text{CCW center}} & \underbrace{(\text{UP}_-,\text{UP}_-)}_{\text{negative saddle}} \\ \underbrace{(\text{DP}_-,\text{DP}_+)}_{\text{CW center}} & \cdots & \underbrace{(\text{UP}_+,\text{UP}_+)}_{\text{positive saddle}} & \underbrace{(\text{DP}_-,\text{DP}_+)}_{\text{CW center}} \\ \vdots & \vdots & \vdots & \vdots \\ \underbrace{(\text{DP}_-,\text{DP}_+)}_{\text{CW center}} & \cdots & \underbrace{(\text{UP}_+,\text{UP}_+)}_{\text{positive saddle}} & \underbrace{(\text{DP}_-,\text{DP}_+)}_{\text{CW center}} \end{bmatrix}_{(2n_{11})\times(2m_{21}+1)}. \tag{1.427}$$

(v_{2d2}) The $((2n_{11})^{\text{th}}\text{LS}, (2m_{21}+1)^{\text{th}}\text{UP})$*-up-parabola lower-saddle appearing bifurcation is from a* $q_1 \times p_2$ *to* $\overline{q}_1 \times \overline{p}_2$ *networks of positive and negative saddles and CCW and CW centers as*

$$\begin{aligned}\bigcup_{u_1=1}^{q_1}\bigcup_{v_1=1}^{p_2}\underbrace{(a_{j_1j_2u_1}, a_{j_2j_1v_1})}_{(n_{j_1u_1},m_{j_2v_1})\text{-XX}} &\rightleftharpoons \underbrace{((2n_{11})^{\text{th}}\text{LS}, (2m_{21}+1)^{\text{th}}\text{UP})}_{((2n_{11}),(2m_{21}+1))\text{-up-parabola lower-saddle}}\\ &\rightleftharpoons \bigcup_{u_2=1}^{\overline{q}_1}\bigcup_{v_2=1}^{\overline{p}_2}\underbrace{(a_{j_1j_2u_2}, a_{j_2j_1v_2})}_{(n_{j_1u_2},m_{j_2v_2})\text{-YY}}.\end{aligned} \tag{1.428}$$

(v_{2d3}) The $((2n_{11})^{\text{th}}\text{LS}, (2m_{21}+1)^{\text{th}}\text{UP})$*-up-parabola lower-saddle switching bifurcation is for the switching of two* $q_1 \times p_2$ *networks of positive and negative saddles and CCW and CW centers as*

$$\bigcup_{u_1=1}^{q_1}\bigcup_{v_1=1}^{p_2}\underbrace{(a_{j_1j_2u_1},a_{j_2j_1v_1})}_{(n_{j_1u_1},m_{j_2v_1})\text{-XX}} \rightleftharpoons \underbrace{((2n_{11})^{\text{th}}\text{LS},(2m_{21}+1)^{\text{th}}\text{UP})}_{((2n_{11}),(2m_{21}+1))\text{-up-parabola lower-saddle}}$$
$$\rightleftharpoons \bigcup_{u_2=1}^{q_1}\bigcup_{v_2=1}^{p_2}\underbrace{(a_{j_1j_2u_2},a_{j_2j_1v_2})}_{(n_{j_1u_2},m_{j_2v_2})\text{-XX}}. \tag{1.429}$$

(v_3) For $n_1 = 2n_{11}+1$ and $m_2 = 2m_{21}$, the equilibrium of $(x^*_{j_2},x^*_{j_1}) = (a_{j_1j_21},a_{j_2j_11})$ has the following bifurcation properties with

$$\begin{gathered}
2n_{11}+1-\sum_{u_1=1}^{p_1} n^{(1)}_{j_1u_1} = 2\sum_{\rho_1=1}^{l_1} n_{\rho_1},\\
2m_{21}-\sum_{s_1=1}^{q_2} m^{(1)}_{j_2v_1} = 2\sum_{\sigma_1=1}^{l_2} m_{\sigma_1};\\
\sum_{u_2=1}^{\bar p_1} n^{(2)}_{j_1u_2} = 2n_{11}+1, \sum_{s_2=1}^{\bar q_2} m^{(2)}_{j_2v_2} = 2m_{21};\\
\sum_{u_i=1}^{p_1} n^{(i)}_{j_1u_i} = 2n_{11}+1, \sum_{s_i=1}^{q_2} m^{(2)}_{j_2v_i} = 2m_{21},\ \text{for } i=1,2;
\end{gathered} \tag{1.430}$$

l_1, l_2 quadratic polynomials are without real roots, and the ρ_1^{th} and σ_1^{th} quadratic polynomials are with powers m_{ρ_1} and n_{σ_1}, respectively.

(v_{3a}) For $a_{j_1j_10}(a_{j_2j_11}-a_{j_1j_11})^{m_1} > 0$ and $a_{j_2j_20} > 0$, there is a $((2n_{11}+1),(2m_{21}))$-up-parabola uppers-saddle equilibrium at $(x^*_{j_2},x^*_{j_1}) = (a_{j_1j_21},a_{j_2j_11})$ as

$$(a_{j_1j_21},a_{j_2j_11}) = \underbrace{((2n_{11}+1)^{\text{th}}\text{UP},(2m_{21})^{\text{th}}\text{US})}_{((2n_{11}+1),(2m_{21}))\text{-up-parabola upper-saddle}}. \tag{1.431}$$

The following three $((2n_{11}+1)^{\text{th}}\text{UP},(2m_{21})^{\text{th}}\text{US})$*-up-parabola upper-saddle appearing and switching bifurcations as follows.*

(v_{3a1}) *The* $((2n_{11}+1)^{\text{th}}\text{UP},(2m_{21})^{\text{th}}\text{US})$*-up-parabola upper-saddle appearing bifurcation is from a* (UP, pF)*-positive up-parabola flow to a* $(2n_{11}+1)\times(2m_{21})$*-network of positive and negative saddles, CCW and CW centers as*

$$\underbrace{(a_{j_1j_21},\dot x_{j_2})}_{(\text{UP,pF})\text{-flow}} \rightleftharpoons \underbrace{((2n_{11}+1)^{\text{th}}\text{UP},(2m_{21})^{\text{th}}\text{US})}_{((2n_{11}+1),(2m_{21}))\text{-up-parabola upper-saddle}}$$
$$\rightleftharpoons \bigcup_{u=1}^{2n_{11}+1}\bigcup_{v=1}^{2m_{21}}\underbrace{(a_{j_1j_2u},a_{j_2j_1v})}_{\text{XX}}, \tag{1.432}$$

where, without 1-dimensional flows,

$$\bigcup_{u=1}^{2n_{11}+1}\bigcup_{v=1}^{2m_{21}}\underbrace{(a_{j_1j_2u},a_{j_2j_1v})}_{\mathrm{XX}} = \left[\begin{array}{cccc} \underbrace{(\mathrm{DP}_-,\mathrm{DP}_+)}_{\text{CW center}} & \cdots & \underbrace{(\mathrm{DP}_-,\mathrm{DP}_+)}_{\text{CW center}} & \underbrace{(\mathrm{UP}_+,\mathrm{UP}_+)}_{\text{positive saddle}} \\ \underbrace{(\mathrm{UP}_-,\mathrm{UP}_-)}_{\text{negative saddle}} & \cdots & \underbrace{(\mathrm{UP}_-,\mathrm{UP}_-)}_{\text{negative saddle}} & \underbrace{(\mathrm{DP}_+,\mathrm{DP}_-)}_{\text{CCW center}} \\ \vdots & \vdots & \vdots & \vdots \\ \underbrace{(\mathrm{DP}_-,\mathrm{DP}_+)}_{\text{CW center}} & \cdots & \underbrace{(\mathrm{DP}_-,\mathrm{DP}_+)}_{\text{CW center}} & \underbrace{(\mathrm{UP}_+,\mathrm{UP}_+)}_{\text{positive saddle}} \end{array}\right]_{(2n_{11}+1)\times(2m_{21})} . \tag{1.433}$$

(v_{3a2}) *The* $((2n_{11}+1)^{\text{th}}\mathrm{UP}, (2m_{21})^{\text{th}}\mathrm{US})$*-up-parabola upper-saddle appearing bifurcation is from a* $q_1 \times p_2$ *to* $\overline{q}_1 \times \overline{p}_2$ *networks of positive and negative saddles and CCW and CW centers as*

$$\begin{aligned}\bigcup_{u_1=1}^{q_1}\bigcup_{v_1=1}^{p_2}\underbrace{(a_{j_1j_2u_1},a_{j_2j_1v_1})}_{(n_{j_1u_1},m_{j_2v_1})\text{-XX}} &\rightleftharpoons \underbrace{((2n_{11}+1)^{\text{th}}\mathrm{UP}, (2m_{21})^{\text{th}}\mathrm{US})}_{((2n_{11}+1),(2m_{21}))\text{-up-parabola upper-saddle}} \\ &\rightleftharpoons \bigcup_{u_2=1}^{\overline{q}_1}\bigcup_{v_2=1}^{\overline{p}_2}\underbrace{(a_{j_1j_2u_2},a_{j_2j_1v_2})}_{(n_{j_1u_2},m_{j_2v_2})\text{-YY}}.\end{aligned} \tag{1.434}$$

(v_{3a3}) *The* $((2n_{11}+1)^{\text{th}}\mathrm{UP}, (2m_{21})^{\text{th}}\mathrm{US})$*-up-parabola upper-saddle switching bifurcation is for the switching of two* $q_1 \times p_2$ *networks of positive and negative saddles and CCW and CW centers as*

$$\begin{aligned}\bigcup_{u_1=1}^{q_1}\bigcup_{v_1=1}^{p_2}\underbrace{(a_{j_1j_2u_1},a_{j_2j_1v_1})}_{(n_{j_1u_1},m_{j_2v_1})\text{-XX}} &\rightleftharpoons \underbrace{((2n_{11}+1)^{\text{th}}\mathrm{UP}, (2m_{21})^{\text{th}}\mathrm{US})}_{((2n_{11}+1),(2m_{21}))\text{-up-parabola upper-saddle}} \\ &\rightleftharpoons \bigcup_{u_2=1}^{q_1}\bigcup_{v_2=1}^{p_2}\underbrace{(a_{j_1j_2u_2},a_{j_2j_1v_2})}_{(n_{j_1u_2},m_{j_2v_2})\text{-XX}}.\end{aligned} \tag{1.435}$$

(v_{3b}) *For* $a_{j_1j_10}(a_{j_2j_11} - a_{j_1j_11})^{m_1} < 0$ *and* $a_{j_2j_20} > 0$*, there is a* $((2n_{11}+1), (2m_{21}))$*-down-parabola upper-saddle equilibrium at* $(x^*_{j_2}, x^*_{j_1}) = (a_{j_1j_21}, a_{j_2j_11})$ *as*

$$(a_{j_1j_21}, a_{j_2j_11}) = \underbrace{((2n_{11}+1)^{\text{th}}\mathrm{DP}, (2m_{21})^{\text{th}}\mathrm{US})}_{((2n_{11}+1),(2m_{21}))\text{-down-parabola upper-saddle}} . \tag{1.436}$$

The following three $((2n_{11}+1)^{\text{th}}\text{DP}, (2m_{21})^{\text{th}}\text{US})$*-down-parabola upper-saddle appearing and switching bifurcations as follows.*

(v_{3b1}) The $((2n_{11}+1)^{\text{th}}\text{DP}, (2m_{21})^{\text{th}}\text{US})$*-down-parabola upper-saddle appearing bifurcation is from a* (DP, pF)*-positive down-parabola flow to a* $(2n_{11}+1)\times(2m_{21})$ *network of positive and negative saddles and CCW and CW centers as*

$$\underbrace{(a_{j_1j_21}, \dot{x}_{j_2})}_{\text{(DP,pF)-flow}} \rightleftharpoons \underbrace{((2n_{11}+1)^{\text{th}}\text{DP}, (2m_{21})^{\text{th}}\text{US})}_{((2n_{11}+1),(2m_{21}))\text{-down-parabola upper-saddle}}$$
$$\rightleftharpoons \bigcup_{u=1}^{2n_{11}+1}\bigcup_{v=1}^{2m_{21}} \underbrace{(a_{j_1j_2u}, a_{j_2j_1v})}_{\text{XX}}, \tag{1.437}$$

where, without 1-dimensional flows,

$$\bigcup_{u=1}^{2n_{11}+1}\bigcup_{v=1}^{2m_{21}} \underbrace{(a_{j_1j_2u}, a_{j_2j_1v})}_{\text{XX}}$$
$$= \begin{bmatrix} \underbrace{(\text{UP}_-,\text{UP}_-)}_{\text{negative saddle}} & \cdots & \underbrace{(\text{UP}_-,\text{UP}_-)}_{\text{negative saddle}} & \underbrace{(\text{DP}_+,\text{DP}_-)}_{\text{CCW center}} \\ \underbrace{(\text{DP}_-,\text{DP}_+)}_{\text{CW center}} & \cdots & \underbrace{(\text{DP}_-,\text{DP}_+)}_{\text{CW center}} & \underbrace{(\text{UP}_+,\text{UP}_+)}_{\text{positive saddle}} \\ \vdots & \vdots & \vdots & \vdots \\ \underbrace{(\text{UP}_-,\text{UP}_-)}_{\text{negative saddle}} & \cdots & \underbrace{(\text{UP}_-,\text{UP}_-)}_{\text{negative saddle}} & \underbrace{(\text{DP}_+,\text{DP}_-)}_{\text{CCW center}} \end{bmatrix}_{(2n_{11}+1)\times(2m_{21})}. \tag{1.438}$$

(v_{3b2}) The $((2n_{11}+1)^{\text{th}}\text{DP}, (2m_{21})^{\text{th}}\text{US})$*-down-parabola upper-saddle appearing bifurcation is from a* $q_1 \times p_2$ *to* $\overline{q}_1 \times \overline{p}_2$ *networks of positive and negative saddles and CCW and CW centers as*

$$\bigcup_{u_1=1}^{q_1}\bigcup_{v_1=1}^{p_2} \underbrace{(a_{j_1j_2u_1}, a_{j_2j_1v_1})}_{(n_{j_1u_1}, m_{j_2v_1})\text{-XX}} \rightleftharpoons \underbrace{((2n_{11}+1)^{\text{th}}\text{DP}, (2m_{21})^{\text{th}}\text{US})}_{((2n_{11}+1),(2m_{21}))\text{-down-parabola upper-saddle}}$$
$$\rightleftharpoons \bigcup_{u_2=1}^{\overline{q}_1}\bigcup_{v_2=1}^{\overline{p}_2} \underbrace{(a_{j_1j_2u_2}, a_{j_2j_1v_2})}_{(n_{j_1u_2}, m_{j_2v_2})\text{-YY}}. \tag{1.439}$$

(v_{3b3}) The $((2n_{11}+1)^{\text{th}}\text{DP}, (2m_{21})^{\text{th}}\text{US})$*-down-parabola upper-saddle switching bifurcation is for the switching of two* $q_1 \times p_2$ *networks of positive and negative saddles and CCW and CW centers as*

$$\bigcup_{u_1=1}^{q_1}\bigcup_{v_1=1}^{p_2}\underbrace{(a_{j_1j_2u_1},a_{j_2j_1v_1})}_{(n_{j_1u_1},m_{j_2v_1})\text{-XX}} \rightleftharpoons \underbrace{((2n_{11}+1)^{\text{th}}\text{DP}, (2m_{21})^{\text{th}}\text{US})}_{((2n_{11}+1),(2m_{21}))\text{-down-parabola upper-saddle}}$$
$$\rightleftharpoons \bigcup_{u_2=1}^{q_1}\bigcup_{v_2=1}^{p_2}\underbrace{(a_{j_1j_2u_2},a_{j_2j_1v_2})}_{(n_{j_1u_2},m_{j_2v_2})\text{-YY}}. \tag{1.440}$$

(v_{3c}) For $a_{j_1j_10}(a_{j_2j_11}-a_{j_1j_11})^{m_1}>0$ and $a_{j_2j_20}<0$, there is a $((2n_{11}+1),(2m_{21}))$-down-parabola lower-saddle equilibrium at $(x^*_{j_2},x^*_{j_1})=(a_{j_1j_21},a_{j_2j_11})$ as

$$(a_{j_1j_21},a_{j_2j_11}) = \underbrace{((2n_{11}+1)^{\text{th}}\text{DP}, (2m_{21})^{\text{th}}\text{LS})}_{((2n_{11}+1),(2m_{21}))\text{-down-parabola lower-saddle}}. \tag{1.441}$$

The following three $((2n_{11}+1)^{\text{th}}\text{DP}, (2m_{21})^{\text{th}}\text{LS})$-down-parabola lower-saddle appearing and switching bifurcations as follows.

(v_{3c1}) The $((2n_{11}+1)^{\text{th}}\text{DP}, (2m_{21})^{\text{th}}\text{LS})$-down-parabola lower-saddle appearing bifurcation is from a (DP, nF)-negative down-parabola flow to a $(2n_{11}+1)\times(2m_{21})$ network of positive and negative saddles and CCW and CW centers as

$$\underbrace{(a_{j_1j_21},\dot{x}_{j_2})}_{\text{(DP,nF)-flow}} \rightleftharpoons \underbrace{((2n_{11}+1)^{\text{th}}\text{DP}, (2m_{21})^{\text{th}}\text{LS})}_{((2n_{11}+1),(2m_{21}))\text{-down-parabola lower-saddle}}$$
$$\rightleftharpoons \bigcup_{u=1}^{2n_{11}+1}\bigcup_{v=1}^{2m_{21}}\underbrace{(a_{j_1j_2u},a_{j_2j_1v})}_{\text{XX}}, \tag{1.442}$$

where, without 1-dimensional flows,

$$\bigcup_{u=1}^{2n_{11}+1}\bigcup_{v=1}^{2m_{21}}\underbrace{(a_{j_1j_2u},a_{j_2j_1v})}_{\text{XX}}$$
$$= \begin{bmatrix} \underbrace{(\text{DP}_-,\text{DP}_+)}_{\text{CW center}} & \cdots & \underbrace{(\text{UP}_+,\text{UP}_+)}_{\text{positive saddle}} & \underbrace{(\text{DP}_-,\text{DP}_+)}_{\text{CW center}} \\ \underbrace{(\text{UP}_-,\text{UP}_-)}_{\text{negative saddle}} & \cdots & \underbrace{(\text{DP}_+,\text{DP}_-)}_{\text{CCW center}} & \underbrace{(\text{UP}_-,\text{UP}_-)}_{\text{negative saddle}} \\ \vdots & \vdots & \vdots & \vdots \\ \underbrace{(\text{DP}_-,\text{DP}_+)}_{\text{CW center}} & \cdots & \underbrace{(\text{UP}_+,\text{UP}_+)}_{\text{positive saddle}} & \underbrace{(\text{DP}_-,\text{DP}_+)}_{\text{CW center}} \end{bmatrix}_{(2n_{11}+1)\times(2m_{21})}. \tag{1.443}$$

(v_{3c2}) The $((2n_{11}+1)^{\text{th}}\text{DP}, (2m_{21})^{\text{th}}\text{LS})$*-down-parabola lower-saddle appearing bifurcation is from a* $q_1 \times p_2$ *to* $\overline{q}_1 \times \overline{p}_2$ *networks of positive and negative saddles and CCW and CW centers as*

$$\begin{aligned}\bigcup_{u_1=1}^{q_1}\bigcup_{v_1=1}^{p_2}\underbrace{(a_{j_1j_2u_1}, a_{j_2j_1v_1})}_{(n_{j_1u_1}, m_{j_2v_1})\text{-XX}} &\rightleftharpoons \underbrace{((2n_{11}+1)^{\text{th}}\text{DP}, (2m_{21})^{\text{th}}\text{LS})}_{((2n_{11}+1),(2m_{21}))\text{-down-parabola lower-saddle}} \\ &\rightleftharpoons \bigcup_{u_2=1}^{\overline{q}_1}\bigcup_{v_2=1}^{\overline{p}_2}\underbrace{(a_{j_1j_2u_2}, a_{j_2j_1v_2})}_{(n_{j_1u_2}, m_{j_2v_2})\text{-YY}}.\end{aligned} \tag{1.444}$$

(v_{3c3}) The $((2n_{11}+1)^{\text{th}}\text{DP}, (2m_{21})^{\text{th}}\text{LS})$*-down-parabola lower-saddle switching bifurcation is for the switching of two* $q_1 \times p_2$ *networks of positive and negative saddles and CCW and CW centers as*

$$\begin{aligned}\bigcup_{u_1=1}^{q_1}\bigcup_{v_1=1}^{p_2}\underbrace{(a_{j_1j_2u_1}, a_{j_2j_1v_1})}_{(n_{j_1u_1}, m_{j_2v_1})\text{-XX}} &\rightleftharpoons \underbrace{((2n_{11}+1)^{\text{th}}\text{DP}, (2m_{21})^{\text{th}}\text{LS})}_{((2n_{11}+1),(2m_{21}))\text{-down-parabola lower-saddle}} \\ &\rightleftharpoons \bigcup_{u_2=1}^{q_1}\bigcup_{v_2=1}^{p_2}\underbrace{(a_{j_1j_2u_2}, a_{j_2j_1v_2})}_{(n_{j_1u_2}, m_{j_2v_2})\text{-YY}}.\end{aligned} \tag{1.445}$$

(v_{3d}) For $a_{j_1j_10}(a_{j_2j_11} - a_{j_1j_11})^{m_1} < 0$ *and* $a_{j_2j_20} < 0$*, there is a* $((2n_{11}+1), (2m_{21}))$*-up-parabola lower-saddle equilibrium at* $(x_{j_2}^*, x_{j_1}^*) = (a_{j_1j_21}, a_{j_2j_11})$ *as*

$$(a_{j_1j_21}, a_{j_2j_11}) = \underbrace{((2n_{11}+1)^{\text{th}}\text{UP}, (2m_{21})^{\text{th}}\text{LS})}_{((2n_{11}+1),(2m_{21}))\text{-up-parabola lower-saddle}}. \tag{1.446}$$

The following three $((2n_{11}+1)^{\text{th}}\text{UP}, (2m_{21})^{\text{th}}\text{LS})$*-up-parabola lower-saddle appearing and switching bifurcations as follows.*

(v_{3d1}) The $((2n_{11}+1)^{\text{th}}\text{UP}, (2m_{21})^{\text{th}}\text{LS})$*-up-parabola lower-saddle appearing bifurcation is from a* (UP, nF)*-negative up-parabola flow to a* $(2n_{11}+1) \times (2m_{21})$*-network of positive and negative saddles, CCW and CW centers as*

$$\begin{aligned}\underbrace{(a_{j_1j_21}, \dot{x}_{j_1})}_{(\text{UP,nF})\text{-flow}} &\rightleftharpoons \underbrace{((2n_{11}+1)^{\text{th}}\text{UP}, (2m_{21})^{\text{th}}\text{LS})}_{((2n_{11}+1),(2m_{21}))\text{-up-parabola lower-saddle}} \\ &\rightleftharpoons \bigcup_{u=1}^{2n_{11}+1}\bigcup_{v=1}^{2m_{21}}\underbrace{(a_{j_1j_2u}, a_{j_2j_1v})}_{\text{XX}},\end{aligned} \tag{1.447}$$

where, without 1-dimensional flows,

$$
\bigcup_{u=1}^{2n_{11}+1}\bigcup_{v=1}^{2m_{21}}\underbrace{(a_{j_1j_2u},a_{j_2j_1v})}_{\text{XX}}
= \begin{bmatrix}
\underbrace{(\text{DP}_+,\text{DP}_-)}_{\text{CCW center}} & \cdots & \underbrace{(\text{DP}_+,\text{DP}_-)}_{\text{CCW center}} & \underbrace{(\text{UP}_-,\text{UP}_-)}_{\text{negative saddle}} \\
\underbrace{(\text{UP}_+,\text{UP}_+)}_{\text{positive saddle}} & \cdots & \underbrace{(\text{UP}_+,\text{UP}_+)}_{\text{positive saddle}} & \underbrace{(\text{DP}_-,\text{DP}_+)}_{\text{CW center}} \\
\vdots & \vdots & \vdots & \vdots \\
\underbrace{(\text{UP}_+,\text{UP}_+)}_{\text{positive saddle}} & \cdots & \underbrace{(\text{UP}_+,\text{UP}_+)}_{\text{positive saddle}} & \underbrace{(\text{DP}_-,\text{DP}_+)}_{\text{CW center}}
\end{bmatrix}_{(2n_{11}+1)\times(2m_{21})}. \tag{1.448}
$$

(v_{3d2}) The $((2n_{11}+1)^{\text{th}}\text{UP}, (2m_{21})^{\text{th}}\text{LS})$*-up-parabola lower-saddle appearing bifurcation is from a* $q_1 \times p_2$ *to* $\overline{q}_1 \times \overline{p}_2$ *networks of positive and negative saddles and CCW and CW centers as*

$$
\begin{aligned}
\bigcup_{u_1=1}^{q_1}\bigcup_{v_1=1}^{p_2}\underbrace{(a_{j_1j_2u_1},a_{j_2j_1v_1})}_{(n_{j_1u_1},m_{j_2v_1})\text{-XX}} &\rightleftharpoons \underbrace{((2n_{11}+1)^{\text{th}}\text{UP}, (2m_{21})^{\text{th}}\text{LS})}_{((2n_{11}+1),(2m_{21}))\text{-up-parabola lower-saddle}} \\
&\rightleftharpoons \bigcup_{u_2=1}^{\overline{q}_1}\bigcup_{v_2=1}^{\overline{p}_2}\underbrace{(a_{j_1j_2u_2},a_{j_2j_1v_2})}_{(n_{j_1u_2},m_{j_2v_2})\text{-YY}}.
\end{aligned} \tag{1.449}
$$

(v_{3d3}) The $((2n_{11}+1)^{\text{th}}\text{UP}, (2m_{21})^{\text{th}}\text{LS})$*-up-parabola lower-saddle switching bifurcation is for the switching of two* $q_1 \times p_2$ *networks of positive and negative saddles and CCW and CW centers as*

$$
\begin{aligned}
\bigcup_{u_1=1}^{q_1}\bigcup_{v_1=1}^{p_2}\underbrace{(a_{j_1j_2u_1},a_{j_2j_1v_1})}_{(n_{j_1u_1},m_{j_2v_1})\text{-XX}} &\rightleftharpoons \underbrace{((2n_{11}+1)^{\text{th}}\text{UP}, (2m_{21})^{\text{th}}\text{LS})}_{((2n_{11}+1),(2m_{21}))\text{-up-parabola lower-saddle}} \\
&\rightleftharpoons \bigcup_{u_2=1}^{q_1}\bigcup_{v_2=1}^{p_2}\underbrace{(a_{j_1j_2u_2},a_{j_2j_1v_2})}_{(n_{j_1u_2},m_{j_2v_2})\text{-XX}}.
\end{aligned} \tag{1.450}
$$

(v_4) For $n_1 = 2n_{11}$ *and* $m_2 = 2m_{21}$*, the equilibrium of* $(x^*_{j_2}, x^*_{j_1}) = (a_{j_1j_21}, a_{j_2j_11})$ *for has the following bifurcation properties with*

$$
2n_{11} - \sum_{u_1=1}^{p_1} n^{(1)}_{j_1u_1} = 2\sum_{\rho_1=1}^{l_1} n_{\rho_1},\ 2m_{21} - \sum_{s_1=1}^{q_2} m^{(1)}_{j_2v_1} = 2\sum_{\sigma_1=1}^{l_2} m_{\sigma_1};
$$

$$\sum_{u_2=1}^{\bar{p}_1} n_{j_1 u_2}^{(2)} = 2n_{11}, \sum_{s_2=1}^{\bar{q}_2} m_{j_2 v_2}^{(2)} = 2m_{21};$$
$$\sum_{u_i=1}^{p_1} n_{j_1 u_i}^{(i)} = 2n_{11}, \sum_{s_i=1}^{q_2} m_{j_2 v_i}^{(2)} = 2m_{21}, \text{ for } i = 1, 2; \tag{1.451}$$

l_1, l_2-quadratic polynomials are without real roots, and the ρ_1^{th} and σ_1^{th}-quadratic polynomials are with powers m_{ρ_1} and n_{σ_1}, respectively.

(v_{4a}) For $a_{j_1j_1 0}(a_{j_2j_1 1} - a_{j_1j_1 1})^{m_1} > 0$ and $a_{j_2j_2 0} > 0$, there is a $((2n_{11}), (2m_{21}))$-double-inflection saddle equilibrium at $(x_{j_2}^*, x_{j_1}^*) = (a_{j_1j_2 1}, a_{j_2j_1 1})$ as

$$(a_{j_1j_2 1}, a_{j_2j_1 1}) = \underbrace{((2n_{11})^{\text{th}}\text{II}_+, (2m_{21})^{\text{th}}\text{II}_+)}_{((2n_{11}),(2m_{21}))\text{-double inflection-saddle}}. \tag{1.452}$$

The following three $((2n_{11})^{\text{th}}\text{II}_+, (2m_{21})^{\text{th}}\text{II}_+)$-double inflection-saddle appearing and switching bifurcations as follows.

(v_{4a1}) The $((2n_{11})^{\text{th}}\text{II}_+, (2m_{21})^{\text{th}}\text{II}_+)$-double inflection-saddle appearing bifurcation is from a (pF, pF)*-flow to a $(2n_{11}) \times (2m_{21})$-network of positive and negative saddles, CCW and CW centers as*

$$\underbrace{(\dot{x}_{j_1}, \dot{x}_{j_2})}_{(\text{pF,pF})\text{-flow}} \rightleftharpoons \underbrace{((2n_{11})^{\text{th}}\text{II}_+, (2m_{21})^{\text{th}}\text{II}_+)}_{((2n_{11}),(2m_{21}))\text{-double inflection-saddle}} \rightleftharpoons \bigcup_{u=1}^{2n_{11}} \bigcup_{v=1}^{2m_{21}} \underbrace{(a_{j_1j_2 u}, a_{j_2j_1 v})}_{\text{XX}}, \tag{1.453}$$

where, without 1-dimensional flows,

$$\bigcup_{u=1}^{2n_{11}} \bigcup_{v=1}^{2m_{21}} \underbrace{(a_{j_1j_2 u}, a_{j_2j_1 v})}_{\text{XX}}$$
$$= \begin{bmatrix} \underbrace{(\text{DP}_-,\text{DP}_+)}_{\text{CW center}} & \cdots & \underbrace{(\text{DP}_-,\text{DP}_+)}_{\text{CW center}} & \underbrace{(\text{UP}_+,\text{UP}_+)}_{\text{positive saddle}} \\ \underbrace{(\text{UP}_-,\text{UP}_-)}_{\text{negative saddle}} & \cdots & \underbrace{(\text{UP}_-,\text{UP}_-)}_{\text{negative saddle}} & \underbrace{(\text{DP}_+,\text{DP}_-)}_{\text{CCW center}} \\ \vdots & \vdots & \vdots & \vdots \\ \underbrace{(\text{UP}_-,\text{UP}_-)}_{\text{negative saddle}} & \cdots & \underbrace{(\text{UP}_-,\text{UP}_-)}_{\text{negative saddle}} & \underbrace{(\text{DP}_+,\text{DP}_-)}_{\text{CCW center}} \end{bmatrix}_{(2n_{11})\times(2m_{21})}. \tag{1.454}$$

(v_{4a2}) The $((2n_{11})^{\text{th}}\text{II}_+, (2m_{21})^{\text{th}}\text{II}_+)$-double inflection-saddle appearing bifurcation is from a $q_1 \times p_2$ to $\overline{q}_1 \times \overline{p}_2$ networks of positive and negative saddles and CCW and CW centers as

$$\bigcup_{u_1=1}^{q_1}\bigcup_{v_1=1}^{p_2}\underbrace{(a_{j_1j_2u_1}, a_{j_2j_1v_1})}_{(n_{j_1u_1},m_{j_2v_1})\text{-XX}} \rightleftharpoons \underbrace{((2n_{11})^{\text{th}}\text{II}_+, (2m_{21})^{\text{th}}\text{II}_+)}_{((2n_{11}),(2m_{21}))\text{-double inflection-saddle}} \rightleftharpoons \bigcup_{u_2=1}^{\overline{q}_1}\bigcup_{v_2=1}^{\overline{p}_2}\underbrace{(a_{j_1j_2u_2}, a_{j_2j_1v_2})}_{(n_{j_1u_2},m_{j_2v_2})\text{-YY}}. \tag{1.455}$$

(v_{4a3}) The $((2n_{11})^{\text{th}}\text{II}_+, (2m_{21})^{\text{th}}\text{II}_+)$-double inflection-saddle switching bifurcation is for the switching of two $q_1 \times p_2$ networks of positive and negative saddles and CCW and CW centers as

$$\bigcup_{u_1=1}^{q_1}\bigcup_{v_1=1}^{p_2}\underbrace{(a_{j_1j_2u_1}, a_{j_2j_1v_1})}_{(n_{j_1u_1},m_{j_2v_1})\text{-XX}} \rightleftharpoons \underbrace{((2n_{11})^{\text{th}}\text{II}_+, (2m_{21})^{\text{th}}\text{II}_+)}_{((2n_{11}),(2m_{21}))\text{-double inflection-saddle}} \rightleftharpoons \bigcup_{u_2=1}^{q_1}\bigcup_{v_2=1}^{p_2}\underbrace{(a_{j_1j_2u_2}, a_{j_2j_1v_2})}_{(n_{j_1u_2},m_{j_2v_2})\text{-XX}}. \tag{1.456}$$

(v_{4b}) For $a_{j_1j_1 0}(a_{j_2j_1 1} - a_{j_1j_1 1})^{m_1} < 0$ and $a_{j_2j_2 0} > 0$, there is a $((2n_{11}), (2m_{21}))$-down-double inflection-saddle equilibrium at $(x^*_{j_2}, x^*_{j_1}) = (a_{j_1j_2 1}, a_{j_2j_1 1})$ as

$$(a_{j_1j_2 1}, a_{j_2j_1 1}) = \underbrace{((2n_{11})^{\text{th}}\text{DI}_+, (2m_{21})^{\text{th}}\text{DI}_-)}_{((2n_{11}),(2m_{21}))\text{-double inflection-saddle}}. \tag{1.457}$$

The following three $((2n_{11})^{\text{th}}\text{DI}_+, (2m_{21})^{\text{th}}\text{DI}_-)$-double inflection-saddle appearing and switching bifurcations as follows.

(v_{4b1}) *The $((2n_{11})^{\text{th}}\text{DI}_+, (2m_{21})^{\text{th}}\text{DI}_-)$-double inflection-saddle appearing bifurcation is from an* (nF, pF)-flow *to a $(2n_{11}+1) \times (2m_{21})$ network of positive and negative saddles and CCW and CW centers as*

$$\underbrace{(\dot{x}_{j_1}, \dot{x}_{j_2})}_{(\text{nF,pF})\text{-flow}} \rightleftharpoons \underbrace{((2n_{11})^{\text{th}}\text{DI}_+, (2m_{21})^{\text{th}}\text{DI}_-)}_{((2n_{11}),(2m_{21}))\text{-double inflection-saddle}} \rightleftharpoons \bigcup_{u=1}^{2n_{11}}\bigcup_{v=1}^{2m_{21}}\underbrace{(a_{j_1j_2u}, a_{j_2j_1v})}_{\text{XX}}, \tag{1.458}$$

where, without 1-dimensional flows,

$$\bigcup_{u=1}^{2n_{11}}\bigcup_{v=1}^{2m_{21}}\underbrace{(a_{j_1j_2u},a_{j_2j_1v})}_{\text{XX}}$$
$$=\begin{bmatrix}\underbrace{(\text{UP}_-,\text{UP}_-)}_{\text{negative saddle}} & \cdots & \underbrace{(\text{UP}_-,\text{UP}_-)}_{\text{negative saddle}} & \underbrace{(\text{DP}_+,\text{DP}_-)}_{\text{CCW center}}\\ \underbrace{(\text{DP}_-,\text{DP}_+)}_{\text{CW center}} & \cdots & \underbrace{(\text{DP}_-,\text{DP}_+)}_{\text{CW center}} & \underbrace{(\text{UP}_+,\text{UP}_+)}_{\text{positive saddle}}\\ \vdots & \vdots & \vdots & \vdots\\ \underbrace{(\text{DP}_-,\text{DP}_+)}_{\text{CW center}} & \cdots & \underbrace{(\text{DP}_-,\text{DP}_+)}_{\text{CW center}} & \underbrace{(\text{UP}_+,\text{UP}_+)}_{\text{positive saddle}}\end{bmatrix}_{(2n_{11})\times(2m_{21})}. \tag{1.459}$$

(v_{4b2}) The $((2n_{11})^{\text{th}}\text{DI}_+,(2m_{21})^{\text{th}}\text{DI}_-)$-double inflection-saddle appearing bifurcation is from a $q_1\times p_2$ to $\overline{q}_1\times\overline{p}_2$ networks of positive and negative saddles and CCW and CW centers as

$$\begin{aligned}\bigcup_{u_1=1}^{q_1}\bigcup_{v_1=1}^{p_2}\underbrace{(a_{j_1j_2u_1},a_{j_2j_1v_1})}_{(n_{j_1u_1},m_{j_2v_1})\text{-XX}} &\rightleftharpoons \underbrace{((2n_{11})^{\text{th}}\text{DI}_+,(2m_{21})^{\text{th}}\text{DI}_-)}_{((2n_{11}),(2m_{21}))\text{-double inflection-saddle}}\\ &\rightleftharpoons \bigcup_{u_2=1}^{\overline{q}_1}\bigcup_{v_2=1}^{\overline{p}_2}\underbrace{(a_{j_1j_2u_2},a_{j_2j_1v_2})}_{(n_{j_1u_2},m_{j_2v_2})\text{-YY}}.\end{aligned} \tag{1.460}$$

(v_{4b3}) The $((2n_{11})^{\text{th}}\text{DI}_+,(2m_{21})^{\text{th}}\text{DI}_-)$-double inflection-saddle switching bifurcation is for the switching of two $q_1\times p_2$ networks of positive and negative saddles and CCW and CW centers as

$$\begin{aligned}\bigcup_{u_1=1}^{q_1}\bigcup_{v_1=1}^{p_2}\underbrace{(a_{j_1j_2u_1},a_{j_2j_1v_1})}_{(n_{j_1u_1},m_{j_2v_1})\text{-XX}} &\rightleftharpoons \underbrace{((2n_{11})^{\text{th}}\text{DI}_+,(2m_{21})^{\text{th}}\text{DI}_-)}_{((2n_{11}),(2m_{21}))\text{-double inflection-saddle}}\\ &\rightleftharpoons \bigcup_{u_2=1}^{q_1}\bigcup_{v_2=1}^{p_2}\underbrace{(a_{j_1j_2u_2},a_{j_2j_1v_2})}_{(n_{j_1u_2},m_{j_2v_2})\text{-YY}}.\end{aligned} \tag{1.461}$$

(v_{4c}) For $a_{j_1j_10}(a_{j_2j_11}-a_{j_1j_11})^{m_1}>0$ and $a_{j_2j_20}<0$, there is a $((2n_{11}),(2m_{21}))$-double inflection saddle equilibrium at $(x^*_{j_2},x^*_{j_1})=(a_{j_1j_21},a_{j_2j_11})$ as

$$(a_{j_1j_21},a_{j_2j_11})=\underbrace{((2n_{11})^{\text{th}}\text{DI}_-,(2m_{21})^{\text{th}}\text{DI}_+)}_{((2n_{11}),(2m_{21}))\text{-double inflection-saddle}}. \tag{1.462}$$

The following three $((2n_{11})^{\text{th}}\text{DI}_{-},(2m_{21})^{\text{th}}\text{DI}_{+})$*-double inflection-saddle appearing and switching bifurcations as follows.*

(v_{4c1})The $((2n_{11})^{\text{th}}\text{DI}_{-},(2m_{21})^{\text{th}}\text{DI}_{+})$*-double inflection-saddle appearing bifurcation is from a* (pF, nF)*-flow to a* $(2n_{11})\times(2m_{21})$ *network of positive and negative saddles and CCW and CW centers as*

$$\underbrace{(\dot{x}_{j_1},\dot{x}_{j_2})}_{\text{(pF,nF)-flow}} \rightleftharpoons \underbrace{((2n_{11})^{\text{th}}\text{DI}_{-},(2m_{21})^{\text{th}}\text{DI}_{+})}_{((2n_{11}),(2m_{21}))\text{-double inflection-saddle}} \rightleftharpoons \bigcup_{u=1}^{2n_{11}}\bigcup_{v=1}^{2m_{21}}\underbrace{(a_{j_1j_2u},a_{j_2j_1v})}_{\text{XX}}, \tag{1.463}$$

where, without 1-dimensional flows,

$$\bigcup_{u=1}^{2n_{11}}\bigcup_{v=1}^{2m_{21}}\underbrace{(a_{j_1j_2u},a_{j_2j_1v})}_{\text{XX}} = \begin{bmatrix} \underbrace{(\text{UP}_{+},\text{UP}_{+})}_{\text{positive saddle}} & \cdots & \underbrace{(\text{UP}_{+},\text{UP}_{+})}_{\text{positive saddle}} & \underbrace{(\text{DP}_{-},\text{DP}_{+})}_{\text{CW center}} \\ \underbrace{(\text{DP}_{+},\text{DP}_{-})}_{\text{CCW center}} & \cdots & \underbrace{(\text{DP}_{+},\text{DP}_{-})}_{\text{CCW center}} & \underbrace{(\text{UP}_{-},\text{UP}_{-})}_{\text{negative saddle}} \\ \vdots & \vdots & \vdots & \vdots \\ \underbrace{(\text{DP}_{+},\text{DP}_{-})}_{\text{CCW center}} & \cdots & \underbrace{(\text{DP}_{+},\text{DP}_{-})}_{\text{CCW center}} & \underbrace{(\text{UP}_{-},\text{UP}_{-})}_{\text{negative saddle}} \end{bmatrix}_{(2n_{11})\times(2m_{21})}. \tag{1.464}$$

(v_{4c2}) The $((2n_{11})^{\text{th}}\text{DI}_{-},(2m_{21})^{\text{th}}\text{DI}_{+})$*-double inflection-saddle appearing bifurcation is from a* $q_1\times p_2$ *to* $\overline{q}_1\times\overline{p}_2$ *networks of positive and negative saddles and CCW and CW centers as*

$$\bigcup_{u_1=1}^{q_1}\bigcup_{v_1=1}^{p_2}\underbrace{(a_{j_1j_2u_1},a_{j_2j_1v_1})}_{(n_{j_1u_1},m_{j_2v_1})\text{-XX}} \rightleftharpoons \underbrace{((2n_{11})^{\text{th}}\text{DI}_{-},(2m_{21})^{\text{th}}\text{DI}_{+})}_{((2n_{11}),(2m_{21}))\text{-double inflection-saddle}} \rightleftharpoons \bigcup_{u_2=1}^{\overline{q}_1}\bigcup_{v_2=1}^{\overline{p}_2}\underbrace{(a_{j_1j_2u_2},a_{j_2j_1v_2})}_{(n_{j_1u_2},m_{j_2v_2})\text{-YY}}. \tag{1.465}$$

(v_{4c3}) The $((2n_{11})^{\text{th}}\text{DI}_{-},(2m_{21})^{\text{th}}\text{DI}_{+})$*-double inflection-saddle switching bifurcation is for the switching of two* $q_1\times p_2$ *networks of positive and negative saddles and CCW and CW centers as*

$$\bigcup_{u_1=1}^{q_1}\bigcup_{v_1=1}^{p_2}\underbrace{(a_{j_1j_2u_1},a_{j_2j_1v_1})}_{(n_{j_1u_1},m_{j_2v_1})\text{-XX}} \rightleftharpoons \underbrace{((2n_{11})^{\text{th}}\text{DI}_-,(2m_{21})^{\text{th}}\text{DI}_+)}_{((2n_{11}),(2m_{21}))\text{-double inflection-saddle}}$$
$$\rightleftharpoons \bigcup_{u_2=1}^{q_1}\bigcup_{v_2=1}^{p_2}\underbrace{(a_{j_1j_2u_2},a_{j_2j_1v_2})}_{(n_{j_1u_2},m_{j_2v_2})\text{-YY}}. \tag{1.466}$$

(v_{4d}) For $a_{j_1j_10}(a_{j_2j_11}-a_{j_1j_11})^{m_1}<0$ and $a_{j_2j_20}<0$, there is a $((2n_{11}),(2m_{21}))$-double inflection-saddle equilibrium at $(x^*_{j_2},x^*_{j_1})=(a_{j_1j_21},a_{j_2j_11})$ as

$$(a_{j_1j_21},a_{j_2j_11})=\underbrace{((2n_{11})^{\text{th}}\text{II}_-,(2m_{21})^{\text{th}}\text{II}_-)}_{((2n_{11}),(2m_{21}))\text{-double inflection-saddle}}. \tag{1.467}$$

The following three $((2n_{11})^{\text{th}}\text{II}_-,(2m_{21})^{\text{th}}\text{II}_-)$-double inflection-saddle appearing and switching bifurcations as follows.

(v_{4d1}) The $((2n_{11})^{\text{th}}\text{II}_-,(2m_{21})^{\text{th}}\text{II}_-)$-double inflection-saddle appearing bifurcation is from an (nF, nF)-flow to a $(2n_{11})\times(2m_{21})$-network of positive and negative saddles, CCW and CW centers as

$$\underbrace{(\dot{x}_{j_1},\dot{x}_{j_2})}_{\text{(nF,nF)-flow}} \rightleftharpoons \underbrace{((2n_{11})^{\text{th}}\text{II}_-,(2m_{21})^{\text{th}}\text{II}_-)}_{((2n_{11}),(2m_{21}))\text{-double inflection-saddle}} \rightleftharpoons \bigcup_{u=1}^{2n_{11}}\bigcup_{v=1}^{2m_{21}}\underbrace{(a_{j_1j_2u},a_{j_2j_1v})}_{\text{XX}}, \tag{1.468}$$

where, without 1-dimensional flows,

$$\bigcup_{u=1}^{2n_{11}}\bigcup_{v=1}^{2m_{21}}\underbrace{(a_{j_1j_2u},a_{j_2j_1v})}_{\text{XX}}$$
$$=\begin{bmatrix}\underbrace{(\text{DP}_+,\text{DP}_-)}_{\text{CCW center}} & \cdots & \underbrace{(\text{DP}_+,\text{DP}_-)}_{\text{CCW center}} & \underbrace{(\text{UP}_-,\text{UP}_-)}_{\text{negative saddle}}\\ \underbrace{(\text{UP}_+,\text{UP}_+)}_{\text{positive saddle}} & \cdots & \underbrace{(\text{UP}_+,\text{UP}_+)}_{\text{positive saddle}} & \underbrace{(\text{DP}_-,\text{DP}_+)}_{\text{CW center}}\\ \vdots & \vdots & \vdots & \vdots\\ \underbrace{(\text{UP}_+,\text{UP}_+)}_{\text{positive saddle}} & \cdots & \underbrace{(\text{UP}_+,\text{UP}_+)}_{\text{positive saddle}} & \underbrace{(\text{DP}_-,\text{DP}_+)}_{\text{CW center}}\end{bmatrix}_{(2n_{11})\times(2m_{21})}. \tag{1.469}$$

(v_{4d2}) The $((2n_{11})^{\text{th}}\text{II}_-,(2m_{21})^{\text{th}}\text{II}_-)$-double inflection-saddle appearing bifurcation is from a $q_1\times p_2$ to $\overline{q}_1\times\overline{p}_2$ networks of positive and negative saddles and CCW and CW centers as

$$\bigcup_{u_1=1}^{q_1}\bigcup_{v_1=1}^{p_2}\underbrace{(a_{j_1j_2u_1},a_{j_2j_1v_1})}_{(n_{j_1u_1},m_{j_2v_1})\text{-XX}} \rightleftharpoons \underbrace{((2n_{11})^{\text{th}}\text{II}_-,(2m_{21})^{\text{th}}\text{II}_-)}_{((2n_{11}),(2m_{21}))\text{-double inflection-saddle}}$$
$$\rightleftharpoons \bigcup_{u_2=1}^{\bar{q}_1}\bigcup_{v_2=1}^{\bar{p}_2}\underbrace{(a_{j_1j_2u_2},a_{j_2j_1v_2})}_{(n_{j_1u_2},m_{j_2v_2})\text{-YY}}. \quad (1.470)$$

(v_{4d3}) The $((2n_{11})^{\text{th}}\text{II}_-,(2m_{21})^{\text{th}}\text{II}_-)$-double inflection-saddle switching bifurcation is for the switching of two $q_1\times p_2$ networks of positive and negative saddles and CCW and CW centers as

$$\bigcup_{u_1=1}^{q_1}\bigcup_{v_1=1}^{p_2}\underbrace{(a_{j_1j_2u_1},a_{j_2j_1v_1})}_{(n_{j_1u_1},m_{j_2v_1})\text{-XX}} \rightleftharpoons \underbrace{((2n_{11})^{\text{th}}\text{II}_-,(2m_{21})^{\text{th}}\text{II}_-)}_{((2n_{11}),(2m_{21}))\text{-double inflection-saddle}}$$
$$\rightleftharpoons \bigcup_{u_2=1}^{q_1}\bigcup_{v_2=1}^{p_2}\underbrace{(a_{j_1j_2u_2},a_{j_2j_1v_2})}_{(n_{j_1u_2},m_{j_2v_2})\text{-XX}}. \quad (1.471)$$

(vi) For $a_{j_2j_21}=a_{j_1j_21}$, there are 64 types of singular infinite-equilibriums at $x^*_{j_2}=a_{j_2j_21}=a_{j_1j_21}$. For $m_2\ge m_1$, 32 types of singular infinite-equilibriums are

$$(a_{j_1j_21},a_{j_2j_11})=\underbrace{([m_1,n_1]^{\text{th}}\text{YY},[m_2-m_1]^{\text{th}}\text{XX})}_{([m_1,n_1],[m_2-m_1])\text{-YYXX}}, \quad (1.472)$$

where

$$\begin{aligned}
&\text{YY}\in\{\text{UP}_-{:}\text{UP}_+,\text{DP}_-{:}\text{DP}_+,\text{DP}_+{:}\text{DP}_-,\text{UP}_+{:}\text{UP}_-\},\\
&\text{XX}\in\{{}_{\text{DI:II}}\text{US},{}_{\text{DI:II}}\text{LS},{}_{\text{II:DI}}\text{US},{}_{\text{II:DI}}\text{LS}\};\\
&\text{YY}\in\{\text{DP}_-{:}\text{UP}_+,\text{UP}_-{:}\text{DP}_+,\text{UP}_+{:}\text{DP}_-,\text{DP}_+{:}\text{UP}_-\},\\
&\text{XX}\in\{{}_{\text{DP}}\text{LS}{:}{}_{\text{UP}}\text{US},{}_{\text{DP}}\text{US}{:}{}_{\text{UP}}\text{LS},{}_{\text{UP}}\text{US}{:}{}_{\text{DP}}\text{LS},{}_{\text{UP}}\text{LS}{:}{}_{\text{DP}}\text{US}\};\\
&\text{YY}\in\{\text{DP}_+{:}\text{UP}_+,\text{UP}_+{:}\text{DP}_+,\text{UP}_-{:}\text{DP}_-,\text{DP}_-{:}\text{UP}_-\},\\
&\text{XX}\in\{{}_{\text{DP}}\text{SI}{:}{}_{\text{UP}}\text{SO},{}_{\text{DP}}\text{SO}{:}{}_{\text{UP}}\text{SI},{}_{\text{UP}}\text{SI}{:}{}_{\text{DP}}\text{SO},{}_{\text{UP}}\text{SO}{:}{}_{\text{DP}}\text{SI}\};\\
&\text{YY}\in\{\text{UP}_+{:}\text{UP}_+,\text{DP}_+{:}\text{DP}_+,\text{DP}_-{:}\text{DP}_-,\text{UP}_-{:}\text{UP}_-\},\\
&\text{XX}\in\{{}_{\text{DI:II}}\text{US},{}_{\text{DI:II}}\text{LS},{}_{\text{II:DI}}\text{US},{}_{\text{II:DI}}\text{LS}\};\\
&\text{YY}\in\{\text{II}_-{:}\text{II}_+,\text{DI}_-{:}\text{DI}_+,\text{DI}_+{:}\text{DI}_-,\text{II}_+{:}\text{II}_-\},\\
&\text{XX}\in\{{}_{\text{II:II}}\text{SO},{}_{\text{DI:DI}}\text{SI}{}_{\text{DI:DI}}\text{SO},{}_{\text{II:II}}\text{SI}\};\\
&\text{YY}\in\{\text{DI}_-{:}\text{II}_+,\text{II}_-{:}\text{DI}_+,\text{II}_+{:}\text{DI}_-,\text{DI}_+{:}\text{II}_-\},\\
&\text{XX}\in\{{}_{\text{UP:UP}}\text{US},{}_{\text{DP:DP}}\text{LS},{}_{\text{DP:DP}}\text{US},{}_{\text{UP:UP}}\text{LS}\};\\
&\text{YY}\in\{\text{DI}_+{:}\text{II}_+,\text{II}_+{:}\text{DI}_+,\text{II}_-{:}\text{DI}_-,\text{DI}_-{:}\text{II}_-\},\\
&\text{XX}\in\{{}_{\text{UP:UP}}\text{SO},{}_{\text{DP:DP}}\text{SI},{}_{\text{DP:DP}}\text{SO},{}_{\text{UP:UP}}\text{SI}\};\\
&\text{YY}\in\{\text{II}_+{:}\text{II}_+,\text{DI}_+{:}\text{DI}_+,\text{DI}_-{:}\text{DI}_-,\text{II}_-{:}\text{II}_-\},\\
&\text{XX}\in\{{}_{\text{II:II}}\text{US},{}_{\text{DI:DI}}\text{LS}{}_{\text{DI:DI}}\text{US},{}_{\text{II:II}}\text{LS}\}.
\end{aligned} \quad (1.473)$$

For $m_2 < m_1$, 32 types of singular infinite-equilibriums are

$$(a_{j_1j_21}, a_{j_2j_11}) = \underbrace{([m_1,n_1]^{\text{th}}\text{YY},[m_1 - m_2]^{\text{th}}\text{ZZ})}_{([m_1,n_1],[m_2-m_1])\text{-YYZZ}}, \tag{1.474}$$

where

$$\begin{aligned}
&\text{YY} \in \{\ \text{UP}_{-}\text{:UP}_{+},\ \text{DP}_{-}\text{:DP}_{+},\ \text{DP}_{+}\text{:DP}_{-},\text{UP}_{+}\text{:UP}_{-}\},\\
&\text{ZZ} \in \{\ {}_{\text{DC:IC}}\text{US},{}_{\text{IC:DC}}\text{LS},{}_{\text{IC:DC}}\text{US},{}_{\text{IC:DC}}\text{LS}\};\\
&\text{YY} \in \{\ \text{DP}_{-}\text{:UP}_{+},\text{UP}_{-}\text{:DP}_{+},\ \text{UP}_{+}\text{:DP}_{-},\text{DP}_{+}\text{:UP}_{-}\},\\
&\text{ZZ} \in \{\ {}_{\text{UA}}\text{LS:}{}_{\text{DA}}\text{US},\ {}_{\text{DA}}\text{US:}{}_{\text{UA}}\text{LS},\ {}_{\text{DA}}\text{US:}{}_{\text{UA}}\text{LS},\ {}_{\text{UA}}\text{LS:}{}_{\text{DA}}\text{US}\};\\
&\text{YY} \in \{\ \text{DP}_{+}\text{:UP}_{+},\ \text{UP}_{+}\text{:DP}_{+},\text{UP}_{-}\text{:DP}_{-},\text{DP}_{-}\text{:UP}_{-}\},\\
&\text{ZZ} \in \{\ {}_{\text{UA}}\text{SI:}{}_{\text{DA}}\text{SO},\ {}_{\text{DA}}\text{SO:}{}_{\text{UA}}\text{SI},\ {}_{\text{DA}}\text{SI:}{}_{\text{UA}}\text{SO},\ {}_{\text{UA}}\text{SO:}{}_{\text{DA}}\text{SI}\};\\
&\text{YY} \in \{\ \text{UP}_{+}\text{:UP}_{+},\ \text{DP}_{+}\text{:DP}_{+},\ \text{DP}_{-}\text{:DP}_{-},\text{UP}_{-}\text{:UP}_{-}\},\\
&\text{ZZ} \in \{\ {}_{\text{DC:IC}}\text{US},{}_{\text{IC:DC}}\text{LS},{}_{\text{IC:DC}}\text{US},{}_{\text{DC:IC}}\text{LS}\};\\
&\text{YY} \in \{\ \text{II}_{-}\text{:II}_{+},\ \text{DI}_{-}\text{:DI}_{+},\ \text{DI}_{+}\text{:DI}_{-},\text{II}_{+}\text{:II}_{-}\},\\
&\text{ZZ} \in \{\ {}_{\text{IC:IC}}\text{SO},{}_{\text{DC:DC}}\text{SI}{}_{\text{DC:DC}}\text{SO},{}_{\text{IC:IC}}\text{SI}\};\\
&\text{YY} \in \{\ \text{DI}_{-}\text{:II}_{+},\ \text{II}_{-}\text{:DI}_{+},\ \text{II}_{+}\text{:DI}_{-},\ \text{DI}_{+}\text{:II}_{-}\},\\
&\text{ZZ} \in \{\ {}_{\text{DA:DA}}\text{US},\ {}_{\text{UA:UA}}\text{LS},\ {}_{\text{UA:UA}}\text{US},\ {}_{\text{DA:DA}}\text{LS}\};\\
&\text{YY} \in \{\ \text{DI}_{+}\text{:II}_{+},\ \text{II}_{+}\text{:DI}_{+},\ \text{II}_{-}\text{:DI}_{-},\ \text{DI}_{-}\text{:II}_{-}\},\\
&\text{ZZ} \in \{\ {}_{\text{DA:DA}}\text{SO},\ {}_{\text{UA:UA}}\text{SI},\ {}_{\text{UA:UA}}\text{SO},\ {}_{\text{DA:DA}}\text{SI}\};\\
&\text{YY} \in \{\ \text{II}_{+}\text{:II}_{+},\ \text{DI}_{+}\text{:DI}_{+},\ \text{DI}_{-}\text{:DI}_{-},\text{II}_{-}\text{:II}_{-}\},\\
&\text{ZZ} \in \{\ {}_{\text{IC:IC}}\text{US},{}_{\text{DC:DC}}\text{LS},{}_{\text{DC:DC}}\text{US},{}_{\text{IC:IC}}\text{LS}\}.
\end{aligned} \tag{1.475}$$

(vi_1) For $m_1 < m_2$, the singularity in the x_{j_1}-direction more than singularity in the x_{j_2}-direction, and the following three $([m_1,n_1]^{\text{th}}\text{YY},[m_2 - m_1]^{\text{th}}\text{XX})$-YYXX *switching bifurcations are given as follows.*

(vi_{1a}) The $([m_1,n_1]^{\text{th}}\text{YY},[m_2 - m_1]^{\text{th}}\text{XX})$-YYXX *switching bifurcation is from a singular equilibrium with a singular 1-dimensional flow switching to a singular 1-dimensional flow with a singular equilibrium.*

$$\left\{\begin{matrix} \underbrace{(a_{j_1j_21}, a_{j_2j_11})}_{(n_1,m_2)\text{-YY}_1} \\ \underbrace{(a_{j_1j_11}, a_{j_1j_21})}_{((m_1,n_1)^{\text{th}}\text{XX}_1,\text{xF})} \end{matrix}\right\} \rightleftharpoons \underbrace{([m_1,n_1]^{\text{th}}\text{YY},[m_2 - m_1]^{\text{th}}\text{XX})}_{([m_1,n_1],[m_2-m_1])\text{-YYXX}} \rightleftharpoons \left\{\begin{matrix} \underbrace{(a_{j_1j_11}, a_{j_1j_21})}_{((m_1,n_1)^{\text{th}}\text{XX}_2,\text{xF})} \\ \underbrace{(a_{j_1j_21}, a_{j_2j_11})}_{(n_1,m_2)\text{-YY}_2} \end{matrix}\right\} \tag{1.476}$$

with

$$\begin{aligned}
&\mathrm{XX}_1, \mathrm{XX}_2 \in \{\text{ DP:UP, UP:DP; UP:UP, DP:DP; }{}_{\mathrm{II:DI}}\mathrm{SO},{}_{\mathrm{DI:II}}\mathrm{SO},\\
&{}_{\mathrm{II:DI}}\mathrm{SI},{}_{\mathrm{DI:II}}\mathrm{SI};\ {}_{\mathrm{II:II}}\mathrm{US},{}_{\mathrm{II:II}}\mathrm{LS},{}_{\mathrm{DI:DI}}\mathrm{US},{}_{\mathrm{DI:DI}}\mathrm{LS}\}\ ;\\
&\mathrm{xF}_1, \mathrm{xF}_2 \in \{\text{ pF, nF}\}\ ;\\
&\mathrm{YY}_1,\mathrm{YY}_2 \in \{\text{ positive saddle, negative saddle, CCW center,}\\
&\text{CW center, up-parabola upper-saddle (lower-saddle),}\\
&\text{down-parabola upper-saddle (lower-saddle)}\\
&\text{double inflection-saddlles}\}\ .
\end{aligned} \tag{1.477}$$

(vi_{1b}) The $([m_1,n_1]^{\text{th}}\mathrm{YY},[m_2-m_1]^{\text{th}}\mathrm{XX})$-YYXX *switching bifurcation is for two* $n_1\times(m_1+m_2)$ *networks of non-singular equilibrium and hyperbolic and hyperbolic-secant flows.*

$$\begin{aligned}
&\bigcup_{r_1=1}^{m_1}\bigcup_{u_1=1}^{n_1}\underbrace{(a_{j_1j_1r_1},a_{j_1j_2u})}_{(\mathrm{XX}_1,\mathrm{xF}_1)}\oplus\bigcup_{u=1}^{n_1}\bigcup_{v_1=1}^{m_2}\underbrace{(a_{j_1j_2u},a_{j_2j_1v_1})}_{\mathrm{YY}_1}\\
&\rightleftharpoons\underbrace{([m_1,n_1]^{\text{th}}\mathrm{YY},[m_2-m_1]^{\text{th}}\mathrm{XX})}_{([m_1,n_1],[m_2-m_1])\text{-YYXX}}\\
&\rightleftharpoons\bigcup_{r_2=1}^{m_1}\bigcup_{u=1}^{n_1}\underbrace{(a_{j_1j_1r_2},a_{j_1j_2u_2})}_{(XX_2,\mathrm{xF}_2)}\oplus\bigcup_{u=1}^{n_1}\bigcup_{v_1=1}^{m_2}\underbrace{(a_{j_1j_2u},a_{j_2j_1v_2})}_{\mathrm{YY}_2},
\end{aligned} \tag{1.478}$$

with

$$\begin{aligned}
&\mathrm{YY}_1,\mathrm{YY}_2 \in \{\text{ positive saddle, negative saddle, CCW center, CW center}\}\ ;\\
&\mathrm{xF}_1, \mathrm{xF}_2 \in \{\text{ pF, nF}\}\ ;\\
&\mathrm{XX}_1, \mathrm{XX}_2 \in \{\text{ hyperbolic flow, hyperbolic-secant flow}\}\ .
\end{aligned} \tag{1.479}$$

(vi_{1c}) The $([m_1,n_1]^{\text{th}}\mathrm{YY},[m_2-m_1]^{\text{th}}\mathrm{XX})$-YYXX *switching bifurcation is for two* $p_1\times(q_1+q_2)$ *networks of singular equilibriums and singular 1-dimensional flows.*

$$\begin{aligned}
&\bigcup_{r_1=1}^{p_1}\bigcup_{u=1}^{q_1}\underbrace{(a_{j_1j_1r_1},a_{j_1j_2u})}_{([m_{r_1},n_u]\mathrm{XX}_1,\mathrm{xF}_1)}\oplus\bigcup_{u=1}^{q_1}\bigcup_{v_1=1}^{p_2}\underbrace{(a_{j_1j_2u},a_{j_2j_1v_1})}_{(n_u,m_{v_1})\text{-}\mathrm{YY}_1}\\
&\rightleftharpoons\underbrace{([m_1,n_1]^{\text{th}}\mathrm{YY},[m_2-m_1]^{\text{th}}\mathrm{XX})}_{([m_1,n_1],[m_2-m_1])\text{-YYXX}}\\
&\rightleftharpoons\bigcup_{r_2=1}^{p_1}\bigcup_{u_2=1}^{q_1}\underbrace{(a_{j_1j_1r_2},a_{j_1j_2u})}_{([m_{r_2},n_u]\mathrm{XX}_1,\mathrm{xF}_2)}\oplus\bigcup_{u=1}^{q_1}\bigcup_{v_2=1}^{p_2}\underbrace{(a_{j_1j_2u},a_{j_2j_1v_2})}_{(n_u,m_{v_2})\text{-}\mathrm{YY}_2}
\end{aligned} \tag{1.480}$$

with

$$\begin{aligned}&XX_1, XX_2 \in \{ \text{DP:UP, UP:DP; UP:UP, DP:DP;}\ {}_{\text{II:DI}}\text{SO},{}_{\text{DI:II}}\text{SO},\\&{}_{\text{II:DI}}\text{SI},{}_{\text{DI:II}}\text{SI};\ {}_{\text{II:II}}\text{US},{}_{\text{II:II}}\text{LS},{}_{\text{DI:DI}}\text{US},{}_{\text{DI:DI}}\text{LS}\} \ ;\\&xF_1, xF_2 \in \{ \text{pF, nF}\} \ ;\\&YY_1,YY_2 \in \{ \text{positive saddle, negative saddle, CCW center,}\\&\text{CW center, up-parabola upper-saddle (lower-saddle),}\\&\text{down-parabola upper-saddle (lower-saddle)}\\&\text{double inflection-saddlles}\} \ .\end{aligned} \tag{1.481}$$

and

$$m_1 = \sum_{r_i=1}^{p_1} m_{j_1 r_i},\ n_1 = \sum_{u_i=1}^{q_1} n_{j_1 u_i},\ m_2 = \sum_{s_i=1}^{q_1} n_{j_2 s_i};\ i = 1, 2. \tag{1.482}$$

(vi_2) For $m_1 < m_2$, the singularity in the x_{j_2}-direction higher than singularity in x_{j_1}-direction, and the following three $([m_1,n_1]^{\text{th}}\text{YY},[m_1 - m_2]^{\text{th}}\text{ZZ})$-YYZZ *switching bifurcations are given as follows.*

(vi_{2a}) The $([m_1,n_1]^{\text{th}}\text{YY},[m_1-m_2]^{\text{th}}\text{ZZ})$-YYZZ *switching bifurcation is for a singular equilibrium with a singular 1-dimensional flow switching to a singular 1-dimensional flow with a singular equilibrium.*

$$\left\{\begin{array}{c}\underbrace{(a_{j_1j_2}1,\ a_{j_2j_1}1)}_{(n_1,m_2)\text{-YY}_1}\\ \underbrace{(a_{j_1j_1}1,\ a_{j_1j_2}1)}_{((m_1,n_1)^{\text{th}}\text{XX}_1,\text{xF})}\end{array}\right\} \rightleftharpoons \underbrace{([m_1,n_1]^{\text{th}}\text{YY},[m_1 - m_2]^{\text{th}}\text{ZZ})}_{([m_1,n_1],[m_2-m_1])\text{-YYZZ}} \rightleftharpoons \left\{\begin{array}{c}\underbrace{(a_{j_1j_1}1,\ a_{j_1j_2}1)}_{((m_1,n_1)^{\text{th}}\text{XX}_2,\text{xF})}\\ \underbrace{(a_{j_1j_2}1,\ a_{j_2j_1}1)}_{(n_1,m_2)\text{-YY}_2}\end{array}\right\} \tag{1.483}$$

with

$$\begin{aligned}&XX_1, XX_2 \in \{ \text{DP:UP, UP:DP; UP:UP, DP:DP;}\ {}_{\text{II:DI}}\text{SO},{}_{\text{DI:II}}\text{SO},\\&{}_{\text{II:DI}}\text{SI},{}_{\text{DI:II}}\text{SI};\ {}_{\text{II:II}}\text{US},{}_{\text{II:II}}\text{LS},{}_{\text{DI:DI}}\text{US},{}_{\text{DI:DI}}\text{LS}\} \ ;\\&xF_1, xF_2 \in \{ \text{pF, nF}\} \ ;\\&YY_1,YY_2 \in \{ \text{positive saddle, negative saddle, CCW center,}\\&\text{CW center, up-parabola upper-saddle (lower-saddle),}\\&\text{down-parabola upper-saddle (lower-saddle)}\\&\text{double inflection-saddlles}\} \ .\end{aligned} \tag{1.484}$$

(vi_{2b}) The $([m_1,n_1]^{\text{th}}\text{YY},[m_1-m_2]^{\text{th}}\text{ZZ})$-YYZZ *switching bifurcation is for two* $m_1 \times (n_1+n_2)$ *networks of non-singular equilibrium and hyperbolic and hyperbolic-secant flows.*

$$\begin{aligned}&\bigcup_{r=1}^{m_1}\bigcup_{u_1=1}^{n_1}\underbrace{(a_{j_1j_1r},a_{j_1j_2u_1})}_{(\mathrm{XX}_1,\mathrm{xF}_1)}\oplus\bigcup_{r=1}^{m_1}\bigcup_{s_1=1}^{n_2}\underbrace{(a_{j_1j_1r},a_{j_2j_2s_1})}_{\mathrm{YY}_1}\\&\rightleftharpoons\underbrace{([m_1,n_1]^{\mathrm{th}}\mathrm{YY},[m_1-m_2]^{\mathrm{th}}\mathrm{ZZ})}_{([m_1,n_1],[m_2-m_1])\text{-YYZZ}}\\&\rightleftharpoons\bigcup_{r=1}^{m_1}\bigcup_{u_2=1}^{n_1}\underbrace{(a_{j_1j_1r},a_{j_1j_2u_1})}_{(XX_2,\mathrm{xF}_2)}\oplus\bigcup_{r=1}^{m_1}\bigcup_{s_2=1}^{n_1}\underbrace{(a_{j_1j_1r},a_{j_2j_2s_2})}_{\mathrm{YY}_2},\end{aligned}\tag{1.485}$$

with

$$\begin{aligned}&\mathrm{YY}_1,\mathrm{YY}_2\in\{\text{ positive and negative saddle, CCW center, CW center}\};\\&\mathrm{xF}_1,\mathrm{xF}_2\in\{\text{ pF, nF}\};\\&\mathrm{XX}_1,\mathrm{XX}_2\in\{\text{ hyperbolic flow, hyperbolic-secant flow}\}.\end{aligned}\tag{1.486}$$

(vi_{2c}) The $([n_2-n_1]^{\mathrm{th}}\mathrm{ZZ},[m_1,n_2]^{\mathrm{th}}\mathrm{YY})$*-YYZZ switching bifurcation is for two* $p_1\times(q_1+q_2)$ *networks of singular equilibriums and singular 1-dimensional flows.*

$$\begin{aligned}&\bigcup_{r_1=1}^{p_1}\bigcup_{u=1}^{q_1}\underbrace{(a_{j_1j_1r},a_{j_1j_2u})}_{([m_{j_1r_1},n_{j_1u}]\mathrm{XX}_1,\mathrm{xF}_1)}\oplus\bigcup_{u=1}^{q_1}\bigcup_{v_1=1}^{p_2}\underbrace{(a_{j_1j_2u},a_{j_2j_1v_1})}_{(n_{j_1u},m_{j_2v_1})\text{-YY}_1}\\&\rightleftharpoons\underbrace{([m_1,n_1]^{\mathrm{th}}\mathrm{YY},[m_1-m_2]^{\mathrm{th}}\mathrm{ZZ})}_{([m_1,n_1],[m_2-m_1])\text{-YYZZ}}\\&\rightleftharpoons\bigcup_{r_2=1}^{p_1}\bigcup_{u=1}^{q_1}\underbrace{(a_{j_1j_1r},a_{j_1j_2u})}_{([m_{j_1r_2},n_{j_1u}]XX_1,\mathrm{xF}_2)}\oplus\bigcup_{u=1}^{q_1}\bigcup_{v_2=1}^{p_2}\underbrace{(a_{j_1j_2u},a_{j_2j_1v_2})}_{(n_{j_1u},m_{j_2v_2})\text{-YY}_2}\end{aligned}\tag{1.487}$$

with

$$\begin{aligned}&\mathrm{XX}_1,\mathrm{XX}_2\in\{\text{ DP:UP, UP:DP; UP:UP, DP:DP; }{}_{\mathrm{II:DI}}\mathrm{SO},{}_{\mathrm{DI:II}}\mathrm{SO},\\&{}_{\mathrm{II:DI}}\mathrm{SI},{}_{\mathrm{DI:II}}\mathrm{SI};\ {}_{\mathrm{II:II}}\mathrm{US},{}_{\mathrm{II:II}}\mathrm{LS},{}_{\mathrm{DI:DI}}\mathrm{US},{}_{\mathrm{DI:DI}}\mathrm{LS}\};\\&\mathrm{xF}_1,\mathrm{xF}_2\in\{\text{ pF, nF}\};\\&\mathrm{YY}_1,\mathrm{YY}_2\in\{\text{ positive saddle, negative saddle, CCW center,}\\&\text{CW center, up-parabola upper-saddle (lower-saddle),}\\&\text{down-parabola upper-saddle (lower-saddle)}\\&\text{double inflection-saddlles}\}.\end{aligned}\tag{1.488}$$

and

$$m_1=\sum_{r_i=1}^{p_1}m_{j_1r_i},\ n_1=\sum_{u_i=1}^{q_1}n_{j_1u_i},\ n_2=\sum_{v_i=1}^{p_2}n_{j_2v_i};\ i=1,2.\tag{1.489}$$

Chapter 2
Proof of Theorem 1.1

In this chapter, Theorem 1.1 is proved. The theorem about the singular equilibriums and 1-dimensional flows with infinite-equilibriums in crossing and product polynomial systems is proved first. The first integral manifolds are obtained for positive and negative saddle, centers, parabola-saddles and inflection-saddle. The theorem about hybrid networks of singular equilibriums and 1-dimensional flows in crossing and product polynomial systems are proved, and the corresponding infinite-equilibriums are discussed. The theorem about hybrid networks of simple equilibriums and 1-dimensional flows with infinite-equilibriums in crossing and product polynomial systems is proved. The theorem about appearing and switching of simple and singular 1-dimensional flows and equilibriums is proved. Finally, the theorem about the switching of networks of simple and singular equilibriums and flows is proved.

2.1 Singular Equilibriums and Infinite-Equilibriums

In this section, the theorem about the singular equilibriums and 1-dimensional flows in the crossing and product polynomial systems will be proved.

2.1.1 Singular Equilibriums and 1-Dimensional Flows

(i) Consider a two-dimensional, nonlinear dynamical system as

$$\begin{aligned}\dot{x}_{j_2} &= a_{j_2 j_2 0}(x_{j_1} - a_{j_2 j_1 1})^{n_2},\\ \dot{x}_{j_1} &= a_{j_1 j_1 0}(x_{j_1} - a_{j_1 j_1 1})^{m_1}(x_{j_2} - a_{j_1 j_2 1})^{n_1}.\end{aligned}$$

A. C. J. Luo, *Two-dimensional Crossing and Product Polynomial Systems*,
https://doi.org/10.1007/978-981-96-5715-5_2

(i_1) The foregoing equation becomes

$$\frac{dx_{j_1}}{dx_{j_2}}=\frac{a_{j_1j_10}(x_{j_1}-a_{j_1j_11})^{m_1}(x_{j_2}-a_{j_1j_21})^{n_1}}{a_{j_2j_20}(x_{j_1}-a_{j_2j_11})^{n_2}},$$

or

$$\frac{(x_{j_1}-a_{j_2j_11})^{m_2}dx_{j_1}}{(x_{j_1}-a_{j_1j_11})^{m_1}}=\frac{a_{j_1j_10}}{a_{j_2j_20}}(x_{j_2}-a_{j_1j_21})^{n_1}dx_{j_2},$$

and

$$\frac{(x_{j_1}-a_{j_2j_11})^{m_2}}{(x_{j_1}-a_{j_1j_11})^{m_1}}=\sum_{\alpha,\beta=0}^{\alpha+\beta=m_2}\frac{m_2!}{\alpha!\beta!}(a_{j_1j_11}-a_{j_2j_11})^{\alpha}(x_{j_1}-a_{j_2j_11})^{\beta-m_1}.$$

With an initial condition (x_{j_10},x_{j_20}) art time $t=t_0$, integration of the foregoing equation gives the first integral manifold as

$$\begin{aligned}&\sum_{\alpha,\beta=0}^{\alpha+\beta=m_2}\frac{m_2!}{\alpha!\beta!}\frac{(a_{j_1j_11}-a_{j_2j_11})^{\alpha}}{\beta-m_1+1}[(x_{j_1}-a_{j_1j_11})^{\beta-m_1+1}-(x_{j_10}-a_{j_1j_11})^{\beta-m_1+1}]\\&=\frac{a_{j_1j_10}}{a_{j_2j_20}}\frac{1}{n_1+1}[(x_{j_2}-a_{j_1j_21})^{n_1+1}-(x_{j_20}-a_{j_1j_21})^{n_1+1}],\\&\text{for } \beta-m_1+1\neq 0;\end{aligned}$$

and

$$\begin{aligned}&\frac{1}{\beta-m_1+1}[(x_{j_1}-a_{j_1j_11})^{\beta-m_1+1}-(x_{j_10}-a_{j_1j_11})^{\beta-m_1+1}]\\&=\ln\frac{|x_{j_1}-a_{j_1j_11}|}{|x_{j_10}-a_{j_1j_11}|},\\&\text{for } \beta-m_1+1=0.\end{aligned}$$

($\mathrm{i}_{1\mathrm{a}}$) At the equilibrium of $x_{j_2}^*=a_{j_1j_21}$ and $\overline{x}_{j_1}\neq a_{j_1j_11},a_{j_2j_11}$ in phase space,

$$\begin{aligned}&\frac{dx_{j_1}}{dx_{j_2}}\Big|_{x_{j_2}^*=a_{j_1j_21}}=\frac{a_{j_1j_10}(\overline{x}_{j_1}-a_{j_1j_11})^{m_1}(x_{j_2}-a_{j_1j_21})^{n_1}}{a_{j_2j_20}(\overline{x}_{j_1}-a_{j_2j_11})^{m_2}}\Big|_{x_{j_2}^*=a_{j_1j_21}}=0,\\&\vdots\\&\frac{d^{n_1}x_{j_1}}{dx_{j_2}^{n_1}}\Big|_{x_{j_2}^*=a_{j_1j_21}}=0.\end{aligned}$$

If

$$\frac{d^{n_1+1}x_{j_1}}{dx_{j_2}^{n_1+1}}\bigg|_{x_{j_2}^*=a_{j_1j_2 1}} = \frac{a_{j_1j_1 0}(\overline{x}_{j_1} - a_{j_1j_1 1})^{m_1} n_1!}{a_{j_2j_2 0}(\overline{x}_{j_1} - a_{j_2j_1 1})^{m_2}} > 0,$$

the $(2n_{11}+1)^{\text{th}}$-order up-parabola flows exists for $n_1 = 2n_{11}+1$, and the $(2n_{11})^{\text{th}}$-order increasing-inflection flow exists for $n_1 = 2n_{11}$. If

$$\frac{d^{n_1+1}x_{j_1}}{dx_{j_2}^{n_1+1}}\bigg|_{x_{j_2}^*=a_{j_1j_2 1}} = \frac{a_{j_1j_1 0}(\overline{x}_{j_1} - a_{j_1j_1 1})^{m_1} n_1!}{a_{j_2j_2 0}(\overline{x}_{j_1} - a_{j_2j_1 1})^{m_2}} < 0,$$

the $(2n_{11}+1)^{\text{th}}$-order down-parabola flows exists for $n_1 = 2n_{11}+1$, and the $(2n_{11})^{\text{th}}$-order decreasing-inflection flow exists for $n_1 = 2n_{11}$.

At the equilibrium of $x_{j_1}^* = a_{j_1j_1 1}$ and $\overline{x}_{j_2} \neq a_{j_1j_2 1}, a_{j_2j_2 1}$, the variational equation is

$$\begin{aligned}\dot{x}_{j_2} &= a_{j_2j_2 0}(a_{j_1j_1 1} - a_{j_2j_1 1})^{m_2},\\ \Delta\dot{x}_{j_1} &= a_{j_1j_1 0}(\overline{x}_{j_2} - a_{j_1j_2 1})^{n_1}(\Delta x_{j_1})^{m_1}.\end{aligned}$$

The flows at $x_{j_1}^* = a_{j_1j_1 1}$ and $\overline{x}_{j_2} \neq a_{j_1j_2 1}$ are the $(2m_{11}+1)^{\text{th}}$-order source and sink for $a_{j_1j_1 0}(\overline{x}_{j_2} - a_{j_1j_2 1})^{n_1} > 0$ and $a_{j_1j_1 0}(\overline{x}_{j_2} - a_{j_1j_2 1})^{n_1} < 0$ for $m_1 = 2m_{11}+1$. The flows at $x_{j_1}^* = a_{j_1j_1 1}$ and $\overline{x}_{j_2} \neq a_{j_1j_2 1}, a_{j_2j_2 1}$, are the $(2m_{11})^{\text{th}}$-order upper-saddle and lower-saddle for $a_{j_1j_1 0}(\overline{x}_{j_2} - a_{j_1j_2 1})^{n_1} > 0$ and $a_{j_1j_1 0}(\overline{x}_{j_2} - a_{j_1j_2 1})^{n_a} < 0$ for $m_1 = 2m_{11}$.

Let $\overline{x}_{j_1} = a_{j_1j_1 1}$ and $\overline{x}_{j_2} = a_{j_1j_2 1} \neq a_{j_2j_2 1}$. Therefore, from the above discussion, there are four cases: (i_{1a1}) $m_1 = 2m_{11}+1$ and $n_1 = 2n_{11}+1$, (i_{1a2}) $m_1 = 2m_{11}$ and $n_1 = 2n_{11}+1$, (i_{1a3}) $m_1 = 2m_{11}+1$ and $n_1 = 2n_{11}$, (i_{1a4}) $m_1 = 2m_{11}$ and $n_1 = 2n_{11}$.

(i_{1a1}) For $m_1=2m_{11}+1$ and $n_1=2n_{11}+1$, the equilibrium of $(x_{j_1}^*, x_{j_2}^*)=(a_{j_1j_1 1}, a_{j_1j_2 1})$ in the x_{j_1}-direction has the following properties as in Eqs. (1.4)–(1.7)

- For $a_{j_1j_1 0} > 0$ and $a_{j_2j_2 0}(a_{j_1j_1 1} - a_{j_2j_1 1})^{m_2} > 0$,

$$(a_{j_1j_1 1}, a_{j_1j_2 1}) = \underbrace{([(2m_{11}+1), (2n_{11}+1)]\text{DP:UP,pF})}_{[(2m_{11}+1),(2n_{11}+1)]\text{-hyperbolic flow }(+)}.$$

The flow at $(x_{j_1}^*, x_{j_2}^*) = (a_{j_1j_1 1}, a_{j_1j_2 1})$ is a ([(2m_{11}+1), (2n_{11} + 1)]DP:UP,pF)-positive hyperbolic flow.

- For $a_{j_1j_1 0} < 0$ and $a_{j_2j_2 0}(a_{j_1j_1 1} - a_{j_2j_1 1})^{m_2} > 0$,

$$(a_{j_1j_1 1}, a_{j_1j_2 1}) = \underbrace{([(2m_{11}+1), (2n_{11}+1)]\text{UP:DP,pF})}_{[(2m_{11}+1),(2n_{11}+1)]\text{-hyperbolic-secant flow }(+)}.$$

The flow at $(x_{j_1}^*, x_{j_2}^*) = (a_{j_1j_1 1}, a_{j_1j_2 1})$ is a ([(2m_{11}+1), (2n_{11} + 1)]UP:DP,pF)-positive hyperbolic-secant flow.

- For $a_{j_1j_1 0} > 0$ and $a_{j_2j_2 0}(a_{j_1j_1 1} - a_{j_2j_1 1})^{m_2} < 0$,

$$(a_{j_1j_11}, a_{j_1j_21}) = \underbrace{([(2m_{11}+1), (2n_{11}+1)]\text{UP:DP,nF})}_{[(2m_{11}+1),(2n_{11}+1)]\text{-hyperbolic-secant flow }(-)}.$$

The flow at $(x_{j_1}^*, x_{j_2}^*) = (a_{j_1j_11}, a_{j_1j_21})$ is a $([(2m_{11}+1), (2n_{11}+1)]$UP:DP,nF)-negative hyperbolic-secant flow.

- For $a_{j_1j_10} < 0$ and $a_{j_2j_20}(a_{j_1j_11} - a_{j_2j_11})^{m_2} < 0$,

$$(a_{j_1j_11}, a_{j_1j_21}) = \underbrace{([(2m_{11}+1), (2n_{11}+1)]\text{DP:UP,nF})}_{[(2m_{11}+1),(2n_{11}+1)]\text{-hyperbolic flow }(-)}.$$

The flow at $(x_{j_1}^*, x_{j_2}^*) = (a_{j_1j_11}, a_{j_1j_21})$ is a $([(2m_{11}+1), (2n_{11}+1)]$DP:UP,pF)-negative hyperbolic flow.

(i_{1a2}) For $m_1 = 2m_{11}$ and $n_1 = 2n_{11} + 1$, the equilibrium of $(x_{j_1}^*, x_{j_2}^*) = (a_{j_1j_11}, a_{j_1j_21})$ for thex_{j_1}-direction has the following properties as in Eqs. (1.8)–(1.11).

- For $a_{j_1j_10} > 0$ and $a_{j_2j_20}(a_{j_1j_11} - a_{j_2j_11})^{m_2} > 0$,

$$(a_{j_1j_11}, a_{j_1j_21}) = \underbrace{([(2m_{11}), (2n_{11}+1)]\text{UP:UP,pF})}_{[(2m_{11}),(2n_{11}+1)]\text{-hyperbolic-secant-to-hyperbolic flow }(+)}.$$

The flow at $(x_{j_1}^*, x_{j_2}^*) = (a_{j_1j_11}, a_{j_1j_21})$ is a $([(2m_{11}), (2n_{11}+1)]$UP:UP,pF)-positive hyperbolic-secant to hyperbolic flow.

- For $a_{j_1j_10} < 0$ and $a_{j_2j_20}(a_{j_1j_11} - a_{j_2j_11})^{m_2} > 0$,

$$(a_{j_1j_11}, a_{j_1j_21}) = \underbrace{([(2m_{11}), (2n_{11}+1)]\text{DP:DP,pF})}_{[(2m_{11}),(2n_{11}+1)]\text{-hyperbolic-to-hyperbolic-secant flow }(+)}.$$

The flow at $(x_{j_1}^*, x_{j_2}^*) = (a_{j_1j_11}, a_{j_1j_21})$ is a $([(2m_{11}), (2n_{11}+1)]$DP:DP,pF)-positive hyperbolic to hyperbolic-secant flow.

- For $a_{j_1j_10} > 0$ and $a_{j_2j_20}(a_{j_1j_11} - a_{j_2j_11})^{m_2} < 0$,

$$(a_{j_1j_11}, a_{j_1j_21}) = \underbrace{([(2m_{11}), (2n_{11}+1)]\text{DP:DP,nF})}_{[(2m_{11}),(2n_{11}+1)]\text{-hyperbolic-to-hyperbolic-secant flow }(-)}.$$

The flow at $(x_{j_1}^*, x_{j_2}^*) = (a_{j_1j_11}, a_{j_1j_21})$ is a $([(2m_{11}), (2n_{11}+1)]$DP:DP,nF)-negative hyperbolic-to-hyperbolic-secant flow.

- For $a_{j_1j_10} < 0$ and $a_{j_2j_20}(a_{j_1j_11} - a_{j_2j_11})^{m_2} < 0$,

$$(a_{j_1j_11}, a_{j_1j_21}) = \underbrace{([(2m_{11}), (2n_{11}+1)]\text{UP:UP,nF})}_{[(2m_{11}),(2n_{11}+1)]\text{-hyperbolic-secant-to-hyperbolic flow }(-)}.$$

The flow at $(x_{j_1}^*, x_{j_2}^*) = (a_{j_1j_11}, a_{j_1j_21})$ is a $([(2m_{11}), (2n_{11}+1)]$UP:UP,nF)-negative hyperbolic-secant-to-hyperbolic flow.

(i_{1a3}) For $m_1 = 2m_{11} + 1$ and $n_1 = 2n_{11}$, the equilibrium of $(x_{j_1}^*, x_{j_2}^*) = (a_{j_1j_11}, a_{j_1j_21})$ for the x_{j_1}-direction has the following properties as in Eqs. (1.12)–(1.15).

- For $a_{j_1j_10} > 0$ and $a_{j_2j_20}(a_{j_1j_11} - a_{j_2j_11})^{m_2} > 0$,

$$(a_{j_1j_11}, a_{j_1j_21}) = \underbrace{([(2m_{11}+1), (2n_{11})]_{\text{DI:II}}\text{SO,pF})}_{[(2m_{11}+1),(2n_{11})]\text{-decreasing-to-increasing inflection source flow }(+)}.$$

The flow at $(x_{j_1}^*, x_{j_2}^*) = (a_{j_1j_11}, a_{j_1j_21})$ is a $([(2m_{11}+1), (2n_{11})]_{\text{DI:II}}\text{SO,pF})$-positive decreasing-to-increasing inflection source flow.

- For $a_{j_1j_10} < 0$ and $a_{j_2j_20}(a_{j_1j_11} - a_{j_2j_11})^{m_2} > 0$,

$$(a_{j_1j_11}, a_{j_1j_21}) = \underbrace{([(2m_{11}+1), (2n_{11})]_{\text{II:DI}}\text{SI,pF})}_{[(2m_{11}+1),(2n_{11})]\text{-increasing-to-decreasing inflection sink flow }(+)}.$$

The flow at $(x_{j_1}^*, x_{j_2}^*) = (a_{j_1j_11}, a_{j_1j_21})$ is a $([(2m_{11}+1), (2n_{11})]_{\text{II:DI}}\text{SI,pF})$-positive increasing-to-decreasing inflection source flow.

- For $a_{j_1j_10} > 0$ and $a_{j_2j_20}(a_{j_1j_11} - a_{j_2j_11})^{m_2} < 0$,

$$(a_{j_1j_11}, a_{j_1j_21}) = \underbrace{([(2m_{11}+1), (2n_{11})]_{\text{II:DI}}\text{SO,nF})}_{[(2m_{11}+1),(2n_{11})]\text{-increasing-to-decreasing inflection source flow }(-)}.$$

The flow at $(x_{j_1}^*, x_{j_2}^*) = (a_{j_1j_11}, a_{j_1j_21})$ is a $([(2m_{11}+1), (2n_{11})]_{\text{II:DI}}\text{SO,nF})$-negative increasing-to-decreasing inflection source flow.

- For $a_{j_1j_10} < 0$ and $a_{j_2j_20}(a_{j_1j_11} - a_{j_2j_11})^{m_2} < 0$,

$$(a_{j_1j_11}, a_{j_1j_21}) = \underbrace{([(2m_{11}+1), (2n_{11})]_{\text{DI:II}}\text{SI,nF})}_{[(2m_{11}+1),(2n_{11})]\text{-decreasing-to-increasing inflection sink flow }(-)}.$$

The flow at $(x_{j_1}^*, x_{j_2}^*) = (a_{j_1j_11}, a_{j_1j_21})$ is a $([(2m_{11}+1), (2n_{11})]_{\text{DI:II}}\text{SI,nF})$-negative decreasing-to-increasing inflection sink flow.

(i_{1a4}) For $m_1 = 2m_{11}$ and $n_1 = 2n_{11}$, the equilibrium of $(x_{j_1}^*, x_{j_2}^*) = (a_{j_1j_11}, a_{j_1j_21})$ for the x_{j_1}-direction has the following properties as in Eqs. (1.16)–(1.19).

- For $a_{j_1j_10} > 0$ and $a_{j_2j_20}(a_{j_1j_11} - a_{j_2j_11})^{m_2} > 0$,

$$(a_{j_1j_11}, a_{j_1j_21}) = \underbrace{([(2m_{11}), (2n_{11})]_{\text{II:II}}\text{US,pF})}_{[(2m_{11}),(2n_{11})]\text{-increasing-inflection upper-saddle flow }(+)}.$$

The flow at $(x_{j_1}^*, x_{j_2}^*) = (a_{j_1j_11}, a_{j_1j_21})$ is a $([(2m_{11}), (2n_{11})]_{\text{II:II}}\text{US,pF})$-positive increasing-inflection upper-saddle flow.

- For $a_{j_1j_10} < 0$ and $a_{j_2j_20}(a_{j_1j_11} - a_{j_2j_11})^{m_2} > 0$,

$$(a_{j_1j_11}, a_{j_1j_21}) = \underbrace{([(2m_{11}), (2n_{11})]_{\text{DI:DI}}\text{LS,pF})}_{[(2m_{11}),(2n_{11})]\text{-decreasing-inflection lower-saddle flow }(+)}.$$

The flow at $(x^*_{j_1}, x^*_{j_2}) = (a_{j_1j_11}, a_{j_1j_21})$ is a $([(2m_1), (2n_1)]_{\text{DI:DI}}\text{LS,pF})$-positive decreasing-inflection lower-saddle flow.

- For $a_{j_1j_10} > 0$ and $a_{j_2j_20}(a_{j_1j_11} - a_{j_2j_11})^{m_2} < 0$,

$$(a_{j_1j_11}, a_{j_1j_21}) = \underbrace{([(2m_{11}), (2n_{11})]_{\text{DI:DI}}\text{US,nF})}_{[(2m_{11}),(2n_{11})]\text{-decreasing-inflection upper-saddle flow }(-)}.$$

The flow at $(x^*_{j_1}, x^*_{j_2}) = (a_{j_1j_11}, a_{j_1j_21})$ is a $([(2m_{11}), (2n_{11})]_{\text{DI:DI}}\text{US,nF})$-negative decreasing-inflection upper-saddle flow.

- For $a_{j_1j_10} < 0$ and $a_{j_2j_20}(a_{j_1j_11} - a_{j_2j_11})^{m_2} < 0$,

$$(a_{j_1j_11}, a_{j_1j_21}) = \underbrace{([(2m_{11}), (2n_{11})]_{\text{II:II}}\text{LS,nF})}_{[(2m_{11}),(2n_{11})]\text{-increasing-inflection lower-saddle flow }(-)}.$$

The flow at $(x^*_{j_1}, x^*_{j_2}) = (a_{j_1j_11}, a_{j_1j_21})$ is a $([(2m_{11}), (2n_{11})]_{\text{II:II}}\text{LS,nF})$-negative increasing-inflection lower-saddle flow.

($\text{i}_{1\text{b}}$) There is an equilibrium for crossing and product equilibrium as

$$(x^*_{j_2}, x^*_{j_1}) = (a_{j_1j_21}, a_{j_2j_11}).$$

Case I: At the equilibrium of $x^*_{j_2} = a_{j_1j_21}$ and $\overline{x}_{j_1} \neq a_{j_1j_11}, a_{j_2j_11}$ in phase space,

$$\frac{dx_{j_1}}{dx_{j_2}}\bigg|_{x^*_{j_2}=a_{j_1j_21}} = \frac{a_{j_1j_10}(\overline{x}_{j_1} - a_{j_1j_11})^{m_1}(x_{j_2} - a_{j_1j_21})^{n_1}}{a_{j_2j_20}(\overline{x}_{j_1} - a_{j_2j_11})^{m_2}}\bigg|_{x^*_{j_2}=a_{j_1j_21}} = 0,$$

$$\vdots$$

$$\frac{d^{n_1}x_{j_1}}{dx^{n_1}_{j_2}}\bigg|_{x^*_{j_2}=a_{j_1j_21}} = 0.$$

If

$$\frac{d^{n_1+1}x_{j_1}}{dx^{n_1+1}_{j_2}}\bigg|_{x^*_{j_2}=a_{j_1j_21}} = \frac{a_{j_1j_10}(\overline{x}_{j_1} - a_{j_1j_11})^{m_1}n_1!}{a_{j_2j_20}(\overline{x}_{j_1} - a_{j_2j_11})^{m_2}} > 0,$$

the $(2n_{11} + 1)^{\text{th}}$-order up-parabola flows exists for $n_1 = 2n_{11} + 1$, and the $(2n_{11})^{\text{th}}$-order increasing-inflection flow exists for $n_1 = 2n_{11}$. If

$$\frac{d^{n_1+1}x_{j_1}}{dx_{j_2}^{n_1+1}}\bigg|_{x_{j_2}^*=a_{j_1j_2 1}} = \frac{a_{j_1j_1 0}(\overline{x}_{j_1}-a_{j_1j_1 1})^{m_1}n_1!}{a_{j_2j_2 0}(\overline{x}_{j_1}-a_{j_2j_1 1})^{m_2}} < 0,$$

the $(2n_{11}+1)^{\text{th}}$-order down-parabola flows exists for $n_1 = 2n_{11}+1$, and the $(2n_{11})^{\text{th}}$-order decreasing-inflection flow exists for $n_1 = 2n_{11}$.

Case II: At the equilibrium of $x_{j_1}^* = a_{j_2j_1 1}$ and $\overline{x}_{j_1} \neq a_{j_1j_1 1}, a_{j_2j_1 1}$ in phase space,

$$\frac{dx_{j_2}}{dx_{j_1}}\bigg|_{x_{j_1}^*=a_{j_2j_1 1}} = \frac{a_{j_2j_2 0}(x_{j_1}-a_{j_2j_1 1})^{m_2}}{a_{j_1j_1 0}(x_{j_1}-a_{j_1j_1 1})^{m_1}(\overline{x}_{j_2}-a_{j_1j_2 1})^{n_1}}\bigg|_{x_{j_1}^*=a_{j_2j_1 1}} = 0,$$

$$\vdots$$

$$\frac{d^{m_2}x_{j_2}}{dx_{j_1}^{m_1}}\bigg|_{x_{j_1}^*=a_{j_2j_1 1}} = 0.$$

If

$$\frac{d^{m_2+1}x_{j_2}}{dx_{j_1}^{m_1+1}}\bigg|_{x_{j_1}^*=a_{j_2j_1 1}} = \frac{a_{j_2j_2 0}m_2!}{a_{j_1j_1 0}(a_{j_2j_1 1}-a_{j_1j_1 1})^{m_1}(\overline{x}_{j_2}-a_{j_1j_2 1})^{n_1}} > 0,$$

the $(2m_{21}+1)^{\text{th}}$-order up-parabola flows exists for $n_2 = 2m_{21}+1$, and the $(2m_{21})^{\text{th}}$-order increasing-inflection flow exists for $m_2 = 2m_{21}$. If

$$\frac{d^{m_2+1}x_{j_2}}{dx_{j_1}^{m_1+1}}\bigg|_{x_{j_1}^*=a_{j_2j_1 1}} = \frac{a_{j_2j_2 0}m_2!}{a_{j_1j_1 0}(a_{j_2j_1 1}-a_{j_1j_1 1})^{m_1}(\overline{x}_{j_2}-a_{j_1j_2 1})^{n_1}} < 0,$$

the $(2m_{21}+1)^{\text{th}}$-order down-parabola flows exists for $n_2 = 2m_{21}+1$, and the $(2m_{21})^{\text{th}}$-order decreasing-inflection flow exists for $m_2 = 2m_{21}$.

Thus there are sixteen types of equilibriums.

(i_{1b1}) For $n_1 = 2n_{11}+1$ and $m_2 = 2m_{21}+1$, the equilibrium of $(x_{j_2}^*, x_{j_1}^*) = (a_{j_1j_2 1}, a_{j_2j_1 1})$ has the following properties as in Eqs. (1.21)–(1.24).

- For $a_{j_1j_1 0}(a_{j_2j_1 1}-a_{j_1j_1 1})^{m_1} > 0$ and $a_{j_2j_2 0} > 0$,

$$(a_{j_1j_2 1}, a_{j_2j_1 1}) = \underbrace{((2n_{11}+1)^{\text{th}}\text{UP}_+, (2m_{21}+1)^{\text{th}}\text{UP}_+)}_{((2n_{12}+1),(2m_{21}+1))\text{-positive saddle}}.$$

  The flow at $(x_{j_2}^*, x_{j_1}^*) = (a_{j_1j_2 1}, a_{j_2j_1 1})$ is a $((2n_{11}+1)^{\text{th}}\text{UP}_+, (2m_{21}+1)^{\text{th}}\text{UP}_+)$-positive saddle.

- For $a_{j_1j_1 0}(a_{j_2j_1 1}-a_{j_1j_1 1})^{m_1} < 0$ and $a_{j_2j_2 0} > 0$,

$$(a_{j_1j_2 1}, a_{j_2j_1 1}) = \underbrace{((2n_{11}+1)^{\text{th}}\text{DP}_+, (2m_{21}+1)^{\text{th}}\text{DP}_-)}_{((2n_{12}+1),(2m_{21}+1))\text{-CCW center}}.$$

The flow at $(x_{j_2}^*, x_{j_1}^*) = (a_{j_1j_2 1}, a_{j_2j_1 1})$ is a $((2n_{11}+1)^{\text{th}}\text{DP}_+, (2m_{21}+1)^{\text{th}}\text{DP}_-)$-counter-clockwise center.

- For $a_{j_1j_1 0}(a_{j_2j_1 1} - a_{j_1j_1 1})^{m_1} > 0$ and $a_{j_2j_2 0} < 0$,

$$(a_{j_1j_2 1}, a_{j_2j_1 1}) = \underbrace{((2n_{11}+1)^{\text{th}}\text{DP}_-, (2m_{21}+1)^{\text{th}}\text{DP}_+)}_{((2n_{11}+1),(2m_{21}+1))\text{-CW center}}.$$

The flow at $(x_{j_2}^*, x_{j_1}^*) = (a_{j_1j_2 1}, a_{j_2j_1 1})$ is a $((2n_{11}+1)^{\text{th}}\text{DP}_-, (2m_{21}+1)^{\text{th}}\text{DP}_+)$-clockwise center.

- For $a_{j_1j_1 0}(a_{j_2j_1 1} - a_{j_1j_1 1})^{m_1} < 0$ and $a_{j_2j_2 0} < 0$,

$$(a_{j_1j_2 1}, a_{j_2j_1 1}) = \underbrace{((2n_{11}+1)^{\text{th}}\text{UP}_-, (2m_{21}+1)^{\text{th}}\text{UP}_-)}_{((2n_{11}+1),(2m_{21}+1))\text{-negative saddle}}.$$

The flow at $(x_{j_2}^*, x_{j_1}^*) = (a_{j_1j_2 1}, a_{j_2j_1 1})$ is a $((2n_{11}+1)^{\text{th}}\text{UP}_-, (2m_{21}+1)^{\text{th}}\text{UP}_-)$-negative saddle.

(i_{1b2}) For $n_1 = 2n_{11}$ and $m_2 = 2m_{21} + 1$,the equilibrium of $(x_{j_2}^*, x_{j_1}^*) = (a_{j_1j_2 1}, a_{j_2j_1 1})$ has the following properties as in Eqs. (1.25)–(1.28).

- For $a_{j_1j_1 0}(a_{j_2j_1 1} - a_{j_1j_1 1})^{m_1} > 0$ and $a_{j_2j_2 0} > 0$,

$$(a_{j_1j_2 1}, a_{j_2j_1 1}) = \underbrace{((2n_{11})^{\text{th}}\text{US},(2m_{21}+1)^{\text{th}}\text{UP})}_{((2n_{11}),(2m_{21}+1))\text{-up-parabola upper-saddle}}.$$

The flow at $(x_{j_2}^*, x_{j_1}^*) = (a_{j_1j_2 1}, a_{j_2j_1 1})$ is a $((2n_{11})^{\text{th}}\text{US},(2m_{21}+1)^{\text{th}}\text{UP})$-up-parabola upper-saddle.

- For $a_{j_1j_1 0}(a_{j_2j_1 1} - a_{j_1j_1 1})^{m_1} < 0$ and $a_{j_2j_2 0} > 0$,

$$(a_{j_1j_2 1}, a_{j_2j_1 1}) = \underbrace{((2n_{11})^{\text{th}}\text{LS},(2m_{21}+1)^{\text{th}}\text{DP})}_{((2n_{11}),(2m_{21}+1))\text{-down-parabola lower-saddle}}.$$

The flow at $(x_{j_2}^*, x_{j_1}^*) = (a_{j_1j_2 1}, a_{j_2j_1 1})$ is a $((2n_{11})^{\text{th}}\text{LS},(2m_{21}+1)^{\text{th}}\text{DP})$-down-parabola lower-saddle.

- For $a_{j_1j_1 0}(a_{j_2j_2 1} - a_{j_1j_2 1})^{n_1} > 0$ and $a_{j_2j_2 0} < 0$,

$$(a_{j_1j_2 1}, a_{j_2j_1 1}) = \underbrace{((2n_{11})^{\text{th}}\text{US},(2m_{21}+1)^{\text{th}}\text{DP})}_{((2n_{11}),(2m_{21}+1))\text{-down-parabola upper-saddle}}.$$

The flow at $(x_{j_2}^*, x_{j_1}^*) = (a_{j_1j_2 1}, a_{j_2j_1 1})$ is a $((2n_{11})^{\text{th}}\text{US},(2m_{21}+1)^{\text{th}}\text{DP})$-down-parabola upper-saddle.

- For $a_{j_1j_10}(a_{j_2j_11}-a_{j_1j_11})^{m_1}<0$ and $a_{j_2j_20}<0$,

$$(a_{j_1j_21},a_{j_2j_11})=\underbrace{((2n_{11})^{\text{th}}\text{LS},(2m_{21}+1)^{\text{th}}\text{UP})}_{((2n_{11}),(2m_{21}+1))\text{-up-parabola lower-saddle}}.$$

The flow at $(x^*_{j_2},x^*_{j_1})=(a_{j_1j_21},a_{j_2j_11})$ is $((2n_{11})^{\text{th}}\text{LS},(2m_{21}+1)^{\text{th}}\text{UP})$-up-parabola lower-saddle.

($\text{i}_{1\text{b}3}$) For $n_1=2n_{11}+1$ and $m_2=2m_{21}$, the equilibrium of $(x^*_{j_2},x^*_{j_1})=(a_{j_1j_21},a_{j_2j_11})$ has the following properties as in Eqs. (1.29)–(1.32).

- For $a_{j_1j_10}(a_{j_2j_11}-a_{j_1j_11})^{m_1}>0$ and $a_{j_2j_20}>0$,

$$(a_{j_1j_21},a_{j_2j_11})=\underbrace{((2n_{11}+1)^{\text{th}}\text{UP},(2m_{21})^{\text{th}}\text{US})}_{((2n_{11}+1),(2m_{21}))\text{-up-parabola upper-saddle}}.$$

The flow at $(x^*_{j_2},x^*_{j_1})=(a_{j_1j_21},a_{j_2j_11})$ is a $((2n_{11}+1)^{\text{th}}\text{UP},(2m_{21})^{\text{th}}\text{US})$-up-parabola upper-saddle.

- For $a_{j_1j_10}(a_{j_2j_11}-a_{j_1j_11})^{m_1}<0$ and $a_{j_2j_20}>0$,

$$(a_{j_1j_21},a_{j_2j_11})=\underbrace{((2n_{11}+1)^{\text{th}}\text{DP},(2m_{21})^{\text{th}}\text{US})}_{((2n_{11}+1),(2m_{21}))\text{-down-parabola upper-saddle}}.$$

The flow at $(x^*_{j_2},x^*_{j_1})=(a_{j_1j_21},a_{j_2j_11})$ is a $((2n_{11}+1)^{\text{th}}\text{DP},(2m_{21})^{\text{th}}\text{US})$-down-parabola upper-saddle.

- For $a_{j_1j_10}(a_{j_2j_11}-a_{j_1j_11})^{m_1}>0$ and $a_{j_2j_20}<0$,

$$(a_{j_1j_21},a_{j_2j_11})=\underbrace{((2n_{11}+1)^{\text{th}}\text{DP},(2m_{21})^{\text{th}}\text{LS})}_{((2n_{11}+1),(2m_{21}))\text{-down-parabola lower-saddle}}.$$

The flow at $(x^*_{j_2},x^*_{j_1})=(a_{j_1j_21},a_{j_2j_11})$ is a $((2n_{11}+1)^{\text{th}}\text{DP},(2m_{21})^{\text{th}}\text{LS})$-down-parabola lower-saddle.

- For $a_{j_1j_10}(a_{j_2j_11}-a_{j_1j_11})^{m_1}<0$ and $a_{j_2j_20}<0$,

$$(a_{j_1j_21},a_{j_2j_11})=\underbrace{((2n_{11}+1)^{\text{th}}\text{UP},(2m_{21})^{\text{th}}\text{LS})}_{((2n_{11}+1),(2m_{21}))\text{-up-parabola lower-saddle}}.$$

The flow at $(x^*_{j_2},x^*_{j_1})=(a_{j_1j_21},a_{j_2j_11})$ is a $((2n_{11}+1)^{\text{th}}\text{UP},(2m_{21})^{\text{th}}\text{LS})$-up-parabola upper-saddle.

(i_{1b4}) For $n_1 = 2n_{11}$ and $m_2 = 2m_{21}$, the equilibrium of $(x_{j_2}^*, x_{j_1}^*) = (a_{j_1j_2 1}, a_{j_2j_1 1})$ has the following properties as in Eqs. (1.33)–(1.36).

- For $a_{j_1j_1 0}(a_{j_2j_1 1} - a_{j_1j_1 1})^{m_1} > 0$ and $a_{j_2j_2 0} > 0$,

$$(a_{j_1j_2 1}, a_{j_2j_1 1}) = \underbrace{((2n_{11})^{\text{th}}\text{II}_+, (2m_{21})^{\text{th}}\text{II}_+)}_{((2n_{11}),(2m_{21}))\text{-double-inflection saddle}}.$$

  The flow at $(x_{j_2}^*, x_{j_1}^*) = (a_{j_1j_2 1}, a_{j_2j_1 1})$ is a $((2n_{11})^{\text{th}}\text{II}_+, (2m_{21})^{\text{th}}\text{II}_+)$-double-inflection saddle.

- For $a_{j_1j_1 0}(a_{j_2j_1 1} - a_{j_1j_1 1})^{m_1} < 0$ and $a_{j_2j_2 0} > 0$,

$$(a_{j_1j_2 1}, a_{j_2j_1 1}) = \underbrace{((2n_{11})^{\text{th}}\text{DI}_+, (2m_{21})^{\text{th}}\text{DI}_-)}_{((2n_{11}),(2m_{21}))\text{-double-inflection saddle}}.$$

  The flow at $(x_{j_2}^*, x_{j_1}^*) = (a_{j_1j_2 1}, a_{j_2j_1 1})$ is a $((2n_{11})^{\text{th}}\text{DI}_+, (2m_{21})^{\text{th}}\text{DI}_-)$-double-inflection saddle.

- For $a_{j_1j_1 0}(a_{j_2j_1 1} - a_{j_1j_1 1})^{m_1} > 0$ and $a_{j_2j_2 0} < 0$,

$$(a_{j_1j_2 1}, a_{j_2j_1 1}) = \underbrace{((2n_{11})^{\text{th}}\text{DI}_-, (2m_{21})^{\text{th}}\text{DI}_+)}_{((2n_{11}),(2m_{21}))\text{-double-inflection saddle}}.$$

  The flow at $(x_{j_2}^*, x_{j_1}^*) = (a_{j_1j_2 1}, a_{j_2j_1 1})$ is a $((2n_{11})^{\text{th}}\text{DI}_-, (2m_{21})^{\text{th}}\text{DI}_+)$-double-inflection saddle.

- For $a_{j_1j_1 0}(a_{j_2j_1 1} - a_{j_1j_1 1})^{m_1} < 0$ and $a_{j_2j_2 0} < 0$,

$$(a_{j_1j_2 1}, a_{j_2j_1 1}) = \underbrace{((2n_{11})^{\text{th}}\text{II}_-, (2m_{21})^{\text{th}}\text{II}_-)}_{((2n_{11}),(2m_{21}))\text{-double-inflection saddle}}.$$

  The flow at $(x_{j_2}^*, x_{j_1}^*) = (a_{j_1j_2 1}, a_{j_2j_1 1})$ is a $((2n_{11})^{\text{th}}\text{II}_-, (2m_{21})^{\text{th}}\text{II}_-)$-double-inflection saddle.

2.1.2 Infinite-Equilibriums

(i_2) For $a_{j_2j_1 1} = a_{j_1j_1 1}$, the foregoing equation becomes

$$\frac{dx_{j_1}}{dx_{j_2}} = \frac{a_{j_1j_1 0}(x_{j_1} - a_{j_1j_1 1})^{m_1 - m_2}(x_{j_2} - a_{j_1j_2 1})^{n_1}}{a_{j_2j_2 0}},$$

or

$$\frac{dx_{j_1}}{(x_{j_1}-a_{j_1j_11})^{m_1-m_2}}=\frac{a_{j_1j_10}}{a_{j_2j_20}}(x_{j_2}-a_{j_1j_21})^{n_1}dx_{j_2}.$$

With an initial condition (x_{j_10}, x_{j_20}) art time $t = t_0$, integration of the foregoing equation gives the first integral manifold as

$$\begin{aligned}&\frac{1}{m_2-m_1+1}[(x_{j_1}-a_{j_1j_11})^{m_2-m_1+1}-(x_{j_10}-a_{j_1j_11})^{m_2-m_1+1}]\\&=\frac{a_{j_1j_10}}{a_{j_2j_20}}\frac{1}{n_1+1}[(x_{j_2}-a_{j_1j_21})^{n_1+1}-(x_{j_20}-a_{j_1j_11})^{n_1+1}],\\&\text{for } m_2-m_1+1\neq 0;\end{aligned}$$

and

$$\begin{aligned}&\frac{1}{m_2-m_1+1}[(x_{j_1}-a_{j_1j_11})^{m_2-m_1+1}-(x_{j_10}-a_{j_1j_11})^{m_2-m_1+1}]=\ln\frac{|x_{j_1}-a_{j_1j_11}|}{|x_{j_10}-a_{j_1j_11}|},\\&\text{for } m_2-m_1+1=0.\end{aligned}$$

(i_{2a}) For $m_2 \geq m_1$, the first integral manifold is

$$\begin{aligned}&\frac{1}{m_2-m_1+1}[(x_{j_1}-a_{j_1j_11})^{m_2-m_1+1}-(x_{j_10}-a_{j_1j_11})^{m_2-m_1+1}]\\&=\frac{a_{j_1j_10}}{a_{j_2j_20}}\frac{1}{n_1+1}[(x_{j_2}-a_{j_1j_21})^{n_1+1}-(x_{j_20}-a_{j_1j_11})^{n_1+1}].\end{aligned}$$

(i_{2a1}) For $x_{j_2}^* = a_{j_2j_21} = a_{j_1j_21}$ and $\overline{x}_{j_1} \neq a_{j_1j_11}$, in phase space,

$$\begin{aligned}&\frac{dx_{j_2}}{dx_{j_1}}\bigg|_{x_{j_1}^*=a_{j_2j_11}}=\frac{a_{j_2j_20}(x_{j_1}-a_{j_1j_11})^{m_2-m_1}}{a_{j_2j_20}(\overline{x}_{j_2}-a_{j_1j_21})^{n_1}}\bigg|_{x_{j_1}^*=a_{j_2j_11}}=0,\\&\vdots\\&\frac{d^{(m_2-m_1)}x_{j_2}}{dx_{j_1}^{(m_2-m_1)}}\bigg|_{x_{j_1}^*=a_{j_2j_11}}=0.\end{aligned}$$

If

$$\frac{d^{(m_2-m_1)+1}x_{j_2}}{dx_{j_1}^{(m_2-m_1)+1}}\bigg|_{x_{j_1}^*=a_{j_2j_11}}=\frac{a_{j_2j_20}(m_2-m_1)!}{a_{j_1j_10}(\overline{x}_{j_2}-a_{j_1j_21})^{n_1}}>0,$$

the flows at $x^*_{j_1} = a_{j_1j_1 1} = a_{j_2j_1 1}$ are $[2(m_{21} - m_{11}) \pm 1]^{\text{th}}$-order up-parabola flow for $(m_1 = 2m_{11} + 1, m_2 = 2m_{21}$ and $m_1 = 2m_{11}, m_2 = 2m_{21} + 1)$, and $[2(m_{21} - m_{11})]^{\text{th}}$-order increasing-inflection flow for $(m_1 = 2m_{11} + 1, m_2 = 2m_{21} + 1$ and $m_1 = 2m_{11}$, $m_2 = 2m_{21})$.

If

$$\frac{d^{(m_2-m_1)+1}x_{j_2}}{dx_{j_1}^{(m_2-m_1)+1}}\bigg|_{x^*_{j_1}=a_{j_2j_1 1}} = \frac{a_{j_2j_2 0}(m_2 - m_1)!}{a_{j_1j_1 0}(\overline{x}_{j_2} - a_{j_1j_2 1})^{n_1}} < 0,$$

the flows at $x^*_{j_1} = a_{j_1j_1 1} = a_{j_2j_1 1}$ are $[2(m_{21} - m_{11}) \pm 1]^{\text{th}}$-order down-parabola flow for $(m_1 = 2m_{11} + 1, m_2 = 2m_{21}$ and $m_1 = 2m_{11}, m_2 = 2m_{21} + 1)$, and $[2(m_{21} - m_{11})]^{\text{th}}$-order decreasing-inflection flow for $(m_1 = 2m_{11} + 1, m_2 = 2m_{21} + 1$ and $m_1 = 2m_{11}$, $m_2 = 2m_{21})$.

The variational equation at $x^*_{j_1} = a_{j_1j_1 1} = a_{j_2j_1 1}$ is

$$\Delta\dot{x}_{j_1} = a_{j_1j_1 0}(\overline{x}_{j_2} - a_{j_1j_2 0})^{n_1}(\Delta x_{j_1})^{m_1}.$$

The flows at $x^*_{j_2} = a_{j_2j_2 1} = a_{j_1j_2 1}$ are $(2m_{11} + 1)^{\text{th}}$ source and sink flows for $a_{j_1j_1 0}(\overline{x}_{j_2} - a_{j_1j_2 0})^{n_1} > 0$ and $a_{j_1j_1 0}(\overline{x}_{j_2} - a_{j_1j_2 0})^{n_1} < 0$, respectively. The flows at $x^*_{j_2} = a_{j_2j_2 1} = a_{j_1j_2 1}$ are $(2m_{11})^{\text{th}}$-upper-saddle and lower-saddle flows for $a_{j_1j_1 0}(\overline{x}_{j_2} - a_{j_1j_2 0})^{n_1} > 0$ and $a_{j_1j_1 0}(\overline{x}_{j_2} - a_{j_1j_2 0})^{n_1} < 0$, respectively.

Thus, based on the above analysis, the flows at $x^*_{j_1} = a_{j_1j_1 1} = a_{j_2j_1 1}$ and on both sides of $x^*_{j_2} = a_{j_1j_2 1}$ are $((2m_{11} + 1)^{\text{th}}\text{XX}, [\cdot]^{\text{th}}\text{YY})$-source and sink flows for $a_{j_1j_1 0}(\overline{x}_{j_2} - a_{j_1j_2 0})^{n_1} > 0$ and $a_{j_1j_1 0}(\overline{x}_{j_2} - a_{j_1j_2 0})^{n_1} < 0$, respectively. The flows at $x^*_{j_1} = a_{j_1j_1 1} = a_{j_2j_1 1}$ and on both sides of $x^*_{j_2} = a_{j_1j_2 1}$ are $((2m_{11})^{\text{th}}\text{XX}, [\cdot]^{\text{th}}\text{YY})$-upper and lower-saddle flows for $a_{j_1j_1 0}(\overline{x}_{j_2} - a_{j_1j_2 0})^{n_1} > 0$ and $a_{j_1j_1 0}(\overline{x}_{j_2} - a_{j_1j_2 0})^{n_1} < 0$, respectively. $[2(m_{21} - m_{11})]^{\text{th}}\text{YY}$ are the $[2(m_{21} - m_{11})]^{\text{th}}$-order increasing and decreasing-inflection flows, and $[2(m_{21} - m_{11}) \pm 1]^{\text{th}}\text{YY}$ are the $[2(m_{21} - m_{11}) \pm 1]^{\text{th}}$-order up and down-parabola flows.

Therefore, there are 16 types of infinite-equilibriums at $x^*_{j_1} = a_{j_1j_1 1} = a_{j_2j_1 1}$.

(i_{2a1a}) For $m_1 = 2m_{11} + 1$ and $m_2 = 2m_{21} + 1$, the infinite-equilibrium of $x^*_{j_1} = a_{j_1j_1 1} = a_{j_2j_1 1}$ with $(\overline{x}_{j_2} \neq a_{j_1j_2 1})$ has the following properties as in Eqs. (1.39)–(1.42).

- For $a_{j_1j_1 0}(\overline{x}_{j_2} - a_{j_1j_2 1})^{n_1} > 0$ and $a_{j_2j_2 0} > 0$,

$$(a_{j_1j_1 1}, \overline{x}_{j_2}) = \underbrace{((2m_{11} + 1)^{\text{th}}\text{SO}, [2(m_{21} - m_{11})]^{\text{th}}\text{II})}_{((2m_{11}+1),[2(m_{21}-m_{11})])\text{-increasing inflection source}}.$$

  The infinite-equilibrium is a $((2m_{11} + 1)^{\text{th}}\text{SO}, [2(m_{21} - m_{11})]^{\text{th}}\text{II})$-increasing-inflection source at $x^*_{j_1} = a_{j_1j_1 1} = a_{j_2j_1 1}$.

- For $a_{j_1j_1 0}(\overline{x}_{j_2} - a_{j_1j_2 1})^{n_1} < 0$ and $a_{j_2j_2 0} > 0$,

$$(a_{j_1j_11}, \overline{x}_{j_2}) = \underbrace{((2m_{11}+1)^{\text{th}}\text{SI}, [2(m_{21}-m_{11})]^{\text{th}}\text{DI})}_{((2m_{11}+1),[2(m_{21}-m_{11})])\text{-decreasing inflection sink}}.$$

The infinite-equilibrium is a $((2m_{11}+1)^{\text{th}}\text{SI}, [2(m_{21}-m_{11})]^{\text{th}}\text{DI})$-decreasing-inflection sink at $x^*_{j_1} = a_{j_1j_11} = a_{j_2j_11}$.

- For $a_{j_1j_10}(\overline{x}_{j_2} - a_{j_1j_21})^{n_1} > 0$ and $a_{j_2j_20} < 0$,

$$(a_{j_1j_11}, \overline{x}_{j_2}) = \underbrace{((2m_{11}+1)^{\text{th}}\text{SO}, [2(m_{21}-m_{11})]^{\text{th}}\text{DI})}_{((2m_{11}+1),[2(m_{21}-m_{11})])\text{-decreasing inflection source}}.$$

The infinite-equilibrium is a $((2m_{11}+1)^{\text{th}}\text{SO}, [2(m_{21}-m_{11})]^{\text{th}}\text{DI})$-decreasing-inflection source at $x^*_{j_1} = a_{j_1j_11} = a_{j_2j_11}$.

- For $a_{j_1j_10}(\overline{x}_{j_2} - a_{j_1j_21})^{n_1} < 0$ and $a_{j_2j_20} < 0$,

$$(a_{j_1j_11}, \overline{x}_{j_2}) = \underbrace{((2m_{11}+1)^{\text{th}}\text{SI}, [2(m_{21}-m_{11})]^{\text{th}}\text{II})}_{((2m_{11}+1),[2(m_{21}-m_{11})])\text{-increasing inflection sink}}.$$

The infinite-equilibrium is a $((2m_{11}+1)^{\text{th}}\text{SI}, [2(m_{21}-m_{11})]^{\text{th}}\text{II})$-increasing-inflection source at $x^*_{j_1} = a_{j_1j_11} = a_{j_2j_11}$.

(i_{2a1b}) For $m_1 = 2m_{11}$ and $m_2 = 2m_{21}+1$, the infinite-equilibrium of $x^*_{j_1} = a_{j_1j_11} = a_{j_2j_11}$ with $(\overline{x}_{j_2} \neq a_{j_1j_21})$ has the following properties as in Eqs. (1.43)–(1.46).

- For $a_{j_1j_10}(\overline{x}_{j_2} - a_{j_1j_21})^{n_1} > 0$ and $a_{j_2j_20} > 0$,

$$(a_{j_1j_11}, \overline{x}_{j_2}) = \underbrace{((2m_{11})^{\text{th}}\text{US}, [2(m_{21}-m_{11})+1]^{\text{th}}\text{UP})}_{((2m_{11}),[2(m_{21}-m_{11})+])\text{-up-parabola upper-saddle}}.$$

The infinite-equilibrium is a $((2m_{11})^{\text{th}}\text{US}, [2(m_{21}-m_{11})+1]^{\text{th}}\text{UP})$-up-parabola upper-saddle at $x^*_{j_1} = a_{j_1j_11} = a_{j_2j_11}$.

- For $a_{j_1j_10}(\overline{x}_{j_2} - a_{j_1j_21})^{n_1} < 0$ and $a_{j_2j_20} > 0$,

$$(a_{j_1j_11}, \overline{x}_{j_2}) = \underbrace{((2m_{11})^{\text{th}}\text{LS}, [2(m_{21}-m_{11})+1]^{\text{th}}\text{DP})}_{((2m_{11}),[2(m_{21}-m_{11})+])\text{-down-parabola lower-saddle}}.$$

The infinite-equilibrium is a $((2m_{11})^{\text{th}}\text{LS}, [2(m_{21}-m_{11})+1]^{\text{th}}\text{DP})$-down-parabola lower-saddle at $x^*_{j_1} = a_{j_1j_11} = a_{j_2j_11}$.

- For $a_{j_1j_10}(\overline{x}_{j_2} - a_{j_1j_21})^{n_1} > 0$ and $a_{j_2j_20} < 0$,

$$(a_{j_1j_11}, \overline{x}_{j_2}) = \underbrace{((2m_{11})^{\text{th}}\text{US}, [2(m_{21}-m_{11})+1]^{\text{th}}\text{DP})}_{((2m_{11}),[2(m_{21}-m_{11})+])\text{-down-parabola upper-saddle}}.$$

The infinite-equilibrium is a $((2m_{11})^{\text{th}}\text{US}, [2(m_{21} - m_{11}) + 1]^{\text{th}}\text{DP})$-down-parabola upper-saddle at $x^*_{j_1} = a_{j_1j_11} = a_{j_2j_11}$.

- For $a_{j_1j_10}(\overline{x}_{j_2} - a_{j_1j_21})^{n_1} < 0$ and $a_{j_2j_20} < 0$,

$$(a_{j_1j_11}, \overline{x}_{j_2}) = \underbrace{((2m_{11})^{\text{th}}\text{LS}, [2(m_{21} - m_{11}) + 1]^{\text{th}}\text{UP})}_{((2m_{11}),[2(m_{21}-m_{11})+])\text{-up-parabola lower-saddle}}.$$

The infinite-equilibrium is a $((2m_{11})^{\text{th}}\text{LS}, [2(m_{21}-m_{11})+1]^{\text{th}}\text{UP})$-up-parabola lower-saddle at $x^*_{j_1} = a_{j_1j_11} = a_{j_2j_11}$.

(i_{2a1c}) For $m_1 = 2m_{11} + 1$ and $m_2 = 2m_{21}$, the infinite-equilibrium of $x^*_{j_1} = a_{j_1j_11} = a_{j_2j_11}$ with $(\overline{x}_{j_2} \neq a_{j_1j_21})$ has the following properties as in Eqs. (1.47)–(1.50).

- For $a_{j_1j_10}(\overline{x}_{j_2} - a_{j_1j_21})^{n_1} > 0$ and $a_{j_2j_20} > 0$,

$$(a_{j_1j_11}, \overline{x}_{j_2}) = \underbrace{((2m_{11} + 1)^{\text{th}}\text{SO}, [2(m_{21} - m_{11}) - 1]^{\text{th}}\text{UP})}_{((2m_{11}+1),[2(m_{21}-m_{11})-1])\text{-up-parabola source}}.$$

The infinite-equilibrium is a $((2m_{11} + 1)^{\text{th}}\text{SO}, [2(m_{21} - m_{11}) - 1]^{\text{th}}\text{UP})$-up-parabola source at $x^*_{j_1} = a_{j_1j_11} = a_{j_2j_11}$.

- For $a_{j_1j_10}(\overline{x}_{j_2} - a_{j_1j_21})^{n_1} < 0$ and $a_{j_2j_20} > 0$,

$$(a_{j_1j_11}, \overline{x}_{j_2}) = \underbrace{((2m_{11} + 1)^{\text{th}}\text{SI}, [2(m_{21} - m_{11}) - 1]^{\text{th}}\text{DP})}_{((2m_{11}),[2(m_{21}-m_{11})-1])\text{-down-parabola sink}}.$$

The infinite-equilibrium is a $((2m_{11} + 1)^{\text{th}}\text{SI}, [2(m_{21} - m_{11}) - 1]^{\text{th}}\text{DP})$-down-parabola sink at $x^*_{j_1} = a_{j_1j_11} = a_{j_2j_11}$.

- For $a_{j_1j_10}(\overline{x}_{j_2} - a_{j_1j_21})^{n_1} > 0$ and $a_{j_2j_20} < 0$,

$$(a_{j_1j_11}, \overline{x}_{j_2}) = \underbrace{((2m_{11} + 1)^{\text{th}}\text{SO}, [2(m_{21} - m_{11}) - 1]^{\text{th}}\text{DP})}_{((2m_{11}),[2(m_{21}-m_{11})-1])\text{-down-parabola source}}.$$

The infinite-equilibrium is a $((2m_{11} + 1)^{\text{th}}\text{SO}, [2(m_{21} - m_{11}) - 1]^{\text{th}}\text{DP})$-down-parabola source at $x^*_{j_1} = a_{j_1j_11} = a_{j_2j_11}$.

- For $a_{j_1j_10}(\overline{x}_{j_2} - a_{j_1j_21})^{n_1} < 0$ and $a_{j_2j_20} < 0$,

$$(a_{j_1j_11}, \overline{x}_{j_2}) = \underbrace{((2m_{11} + 1)^{\text{th}}\text{SI}, [2(m_{21} - m_{11}) - 1]^{\text{th}}\text{UP})}_{((2m_{11}+1),[2(m_{21}-m_{11})-1])\text{-up-parabola sink}}.$$

The infinite-equilibrium is a $((2m_{11} + 1)^{\text{th}}\text{SI}, [2(m_{21} - m_{11}) - 1]^{\text{th}}\text{UP})$-up-parabola sink at $x^*_{j_1} = a_{j_1j_11} = a_{j_2j_11}$.

(i_{2a1d}) For $m_1 = 2m_{11}$ and $m_2 = 2m_{21}$, the infinite-equilibrium of $x_{j_1}^* = a_{j_1 j_1 1} = a_{j_2 j_1 1}$ with ($\overline{x}_{j_2} \neq a_{j_1 j_2 1}$) has the following properties as in Eqs. (1.51)–(1.54).

- For $a_{j_1 j_1 0}(\overline{x}_{j_2} - a_{j_1 j_2 1})^{n_1} > 0$ and $a_{j_2 j_2 0} > 0$,

$$(a_{j_1 j_1 1}, \overline{x}_{j_2}) = \underbrace{((2m_{11})^{\text{th}}\text{US}, [2(m_{21} - m_{11})]^{\text{th}}\text{II})}_{((2m_{11}),[2(m_{21}-m_{11})])\text{-increasing-inflection upper-saddle}} .$$

The infinite-equilibrium is a $((2m_{11})^{\text{th}}\text{US}, [2(m_{21} - m_{11})]^{\text{th}}\text{II})$-increasing-inflection upper-saddle at $x_{j_1}^* = a_{j_1 j_1 1} = a_{j_2 j_1 1}$.

- For $a_{j_1 j_1 0}(\overline{x}_{j_2} - a_{j_1 j_2 1})^{n_1} < 0$ and $a_{j_2 j_2 0} > 0$,

$$(a_{j_1 j_1 1}, \overline{x}_{j_2}) = \underbrace{((2m_{11})^{\text{th}}\text{LS}, [2(m_{21} - m_{11})]^{\text{th}}\text{DI})}_{((2m_{11}),[2(m_{21}-m_{11})])\text{-decreasing-inflection lower-saddle}} .$$

The infinite-equilibrium is a $((2m_{11})^{\text{th}}\text{LS}, [2(m_{21} - m_{11})]^{\text{th}}\text{DI})$-decreasing-inflection lower-saddle at $x_{j_1}^* = a_{j_1 j_1 1} = a_{j_2 j_1 1}$.

- For $a_{j_1 j_1 0}(\overline{x}_{j_2} - a_{j_1 j_2 1})^{n_1} > 0$ and $a_{j_2 j_2 0} < 0$,

$$(a_{j_1 j_1 1}, \overline{x}_{j_2}) = \underbrace{((2m_{11})^{\text{th}}\text{US}, [2(m_{21} - m_{11})]^{\text{th}}\text{DI})}_{((2m_{11}),[2(m_{21}-m_{11})])\text{-decreasing-inflection upper-saddle}} .$$

The infinite-equilibrium is a $((2m_{11})^{\text{th}}\text{US}, [2(m_{21} - m_{11})]^{\text{th}}\text{DI})$-decreasing-inflection upper-saddle at $x_{j_1}^* = a_{j_1 j_1 1} = a_{j_2 j_1 1}$.

- For $a_{j_1 j_1 0}(\overline{x}_{j_2} - a_{j_1 j_2 1})^{n_1} < 0$ and $a_{j_2 j_2 0} < 0$,

$$(a_{j_1 j_1 1}, \overline{x}_{j_2}) = \underbrace{((2m_{11})^{\text{th}}\text{LS}, [2(m_{21} - m_{11})]^{\text{th}}\text{II})}_{((2m_{11}),[2(m_{21}-m_{11})])\text{-increasing-inflection lower-saddle}} .$$

The infinite-equilibrium is a $((2m_{11})^{\text{th}}\text{LS}, [2(m_{21} - m_{11})]^{\text{th}}\text{II})$-increasing-inflection lower-saddle at $x_{j_1}^* = a_{j_1 j_1 1} = a_{j_2 j_1 1}$.

(i_{2a2}) At $x_{j_2}^* = a_{j_1 j_2 1}$ and $\overline{x}_{j_1} \neq a_{j_1 j_1 1}, a_{j_2 j_1 1}$, in phase space,

$$\begin{aligned}
\frac{dx_{j_1}}{dx_{j_2}}\bigg|_{x_{j_2}^* = a_{j_1 j_2 1}} &= \frac{a_{j_1 j_1 0}}{a_{j_2 j_2 0}} \frac{(\overline{x}_{j_1} - a_{j_1 j_1 1})^{m_1} (x_{j_2} - a_{j_1 j_2 1})^{n_1}}{(\overline{x}_{j_1} - a_{j_2 j_1 1})^{m_2}}\bigg|_{x_{j_2}^* = a_{j_1 j_2 1}} = 0,\\
&\vdots\\
\frac{d^{n_1} x_{j_1}}{dx_{j_2}^{n_1}}\bigg|_{x_{j_2}^* = a_{j_1 j_2 1}} &= 0.
\end{aligned}$$

If

$$\left.\frac{d^{n_1+1}x_{j_1}}{dx_{j_2}^{n_1+1}}\right|_{x_{j_2}^*=a_{j_1j_21}} = \frac{a_{j_1j_10}}{a_{j_2j_20}} \frac{(\overline{x}_{j_1} - a_{j_1j_11})^{m_1} n_1!}{(\overline{x}_{j_1} - a_{j_2j_11})^{m_2}} > 0,$$

the $(2n_{11}+1)^{\text{th}}$-order up-parabola flows exists for $n_1 = 2n_{11}+1$, and the $(2n_{11})^{\text{th}}$-order increasing-inflection flow exists for $n_1 = 2n_{11}$. If

$$\left.\frac{d^{n_1+1}x_{j_1}}{dx_{j_2}^{n_1+1}}\right|_{x_{j_2}^*=a_{j_1j_21}} = \frac{a_{j_1j_10}}{a_{j_2j_20}} \frac{(\overline{x}_{j_1} - a_{j_1j_11})^{m_1} n_1!}{(\overline{x}_{j_1} - a_{j_2j_11})^{m_2}} < 0,$$

the $(2n_{11}+1)^{\text{th}}$-order down-parabola flows exists for $n_1 = 2n_{11}+1$, and the $(2n_{11})^{\text{th}}$-order decreasing-inflection flow exists for $n_1 = 2n_{11}$.

At $\overline{x}_{j_1} \neq a_{j_1j_11}, a_{j_2j_11}$, let

$$\dot{x}_{j_2} = a_{j_2j_20}(\overline{x}_{j_1} - a_{j_2j_1v})^{m_2}.$$

If $a_{j_2j_20}(\overline{x}_{j_1} - a_{j_2j_1v})^{m_2} > 0$ and $a_{j_2j_20}(\overline{x}_{j_1} - a_{j_2j_1v})^{m_2} < 0$, the parabola and inflection flows are positive and negative in the x_{j_2}-direction, respectively.

Therefore, from the above analysis and the case of (ii_{1b1}) at $\overline{x}_{j_1} = a_{j_2j_11}$, there are 32 types of equilibrium $(x_{j_1}^*, x_{j_2}^*) = (a_{j_1j_11}, a_{j_1j_21})$ on the infinite-equilibrium of $x_{j_1}^* = a_{j_1j_11} = a_{j_2j_11}$.

(i_{2a2a}) For $m_1 = 2m_{11}+1$, $n_1 = 2n_{11}+1$ and $m_2 = 2m_{21}+1$, the equilibrium of $(x_{j_2}^*, x_{j_1}^*) = (a_{j_1j_21}, a_{j_2j_11})$ on the infinite-equilibrium of $x_{j_1}^* = a_{j_1j_11} = a_{j_2j_11}$ has the following properties as in Eqs. (1.55)–(1.58).

- For $a_{j_1j_10} > 0$ and $a_{j_2j_20} > 0$,

$$(a_{j_1j_21}, a_{j_2j_11}) = \underbrace{([(2m_{11}+1),(2n_{11}+1)]^{\text{th}}\text{UP}_{-}\text{:UP}_{+},[2(m_{21}-m_{11})]^{\text{th}}{}_{\text{DI:II}}\text{US})}_{([(2m_{11}+1),(2n_{11}+1)],[2(m_{21}-m_{11})])\text{-inflection up-parabola upper-saddle}}.$$

The equilibrium is a $([(2m_{11}+1),(2n_{11}+1)]^{\text{th}}\text{UP}_{-}\text{:UP}_{+},[2(m_{21}-m_{11})]^{\text{th}}{}_{\text{DI:II}}\text{US})$-inflection up-parabola upper-saddle at $(x_{j_2}^*, x_{j_1}^*) = (a_{j_1j_21}, a_{j_2j_11})$.

- For $a_{j_1j_10} < 0$ and $a_{j_2j_20} > 0$,

$$(a_{j_1j_21}, a_{j_2j_11}) = \underbrace{([(2m_{11}+1),(2n_{11}+1)]^{\text{th}}\text{DP}_{-}\text{:DP}_{+},[2(m_{21}-m_{11})]^{\text{th}}{}_{\text{II:DI}}\text{LS})}_{([(2m_{11}+1),(2n_{11}+1)],[2(m_{21}-m_{11})])\text{-inflection down-parabola lower-saddle}}.$$

The equilibrium is a $([(2m_{11}+1),(2n_{11}+1)]^{\text{th}}\text{DP}_{-}\text{:DP}_{+},[2(m_{21}-m_{11})]^{\text{th}}{}_{\text{II:DI}}\text{LS})$-inflection down-parabola lower-saddle at $(x_{j_2}^*, x_{j_1}^*) = (a_{j_1j_21}, a_{j_2j_11})$.

- For $a_{j_1j_10} > 0$ and $a_{j_2j_20} < 0$,

$$
\begin{aligned}
&(a_{j_1j_2 1}, a_{j_2j_1 1})\\
&= \underbrace{([(2m_{11}+1),(2n_{11}+1)]^{\text{th}}\text{DP}_{+}\text{:DP}_{-},[2(m_{21}-m_{11})]^{\text{th}}{}_{\text{II:DI}}\text{US})}_{([(2m_{11}+1),(2n_{11}+1)],[2(m_{21}-m_{11})])\text{-inflection down-parabola upper-saddle}}.
\end{aligned}
$$

The equilibrium is a $([(2m_{11}+1),(2n_{11}+1)]^{\text{th}}\text{DP}_{+}\text{:DP}_{-},[2(m_{21}-m_{11})]^{\text{th}}{}_{\text{II:DI}}\text{US})$-inflection down-parabola upper-saddle at $(x^*_{j_2}, x^*_{j_1}) = (a_{j_1j_2 1}, a_{j_2j_1 1})$.

- For $a_{j_1j_1 0} < 0$ and $a_{j_2j_2 0} < 0$,

$$
\begin{aligned}
&(a_{j_1j_2 1}, a_{j_2j_1 1})\\
&= \underbrace{([(2m_{11}+1),(2n_{11}+1)]^{\text{th}}\text{UP}_{+}\text{:UP}_{-},[2(m_{21}-m_{11})]^{\text{th}}{}_{\text{DI:II}}\text{LS})}_{([(2m_{11}+1),(2n_{11}+1)],[2(m_{21}-m_{11})])\text{-inflection up-parabola lower-saddle}}.
\end{aligned}
$$

The equilibrium is a $([(2m_{11}+1),(2n_{11}+1)]^{\text{th}}\text{UP}_{+}\text{:UP}_{-},[2(m_{21}-m_{11})]^{\text{th}}{}_{\text{DI:II}}\text{LS})$- inflection up-parabola lower-saddle at $(x^*_{j_2}, x^*_{j_1}) = (a_{j_1j_2 1}, a_{j_2j_1 1})$.

(i_{2a2b}) For $m_1 = 2m_{11}$, $n_1 = 2n_{11} + 1$ and $m_2 = 2m_{21} + 1$, the equilibrium of $(x^*_{j_2}, x^*_{j_1}) = (a_{j_1j_2 1}, a_{j_2j_1 1})$ on the infinite-equilibrium of $x^*_{j_1} = a_{j_1j_1 1} = a_{j_2j_1 1}$ has the following properties as in Eqs. (1.59)–(1.62).

- For $a_{j_1j_1 0} > 0$ and $a_{j_2j_2 0} > 0$,

$$
\begin{aligned}
&(a_{j_1j_2 1}, a_{j_2j_1 1})\\
&= \underbrace{([(2m_{11}),(2n_{11}+1)]^{\text{th}}\text{DP}_{-}\text{:UP}_{+},[2(m_{21}-m_{11})+1]^{\text{th}}{}_{\text{DP}}\text{LS}:{}_{\text{UP}}\text{US})}_{([(2m_{11}),(2n_{11}+1)],[2(m_{21}-m_{11})+1])\text{-positive hyperbolic parabola lower-to-upper saddle}}.
\end{aligned}
$$

The equilibrium is a $([(2m_{11}),(2n_{11}+1)]^{\text{th}}\text{DP}_{-}\text{:UP}_{+},[2(m_{21}-m_{11})+1]^{\text{th}}{}_{\text{DP}}\text{LS:}{}_{\text{UP}}\text{US})$-positive hyperbolic parabola lower-to-upper saddle at $(x^*_{j_2}, x^*_{j_1}) = (a_{j_1j_2 1}, a_{j_2j_1 1})$.

- For $a_{j_1j_1 0} < 0$ and $a_{j_2j_2 0} > 0$,

$$
\begin{aligned}
&(a_{j_1j_2 1}, a_{j_2j_1 1})\\
&= \underbrace{([(2m_{11}),(2n_{11}+1)]^{\text{th}}\text{UP}_{-}\text{:DP}_{+},[2(m_{21}-m_{11})+1]^{\text{th}}{}_{\text{UP}}\text{US}:{}_{\text{DP}}\text{LS})}_{([(2m_{11}),(2n_{11}+1)],[2(m_{21}-m_{11})+1])\text{-CCW circular parabola upper-to-lower saddle}}.
\end{aligned}
$$

The equilibrium is a $([(2m_{11}),(2n_{11}+1)]^{\text{th}}\text{UP}_{-}\text{:DP}_{+},[2(m_{21}-m_{11})+1]^{\text{th}}{}_{\text{UP}}\text{US:}{}_{\text{DP}}\text{LS})$-CCW circular parabola upper-to-lower saddle at $(x^*_{j_2}, x^*_{j_1}) = (a_{j_1j_2 1}, a_{j_2j_1 1})$.

- For $a_{j_1j_1 0} > 0$ and $a_{j_2j_2 0} < 0$,

$$
\begin{aligned}
&(a_{j_1j_2 1}, a_{j_2j_1 1})\\
&= \underbrace{([(2m_{11}),(2n_{11}+1)]^{\text{th}}\text{UP}_{+}\text{:DP}_{-},[2(m_{21}-m_{11})+1]^{\text{th}}{}_{\text{UP}}\text{LS}:{}_{\text{DP}}\text{US})}_{([(2m_{11}),(2n_{11}+1)],[2(m_{21}-m_{11})+1])\text{-CW circular parabola lower-to-upper saddle}}.
\end{aligned}
$$

The equilibrium is a $([(2m_{11}),(2n_{11}+1)]^{\text{th}}\text{UP}_{+}\text{:DP}_{-},[2(m_{21}-m_{11})+1]^{\text{th}}{}_{\text{UP}}\text{LS:}{}_{\text{DP}}\text{US})$-CW circular parabola lower-to-upper saddle at $(x^*_{j_2}, x^*_{j_1}) = (a_{j_1j_21}, a_{j_2j_11})$.

- For $a_{j_1j_10} < 0$ and $a_{j_2j_20} < 0$,

$$(a_{j_1j_21}, a_{j_2j_11}) = \underbrace{([(2m_{11}),(2n_{11}+1)]^{\text{th}}\text{DP}_{+}\text{:UP}_{-},[2(m_{21}-m_{11})+1]^{\text{th}}{}_{\text{DP}}\text{US}:{}_{\text{UP}}\text{LS})}_{([(2m_{11}),(2n_{11}+1)],[2(m_{21}-m_{11})+1])\text{-negative hyperbolic parabola upper-to-lower saddle}}.$$

The equilibrium is a $([(2m_{11}),(2n_{11}+1)]^{\text{th}}\text{DP}_{-}\text{:UP}_{+},[2(m_{21}-m_{11})+1]^{\text{th}}{}_{\text{DP}}\text{US:}{}_{\text{UP}}\text{LS})$-negative hyperbolic parabola upper-to-lower saddle at $(x^*_{j_2}, x^*_{j_1}) = (a_{j_1j_21}, a_{j_2j_11})$.

(i_{2a2c}) For $m_1 = 2m_{11}+1$, $n_1 = 2n_{11}+1$ and $m_2 = 2m_{21}$, the equilibrium of $(x^*_{j_2}, x^*_{j_1}) = (a_{j_1j_21}, a_{j_2j_11})$ on the infinite-equilibrium of $x^*_{j_1} = a_{j_1j_11} = a_{j_2j_11}$ has the following properties as in Eqs. (1.63)–(1.66).

- For $a_{j_1j_10} > 0$ and $a_{j_2j_20} > 0$,

$$(a_{j_1j_21}, a_{j_2j_11}) = \underbrace{([(2m_{11}+1),(2n_{11}+1)]^{\text{th}}\text{DP}_{+}\text{:UP}_{+},[2(m_{21}-m_{11})-1]^{\text{th}}{}_{\text{DP}}\text{SI}:{}_{\text{UP}}\text{SO})}_{([(2m_{11}+1),(2n_{11}+1)],[2(m_{21}-m_{11})-1])\text{-hyperbolic parabola sink-to-source}}.$$

The equilibrium is a $([(2m_{11}+1),(2n_{11}+1)]^{\text{th}}\text{DP}_{+}\text{:UP}_{+},[2(m_{21}-m_{11})-1]^{\text{th}}{}_{\text{DP}}\text{SI}:{}_{\text{UP}}\text{SO})$-hyperbolic parabola sink-to-source at $(x^*_{j_2}, x^*_{j_1}) = (a_{j_1j_21}, a_{j_2j_11})$.

- For $a_{j_1j_10} < 0$ and $a_{j_2j_20} > 0$,

$$(a_{j_1j_21}, a_{j_2j_11}) = \underbrace{([(2m_{11}+1),(2n_{11}+1)]^{\text{th}}\text{UP}_{+}\text{:DP}_{+},[2(m_{21}-m_{11})-1]^{\text{th}}{}_{\text{UP}}\text{SO}:{}_{\text{DP}}\text{SI})}_{([(2m_{11}+1),(2n_{11}+1)],[2(m_{21}-m_{11})-1])\text{-circular parabola source-to-sink}}.$$

The equilibrium is a $([(2m_{11}+1),(2n_{11}+1)]^{\text{th}}\text{UP}_{+}\text{:DP}_{+},[2(m_{21}-m_{11})-1]^{\text{th}}{}_{\text{UP}}\text{SO}:{}_{\text{DP}}\text{SI})$-circular parabola source-to-sink at $(x^*_{j_2}, x^*_{j_1}) = (a_{j_1j_21}, a_{j_2j_11})$.

- For $a_{j_1j_10} > 0$ and $a_{j_2j_20} < 0$,

$$(a_{j_1j_21}, a_{j_2j_11}) = \underbrace{([(2m_{11}+1),(2n_{11}+1)]^{\text{th}}\text{UP}_{-}\text{:DP}_{-},[2(m_{21}-m_{11})-1]^{\text{th}}{}_{\text{UP}}\text{SI}:{}_{\text{DP}}\text{SO})}_{([(2m_{11}+1),(2n_{11}+1)],[2(m_{21}-m_{11})-1])\text{-circular parabola sink-to-source}}.$$

The equilibrium is a $([(2m_{11}+1),(2n_{11}+1)]^{\text{th}}\text{UP}_{-}\text{:DP}_{-},[2(m_{21}-m_{11})-1]^{\text{th}}{}_{\text{UP}}\text{SI}:{}_{\text{DP}}\text{SO})$-circular parabola sink-to-source at $(x^*_{j_2}, x^*_{j_1}) = (a_{j_1j_21}, a_{j_2j_11})$.

- For $a_{j_1j_10} < 0$ and $a_{j_2j_20} < 0$,

$$(a_{j_1j_21}, a_{j_2j_11}) = \underbrace{([(2m_{11}+1),(2n_{11}+1)]^{\text{th}}\text{DP}_-\text{:UP}_-,[2(m_{21}-m_{11})-1]^{\text{th}}{}_{\text{DP}}\text{SO}:{}_{\text{UP}}\text{SI})}_{([(2m_{11}+1),(2n_{11}+1)],[2(m_{21}-m_{11})-1])\text{-hyperbolic parabola source-to-sink}}.$$

The equilibrium is a $([(2m_{11}+1),(2n_{11}+1)]^{\text{th}}\text{DP}_-\text{:UP}_-,[2(m_{21}-m_{11})-1]^{\text{th}}{}_{\text{DP}}\text{SO}:{}_{\text{UP}}\text{SI})$-hyperbolic parabola source-to-sink at $(x^*_{j_2}, x^*_{j_1}) = (a_{j_1j_21}, a_{j_2j_11})$.

(i_{2a2d}) For $m_1 = 2m_{11}$, $n_1 = 2n_{11}+1$ and $m_2 = 2m_{21}$, the equilibrium of $(x^*_{j_2}, x^*_{j_1}) = (a_{j_1j_21}, a_{j_2j_11})$ on the infinite-equilibrium of $x^*_{j_1} = a_{j_1j_11} = a_{j_2j_11}$ has the following properties as in Eqs. (1.67)–(1.70).

- For $a_{j_1j_10} > 0$ and $a_{j_2j_20} > 0$,

$$(a_{j_1j_21}, a_{j_2j_11}) = \underbrace{([(2m_{11}),(2n_{11}+1)]^{\text{th}}\text{UP}_+\text{:UP}_+,[2(m_{21}-m_{11})]^{\text{th}}{}_{\text{DI:II}}\text{US})}_{([(2m_{11}),(2n_{11}+1)],[2(m_{21}-m_{11})])\text{-inflection up-parabola upper-saddle}}.$$

The equilibrium is a $([(2m_{11}),(2n_{11}+1)]^{\text{th}}\text{UP}_+\text{:UP}_+,[2(m_{21}-m_{11})]^{\text{th}}{}_{\text{DI:II}}\text{US})$-inflection up-parabola upper-saddle at $(x^*_{j_2}, x^*_{j_1}) = (a_{j_1j_21}, a_{j_2j_11})$.

- For $a_{j_1j_10} < 0$ and $a_{j_2j_20} > 0$,

$$(a_{j_1j_21}, a_{j_2j_11}) = \underbrace{([(2m_{11}),(2n_{11}+1)]^{\text{th}}\text{DP}_+\text{:DP}_+,[2(m_{21}-m_{11})]^{\text{th}}{}_{\text{II:DI}}\text{US})}_{([(2m_{11}),(2n_{11}+1)],[2(m_{21}-m_{11})])\text{-inflection down-parabola upper-saddle}}.$$

The equilibrium is a $([(2m_{11}),(2n_{11}+1)]^{\text{th}}\text{DP}_+\text{:DP}_+,[2(m_{21}-m_{11})]^{\text{th}}{}_{\text{DI:II}}\text{US})$-inflection down-parabola upper-saddle at $(x^*_{j_2}, x^*_{j_1}) = (a_{j_1j_21}, a_{j_2j_11})$.

- For $a_{j_1j_10} > 0$ and $a_{j_2j_20} < 0$,

$$(a_{j_1j_21}, a_{j_2j_11}) = \underbrace{([(2m_{11}),(2n_{11}+1)]^{\text{th}}\text{DP}_-\text{:DP}_-,[2(m_{21}-m_{11})]^{\text{th}}{}_{\text{II:DI}}\text{LS})}_{([(2m_{11}),(2n_{11}+1)],[2(m_{21}-m_{11})])\text{-inflection down-parabola lower-saddle}}.$$

The equilibrium is a $([(2m_{11}),(2n_{11}+1)]^{\text{th}}\text{DP}_-\text{:DP}_-,[2(m_{21}-m_{11})]^{\text{th}}{}_{\text{DI:II}}\text{LS})$-inflection down-parabola lower-saddle at $(x^*_{j_2}, x^*_{j_1}) = (a_{j_1j_21}, a_{j_2j_11})$.

- For $a_{j_1j_10} < 0$ and $a_{j_2j_20} < 0$,

$$(a_{j_1j_21}, a_{j_2j_11}) = \underbrace{([(2m_{11}),(2n_{11}+1)]^{\text{th}}\text{UP}_-\text{:UP}_-,[2(m_{21}-m_{11})]^{\text{th}}{}_{\text{DI:II}}\text{LS})}_{([(2m_{11}),(2n_{11}+1)],[2(m_{21}-m_{11})])\text{-inflection up-parabola lower-saddle}}.$$

The equilibrium is a $([(2m_{11}),(2n_{11}+1)]^{\text{th}}\text{UP}_{+}\text{:UP}_{+},[2(m_{21}-m_{11})]^{\text{th}}{}_{\text{DI:II}}\text{US})$-inflection up-parabola lower-saddle at $(x_{j_2}^{*}, x_{j_1}^{*}) = (a_{j_1 j_2 1}, a_{j_2 j_1 1})$.

(i_{2a2e}) For $m_1 = 2m_{11} + 1$, $n_1 = 2n_{11}$ and $m_2 = 2m_{21} + 1$, the equilibrium of $(x_{j_2}^{*}, x_{j_1}^{*}) = (a_{j_1 j_2 1}, a_{j_2 j_1 1})$ on the infinite-equilibrium of $x_{j_1}^{*} = a_{j_1 j_1 1} = a_{j_2 j_1 1}$ has the following properties as in Eqs. (1.71)–(1.74).

- For $a_{j_1 j_1 0} > 0$ and $a_{j_2 j_2 0} > 0$,

$$(a_{j_1 j_2 1}, a_{j_2 j_1 1}) = \underbrace{([(2m_{11}+1),(2n_{11})]^{\text{th}}\text{II}_{-}\text{:II}_{+},[2(m_{21}-m_{11})]^{\text{th}}{}_{\text{II:II}}\text{SO})}_{([(2m_{11}+1),(2n_{11})],[2(m_{21}-m_{11})])\text{-increasing-inflection source}}.$$

  The equilibrium is a $([(2m_{11}+1),(2n_{11})]^{\text{th}}\text{II}_{-}\text{:II}_{+},[2(m_{21}-m_{11})]^{\text{th}}{}_{\text{II:II}}\text{SO})$-increasing-inflection source at $(x_{j_2}^{*}, x_{j_1}^{*}) = (a_{j_1 j_2 1}, a_{j_2 j_1 1})$.

- For $a_{j_1 j_1 0} < 0$ and $a_{j_2 j_2 0} > 0$,

$$(a_{j_1 j_2 1}, a_{j_2 j_1 1}) = \underbrace{([(2m_{11}+1),(2n_{11})]^{\text{th}}\text{DI}_{-}\text{:DI}_{+},[2(m_{21}-m_{11})]^{\text{th}}{}_{\text{DI:DI}}\text{SI})}_{([(2m_{11}+1),(2n_{11})],[2(m_{21}-m_{11})])\text{-decreasing-inflection sink}}.$$

  The equilibrium is a $([(2m_{11}+1),(2n_{11})]^{\text{th}}\text{DI}_{-}\text{:DI}_{+},[2(m_{21}-m_{11})]^{\text{th}}{}_{\text{DI:DI}}\text{SI})$-decreasing-inflection sink at $(x_{j_2}^{*}, x_{j_1}^{*}) = (a_{j_1 j_2 1}, a_{j_2 j_1 1})$.

- For $a_{j_1 j_1 0} > 0$ and $a_{j_2 j_2 0} < 0$,

$$(a_{j_1 j_2 1}, a_{j_2 j_1 1}) = \underbrace{([(2m_{11}+1),(2n_{11})]^{\text{th}}\text{DI}_{+}\text{:DI}_{-},[2(m_{21}-m_{11})]^{\text{th}}{}_{\text{DI:DI}}\text{SO})}_{([(2m_{11}+1),(2n_{11})],[2(m_{21}-m_{11})])\text{-decreasing-inflection source}}.$$

  The equilibrium is a $([(2m_{11}+1),(2n_{11})]^{\text{th}}\text{DI}_{+}\text{:DI}_{-},[2(m_{21}-m_{11})]^{\text{th}}{}_{\text{DI:DI}}\text{SO})$-decreasing-inflection source at $(x_{j_2}^{*}, x_{j_1}^{*}) = (a_{j_1 j_2 1}, a_{j_2 j_1 1})$.

- For $a_{j_1 j_1 0} < 0$ and $a_{j_2 j_2 0} < 0$,

$$(a_{j_1 j_2 1}, a_{j_2 j_1 1}) = \underbrace{([(2m_{11}+1),(2n_{11})]^{\text{th}}\text{II}_{+}\text{:II}_{-},[2(m_{21}-m_{11})]^{\text{th}}{}_{\text{II:II}}\text{SI})}_{([(2m_{11}+1),(2n_{11})],[2(m_{21}-m_{11})])\text{-increasing-inflection sink}}.$$

  The equilibrium is a $([(2m_{11}+1),(2n_{11})]^{\text{th}}\text{II}_{+}\text{:II}_{-},[2(m_{21}-m_{11})]^{\text{th}}{}_{\text{II:II}}\text{SI})$-increasing-inflection sink at $(x_{j_2}^{*}, x_{j_1}^{*}) = (a_{j_1 j_2 1}, a_{j_2 j_1 1})$.

(i_{2a2f}) For $m_1 = 2m_{11}$, $n_1 = 2n_{11}$ and $m_2 = 2m_{21} + 1$, the equilibrium of $(x_{j_2}^{*}, x_{j_1}^{*}) = (a_{j_1 j_2 1}, a_{j_2 j_1 1})$ on the infinite-equilibrium of $x_{j_1}^{*} = a_{j_1 j_1 1} = a_{j_2 j_1 1}$ has the following properties as in Eqs. (1.75)–(1.78).

- For $a_{j_1 j_1 0} > 0$ and $a_{j_2 j_2 0} > 0$,

$$(a_{j_1 j_2 1}, a_{j_2 j_1 1}) = \underbrace{([(2m_{11}),(2n_{11})]^{\text{th}}\text{DI}_{-}\text{:II}_{+},[2(m_{21}-m_{11})+1]^{\text{th}}{}_{\text{UP:UP}}\text{US})}_{([(2m_{11}),(2n_{11})],[2(m_{21}-m_{11})+1])\text{-double up-parabola upper-saddle}}.$$

The equilibrium is a $([(2m_{11}),(2n_{11})]^{\text{th}}\text{DI}_{-}\text{:DI}_{+},[2(m_{21}-m_{11})+1]^{\text{th}}{}_{\text{UP:UP}}\text{US})$-double up-parabola upper-saddle at $(x^*_{j_2}, x^*_{j_1}) = (a_{j_1j_2 1}, a_{j_2j_1 1})$.

- For $a_{j_1j_1 0} < 0$ and $a_{j_2j_2 0} > 0$,

$$(a_{j_1j_2 1}, a_{j_2j_1 1}) = \underbrace{([(2m_{11}),(2n_{11})]^{\text{th}}\text{II}_{-}\text{:DI}_{+},[2(m_{21}-m_{11})+1]^{\text{th}}{}_{\text{DP:DP}}\text{LS})}_{([(2m_{11}),(2n_{11})],[2(m_{21}-m_{11})+1])\text{-double down-parabola lower-saddle}}.$$

The equilibrium is a $([(2m_{11}),(2n_{11})]^{\text{th}}\text{II}_{-}\text{:DI}_{+},[2(m_{21}-m_{11})+1]^{\text{th}}{}_{\text{DP:DP}}\text{LS})$-double down-parabola lower-saddle at $(x^*_{j_2}, x^*_{j_1}) = (a_{j_1j_2 1}, a_{j_2j_1 1})$.

- For $a_{j_1j_1 0} > 0$ and $a_{j_2j_2 0} < 0$,

$$(a_{j_1j_2 1}, a_{j_2j_1 1}) = \underbrace{([(2m_{11}),(2n_{11})]^{\text{th}}\text{II}_{+}\text{:DI}_{-},[2(m_{21}-m_{11})+1]^{\text{th}}{}_{\text{DP:DP}}\text{US})}_{([(2m_{11}),(2n_{11})],[2(m_{21}-m_{11})+1])\text{-double down-parabola upper-saddle}}.$$

The equilibrium is a $([(2m_{11}),(2n_{11})]^{\text{th}}\text{II}_{+}\text{:DI}_{-},[2(m_{21}-m_{11})+1]^{\text{th}}{}_{\text{DP:DP}}\text{US})$-double down-parabola upper-saddle at $(x^*_{j_2}, x^*_{j_1}) = (a_{j_1j_2 1}, a_{j_2j_1 1})$.

- For $a_{j_1j_1 0} < 0$ and $a_{j_2j_2 0} < 0$,

$$(a_{j_1j_2 1}, a_{j_2j_1 1}) = \underbrace{([(2m_{11}),(2n_{11})]^{\text{th}}\text{DI}_{+}\text{:II}_{-},[2(m_{21}-m_{11})+1]^{\text{th}}{}_{\text{UP:UP}}\text{LS})}_{([(2m_{11}),(2n_{11})],[2(m_{21}-m_{11})+1])\text{-double up-parabola lower-saddle}}.$$

The equilibrium is a $([(2m_{11}),(2n_{11})]^{\text{th}}\text{DI}_{+}\text{:II}_{-},[2(m_{21}-m_{11})+1]^{\text{th}}{}_{\text{UP:UP}}\text{LS})$-double up-parabola lower-saddle at $(x^*_{j_2}, x^*_{j_1}) = (a_{j_1j_2 1}, a_{j_2j_1 1})$.

$(\text{i}_{2\text{a}2\text{g}})$ For $m_1 = 2m_{11}+1$, $n_1 = 2n_{11}$ and $m_2 = 2m_{21}$, the equilibrium of $(x^*_{j_2}, x^*_{j_1}) = (a_{j_1j_2 1}, a_{j_2j_1 1})$ on the infinite-equilibrium of $x^*_{j_1} = a_{j_1j_1 1} = a_{j_2j_1 1}$ has the following properties as in Eqs. (1.79)–(1.82).

- For $a_{j_1j_1 0} > 0$ and $a_{j_2j_2 0} > 0$,

$$(a_{j_1j_2 1}, a_{j_2j_1 1}) = \underbrace{([(2m_{11}+1),(2n_{11})]^{\text{th}}\text{DI}_{+}\text{:II}_{+},[2(m_{21}-m_{11})-1]^{\text{th}}{}_{\text{UP:UP}}\text{SO})}_{([(2m_{11}+1),(2n_{11})],[2(m_{21}-m_{11})-1])\text{-double up-parabola source}}.$$

The equilibrium is a $([(2m_{11}+1),(2n_{11})]^{\text{th}}\text{DI}_{+}\text{:II}_{+},[2(m_{21}-m_{11})-1]^{\text{th}}{}_{\text{UP:UP}}\text{SO})$-double up-parabola source at $(x^*_{j_2}, x^*_{j_1}) = (a_{j_1j_2 1}, a_{j_2j_1 1})$.

- For $a_{j_1j_1 0} < 0$ and $a_{j_2j_2 0} > 0$,

$$(a_{j_1j_2 1}, a_{j_2j_1 1}) = \underbrace{([(2m_{11}+1),(2n_{11})]^{\text{th}}\text{II}_{+}\text{:DI}_{+},[2(m_{21}-m_{11})-1]^{\text{th}}{}_{\text{DP:DP}}\text{SI})}_{([(2m_{11}+1),(2n_{11})],[2(m_{21}-m_{11})-1])\text{-double down-parabola sink}}.$$

The equilibrium is a $([(2m_{11}+1),(2n_{11})]^{\text{th}}\text{II}_{+}\text{:DI}_{+},[2(m_{21}-m_{11})-1]^{\text{th}}{}_{\text{DP:DP}}\text{SI})$-double down-parabola sink at $(x^*_{j_2}, x^*_{j_1}) = (a_{j_1j_2 1}, a_{j_2j_1 1})$.

- For $a_{j_1j_10} > 0$ and $a_{j_2j_20} < 0$,

$$(a_{j_1j_21}, a_{j_2j_11}) = \underbrace{([(2m_{11}+1),(2n_{11})]^{\text{th}}\text{II}_{-}\text{:DI}_{-},[2(m_{21}-m_{11})-1]^{\text{th}}{}_{\text{DP:DP}}\text{SO})}_{([(2m_{11}+1),(2n_{11})],[2(m_{21}-m_{11})+1])\text{-double down-parabola source}}.$$

The equilibrium is a $([(2m_{11}+1),(2n_{11})]^{\text{th}}\text{II}_{-}\text{:DI}_{-},[2(m_{21}-m_{11})-1]^{\text{th}}{}_{\text{DP:DP}}\text{SO})$-double down-parabola source at $(x^*_{j_2}, x^*_{j_1}) = (a_{j_1j_21}, a_{j_2j_11})$.

- For $a_{j_1j_10} < 0$ and $a_{j_2j_20} < 0$,

$$(a_{j_1j_21}, a_{j_2j_11}) = \underbrace{([(2m_{11}+1),(2n_{11})]^{\text{th}}\text{DI}_{-}\text{:II}_{-},[2(m_{21}-m_{11})-1]^{\text{th}}{}_{\text{UP:UP}}\text{SI})}_{([(2m_{11}+1),(2n_{11})],[2(m_{21}-m_{11})+1])\text{-double up-parabola sink}}. \tag{1.82}$$

The equilibrium is a $([(2m_{11}+1),(2n_{11})]^{\text{th}}\text{DI}_{-}\text{:II}_{-},[2(m_{21}-m_{11})-1]^{\text{th}}{}_{\text{UP:UP}}\text{SI})$-double up-parabola sink at $(x^*_{j_2}, x^*_{j_1}) = (a_{j_1j_21}, a_{j_2j_11})$.

(i_{2a2h}) For $m_1 = 2m_{11}, n_1 = 2n_{11}$ and $m_2 = 2m_{21}$, the equilibrium of $(x^*_{j_2}, x^*_{j_1}) = (a_{j_1j_21}, a_{j_2j_11})$ on the infinite-equilibrium of $x^*_{j_1} = a_{j_1j_11} = a_{j_2j_11}$ has the following properties as in Eqs. (1.83)–(1.86).

- For $a_{j_1j_10} > 0$ and $a_{j_2j_20} > 0$,

$$(a_{j_1j_21}, a_{j_2j_11}) = \underbrace{([(2m_{11}),(2n_{11})]^{\text{th}}\text{II}_{+}\text{:II}_{+},[2(m_{21}-m_{11})]^{\text{th}}{}_{\text{II:II}}\text{US})}_{([(2m_{11}),(2n_{11})],[2(m_{21}-m_{11})])\text{-increasing-inflection upper-saddle}}.$$

The equilibrium is a $([(2m_{11}),(2n_{11})]^{\text{th}}\text{II}_{+}\text{:II}_{+},[2(m_{21}-m_{11})]^{\text{th}}{}_{\text{II:II}}\text{US})$-increasing-inflection upper-saddle at $(x^*_{j_2}, x^*_{j_1}) = (a_{j_1j_21}, a_{j_2j_11})$.

- For $a_{j_1j_10} < 0$ and $a_{j_2j_20} > 0$,

$$(a_{j_1j_21}, a_{j_2j_11}) = \underbrace{([(2m_{11}),(2n_{11})]^{\text{th}}\text{DI}_{+}\text{:DI}_{+},[2(m_{21}-m_{11})]^{\text{th}}{}_{\text{DI:DI}}\text{LS})}_{([(2m_{11}),(2n_{11})],[2(m_{21}-m_{11})])\text{-decreasing-inflection lower-saddle}}.$$

The equilibrium is a $([(2m_{11}),(2n_{11})]^{\text{th}}\text{DI}_{+}\text{:DI}_{+},[2(m_{21}-m_{11})]^{\text{th}}{}_{\text{DI:DI}}\text{LS})$-decreasing-inflection lower-saddle at $(x^*_{j_2}, x^*_{j_1}) = (a_{j_1j_21}, a_{j_2j_11})$.

- For $a_{j_1j_10} > 0$ and $a_{j_2j_20} < 0$,

$$(a_{j_1j_21}, a_{j_2j_11}) = \underbrace{([(2m_{11}),(2n_{11})]^{\text{th}}\text{DI}_{-}\text{:DI}_{-},[2(m_{21}-m_{11})]^{\text{th}}{}_{\text{DI:DI}}\text{US})}_{([(2m_{11}),(2n_{11})],[2(m_{21}-m_{11})])\text{-decreasing-inflection upper-saddle}}.$$

The equilibrium is a $([(2m_{11}),(2n_{11})]^{\text{th}}\text{DI}_{-}\text{:DI}_{-},[2(m_{21}-m_{11})]^{\text{th}}{}_{\text{DI:DI}}\text{US})$-decreasing-inflection upper-saddle at $(x^*_{j_2}, x^*_{j_1}) = (a_{j_1j_21}, a_{j_2j_11})$.

- For $a_{j_1j_10} < 0$ and $a_{j_2j_20} < 0$,

$$(a_{j_1j_21}, a_{j_2j_11}) = \underbrace{([(2m_{11}),(2n_{11})]^{\text{th}}\text{II}_{-}\text{:II}_{-},[2(m_{21}-m_{11})]^{\text{th}}{}_{\text{II:II}}\text{LS})}_{([(2m_{11}),(2n_{11})],[2(m_{21}-m_{11})])\text{-increasing-inflection lower-saddle}}.$$

The equilibrium is a $([(2m_{11}),(2n_{11})]^{\text{th}}\text{II}_{-}\text{:II}_{-},[2(m_{21}-m_{11})]^{\text{th}}{}_{\text{II:II}}\text{LS})$-increasing-inflection lower-saddle at $(x_{j_2}^*, x_{j_1}^*) = (a_{j_1j_21}, a_{j_2j_11})$.

(i_{2b}) For $n_1 < n_2$, the first integral manifold is

$$-\frac{1}{m_1-m_2-1}\Big[\frac{1}{(x_{j_1}-a_{j_1j_11})^{m_1-m_2+1}} - \frac{1}{(x_{j_10}-a_{j_1j_11})^{m_1-m_2+1}}\Big]$$
$$= \frac{a_{j_1j_10}}{a_{j_2j_20}}\frac{1}{n_1+1}[(x_{j_2}-a_{j_1j_21})^{n_1+1} - (x_{j_20}-a_{j_1j_11})^{n_1+1}],$$

for $m_2 - m_1 + 1 \neq 0$;

$$\frac{1}{m_2-m_1+1}[(x_{j_1}-a_{j_1j_11})^{m_2-m_1+1} - (x_{j_10}-a_{j_1j_11})^{m_2-m_1+1}]$$
$$= \ln\frac{|x_{j_1}-a_{j_1j_11}|}{|x_{j_10}-a_{j_1j_11}|},$$

for $m_2 - m_1 + 1 = 0$.

For $x_{j_1}^* = a_{j_1j_11} = a_{j_2j_11}$ and $\overline{x}_{j_2} \neq a_{j_1j_21}$, if

$$\frac{dx_{j_2}}{dx_{j_2}}\Big|_{x_{j_1}^*=a_{j_1j_11}\pm\varepsilon} = \lim_{\varepsilon\to 0}\frac{a_{j_2j_20}}{a_{j_1j_10}(\pm\varepsilon)^{m_1-m_2}(\overline{x}_{j_2}-a_{j_1j_21})^{n_1}} = +\infty,$$

the flow at $x_{j_1}^* = a_{j_1j_11} = a_{j_2j_11}$ is a $[2(m_{11}-m_{21})]^{\text{th}}$-order increasing-concave flow for $m_1 = 2m_{11}+1, m_2 = 2m_{21}+1$ and $m_1 = 2m_{11}, m_2 = 2m_{21}$, if

$$\frac{dx_{j_2}}{dx_{j_2}}\Big|_{x_{j_1}^*=a_{j_1j_11}\pm\varepsilon} = \lim_{\varepsilon\to 0}\frac{a_{j_2j_20}}{a_{j_1j_10}(\pm\varepsilon)^{m_1-m_2}(\overline{x}_{j_2}-a_{j_1j_21})^{n_1}} = -\infty,$$

the flow at $x_{j_2}^* = a_{j_2j_21} = a_{j_1j_21}$ is a $[2(m_{11}-m_{21})]^{\text{th}}$-order decreasing-concave flow for $m_1 = 2m_{11}+1, m_2 = 2m_{21}+1$ and $m_1 = 2m_{11}, m_2 = 2m_{21}$.

For $x_{j_1}^* = a_{j_1j_11} = a_{j_2j_11}$ and $\overline{x}_{j_2} \neq a_{j_1j_21}$, if

$$\frac{dx_{j_2}}{dx_{j_2}}\Big|_{x_{j_1}^*=a_{j_1j_11}\pm\varepsilon} = \lim_{\varepsilon\to 0}\frac{a_{j_2j_20}}{a_{j_1j_10}(\pm\varepsilon)^{m_1-m_2}(\overline{x}_{j_2}-a_{j_1j_21})^{n_1}} = \pm\infty,$$

the flow at $x_{j_1}^* = a_{j_1j_11} = a_{j_2j_11}$ is a $[2(m_{11}-m_{21})\pm 1]^{\text{th}}$-order up-asymptote flow for $m_1 = 2m_{11}, m_2 = 2m_{21}+1$ and $m_1 = 2m_{11}+1, m_2 = 2m_{21}$, if

$$\frac{dx_{j_2}}{dx_{j_2}}\bigg|_{x_{j_1}^*=a_{j_1j_11}\pm\varepsilon}=\lim_{\varepsilon\to 0}\frac{a_{j_2j_20}}{a_{j_1j_10}(\pm\varepsilon)^{m_1-m_2}(\overline{x}_{j_2}-a_{j_1j_21})^{n_1}}=\mp\infty,$$

the flow at $x_{j_1}^*=a_{j_1j_11}=a_{j_2j_11}$ is a $[2(m_{11}-m_{21})\pm 1]^{\text{th}}$-order down-asymptote flow for $m_1=2m_{11}, m_2=2m_{21}+1$ and $m_1=2m_{11}+1, m_2=2m_{21}$.

The variational equation at $x_{j_1}^*=a_{j_1j_11}=a_{j_2j_11}$ is

$$\Delta\dot{x}_{j_1}=a_{j_1j_10}(\overline{x}_{j_2}-a_{j_1j_21})^{n_1}(\Delta x_{j_1})^{m_1}.$$

The flows at $x_{j_1}^*=a_{j_1j_11}=a_{j_2j_11}$ and $\overline{x}_{j_2}\neq a_{j_1j_21}$ are $(2m_{11}+1)^{\text{th}}$-source and sink flows for $a_{j_1j_10}(\overline{x}_{j_2}-a_{j_1j_21})^{n_1}>0$ and $a_{j_1j_10}(\overline{x}_{j_2}-a_{j_1j_21})^{n_1}<0$, respectively. The flows at $x_{j_1}^*=a_{j_1j_11}=a_{j_2j_11}$ and $\overline{x}_{j_2}\neq a_{j_1j_21}$ are $(2m_{11})^{\text{th}}$-upper saddle and lower-saddle flows for $a_{j_1j_10}(\overline{x}_{j_2}-a_{j_1j_21})^{n_1}>0$ and $a_{j_1j_10}(\overline{x}_{j_2}-a_{j_1j_21})^{n_1}<0$, respectively.

Thus, based on the above analysis, the flows at $x_{j_1}^*=a_{j_1j_11}=a_{j_2j_11}$ and on both sides of $x_{j_2}^*=a_{j_1j_21}$ are $((2m_{11}+1)^{\text{th}}\text{XX}, [\cdot]^{\text{th}}\text{YY})$-source and sink flows for $a_{j_1j_10}(\overline{x}_{j_2}-a_{j_1j_20})^{n_1}>0$ and $a_{j_1j_10}(\overline{x}_{j_2}-a_{j_1j_20})^{n_1}<0$, respectively. The flows at $x_{j_1}^*=a_{j_1j_11}=a_{j_2j_11}$ and on both sides of $x_{j_2}^*=a_{j_1j_21}$ are $((2m_{11})^{\text{th}}\text{XX}, [\cdot]^{\text{th}}\text{YY})$-upper and lower-saddle flows for $a_{j_1j_10}(\overline{x}_{j_2}-a_{j_1j_20})^{n_1}>0$ and $a_{j_1j_10}(\overline{x}_{j_2}-a_{j_1j_20})^{n_1}<0$, respectively. $[2(m_{11}-m_{21})]^{\text{th}}\text{YY}$ are the $[2(m_{11}-m_{21})]^{\text{th}}$-order increasing and decreasing concave flows, and $[2(m_{21}-m_{11})\pm 1]^{\text{th}}\text{YY}$ are the $[2(m_{21}-m_{11})\pm 1]^{\text{th}}$-order up and down asymptote flows.

(i_{2b1}) There are 16 types of infinite-equilibriums at $x_{j_1}^*=a_{j_1j_11}=a_{j_2j_11}$.

(i_{2b1a}) For $m_1=2m_{11}+1$ and $m_2=2m_{21}+1$, the infinite-equilibrium of $x_{j_1}^*=a_{j_1j_11}=a_{j_2j_11}$ and $\overline{x}_{j_2}\neq a_{j_1j_21}$ has the following properties as in Eqs. (1.88)–(1.91).

- For $a_{j_1j_10}(\overline{x}_{j_2}-a_{j_1j_21})^{n_1}>0$ and $a_{j_2j_20}>0$,

$$(a_{j_1j_11},\overline{x}_{j_2})=\underbrace{((2m_{11}+1)^{\text{th}}\text{SO},[2(m_{11}-m_{21})]^{\text{th}}\text{IC})}_{((2m_{11}+1),[2(m_{11}-m_{21})])\text{-increasing concave source}}.$$

  The infinite-equilibrium is a $((2m_{11}+1)^{\text{th}}\text{SO},[2(m_{11}-m_{21})]^{\text{th}}\text{IC})$-increasing-concave source at $x_{j_1}^*=a_{j_1j_11}=a_{j_2j_11}$.

- For $a_{j_1j_10}(\overline{x}_{j_2}-a_{j_1j_21})^{n_1}<0$ and $a_{j_2j_20}>0$,

$$(a_{j_1j_11},\overline{x}_{j_2})=\underbrace{((2m_{11}+1)^{\text{th}}\text{SI},[2(m_{11}-m_{21})]^{\text{th}}\text{DC})}_{((2m_{11}+1),[2(m_{11}-m_{21})])\text{-decreasing concave sink}}.$$

  The infinite-equilibrium is a $((2m_{11}+1)^{\text{th}}\text{SI},[2(m_{11}-m_{21})]^{\text{th}}\text{DC})$-decreasing-concave sink at $x_{j_1}^*=a_{j_1j_11}=a_{j_2j_11}$.

- For $a_{j_1j_10}(\overline{x}_{j_2}-a_{j_1j_21})^{n_1}>0$ and $a_{j_2j_20}<0$,

$$(a_{j_1j_11},\overline{x}_{j_2})=\underbrace{((2m_{11}+1)^{\text{th}}\text{SO},[2(m_{11}-m_{21})]^{\text{th}}\text{DC})}_{((2m_{11}+1),[2(m_{11}-m_{21})])\text{-decreasing concave source}}.$$

The infinite-equilibrium is a $((2m_{11}+1)^{\text{th}}\text{SO},[2(m_{11}-m_{21})]^{\text{th}}\text{DC})$-decreasing-concave source at $x_{j_1}^* = a_{j_1j_11} = a_{j_2j_11}$.

- For $a_{j_1j_10}(\overline{x}_{j_2} - a_{j_1j_21})^{n_1} < 0$ and $a_{j_2j_20} < 0$,

$$(a_{j_1j_11}, \overline{x}_{j_2}) = \underbrace{((2m_{11}+1)^{\text{th}}\text{SI},[2(m_{11}-m_{21})]^{\text{th}}\text{IC})}_{((2m_{11}+1),[2(m_{11}-m_{21})])\text{increasing concave sink}}.$$

The infinite-equilibrium is a $((2m_{11}+1)^{\text{th}}\text{SI},[2(m_{11}-m_{21})]^{\text{th}}\text{IC})$-increasing-concave sink at $x_{j_1}^* = a_{j_1j_11} = a_{j_2j_11}$.

(i_{2b1b}) For $m_1 = 2m_{11}$ and $m_2 = 2m_{21}+1$, the infinite-equilibrium of $x_{j_1}^* = a_{j_1j_11} = a_{j_2j_11}$ with $\overline{x}_{j_2} \neq a_{j_1j_21}$ has the following properties as in Eqs. (1.92)–(1.95).

- For $a_{j_1j_10}(\overline{x}_{j_2} - a_{j_1j_21})^{n_1} > 0$ and $a_{j_2j_20} > 0$,

$$(a_{j_1j_11}, \overline{x}_{j_2}) = \underbrace{((2m_{11})^{\text{th}}\text{US},[2(m_{11}-m_{21})-1]^{\text{th}}\text{DA})}_{((2m_{11}),[2(m_{11}-m_{21})-1])\text{-down-asymptote upper-saddle}}.$$

The infinite-equilibrium is a $((2m_{11})^{\text{th}}\text{US},[2(m_{11}-m_{21})-1]^{\text{th}}\text{DA})$-down-asymptote upper-saddle at $x_{j_1}^* = a_{j_1j_11} = a_{j_2j_11}$.

- For $a_{j_1j_10}(\overline{x}_{j_2} - a_{j_1j_21})^{n_1} < 0$ and $a_{j_2j_20} > 0$,

$$(a_{j_1j_11}, \overline{x}_{j_2}) = \underbrace{((2m_{11})^{\text{th}}\text{LS},[2(m_{11}-m_{21})-1]^{\text{th}}\text{UA})}_{((2m_{11}),[2(m_{11}-m_{21})-1])\text{-up-asymptote lower-saddle}}.$$

The infinite-equilibrium is a $((2m_{11})^{\text{th}}\text{LS},[2(m_{11}-m_{21})-1]^{\text{th}}\text{UA})$-up-asymptote lower-saddle at $x_{j_1}^* = a_{j_1j_11} = a_{j_2j_11}$.

- For $a_{j_1j_10}(\overline{x}_{j_2} - a_{j_1j_21})^{n_1} > 0$ and $a_{j_2j_20} < 0$,

$$(a_{j_1j_11}, \overline{x}_{j_2}) = \underbrace{((2m_{11})^{\text{th}}\text{US},[2(m_{11}-m_{21})-1]^{\text{th}}\text{UA})}_{((2m_{11}),[2(m_{11}-m_{21})-1])\text{-up-asymptote upper-saddle}}.$$

The infinite-equilibrium is a $((2m_{11})^{\text{th}}\text{US},[2(m_{11}-m_{21})-1]^{\text{th}}\text{UA})$-up-asymptote upper-saddle at $x_{j_1}^* = a_{j_1j_11} = a_{j_2j_11}$.

- For $a_{j_1j_10}(\overline{x}_{j_2} - a_{j_1j_21})^{n_1} < 0$ and $a_{j_2j_20} < 0$,

$$(a_{j_1j_11}, \overline{x}_{j_2}) = \underbrace{((2m_{11})^{\text{th}}\text{LS},[2(m_{11}-m_{21})-1]^{\text{th}}\text{DA})}_{((2m_{11}),[2(m_{11}-m_{21})-1])\text{-down-asymptote lower-saddle}}.$$

The infinite-equilibrium is a $((2m_{11})^{\text{th}}\text{LS},[2(m_{11}-m_{21})-1]^{\text{th}}\text{DA})$-down-asymptote lower-saddle at $x_{j_1}^* = a_{j_1j_11} = a_{j_2j_11}$.

(i_{2b1c}) For $m_1 = 2m_{11}+1$ and $m_2 = 2m_{21}$, the infinite-equilibrium of $x_{j_1}^* = a_{j_1j_11} = a_{j_2j_11}$ and $\overline{x}_{j_2} \neq a_{j_1j_21}$ has the following properties as in Eqs. (1.96)–(1.99).

- For $a_{j_1j_1 0}(\overline{x}_{j_2} - a_{j_1j_2 1})^{n_1} > 0$ and $a_{j_2j_2 0} > 0$,

$$(a_{j_1j_1 1}, \overline{x}_{j_2}) = \underbrace{((2m_{11}+1)^{\text{th}}\text{SO},[2(m_{11}-m_{21})+1]^{\text{th}}\text{DA})}_{((2m_{11}+1),[2(m_{11}-m_{21})+1])\text{-down-asymptote source}}.$$

The infinite-equilibrium is a $((2m_{11}+1)^{\text{th}}\text{SO},[2(m_{11}-m_{21})+1]^{\text{th}}\text{DA})$-down-asymptote source at $x^*_{j_1} = a_{j_1j_1 1} = a_{j_2j_1 1}$.

- For $a_{j_1j_1 0}(\overline{x}_{j_2} - a_{j_1j_2 1})^{n_1} < 0$ and $a_{j_2j_2 0} > 0$,

$$(a_{j_1j_1 1}, \overline{x}_{j_2}) = \underbrace{((2m_{11}+1)^{\text{th}}\text{SI},[2(m_{11}-m_{21})+1]^{\text{th}}\text{UA})}_{((2m_{11}+1),[2(m_{11}-m_{21})+1])\text{-up-asymptote sink}}.$$

The infinite-equilibrium is a $((2m_{11}+1)^{\text{th}}\text{SI},[2(m_{11}-m_{21})+1]^{\text{th}}\text{UA})$-up-asymptote sink at $x^*_{j_1} = a_{j_1j_1 1} = a_{j_2j_1 1}$.

- For $a_{j_1j_1 0}(\overline{x}_{j_2} - a_{j_1j_2 1})^{n_1} > 0$ and $a_{j_2j_2 0} < 0$,

$$(a_{j_1j_1 1}, \overline{x}_{j_2}) = \underbrace{((2m_{11}+1)^{\text{th}}\text{SO},[2(m_{11}-m_{21})+1]^{\text{th}}\text{UA})}_{((2m_{11}+1),[2(m_{11}-m_{21})+1])\text{-up-asymptote source}}.$$

The infinite-equilibrium is a $((2m_{11}+1)^{\text{th}}\text{SO},[2(m_{11}-m_{21})+1]^{\text{th}}\text{UA})$-up-asymptote source at $x^*_{j_1} = a_{j_1j_1 1} = a_{j_2j_1 1}$.

- For $a_{j_1j_1 0}(\overline{x}_{j_2} - a_{j_1j_2 1})^{n_1} < 0$ and $a_{j_2j_2 0} < 0$,

$$(a_{j_1j_1 1}, \overline{x}_{j_2}) = \underbrace{((2m_{11}+1)^{\text{th}}\text{SI},[2(m_{11}-m_{21})+1]^{\text{th}}\text{DA})}_{((2m_{11}+1),[2(m_{11}-m_{21})+1])\text{-down-asymptote sink}}.$$

The infinite-equilibrium is a $((2m_{11}+1)^{\text{th}}\text{SI},[2(m_{11}-m_{21})+1]^{\text{th}}\text{DA})$-down-asymptote lower-saddle at $x^*_{j_1} = a_{j_1j_1 1} = a_{j_2j_1 1}$.

($\text{i}_{2\text{b}1\text{d}}$) For $m_1 = 2m_{11}$ and $m_2 = 2m_{21}$, the infinite-equilibrium of $x^*_{j_1} = a_{j_1j_1 1} = a_{j_2j_1 1}$ with $\overline{x}_{j_1} \neq a_{j_1j_1 1}$ has the following properties as in Eqs. (1.100)–(1.103).

- For $a_{j_1j_1 0}(\overline{x}_{j_2} - a_{j_1j_2 1})^{n_1} > 0$ and $a_{j_2j_2 0} > 0$,

$$(a_{j_1j_1 1}, \overline{x}_{j_2}) = \underbrace{((2m_{11})^{\text{th}}\text{US},[2(m_{11}-m_{21})]^{\text{th}}\text{IC})}_{((2m_{11}),[2(m_{11}-m_{21})])\text{-increasing-concave upper-saddle}}.$$

The infinite-equilibrium is a $((2m_{11})^{\text{th}}\text{US},[2(m_{11}-m_{21})]^{\text{th}}\text{IC})$-increasing-concave upper-saddle at $x^*_{j_1} = a_{j_1j_1 1} = a_{j_2j_1 1}$.

- For $a_{j_1j_1 0}(\overline{x}_{j_2} - a_{j_1j_2 1})^{n_1} < 0$ and $a_{j_2j_2 0} > 0$,

$$(a_{j_1j_1 1}, \overline{x}_{j_2}) = \underbrace{((2m_{11})^{\text{th}}\text{LS},[2(m_{11}-m_{21})]^{\text{th}}\text{DC})}_{((2m_{11}),[2(m_{11}-m_{21})])\text{-decreasing-concave lower-saddle}}.$$

The infinite-equilibrium is a $((2m_{11})^{\text{th}}\text{LS},[2(m_{11}-m_{21})]^{\text{th}}\text{DC})$-decreasing-concave lower-saddle at $x_{j_1}^*=a_{j_1j_11}=a_{j_2j_11}$.

- For $a_{j_1j_10}(\overline{x}_{j_2}-a_{j_1j_21})^{n_1}>0$ and $a_{j_2j_20}<0$,

$$(a_{j_1j_11},\overline{x}_{j_2})=\underbrace{((2m_{11})^{\text{th}}\text{US},[2(m_{11}-m_{21})]^{\text{th}}\text{DC})}_{((2m_{11}),[2(m_{11}-m_{21})])\text{-decreasing-concave upper-saddle}}.$$

The infinite-equilibrium is a $((2m_{11})^{\text{th}}\text{US},[2(m_{11}-m_{21})]^{\text{th}}\text{DC})$-decreasing-concave upper-saddle at $x_{j_1}^*=a_{j_1j_11}=a_{j_2j_11}$.

- For $a_{j_1j_10}(\overline{x}_{j_2}-a_{j_1j_21})^{n_1}<0$ and $a_{j_2j_20}<0$,

$$(a_{j_1j_11},\overline{x}_{j_2})=\underbrace{((2m_{11})^{\text{th}}\text{LS},[2(m_{11}-m_{21})]^{\text{th}}\text{IC})}_{((2m_{11}),[2(m_{11}-m_{21})])\text{-increasing-concave lower-saddle}}.$$

The infinite-equilibrium is a $((2m_{11})^{\text{th}}\text{LS},[2(m_{11}-m_{21})]^{\text{th}}\text{IC})$-increasing-concave lower-saddle at $x_{j_1}^*=a_{j_1j_11}=a_{j_2j_11}$.

(i_{2b2}) At $x_{j_2}^*=a_{j_1j_21}$ and $\overline{x}_{j_1}\neq a_{j_1j_11},a_{j_2j_11}$, in phase space,

$$\frac{dx_{j_1}}{dx_{j_2}}\bigg|_{x_{j_2}^*=a_{j_1j_21}}=\frac{a_{j_1j_10}}{a_{j_2j_20}}\frac{(\overline{x}_{j_1}-a_{j_1j_11})^{m_1}(x_{j_2}-a_{j_1j_21})^{n_1}}{(\overline{x}_{j_1}-a_{j_2j_11})^{m_2}}\bigg|_{x_{j_2}^*=a_{j_1j_21}}=0,$$

$$\vdots$$

$$\frac{d^{n_1}x_{j_1}}{dx_{j_2}^{n_1}}\bigg|_{x_{j_2}^*=a_{j_1j_21}}=0.$$

If

$$\frac{d^{n_{j_1u_1}+1}x_{j_1}}{dx_{j_2}^{n_{j_1u_1}+!}}\bigg|_{x_{j_2}^*=a_{j_1j_2u_1}}=\frac{a_{j_1j_10}}{a_{j_2j_20}}\frac{(\overline{x}_{j_1}-a_{j_1j_11})^{m_1}n_1!}{(\overline{x}_{j_1}-a_{j_2j_11})^{m_2}}>0,$$

the $(2n_{11}+1)^{\text{th}}$-order up-parabola flows exists for $n_1=2n_{11}+1$, and the $(2n_{11})^{\text{th}}$-order increasing-inflection flow exists for $n_1=2n_{11}$. If

$$\frac{d^{n_{j_1u_1}+1}x_{j_1}}{dx_{j_2}^{n_{j_1u_1}+!}}\bigg|_{x_{j_2}^*=a_{j_1j_2u_1}}=\frac{a_{j_1j_10}}{a_{j_2j_20}}\frac{(\overline{x}_{j_1}-a_{j_1j_11})^{m_1}n_1!}{(\overline{x}_{j_1}-a_{j_2j_11})^{m_2}}<0,$$

the $(2n_{11}+1)^{\text{th}}$-order down-parabola flows exists for $n_1=2n_{11}+1$, and the $(2n_{11})^{\text{th}}$-order decreasing-inflection flow exists for $n_1=2n_{11}$. If

At $\overline{x}_{j_1}\neq a_{j_1j_11},a_{j_2j_11}$, let

$$\dot{x}_{j_2} = a_{j_2j_20}(\overline{x}_{j_1} - a_{j_2j_11})^{m_2}.$$

If $a_{j_2j_20}(\overline{x}_{j_1} - a_{j_2j_11})^{m_2} > 0$ and $a_{j_2j_20}(\overline{x}_{j_1} - a_{j_2j_11})^{m_2} < 0$, the parabola and inflection flows are positive and negative in the x_{j_2}-direction, respectively.

Therefore, from the above analysis and the case of ($\text{ii}_{1\text{b}1}$) at $\overline{x}_{j_1} = a_{j_2j_11}$, there are 32 types of equilibrium $(x^*_{j_2}, x^*_{j_1}) = (a_{j_1j_21}, a_{j_2j_11})$ on the infinite-equilibrium of $x^*_{j_1} = a_{j_1j_11} = a_{j_2j_11}$.

($\text{i}_{2\text{b}2\text{a}}$) For $m_1 = 2m_{11} + 1$, $n_1 = 2n_{11} + 1$, and $m_2 = 2m_{21} + 1$, the equilibriums of $(x^*_{j_2}, x^*_{j_1}) = (a_{j_1j_21}, a_{j_2j_11})$ on the equilibrium of $x^*_{j_1} = a_{j_1j_11} = a_{j_2j_11}$ have the following properties as in Eqs. (1.104)–(1.107).

- For $a_{j_1j_10} > 0$ and $a_{j_2j_20} > 0$,

$$(a_{j_1j_21}, a_{j_2j_11}. \\ = \underbrace{([(2m_{11}+1),(2n_{11}+1)]^{\text{th}}\text{UP}_{-}\text{:UP}_{+},[2(m_{11}-m_{21})]^{\text{th}}{}_{\text{DC:IC}}\text{US})}_{([(2m_{11}+1),(2n_{11}+1)],[2(m_{11}-m_{21})])\text{-concave up-parabola upper-saddle}}.$$

The equilibrium is a $([(2m_{11}+1),(2n_{11}+1)]^{\text{th}}\text{UP}_{-}\text{:UP}_{+},[2(m_{11}-m_{21})]^{\text{th}}{}_{\text{DC:IC}}\text{US})$-concave up-parabola upper-saddle at $(x^*_{j_2}, x^*_{j_1}) = (a_{j_1j_21}, a_{j_2j_11})$.

- For $a_{j_1j_10} < 0$ and $a_{j_2j_20} > 0$,

$$(a_{j_1j_21}, a_{j_2j_11}) \\ = \underbrace{([(2m_{11}+1),(2n_{11}+1)]^{\text{th}}\text{DP}_{-}\text{:DP}_{+},[2(m_{11}-m_{21})]^{\text{th}}{}_{\text{IC:DC}}\text{LS})}_{([(2m_{11}+1),(2n_{11}+1)],[2(m_{11}-m_{21})])\text{-concave down-parabola lower-saddle}}.$$

The equilibrium is a $([(2m_{11}+1),(2n_{11}+1)]^{\text{th}}\text{DP}_{-}\text{:DP}_{+},[2(m_{11}-m_{21})]^{\text{th}}{}_{\text{IC:DC}}\text{LS})$-concave down-parabola lower-saddle at $(x^*_{j_2}, x^*_{j_1}) = (a_{j_1j_21}, a_{j_2j_11})$.

- For $a_{j_1j_10} > 0$ and $a_{j_2j_20} < 0$,

$$(a_{j_1j_21}, a_{j_2j_11}) \\ = \underbrace{([(2m_{11}+1),(2n_{11}+1)]^{\text{th}}\text{DP}_{+}\text{:DP}_{+},[2(m_{11}-m_{21})]^{\text{th}}{}_{\text{IC:DC}}\text{US})}_{([(2m_{11}+1),(2n_{11}+1)],[2(m_{11}-m_{21})])\text{-concave down-parabola upper-saddle}}.$$

The equilibrium is a $([(2m_{11}+1),(2n_{11}+1)]^{\text{th}}\text{DP}_{+}\text{:DP}_{+},[2(m_{11}-m_{21})]^{\text{th}}{}_{\text{IC:DC}}\text{US})$-concave down-parabola upper-saddle at $(x^*_{j_2}, x^*_{j_1}) = (a_{j_1j_21}, a_{j_2j_11})$.

- For $a_{j_1j_10} < 0$ and $a_{j_2j_20} < 0$,

$$(a_{j_1j_21}, a_{j_2j_11}) \\ = \underbrace{([(2m_{11}+1),(2n_{11}+1)]^{\text{th}}\text{UP}_{+}\text{:UP}_{-},[2(m_{11}-m_{21})]^{\text{th}}{}_{\text{DC:IC}}\text{LS})}_{([(2m_{11}+1),(2n_{11}+1)],[2(m_{11}-m_{21})])\text{-concave up-parabola lower-saddle}}.$$

The equilibrium is a $([(2m_{11}+1),(2n_{11}+1)]^{\text{th}}\text{UP}_{+}\text{:UP}_{-},[2(m_{11}-m_{21})]^{\text{th}}{}_{\text{DC:IC}}\text{LS})$- concave up-parabola lower-saddle at $(x_{j_2}^*, x_{j_1}^*) = (a_{j_1j_2 1}, a_{j_2j_1 1})$.

(i_{2b2b}) For $m_1 = 2m_{11}, n_1 = 2n_{11}+1$ and $m_2 = 2m_{21}+1$, the equilibriums of $(x_{j_2}^*, x_{j_1}^*) = (a_{j_1j_2 1}, a_{j_2j_1 1})$ on the infinite-equilibrium of $x_{j_1}^* = a_{j_1j_1 1} = a_{j_2j_1 1}$ have the following properties as in Eqs. (1.108)–(1.111).

- For $a_{j_1j_1 0} > 0$ and $a_{j_2j_2 0} > 0$,

$$(a_{j_1j_1 1}, a_{j_2j_2 1}) = \underbrace{([(2m_{11}),(2n_{11}+1)]^{\text{th}}\text{DP}_{-}\text{:UP}_{+},[2(m_{11}-m_{21})-1]^{\text{th}}{}_{\text{UA}}\text{LS:}{}_{\text{DA}}\text{US})}_{([(2m_{11}),(2n_{11}+1)],[2(m_{11}-m_{21})-1])\text{-hyperbolic asymptote lower-to-upper saddle}}.$$

The equilibrium is a $([(2m_{11}),(2n_{11}+1)]^{\text{th}}\text{DP}_{-}\text{:UP}_{+},[2(m_{11}-m_{21})-1]^{\text{th}}{}_{\text{UA}}\text{LS:}{}_{\text{DA}}\text{US})$-hyperbolic asymptote lower-to-upper saddle at $(x_{j_2}^*, x_{j_1}^*) = (a_{j_1j_2 1}, a_{j_2j_1 1})$.

- For $a_{j_1j_1 0} < 0$ and $a_{j_2j_2 0} > 0$,

$$(a_{j_1j_1 1}, a_{j_2j_2 1}) = \underbrace{([(2m_{11}),(2n_{11}+1)]^{\text{th}}\text{UP}_{-}\text{:DP}_{+},[2(m_{11}-m_{21})-1]^{\text{th}}{}_{\text{DA}}\text{US:}{}_{\text{UA}}\text{LS})}_{([(2m_{11}),(2n_{11}+1)],[2(m_{11}-m_{21})-1])\text{-hyperbolic-secant asymptote upper-to-lower saddle}}.$$

The equilibrium is a $([(2m_{11}),(2n_{11}+1)]^{\text{th}}\text{UP}_{-}\text{:DP}_{+}, [2(m_{11}-m_{21})-1]^{\text{th}}{}_{\text{DA}}\text{US:}{}_{\text{UA}}\text{LS})$-hyperbolic-secant asymptote upper-to-lower saddle at $(x_{j_2}^*, x_{j_1}^*) = (a_{j_1j_2 1}, a_{j_2j_1 1})$.

- For $a_{j_1j_1 0} > 0$ and $a_{j_2j_2 0} < 0$,

$$(a_{j_1j_1 1}, a_{j_2j_2 1}) = \underbrace{([(2m_{11}),(2n_{11}+1)]^{\text{th}}\text{UP}_{+}\text{:DP}_{-},[2(m_{11}-m_{21})-1]^{\text{th}}{}_{\text{DA}}\text{LS:}{}_{\text{UA}}\text{US})}_{([(2m_{11}),(2n_{11}+1)],[2(m_{11}-m_{21})-1])\text{-hyperbolic-secant asymptote lower-to-upper saddle}}.$$

The equilibrium is a $([(2m_{11}),(2n_{11}+1)]^{\text{th}}\text{UP}_{+}\text{:DP}_{-},[2(m_{11}-m_{21})-1]^{\text{th}}{}_{\text{DA}}\text{LS:}{}_{\text{UA}}\text{US})$-hyperbolic-secant asymptote lower-to-upper saddle at $(x_{j_2}^*, x_{j_1}^*) = (a_{j_1j_2 1}, a_{j_2j_1 1})$.

- For $a_{j_1j_1 0} < 0$ and $a_{j_2j_2 0} < 0$,

$$(a_{j_1j_1 1}, a_{j_2j_2 1}) = \underbrace{([(2m_{11}),(2n_{11}+1)]^{\text{th}}\text{DP}_{+}\text{:UP}_{-},[2(m_{11}-m_{21})-1]^{\text{th}}{}_{\text{UA}}\text{US:}{}_{\text{DA}}\text{LS})}_{([(2m_{11}),(2n_{11}+1)],[2(m_{11}-m_{21})-1])\text{-hyperbolic asymptote upper-to-lower saddle}}.$$

The equilibrium is a $([(2m_{11}),(2n_{11}+1)]^{\text{th}}\text{DP}_{+}\text{:UP}_{-},[2(m_{11}-m_{21})-1]^{\text{th}}{}_{\text{UA}}\text{US:}{}_{\text{DA}}\text{LS})$-hyperbolic asymptote upper-to-lower saddle at $(x_{j_2}^*, x_{j_1}^*) = (a_{j_1j_2 1}, a_{j_2j_1 1})$.

(i_{2b2c}) For $m_1 = 2m_{11}+1, n_1 = 2n_{11}+1$, and $m_2 = 2m_{21}$, the equilibriums of $(x^*_{j_2}, x^*_{j_1}) = (a_{j_1j_2 1}, a_{j_2j_1 1})$ on the infinite-equilibrium of $x^*_{j_1} = a_{j_1j_1 1} = a_{j_2j_1 1}$ have the following properties as in Eqs. (1.112)–(1.115).

- For $a_{j_1j_1 0} > 0$ and $a_{j_2j_2 0} > 0$,

$$(a_{j_1j_1 1}, a_{j_2j_2 1}) = \underbrace{([(2m_{11}+1),(2n_{11}+1)]^{\text{th}}\text{DP}_+\text{:UP}_+,[2(m_{11}-m_{21})+1]^{\text{th}}{}_{\text{UA}}\text{SI:}{}_{\text{DA}}\text{SO})}_{([(2m_{11}+1),(2n_{11}+1)],[2(m_{11}-m_{21})+1])\text{-hyperbolic asymptote sink-to-source}}.$$

The equilibrium is a $([(2m_{11}+1),(2n_{11}+1)]^{\text{th}}\text{DP}_+\text{:UP}_+,[2(m_{11}-m_{21})+1]^{\text{th}}{}_{\text{UA}}\text{SI:}{}_{\text{DA}}\text{SO})$-hyperbolic-asymptote sink-to-source at $(x^*_{j_2}, x^*_{j_1}) = (a_{j_1j_2 1}, a_{j_2j_1 1})$.

- For $a_{j_1j_1 0} < 0$ and $a_{j_2j_2 0} > 0$,

$$(a_{j_1j_1 1}, a_{j_2j_2 1}) = \underbrace{([(2m_{11}+1),(2n_{11}+1)]^{\text{th}}\text{UP}_+\text{:DP}_+,[2(m_{11}-m_{21})+1]^{\text{th}}{}_{\text{DA}}\text{SO:}{}_{\text{UA}}\text{SI})}_{([(2m_{11}+1),(2n_{11}+1)],[2(m_{11}-m_{21})+1])\text{-hyperbolic-secant asymptote source-to-sink}}.$$

The equilibrium is a $([(2m_{11}+1),(2n_{11}+1)]^{\text{th}}\text{UP}_+\text{:DP}_+,[2(m_{11}-m_{21})+1]^{\text{th}}{}_{\text{DA}}\text{SO:}{}_{\text{UA}}\text{SI})$ -hyperbolic-secant asymptote source-to-sink at $(x^*_{j_2}, x^*_{j_1})=(a_{j_1j_2 1}, a_{j_2j_1 1})$.

- For $a_{j_1j_1 0} > 0$ and $a_{j_2j_2 0} < 0$,

$$(a_{j_1j_1 1}, a_{j_2j_2 1}) = \underbrace{([(2m_{11}+1),(2n_{11}+1)]^{\text{th}}\text{UP}_-\text{:DP}_-,[2(m_{11}-m_{21})+1]^{\text{th}}{}_{\text{DA}}\text{SI:}{}_{\text{UA}}\text{SO})}_{([(2m_{11}+1),(2n_{11}+1)],[2(m_{11}-m_{21})+1])\text{-hyperbolic-secant asymptote sink-to-source}}.$$

The equilibrium is a $([(2m_{11}+1),(2n_{11}+1)]^{\text{th}}\text{UP}_-\text{:DP}_-,[2(m_{11}-m_{21})+1]^{\text{th}}{}_{\text{DA}}\text{SI:}{}_{\text{UA}}\text{SO})$ -hyperbolic-secant asymptote sink-to-source at $(x^*_{j_2}, x^*_{j_1})=(a_{j_1j_2 1}, a_{j_2j_1 1})$.

- For $a_{j_1j_1 0} < 0$ and $a_{j_2j_2 0} < 0$,

$$(a_{j_1j_1 1}, a_{j_2j_2 1}) = \underbrace{([(2m_{11}+1),(2n_{11}+1)]^{\text{th}}\text{DP}_-\text{:UP}_-,[2(m_{11}-m_{21})+1]^{\text{th}}{}_{\text{UA}}\text{SO:}{}_{\text{DA}}\text{SI})}_{([(2m_{11}+1),(2n_{11}+1)],[2(m_{11}-m_{21})+1])\text{-hyperbolic asymptote source-to-sink}}.$$

The equilibrium is a $([(2m_{11}+1),(2n_{11}+1)]^{\text{th}}\text{SI}, [2(m_{11}-m_{21})+1]^{\text{th}}\text{UA:DA})$ hyperbolic-asymptote lower-saddle at $(x^*_{j_2}, x^*_{j_1}) = (a_{j_1j_2 1}, a_{j_2j_1 1})$.

(i_{2a2d}) For $m_1 = 2m_{11}, n_1 = 2n_{11}+1$ and $m_2 = 2m_{21}$, the equilibriums of $(x^*_{j_2}, x^*_{j_1}) = (a_{j_1j_2 1}, a_{j_2j_1 1})$ on the infinite-equilibrium of $x^*_{j_2} = a_{j_2j_2 1} = a_{j_1j_2 1}$ have the following properties as in Eqs. (1.116)–(1.119).

- For $a_{j_1j_10} > 0$ and $a_{j_2j_20} > 0$,

$$(a_{j_1j_21}, a_{j_2j_11}) = \underbrace{([(2m_{11}),(2n_{11}+1)]^{\text{th}}\text{UP}_{+}\text{:UP}_{+},[2(m_{11}-m_{21})]^{\text{th}}{}_{\text{DC:IC}}\text{US})}_{([(2m_{11}),(2n_{11}+1)],[2(m_{11}-m_{21})])\text{-positve concave up-parabola upper-saddle}}.$$

The equilibrium is a $([(2m_{11}),(2n_{11}+1)]^{\text{th}}\text{UP}_{+}\text{:UP}_{+},[2(m_{11}-m_{21})]^{\text{th}}{}_{\text{DC:IC}}\text{US})$ -positive-concave up-parabola upper-saddle at $(x^*_{j_2}, x^*_{j_1})=(a_{j_1j_21}, a_{j_2j_11})$.

- For $a_{j_1j_10} < 0$ and $a_{j_2j_20} > 0$,

$$(a_{j_1j_21}, a_{j_2j_11}) = \underbrace{([(2m_{11}),(2n_{11}+1)]^{\text{th}}\text{DP}_{+}\text{:DP}_{+},[2(m_{11}-m_{21})]^{\text{th}}{}_{\text{IC:DC}}\text{US})}_{([(2m_{11}),(2n_{11}+1)],[2(m_{11}-m_{21})])\text{-positve concave down-parabola upper-saddle}}.$$

The equilibrium is a $([(2m_{11}),(2n_{11}+1)]^{\text{th}}\text{DP}_{+}\text{:DP}_{+},[2(m_{11}-m_{21})]^{\text{th}}{}_{\text{IC:DC}}\text{US})$ -positive-concave down-parabola upper-saddle at $(x^*_{j_2}, x^*_{j_1})=(a_{j_1j_21},a_{j_2j_11})$.

- For $a_{j_1j_10} > 0$ and $a_{j_2j_20} < 0$,

$$(a_{j_1j_21}, a_{j_2j_11}) = \underbrace{([(2m_{11}),(2n_{11}+1)]^{\text{th}}\text{DP}_{-}\text{:DP}_{-},[2(m_{11}-m_{21})]^{\text{th}}{}_{\text{IC:DC}}\text{LS})}_{([(2m_{11}),(2n_{11}+1)],[2(m_{11}-m_{21})])\text{-negative concave down-parabola lower-saddle}}.$$

The equilibrium is a $([(2m_{11}),(2n_{11}+1)]^{\text{th}}\text{DP}_{-}\text{:DP}_{-},[2(m_{11}-m_{21})]^{\text{th}}{}_{\text{IC:DC}}\text{LS})$-negative-concave down-parabola lower-saddle at $(x^*_{j_2}, x^*_{j_1}) = (a_{j_1j_21}, a_{j_2j_11})$.

- For $a_{j_1j_10} < 0$ and $a_{j_2j_20} < 0$,

$$(a_{j_1j_21}, a_{j_2j_11}) = \underbrace{([(2m_{11}),(2n_{11}+1)]^{\text{th}}\text{UP}_{-}\text{:UP}_{-},[2(m_{11}-m_{21})]^{\text{th}}{}_{\text{DC:IC}}\text{LS})}_{([(2m_{11}),(2n_{11}+1)],[2(m_{11}-m_{21})])\text{-negative concave up-parabola lower-saddle}}.$$

The equilibrium is a $([(2m_{11}),(2n_{11}+1)]^{\text{th}}\text{UP}_{-}\text{:UP}_{-},[2(m_{11}-m_{21})]^{\text{th}}{}_{\text{DC:IC}}\text{LS})$-negative-concave up-parabola lower-saddle at $(x^*_{j_2}, x^*_{j_1}) = (a_{j_1j_21}, a_{j_2j_11})$.

(i_{2b2e}) For $m_1 = 2m_{11} + 1$, $n_1 = 2n_{11}$ and $m_2 = 2m_{21} + 1$, the equilibrium of $(x^*_{j_2}, x^*_{j_1}) = (a_{j_1j_21}, a_{j_2j_11})$ on the infinite-equilibrium of $x^*_{j_1} = a_{j_1j_11} = a_{j_2j_11}$ have the following properties as in Eqs. (1.120)–(1.123).

- For $a_{j_1j_10} > 0$ and $a_{j_2j_20} > 0$,

$$(a_{j_1j_21}, a_{j_2j_11}) = \underbrace{([(2m_{11}+1),(2n_{11})]^{\text{th}}\text{II}_{-}\text{:II}_{+},[2(m_{11}-m_{21})]^{\text{th}}{}_{\text{IC:IC}}\text{SO})}_{([(2m_{11}+1),(2n_{11})],[2(m_{11}-m_{21})])\text{-increasing-concave inflection-source}}.$$

The equilibrium is a $([(2m_{11}+1),(2n_{11})]^{\text{th}}\text{II}_{-}\text{:II}_{+},[2(m_{11}-m_{21})]^{\text{th}}{}_{\text{IC:IC}}\text{SO})$-increasing-concave inflection-source at $(x_{j_2}^*, x_{j_1}^*) = (a_{j_1j_21}, a_{j_2j_11})$.

- For $a_{j_1j_10} < 0$ and $a_{j_2j_20} > 0$,

$$(a_{j_1j_21}, a_{j_2j_11}) = \underbrace{([(2m_{11}+1),(2n_{11})]^{\text{th}}\text{DI}_{-}\text{:DI}_{+},[2(m_{11}-m_{21})]^{\text{th}}{}_{\text{DC:DC}}\text{SI})}_{([(2m_{11}+1),(2n_{11})],[2(m_{11}-m_{21})])\text{-decreasing-concave inflection-sink}}.$$

The equilibrium is a $([(2m_{11}+1),(2n_{11})]^{\text{th}}\text{DI}_{-}\text{:DI}_{+},[2(m_{11}-m_{21})]^{\text{th}}{}_{\text{DC:DC}}\text{SI})$-decreasing-concave inflection-sink at $(x_{j_2}^*, x_{j_1}^*) = (a_{j_1j_21}, a_{j_2j_11})$.

- For $a_{j_1j_10} > 0$ and $a_{j_2j_20} < 0$,

$$(a_{j_1j_21}, a_{j_2j_11}) = \underbrace{([(2m_{11}+1),(2n_{11})]^{\text{th}}\text{DI}_{+}\text{:DI}_{-},[2(m_{11}-m_{21})]^{\text{th}}{}_{\text{DC:DC}}\text{SO})}_{([(2m_{11}+1),(2n_{11})],[2(m_{11}-m_{21})])\text{-decreasing-concave inflection-source}}.$$

The equilibrium is a $([(2m_{11}+1),(2n_{11})]^{\text{th}}\text{DI}_{+}\text{:DI}_{-},[2(m_{11}-m_{21})]^{\text{th}}{}_{\text{DC:DC}}\text{SO})$-decreasing concave inflection-source at $(x_{j_2}^*, x_{j_1}^*) = (a_{j_1j_21}, a_{j_2j_11})$.

- For $a_{j_1j_10} < 0$ and $a_{j_2j_20} < 0$,

$$(a_{j_1j_21}, a_{j_2j_11}) = \underbrace{([(2m_{11}+1),(2n_{11})]^{\text{th}}\text{II}_{+}\text{:II}_{-},[2(m_{11}-m_{21})]^{\text{th}}{}_{\text{IC:IC}}\text{SI})}_{([(2m_{11}+1),(2n_{11})],[2(m_{11}-m_{21})])\text{-increasing-concave inflection-sink}}.$$

The equilibrium is a $([(2m_{11}+1),(2n_{11})]^{\text{th}}\text{II}_{+}\text{:II}_{-},[2(m_{11}-m_{21})]^{\text{th}}{}_{\text{IC:IC}}\text{SI})$-increasing-concave inflection-sink at $(x_{j_2}^*, x_{j_1}^*) = (a_{j_1j_21}, a_{j_2j_11})$.

(i_{2b2f}) For $m_1 = 2m_{11}$, $n_1 = 2n_{11}$, and $m_2 = 2m_{21}+1$, the equilibrium of $(x_{j_2}^*, x_{j_1}^*) = (a_{j_1j_21}, a_{j_2j_11})$ on the infinite-equilibrium of $x_{j_1}^* = a_{j_1j_11} = a_{j_2j_11}$ have the following properties as in Eqs. (1.124)–(1.127).

- For $a_{j_1j_10} > 0$ and $a_{j_2j_20} > 0$,

$$(a_{j_1j_21}, a_{j_2j_11}) = \underbrace{([(2m_{11}),(2n_{11})]^{\text{th}}\text{DI}_{-}\text{:II}_{+},[2(m_{11}-m_{21})-1]^{\text{th}}{}_{\text{DA:DA}}\text{US})}_{([(2m_{11}),(2n_{11})],[2(m_{11}-m_{21})-1])\text{-inflection down-asymptote upper-saddle}}.$$

The equilibrium is a $([(2m_{11}),(2n_{11})]^{\text{th}}\text{DI}_{-}\text{:II}_{+},[2(m_{11}-m_{21})-1]^{\text{th}}{}_{\text{DA:DA}}\text{US})$-inflection down-asymptote upper-saddle at $(x^*_{j_1}, x^*_{j_2}) = (a_{j_1j_11}, a_{j_1j_21})$.

- For $a_{j_1j_10} < 0$ and $a_{j_2j_20} > 0$,

$$
\begin{aligned}
&(a_{j_1j_21}, a_{j_2j_11})\\
&\quad = \underbrace{([(2m_{11}),(2n_{11})]^{\text{th}}\text{II}_{-}\text{:DI}_{+},[2(m_{11}-m_{21})-1]^{\text{th}}{}_{\text{UA:UA}}\text{LS})}_{([(2m_{11}),(2n_{11})],[2(m_{11}-m_{21})-1])\text{-inflection up-asymptote lower-saddle}}.
\end{aligned}
$$

The equilibrium is a $([(2m_{11}),(2n_{11})]^{\text{th}}\text{II}_{-}\text{:DI}_{+},[2(m_{11}-m_{21})-1]^{\text{th}}{}_{\text{DA:DA}}\text{LS})$-inflection up-asymptote lower-saddle at $(x^*_{j_2}, x^*_{j_1}) = (a_{j_1j_21}, a_{j_2j_11})$.

- For $a_{j_1j_10} > 0$ and $a_{j_2j_20} < 0$,

$$
\begin{aligned}
&(a_{j_1j_21}, a_{j_2j_11})\\
&\quad = \underbrace{([(2m_{11}),(2n_{11})]^{\text{th}}\text{II}_{+}\text{:DI}_{-},[2(m_{11}-m_{21})-1]^{\text{th}}{}_{\text{UA:UA}}\text{US})}_{([(2m_{11}),(2n_{11})],[2(m_{11}-m_{21})-1])\text{-inflection up-asymptote upper-saddle}}.
\end{aligned}
$$

The equilibrium is a $([(2m_{11}),(2n_{11})]^{\text{th}}\text{II}_{+}\text{:DI}_{-},[2(m_{11}-m_{21})-1]^{\text{th}}{}_{\text{UA:UA}}\text{US})$-inflection up-asymptote upper-saddle at $(x^*_{j_2}, x^*_{j_1}) = (a_{j_1j_21}, a_{j_2j_11})$.

- For $a_{j_1j_10} < 0$ and $a_{j_2j_20} < 0$,

$$
\begin{aligned}
&(a_{j_1j_21}, a_{j_2j_11})\\
&\quad = \underbrace{([(2m_{11}),(2n_{11})]^{\text{th}}\text{DI}_{+}\text{:II}_{-},[2(m_{11}-m_{21})-1]^{\text{th}}{}_{\text{DA:DA}}\text{LS})}_{([(2m_{11}),(2n_{11})],[2(m_{11}-m_{21})-1])\text{-inflection down-asymptote lower-saddle}}.
\end{aligned}
$$

The equilibrium is a $([(2m_{11}),(2n_{11})]^{\text{th}}\text{DI}_{+}\text{:II}_{-},[2(m_{11}-m_{21})-1]^{\text{th}}{}_{\text{DA:DA}}\text{LS})$-inflection down-asymptote lower-saddle at $(x^*_{j_2}, x^*_{j_1}) = (a_{j_1j_21}, a_{j_2j_11})$.

$(\text{i}_{2\text{b}2\text{g}})$ For $m_1 = 2m_{11}+1$, $n_1 = 2n_{11}$, and $m_2 = 2m_{21}$, the equilibrium $(x^*_{j_2}, x^*_{j_1}) = (a_{j_1j_21}, a_{j_2j_11})$ on the infinite-equilibrium of $x^*_{j_1} = a_{j_1j_11} = a_{j_2j_11}$ have the following properties as in Eqs. (1.128)–(1.131).

- For $a_{j_1j_10} > 0$ and $a_{j_2j_20} > 0$,

$$
\begin{aligned}
&(a_{j_1j_21}, a_{j_2j_11})\\
&\quad = \underbrace{([(2m_{11}+1),(2n_{11})]^{\text{th}}\text{DI}_{+}\text{:II}_{+},[2(m_{11}-m_{21})+1]^{\text{th}}{}_{\text{DA:DA}}\text{SO})}_{([(2m_{11}+1),(2n_{11})],[2(m_{11}-m_{21})+1])\text{-inflection down-asymptote source}}.
\end{aligned}
$$

The equilibrium is a $([(2m_{11}+1),(2n_{11})]^{\text{th}}\text{DI}_{+}\text{:II}_{+},[2(m_{11}-m_{21})+1]^{\text{th}}{}_{\text{DA:DA}}\text{SO})$-inflection down-asymptote source at $(x^*_{j_2}, x^*_{j_1}) = (a_{j_1j_21}, a_{j_2j_11})$.

- For $a_{j_1 j_1 0} < 0$ and $a_{j_2 j_2 0} > 0$,

$$(a_{j_1 j_2 1}, a_{j_2 j_1 1}) = \underbrace{([(2m_{11}+1),(2n_{11})]^{\text{th}}\text{II}_+:\text{DI}_+,[2(m_{11}-m_{21})+1]^{\text{th}}{}_{\text{UA:UA}}\text{SI})}_{([(2m_{11}+1),(2n_{11})],[2(m_{11}-m_{21})+1])\text{-inflection up-asymptote sink}}.$$

The equilibrium is a $([(2m_{11}+1),(2n_{11})]^{\text{th}}\text{II}_+:\text{DI}_+,[2(m_{11}-m_{21})+1]^{\text{th}}{}_{\text{UA:UA}}\text{SI})$-inflection up-asymptote sink at $(x_{j_2}^*, x_{j_1}^*) = (a_{j_1 j_2 1}, a_{j_2 j_1 1})$.

- For $a_{j_1 j_1 0} > 0$ and $a_{j_2 j_2 0} < 0$,

$$(a_{j_1 j_2 1}, a_{j_2 j_1 1}) = \underbrace{([(2m_{11}+1),(2n_{11})]^{\text{th}}\text{II}_-:\text{DI}_-,[2(m_{11}-m_{21})+1]^{\text{th}}{}_{\text{UA:UA}}\text{SO})}_{([(2m_{11}+1),(2n_{11})],[2(m_{11}-m_{21})+1])\text{-inflection up-asymptote source}}.$$

The equilibrium is a $([(2m_{11}+1),(2n_{11})]^{\text{th}}\text{II}_-:\text{DI}_-,[2(m_{11}-m_{21})+1]^{\text{th}}{}_{\text{UA:UA}}\text{SO})$-inflection up-asymptote source at $(x_{j_2}^*, x_{j_1}^*) = (a_{j_1 j_2 1}, a_{j_2 j_1 1})$.

- For $a_{j_1 j_1 0} < 0$ and $a_{j_2 j_2 0} < 0$,

$$(a_{j_1 j_2 1}, a_{j_2 j_1 1}) = \underbrace{([(2m_{11}+1),(2n_{11})]^{\text{th}}\text{DI}_-:\text{II}_-,[2(m_{11}-m_{21})+1]^{\text{th}}{}_{\text{DA:DA}}\text{SI})}_{([(2m_{11}+1),(2n_{11})],[2(m_{11}-m_{21})+1])\text{-inflection down-asymptote sink}}.$$

The equilibrium is a $([(2m_{11}+1),(2n_{11})]^{\text{th}}\text{DI}_-:\text{II}_-,[2(m_{11}-m_{21})+1]^{\text{th}}{}_{\text{DA:DA}}\text{SI})$-inflection down-asymptote sink at $(x_{j_2}^*, x_{j_1}^*) = (a_{j_1 j_2 1}, a_{j_2 j_1 1})$.

(i_{2b2h}) For $m_1 = 2m_{11}$, $n_1 = 2n_{11}$, and $m_2 = 2m_{21}$, the equilibrium of $(x_{j_2}^*, x_{j_1}^*) = (a_{j_1 j_2 1}, a_{j_2 j_1 1})$ on the infinite-equilibrium of $x_{j_1}^* = a_{j_1 j_1 1} = a_{j_2 j_1 1}$ have the following properties as in Eqs. (1.132)–(1.135).

- For $a_{j_1 j_1 0} > 0$ and $a_{j_2 j_2 0} > 0$,

$$(a_{j_1 j_2 1}, a_{j_2 j_1 1}) = \underbrace{([(2m_{11}),(2n_{11})]^{\text{th}}\text{II}_+:\text{II}_+,[2(m_{11}-m_{21})]^{\text{th}}{}_{\text{IC:IC}}\text{US})}_{([(2m_{11}+1),(2n_{11})],[2(m_{11}-m_{21})])\text{-increasing-concave inflection-upper-saddle}}.$$

The equilibrium is a $([(2m_{11}),(2n_{11})]^{\text{th}}\text{II}_+:\text{II}_+,[2(m_{11}-m_{21})]^{\text{th}}{}_{\text{IC:IC}}\text{US})$-increasing-concave inflection-upper-saddle at $(x_{j_2}^*, x_{j_1}^*) = (a_{j_1 j_2 1}, a_{j_2 j_1 1})$.

- For $a_{j_1 j_1 0} < 0$ and $a_{j_2 j_2 0} > 0$,

$$(a_{j_1 j_2 1}, a_{j_2 j_1 1}) = \underbrace{([(2m_{11}),(2n_{11})]^{\text{th}}\text{DI}_+:\text{DI}_+,[2(m_{11}-m_{21})]^{\text{th}}{}_{\text{DC:DC}}\text{LS})}_{([(2m_{11}),(2n_{11})],[2(m_{11}-m_{21})])\text{-decreasing-concave inflection-lower-saddle}}.$$

The equilibrium is a $([(2m_{11}),(2n_{11})]^{\text{th}}\text{DI}_{+}{:}\text{DI}_{+},[2(m_{11}-m_{21})]^{\text{th}}{}_{\text{DC:DC}}\text{LS})$-decreasing-concave inflection-lower-saddle at $(x_{j_2}^*, x_{j_1}^*) = (a_{j_1j_21}, a_{j_2j_11})$.

- For $a_{j_1j_10} > 0$ and $a_{j_2j_20} < 0$,

$$
\begin{aligned}
&(a_{j_1j_21}, a_{j_2j_11}) \\
&\quad = \underbrace{([(2m_{11}),(2n_{11})]^{\text{th}}\text{DI}_{-}{:}\text{DI}_{-},[2(m_{11}-m_{21})]^{\text{th}}{}_{\text{DC:DC}}\text{US})}_{([(2m_{11}),(2n_{11})],[2(m_{11}-m_{21})])\text{-decreasing-concave inflection-upper-saddle}}.
\end{aligned}
$$

The equilibrium is a $([(2m_{11}),(2n_{11})]^{\text{th}}\text{DI}_{-}{:}\text{DI}_{-},[2(m_{11}-m_{21})]^{\text{th}}{}_{\text{DC:DC}}\text{US})$-decreasing-concave inflection-lower-saddle at $(x_{j_2}^*, x_{j_1}^*) = (a_{j_1j_21}, a_{j_2j_11})$.

- For $a_{j_1j_10} < 0$ and $a_{j_2j_20} < 0$,

$$
\begin{aligned}
&(a_{j_1j_21}, a_{j_2j_11}) \\
&\quad = \underbrace{([(2m_{11}),(2n_{11})]^{\text{th}}\text{II}_{-}{:}\text{II}_{-},[2(m_{11}-m_{21})]^{\text{th}}{}_{\text{IC:IC}}\text{LS})}_{([(2m_{11}),(2n_{11})],[2(m_{11}-m_{21})])\text{-increasing-concave inflection-lower-saddle}}.
\end{aligned}
$$

The equilibrium is a $([(2m_{11}),(2n_{11})]^{\text{th}}\text{II}_{-}{:}\text{II}_{-},[2(m_{11}-m_{21})]^{\text{th}}{}_{\text{IC:IC}}\text{LS})$-increasing-concave inflection-lower-saddle at $(x_{j_2}^*, x_{j_1}^*) = (a_{j_1j_21}, a_{j_2j_11})$.

2.2 Hybrid Networks of Singular Equilibriums and Flows

In this section, theorem about the hybrid networks of singular equilibriums and flows will be proved, and the corresponding infinite-equilibriums will be discussed.

(ii) Consider

$$
\begin{aligned}
\dot{x}_{j_2} &= a_{j_2j_20}\prod_{v=1}^{p_2}(x_{j_1} - a_{j_2j_1v})^{m_{j_2s}} \\
\dot{x}_{j_1} &= a_{j_1j_10}\prod_{r=1}^{p_1}(x_{j_1} - a_{j_1j_1r})^{m_{j_1r}}\prod_{u=1}^{q_1}(x_{j_2} - a_{j_1j_2u})^{n_{j_1u}},
\end{aligned}
$$

where

$$
\sum_{r=1}^{p_1} m_{j_1r} = m_1, \sum_{u=1}^{q_1} n_{j_2u} = n_1; \sum_{v=1}^{p_2} n_{j_2v} = m_2.
$$

2.2.1 Hybrid Networks of Singular Equilibriums and Flows

(ii_1) For $a_{j_1j_1v} \neq a_{j_2j_2s}$ $(u = 1, 2, \ldots, p_1; v = 1, 2, \ldots, p_2)$, consider

$$\frac{dx_{j_1}}{dx_{j_2}} = \frac{a_{j_1j_10}}{a_{j_2j_20}} \frac{\prod_{r=1}^{p_1} (x_{j_1} - a_{j_1j_1r})^{m_{j_1r}} \prod_{u=1}^{q_1} (x_{j_2} - a_{j_1j_2u})^{n_{j_1u}}}{\prod_{s=1}^{p_2} (x_{j_1} - a_{j_2j_1v})^{m_{j_2v}}}.$$

The deformed equation is

$$\frac{\prod_{v=1}^{p_2} (x_{j_1} - a_{j_2j_1v})^{m_{j_2v}} dx_{j_1}}{\prod_{r=1}^{p_1} (x_{j_1} - a_{j_1j_1r})^{m_{j_1r}}} = \frac{a_{j_1j_10}}{a_{j_2j_20}} \prod_{u=1}^{q_1} (x_{j_2} - a_{j_1j_2u})^{n_{j_1u}} dx_{j_2},$$

The expansion of the above equation gives

$$\begin{aligned}
&\sum_{v_0=0}^{m_2} \frac{\sum\limits_{t_1,l_1=0}^{t_1+l_1=m_{j_2v_1}} \cdots \sum\limits_{t_{p_2},l_{p_2}=0}^{t_{p_2}+l_{p_2}=m_{j_2v_{p_2}}} \prod\limits_{k=1}^{p_2} \frac{m_{j_1v_k}!}{t_k!l_k!}(a_{j_1j_1r_1} - a_{j_2j_1k})^{t_k} \delta_{\beta_0}^{\Sigma_{r=1}^{p_2} l_r} (x_{j_1} - a_{j_1j_1r_1})^{v_0}}{(x_{j_1} - a_{j_1j_1r_1})^{m_{j_1r_1}} \prod\limits_{r_2=1, r_2\neq r_1}^{p_1} (x_{j_1} - a_{j_1j_1r_2})^{m_{j_1r_2}}} dx_{j_1} \\
&= \frac{a_{j_1j_10}}{a_{j_2j_20}} [(x_{j_2} - a_{j_1j_2u_1})^{n_1} + \sum_{\alpha=1}^{n_1 - n_{j_1u_1}} b_{j_1j_2\alpha} (x_{j_1} - a_{j_1j_2u_1})^{n_1-\alpha}] dx_{j_2},
\end{aligned}$$

where

$$\begin{aligned}
b_{j_1j_2\alpha} &= \sum_{l_{q_1}, t_{q_1}=0}^{l_{q_1}+t_{q_1}=n_{j_1u_{q_1}}} \cdots \sum_{l_2,t_2=0}^{l_2+t_2=n_{j_1u_2}} [\prod_{k=2}^{q_1} \frac{n_{j_1u_k}!}{l_k!t_k!} (a_{j_1j_2u_1} - a_{j_1j_2u_k})^{l_k}] \delta_{t_2+\cdots+t_{q_1}}^{n_1-\alpha}; \\
b_{j_1j_20} &= 1.
\end{aligned}$$

Let

$$b_{j_2j_1v_0} = \sum_{t_1,l_1=0}^{t_1+l_1=m_{j_2v_1}} \cdots \sum_{t_{p_2},l_{p_2}=0}^{t_{p_2}+l_{p_2}=m_{j_1v_{p_2}}} \prod_{k=1}^{p_2} \frac{m_{j_1v_k}!}{t_k!l_k!} (a_{j_1j_1r_1} - a_{j_2j_1v_k})^{t_k} \delta_{v_0}^{\Sigma_{r=1}^{p_2} l_r}.$$

We have

$$\frac{\sum_{\nu_0=0}^{m_1-1} b_{j_2j_1\nu_0}(x_{j_1}-a_{j_1j_1r_1})^{\nu_0}+\sum_{\nu_0=m_1}^{m_2} b_{j_2j_1\nu_0}(x_{j_1}-a_{j_1j_1r_1})^{\nu_0}}{(x_{j_1}-a_{j_1j_1r_1})^{m_{j_1r_1}}\prod_{r_2=1,r_2\neq r_1}^{p_1}(x_{j_1}-a_{j_1j_1r_2})^{m_{j_1r_2}}}dx_{j_1}$$

$$=\frac{a_{j_1j_10}}{a_{j_2j_20}}[(x_{j_2}-a_{j_1j_2u_1})^{n_1}+\sum_{\alpha=1}^{n_1-n_{r_1}} b_{j_1j_2\alpha}(x_{j_1}-a_{j_1j_2u_1})^{n_1-\alpha}]dx_{j_2}.$$

The first part is given by

$$\frac{\sum_{\nu_0=0}^{m_1-1} b_{j_2j_1\nu_0}(x_{j_1}-a_{j_1j_1r_1})^{\nu_0}}{(x_{j_1}-a_{j_1j_1r_1})^{m_{j_1r_1}}\prod_{r_2=1,r_2\neq r_1}^{p_1}(x_{j_1}-a_{j_1j_1r_2})^{m_{j_1r_2}}}$$

$$=\sum_{\nu_0=0}^{m_1-1} b_{j_2j_1\nu_0}\frac{(x_{j_1}-a_{j_1j_1r_1})^{\nu_0-n_{j_2s_1}}}{\prod_{r_2=1,r_2\neq r_1}^{p_1}(x_{j_1}-a_{j_1j_1r_2})^{m_{j_1r_2}}}$$

$$=\sum_{\nu_0=0}^{m_{j_1r_1}-1} b_{j_2j_1\nu_0}\sum_{j=1}^{p_1}\sum_{l_j=1}^{\overline{m}_{j_2r_j}} A^{(r_1)}_{l_j\overline{m}_{j_1r_j}}\frac{1}{(x_{j_1}-a_{j_1j_1r_j})^{l_j}}$$

$$+\sum_{\nu_0=m_{j_1r_1}}^{m_1-1} b_{j_2j_1\nu_0}\sum_{j=2}^{p_1}\sum_{l_j=1}^{\overline{m}_{j_2r_j}} A^{(r_1)}_{l_j\overline{m}_{j_1r_j}}\frac{1}{(x_{j_1}-a_{j_1j_1r_j})^{l_j}};$$

where

$$A_{l_\rho\overline{m}^{(r_1)}_{j_1r_\rho}}=\frac{1}{(\overline{m}_{j_1r_\rho}-l_\rho)!}\frac{d^{\overline{m}_{j_1r_\rho}-l_\rho}}{dx_{j_1}^{\overline{m}_{j_1r_\rho}-l_\rho}}\frac{1}{\prod_{\sigma=1,\sigma\neq\rho}^{p_1}(x_{j_1}-a_{j_1j_1r_\sigma})^{m_{j_ir_\sigma}}}\Bigg|_{x^*_{j1}=a_{j_ij_1r_\rho}}$$

$$(l_\rho=1,2,\ldots,\overline{m}_{j_1r_\rho},\ \rho=1,2,\ldots,p_1),$$

$$A^{(r_1)}_{\overline{m}_{j_1r_\rho}\overline{m}_{jr_\rho}}=\frac{1}{\prod_{\sigma=1,\sigma\neq\rho}(a_{j_1j_1r_\rho}-a_{j_1j_1r_\sigma})^{m_{j_1r_\rho}}},$$

$$B^{(r_1)}_{l_\rho\overline{m}_{j_1r_\rho}}=\frac{1}{(\overline{m}_{j_1r_\rho}-l_\rho)!}\frac{d^{\overline{m}_{j_1r_\rho}-l_\rho}}{dx_{j_1}^{\overline{m}_{j_1r_\rho}-l_\rho}}\frac{1}{\prod_{\sigma=2,\sigma\neq\rho}^{p_1}(x_{j_1}-a_{j_1j_1r_\sigma})^{m_{j_1r_\sigma}}}\Bigg|_{x^*_{j1}=a_{j_1j_1r_\rho}}$$

$$(l_\rho=1,2,\ldots,\overline{m}_{j_1r_\rho},\ \rho=2,3,\ldots,p_1),$$

$$B^{(r_1)}_{\overline{m}_{j_1r_\rho}\overline{m}_{j_1r_\rho}}=\frac{1}{\prod_{\sigma=2,\sigma\neq\rho}^{p_1}(a_{j_1j_1r_\rho}-a_{j_1j_1r_\sigma})^{m_{j_1r_\sigma}}}$$

$$\{\overline{m}_{j_1r_\rho}|\rho=1,2,\ldots,p_1\}=\{m_{j_1r_1}-\beta_0,m_{j_1r_2},\ldots,\mathrm{m}_{j_1r_{p1}}\},$$

The second part is deformed by

$$\frac{\sum_{\nu_0=m_1}^{m_2} b_{j_2 j_1 \nu_0}(x_{j_1}-a_{j_1 j_1 r_1})^{\nu_0}}{(x_{j_1}-a_{j_1 j_1 r_1})^{m_{j_1 r_1}} \prod_{r_2=1, r_2\neq r_1}^{p_1} (x_{j_1}-a_{j_1 j_1 r_2})^{m_{j_1 r_2}}}$$

$$= \sum_{\nu_0=m_1}^{m_2} b_{j_2 j_1 \nu_0} \frac{(x_{j_1}-a_{j_1 j_1 r_1})^{\nu_0 - m_{j_1 r_1}}}{(x_{j_1}-a_{j_1 j_1 r_2})^{m_{j_1 r_2}} \prod_{r_3=1, r_3\neq r_1, r_2}^{p_1} (x_{j_1}-a_{j_1 j_1 r_3})^{m_{j_1 r_3}}}$$

$$= \sum_{\nu_0=m_1}^{m_2} b_{j_2 j_1 \nu_0} \Bigg[\sum_{\substack{\mu_1 \in [0, \nu_0 - m_1] \\ \nu_1 \in [m_1 - m_{j_1 r_1}, \nu_0 - m_{j_1 r_1}]}}^{\mu_1+\nu_1=\nu_0-m_{j_1 r_1}} \frac{(\nu_0 - m_{j_1 r_1})!}{\mu_1! \nu_1!} \frac{(a_{j_1 j_1 r_2}-a_{j_1 j_1 r_1})^{\mu_1}(x_{j_1}-a_{j_1 j_1 r_2})^{\nu_1 - m_{j_1 r_2}}}{\prod_{r_3=1, r_3\neq r_1, r_2}^{p_1} (x_{j_1}-a_{j_1 j_1 r_3})^{m_{j_1 r_3}}}$$

$$+ \sum_{\substack{\mu_1 \in [\nu_0 - \Sigma_{i=1}^2 m_{j_1 r_i}, \nu_0 - m_{j_1 r_1}) \\ \nu_1 \in [0, m_{j_1 r_2})}}^{\mu_1+\nu_1=\nu_0-m_{j_1 r_1}} \frac{(\nu_0 - m_{j_1 r_1})!}{\mu_1! \nu_1!} \frac{(a_{j_1 j_1 r_2}-a_{j_1 j_1 r_1})^{\mu_1}}{(x_{j_1}-a_{j_1 j_1 r_2})^{m_{j_1 r_2}-\nu_1} \prod_{r_3=1, r_3\neq r_1, r_2}^{p_1} (x_{j_1}-a_{j_1 j_1 r_3})^{m_{j_1 r_3}}}$$

$$+ \sum_{\substack{\mu_1 \in (\nu_0 - m_1, \nu_0 - \Sigma_{i=1}^2 m_{j_1 r_i}] \\ \nu_1 \in [m_{j_1 r_2}, m_1 - m_{j_1 r_1})}}^{\mu_1+\nu_1=\nu_0-m_{j_1 r_1}} \frac{(\nu_0 - m_{j_1 r_1})!}{\mu_1! \nu_1!} \frac{(a_{j_1 j_1 r_2}-a_{j_1 j_1 r_1})^{\mu_1}(x_{j_1}-a_{j_1 j_1 r_2})^{\nu_1 - m_{j_1 r_2}}}{\prod_{r_3=1, r_3\neq r_1, r_2}^{p_1} (x_{j_1}-a_{j_1 j_1 r_3})^{m_{j_1 r_3}}} \Bigg]$$

Thus

$$\frac{\sum_{\nu_0=m_1}^{m_2} b_{j_2 j_1 \nu_0}(x_{j_1}-a_{j_1 j_1 r_1})^{\nu_0}}{(x_{j_1}-a_{j_1 j_1 r_1})^{m_{j_1 r_1}} \prod_{r_2=1, r_2\neq r_1}^{p_1} (x_{j_1}-a_{j_1 j_1 r_2})^{m_{j_1 r_2}}}$$

$$= \sum_{\nu_0=m_1}^{m_2} b_{j_2 j_1 \nu_0} \times \Bigg[\sum_{\substack{\mu_1 \in [0, \nu_0 - m_1] \\ \nu_1 \in [m_1 - m_{j_1 r_1}, \nu_0 - m_{j_1 r_1}]}}^{\mu_1+\nu_1=\nu_0-m_{j_1 r_1}} \frac{(\nu_0 - m_{j_1 r_1})!}{\mu_1! \nu_1!} (a_{j_1 j_1 r_2}-a_{j_1 j_1 r_1})^{\mu_1}$$

$$\times \frac{(x_{j_1}-a_{j_1 j_1 r_2})^{\nu_1 - m_{j_1 r_2}}}{\prod_{r_3=1, r_3\neq r_1, r_2}^{p_1} (x_{j_1}-a_{j_1 j_1 r_3})^{m_{j_1 r_3}}}$$

$$+ \sum_{\substack{\mu_1 \in [\nu_0 - \Sigma_{i=1}^2 m_{j_1 r_i}, \nu_0 - m_{j_1 r_1}) \\ \nu_1 \in [0, m_{j_1 r_2})}}^{\mu_1+\nu_1=\nu_0-m_{j_1 r_1}} \frac{(\nu_0 - m_{j_1 r_1})!}{\mu_1! \nu_1!} (a_{j_1 j_1 r_2}-a_{j_1 j_1 r_1})^{\mu_1}$$

$$\times \sum_{\rho=2}^{p_1} \sum_{l_2=1}^{\overline{m}_{r_\rho}} A_{l_2 \overline{m}_{s_\rho}}^{(r_2)} \frac{1}{(x_{j_1}-a_{j_1 j_1 r_\rho})^{l_2}}$$

$$+\sum_{\substack{\mu_1\in[\nu_0-\Sigma_{i=1}^{2}m_{j_1r_i},\nu_0-m_{j_1r_1})\\ \nu_1\in[0,m_{j_1r_2})}}^{\mu_1+\nu_1=\nu_0-m_{j_1r_1}}\frac{(\nu_0-m_{j_1r_1})!}{\mu_1!\nu_1!}(a_{j_1j_1r_2}-a_{j_1j_1r_1})^{\mu_1}$$

$$\left.\sum_{\rho=3}^{q_2}\sum_{l_2=1}^{\overline{m}_{r_\rho}}B_{l_2\overline{n}_{s_\rho}}^{(r_2)}\frac{1}{(x_{j_1}-a_{j_1j_1r_\rho})^{l_2}}\right].$$

where

$$A_{l_\rho\overline{m}_{j_1l_\rho}}^{(r_2)}=\frac{1}{(\overline{n}_{j_2s_\rho}-l_\rho)!}\frac{d^{\overline{n}_{2j_2s_\rho}-l_\rho}}{dx_{j_2j_\rho}^{\overline{n}_{2j_2s_\rho}-l_\rho}}\frac{1}{\prod_{\sigma=2,\sigma\neq\rho}^{q_2}(x_{j_2}-a_{j_1j_1s_\sigma})^{n_{2j_2\sigma}}}\Bigg|_{x_{j2}^*=a_{jj_2j_\rho}}$$

$$(l_\rho=1,2,\ldots,\overline{n}_{j_2s_\rho},\ \rho=2,3,\ldots,q_2),$$

$$A_{\overline{n}_{j1}l_\rho\overline{n}_{j_2l_\rho}}^{(r_2)}=\frac{1}{\prod_{\sigma=2,\sigma\neq\rho}^{q_2}(a_{j_2j_2s_\rho}-a_{j_sj_ss_\sigma})^{n_{jj_2s_\sigma}}},$$

$$B_{l_\rho\overline{n}_{2j_\rho}}^{(r_2)}=\frac{1}{(\overline{n}_{j_2s_\rho}-l_\rho)!}\frac{d^{\overline{n}_{2j_2}-l_\rho}}{dx_{j_2}^{\overline{n}_{2j_2}-l_\rho}}\frac{(x_{j_2}-a_{j_2j_2s_2})^{\beta_1-\sum_{i-1}^{2}n_{jj_2}}}{\prod_{\sigma=3,\sigma\neq\rho}^{q_2}(x_{j_2}-a_{j_1j_1s_\sigma})^{n_{jj_2}\sigma}}\Bigg|_{x_{j2}^*=a_{jj_2j_\rho}}$$

$$(l_\rho=1,2,\ldots,\overline{n}_{(2)s_\rho},\ \rho=3,4,\ldots,q_2),$$

$$B_{\overline{n}_{j2}s_\rho\overline{n}_{j2}}^{(s_2)}=\frac{(a_{j_2j_2\rho_\rho}-a_{j_2j_2s_2})^{\beta_1-\sum_{i-1}^{2}n_{j2s_i}}}{\prod_{\sigma=3,\sigma\neq\rho}^{q_2}(a_{j_2j_2s_\rho}-a_{j_1j_1r_\sigma})^{m_{j1s_\sigma}}};$$

$$\{\overline{m}_{j_1r_\rho}|\rho=2,3,\ldots,q_2\}=\{\sum_{i=1}^{2}m_{j_1r_i}-\beta_1,m_{j_1r_3},\ldots,m_{j_1r_{g_2}}\},$$

Repeating the same procedure, the following relations are obtained.

$$\frac{(x_{j_1}-a_{j_1j_1r_2})^{\nu_1-m_{j_1r_2}}}{\prod\limits_{r_3=1,r_3\neq r_1,r_2}^{p_1}(x_{j_1}-a_{j_1j_1r_3})^{m_{j_1r_3}}}$$

$$= \sum_{\substack{\mu_2\in[0,\nu_0-m_1]\\ \nu_2\in[m_1-\Sigma_{i=1}^2 m_{j_1r_i},\nu_0-\Sigma_{i=1}^2 m_{j_1r_i}]}}^{\mu_2+\nu_2=\nu_0-\Sigma_{i=1}^2 m_{j_1r_i}} \prod_{k=1}^{2}\frac{(\nu_{k-1}-\Sigma_{i=1}^k m_{j_1r_i})!}{\mu_k!\nu_k!}(a_{j_1j_1r_{k+1}}-a_{j_1j_1r_k})^{\mu_k}$$

$$\times\frac{(x_{j_1}-a_{j_1j_2s_1})^{\nu_2-m_{j_1r_2}}}{\prod\limits_{s_3=1,s_3\neq s_1,s_2}^{q_2}(x_{j_2}-a_{j_2j_2s_3})^{n_{j_2s_3}}}$$

$$+\sum_{\substack{\mu_2\in[\nu_0-\Sigma_{i=1}^3 m_{j_1r_i},\nu_0-\Sigma_{i=1}^2 m_{j_1r_i})\\ \beta_1\in[0,n_{j_2s_3})}}^{\mu_2+\nu_2=\nu_0-\Sigma_{i=1}^2 m_{j_1r_i}} \prod_{k=1}^{2}\frac{(\nu_{k-1}-\Sigma_{i=1}^k m_{j_1r_i})!}{\mu_k!\nu_k!}(a_{j_1j_1r_{k+1}}-a_{j_1j_1r_k})^{\mu_k}$$

$$\times\sum_{\rho=3}^{p_1}\sum_{l_2=1}^{\overline{m}_{r_\rho}}\frac{A^{(r_3)}_{l_2\overline{m}_{j_1r_\rho}}}{(x_{j_1}-a_{j_1j_1r_\rho})^{l_2}}$$

$$+\sum_{\substack{\mu_2\in(\nu_0-m_1,\beta_0-\Sigma_{i=1}^3 m_{j_1r_i}]\\ \nu_2\in[m_{j_1r_3},m_1-\Sigma_{i=1}^2 m_{j_1r_i})}}^{\mu_2+\nu_2=\nu_0-\Sigma_{i=1}^2 m_{j_1r_i}} \prod_{k=1}^{2}\frac{(\nu_{k-1}-\Sigma_{i=1}^k m_{j_1r_i})!}{\mu_k!\nu_k!}(a_{j_1j_1r_{k+1}}-a_{j_1j_1r_k})^{\mu_k}$$

$$\times\sum_{\rho=4}^{p_1}\sum_{l_2=1}^{\overline{m}_{r_\rho}}\frac{B^{(r_3)}_{l_2\overline{m}_{j_1r_\rho}}}{(x_{j_1}-a_{j_1j_1r_\rho})^{l_2}}.$$

and

$$A^{(r_3)}_{l_\rho\overline{m}_{j_1r_\rho}} = \frac{1}{(\overline{m}_{j_1r_\rho}-l_\rho)!}\frac{d^{\overline{m}_{j_1r_\rho}-l_\rho}}{dx_{j_1}^{\overline{m}_{j_1r_\rho}-l_\rho}}\frac{1}{\prod\limits_{\sigma=3,\sigma\neq\rho}^{p_1}(x_{j_1}-a_{j_1j_1r_\sigma})^{m_{j_1r_\sigma}}}\Big|_{x^*_{j_1}=a_{j_1j_1r_\rho}}$$

$$(l_\rho=1,2,\ldots,\overline{m}_{j_1r_\rho},\ \rho=3,4,\ldots,p_1),$$

$$A^{(r_3)}_{l_\rho\overline{m}_{jr_\rho}} = \frac{1}{\prod\limits_{\sigma=3,\sigma\neq\rho}^{p_1}(a_{j_1j_1r_\rho}-a_{j_1j_1r_\sigma})^{m_{j_1r_\sigma}}},$$

$$B^{(r_3)}_{l_\rho\overline{m}_{j_1r_\rho}} = \frac{1}{(\overline{n}_{j_2s_\rho}-l_\rho)!}\frac{d^{\overline{m}_{j_1r_\rho}-l_\rho}}{dx_{j_2}^{\overline{m}_{j_1r_\rho}-l_\rho}}\frac{(x_{j_1}-a_{j_1j_1r_3})^{\nu_1-\sum_{i=1}^{2}m_{jr_1}}}{\prod\limits_{\sigma=4,\sigma\neq\rho}^{p_1}(x_{j_1}-a_{j_1j_1r_\sigma})^{m_{j1r_\sigma}}}\Big|_{x^*_{j1}=a_{j_1j_1r_\rho}}$$

$$(l_\rho=1,2,\ldots,\overline{m}_{j_1r_\rho},\ \rho=4,5,\ldots,p_1),$$

$$B^{(r_3)}_{\overline{m}_{j_1}r_\rho\overline{m}_{j_1}r_\rho} = \frac{(a_{j_1j_1r_\rho}-a_{j_1j_1r_3})^{\nu_1-\sum_{i=1}^{2}m_{jr_1}}}{\prod\limits_{\sigma=4,\sigma\neq\rho}^{p_1}(a_{j_1j_1r_\rho}-a_{j_1j_1r_\sigma})^{m_{jj_1r_\sigma}}};$$

$$\{\overline{m}_{j_1 r_\rho} | \rho = 3, 4, \ldots, p_1\} = \{\sum_{i=1}^{2} m_{j_1 r_i} - \nu_1, m_{j_1 r_3}, \ldots, m_{j_1 r_1}\},$$

Continuing the same procedure until r_{p-1}, the following relation is obtained as

$$\sum_{\nu_0=m_1}^{m_2} b_{j_2 j_1 \nu_0} \sum_{\substack{\mu_1 \in [0, \nu_0 - m_1], \\ \nu_1 \in [m_1 - m_{j_1 r_1}, \nu_0 - m_{j_1 r_1}]}}^{\mu_1+\nu_1=\nu_0-m_{j_1 r_1}} \sum_{\substack{\mu_2 \in [0, \nu_1 - m_2] \\ \nu_2 \in [m_1 - \Sigma_{i=1}^{2} m_{j_1 r_i}, \nu_1 - \Sigma_{i=1}^{2} m_{j_1 r_i}]}}^{\mu_2+\nu_2=\nu_1-\Sigma_{i=1}^{2} m_{j_1 r_i}} \cdots$$

$$\sum_{\substack{\mu_{p_1-1} \in [0, \nu_{p_1-2} - m_1] \\ \nu_{p_1-1} \in [m_1 - \Sigma_{i=1}^{p_1-1} m_{j_1 r_i}, \nu_{p_1-2} - \Sigma_{i=1}^{p_1-1} m_{j_1 r_i}]}}^{\mu_{q_2-1}+\nu_{q_2-1}=\nu_{p_1-2}-\Sigma_{i=1}^{p_1-1} m_{j_1 r_i}}$$

$$\prod_{j=1}^{p_1-1} \frac{(\nu_{j-1} - \Sigma_{i=1}^{j} m_{j_1 r_i})!}{\mu_j! \nu_j!} (a_{j_1 j_1 r_{j+1}} - a_{j_1 j_1 r_j})^{\mu_j} (x_{j_1} - a_{j_1 j_1 r_{p_1}})^{\nu_{p_1-1} - m_{j_1 r_{p_1}}} dx_{j_1}$$

$$+ \sum_{\nu_0=m_1}^{m_2} b_{j_2 j_1 \nu_0} \sum_{k=2}^{p_1-1} \Bigg[\sum_{\substack{\mu_1 \in [0, \nu_0 - m_1), \\ \nu_1 \in (m_1 - m_{j_1 r_1}, \nu_0 - m_{j_1 r_1}]}}^{\mu_1+\nu_1=\nu_0-m_{j_1 r_1}} \sum_{\substack{\mu_2 \in [0, \nu_1 - m_1] \\ \nu_2 \in (m_1 - \Sigma_{i=1}^{2} m_{j_1 r_i}, \nu_1 - \Sigma_{i=1}^{2} m_{j_1 r_i})}}^{\mu_2+\nu_2=\nu_1-\Sigma_{i=1}^{2} m_{j_1 r_i}} \cdots$$

$$\Bigg(\sum_{\substack{\mu_{k-1} \in (\nu_{k-2} - m_1 + m_{j r_k}, \nu_{k-2} - \Sigma_{k-1}^{k-1} m_{j r_i}] \\ \nu_{k-1} \in [0, m_1 - \Sigma_{i-1}^{k} m_{j r_i}^{k})}}^{\mu_{k-1}+\nu_{k-1}=\nu_{k-2}-\Sigma_{i-1}^{k-1} m_{j r_j}} \prod_{j=1}^{k-1} \frac{(\nu_{j-1} - \Sigma_{i=1}^{j} m_{j_1 r_i})!}{\mu_j! \nu_j!} (a_{j_1 j_j r_{j+1}} - a_{j_1 j_1 r_j})^{\mu_j}$$

$$\times \sum_{\rho=k}^{p_1} \sum_{l_\rho=1}^{m_{j_1 r_\rho}} A_{l_\rho m_{j_1 r_\rho}}^{(r_k)} \frac{dx_{j_1}}{(x_{j_1} - a_{j_1 j_1 r_\rho})^{l_\rho}}$$

$$+ \sum_{\substack{\mu_{k-1} \in (\nu_{k-2} - m_1, \nu_{k-2} - m_1 + m_{j_1 r_k}] \\ \nu_{k-1} \in [m_1 - \Sigma_{i=1}^{k} m_{j_1 r_i}, m_1 - \Sigma_{i=1}^{k-1} m_{j_1 r_i})}}^{\mu_{k-1}+\nu_{k-1}=\nu_{k-2}-\Sigma_{i=1}^{k-1} m_{j_1 r_i}} \prod_{j=1}^{k-1} \frac{(\nu_{j-1} - \Sigma_{i=1}^{j} m_{j_1 r_i})!}{\mu_j! \nu_j!} (a_{j_1 j_1 r_{j+1}} - a_{j_1 j_1 r_j})^{\mu_j}$$

$$\times \sum_{\rho=k+1}^{p_1} \sum_{l_\rho=1}^{\overline{m}_{j_1 r_\rho}} B_{l_\rho \overline{m}_{j_1 r_\rho}}^{(r_k)} \frac{dx_{j_1}}{(x_{j_1} - a_{j_1 j_1 r_\rho})^{l_\rho}} \Bigg) \Bigg]$$

$$+ \Bigg(\sum_{\nu_0=0}^{m_{j_1 s_1}-1} b_{j_2 j_1 \nu_0} \sum_{\rho=1}^{p_1} \sum_{l_\rho=1}^{\overline{m}_{j_1 r_\rho}} A_{l_\rho \overline{m}_{j_1 r_\rho}}^{(r_1)} \frac{dx_{j_1}}{(x_{j_1} - a_{j_1 j_1 r_\rho})^{l_\rho}}$$

$$+ \sum_{\nu_0=m_{j_1 r_1}}^{m_1-1} b_{j_2 j_1 \nu_0} \sum_{\rho=2}^{p_1} \sum_{l_\rho=1}^{\overline{m}_{j_1 r_\rho}} B_{l_\rho \overline{m}_{j_1 r_\rho}}^{(r_1)} \frac{dx_{j_1}}{(x_{j_1} - a_{j_1 j_1 r_\rho})^{l_\rho}} \Bigg) \Bigg\}$$

$$= \frac{a_{j_1j_10}}{a_{j_2j_20}}[(x_{j_2} - a_{j_1j_2u_1})^{n_1} + \sum_{\alpha=1}^{n_1-n_{r_1}} b_{j_1j_2\alpha}(x_{j_1} - a_{j_1j_2u_1})^{n_1-\alpha}]dx_{j_2}.$$

With an initial condition (x_{j_10}, x_{j_20}) at time $t = t_0$, the integration of the foregoing equation gives the first integral manifold as

$$\sum_{\nu_0=m_1}^{m_2} b_{j_2j_1\nu_0} \sum_{\substack{\mu_1+\nu_1=\nu_0-m_{j_1r_1} \\ \mu_1\in[0,\nu_0-m_1], \\ \nu_1\in[m_1-m_{j_1r_1},\nu_0-m_{j_1r_1}]}} \sum_{\substack{\mu_2+\nu_2=\nu_1-\Sigma_{i=1}^2 m_{j_1r_i} \\ \mu_2\in[0,\nu_1-m_2] \\ \nu_2\in[m_1-\Sigma_{i=1}^2 m_{j_1r_i},\nu_1-\Sigma_{i=1}^2 m_{j_1r_i}]}} \cdots$$

$$\sum_{\substack{\mu_{q_2-1}+\nu_{q_2-1}=\nu_{p_1-2}-\Sigma_{i=1}^{p_1-1} m_{j_1r_i} \\ \mu_{p_1-1}\in[0,\nu_{p_1-2}-m_1] \\ \nu_{p_1-1}\in[m_1-\Sigma_{i=1}^{p_1-1} m_{j_1r_i},\nu_{p_1-2}-\Sigma_{i=1}^{p_1-1} m_{j_1r_i}]}} \prod_{j=1}^{p_1-1} \frac{(\nu_{j-1} - \Sigma_{i=1}^{j} m_{j_1r_i})!}{\mu_j!\nu_j!}(a_{j_1j_1r_{j+1}} - a_{j_1j_1r_j})^{\mu_j}$$

$$\times \frac{1}{\nu_{p_1-1} - m_{j_1r_{p_1}} + 1}\left[(x_{j_1} - a_{j_1j_1r_{p_1}})^{\nu_{p_1-1}-m_{j_1r_{p_1}}+1} - (x_{j_10} - a_{j_1j_1r_{p_1}})^{\nu_{p_1-1}-m_{j_1r_{p_1}}+1}\right]$$

$$- \sum_{\nu_0=m_1}^{m_2} b_{j_2j_1\nu_0} \sum_{k=2}^{p_1-1} \left(\sum_{\substack{\mu_1+\nu_1=\nu_0-m_{j_1r_1} \\ \mu_1\in[0,\nu_0-m_1), \\ \nu_1\in(m_1-m_{j_1r_1},\nu_0-m_{j_1r_1}]}} \sum_{\substack{\mu_2+\nu_2=\nu_1-\Sigma_{i=1}^2 m_{j_1r_i} \\ \mu_2\in[0,\nu_1-m_1] \\ \nu_2\in(m_1-\Sigma_{i=1}^2 m_{j_1r_i},\nu_1-\Sigma_{i=1}^2 m_{j_1r_i})}} \cdots \right.$$

$$\left\{ \sum_{\substack{\mu_{k-1}+\nu_{k-1}=\nu_{k-2}-\Sigma_{i=1}^{k-1} m_{j_1r_i} \\ \mu_{k-1}\in(\nu_{k-2}-m_1+m_{j_1r_k},\nu_{k-2}-\Sigma_{i=1}^{k-1} m_{j_1r_i}] \\ \nu_{k-1}\in[0,m_1-\Sigma_{i=1}^{k} m_{j_1r_i})}} \prod_{j=1}^{k-1} \frac{(\nu_{j-1} - \Sigma_{i=1}^{j} m_{j_1r_i})!}{\mu_j!\nu_j!}(a_{j_1j_1r_{j+1}} - a_{j_1j_1r_j})^{\mu_j} \right.$$

$$\times \sum_{\rho=k}^{p_1} \sum_{l_\rho=1}^{m_{j_1r_\rho}} A^{(r_k)}_{l_\rho m_{j_1r_\rho}} \frac{1}{l_\rho - 1}\left[\frac{1}{(x_{j_1} - a_{j_1j_1r_\rho})^{l_\rho-1}} - \frac{1}{(x_{j_10} - a_{j_1j_1r_\rho})^{l_\rho-1}}\right]$$

$$+ \sum_{\substack{\mu_{k-1}+\nu_{k-1}=\nu_{k-2}-\Sigma_{i=1}^{k-1} m_{j_1r_i} \\ \mu_{k-1}\in(\nu_{k-2}-m_1,\nu_{k-2}-m_1+m_{j_1r_k}] \\ \nu_{k-1}\in[m_1-\Sigma_{i=1}^{k} m_{j_1r_i},m_1-\Sigma_{i=1}^{k-1} m_{j_1r_i})}} \prod_{j=1}^{k-1} \frac{(\nu_{j-1} - \Sigma_{i=1}^{j} m_{j_1r_i})!}{\mu_j!\nu_j!}(a_{j_1j_1r_{j+1}} - a_{j_1j_1r_j})^{\mu_j}$$

$$\left.\left. \times \sum_{\rho=k+1}^{p_1} \sum_{l_\rho=1}^{\overline{m}_{j_1r_\rho}} B^{(r_k)}_{l_\rho\bar{m}_{j_1r_\rho}} \frac{1}{l_\rho - 1}\left[\frac{1}{(x_{j_1} - a_{j_1j_1r_\rho})^{l_\rho-1}} - \frac{1}{(x_{j_10} - a_{j_1j_1r_\rho})^{l_\rho-1}}\right]\right\}\right)$$

$$+\left\{ \sum_{\nu_0=0}^{m_{j_1r_1}-1} b_{j_2j_1\nu_0} \sum_{\rho=1}^{p_1} \sum_{l_\rho=1}^{\bar{m}_{j_1r_\rho}} A^{(r_1)}_{l_\rho\bar{m}_{j_1r_\rho}} \frac{1}{l_\rho - 1}\left[\frac{1}{(x_{j_1} - a_{j_1j_1r_\rho})^{l_\rho-1}} - \frac{1}{(x_{j_10} - a_{j_1j_1r_\rho})^{l_\rho-1}}\right]\right.$$

$$\left. + \sum_{\nu_0=m_{j_1r_1}}^{m_1-1} b_{j_2j_1\nu_0} \sum_{\rho=2}^{p_1} \sum_{l_\rho=1}^{\bar{m}_{j_1r_\rho}} B^{(r_1)}_{l_\rho\bar{m}_{j_1r_\rho}} \frac{1}{l_\rho - 1}\left[\frac{1}{(x_{j_1} - a_{j_1j_1r_\rho})^{l_\rho-1}} - \frac{1}{(x_{j_10} - a_{j_1j_1r_\rho})^{l_\rho-1}}\right]\right\}$$

$$= \frac{a_{j_1j_10}}{a_{j_2j_20}} \frac{1}{n_1 + 1}\left\{\left[(x_{j_2} - a_{j_1j_2u_1})^{n_1+1} - (x_{j_20} - a_{j_1j_2u_1})^{n_1+1}\right]\right.$$

$$\left. + \sum\nolimits_{\alpha=1}^{n_1-n_{r_1}} b_{j_1j_2\alpha} \frac{1}{n_1 - \alpha + 1}\left[(x_{j_2} - a_{j_1j_2u_1})^{n_1-\alpha+1} - (x_{j_{20}} - a_{j_1j_2u_1})^{n_1-\alpha+1}\right]\right\}.$$

where

$$A^{(r_k)}_{l_\rho \bar{m}_{j_1 r_\rho}} = \frac{1}{(\bar{m}_{j_1 r_\rho} - l_\rho)!} \frac{d^{\bar{m}_{j_1 r_\rho} - l_\rho}}{dx_{j_1}^{\bar{m}_{j_1 r_\rho} - l_\rho}} \frac{1}{\prod_{\sigma=k,\sigma\neq\rho}^{p_1} (x_{j_1} - a_{j_1 j_1 r_\sigma})^{m_{j_1 r_\sigma}}} \Big|_{x^*_{j_1} = a_{j_1 j_1 r_\rho}}$$

$$(l_\rho = 1, 2, \ldots, \bar{m}_{j_1 r_\rho}, \rho = k, k+1, \ldots, p_1),$$

$$A^{(r_k)}_{\bar{m}_{j_1 r_\rho} \bar{m}_{j_1 r_\rho}} = \frac{1}{\prod_{\sigma=k,\sigma\neq\rho}^{p_1} (a_{j_1 j_1 r_\rho} - a_{j_1 j_1 r_\sigma})^{m_{j_1 r_\sigma}}},$$

$$B^{(r_k)}_{l_\rho \bar{m}_{j_1 r_\rho}} = \frac{1}{(\bar{m}_{j_1 r_\rho} - l_\rho)!} \frac{d^{\bar{m}_{j_1 r_\rho} - l_\rho}}{dx_{j_1}^{\bar{m}_{j_1 r_\rho} - l_\rho}} \frac{(x_{j_1} - a_{j_1 j_1 r_k})^{\nu_{k-1} - \sum_{i=1}^{k} m_{j_1 r_i}}}{\prod_{\sigma=k+1,\sigma\neq\rho}^{p_1} (x_{j_1} - a_{j_1 j_1 r_\sigma})^{m_{j_1 r_\sigma}}} \Big|_{x^*_{j_1} = a_{j_1 j_1 r_\rho}}$$

$$(l_\rho = 1, 2, \cdots, \bar{m}_{j_1 r_\rho}, \rho = k+1, \cdots, p_1),$$

$$B^{(r_k)}_{\bar{m}_{j_1 r_\rho} \bar{m}_{j_1 r_\rho}} = \frac{(x_{j_1} - a_{j_1 j_1 r_k})^{\nu_{k-1} - \sum_{i=1}^{k} m_{j_1 r_i}}}{\prod_{\sigma=k+1,\sigma\neq\rho}^{p_1} (a_{j_1 j_1 r_\rho} - a_{j_1 j_1 r_\sigma})^{m_{j_1 r_\sigma}}},$$

$$\{\bar{m}_{j_1 r_\rho} | \rho = k, k+1, \cdots, p_1\} = \{\sum_{i=1}^{k} m_{j_1 r_i} - \nu_{k-1}, m_{j_1 r_{k+1}}, \cdots, m_{j_1 r_{p_1}}\},$$

for $k \in \{1, 2, \cdots, p_1 - 1\}$;

and

$$b_{j_2 j_1 \nu_0} = \sum_{t_1, l_1 = 0}^{t_1 + l_1 = n_{j_1 u_1}} \cdots \sum_{t_{p_2}, l_{p_2} = 0}^{t_{p_2} + l_{p_2} = m_{j_2 \nu_{p_2}}} \prod_{k=1}^{p_2} \frac{n_{j_1 u_k}!}{t_k! l_k!} (a_{j_1 j_1 r_1} - a_{j_2 j_1 k})^{t_k} \delta_{\nu_0}^{\Sigma_{k=1}^{p_2} l_k},$$

$$\frac{1}{\nu_{p_1 - 1} - m_{j_1 p_1} + 1} \left[(x_{j_1} - a_{j_1 j_1 r_{p_1}})^{\nu_{p_1 - 1} - m_{j_1 p_1} + 1} - (x_{j_1 0} - a_{j_1 j_1 r_{p_1}})^{\nu_{p_1 - 1} - m_{j_1 p_1} + 1} \right]$$

$$= \ln \frac{|x_{j_1} - a_{j_1 j_1 r_{p_1}}|}{|x_{j_1 0} - a_{j_1 j_1 r_{p_1}}|},$$

for $\nu_{p_1 - 1} - m_{j_1 p_1} + 1 = 0$;

$$\lim_{l_k \to 1} \frac{1}{l_k - 1} \left[\frac{1}{(x_{j_1} - a_{j_1 j_1 r_k})^{l_k - 1}} - \frac{1}{(x_{j_1 0} - a_{j_1 j_1 r_k})^{l_k - 1}} \right] = -\ln \frac{|x_{j_1} - a_{j_1 j_1 r_k}|}{|x_{j_1 0} - a_{j_1 j_1 r_k}|},$$

for $l_k = 1$ $(k = 1, 2, \ldots, p_1 - 1)$.

The network matrices of singular equilibriums and 1-dimensional flows are defined as

$$\bigcup_{\eta=j_1}^{j_2} \bigcup_{u=1}^{q_1} \bigcup_{l=1}^{p_1+p_2} (c_{\eta j_1 l}, a_{j_1 j_2 u}) = \bigcup_{r=1}^{p_1} \bigcup_{u=1}^{q_1} \underbrace{(a_{j_1 j_1 r}, a_{j_1 j_2 u})}_{([m_{j_1 r}, n_{j_1 u}] - XX, xF)} \oplus \bigcup_{u=1}^{q_1} \bigcup_{v=1}^{p_2} \underbrace{(a_{j_1 j_2 u}, a_{j_2 j_1 v})}_{(n_{j_1 u}, m_{j_2 v}) - XX},$$

$$\{c_{\eta j_1 l} | \eta \in \{j_1, j_2\}, l = 1, 2, \ldots, p_1 + p_2\}$$

$$= \text{sort}\{a_{j_1 j_1 r}, a_{j_2 j_1 v} | r = 1, 2, \ldots p_1; v = 1, 2, \ldots, p_2\},$$

$$c_{\eta_1 j_1 l_1} < c_{\eta_2 j_1 l_2} \text{ for } l_1 < l_2, \eta_1; \eta_2 \in \{j_1, j_2\};$$

where

$$\bigcup_{r=1}^{p_1} \bigcup_{u=1}^{q_1} \underbrace{(a_{j_1 j_1 r}, a_{j_1 j_2 u})}_{([m_{j_1 r}, n_{j_1 u}]\text{-XX,xF})}$$

$$= \begin{bmatrix} \underbrace{(a_{j_1 j_1 1}, a_{j_1 j_2 q_1})}_{([m_{j_1 1}, n_{j_1 q_1}]\text{XX,xF})} & \cdots & \underbrace{(a_{j_1 j_1 (p_1-1)}, a_{j_1 j_2 q_1})}_{([m_{1(p_1-1)}, n_{j_1 q_1}]\text{XX,xF})} & \underbrace{(a_{j_1 j_1 p_1}, a_{j_1 j_2 q_1})}_{([m_{j_1 p_1}, n_{j_1 q_1}]\text{XX,xF})} \\ \underbrace{(a_{j_1 j_1 1}, a_{j_1 j_2 (q_1-1)})}_{([m_{j_1 1}, n_{j_1 (q_1-1)}]\text{XX,xF})} & \cdots & \underbrace{(a_{j_1 j_1 (p_1-1)}, a_{j_1 j_2 (q_1-1})}_{([m_{j_1 (p_1-1)}, n_{j_1 (q_1-1)}]\text{XX,xF})} & \underbrace{(a_{j_1 j_1 p_1}, a_{j_1 j_2 (q_1-1)})}_{([m_{j_1 p_1}, n_{j_1 (q_1-1)}]\text{XX,xF})} \\ \vdots & \vdots & \vdots & \vdots \\ \underbrace{(a_{j_1 j_1 1}, a_{j_1 j_2 1})}_{([m_{j_1 1}, n_{j_1 1}]\text{XX,xF})} & \cdots & \underbrace{(a_{j_1 j_1 (p_1-1)}, a_{j_1 j_2 1})}_{([m_{j_1 (p_1-1)}, n_{j_1 1}]\text{XX,xF})} & \underbrace{(a_{j_1 j_1 p_1}, a_{j_1 j_2 1})}_{([m_{j_1 p_1}, n_{j_1 1}]\text{XX,xF})} \end{bmatrix}_{q_1 \times p_1},$$

and

$$\bigcup_{u=1}^{q_1} \bigcup_{v=1}^{p_2} \underbrace{(a_{j_1 j_2 u}, a_{j_2 j_1 v})}_{(n_{j_1 u}, m_{j_2 v})\text{-XX}}$$

$$= \begin{bmatrix} \underbrace{(a_{j_1 j_2 q_1}, a_{j_2 j_1 1})}_{(n_{j_1 q_1}, m_{j_2 (p_2-1)})\text{-YY}} & \cdots & \underbrace{(a_{j_1 j_2 q_1}, a_{j_2 j_1 (p_2-1)})}_{(n_{j_1 q_1}, m_{j_2 (p_2-1)})\text{-YY}} & \underbrace{(a_{j_1 j_2 q_1}, a_{j_2 j_1 p_2})}_{(n_{j_1 q_1}, m_{j_2 p_2})\text{-YY}} \\ \underbrace{(a_{j_1 j_2 (q_1-1)}, a_{j_2 j_1 1})}_{(n_{j_1 q_1}, m_{j_2 (p_2-1)})\text{-YY}} & \cdots & \underbrace{(a_{j_1 j_2 (q_1-1)}, a_{j_2 j_1 (p_2-1)})}_{(n_{j_1 q_1}, m_{j_2 (p_2-1)})\text{-YY}} & \underbrace{(a_{j_1 j_2 (q_1-1)}, a_{j_2 j_1 p_2})}_{(n_{j_2 (q_1-1)}, m_{j_2 p_2})\text{-YY}} \\ \vdots & \vdots & \vdots & \vdots \\ \underbrace{(a_{j_1 j_2 1}, a_{j_2 j_1 1})}_{(n_{j_2 1}, m_{j_2 1})\text{-YY}} & \cdots & \underbrace{(a_{j_1 j_2 1}, a_{j_2 j_1 (p_2-1)})}_{(n_{j_1 1}, m_{j_2 p_2})\text{-YY}} & \underbrace{(a_{j_1 j_2 1}, a_{j_2 j_1 p_2})}_{(n_{j_1 1}, m_{j_2 p_2})\text{-YY}} \end{bmatrix}_{q_1 \times p_2}.$$

(ii_{1a}) There is an equilibrium for 1-dimensional flow as

$$(x_{j_1}^*, x_{j_2}^*) = (a_{j_1 j_1 r_1}, a_{j_1 j_2 u_1}).$$

For $a_{j_1 j_1 r} \neq a_{j_2 j_1 v}$ ($r = 1, 2, \ldots, p_1$; $v = 1, 2, \ldots, p_2$), consider

$$\frac{dx_{j_1}}{dx_{j_2}} = \frac{a_{j_1 j_1 0}}{a_{j_2 j_2 0}} \frac{\prod_{r=1}^{p_1} (x_{j_1} - a_{j_1 j_1 r})^{m_{j_1 r}} \prod_{u=1}^{q_1} (x_{j_2} - a_{j_1 j_2 u})^{n_{j_1 u}}}{\prod_{v=1}^{p_2} (x_{j_1} - a_{j_2 j_1 v})^{m_{j_2 v}}}.$$

At $x^*_{j_2} = a_{j_1j_2u_1}$ and $\overline{x}_{j_1} \neq a_{j_1j_1r}, a_{j_2j_1v}(r = 1, 2, \ldots, p_1; v = 1, 2, \ldots, p_2)$, in phase space,

$$\frac{dx_{j_1}}{dx_{j_2}}\bigg|_{x^*_{j_2}=a_{j_1j_2u_1}} = \frac{a_{j_1j_10}}{a_{j_2j_20}} \frac{\prod_{r=1}^{p_1} (\overline{x}_{j_1} - a_{j_1j_1r})^{m_{j_1r}} \prod_{u_2=1,u_2\neq u_1}^{q_1} (x_{j_2} - a_{j_1j_2u_2})^{n_{j_1u_2}}}{\prod_{v=1}^{p_2} (\overline{x}_{j_1} - a_{j_2j_1v})^{m_{j_2v}}}$$

$$\times (x_{j_2} - a_{j_1j_2u_1})^{n_{j_1u_1}}\bigg|_{x^*_{j_2}=a_{j_1j_2u_1}} = 0,$$

$$\vdots$$

$$\frac{d^{n_{j_1u_1}} x_{j_1}}{dx_{j_2}^{n_{j_1u_1}}}\bigg|_{x^*_{j_2}=a_{j_1j_2u_1}} = 0.$$

If

$$\frac{d^{n_{j_1u_1}+1} x_{j_1}}{dx_{j_2}^{n_{j_1u_1}+!}}\bigg|_{x^*_{j_2}=a_{j_1j_2u_1}}$$

$$= \frac{a_{j_1j_10}}{a_{j_2j_20}} \frac{\prod_{r=1}^{p_1} (\overline{x}_{j_1} - a_{j_1j_1r})^{m_{j_1r}} \prod_{u_2=1,u_2\neq u_1}^{q_1} (a_{j_1j_2u_1} - a_{j_1j_2u_2})^{n_{j_1u_2}} n_{j_1u_1}!}{\prod_{v=1}^{p_2} (\overline{x}_{j_1} - a_{j_2j_1v})^{m_{j_2v}}} > 0,$$

the $(2n_{j_1u_1 1} + 1)^{\text{th}}$-order up-parabola flows exists for $n_{j_1u_1} = 2n_{j_1u_1 1} + 1$, and the $(2n_{j_1u_1 1})^{\text{th}}$-order increasing-inflection flow exists for $n_{j_1u_1} = 2n_{j_1u_1 1}$. If

$$\frac{d^{n_{j_1u_1}+1} x_{j_1}}{dx_{j_2}^{n_{j_1u_1}+1}}\bigg|_{x^*_{j_2}=a_{j_1j_2u_1}}$$

$$= \frac{a_{j_1j_10}}{a_{j_2j_20}} \frac{\prod_{r=1}^{p_1} (\overline{x}_{j_1} - a_{j_1j_1r})^{m_{j_1r}} \prod_{u_2=1,u_2\neq u_1}^{q_1} (a_{j_1j_2u_1} - a_{j_1j_2u_2})^{n_{j_1u_2}} n_{j_1u_1}!}{\prod_{v=1}^{p_2} (\overline{x}_{j_1} - a_{j_2j_1v})^{m_{j_2v}}} < 0,$$

the $(2n_{j_1u_1 1} + 1)^{\text{th}}$-order down-parabola flows exists for $n_{j_1u_1} = 2n_{j_1u_1 1} + 1$, and the $(2n_{j_1u_1 1})^{\text{th}}$-order decreasing-inflection flow exists for $n_{j_1u_1} = 2n_{j_1u_1 1}$.

At $\overline{x}_{j_1} \neq a_{j_1j_1r}, a_{j_2j_1v}(r = 1, 2, \ldots, p_1; v = 1, 2, \ldots, p_2)$, let

$$\dot{x}_{j_2} = a_{j_2j_20} \prod_{v=1}^{p_2} (\overline{x}_{j_1} - a_{j_2j_1v})^{m_{j_2v}}.$$

If $a_{j_2j_20}\prod_{v=1}^{p_2}(\overline{x}_{j_1}-a_{j_2j_1v})^{m_{j_2v}}>0$ and $a_{j_2j_20}\prod_{v=1}^{p_2}(\overline{x}_{j_1}-a_{j_2j_1v})^{m_{j_2v}}<0$, the parabola and inflection flows are positive and negative in the x_{j_2}-direction, respectively.

At the equilibrium of $x^*_{j_1}=a_{j_1j_1r_1}$ and $\overline{x}_{j_2}\neq a_{j_1j_2u}, a_{j_2j_2s}$, the variational equation is

$$\dot{x}_{j_2}=a_{j_2j_20}\prod_{s=1}^{q_2}(a_{j_1j_1r_1}-a_{j_2j_1v})^{m_{j_2v}}$$

$$\Delta\dot{x}_{j_1}=a_{j_1j_10}\prod_{r_2=1,r_2\neq r_1}^{p_1}(a_{j_1j_1r_1}-a_{j_1j_1r_2})^{m_{j_1r_2}}\prod_{u=1}^{q_1}(\overline{x}_{j_2}-a_{j_1j_2u})^{n_{j_1u}}(\Delta x_{j_1})^{m_{j_1r_1}}.$$

For

$$a_{j_1j_10}\prod_{r_2=1,r_2\neq r_1}^{p_1}(a_{j_1j_1r_1}-a_{j_1j_1r_2})^{m_{j_1r_2}}\prod_{u=1}^{q_1}(\overline{x}_{j_2}-a_{j_1j_2u})^{n_{j_1u}}>0 \text{ and}$$
$$a_{j_1j_10}\prod_{r_2=1,r_2\neq r_1}^{p_1}(a_{j_1j_1r_1}-a_{j_1j_1r_2})^{m_{j_1r_2}}\prod_{u=1}^{q_1}(\overline{x}_{j_2}-a_{j_1j_2u})^{n_{j_1u}}<0,$$

the flows at $x^*_{j_1}=a_{j_1j_1r_1}$ and $\overline{x}_{j_2}\neq a_{j_1j_2u}, a_{j_2j_2s}$ are the $(2m_{j_1r_1}1+1)^{\text{th}}$-order source and sink for $m_{j_1r_1}=2m_{j_1r_1}1+1$; and the flows at $x^*_{j_1}=a_{j_1j_1r_1}$ and $\overline{x}_{j_2}\neq a_{j_1j_2u}, a_{j_2j_2s}$, are the $(2m_{j_1r_1}1)^{\text{th}}$-order upper-saddle and lower-saddle for $m_{j_1r_1}=2m_{j_1r_1}1$.

Let $\overline{x}_{j_1}=a_{j_1j_1r_1}\neq a_{j_2j_1v}$ and $\overline{x}_{j_2}=a_{j_1j_2u_1}$. Therefore, from the above discussion, there are four cases: (ii_{1a1}) $m_{j_1r_1}=2m_{j_1r_1}1+1$ and $n_{j_1u_1}=2n_{j_1u_1}1+1$, ($\text{ii}_{1a2}$) $m_{j_1r_1}=2m_{j_1r_1}1$ and $n_{j_1u_1}=2n_{j_1u_1}1+1$, ($\text{ii}_{1a3}$) $m_{j_1r_1}=2m_{j_1r_1}1+1$ and $n_{j_1u_1}=2n_{j_1u_1}1$, (ii_{1a4}) $m_{j_1r_1}=2m_{j_1r_1}1$ and $n_{j_1u_1}=2n_{j_1u_1}1$.

(ii_{1a1}) For $m_{j_1r_1}=2m_{j_1r_1}1+1$ and $n_{j_1u_1}=2n_{j_1u_1}1+1$, the equilibrium of $(x^*_{j_1},x^*_{j_2})=(a_{j_1j_1r_1},a_{j_1j_2u_1})$ has the following properties as in Eqs. (1.145)–(1.148).

- For $a_{j_1j_10}\prod_{r_2=1,r_2\neq r_1}^{p_1}(a_{j_1j_1r_1}-a_{j_1j_1r_2})^{m_{j_1r_2}}\prod_{u_2=1,u_2\neq u_1}^{q_1}(a_{j_1j_2u_1}-a_{j_1j_2u_2})^{n_{j_1u_2}}>0$ and $a_{j_2j_20}\prod_{v=1}^{p_2}(a_{j_1j_1r_1}-a_{j_2j_1v})^{m_{j_2v}}>0$,

$$(a_{j_1j_1r_1},a_{j_1j_2u_1})=\underbrace{([(2m_{j_1r_1}1{+}1),(2n_{j_1u_1}1+1)]\text{DP:UP,pF})}_{[(2m_{j_1r_1}1+1),(2n_{j_1u_1}1+1)]\text{-hyperbolic flow }(+)}$$

  The flow at $(x^*_{j_1},x^*_{j_2})=(a_{j_1j_1r_1},a_{j_1j_2u_1})$ is a $([(2m_{j_1r_1}1{+}1),(2n_{j_1u_1}1+1)]\text{DP:UP,pF})$-positive hyperbolic flow .

- For $a_{j_1j_10}\prod_{r_2=1,r_2\neq r_1}^{p_1}(a_{j_1j_1r_1}-a_{j_1j_1r_2})^{m_{j_1r_2}}\prod_{u_2=1,u_2\neq u_1}^{q_1}(a_{j_1j_2u_1}-a_{j_1j_2u_2})^{n_{j_1u_2}}<0$ and $a_{j_2j_20}\prod_{v=1}^{p_2}(a_{j_1j_1r_1}-a_{j_2j_1v})^{m_{j_2v}}>0$,

$$(a_{j_1j_1r_1}, a_{j_1j_2u_1}) = \underbrace{([(2m_{j_1r_1}1+1), (2n_{j_1u_1}1 + 1)]\text{UP:DP,pF})}_{[(2m_{j_1r_1}1+1),(2n_{j_1u_1}1+1)]\text{-hyperbolic-secant flow }(+)}$$

The flow at $(x_{j_1}^*, x_{j_2}^*)=(a_{j_1j_1r_1}, a_{j_1j_2u_1})$ is a ([$(2m_{j_1r_1}1+1), (2n_{j_1u_1}1 + 1)$]UP:DP,pF)-positive hyperbolic-secant flow .

- For $a_{j_1j_10}\prod_{r_2=1,r_2\neq r_1}^{p_1}(a_{j_1j_1r_1}-a_{j_1j_1r_2})^{m_{j_1r_2}}\prod_{u_2=1,u_2\neq u_1}^{q_1}(a_{j_1j_2u_1}-a_{j_1j_2u_2})^{n_{j_1u_2}} > 0$ and $a_{j_2j_20}\prod_{v=1}^{p_2}(a_{j_1j_1r_1}-a_{j_2j_1v})^{m_{j_2v}} < 0$,

$$(a_{j_1j_1r_1}, a_{j_1j_2u_1}) = \underbrace{([(2m_{j_1r_1}1+1), (2n_{j_1u_1}1 + 1)]\text{UP:DP,nF})}_{[(2m_{j_1r_1}1+1),(2n_{j_1u_1}1+1)]\text{-hyperbolic-secant flow }(-)}$$

The flow at $(x_{j_1}^*, x_{j_2}^*)=(a_{j_1j_1r_1}, a_{j_1j_2u_1})$ is a ([$(2m_{j_1r_1}1+1), (2n_{j_1u_1}1 + 1)$]UP:DP,nF)-negative hyperbolic-secant flow .

- For $a_{j_1j_10}\prod_{r_2=1,r_2\neq r_1}^{p_1}(a_{j_1j_1r_1}-a_{j_1j_1r_2})^{m_{j_1r_2}}\prod_{u_2=1,u_2\neq u_1}^{q_1}(a_{j_1j_2u_1}-a_{j_1j_2u_2})^{n_{j_1u_2}} < 0$ and $a_{j_2j_20}\prod_{v=1}^{p_2}(a_{j_1j_1r_1}-a_{j_2j_1v})^{m_{j_2v}} < 0$,

$$(a_{j_1j_1r_1}, a_{j_1j_2u_1}) = \underbrace{([(2m_{j_1r_1}1+1), (2n_{j_1u_1}1 + 1)]\text{DP:UP,nF})}_{[(2m_{j_1r_1}1+1),(2n_{j_1u_1}1+1)]\text{-hyperbolic flow }(-)}$$

The flow at $(x_{j_1}^*, x_{j_2}^*)=(a_{j_1j_1r_1}, a_{j_1j_2u_1})$ is a ([$(2m_{j_1r_1}1+1), (2n_{j_1u_1}1 + 1)$]DP:UP,pF)-negative hyperbolic flow .

(ii_{1a2}) For $m_{j_1r_1} = 2m_{j_1r_1}1$ and $n_{j_1u_1} = 2n_{j_1u_1}1 + 1$, the equilibrium of $(x_{j_1}^*, x_{j_2}^*) = (a_{j_1j_1r_1}, a_{j_1j_2u_1})$ for the x_{j_1}-direction has the following properties as in Eq. (1.149)-(1.152).

- For $a_{j_1j_10}\prod_{r_2=1,r_2\neq r_1}^{p_1}(a_{j_1j_1r_1}-a_{j_1j_1r_2})^{m_{j_1r_2}}\prod_{u_2=1,u_2\neq u_1}^{q_1}(a_{j_1j_2u_1}-a_{j_1j_2u_2})^{n_{j_1u_2}} > 0$ and $a_{j_2j_20}\prod_{v=1}^{p_2}(a_{j_1j_1r_1}-a_{j_2j_1v})^{m_{j_2v}} > 0$,

$$(a_{j_1j_1r_1}, a_{j_1j_2u_1}) = \underbrace{([(2m_{j_1r_1}1), (2n_{j_1u_1}1 + 1)]\text{UP:UP,pF})}_{[(2m_{j_1r_1}1),(2n_{j_1u_1}1+1)]\text{-hyperbolic-secant-to-hyperbolic flow }(+)}.$$

The flow at $(x_{j_1}^*, x_{j_2}^*)=(a_{j_1j_1r_1}, a_{j_1j_2u_1})$ is a ([$(2m_{j_1r_1}1), (2n_{j_1u_1}1 + 1)$]UP:UP,pF)-positive hyperbolic-secant to hyperbolic flow .

- For $a_{j_1j_10}\prod_{r_2=1,r_2\neq r_1}^{p_1}(a_{j_1j_1r_1}-a_{j_1j_1r_2})^{m_{j_1r_2}}\prod_{u_2=1,u_2\neq u_1}^{q_1}(a_{j_1j_2u_1}-a_{j_1j_2u_2})^{n_{j_1u_2}}<0$ and $a_{j_2j_20}\prod_{v=1}^{p_2}(a_{j_1j_1r_1}-a_{j_2j_1v})^{m_{j_2v}}>0$,

$$(a_{j_1j_1r_1},a_{j_1j_2u_1})=\underbrace{([(2m_{j_1r_1}1),(2n_{j_1u_1}1+1)]\text{DP:DP,pF})}_{[(2m_{j_1r_1}1),(2n_{j_1u_1}1+1)]\text{-hyperbolic-to-hyperbolic-secant flow }(+)}.$$

The flow at $(x^*_{j_1},x^*_{j_2})=(a_{j_1j_1r_1},a_{j_1j_2u_1})$ is a $([(2m_{j_1r_1}1),(2n_{j_1u_1}1+1)]$DP:DP,pF)-positive hyperbolic to hyperbolic-secant flow .

- For $a_{j_1j_10}\prod_{r_2=1,r_2\neq r_1}^{p_1}(a_{j_1j_1r_1}-a_{j_1j_1r_2})^{m_{j_1r_2}}\prod_{u_2=1,u_2\neq u_1}^{q_1}(a_{j_1j_2u_1}-a_{j_1j_2u_2})^{n_{j_1u_2}}>0$ and $a_{j_2j_20}\prod_{v=1}^{p_2}(a_{j_1j_1r_1}-a_{j_2j_1v})^{m_{j_2v}}<0$,

$$(a_{j_1j_1r_1},a_{j_1j_2u_1})=\underbrace{([(2m_{j_1r_1}1),(2n_{j_1u_1}1+1)]\text{DP:DP,nF})}_{[(2m_{j_1r_1}1),(2n_{j_1u_1}1+1)]\text{-hyperbolic-to-hyperbolic-secant flow }(-)}.$$

The flow at $(x^*_{j_1},x^*_{j_2})=(a_{j_1j_1r_1},a_{j_1j_2u_1})$ is a $([(2m_{j_1r_1}1),(2n_{j_1u_1}1+1)]$DP:DP,nF)-negative hyperbolic-to-hyperbolic-secant flow .

- For $a_{j_1j_10}\prod_{r_2=1,r_2\neq r_1}^{p_1}(a_{j_1j_1r_1}-a_{j_1j_1r_2})^{m_{j_1r_2}}\prod_{u_2=1,u_2\neq u_1}^{q_1}(a_{j_1j_2u_1}-a_{j_1j_2u_2})^{n_{j_1u_2}}<0$ and $a_{j_2j_20}\prod_{v=1}^{p_2}(a_{j_1j_1r_1}-a_{j_2j_1v})^{m_{j_2v}}<0$,

$$(a_{j_1j_1r_1},a_{j_1j_2u_1})=\underbrace{([(2m_{j_1r_1}1),(2n_{j_1u_1}1+1)]\text{UP:UP,nF})}_{[(2m_{j_1r_1}1),(2n_{j_1u_1}1+1)]\text{-hyperbolic-secant-to-hyperbolic flow }(-)}.$$

The flow at $(x^*_{j_1},x^*_{j_2})=(a_{j_1j_1r_1},a_{j_1j_2u_1})$ is a $([(2m_{j_1r_1}1),(2n_{j_1u_1}1+1)]$UP:UP,nF)-negative hyperbolic-secant-to-hyperbolic flow .

(ii_{1a3}) For $m_{j_1r_1}=2m_{j_1r_1}1+1$ and $n_{j_1u_1}=2n_{j_1u_1}1$, the equilibrium of $(x^*_{j_1},x^*_{j_2})=(a_{j_1j_1r_1},a_{j_1j_2u_1})$ for the x_{j_1}-direction has the following properties as in Eqs. (1.153)–(1.156).

- For $a_{j_1j_10}\prod_{r_2=1,r_2\neq r_1}^{p_1}(a_{j_1j_1r_1}-a_{j_1j_1r_2})^{m_{j_1r_2}}\prod_{u_2=1,u_2\neq u_1}^{q_1}(a_{j_1j_2u_1}-a_{j_1j_2u_2})^{n_{j_1u_2}}>0$ and $a_{j_2j_20}\prod_{v=1}^{p_2}(a_{j_1j_1r_1}-a_{j_2j_1v})^{m_{j_2v}}>0$,

$$(a_{j_1j_1r_1},a_{j_1j_2u_1})=\underbrace{([(2m_{j_1r_1}1+1),(2n_{j_1u_1}1)]_{\text{DI:II}}\text{SO,pF})}_{[(2m_{j_1r_1}1+1),(2n_{j_1u_1}1)]\text{-decreasing-to-increasing inflection source flow }(+)}.$$

The flow at $(x_{j_1}^*, x_{j_2}^*) = (a_{j_1j_1r_1}, a_{j_1j_2u_1})$ is a $([(2m_{j_1r_1}1{+}1), (2n_{j_1u_1}1)]_{\mathrm{DI:II}}\mathrm{SO,pF})$-positive decreasing-to-increasing inflection source flow .

- For $a_{j_1j_10} \prod_{r_2=1, r_2\neq r_1}^{p_1} (a_{j_1j_1r_1} - a_{j_1j_1r_2})^{m_{j_1r_2}} \prod_{u_2=1, u_2\neq u_1}^{q_1} (a_{j_1j_2u_1} - a_{j_1j_2u_2})^{n_{j_1u_2}} < 0$ and $a_{j_2j_20} \prod_{v=1}^{p_2} (a_{j_1j_1r_1} - a_{j_2j_1v})^{m_{j_2v}} > 0$,

$$(a_{j_1j_1r_1}, a_{j_1j_2u_1}) = \underbrace{([(2m_{j_1r_1}1{+}1), (2n_{j_1u_1}1)]_{\mathrm{II:DI}}\mathrm{SI,pF})}_{[(2m_{j_1r_1}1+1),(2n_{j_1u_1}1)]\text{-increasing-to-decreasing inflection sink flow } (+)}.$$

The flow at $(x_{j_1}^*, x_{j_2}^*) = (a_{j_1j_1r_1}, a_{j_1j_2u_1})$ is a $([(2m_{j_1r_1}1{+}1), (2n_{j_1u_1}1)]_{\mathrm{II:DI}}\mathrm{SI,pF})$-positive increasing-to-decreasing inflection source flow .

- For $a_{j_1j_10} \prod_{r_2=1, r_2\neq r_1}^{p_1} (a_{j_1j_1r_1} - a_{j_1j_1r_2})^{m_{j_1r_2}} \prod_{u_2=1, u_2\neq u_1}^{q_1} (a_{j_1j_2u_1} - a_{j_1j_2u_2})^{n_{j_1u_2}} > 0$ and $a_{j_2j_20} \prod_{v=1}^{p_2} (a_{j_1j_1r_1} - a_{j_2j_1v})^{m_{j_2v}} < 0$,

$$(a_{j_1j_1r_1}, a_{j_1j_2u_1}) = \underbrace{([(2m_{j_1r_1}1{+}1), (2n_{j_1u_1}1)]_{\mathrm{II:DI}}\mathrm{SO,nF})}_{[(2m_{j_1r_1}1+1),(2n_{j_1u_1}1)]\text{-increasing-to-decreasing inflection source flow } (-)}.$$

The flow at $(x_{j_1}^*, x_{j_2}^*) = (a_{j_1j_1r_1}, a_{j_1j_2u_1})$ is a $([(2m_{j_1r_1}1{+}1), (2n_{j_1u_1}1)]_{\mathrm{II:DI}}\mathrm{SO,nF})$-negative increasing-to-decreasing inflection source flow .

- For $a_{j_1j_10} \prod_{r_2=1, r_2\neq r_1}^{p_1} (a_{j_1j_1r_1} - a_{j_1j_1r_2})^{m_{j_1r_2}} \prod_{u_2=1, u_2\neq u_1}^{q_1} (a_{j_1j_2u_1} - a_{j_1j_2u_2})^{n_{j_1u_2}} < 0$ and $a_{j_2j_20} \prod_{v=1}^{p_2} (a_{j_1j_1r_1} - a_{j_2j_1v})^{m_{j_2v}} < 0$,

$$(a_{j_1j_1r_1}, a_{j_1j_2u_1}) = \underbrace{([(2m_{j_1r_1}1{+}1), (2n_{j_1u_1}1)]_{\mathrm{DI:II}}\mathrm{SI,nF})}_{[(2m_{j_1r_1}1+1),(2n_{j_1u_1}1)]\text{-decreasing-to-increasing inflection sink flow } (-)}.$$

The flow at $(x_{j_1}^*, x_{j_2}^*) = (a_{j_1j_1r_1}, a_{j_1j_2u_1})$ is a $([(2m_{j_1r_1}1{+}1), (2n_{j_1u_1}1)]_{\mathrm{DI:II}}\mathrm{SI,nF})$-negative decreasing-to-increasing inflection sink flow .

(ii_{1a4}) For $m_{j_1r_1} = 2m_{j_1r_1}1$ and $n_{j_1u_1} = 2n_{j_1u_1}1$, the equilibrium of $(x_{j_1}^*, x_{j_2}^*) = (a_{j_1j_1r_1}, a_{j_1j_2u_1})$ for the x_{j_1}-direction has the following properties as in Eqs. (1.157)–(1.160).

- For $a_{j_1j_10} \prod_{r_2=1, r_2\neq r_1}^{p_1} (a_{j_1j_1r_1} - a_{j_1j_1r_2})^{m_{j_1r_2}} \prod_{u_2=1, u_2\neq u_1}^{q_1} (a_{j_1j_2u_1} - a_{j_1j_2u_2})^{n_{j_1u_2}} > 0$ and $a_{j_2j_20} \prod_{v=1}^{p_2} (a_{j_1j_1r_1} - a_{j_2j_1v})^{m_{j_2v}} > 0$,

$$(a_{j_1j_1r_1}, a_{j_1j_2u_1}) = \underbrace{([(2m_{j_1r_1}1), (2n_{j_1u_1}1)]_{\text{II:II}}\text{US,pF})}_{[(2m_{j_1r_1}1),(2n_{j_1u_1}1)]\text{-increasing-inflection upper-saddle flow }(+)}.$$

The flow at $(x^*_{j_1}, x^*_{j_2}) = (a_{j_1j_1r_1}, a_{j_1j_2u_1})$ is a $([(2m_{j_1r_1}1), (2n_{j_1u_1}1)]_{\text{II:II}}\text{US,pF})$-positive increasing-inflection upper-saddle flow .

- For $a_{j_1j_10} \prod_{r_2=1, r_2\neq r_1}^{p_1} (a_{j_1j_1r_1} - a_{j_1j_1r_2})^{m_{j_1r_2}} \prod_{u_2=1, u_2\neq u_1}^{q_1} (a_{j_1j_2u_1} - a_{j_1j_2u_2})^{n_{j_1u_2}} < 0$ and $a_{j_2j_20} \prod_{v=1}^{p_2} (a_{j_1j_1r_1} - a_{j_2j_1v})^{m_{j_2v}} > 0$,

$$(a_{j_1j_1r_1}, a_{j_1j_2u_1}) = \underbrace{([(2m_{j_1r_1}1), (2n_{j_1u_1}1)]_{\text{DI:DI}}\text{LS,pF})}_{[(2m_{j_1r_1}1),(2n_{j_1u_1}1)]\text{-decreasing-inflection lower-saddle flow }(+)}.$$

The flow at $(x^*_{j_1}, x^*_{j_2}) = (a_{j_1j_1r_1}, a_{j_1j_2u_1})$ is a $([(2m_{j_1r_1}1), (2n_{j_1u_1}1)]_{\text{DI:DI}}\text{LS,pF})$-positive decreasing-inflection lower-saddle flow .

- For $a_{j_1j_10} \prod_{r_2=1, r_2\neq r_1}^{p_1} (a_{j_1j_1r_1} - a_{j_1j_1r_2})^{m_{j_1r_2}} \prod_{u_2=1, u_2\neq u_1}^{q_1} (a_{j_1j_2u_1} - a_{j_1j_2u_2})^{n_{j_1u_2}} > 0$ and $a_{j_2j_20} \prod_{v=1}^{p_2} (a_{j_1j_1r_1} - a_{j_2j_1v})^{m_{j_2v}} < 0$,

$$(a_{j_1j_1r_1}, a_{j_1j_2u_1}) = \underbrace{([(2m_{j_1r_1}1), (2n_{j_1u_1}1)]_{\text{DI:DI}}\text{US,nF})}_{[(2m_{j_1r_1}1),(2n_{j_1u_1}1)]\text{-decreasing-inflection upper-saddle flow }(-)}.$$

The flow at $(x^*_{j_1}, x^*_{j_2}) = (a_{j_1j_1r_1}, a_{j_1j_2u_1})$ is a $([(2m_{j_1r_1}1), (2n_{j_1u_1}1)]_{\text{DI:DI}}\text{US,nF})$-negative decreasing-inflection upper-saddle flow .

- For $a_{j_1j_10} \prod_{r_2=1, r_2\neq r_1}^{p_1} (a_{j_1j_1r_1} - a_{j_1j_1r_2})^{m_{j_1r_2}} \prod_{u_2=1, u_2\neq u_1}^{q_1} (a_{j_1j_2u_1} - a_{j_1j_2u_2})^{n_{j_1u_2}} < 0$ and $a_{j_2j_20} \prod_{v=1}^{p_2} (a_{j_1j_1r_1} - a_{j_2j_1v})^{m_{j_2v}} < 0$,

$$(a_{j_1j_1r_1}, a_{j_1j_2u_1}) = \underbrace{([(2m_{j_1r_1}1), (2n_{j_1u_1}1)]_{\text{II:II}}\text{LS,nF})}_{[(2m_{j_1r_1}1),(2n_{j_1u_1}1)]\text{-increasing-inflection lower-saddle flow }(-)}.$$

The flow at $(x^*_{j_1}, x^*_{j_2}) = (a_{j_1j_1r_1}, a_{j_1j_2u_1})$ is a $([(2m_{j_1r_1}1), (2n_{j_1u_1}1)]_{\text{II:II}}\text{LS,nF})$-negative increasing-inflection lower-saddle flow .

($\text{ii}_{1\text{b}}$) There is an equilibrium for equilibrium as

$$(x^*_{j_2}, x^*_{j_1}) = (a_{j_1j_2u_1}, a_{j_2j_1v_1}).$$

For $a_{j_2j_1v} \neq a_{j_1j_1r}$ $(r = 1, 2, \ldots, p_1; v = 1, 2, \ldots, p_2)$, consider

$$\frac{dx_{j_2}}{dx_{j_1}}=\frac{a_{j_2j_20}}{a_{j_1j_10}}\frac{\prod_{v=1}^{p_2}(x_{j_1}-a_{j_2j_1v})^{m_{j_2v}}}{\prod_{r=1}^{p_1}(x_{j_1}-a_{j_1j_1r})^{m_{j_1r}}\prod_{u=1}^{q_1}(x_{j_2}-a_{j_1j_2u})^{n_{j_1u}}}.$$

At the equilibrium of $x_{j_2}^*=a_{j_2j_1v_1}$ and $\overline{x}_{j_1}\neq a_{j_1j_1r}$, in phase space,

$$\begin{aligned}\left.\frac{dx_{j_2}}{dx_{j_1}}\right|_{x_{j_1}^*=a_{j_2j_1v_1}}&=\frac{a_{j_2j_20}}{a_{j_1j_10}}\left.\frac{(x_{j_1}-a_{j_2j_1v_1})^{m_{j_2v_1}}\prod_{v_2=1,v_2\neq v_1}^{p_2}(x_{j_1}-a_{j_2j_1v_2})^{m_{j_2v_2}}}{\prod_{r=1}^{p_1}(x_{j_1}-a_{j_1j_1r})^{m_{j_1r}}\prod_{u=1}^{q_1}(\overline{x}_{j_2}-a_{j_1j_2u})^{n_{j_1u}}}\right|_{x_{j_1}^*=a_{j_2j_1v_1}}\\&=0,\\&\vdots\\\left.\frac{d^{m_{j_2v_1}}x_{j_2}}{dx_{j_1}^{m_{j_2v_1}}}\right|_{x_{j_1}^*=a_{j_2j_1v_1}}&=0.\end{aligned}$$

If

$$\left.\frac{d^{m_{j_1u_1}+1}x_{j_2}}{dx_{j_1}^{m_{j_1u_1}+1}}\right|_{x_{j_1}^*=a_{j_2j_1v_1}}=\frac{a_{j_2j_20}}{a_{j_1j_10}}\frac{\prod_{v_2=1,v_2\neq v_1}^{p_2}(a_{j_2j_1v_1}-a_{j_2j_1v_2})^{m_{j_2v_2}}m_{j_2v_1}!}{\prod_{r=1}^{p_1}(a_{j_2j_1v_1}-a_{j_1j_1r})^{m_{j_1r}}\prod_{u=1}^{q_1}(\overline{x}_{j_2}-a_{j_1j_2u})^{n_{j_1u}}}>0,$$

the $(2m_{j_2v_1 1}+1)^{\text{th}}$-order up-parabola flows exists for $m_{j_2v_1}=2m_{j_2v_1 1}+1$, and the $(2m_{j_2v_1 1})^{\text{th}}$-order increasing-inflection flow exists for $m_{j_2v_1}=2m_{j_2v_1 1}$. If

$$\left.\frac{d^{m_{j_1u_1}+1}x_{j_2}}{dx_{j_1}^{m_{j_1u_1}+1}}\right|_{x_{j_1}^*=a_{j_2j_1v_1}}=\frac{a_{j_2j_20}}{a_{j_1j_10}}\frac{\prod_{v_2=1,v_2\neq v_1}^{p_2}(a_{j_2j_1v_1}-a_{j_2j_1v_2})^{m_{j_2v_2}}m_{j_2v_1}!}{\prod_{r=1}^{p_1}(a_{j_2j_1v_1}-a_{j_1j_1r})^{m_{j_1r}}\prod_{u=1}^{q_1}(\overline{x}_{j_2}-a_{j_1j_2u})^{n_{j_1u}}}<0,$$

the $(2m_{j_2v_1 1}+1)^{\text{th}}$-order down-parabola flows exists for $m_{j_2v_1}=2m_{j_2v_1 1}+1$, and the $(2m_{j_2v_1 1})^{\text{th}}$-order decreasing-inflection flow exists for $m_{j_2v_1}=2m_{j_2v_1 1}$.

Let

$$\dot{x}^*_{j_1} = a_{j_1j_10}\prod_{r=1}^{p_1}(a_{j_2j_1v_1} - a_{j_1j_1r})^{m_{j_1r}}\prod_{u=1}^{q_1}(\overline{x}_{j_2} - a_{j_1j_2u})^{n_{j_1u}}.$$

If $\dot{x}^*_{j_1} > 0$ and $\dot{x}^*_{j_1} < 0$, the parabola and inflection flows are positive and negative in the x_{j_1}-direction, respectively.

Thus, from the above analysis and the proof of the first part of the case (ii$_{1a}$), the equilibriums of $(x^*_{j_2}, x^*_{j_1}) = (a_{j_1j_2u_1}, a_{j_2j_1v_1})$ have the 16 types of singular properties.

(ii$_{1b1}$) For $n_{j_1u_1} = 2n_{j_1u_1}1 + 1$ and $m_{j_2v_1} = 2m_{j_2v_1}1 + 1$, the equilibrium of $(x^*_{j_2}, x^*_{j_1}) = (a_{j_1j_2u_1}, a_{j_2j_1v_1})$ has the following properties as in Eqs. (1.162)–(1.165)

- For $a_{j_1j_10}\prod_{r=1}^{p_1}(a_{j_2j_1v_1} - a_{j_1j_1r})^{m_{j_1r}}\prod_{u_2=1,u_2\neq u_1}^{q_1}(a_{j_1j_2u_1} - a_{j_1j_2u_2})^{n_{j_1u_2}} > 0$ and $a_{j_2j_20}\prod_{v_2=1,v_2\neq v_1}^{p_2}(a_{j_2j_1v_1} - a_{j_2j_1v_2})^{m_{j_2v_2}} > 0$,

$$(a_{j_1j_2u_1}, a_{j_2j_1v_1}) = \underbrace{((2n_{j_1u_1}1 + 1)^{\text{th}}\text{UP}_+, (2m_{j_2v_1}1 + 1)^{\text{th}}\text{UP}_+)}_{((2n_{j_1u_1}1+1),(2m_{j_2v_1}1+1))\text{-positive saddle}}.$$

  The flow at $(x^*_{j_2}, x^*_{j_1}) = (a_{j_1j_2u_1}, a_{j_2j_1v_1})$ is a $((2n_{j_1u_1}1+1)^{\text{th}}\text{UP}_+, (2m_{j_2v_1}1+1)^{\text{th}}\text{UP}_+)$-positive saddle .

- For $a_{j_1j_10}\prod_{r=1}^{p_1}(a_{j_2j_1v_1} - a_{j_1j_1r})^{m_{j_1r}}\prod_{u_2=1,u_2\neq u_1}^{q_1}(a_{j_1j_2u_1} - a_{j_1j_2u_2})^{n_{j_1u_2}} < 0$ and $a_{j_2j_20}\prod_{v_2=1,v_2\neq v_1}^{p_2}(a_{j_2j_1v_1} - a_{j_2j_1v_2})^{m_{j_2v_2}} > 0$,

$$(a_{j_1j_2u_1}, a_{j_2j_1v_1}) = \underbrace{((2n_{j_1u_1}1 + 1)^{\text{th}}\text{DP}_+, (2m_{j_2v_1}1 + 1)^{\text{th}}\text{DP}_-)}_{((2n_{j_1u_1}1+1),(2m_{j_2v_1}1+1))\text{-CCW center}}.$$

  The flow at $(x^*_{j_2}, x^*_{j_1}) = (a_{j_1j_2u_1}, a_{j_2j_1v_1})$ is a $((2n_{j_1u_1}1+1)^{\text{th}}\text{DP}_+, (2m_{j_2v_1}1+1)^{\text{th}}\text{DP}_-)$-counter-clockwise center .

- For $a_{j_1j_10}\prod_{r=1}^{p_1}(a_{j_2j_1v_1} - a_{j_1j_1r})^{m_{j_1r}}\prod_{u_2=1,u_2\neq u_1}^{q_1}(a_{j_1j_2u_1} - a_{j_1j_2u_2})^{n_{j_1u_2}} > 0$ and $a_{j_2j_20}\prod_{v_2=1,v_2\neq v_1}^{p_2}(a_{j_2j_1v_1} - a_{j_2j_1v_2})^{m_{j_2v_2}} < 0$,

$$(a_{j_1j_2u_1}, a_{j_2j_1v_1}) = \underbrace{((2n_{j_1u_1}1 + 1)^{\text{th}}\text{DP}_-, (2m_{j_2v_1}1 + 1)^{\text{th}}\text{DP}_+)}_{((2n_{j_1u_1}1+1),(2m_{j_2v_1}1+1))\text{-CW center}}.$$

  The flow at $(x^*_{j_2}, x^*_{j_1}) = (a_{j_1j_2u_1}, a_{j_2j_1v_1})$ is a $((2n_{j_1u_1}1+1)^{\text{th}}\text{DP}_-, (2m_{j_2v_1}1+1)^{\text{th}}\text{DP}_+)$-clockwise center .

- For $a_{j_1j_10}\prod_{r=1}^{p_1}(a_{j_2j_1v_1}-a_{j_1j_1r})^{m_{j_1r}}\prod_{u_2=1,u_2\neq u_1}^{q_1}(a_{j_1j_2u_1}-a_{j_1j_2u_2})^{n_{j_1u_2}}<0$ and $a_{j_2j_20}\prod_{v_2=1,v_2\neq v_1}^{p_2}(a_{j_2j_1v_1}-a_{j_2j_1v_2})^{m_{j_2v_2}}<0$,

$$(a_{j_1j_2u_1},a_{j_2j_1v_1})=\underbrace{((2n_{j_1u_1}1+1)^{\text{th}}\text{UP}_-,(2m_{j_2v_1}1+1)^{\text{th}}\text{UP}_-)}_{((2n_{j_1u_1}1+1),(2m_{j_2v_1}1+1))\text{-negative saddle}}.$$

The flow at $(x_{j_2}^*,x_{j_1}^*)=(a_{j_1j_2u_1},a_{j_2j_1v_1})$ is a $((2n_{j_1u_1}1+1)^{\text{th}}\text{UP}_-,(2m_{j_2v_1}1+1)^{\text{th}}\text{UP}_-)$-negative saddle .

(ii$_{1b2}$) For $n_{j_1u_1}=2n_{j_1u_1}1$ and $m_{j_2v_1}=2m_{j_2v_1}1+1$, the equilibrium of $(x_{j_2}^*,x_{j_1}^*)=(a_{j_1j_2u_1},a_{j_2j_1v_1})$ has the following properties as in Eqs. (1.166)–(1.169).

- For $a_{j_1j_10}\prod_{r=1}^{p_1}(a_{j_2j_1v_1}-a_{j_1j_1r})^{m_{j_1r}}\prod_{u_2=1,u_2\neq u_1}^{q_1}(a_{j_1j_2u_1}-a_{j_1j_2u_2})^{n_{j_1u_2}}>0$ and $a_{j_2j_20}\prod_{v_2=1,v_2\neq v_1}^{p_2}(a_{j_2j_1v_1}-a_{j_2j_1v_2})^{m_{j_2v_2}}>0$,

$$(a_{j_1j_2u_1},a_{j_2j_1v_1})=\underbrace{((2n_{j_1u_1}1)^{\text{th}}\text{US},(2m_{j_2v_1}1+1)^{\text{th}}\text{UP})}_{((2n_{j_1u_1}1),(2m_{j_2v_1}1+1))\text{-up-parabola upper-saddle}}.$$

The flow at $(x_{j_2}^*,x_{j_1}^*)=(a_{j_1j_2u_1},a_{j_2j_1v_1})$ is a $((2n_{j_1u_1}1)^{\text{th}}\text{US},(2m_{j_2v_1}1+1)^{\text{th}}\text{UP})$ -up-parabola upper-saddle .

- For $a_{j_1j_10}\prod_{r=1}^{p_1}(a_{j_2j_1v_1}-a_{j_1j_1r})^{m_{j_1r}}\prod_{u_2=1,u_2\neq u_1}^{q_1}(a_{j_1j_2u_1}-a_{j_1j_2u_2})^{n_{j_1u_2}}<0$ and $a_{j_2j_20}\prod_{v_2=1,v_2\neq v_1}^{p_2}(a_{j_2j_1v_1}-a_{j_2j_1v_2})^{m_{j_2v_2}}>0$,

$$(a_{j_1j_2u_1},a_{j_2j_1v_1})=\underbrace{((2n_{j_1u_1}1)^{\text{th}}\text{LS},(2m_{j_2v_1}1+1)^{\text{th}}\text{DP})}_{((2n_{j_1u_1}1),(2m_{j_2v_1}1+1))\text{-down-parabola lower-saddle}}.$$

The flow at $(x_{j_2}^*,x_{j_1}^*)=(a_{j_1j_2u_1},a_{j_2j_1v_1})$ is a $((2n_{j_1u_1}1)^{\text{th}}\text{LS},(2m_{j_2v_1}1+1)^{\text{th}}\text{DP})$ -down-parabola lower-saddle.

- For $a_{j_1j_10}\prod_{r=1}^{p_1}(a_{j_2j_1v_1}-a_{j_1j_1r})^{m_{j_1r}}\prod_{u_2=1,u_2\neq u_1}^{q_1}(a_{j_1j_2u_1}-a_{j_1j_2u_2})^{n_{j_1u_2}}>0$ and $a_{j_2j_20}\prod_{v_2=1,v_2\neq v_1}^{p_2}(a_{j_2j_1v_1}-a_{j_2j_1v_2})^{m_{j_2v_2}}<0$,

$$(a_{j_1j_2u_1},a_{j_2j_1v_1})=\underbrace{((2n_{j_1u_1}1)^{\text{th}}\text{US},(2m_{j_2v_1}1+1)^{\text{th}}\text{DP})}_{((2n_{j_1u_1}1),(2m_{j_2v_1}1+1))\text{-down-parabola upper-saddle}}.$$

The flow at $(x_{j_2}^*, x_{j_1}^*) = (a_{j_1j_2u_1}, a_{j_2j_1v_1})$ is a $((2n_{j_1u_1}1)^{\text{th}}\text{US},(2m_{j_2v_1}1+1)^{\text{th}}\text{DP})$ -down-parabola upper-saddle .

- For $a_{j_1j_10}\prod_{r=1}^{p_1}(a_{j_2j_1v_1}-a_{j_1j_1r})^{m_{j_1r}}\prod_{u_2=1,u_2\neq u_1}^{q_1}(a_{j_1j_2u_1}-a_{j_1j_2u_2})^{n_{j_1u_2}}<0$ and $a_{j_2j_20}\prod_{v_2=1,v_2\neq v_1}^{p_2}(a_{j_2j_1v_1}-a_{j_2j_1v_2})^{m_{j_2v_2}}<0$,

$$(a_{j_1j_2u_1}, a_{j_2j_1v_1}) = \underbrace{((2n_{j_1u_1}1)^{\text{th}}\text{LS},(2m_{j_2v_1}1+1)^{\text{th}}\text{UP})}_{((2n_{j_1u_1}1),(2m_{j_2v_1}1+1))\text{-up-parabola lower-saddle}}.$$

The flow at $(x_{j_2}^*, x_{j_1}^*) = (a_{j_1j_2u_1}, a_{j_2j_1v_1})$ is a $((2n_{j_1u_1}1)^{\text{th}}\ \text{LS},(2m_{j_2v_1}1+1)^{\text{th}}\text{UP})$ -up-parabola lower-saddle .

(ii_{1b3}) For $n_{j_1u_1} = 2n_{j_1u_1}1+1$ and $m_{j_2v_1} = 2m_{j_2v_1}1$, the equilibrium of $(x_{j_2}^*, x_{j_1}^*) = (a_{j_1j_2u_1}, a_{j_2j_1v_1})$ has the following properties as in Eqs. (1.170)–(1.173).

- For $a_{j_1j_10}\prod_{r=1}^{p_1}(a_{j_2j_1v_1}-a_{j_1j_1r})^{m_{j_1r}}\prod_{u_2=1,u_2\neq u_1}^{q_1}(a_{j_1j_2u_1}-a_{j_1j_2u_2})^{n_{j_1u_2}}>0$ and $a_{j_2j_20}\prod_{v_2=1,v_2\neq v_1}^{p_2}(a_{j_2j_1v_1}-a_{j_2j_1v_2})^{m_{j_2v_2}}>0$,

$$(a_{j_1j_2u_1}, a_{j_2j_1v_1}) = \underbrace{((2n_{j_1u_1}1+1)^{\text{th}}\text{UP},(2m_{j_2v_1}1)^{\text{th}}\text{US})}_{((2n_{j_1u_1}1+1),(2m_{j_2v_1}1))\text{-up-parabola upper-saddle}}.$$

The flow at $(x_{j_2}^*, x_{j_1}^*) = (a_{j_1j_2u_1}, a_{j_2j_1v_1})$ is a $((2n_{j_1u_1}1+1)^{\text{th}}\text{UP},(2m_{j_2v_1}1)^{\text{th}}\text{US})$ -up-parabola upper-saddle .

- For $a_{j_1j_10}\prod_{r=1}^{p_1}(a_{j_2j_1v_1}-a_{j_1j_1r})^{m_{j_1r}}\prod_{u_2=1,u_2\neq u_1}^{q_1}(a_{j_1j_2u_1}-a_{j_1j_2u_2})^{n_{j_1u_2}}<0$ and $a_{j_2j_20}\prod_{v_2=1,v_2\neq v_1}^{p_2}(a_{j_2j_1v_1}-a_{j_2j_1v_2})^{m_{j_2v_2}}>0$,

$$(a_{j_1j_2u_1}, a_{j_2j_1v_1}) = \underbrace{((2n_{j_1u_1}1+1)^{\text{th}}\text{DP},(2m_{j_2v_1}1)^{\text{th}}\text{US})}_{((2n_{j_1u_1}1+1),(2m_{j_2v_1}1))\text{-down-parabola upper-saddle}}.$$

The flow at $(x_{j_2}^*, x_{j_1}^*) = (a_{j_1j_2u_1}, a_{j_2j_1v_1})$ is a $((2n_{j_1u_1}1+1)^{\text{th}}\text{DP},(2m_{j_2v_1}1)^{\text{th}}\text{US})$ -down-parabola upper-saddle .

- For $a_{j_1j_10}\prod_{r=1}^{p_1}(a_{j_2j_1v_1}-a_{j_1j_1r})^{m_{j_1r}}\prod_{u_2=1,u_2\neq u_1}^{q_1}(a_{j_1j_2u_1}-a_{j_1j_2u_2})^{n_{j_1u_2}}>0$ and $a_{j_2j_20}\prod_{v_2=1,v_2\neq v_1}^{p_2}(a_{j_2j_1v_1}-a_{j_2j_1v_2})^{m_{j_2v_2}}<0$,

$$(a_{j_1j_2u_1}, a_{j_2j_1v_1}) = \underbrace{((2n_{j_1u_1}1+1)^{\text{th}}\text{DP},(2m_{j_2v_1}1)^{\text{th}}\text{LS})}_{((2n_{j_1u_1}1+1),(2m_{j_2v_1}1))\text{-down-parabola lower-saddle}} .$$

The flow at $(x_{j_2}^*, x_{j_1}^*) = (a_{j_1j_2u_1}, a_{j_2j_1v_1})$ is a $((2n_{j_1u_1}1+1)^{\text{th}}\text{DP},(2m_{j_2v_1}1)^{\text{th}}\text{LS})$ -down-parabola lower-saddle .

- For $a_{j_1j_10}\prod_{r=1}^{p_1}(a_{j_2j_1v_1}-a_{j_1j_1r})^{m_{j_1r}}\prod_{u_2=1,u_2\neq u_1}^{q_1}(a_{j_1j_2u_1}-a_{j_1j_2u_2})^{n_{j_1u_2}}<0$ and $a_{j_2j_20}\prod_{v_2=1,v_2\neq v_1}^{p_2}(a_{j_2j_1v_1}-a_{j_2j_1v_2})^{m_{j_2v_2}}<0$,

$$(a_{j_1j_2u_1}, a_{j_2j_1v_1}) = \underbrace{((2n_{j_1u_1}1+1)^{\text{th}}\text{UP},(2m_{j_2v_1}1)^{\text{th}}\text{LS})}_{((2n_{j_1u_1}1+1),(2m_{j_2v_1}1))\text{-up-parabola lower-saddle}} .$$

The flow at $(x_{j_2}^*, x_{j_1}^*) = (a_{j_1j_2u_1}, a_{j_2j_1v_1})$ is a $((2n_{j_1u_1}1+1)^{\text{th}}\text{UP},(2m_{j_2v_1}1)^{\text{th}}\text{LS})$-up-parabola lower-saddle .

(ii_{1b4}) For $n_{j_1u_1} = 2n_{j_1u_1}1$ and $m_{j_2v_1} = 2m_{j_2v_1}1$, the equilibrium of $(x_{j_2}^*, x_{j_1}^*) = (a_{j_1j_2u_1}, a_{j_2j_1v_1})$ has the following properties as in Eqs. (1.174)–(1.177).

- For $a_{j_1j_10}\prod_{r=1}^{p_1}(a_{j_2j_1v_1}-a_{j_1j_1r})^{m_{j_1r}}\prod_{u_2=1,u_2\neq u_1}^{q_1}(a_{j_1j_2u_1}-a_{j_1j_2u_2})^{n_{j_1u_2}}>0$ and $a_{j_2j_20}\prod_{v_2=1,v_2\neq v_1}^{p_2}(a_{j_2j_1v_1}-a_{j_2j_1v_2})^{m_{j_2v_2}}>0$,

$$(a_{j_1j_2u_1}, a_{j_2j_1v_1}) = \underbrace{((2n_{j_1u_1}1)^{\text{th}}\text{II}_+, (2m_{j_2v_1}1)^{\text{th}}\text{II}_+)}_{((2n_{j_1u_1}1),(2m_{j_2v_1}1))\text{-double inflection-saddle}} .$$

The flow at $(x_{j_2}^*, x_{j_1}^*) = (a_{j_1j_2u_1}, a_{j_2j_1v_1})$ is a $((2n_{j_1u_1}1)^{\text{th}}\text{II}_+, (2m_{j_2v_1}1)^{\text{th}}\text{II}_+)$-double inflection-saddle .

- For $a_{j_1j_10}\prod_{r=1}^{p_1}(a_{j_2j_1v_1}-a_{j_1j_1r})^{m_{j_1r}}\prod_{u_2=1,u_2\neq u_1}^{q_1}(a_{j_1j_2u_1}-a_{j_1j_2u_2})^{n_{j_1u_2}}<0$ and $a_{j_2j_20}\prod_{v_2=1,v_2\neq v_1}^{p_2}(a_{j_2j_1v_1}-a_{j_2j_1v_2})^{m_{j_2v_2}}>0$,

$$(a_{j_1j_2u_1}, a_{j_2j_1v_1}) = \underbrace{((2n_{j_1u_1}1)^{\text{th}}\text{DI}_+, (2m_{j_2v_1}1)^{\text{th}}\text{DI}_-)}_{((2n_{j_1u_1}1),(2m_{j_2v_1}1))\text{-double inflection-saddle}} .$$

The flow at $(x_{j_2}^*, x_{j_1}^*) = (a_{j_1j_2u_1}, a_{j_2j_1v_1})$ is a $((2n_{j_1u_1}1)^{\text{th}}\ \text{DI}_+, (2m_{j_2v_1}1)^{\text{th}}\text{DI}_-)$-double inflection-saddle .

- For $a_{j_1j_10}\prod_{r=1}^{p_1}(a_{j_2j_1v_1}-a_{j_1j_1r})^{m_{j_1r}}\prod_{u_2=1,u_2\neq u_1}^{q_1}(a_{j_1j_2u_1}-a_{j_1j_2u_2})^{n_{j_1u_2}}>0$ and $a_{j_2j_20}\prod_{v_2=1,v_2\neq v_1}^{p_2}(a_{j_2j_1v_1}-a_{j_2j_1v_2})^{m_{j_2v_2}}<0$,

$$(a_{j_1j_2u_1},a_{j_2j_1v_1})=\underbrace{((2n_{j_1u_1}1)^{\text{th}}\text{DI}_-,(2m_{j_2v_1}1)^{\text{th}}\text{DI}_+)}_{((2n_{j_1u_1}1),(2m_{j_2v_1}1))\text{-double inflection-saddle}}.$$

The flow at $(x_{j_2}^*,x_{j_1}^*)=(a_{j_1j_2u_1},a_{j_2j_1v_1})$ is a $((2n_{j_1u_1}1)^{\text{th}}$ DI_-, $(2m_{j_2v_1}1)^{\text{th}}\text{DI}_+)$-double- inflection-saddle .

- For $a_{j_1j_10}\prod_{r=1}^{p_1}(a_{j_2j_1v_1}-a_{j_1j_1r})^{m_{j_1r}}\prod_{u_2=1,u_2\neq u_1}^{q_1}(a_{j_1j_2u_1}-a_{j_1j_2u_2})^{n_{j_1u_2}}<0$ and $a_{j_2j_20}\prod_{v_2=1,v_2\neq v_1}^{p_2}(a_{j_2j_1v_1}-a_{j_2j_1v_2})^{m_{j_2v_2}}<0$,

$$(a_{j_1j_2u_1},a_{j_2j_1v_1})=\underbrace{((2n_{j_1u_1}1)^{\text{th}}\text{II}_-,(2m_{j_2v_1}1)^{\text{th}}\text{II}_-)}_{((2n_{j_1u_1}1),(2m_{j_2v_1}1))\text{-double inflection-saddle}}.$$

The flow at $(x_{j_2}^*,x_{j_1}^*)=(a_{j_1j_2u_1},a_{j_2j_1v_1})$ is a $((2n_{j_1u_1}1)^{\text{th}}\text{II}_-,(2m_{j_2v_1}1)^{\text{th}}\text{II}_-)$-double inflection-saddle .

2.2.2 Infinite-Equilibriums in the Hybrid Networks

(ii$_2$) For $a_{j_1j_1r_1}=a_{j_2j_1v_1}$ $(r_1\in\{1,2,\ldots,p_1\};\ v_1\in\{1,2,\ldots,p_2\})$

$$\begin{aligned}\dot{x}_{j_2}&=a_{j_2j_20}(x_{j_1}-a_{j_2j_1v_1})^{m_{j_1v_1}}\prod_{v_1=1,v_2\neq v_1}^{p_2}(x_{j_1}-a_{j_2j_1v_2})^{m_{j_1v_2}},\\ \dot{x}_{j_1}&=a_{j_1j_10}(x_{j_1}-a_{j_1j_1r_1})^{m_{j_1r_1}}\prod_{r_2=1,r_2\neq r_1}^{p_1}(x_{j_1}-a_{j_1j_1r_2})^{m_{j_1r_2}}\prod_{u=1}^{q_1}(x_{j_2}-a_{j_1j_2u})^{n_{j_1u}}.\end{aligned}$$

Consider

$$\frac{dx_{j_2}}{dx_{j_1}}=\frac{a_{j_2j_20}}{a_{j_1j_10}}\frac{(x_{j_1}-a_{j_2j_1v_1})^{m_{j_1v_1}}\prod_{v_1=1,v_2\neq v_1}^{p_2}(x_{j_1}-a_{j_2j_1v_2})^{m_{j_1v_2}}}{(x_{j_1}-a_{j_1j_1r_1})^{m_{j_1r_1}}\prod_{r_2=1,r_2\neq r_1}^{p_1}(x_{j_1}-a_{j_1j_1r_2})^{m_{j_1r_2}}\prod_{u=1}^{q_1}(x_{j_2}-a_{j_1j_2u})^{n_{j_1u}}}.$$

The deformed equation is

$$\frac{(x_{j_1}-a_{j_2j_1v_1})^{m_{j_1v_1}}\prod_{v_1=1,v_2\neq v_1}^{p_2}(x_{j_1}-a_{j_2j_1v_2})^{m_{j_1v_2}}}{(x_{j_1}-a_{j_1j_1r_1})^{m_{j_1r_1}}\prod_{r_2=1,r_2\neq r_1}^{p_1}(x_{j_1}-a_{j_1j_1r_2})^{m_{j_1r_2}}}dx_{j_1}$$
$$=\frac{a_{j_1j_10}}{a_{j_2j_20}}\prod_{u=1}^{q_1}(x_{j_2}-a_{j_1j_2u})^{n_{j_1u}}dx_{j_2},$$

The expansion of the above equation gives

$$\sum_{v_0=m_{j_2v_1}}^{m_2}(x_{j_1}-a_{j_1j_1r_1})^{v_0}$$
$$\times\frac{\sum_{t_2,l_2=0}^{t_2+l_2=m_{j_2v_2}}\cdots\sum_{t_{p_2},l_{p_2}=0}^{t_{p_2}+l_{p_2}=m_{j_2v_{p_2}}}\prod_{k=2}^{p_2}\frac{m_{j_2v_k}!}{t_k!l_k!}(a_{j_1j_1r_1}-a_{j_2j_1k})^{t_k}\delta_{v_0}^{m_{j_2v_1}+\Sigma_{k=2}^{p_2}l_k}}{(x_{j_1}-a_{j_1j_1r_1})^{m_{j_1r_1}}\prod_{r_2=1,r_2\neq r_1}^{p_1}(x_{j_1}-a_{j_1j_1r_2})^{m_{j_1r_2}}}dx_{j_1}$$
$$=\frac{a_{j_1j_10}}{a_{j_2j_20}}\prod_{u=1}^{q_1}(x_{j_2}-a_{j_1j_2u})^{n_{j_1u}}dx_{j_2},$$

Let

$$b_{j_2j_1v_0}=\sum_{t_2,l_2=0}^{t_2+l_2=m_{j_2v_2}}\cdots\sum_{t_{p_2},l_{p_2}=0}^{t_{p_2}+l_{p_2}=m_{j_2v_{p_2}}}\prod_{k=2}^{p_2}\frac{m_{j_2v_k}!}{t_k!l_k!}(a_{j_1j_1r_1}-a_{j_2j_1k})^{t_k}\delta_{v_0}^{m_{j_2v_1}+\Sigma_{k=2}^{p_2}l_k}.$$

We have

$$\frac{\sum_{v_0=m_{j_2v_1}}^{m_1-1}b_{j_2j_1v_0}(x_{j_1}-a_{j_1j_1r_1})^{v_0}+\sum_{v_0=m_1}^{m_2}b_{j_2j_1v_0}(x_{j_1}-a_{j_1j_1r_1})^{v_0}}{(x_{j_1}-a_{j_1j_1r_1})^{m_{j_1r_1}}\prod_{r_2=1,r_2\neq r_1}^{p_1}(x_{j_1}-a_{j_1j_1r_2})^{m_{j_1r_2}}}dx_{j_1}$$
$$=\frac{a_{j_1j_10}}{a_{j_2j_20}}\prod_{u=1}^{q_1}(x_{j_2}-a_{j_1j_2u})^{n_{j_1u}}dx_{j_2},$$

For $n_{j_1u_1}<n_{j_2s_1}$, the first part is given by

$$\frac{\sum_{v_0=m_{j_2v_1}}^{m_1-1}b_{j_2j_1v_0}(x_{j_1}-a_{j_1j_1r_1})^{v_0}}{(x_{j_1}-a_{j_1j_1r_1})^{m_{j_1r_1}}\prod_{r_2=1,r_2\neq r_1}^{p_1}(x_{j_1}-a_{j_1j_1r_2})^{m_{j_1r_2}}}$$

$$= \sum_{\nu_0=m_{j_2\nu_1}}^{m_1-1} b_{j_2j_1\nu_0} \frac{(x_{j_1} - a_{j_1j_1r_1})^{\nu_0-m_{j_1r_1}}}{\prod_{r_2=1,r_2\neq r_1}^{p_1} (x_{j_1} - a_{j_1j_1r_2})^{m_{j_1r_2}}}$$

$$= \sum_{\nu_0=m_{j_2\nu_1}}^{m_{j_1r_1}-1} b_{j_2j_1\nu_0} \sum_{j=1}^{p_1} \sum_{l_j=1}^{\overline{m}_{j_1r_j}} A^{(r_1)}_{l_j\overline{m}_{j_1r_j}} \frac{1}{(x_{j_1} - a_{j_1j_1r_j})^{l_j}}$$

$$+ \sum_{\nu_0=m_{j_2r_1}}^{m_1-1} b_{j_2j_1\nu_0} \sum_{j=2}^{p_1} \sum_{l_j=1}^{\overline{m}_{j_1r_j}} B^{(r_1)}_{l_j\overline{m}_{j_1r_j}} \frac{1}{(x_{j_1} - a_{j_1j_1r_j})^{l_j}};$$

where

$$A^{(r_1)}_{l_\rho\overline{m}_{j1r_\rho}} = \frac{1}{(\overline{m}_{j_1r_\rho} - l_\rho)!} \frac{d^{\overline{m}_{j1r_\rho}-l_\rho}}{dx_{j_1}^{\overline{m}_{j1r_\rho}-l_\rho}} \frac{1}{\prod_{\sigma=1,\sigma\neq\rho}^{p_1} (x_{j_1} - a_{j_1j_1r_\sigma})^{m_{j1r_\sigma}}} |x^*_{j_1} = a_{j1jr_\rho}$$

$$(l_\rho = 1, 2, \ldots, \overline{m}_{j_1r_\rho}, \rho = 1, 2, \ldots, p_1),$$

$$A^{(r_1)}_{\overline{m}_{j_1r_\rho}\overline{m}_{j_1r_\rho}} = \frac{1}{\prod_{\sigma=1,\sigma\neq\rho} (a_{j_1j_1r_\rho} - a_{j_1j_1r_\sigma})^{m_{j1r_\sigma}}},$$

$$B^{(r_1)}_{l_\rho\overline{m}_{j1r_\rho}} = \frac{1}{(\overline{m}_{j_1r_\rho} - l_\rho)!} \frac{d^{\overline{m}_{j1r_\rho}-l_\rho}}{dx_{j_1}^{\overline{m}_{jr_\rho}-l_\rho}} \frac{(x_{j_2} - a_{j_2j_2s_1})^{\nu_0-m_{j11}}}{\prod_{\sigma=1,\sigma\neq\rho}^{p_1} (x_{j_1} - a_{j_1j_1r_\sigma})^{m_{j1r_\sigma}}}\Bigg|_{x^*_{j1}=a_{j1\{r_\rho}}$$

$$(l_\rho = 1, 2, \ldots, \overline{m}_{j_1r_\rho}, \rho = 2, 3, \ldots, p_1),$$

$$B^{(r_1)}_{\overline{m}_{j_1r_\rho}\overline{m}_{j_1r_\rho}} = \frac{(a_{j_1j_1r_\rho} - a_{j_1j_1r_1})^{\nu_0-m_{jj_1}}}{\prod_{\sigma=1,\sigma\neq\rho}^{p_1} (a_{j_1j_1r_\rho} - a_{j_1j_1r_\sigma})^{m_{jlr_\sigma}}}$$

$$\{\overline{m}_{j_1r_\rho} | \rho = 1, 2, \ldots, p_1\} = \{m_{j_1r_1} - \nu_0, \mathrm{m}_{j_1r_2}, \ldots, m_{j_1r_{p_1}}\}.$$

For $m_{j_2\nu_1} = m_{j_1r_1}$, the first part is given by

$$\frac{\sum_{\nu_0=m_{j_2\nu_1}}^{m_1-1} b_{j_2j_1\nu_0}(x_{j_1} - a_{j_1j_1r_1})^{\nu_0}}{(x_{j_1} - a_{j_1j_1r_1})^{m_{j_1r_1}} \prod_{r_2=1,r_2\neq r_1}^{p_1} (x_{j_1} - a_{j_1j_1r_2})^{m_{j_1r_2}}}$$

$$= \sum_{\nu_0=m_{j_2\nu_1}}^{m_1-1} b_{j_2j_1\nu_0} \frac{(x_{j_1} - a_{j_1j_1r_1})^{\nu_0-m_{j_1r_1}}}{\prod_{r_2=1,r_2\neq r_1}^{p_1} (x_{j_1} - a_{j_1j_1r_2})^{m_{j_1r_2}}}$$

$$= \sum_{\nu_0=m_{j_1r_1}}^{m_1-1} b_{j_2j_1\nu_0} \sum_{j=2}^{p_1} \sum_{l_j=1}^{\overline{m}_{j_1r_j}} B_{l_j\overline{m}_{j_1r_j}}^{(r_1)} \frac{1}{(x_{j_1}-a_{j_1j_1r_j})^{l_j}};$$

For $m_{j_2\nu_1} > m_{j_1r_1}$, the first part is given by

$$\frac{\sum_{\nu_0=m_{j_2\nu_1}}^{m_1-1} b_{j_2j_1\nu_0}(x_{j_1}-a_{j_1j_1r_1})^{\nu_0}}{(x_{j_1}-a_{j_1j_1r_1})^{m_{j_1r_1}} \prod_{r_2=1,r_2\neq r_1}^{p_1} (x_{j_1}-a_{j_1j_1r_2})^{m_{j_1r_2}}}$$

$$= \sum_{\nu_0=m_{j_2\nu_1}}^{m_1-1} b_{j_2j_1\nu_0} \frac{(x_{j_1}-a_{j_1j_1r_1})^{\nu_0-m_{j_1r_1}}}{\prod_{r_2=1,r_2\neq r_1}^{p_1} (x_{j_1}-a_{j_1j_1r_2})^{m_{j_1r_2}}}$$

$$= \sum_{\nu_0=m_{j_1r_1}}^{m_1-1} b_{j_2j_1\nu_0} \sum_{j=2}^{p_1} \sum_{l_j=1}^{\overline{m}_{j_1r_j}} B_{l_j\overline{m}_{j_1r_j}}^{(r_1)} \frac{1}{(x_{j_1}-a_{j_1j_1r_j})^{l_j}};$$

Following the proof of Eqs. (1.105)–(1.109) for $a_{j_1j_1r_1} \neq a_{j_2j_1r_1}$, with the above equations, the first integral manifold for $a_{j_1j_1r_1} = a_{j_2j_1r_1}$ is given by

$$\sum_{\nu_0=m_1}^{m_2} b_{j_2j_1\nu_0} \sum_{\substack{\mu_1\in[0,\nu_0-m_1],\\ \nu_1\in[m_1-m_{j_1r_1},\beta_0-m_{j_1r_1}]}}^{\mu_1+\nu_1=\nu_0-m_{j_1s_1}} \sum_{\substack{\mu_2\in[0,\nu_1-m_1]\\ \nu_2\in[m_1-\Sigma_{i=1}^2 m_{j_1r_i},\nu_1-\Sigma_{i=1}^2 m_{j_1r_i}]}}^{\mu_2+\nu_2=\nu_1-\Sigma_{i=1}^2 m_{j_1r_i}} \cdots$$

$$\sum_{\substack{\mu_{p_2-1}\in[0,\nu_{q_2-2}-m_1]\\ \nu_{p_2-1}\in[m_1-\Sigma_{i=1}^{p_2-1} m_{j_1r_i},\nu_{p-2}-\Sigma_{i=1}^{q_2-1} m_{j_1r_i}]}}^{\mu_{p_2-1}+\nu_{p_2-1}=\nu_{p_2-2}-\Sigma_{i=1}^{p_2-1} m_{j_1r_i}} \prod_{j=1}^{p_2-1} \frac{(\nu_{j-1}-\Sigma_{i=1}^{j} m_{j_1r_i})!}{\mu_j!\nu_j!}(a_{j_1j_1r_{j+1}}-a_{j_1j_1r_j})^{\mu_j}$$

$$\times \frac{1}{\nu_{p_2-1}-m_{j_1r_{p_2}}+1}\left[(x_{j_1}-a_{j_1j_1r_\rho})^{\nu_{p_2-1}-m_{j_1r_{p_2}}+1}-(x_{j_10}-a_{j_1j_1r_\rho})^{\nu_{p_2-1}-m_{j_1r_{p_2}}+1}\right]$$

$$-\sum_{\beta_0=p_2}^{p_1} b_{j_2j_1\nu_0} \sum_{k=2}^{p_2-1}\Bigg(\sum_{\substack{\alpha_1\in[0,\beta_0-n_2),\\ \beta_1\in(n_2-n_{j_2s_1},\beta_0-n_{j_2s_1}]}}^{\alpha_1+\beta_1=\beta_0-n_{j_2s_1}} \sum_{\substack{\alpha_2\in[0,\beta_1-n_2]\\ \beta_2\in(n_2-\Sigma_{i=1}^2 n_{j_2s_i},\beta_1-\Sigma_{i=1}^2 n_{j_2s_i})}}^{\alpha_2+\beta_2=\beta_1-\Sigma_{i=1}^2 n_{j_2s_i}} \cdots$$

$$\times\left\{ \sum_{\substack{\mu_{k-1}\in(\nu_{k-2}-m_1+m_{j_1r_k},\nu_{k-2}-\Sigma_{i=1}^{k-1} m_{j_1r_i}]\\ \nu_{k-1}\in[0,m_1-\Sigma_{i=1}^{k} m_{j_1r_i})}}^{\mu_{k-1}+\nu_{k-1}=\nu_{k-2}-\Sigma_{i=1}^{k-1} m_{j_1r_i}} \prod_{j=1}^{k-1} \frac{(\nu_{j-1}-\Sigma_{i=1}^{j} m_{j_1r_i})!}{\mu_j!\nu_j!}(a_{j_1j_1r_{j+1}}-a_{j_1j_1r_j})^{\mu_j}\right.$$

$$\times \sum_{\rho=k}^{p_2} \sum_{l_\rho=1}^{m_{j_1r_\rho}} \frac{A_{l_\rho m_{j_1r_\rho}}^{(r_k)}}{l_\rho-1}\left[\frac{1}{(x_{j_1}-a_{j_1j_1r_\rho})^{l_\rho-1}}-\frac{1}{(x_{j_20}-a_{j_2j_2s_\rho})^{l_\rho-1}}\right]$$

$$+\sum_{\substack{\mu_{k-1}\in(\nu_{k-2}-m_1,\nu_{k-2}-m_1+m_{j_1r_k}]\\ \nu_{k-1}\in[m_1-\Sigma_{i=1}^{k}m_{j_1r_i},m_1-\Sigma_{i=1}^{k-1}m_{j_1r_i})}}^{\mu_{k-1}+\nu_{k-1}=\nu_{k-2}-\Sigma_{i=1}^{k-1}m_{j_1r_i}}\prod_{j=1}^{k-1}\frac{(\nu_{j-1}-\Sigma_{i=1}^{j}m_{j_1r_i})!}{\mu_j!\nu_j!}(a_{j_2j_2s_{j+1}}-a_{j_2j_2s_j})^{\alpha_j}$$

$$\times\sum_{\rho=k+1}^{p_2}\sum_{l_\rho=1}^{\bar{m}_{j_1r_\rho}}\frac{B_{l_\rho\bar{m}_{j_1r_\rho}}^{(r_k)}}{l_\rho-1}\left[\frac{1}{(x_{j_1}-a_{j_1j_1r_\rho})^{l_\rho-1}}-\frac{1}{(x_{j_10}-a_{j_1j_1r_\rho})^{l_\rho-1}}\right]\Bigg\}\Bigg)+\Pi(\nu_1)$$

$$=\frac{a_{j_1j_10}}{a_{j_2j_20}}\Bigg\{\frac{1}{n_1+1}\left[(x_{j_2}-a_{j_1j_2u_1})^{n_1+1}-(x_{j_20}-a_{j_1j_2u_1})^{n_1+1}\right]$$

$$+\sum\nolimits_{\alpha=1}^{n_1-n_{r_1}}\frac{1}{n_1-\alpha+1}b_{j_1j_2\alpha}\left[(x_{j_1}-a_{j_1j_2u_1})^{n_1-\alpha+1}-(x_{j_10}-a_{j_1j_2u_1})^{n_1-\alpha+1}\right]\Bigg\},$$

where

$$\Pi(\nu_1)=-\Bigg(\sum_{\beta_0=0}^{m_{j_1r_1}-1}b_{j_2j_1\nu_0}\sum_{\rho=1}^{p_1}\sum_{l_\rho=1}^{\bar{m}_{j_1r_\rho}}\frac{A_{l_\rho\bar{m}_{j_1r_\rho}}^{(r_1)}}{l_\rho-1}\left[\frac{1}{(x_{j_1}-a_{j_1j_1r_\rho})^{l_\rho-1}}-\frac{1}{(x_{j_10}-a_{j_1j_1r_\rho})^{l_\rho-1}}\right]$$

$$+\sum_{\beta_0=m_{j_1r_1}}^{m_1-1}b_{j_1j_2\nu_0}\sum_{\rho=2}^{p_1}\sum_{l_\rho=1}^{\bar{m}_{j_1r_\rho}}\frac{B_{l_\rho\bar{m}_{j_1r_\rho}}^{(r_1)}}{l_\rho-1}\left[\frac{1}{(x_{j_1}-a_{j_1j_1r_\rho})^{l_\rho-1}}-\frac{1}{(x_{j_10}-a_{j_1j_1r_\rho})^{l_\rho-1}}\right]\Bigg)$$

for $m_{j_2\nu_1}<m_{j_1r_1}$;

$$\Pi(\nu_1)=-\sum_{\nu_0=m_{j_1r_1}}^{m_1-1}b_{j_2j_1\nu_0}\sum_{\rho=2}^{p_1}\sum_{l_\rho=1}^{\bar{m}_{j_1r_\rho}}\frac{B_{l_\rho\bar{m}_{j_1r_\rho}}^{(r_1)}}{l_\rho-1}\left[\frac{1}{(x_{j_1}-a_{j_1j_1r_\rho})^{l_\rho-1}}-\frac{1}{(x_{j_10}-a_{j_1j_1r_\rho})^{l_\rho-1}}\right],$$

for $m_{j_2\nu_1}=m_{j_1r_1}$;

$$\Pi(\nu_1)=-\sum_{\nu_0=m_{j_2\nu_1}}^{m_1-1}b_{j_2j_1\nu_0}\sum_{\rho=2}^{p_1}\sum_{l_\rho=1}^{\bar{m}_{j_1r_\rho}}\frac{B_{l_\rho\bar{m}_{j_1r_\rho}}^{(r_1)}}{l_\rho-1}\left[\frac{1}{(x_{j_1}-a_{j_1j_1r_\rho})^{l_\rho-1}}-\frac{1}{(x_{j_10}-a_{j_1j_1r_\rho})^{l_\rho-1}}\right],$$

for $m_{j_2\nu_1}>m_{j_1r_1}$.

and

$$b_{j_2j_1\nu_0}=\sum_{t_2,l_2=0}^{t_2+l_2=m_{j_2\nu_2}}\cdots\sum_{t_{p_2},l_{p_2}=0}^{t_{p_2}+l_{p_2}=m_{j_2\nu_{p_2}}}\prod_{k=2}^{p_2}\frac{m_{j_2\nu_k}!}{t_k!l_k!}(a_{j_1j_1r_1}-a_{j_2j_1\nu_k})^{t_k}\delta_{\beta_0}^{\Sigma_{k=2}^{p_2}l_k+m_{j_2\nu_1}},$$

$$\frac{1}{\beta_{p_1-1}-m_{j_1p_1}+1}\left[(x_{j_1}-a_{j_1j_1r_{p_1}})^{\beta_{p_1-1}-m_{j_1p_1}+1}-(x_{j_10}-a_{j_1j_1r_{p_1}})^{\beta_{p_1-1}-m_{j_1p_1}+1}\right]$$

$$=\ln\frac{|x_{j_1}-a_{j_1j_1r_{p_1}}|}{|x_{j_10}-a_{j_1j_1r_{p_1}}|},\ \text{for }\beta_{p_1-1}-m_{j_1p_1}+1=0;$$

$$\frac{1}{l_k-1}\left[\frac{1}{(x_{j_1}-a_{j_1j_1r_k})^{l_k-1}}-\frac{1}{(x_{j_10}-a_{j_1j_1r_k})^{l_k-1}}\right]=-\ln\frac{|x_{j_1}-a_{j_1j_1r_k}|}{|x_{j_10}-a_{j_1j_1r_k}|},$$
for $l_k=1$ $(k=1,2,\cdots,p_1-1)$.

Define a network of hybrid singular 1-dimensional flow and singular equilibriums as

$$\begin{aligned}\text{NM}=&\underbrace{(a_{j_1j_2u_1},a_{j_2j_1v_1})}_{([m_{j_1r_1},n_{j_1u_1}]^{\text{th}}\text{YY},[(m_{j_2v_1}-m_{j_1r_1})]^{\text{th}}\text{XX})}\oplus\bigcup_{u_2=1,u_2\neq u_1}^{q_1}\underbrace{(a_{j_1j_2u_2},a_{j_2j_1v_1})}_{([m_{j_1r_1},n_{j_1u_2}]^{\text{th}}\text{YY},[(m_{j_2v_1}-m_{j_1r_1})]^{\text{th}}\text{XX})}\\&\oplus\bigcup_{u=1}^{q_1}\bigcup_{r_2=1,r_2\neq r_1}^{p_1}\underbrace{(a_{j_1j_1r_2},a_{j_1j_2u})}_{([m_{j_1r_2},n_{j_1u}]XX,xx)}\oplus\bigcup_{u=1}^{q_1}\bigcup_{v_1=1,v_2\neq v_1}^{p_2}\underbrace{(a_{j_1j_2u},a_{j_2j_1v_2})}_{(n_{j_1u},m_{j_2v_2})^{\text{th}}\text{-UU}}.\end{aligned}$$

(ii$_{2a}$) For $m_{j_1r_1}\le m_{j_2v_1}$,at $x_{j_1}^*=a_{j_1j_1r_1}=a_{j_2j_1v_1}$ and $\overline{x}_{j_2}\neq a_{j_1j_2u}$, in phase space,

$$\begin{aligned}&\left.\frac{dx_{j_2}}{dx_{j_1}}\right|_{x_{j_1}^*=a_{j_2j_1v_1}}\\&=\left.\frac{a_{j_2j_20}(x_{j_1}-a_{j_2j_1v_1})^{m_{j_2v_1}-m_{j_1r_1}}\prod_{v_2=1,v_2\neq v_1}^{p_2}(x_{j_1}-a_{j_2j_1v_2})^{m_{j_2v_2}}}{a_{j_1j_10}\prod_{r=1}^{p_1}(x_{j_1}-a_{j_1j_1r})^{m_{j_1r}}\prod_{u=1}^{q_1}(\overline{x}_{j_2}-a_{j_1j_2u})^{n_{j_1u}}}\right|_{x_{j_1}^*=a_{j_2j_1v_1}}\\&=0,\\&\vdots\\&\left.\frac{d^{m_{j_2v_1}-m_{j_1r_1}}x_{j_2}}{dx_{j_1}^{m_{j_2v_1}-m_{j_1r_1}}}\right|_{x_{j_1}^*=a_{j_2j_1v_1}}=0.\end{aligned}$$

If

$$\begin{aligned}&\left.\frac{d^{(m_{j_2v_1}-m_{j_1r_1})+1}x_{j_2}}{dx_{j_1}^{(m_{j_2v_1}-m_{j_1r_1})+1}}\right|_{x_{j_1}^*=a_{j_2j_1v_1}}\\&=\frac{a_{j_2j_20}(m_{j_2v_1}-m_{j_1r_1})!\prod_{v_2=1,v_2\neq v_1}^{p_2}(a_{j_2j_1v_1}-a_{j_2j_1v_2})^{m_{j_2v_2}}}{a_{j_1j_10}\prod_{r=1}^{p_1}(a_{j_2j_1v_1}-a_{j_1j_1r})^{m_{j_1r}}\prod_{u=1}^{q_1}(\overline{x}_{j_2}-a_{j_1j_2u})^{n_{j_1u}}}>0,\end{aligned}$$

the flows at $x_{j_1}^*=a_{j_1j_1r_1}=a_{j_2j_1v_1}$ are $[2(m_{j_2v_11}-m_{j_1r_11})\pm1]^{\text{th}}$-order up-parabola flow for ($m_{j_2v_1}=2m_{j_2v_11}+1$, $m_{j_1r_1}=2m_{j_1r_11}$ and $m_{j_2v_1}=2m_{j_2v_11}$, $m_{j_1r_1}=2m_{j_1r_11}+1$), and $[2(n_{j_1u_11}-n_{j_2s_11})]^{\text{th}}$-order increasing-inflection flow for ($m_{j_2v_1}=2m_{j_2v_11}+1$,

$m_{j_1r_1} = 2m_{j_1r_1}1 + 1$ and $m_{j_2v_1} = 2m_{j_2v_1}1$, $m_{j_1r_1} = 2m_{j_1r_1}1$). If

$$\frac{d^{(m_{j_2v_1}-m_{j_1r_1})+1}x_{j_2}}{dx_{j_1}^{(m_{j_2v_1}-m_{j_1r_1})+1}}\Big|_{x_{j_1}^*=a_{j_2j_1v_1}}$$
$$= \frac{a_{j_2j_20}(m_{j_2v_1}-m_{j_1r_1})! \prod_{v_2=1,v_2\neq v_1}^{p_2}(a_{j_2j_1v_1}-a_{j_2j_1v_2})^{m_{j_2v_2}}}{a_{j_1j_10}\prod_{r_2=1,r_2\neq r_1}^{p_1}(a_{j_2j_1v_1}-a_{j_1j_1r})^{m_{j_1r}}\prod_{u=1}^{q_1}(\overline{x}_{j_2}-a_{j_1j_2u})^{n_{j_1u}}} < 0,$$

the flows at $x_{j_1}^* = a_{j_1j_1r_1} = a_{j_2j_1v_1}$ are $[2(m_{j_2v_1}1 - m_{j_1r_1}1) \pm 1]^{\text{th}}$-order down-parabola flow for ($m_{j_2v_1} = 2m_{j_2v_1}1+1$, $m_{j_1r_1} = 2m_{j_1r_1}1$ and $m_{j_2v_1} = 2m_{j_2v_1}1$, $m_{j_1r_1} = 2m_{j_1r_1}1+1$), and $[2(n_{j_1u_1}1 - n_{j_2s_1}1)]^{\text{th}}$-order decreasing-inflection flow for ($m_{j_2v_1} = 2m_{j_2v_1}1 + 1$, $m_{j_1r_1} = 2m_{j_1r_1}1 + 1$ and $m_{j_2v_1} = 2m_{j_2v_1}1$, $m_{j_1r_1} = 2m_{j_1r_1}1$).

The variational equation at $x_{j_1}^* = a_{j_1j_1r_1} = a_{j_2j_1v_1}$ is

$$\Delta\dot{x}_{j_1} = a_{j_1j_10}\prod_{r_2=1,r_2\neq r_1}^{p_1}(a_{j_2j_1v_1}-a_{j_1j_1r_2})^{m_{j_1r_2}}\prod_{u=1}^{q_1}(\overline{x}_{j_2}-a_{j_1j_2u})^{n_{j_1u}}(\Delta x_{j_1})^{m_{j_1r_1}}.$$

For

$$a_{j_1j_10}\prod_{r_2=1,r_2\neq r_1}^{p_1}(a_{j_2j_1v_1}-a_{j_1j_1r_2})^{m_{j_1r_2}}\prod_{u=1}^{q_1}(\overline{x}_{j_2}-a_{j_1j_2u})^{n_{j_1u}} > 0 \text{ and}$$
$$a_{j_1j_10}\prod_{r_2=1,r_2\neq r_1}^{p_1}(a_{j_2j_1v_1}-a_{j_1j_1r_2})^{m_{j_1r_2}}\prod_{u=1}^{q_1}(\overline{x}_{j_2}-a_{j_1j_2u})^{n_{j_1u}} < 0,$$

the flows at $x_{j_1}^* = a_{j_1j_1r_1} = a_{j_2j_1v_1}$ are $(2m_{j_1r_1}1 + 1)^{\text{th}}$-source and sink flows for $m_{j_1r_1} = 2m_{j_1r_1}1 + 1$, and $(2m_{j_1r_1}1)^{\text{th}}$-upper-saddle and lower-saddle flows for $m_{j_1r_1} = 2m_{j_1r_1}1$.

Thus, based on the above analysis, Thus, based on the above analysis, the flows at $x_{j_1}^* = a_{j_1j_1}1 = a_{j_2j_1}1$ and on both sides of $x_{j_2}^* = a_{j_1j_2}1$ are $((2m_{j_1r_1}1+1)^{\text{th}}\text{XX}, [\cdot]^{\text{th}}\text{YY})$-source and sink flows for

$$a_{j_1j_10}\prod_{r_2=1,r_2\neq r_1}^{p_1}(a_{j_2j_1v_1}-a_{j_1j_1r_2})^{m_{j_1r_2}}\prod_{u=1}^{q_1}(\overline{x}_{j_2}-a_{j_1j_2u})^{n_{j_1u}} > 0 \text{ and}$$
$$a_{j_1j_10}\prod_{r_2=1,r_2\neq r_1}^{p_1}(a_{j_2j_1v_1}-a_{j_1j_1r_2})^{m_{j_1r_2}}\prod_{u=1}^{q_1}(\overline{x}_{j_2}-a_{j_1j_2u})^{n_{j_1u}} < 0,$$

respectively. The flows at $x_{j_1}^* = a_{j_1j_1}1 = a_{j_2j_1}1$ and on both sides of $x_{j_2}^* = a_{j_1j_2}1$ are $((2m_{11})^{\text{th}}\text{XX}, [\cdot]^{\text{th}}\text{YY})$-upper and lower-saddle flows for

$$a_{j_1j_10}\prod_{r_2=1,r_2\neq r_1}^{p_1}(a_{j_2j_1v_1}-a_{j_1j_1r_2})^{m_{j_1r_2}}\prod_{u=1}^{q_1}(\overline{x}_{j_2}-a_{j_1j_2u})^{n_{j_1u}}>0 \text{ and}$$

$$a_{j_1j_10}\prod_{r_2=1,r_2\neq r_1}^{p_1}(a_{j_2j_1v_1}-a_{j_1j_1r_2})^{m_{j_1r_2}}\prod_{u=1}^{q_1}(\overline{x}_{j_2}-a_{j_1j_2u})^{n_{j_1u}}<0,$$

respectively. $[2(m_{j_2v_1}1-m_{j_1r_1}1)]^{\text{th}}$YY are the $[2(m_{j_2v_1}1-m_{j_1r_1}1)]^{\text{th}}$-order up and down parabola flows, and $[2(m_{j_2v_1}1-m_{j_1r_1}1)\pm 1]^{\text{th}}$YY are the $[2(m_{j_2v_1}1-m_{j_1r_1}1)\pm 1]^{\text{th}}$-order increasing and decreasing inflection flows.

(ii$_{2a1}$) For $m_{j_1r_1}\leq m_{j_2v_1}$, 16 types of infinite-equilibriums of $x_{j_1}^*=a_{j_1j_1r_1}=a_{j_2j_1v_1}$ exist.

(ii$_{2a1a}$) For $m_{j_1r_1}=2m_{j_1r_1}1+1$ and $m_{j_2v_1}=2m_{j_1v_1}1+1$, the infinite-equilibrium of $x_{j_1}^*=a_{j_2j_1v_1}=a_{j_1j_1r_1}$ with $\overline{x}_{j_2}\neq a_{j_1j_2u_1}$ has the following properties as in Eqs. (1.183)–(1.186).

- For $a_{j_1j_10}\prod_{r_2=1,r_2\neq r_1}^{p_1}(a_{j_1j_1r_1}-a_{j_1j_1r_2})^{m_{j_1r_2}}\prod_{u=1}^{q_1}(\overline{x}_{j_2}-a_{j_1j_2u})^{n_{j_1u}}>0$ and $a_{j_2j_20}\prod_{v_2=1,v_2\neq v_1}^{p_2}(a_{j_1j_1r_1}-a_{j_2j_1v_2})^{m_{j_2v_2}}>0$,

$$(a_{j_1j_1r_1},\overline{x}_{j_2})=\underbrace{((2m_{j_1r_1}1+1)^{\text{th}}\text{SO},[2(m_{j_2v_1}1-m_{j_1r_1}1)]^{\text{th}}\text{II})}_{((2m_{j_1r_1}1+1),[2(m_{j_2v_1}1-m_{j_1r_1}1)])\text{-increasing-inflection source}}.$$

The infinite-equilibrium is a $((2m_{j_1r_1}1+1)^{\text{th}}\text{SO},[2(m_{j_2v_1}1-m_{j_1r_1}1)]^{\text{th}}\text{II})$-increasing-inflection source at $x_{j_1}^*=a_{j_2j_1v_1}=a_{j_1j_1r_1}$.

- For $a_{j_1j_10}\prod_{r_2=1,r_2\neq r_1}^{p_1}(a_{j_1j_1r_1}-a_{j_1j_1r_2})^{m_{j_1r_2}}\prod_{u=1}^{q_1}(\overline{x}_{j_2}-a_{j_1j_2u})^{n_{j_1u}}<0$ and $a_{j_2j_20}\prod_{v_2=1,v_2\neq v_1}^{p_2}(a_{j_1j_1r_1}-a_{j_2j_1v_2})^{m_{j_2v_2}}>0$,

$$(a_{j_1j_1r_1},\overline{x}_{j_2})=\underbrace{((2m_{j_1r_1}1+1)^{\text{th}}\text{SI},[2(m_{j_2v_1}1-m_{j_1r_1}1)]^{\text{th}}\text{DI})}_{((2m_{j_1r_1}1+1),[2(m_{j_2v_1}1-m_{j_1r_1}1)])\text{-decreasing-inflection sink}}.$$

The infinite-equilibrium is a $((2m_{j_1r_1}1+1)^{\text{th}}\text{SI},[2(m_{j_2v_1}1-m_{j_1r_1}1)]^{\text{th}}\text{DI})$-decreasing-inflection sink at $x_{j_1}^*=a_{j_2j_1v_1}=a_{j_1j_1r_1}$.

- For $a_{j_1j_10}\prod_{r_2=1,r_2\neq r_1}^{p_1}(a_{j_1j_1r_1}-a_{j_1j_1r_2})^{m_{j_1r_2}}\prod_{u=1}^{q_1}(\overline{x}_{j_2}-a_{j_1j_2u})^{n_{j_1u}}>0$ and $a_{j_2j_20}\prod_{v_2=1,v_2\neq v_1}^{p_2}(a_{j_2j_1v_1}-a_{j_2j_1v_2})^{m_{j_2v_2}}<0$,

$$(a_{j_1j_1r_1}, \overline{x}_{j_2}) = \underbrace{((2m_{j_1r_1}1 + 1)^{\text{th}}\text{SO},[2(m_{j_2v_1}1 - m_{j_1r_1}1)]^{\text{th}}\text{DI})}_{((2m_{j_1r_1}1+1),[2(m_{j_2v_1}1-m_{j_1r_1}1)])\text{-decreasing-inflection source}}.$$

The infinite-equilibrium is a $((2m_{j_1r_1}1 + 1)^{\text{th}}\text{SO},[2(m_{j_2v_1}1 - m_{j_1r_1}1)]^{\text{th}}\text{DI})$-decreasing-inflection sink at $x_{j_1}^* = a_{j_2j_1v_1} = a_{j_1j_1r_1}$.

- For $a_{j_1j_10}\prod_{r_2=1,r_2\neq r_1}^{p_1}(a_{j_1j_1r_1} - a_{j_1j_1r_2})^{m_{j_1r_2}}\prod_{u=1}^{q_1}(\overline{x}_{j_2} - a_{j_1j_2u})^{n_{j_1u}} < 0$ and $a_{j_2j_20}\prod_{v_2=1,v_2\neq v_1}^{p_2}(a_{j_2j_1v_1} - a_{j_2j_1v_2})^{m_{j_2v_2}} < 0$,

$$(a_{j_1j_1r_1}, \overline{x}_{j_2}) = \underbrace{((2m_{j_1r_1}1 + 1)^{\text{th}}\text{SI},[2(m_{j_2v_1}1 - m_{j_1r_1}1)]^{\text{th}}\text{II})}_{((2m_{j_1r_1}1+1),[2(m_{j_2v_1}1-m_{j_1r_1}1)])\text{-increasing-inflection sink}}.$$

The infinite-equilibrium is a $((2m_{j_1r_1}1 + 1)^{\text{th}}\text{SI},[2(m_{j_2v_1}1 - m_{j_1r_1}1)]^{\text{th}}\text{II})$-increasing-inflection sink at $x_{j_1}^* = a_{j_2j_1v_1} = a_{j_1j_1r_1}$.

($\text{ii}_{2\text{a1b}}$) For $m_{j_1r_1} = 2m_{j_1r_1}1$ and $m_{j_2v_1} = 2m_{j_1v_1}1 + 1$, the infinite-equilibrium of $x_{j_1}^* = a_{j_2j_1v_1} = a_{j_1j_1r_1}$ with $\overline{x}_{j_2} \neq a_{j_1j_2u_1}$ has the following properties as in Eqs. (1.187)–(1.190).

- For $a_{j_1j_10}\prod_{r_2=1,r_2\neq r_1}^{p_1}(a_{j_1j_1r_1} - a_{j_1j_1r_2})^{m_{j_1r_2}}\prod_{u=1}^{q_1}(\overline{x}_{j_2} - a_{j_1j_2u})^{n_{j_1u}} > 0$ and $a_{j_2j_20}\prod_{v_2=1,v_2\neq v_1}^{p_2}(a_{j_1j_1r_1} - a_{j_2j_1v_2})^{m_{j_2v_2}} > 0$,

$$(a_{j_1j_1r_1}, \overline{x}_{j_2}) = \underbrace{((2m_{j_1r_1}1)^{\text{th}}\text{US},[2(m_{j_2v_1}1 - m_{j_1r_1}1) + 1]^{\text{th}}\text{UP})}_{((2m_{j_1r_1}1),[2(m_{j_2v_1}1-m_{j_1r_1}1)+1])\text{-up-parabola upper-saddle}}.$$

The infinite-equilibrium is a $((2m_{j_1r_1}1)^{\text{th}}\text{US},[2(m_{j_2v_1}1-m_{j_1r_1}1) + 1]^{\text{th}}\text{UP})$-up-parabola upper-saddle at $x_{j_1}^* = a_{j_2j_1v_1} = a_{j_1j_1r_1}$.

- For $a_{j_1j_10}\prod_{r_2=1,r_2\neq r_1}^{p_1}(a_{j_1j_1r_1} - a_{j_1j_1r_2})^{m_{j_1r_2}}\prod_{u=1}^{q_1}(\overline{x}_{j_2} - a_{j_1j_2u})^{n_{j_1u}} < 0$ and $a_{j_2j_20}\prod_{v_2=1,v_2\neq v_1}^{p_2}(a_{j_2j_1v_1} - a_{j_2j_1v_2})^{m_{j_2v_2}} > 0$,

$$(a_{j_1j_1r_1}, \overline{x}_{j_2}) = \underbrace{((2m_{j_1r_1}1)^{\text{th}}\text{LS},[2(m_{j_2v_1}1 - m_{j_1r_1}1) + 1]^{\text{th}}\text{DP})}_{((2m_{j_1r_1}1),[2(m_{j_2v_1}1-m_{j_1r_1}1)+1])\text{-down-parabola lower-saddle}}.$$

The infinite-equilibrium is a $((2m_{j_1r_1}1)^{\text{th}}\text{LS},[2(m_{j_2v_1}1 - m_{j_1r_1}1) + 1]^{\text{th}}\text{DP})$-down-parabola lower-saddle at $x_{j_1}^* = a_{j_2j_1v_1} = a_{j_1j_1r_1}$.

- For $a_{j_1j_10}\prod_{r_2=1,r_2\neq r_1}^{p_1}(a_{j_1j_1r_1}-a_{j_1j_1r_2})^{m_{j_1r_2}}\prod_{u=1}^{q_1}(\overline{x}_{j_2}-a_{j_1j_2u})^{n_{j_1u}}>0$ and $a_{j_2j_20}\prod_{v_2=1,v_2\neq v_1}^{p_2}(a_{j_2j_1v_1}-a_{j_2j_1v_2})^{m_{j_2v_2}}<0$,

$$(a_{j_1j_1r_1},\overline{x}_{j_2})=\underbrace{((2m_{j_1r_11})^{\text{th}}\text{US},[2(m_{j_2v_11}-m_{j_1r_11})+1]^{\text{th}}\text{DP})}_{((2m_{j_1r_11}),[2(m_{j_2v_11}-m_{j_1r_11})+1])\text{-down-parabola upper-saddle}}.$$

The infinite-equilibrium is a $((2m_{j_1r_11})^{\text{th}}\text{US},[2(m_{j_2v_11}-m_{j_1r_11})+1]^{\text{th}}\text{DP})$-down-parabola upper-saddle at $x_{j_1}^*=a_{j_2j_1v_1}=a_{j_1j_1r_1}$.

- For $a_{j_1j_10}\prod_{r_2=1,r_2\neq r_1}^{p_1}(a_{j_1j_1r_1}-a_{j_1j_1r_2})^{m_{j_1r_2}}\prod_{u=1}^{q_1}(\overline{x}_{j_2}-a_{j_1j_2u})^{n_{j_1u}}<0$ and $a_{j_2j_20}\prod_{v_2=1,v_2\neq v_1}^{p_2}(a_{j_2j_1v_1}-a_{j_2j_1v_2})^{m_{j_2v_2}}<0$,

$$(a_{j_1j_1r_1},\overline{x}_{j_2})=\underbrace{((2m_{j_1r_11})^{\text{th}}\text{LS},[2(m_{j_2v_11}-m_{j_1r_11})+1]^{\text{th}}\text{UP})}_{((2m_{j_1r_11}),[2(m_{j_2v_11}-m_{j_1r_11})+1])\text{-up-parabola lower-saddle}}.$$

The infinite-equilibrium is a $((2m_{j_1r_11})^{\text{th}}\text{LS},[2(m_{j_2v_11}-m_{j_1r_11})+1]^{\text{th}}\text{UP})$-up-parabola lower-saddle at $x_{j_1}^*=a_{j_2j_1v_1}=a_{j_1j_1r_1}$.

(ii$_{2a1c}$) For $m_{j_1r_1}=2m_{j_1r_11}+1$ and $m_{j_2v_1}=2m_{j_1v_11}$, the infinite-equilibrium of $x_{j_1}^*=a_{j_2j_1v_1}=a_{j_1j_1r_1}$ with $\overline{x}_{j_2}\neq a_{j_1j_2u_1}$ has the following properties as in Eqs. (1.191)–(1.194).

- For $a_{j_1j_10}\prod_{r_2=1,r_2\neq r_1}^{p_1}(a_{j_1j_1r_1}-a_{j_1j_1r_2})^{m_{j_1r_2}}\prod_{u=1}^{q_1}(\overline{x}_{j_2}-a_{j_1j_2u})^{n_{j_1u}}>0$ and $a_{j_2j_20}\prod_{v_2=1,v_2\neq v_1}^{p_2}(a_{j_2j_1v_1}-a_{j_2j_1v_2})^{m_{j_2v_2}}>0$,

$$(a_{j_1j_1r_1},\overline{x}_{j_2})=\underbrace{((2m_{j_1r_11}+1)^{\text{th}}\text{SO},[2(m_{j_2v_11}-m_{j_1r_11})-1]^{\text{th}}\text{UP})}_{((2m_{j_1r_11}+1),[2(m_{j_2v_11}-m_{j_1r_11})-1])\text{-up-parabola source}}.$$

The infinite-equilibrium is a $((2m_{j_1r_11}+1)^{\text{th}}\text{SO},[2(m_{j_2v_11}-m_{j_1r_11})-1]^{\text{th}}\text{UP})$-up-parabola source at $x_{j_1}^*=a_{j_2j_1v_1}=a_{j_1j_1r_1}$.

- For $a_{j_1j_10}\prod_{r_2=1,r_2\neq r_1}^{p_1}(a_{j_1j_1r_1}-a_{j_1j_1r_2})^{m_{j_1r_2}}\prod_{u=1}^{q_1}(\overline{x}_{j_2}-a_{j_1j_2u})^{n_{j_1u}}<0$ and $a_{j_2j_20}\prod_{v_2=1,v_2\neq v_1}^{p_2}(a_{j_2j_1v_1}-a_{j_2j_1v_2})^{m_{j_2v_2}}>0$,

$$(a_{j_1j_1r_1},\overline{x}_{j_2})=\underbrace{((2m_{j_1r_11}+1)^{\text{th}}\text{SI},[2(m_{j_2v_11}-m_{j_1r_11})-1]^{\text{th}}\text{DP})}_{((2m_{j_1r_11}+1),[2(m_{j_2v_11}-m_{j_1r_11})-1])\text{-down-parabola sink}}.$$

The infinite-equilibrium is a $((2m_{j_1r_1 1}+1)^{\text{th}}\text{SI},[2(m_{j_2v_1 1}-m_{j_1r_1 1})-1]^{\text{th}}$ DP)-down-parabola sink at $x_{j_1}^*=a_{j_2j_1v_1}=a_{j_1j_1r_1}$.

- For $a_{j_1j_10}\prod_{r_2=1,r_2\neq r_1}^{p_1}(a_{j_1j_1r_1}-a_{j_1j_1r_2})^{m_{j_1r_2}}\prod_{u=1}^{q_1}(\overline{x}_{j_2}-a_{j_1j_2u})^{n_{j_1u}}>0$ and $a_{j_2j_20}\prod_{v_2=1,v_2\neq v_1}^{p_2}(a_{j_2j_1v_1}-a_{j_2j_1v_2})^{m_{j_2v_2}}<0$,

$$(a_{j_1j_1r_1},\overline{x}_{j_2})=\underbrace{((2m_{j_1r_1 1}+1)^{\text{th}}\text{SO},[2(m_{j_2v_1 1}-m_{j_1r_1 1})-1]^{\text{th}}\text{DP})}_{((2m_{j_1r_1 1}+1),[2(m_{j_2v_1 1}-m_{j_1r_1 1})-1])\text{-down-parabola source}}.$$

The infinite-equilibrium is a $((2m_{j_1r_1 1}+1)^{\text{th}}\text{SO},[2(m_{j_2v_1 1}-m_{j_1r_1 1})-1]^{\text{th}}$ DP)-down-parabola source at $x_{j_1}^*=a_{j_2j_1v_1}=a_{j_1j_1r_1}$.

- For $a_{j_1j_10}\prod_{r_2=1,r_2\neq r_1}^{p_1}(a_{j_1j_1r_1}-a_{j_1j_1r_2})^{m_{j_1r_2}}\prod_{u=1}^{q_1}(\overline{x}_{j_2}-a_{j_1j_2u})^{n_{j_1u}}<0$ and $a_{j_2j_20}\prod_{v_2=1,v_2\neq v_1}^{p_2}(a_{j_2j_1v_1}-a_{j_2j_1v_2})^{m_{j_2v_2}}<0$,

$$(a_{j_1j_1r_1},\overline{x}_{j_2})=\underbrace{((2m_{j_1r_1 1}+1)^{\text{th}}\text{SI},[2(m_{j_2v_1 1}-m_{j_1r_1 1})-1]^{\text{th}}\text{UP})}_{((2m_{j_1r_1 1}+1),[2(m_{j_2v_1 1}-m_{j_1r_1 1})-1])\text{-up-parabola sink}}.$$

The infinite-equilibrium is a $((2m_{j_1r_1 1}+1)^{\text{th}}\text{SI},[2(m_{j_2v_1 1}-m_{j_1r_1 1})-1]^{\text{th}}$ UP)-up-parabola sink at $x_{j_1}^*=a_{j_2j_1v_1}=a_{j_1j_1r_1}$.

($\text{ii}_{2\text{a1d}}$) For $m_{j_1r_1}=2m_{j_1r_1 1}$ and $m_{j_2v_1}=2m_{j_1v_1 1}$, the infinite-equilibrium of $x_{j_1}^*=a_{j_2j_1v_1}=a_{j_1j_1r_1}$ with $\overline{x}_{j_2}\neq a_{j_1j_2u_1}$ has the following properties as in Eqs. (1.195)–(1.198).

- For $a_{j_1j_10}\prod_{r_2=1,r_2\neq r_1}^{p_1}(a_{j_1j_1r_1}-a_{j_1j_1r_2})^{m_{j_1r_2}}\prod_{u=1}^{q_1}(\overline{x}_{j_2}-a_{j_1j_2u})^{n_{j_1u}}>0$ and $a_{j_2j_20}\prod_{v_2=1,v_2\neq v_1}^{p_2}(a_{j_2j_1v_1}-a_{j_2j_1v_2})^{m_{j_2v_2}}>0$,

$$(a_{j_1j_1r_1},\overline{x}_{j_2})=\underbrace{((2m_{j_1r_1 1})^{\text{th}}\text{US},[2(m_{j_2v_1 1}-m_{j_1r_1 1})]^{\text{th}}\text{II})}_{((2m_{j_1r_1 1}),[2(m_{j_2v_1 1}-m_{j_1r_1 1})])\text{-increasing-inflection upper-saddle}}.$$

The infinite-equilibrium is a $((2m_{j_1r_1 1})^{\text{th}}\text{US},[2(m_{j_2v_1 1}-m_{j_1r_1 1})]^{\text{th}}\text{II})$-increasing-inflection upper-saddle at $x_{j_1}^*=a_{j_2j_1v_1}=a_{j_1j_1r_1}$.

- For $a_{j_1j_10}\prod_{r_2=1,r_2\neq r_1}^{p_1}(a_{j_1j_1r_1}-a_{j_1j_1r_2})^{m_{j_1r_2}}\prod_{u=1}^{q_1}(\overline{x}_{j_2}-a_{j_1j_2u})^{n_{j_1u}}<0$ and $a_{j_2j_20}\prod_{v_2=1,v_2\neq v_1}^{p_2}(a_{j_2j_1v_1}-a_{j_2j_1v_2})^{m_{j_2v_2}}>0$,

$$(a_{j_1j_1r_1}, \overline{x}_{j_2}) = \underbrace{((2m_{j_1r_1}1)^{\text{th}}\text{LS},[2(m_{j_2v_1}1 - m_{j_1r_1}1)]^{\text{th}}\text{DI})}_{((2m_{j_1r_1}1),[2(m_{j_2v_1}1-m_{j_1r_1}1)])\text{-decreasing-inflection lower-saddle}}.$$

The infinite-equilibrium is a $((2m_{j_1r_1}1)^{\text{th}}\text{LS},[2(m_{j_2v_1}1 - m_{j_1r_1}1)]^{\text{th}}\text{DI})$-decreasing-inflection lower-saddle at $x^*_{j_1} = a_{j_2j_1v_1} = a_{j_1j_1r_1}$.

- For $a_{j_1j_10} \prod_{r_2=1,r_2\neq r_1}^{p_1} (a_{j_1j_1r_1} - a_{j_1j_1r_2})^{m_{j_1r_2}} \prod_{u=1}^{q_1} (\overline{x}_{j_2} - a_{j_1j_2u})^{n_{j_1u}} > 0$ and $a_{j_2j_20} \prod_{v_2=1,v_2\neq v_1}^{p_2} (a_{j_2j_1v_1} - a_{j_2j_1v_2})^{m_{j_2v_2}} < 0,$

$$(a_{j_1j_1r_1}, \overline{x}_{j_2}) = \underbrace{((2m_{j_1r_1}1)^{\text{th}}\text{US},[2(m_{j_2v_1}1 - m_{j_1r_1}1)]^{\text{th}}\text{DI})}_{((2m_{j_1r_1}1),[2(m_{j_2v_1}1-m_{j_1r_1}1)])\text{-decreasing-inflection upper-saddle}}.$$

The infinite-equilibrium is a $((2m_{j_1r_1}1)^{\text{th}}\text{US},[2(m_{j_2v_1}1-m_{j_1r_1}1)]^{\text{th}}\text{DI})$-decreasing-inflection upper-saddle at $x^*_{j_1} = a_{j_2j_1v_1} = a_{j_1j_1r_1}$.

- For $a_{j_1j_10} \prod_{r_2=1,r_2\neq r_1}^{p_1} (a_{j_1j_1r_1} - a_{j_1j_1r_2})^{m_{j_1r_2}} \prod_{u=1}^{q_1} (\overline{x}_{j_2} - a_{j_1j_2u})^{n_{j_1u}} < 0$ and $a_{j_2j_20} \prod_{v_2=1,v_2\neq v_1}^{p_2} (a_{j_2j_1v_1} - a_{j_2j_1v_2})^{m_{j_2v_2}} < 0,$

$$(a_{j_1j_1r_1}, \overline{x}_{j_2}) = \underbrace{((2m_{j_1r_1}1)^{\text{th}}\text{LS},[2(m_{j_2v_1}1 - m_{j_1r_1}1)]^{\text{th}}\text{II})}_{((2m_{j_1r_1}1),[2(m_{j_2v_1}1-m_{j_1r_1}1)])\text{-increasing-inflection lower-saddle}}.$$

The infinite-equilibrium is a $((2m_{j_1r_1}1)^{\text{th}}\text{SI},[2(m_{j_2v_1}1-m_{j_1r_1}1)]^{\text{th}}\text{II})$-increasing-inflection lower-saddle at $x^*_{j_1} = a_{j_2j_1v_1} = a_{j_1j_1r_1}$.

($\text{ii}_{2\text{a}2}$) At $x^*_{j_2} = a_{j_1j_2u_1}$ and $\overline{x}_{j_1} \neq a_{j_1j_1r}, a_{j_2j_1v}(r = 1, 2, \ldots, p_1; v = 1, 2, \ldots, p_2)$, in phase space,

$$\begin{aligned}\left.\frac{dx_{j_1}}{dx_{j_2}}\right|_{x^*_{j_2}=a_{j_1j_2u_1}} &= \frac{a_{j_1j_10}}{a_{j_2j_20}} \frac{\prod_{r=1}^{p_1} (\overline{x}_{j_1} - a_{j_1j_1r})^{m_{j_1r}} \prod_{u_2=1,u_2\neq u_1}^{q_1} (x_{j_2} - a_{j_1j_2u_2})^{n_{j_1u_2}}}{\prod_{v=1}^{p_2} (\overline{x}_{j_1} - a_{j_2j_1v})^{m_{j_2v}}} \\ &\quad \times \left.(x_{j_2} - a_{j_1j_2u_1})^{n_{j_1u_1}}\right|_{x^*_{j_2}=a_{j_1j_2u_1}} = 0, \\ &\vdots \\ \left.\frac{d^{n_{j_1u_1}}x_{j_1}}{dx_{j_2}^{n_{j_1u_1}}}\right|_{x^*_{j_2}=a_{j_1j_2u_1}} &= 0.\end{aligned}$$

If

$$\left.\frac{d^{n_{j_1u_1}+1}x_{j_1}}{dx_{j_2}^{n_{j_1u_1}+!}}\right|_{x_{j_2}^*=a_{j_1j_2u_1}} = \frac{a_{j_1j_10}}{a_{j_2j_20}} \frac{\prod_{r=1}^{p_1}(\overline{x}_{j_1}-a_{j_1j_1r})^{m_{j_1r}} \prod_{u_2=1,u_2\neq u_1}^{q_1}(a_{j_1j_2u_1}-a_{j_1j_2u_2})^{n_{j_1u_2}} n_{j_1u_1}!}{\prod_{v=1}^{p_2}(\overline{x}_{j_1}-a_{j_2j_1v})^{m_{j_2v}}} > 0,$$

the $(2n_{j_1u_11}+1)^{\text{th}}$-order up-parabola flows exists for $n_{j_1u_1}=2n_{j_1u_11}+1$, and the $(2n_{j_1u_11})^{\text{th}}$-order increasing-inflection flow exists for $n_{j_1u_1}=2n_{j_1u_11}$. If

$$\left.\frac{d^{n_{j_1u_1}+1}x_{j_1}}{dx_{j_2}^{n_{j_1u_1}+1}}\right|_{x_{j_2}^*=a_{j_1j_2u_1}} = \frac{a_{j_1j_10}}{a_{j_2j_20}} \frac{\prod_{r=1}^{p_1}(\overline{x}_{j_1}-a_{j_1j_1r})^{m_{j_1r}} \prod_{u_2=1,u_2\neq u_1}^{q_1}(a_{j_1j_2u_1}-a_{j_1j_2u_2})^{n_{j_1u_2}} n_{j_1u_1}!}{\prod_{v=1}^{p_2}(\overline{x}_{j_1}-a_{j_2j_1v})^{m_{j_2v}}} < 0,$$

the $(2n_{j_1u_11}+1)^{\text{th}}$-order down-parabola flows exists for $n_{j_1u_1}=2n_{j_1u_11}+1$, and the $(2n_{j_1u_11})^{\text{th}}$-order decreasing-inflection flow exists for $n_{j_1u_1}=2n_{j_1u_11}$.

At $\overline{x}_{j_1}\neq a_{j_1j_1r}, a_{j_2j_1v} (r=1,2,\ldots,p_1; v=1,2,\ldots,p_2)$, let

$$\dot{x}_{j_2}=a_{j_2j_20}\prod_{v=1}^{p_2}(\overline{x}_{j_1}-a_{j_2j_1v})^{m_{j_2v}}.$$

If $a_{j_2j_20}\prod_{v=1}^{p_2}(\overline{x}_{j_1}-a_{j_2j_1v})^{m_{j_2v}}>0$ and $a_{j_2j_20}\prod_{v=1}^{p_2}(\overline{x}_{j_1}-a_{j_2j_1v})^{m_{j_2v}}<0$, the parabola and inflection flows are positive and negative in the x_{j_2}-direction, respectively.

From the above analysis and case ($\text{ii}_{2\text{a}1}$), there are 32 types of equilibrium $(x_{j_2}^*, x_{j_1}^*)=(a_{j_1j_2u_1}, a_{j_2j_1v_1})$ on the infinite-equilibrium of $x_{j_1}^*=a_{j_1j_1r_1}=a_{j_2j_1v_1}$.

($\text{ii}_{2\text{a}2\text{a}}$) For $m_{j_1r_1}=2m_{j_1r_11}+1$, $n_{j_1u_1}=2n_{j_1u_11}+1$ and $m_{j_2v_1}=2m_{j_1v_11}+1$, the equilibrium of $(x_{j_2}^*, x_{j_1}^*)=(a_{j_1j_2u_1}, a_{j_2j_1v_1})$ on the infinite-equilibrium $x_{j_1}^*=a_{j_2j_1v_1}=a_{j_1j_1r_1}$ has the following properties as in Eqs. (1.199)–(1.202).

- For $a_{j_1j_10}\prod_{r_2=1,r_2\neq r_1}^{p_1}(a_{j_1j_1r_1}-a_{j_1j_1r_2})^{m_{j_1r_2}}\prod_{u_2=1,u_2\neq u_1}^{q_1}(a_{j_1j_2u_1}-a_{j_1j_2u_2})^{n_{j_1u_2}}>0$ and $a_{j_2j_20}\prod_{v_2=1,v_2\neq v_1}^{p_2}(a_{j_2j_1v_1}-a_{j_2j_1v_2})^{m_{j_2v_2}}>0$,

$$(a_{j_1j_2u_1}, a_{j_2j_1v_1}) = \underbrace{([(2m_{j_1r_11}+1),(2n_{j_1u_11}+1)]^{\text{th}}\text{UP}_-\text{:UP}_+,[2(m_{j_2v_11}-m_{j_1r_11})]^{\text{th}}{}_{\text{DI:II}}\text{US})}_{([(2m_{j_1r_11}+1),(2n_{j_1u_11}+1)],[2(m_{j_2v_11}-m_{j_1r_11})])\text{-inflection up-parabola upper-saddle}}.$$

The equilibrium is a $([(2m_{j_1r_1}1+1),(2n_{j_1u_1}1+1)]^{\text{th}}\text{UP}_-{:}\text{UP}_+,[2(m_{j_2v_1}1-m_{j_1r_1}1)]^{\text{th}}{}_{\text{DI:II}}\text{US})$ -inflection up-parabola upper-saddle at $(x^*_{j_2},x^*_{j_1})=(a_{j_1j_2u_1},a_{j_2j_1v_1})$.

- For $a_{j_1j_10}\prod_{r_2=1,r_2\neq r_1}^{p_1}(a_{j_1j_1r_1}-a_{j_1j_1r_2})^{m_{j_1r_2}}\prod_{u_2=1,u_2\neq u_1}^{q_1}(a_{j_1j_2u_1}-a_{j_1j_2u_2})^{n_{j_1u_2}}<0$ and $a_{j_2j_20}\prod_{v_2=1,v_2\neq v_1}^{p_2}(a_{j_2j_1v_1}-a_{j_2j_1v_2})^{m_{j_2v_2}}>0$,

$$\begin{aligned}&(a_{j_1j_2u_1},a_{j_2j_1v_1})\\&=\underbrace{([(2m_{j_1r_1}1+1),(2n_{j_1u_1}1+1)]^{\text{th}}\text{DP}_-{:}\text{DP}_+,[2(m_{j_2v_1}1-m_{j_1r_1}1)]^{\text{th}}{}_{\text{II:DI}}\text{LS})}_{([(2m_{j_1r_1}1+1),(2n_{j_1u_1}1+1)],[2(m_{j_2v_1}1-m_{j_1r_1}1)])\text{-inflection down-parabola lower-saddle}}.\end{aligned}$$

The equilibrium is a $([(2m_{j_1r_1}1+1),(2n_{j_1u_1}1+1)]^{\text{th}}\text{DP}_-{:}\text{DP}_+,[2(m_{j_2v_1}1-m_{j_1r_1}1)]^{\text{th}}{}_{\text{II:DI}}\text{LS})$ -inflection down-parabola lower-saddle at $(x^*_{j_2},x^*_{j_1})=(a_{j_1j_2u_1},a_{j_2j_1v_1})$.

- For $a_{j_1j_10}\prod_{r_2=1,r_2\neq r_1}^{p_1}(a_{j_1j_1r_1}-a_{j_1j_1r_2})^{m_{j_1r_2}}\prod_{u_2=1,u_2\neq u_1}^{q_1}(a_{j_1j_2u_1}-a_{j_1j_2u_2})^{n_{j_1u_2}}>0$ and $a_{j_2j_20}\prod_{v_2=1,v_2\neq v_1}^{p_2}(a_{j_2j_1v_1}-a_{j_2j_1v_2})^{m_{j_2v_2}}<0$,

$$\begin{aligned}&(a_{j_1j_2u_1},a_{j_2j_1v_1})\\&=\underbrace{([(2m_{j_1r_1}1+1),(2n_{j_1u_1}1+1)]^{\text{th}}\text{DP}_+{:}\text{DP}_-,[2(m_{j_2v_1}1-m_{j_1r_1}1)]^{\text{th}}{}_{\text{II:DI}}\text{US})}_{([(2m_{j_1r_1}1+1),(2n_{j_1u_1}1+1)],[2(m_{j_2v_1}1-m_{j_1r_1}1)])\text{-inflection down-parabola upper-saddle}}.\end{aligned}$$

The equilibrium is a $([(2m_{j_1r_1}1+1),(2n_{j_1u_1}1+1)]^{\text{th}}\text{DP}_+{:}\text{DP}_-,[2(m_{j_2v_1}1-m_{j_1r_1}1)]^{\text{th}}{}_{\text{II:DI}}\text{US})$-inflection down-parabola upper-saddle at $(x^*_{j_2},x^*_{j_1})=(a_{j_1j_2u_1},a_{j_2j_1v_1})$.

- For $a_{j_1j_10}\prod_{r_2=1,r_2\neq r_1}^{p_1}(a_{j_1j_1r_1}-a_{j_1j_1r_2})^{m_{j_1r_2}}\prod_{u_2=1,u_2\neq u_1}^{q_1}(a_{j_1j_2u_1}-a_{j_1j_2u_2})^{n_{j_1u_2}}<0$ and $a_{j_2j_20}\prod_{v_2=1,v_2\neq v_1}^{p_2}(a_{j_2j_1v_1}-a_{j_2j_1v_2})^{m_{j_2v_2}}<0$,

$$\begin{aligned}&(a_{j_1j_2u_1},a_{j_2j_1v_1})\\&=\underbrace{([(2m_{j_1r_1}1+1),(2n_{j_1u_1}1+1)]^{\text{th}}\text{UP}_-{:}\text{UP}_-,[2(m_{j_2v_1}1-m_{j_1r_1}1)]^{\text{th}}{}_{\text{DI:II}}\text{LS})}_{([(2m_{j_1r_1}1+1),(2n_{j_1u_1}1+1)],[2(m_{j_2v_1}1-m_{j_1r_1}1)])\text{-inlfection up-parabola lower-saddle}}.\end{aligned}$$

The equilibrium is a $([(2m_{j_1r_1}1+1),(2n_{j_1u_1}1+1)]^{\text{th}}\text{UP}_-{:}\text{UP}_-,[2(m_{j_2v_1}1-m_{j_1r_1}1)]^{\text{th}}{}_{\text{DI:II}}\text{LS})$ -inflection up-parabola lower-saddle at $(x^*_{j_2},x^*_{j_1})=(a_{j_1j_2u_1},a_{j_2j_1v_1})$.

(ii$_{2a2b}$) For $m_{j_1r_1}=2m_{j_1r_1}1$, $n_{j_1u_1}=2n_{j_1u_1}1+1$ and $m_{j_2v_1}=2m_{j_1v_1}1+1$, the equilibrium of $(x^*_{j_2},x^*_{j_1})=(a_{j_1j_2u_1},a_{j_2j_1v_1})$ on the infinite-equilibrium $x^*_{j_1}=a_{j_2j_1v_1}=a_{j_1j_1r_1}$ has the following properties as in Eqs. (1.203)–(1.206).

- For $a_{j_1j_10}\prod_{r_2=1,r_2\neq r_1}^{p_1}(a_{j_1j_1r_1}-a_{j_1j_1r_2})^{m_{j_1r_2}}\prod_{u_2=1,u_2\neq u_1}^{q_1}(a_{j_1j_2u_1}-a_{j_1j_2u_2})^{n_{j_1u_2}}>0$ and $a_{j_2j_20}\prod_{v_2=1,v_2\neq v_1}^{p_2}(a_{j_2j_1v_1}-a_{j_2j_1v_2})^{m_{j_2v_2}}>0$,

$$(a_{j_1j_2u_1},a_{j_2j_1v_1}) = \underbrace{([(2m_{j_1r_11}),(2n_{j_1u_11}+1)]^{\text{th}}\text{DP}_-\text{:UP}_+,[2(m_{j_2v_11}-m_{j_1r_11})+1]^{\text{th}}_{\text{DP}}\text{LS:}_{\text{UP}}\text{US})}_{([2m_{j_1r_11}+1),(2n_{j_1u_11}+1)],[2(m_{j_2v_11}-m_{j_1r_11})+1])\text{-positive hyperbolic lower-to-upper saddle}}.$$

The equilibrium is a $[(2m_{j_1r_11}),(2n_{j_1u_11}+1)]^{\text{th}}\text{DP}_-\text{:UP}_+,[2(m_{j_2v_11}-m_{j_1r_11})+1]^{\text{th}}$ $_{\text{DP}}\text{LS:}_{\text{UP}}\text{US}$)-positive hyperbolic lower-to-upper saddle at $(x^*_{j_2},x^*_{j_1})=(a_{j_1j_2u_1},a_{j_2j_1v_1})$.

- For $a_{j_1j_10}\prod_{r_2=1,r_2\neq r_1}^{p_1}(a_{j_1j_1r_1}-a_{j_1j_1r_2})^{m_{j_1r_2}}\prod_{u_2=1,u_2\neq u_1}^{q_1}(a_{j_1j_2u_1}-a_{j_1j_2u_2})^{n_{j_1u_2}}<0$ and $a_{j_2j_20}\prod_{v_2=1,v_2\neq v_1}^{p_2}(a_{j_2j_1v_1}-a_{j_2j_1v_2})^{m_{j_2v_2}}>0$,

$$(a_{j_1j_2u_1},a_{j_2j_1v_1}) = \underbrace{([(2m_{j_1r_11}),(2n_{j_1u_11}+1)]^{\text{th}}\text{UP}_-\text{:DP}_+,[2(m_{j_2v_11}-m_{j_1r_11})+1]^{\text{th}}_{\text{UP}}\text{US:}_{\text{DP}}\text{LS})}_{([2m_{j_1r_11}+1),(2n_{j_1u_11}+1)],[2(m_{j_2v_11}-m_{j_1r_11})+1])\text{-CCW circular upper-to-lower saddle}}.$$

The equilibrium is a $[(2m_{j_1r_11}),(2n_{j_1u_11}+1)]^{\text{th}}\text{UP}_-\text{:DP}_+,[2(m_{j_2v_11}-m_{j_1r_11})+1]^{\text{th}}$ $_{\text{UP}}\text{US:}_{\text{DP}}\text{LS}$) -CCW circular upper-to-lower saddle at $(x^*_{j_2},x^*_{j_1})=(a_{j_1j_2u_1},a_{j_2j_1v_1})$.

- For $a_{j_1j_10}\prod_{r_2=1,r_2\neq r_1}^{p_1}(a_{j_1j_1r_1}-a_{j_1j_1r_2})^{m_{j_1r_2}}\prod_{u_2=1,u_2\neq u_1}^{q_1}(a_{j_1j_2u_1}-a_{j_1j_2u_2})^{n_{j_1u_2}}>0$ and $a_{j_2j_20}\prod_{v_2=1,v_2\neq v_1}^{p_2}(a_{j_2j_1v_1}-a_{j_2j_1v_2})^{m_{j_2v_2}}<0$,

$$(a_{j_1j_2u_1},a_{j_2j_1v_1}) = \underbrace{([(2m_{j_1r_11}),(2n_{j_1u_11}+1)]^{\text{th}}\text{UP}_+\text{:DP}_-,[2(m_{j_2v_11}-m_{j_1r_11})+1]^{\text{th}}_{\text{UP}}\text{LS:}_{\text{DP}}\text{US})}_{([2m_{j_1r_11}+1),(2n_{j_1u_11}+1)],[2(m_{j_2v_11}-m_{j_1r_11})-1])\text{-CW circular lower-to-upper saddle}}.$$

The equilibrium is a $[(2m_{j_1r_11}),(2n_{j_1u_11}+1)]^{\text{th}}\text{UP}_-\text{:DP}_+,[2(m_{j_2v_11}-m_{j_1r_11})+1]^{\text{th}}$ $_{\text{UP}}\text{US:}_{\text{DP}}\text{LS}$) -CW circular lower-to-upper saddle at $(x^*_{j_2},x^*_{j_1})=(a_{j_1j_2u_1},a_{j_2j_1v_1})$.

- For $a_{j_1j_10}\prod_{r_2=1,r_2\neq r_1}^{p_1}(a_{j_1j_1r_1}-a_{j_1j_1r_2})^{m_{j_1r_2}}\prod_{u_2=1,u_2\neq u_1}^{q_1}(a_{j_1j_2u_1}-a_{j_1j_2u_2})^{n_{j_1u_2}}<0$ and $a_{j_2j_20}\prod_{v_2=1,v_2\neq v_1}^{p_2}(a_{j_2j_1v_1}-a_{j_2j_1v_2})^{m_{j_2v_2}}<0$,

$$(a_{j_1j_2u_1}, a_{j_2j_1v_1}) = \underbrace{([(2m_{j_1r_1}1),(2n_{j_1u_1}1+1)]^{\text{th}}\text{DP}_{+}\text{:UP}_{-},[2(m_{j_2v_1}1-m_{j_1r_1}1)+1]^{\text{th}}_{\text{DP}}\text{US:}_{\text{UP}}\text{LS})}_{([2m_{j_1r_1}1+1),(2n_{j_1u_1}1+1)],[2(m_{j_2v_1}1-m_{j_1r_1}1)+1])\text{-negative hyperbolic upper-to-lower saddle}}.$$

The equilibrium is a $[(2m_{j_1r_1}1),(2n_{j_1u_1}1+1)]^{\text{th}}\text{DP}_{-}\text{:UP}_{+},[2(m_{j_2v_1}1-m_{j_1r_1}1)+1]^{\text{th}}$ $_{\text{DP}}\text{LS:}_{\text{UP}}\text{US})$ -negative hyperbolic upper-to-lower saddle at $(x^*_{j_2}, x^*_{j_1})=(a_{j_1j_2u_1}, a_{j_2j_1v_1})$.

($\text{ii}_{2\text{a}2\text{c}}$) For $m_{j_1r_1}=2m_{j_1r_1}1+1$, $n_{j_1u_1}=2n_{j_1u_1}1+1$, and $m_{j_2v_1}=2m_{j_1v_1}1$, the equilibrium of $(x^*_{j_1}, x^*_{j_2})=(a_{j_1j_1r_1}, a_{j_1j_2u_1})$ on the infinite-equilibrium $x^*_{j_1}=a_{j_2j_1v_1}=a_{j_1j_1r_1}$ has the following properties as in Eqs. (1.207)–(1.210).

- For $a_{j_1j_10}\prod_{r_2=1,r_2\neq r_1}^{p_1}(a_{j_1j_1r_1}-a_{j_1j_1r_2})^{m_{j_1r_2}}\prod_{u_2=1,u_2\neq u_1}^{q_1}(a_{j_1j_2u_1}-a_{j_1j_2u_2})^{n_{j_1u_2}}>0$ and $a_{j_2j_20}\prod_{v_2=1,v_2\neq v_1}^{p_2}(a_{j_2j_1v_1}-a_{j_2j_1v_2})^{m_{j_2v_2}}>0$,

$$(a_{j_1j_2u_1}, a_{j_2j_1v_1}) = \underbrace{([(2m_{j_1r_1}1+1),(2n_{j_1u_1}1+1)]^{\text{th}}\text{DP}_{+}\text{:UP}_{+},[2(m_{j_2v_1}1-m_{j_1r_1}1)-1]^{\text{th}}_{\text{DP}}\text{SI:}_{\text{UP}}\text{SO})}_{([2m_{j_1r_1}1+1),(2n_{j_1u_1}1+1)],[2(m_{j_2v_1}1-m_{j_1r_1}1)-1])\text{-hyperbolic parabola sink-to-source}}.$$

The equilibrium is a $[(2m_{j_1r_1}1),(2n_{j_1u_1}1+1)]^{\text{th}}\text{DP}_{-}\text{:UP}_{+},[2(m_{j_2v_1}1-m_{j_1r_1}1)-1]^{\text{th}}$ $_{\text{DP}}\text{SI:}_{\text{UP}}\text{SO})$ -hyperbolic parabola sink-to-source at $(x^*_{j_2}, x^*_{j_1})=(a_{j_1j_2u_1}, a_{j_2j_1v_1})$.

- For $a_{j_1j_10}\prod_{r_2=1,r_2\neq r_1}^{p_1}(a_{j_1j_1r_1}-a_{j_1j_1r_2})^{m_{j_1r_2}}\prod_{u_2=1,u_2\neq u_1}^{q_1}(a_{j_1j_2u_1}-a_{j_1j_2u_2})^{n_{j_1u_2}}<0$ and $a_{j_2j_20}\prod_{v_2=1,v_2\neq v_1}^{p_2}(a_{j_2j_1v_1}-a_{j_2j_1v_2})^{m_{j_2v_2}}>0$,

$$(a_{j_1j_2u_1}, a_{j_2j_1v_1}) = \underbrace{([(2m_{j_1r_1}1+1),(2n_{j_1u_1}1+1)]^{\text{th}}\text{UP}_{+}\text{:DP}_{+},[2(m_{j_2v_1}1-m_{j_1r_1}1)-1]^{\text{th}}_{\text{UP}}\text{SO:}_{\text{DP}}\text{SI})}_{([2m_{j_1r_1}1+1),(2n_{j_1u_1}1+1)],[2(m_{j_2v_1}1-m_{j_1r_1}1)-1])\text{-circular parabola source-to-sink}}.$$

The equilibrium is a $[(2m_{j_1r_1}1),(2n_{j_1u_1}1+1)]^{\text{th}}\text{DP}_{-}\text{:UP}_{+},[2(m_{j_2v_1}1-m_{j_1r_1}1)-1]^{\text{th}}$ $_{\text{UP}}\text{SO:}_{\text{UP}}\text{SI})$ -circular parabola source-to-sink at $(x^*_{j_2}, x^*_{j_1})=(a_{j_1j_2u_1}, a_{j_2j_1v_1})$.

- For $a_{j_1j_10}\prod_{r_2=1,r_2\neq r_1}^{p_1}(a_{j_1j_1r_1}-a_{j_1j_1r_2})^{m_{j_1r_2}}\prod_{u_2=1,u_2\neq u_1}^{q_1}(a_{j_1j_2u_1}-a_{j_1j_2u_2})^{n_{j_1u_2}}>0$ and $a_{j_2j_20}\prod_{v_2=1,v_2\neq v_1}^{p_2}(a_{j_2j_1v_1}-a_{j_2j_1v_2})^{m_{j_2v_2}}<0$,

$$(a_{j_1j_2u_1}, a_{j_2j_1v_1}) = \underbrace{([(2m_{j_1r_1}1+1),(2n_{j_1u_1}1+1)]^{\text{th}}\text{UP}_{-}\text{:DP}_{-},[2(m_{j_2v_1}1-m_{j_1r_1}1)-1]^{\text{th}}_{\text{UP}}\text{SI:}_{\text{DP}}\text{SO})}_{([2m_{j_1r_1}1+1),(2n_{j_1u_1}1+1)],[2(m_{j_2v_1}1-m_{j_1r_1}1)-1])\text{-circular parabola sink-to-source}}.$$

The equilibrium is a $[(2m_{j_1r_1}1),(2n_{j_1u_1}1+1)]^{\text{th}}\text{DP}_{-}\text{:UP}_{+},[2(m_{j_2v_1}1-m_{j_1r_1}1)-1]^{\text{th}}$ $_{\text{UP}}\text{SI:}_{\text{UP}}\text{SO})$ -circular parabola sink-to-source at $(x^*_{j_2}, x^*_{j_1})=(a_{j_1j_2u_1}, a_{j_2j_1v_1})$.

- For $a_{j_1j_10}\prod_{r_2=1,r_2\neq r_1}^{p_1}(a_{j_1j_1r_1}-a_{j_1j_1r_2})^{m_{j_1r_2}}\prod_{u_2=1,u_2\neq u_1}^{q_1}(a_{j_1j_2u_1}-a_{j_1j_2u_2})^{n_{j_1u_2}}<0$ and $a_{j_2j_20}\prod_{v_2=1,v_2\neq v_1}^{p_2}(a_{j_2j_1v_1}-a_{j_2j_1v_2})^{m_{j_2v_2}}<0$,

$$(a_{j_1j_2u_1}, a_{j_2j_1v_1}) = \underbrace{([(2m_{j_1r_1}1+1),(2n_{j_1u_1}1+1)]^{\text{th}}\text{DP}_{-}\text{:UP}_{-},[2(m_{j_2v_1}1-m_{j_1r_1}1)-1]^{\text{th}}_{\text{DP}}\text{SO:}_{\text{UP}}\text{SI})}_{([2m_{j_1r_1}1+1),(2n_{j_1u_1}1+1)],[2(m_{j_2v_1}1-m_{j_1r_1}1)-1])\text{-hyperbolic parabola source-to-sink}}.$$

The equilibrium is a $[(2m_{j_1r_1}1),(2n_{j_1u_1}1+1)]^{\text{th}}\text{DP}_{-}\text{:UP}_{-},[2(m_{j_2v_1}1-m_{j_1r_1}1)-1]^{\text{th}}$ $_{\text{DP}}\text{SO:}_{\text{UP}}\text{SI})$ -hyperbolic parabola source-to-sink at $(x^*_{j_2}, x^*_{j_1})=(a_{j_1j_2u_1}, a_{j_2j_1v_1})$.

(ii_{2a2d}) For $m_{j_1r_1}=2m_{j_1r_1}1$, $n_{j_1u_1}=2n_{j_1u_1}1+1$, and $m_{j_2v_1}=2m_{j_1v_1}1$, the equilibrium of $(x^*_{j_2}, x^*_{j_1})=(a_{j_1j_2u_1}, a_{j_2j_1v_1})$ on the infinite-equilibrium $x^*_{j_1}=a_{j_2j_1v_1}=a_{j_1j_1r_1}$ has the following properties as in Eqs. (1.211)–(1.214).

- For $a_{j_1j_10}\prod_{r_2=1,r_2\neq r_1}^{p_1}(a_{j_1j_1r_1}-a_{j_1j_1r_2})^{m_{j_1r_2}}\prod_{u_2=1,u_2\neq u_1}^{q_1}(a_{j_1j_2u_1}-a_{j_1j_2u_2})^{n_{j_1u_2}}>0$ and $a_{j_2j_20}\prod_{v_2=1,v_2\neq v_1}^{p_2}(a_{j_2j_1v_1}-a_{j_2j_1v_2})^{m_{j_2v_2}}>0$,

$$(a_{j_1j_2u_1}, a_{j_2j_1v_1}) = \underbrace{([(2m_{j_1r_1}1),(2n_{j_1u_1}1+1)]^{\text{th}}\text{UP}_{+}\text{:UP}_{+},[2(m_{j_2v_1}1-m_{j_1r_1}1)]^{\text{th}}{}_{\text{DI:II}}\text{US})}_{([(2m_{j_1r_1}1),(2n_{j_1u_1}1+1)],[2(m_{j_2v_1}1-m_{j_1r_1}1)])\text{-inflection up-parabola upper-saddle}}.$$

The equilibrium is a $([(2m_{j_1r_1}1),(2n_{j_1u_1}1+1)]^{\text{th}}\text{UP}_{+}\text{:UP}_{+},[2(m_{j_2v_1}1-m_{j_1r_1}1)]^{\text{th}}$ $_{\text{DI:II}}\text{US})$-inflection up-parabola upper-saddle at $(x^*_{j_2}, x^*_{j_1})=(a_{j_1j_2u_1}, a_{j_2j_1v_1})$.

- For $a_{j_1j_10}\prod_{r_2=1,r_2\neq r_1}^{p_1}(a_{j_1j_1r_1}-a_{j_1j_1r_2})^{m_{j_1r_2}}\prod_{u_2=1,u_2\neq u_1}^{q_1}(a_{j_1j_2u_1}-a_{j_1j_2u_2})^{n_{j_1u_2}}<0$ and $a_{j_2j_20}\prod_{v_2=1,v_2\neq v_1}^{p_2}(a_{j_2j_1v_1}-a_{j_2j_1v_2})^{m_{j_2v_2}}>0$,

$$(a_{j_1j_2u_1}, a_{j_2j_1v_1}) = \underbrace{([(2m_{j_1r_1}1),(2n_{j_1u_1}1+1)]^{\text{th}}\text{DP}_{+}\text{:DP}_{+},[2(m_{j_2v_1}1-m_{j_1r_1}1)]^{\text{th}}{}_{\text{II:DI}}\text{US})}_{([(2m_{j_1r_1}1),(2n_{j_1u_1}1+1)],[2(m_{j_2v_1}1-m_{j_1r_1}1)])\text{-inflection up-parabola upper-saddle}}.$$

The equilibrium is a $([(2m_{j_1r_1}1),(2n_{j_1u_1}1+1)]^{\text{th}}\text{DP}_+\text{:DP}_+,[2(m_{j_2v_1}1-m_{j_1r_1}1)]^{\text{th}}{}_{\text{II:DI}}\text{US})$-inflection down-parabola upper-saddle at $(x_{j_2}^*, x_{j_1}^*) = (a_{j_1j_2u_1}, a_{j_2j_1v_1})$.

- For $a_{j_1j_10}\prod_{r_2=1,r_2\neq r_1}^{p_1}(a_{j_1j_1r_1}-a_{j_1j_1r_2})^{m_{j_1r_2}}\prod_{u_2=1,u_2\neq u_1}^{q_1}(a_{j_1j_2u_1}-a_{j_1j_2u_2})^{n_{j_1u_2}}>0$ and $a_{j_2j_20}\prod_{v_2=1,v_2\neq v_1}^{p_2}(a_{j_2j_1v_1}-a_{j_2j_1v_2})^{m_{j_2v_2}}<0$,

$$(a_{j_1j_2u_1}, a_{j_2j_1v_1}) = \underbrace{([(2m_{j_1r_1}1),(2n_{j_1u_1}1+1)]^{\text{th}}\text{DP}_-\text{:DP}_-,[2(m_{j_2v_1}1-m_{j_1r_1}1)]^{\text{th}}{}_{\text{II:DI}}\text{LS})}_{([(2m_{j_1r_1}1),(2n_{j_1u_1}1+1)],[2(m_{j_2v_1}1-m_{j_1r_1}1)])\text{-inflection down-parabola lower-saddle}}.$$

The equilibrium is a $([(2m_{j_1r_1}1),(2n_{j_1u_1}1+1)]^{\text{th}}\text{DP}_-\text{:DP}_-,[2(m_{j_2v_1}1-m_{j_1r_1}1)]^{\text{th}}{}_{\text{II:DI}}\text{LS})$-inflection down-parabola lower-saddle at $(x_{j_2}^*, x_{j_1}^*) = (a_{j_1j_2u_1}, a_{j_2j_1v_1})$.

- For $a_{j_1j_10}\prod_{r_2=1,r_2\neq r_1}^{p_1}(a_{j_1j_1r_1}-a_{j_1j_1r_2})^{m_{j_1r_2}}\prod_{u_2=1,u_2\neq u_1}^{q_1}(a_{j_1j_2u_1}-a_{j_1j_2u_2})^{n_{j_1u_2}}<0$ and $a_{j_2j_20}\prod_{v_2=1,v_2\neq v_1}^{p_2}(a_{j_2j_1v_1}-a_{j_2j_1v_2})^{m_{j_2v_2}}<0$,

$$(a_{j_1j_2u_1}, a_{j_2j_1v_1}) = \underbrace{([(2m_{j_1r_1}1),(2n_{j_1u_1}1+1)]^{\text{th}}\text{UP}_-\text{:UP}_-,[2(m_{j_2v_1}1-m_{j_1r_1}1)]^{\text{th}}{}_{\text{DI:II}}\text{LS})}_{([(2m_{j_1r_1}1),(2n_{j_1u_1}1+1)],[2(m_{j_2v_1}1-m_{j_1r_1}1)])\text{-inflection up-parabola lower-saddle}}.$$

The equilibrium is a $([(2m_{j_11_11}),(2n_{j_1l_1}1+1)]^{\text{th}}\text{UP}_+:\text{UP}_+,[2(m_{j_2h_1}1-m_{j_1r_1})]^{\text{th}}{}_{\text{Di-II}}\text{LS})$-inflection up-parabola lower-saddle at $(x_{j_2}^*, x_{j_1}^*)=(a_{j_1j_2l_1}, a_{j_2j_1v_1})$.

($\text{ii}_{2\text{a}2\text{e}}$) For $m_{j_1r_1}=2m_{j_1r_1}1+1$, $n_{j_1u_1}=2n_{j_1u_1}1$, and $m_{j_2v_1}=2m_{j_1v_1}1+1$, the equilibrium of $(x_{j_2}^*, x_{j_1}^*) = (a_{j_1j_2u_1}, a_{j_2j_1v_1})$ on the infinite-equilibrium $x_{j_1}^* = a_{j_2j_1v_1} = a_{j_1j_1r_1}$ has the following properties as in Eqs. (1.215)–(1.218).

- For $a_{j_1j_10}\prod_{r_2=1,r_2\neq r_1}^{p_1}(a_{j_1j_1r_1}-a_{j_1j_1r_2})^{m_{j_1r_2}}\prod_{u_2=1,u_2\neq u_1}^{q_1}(a_{j_1j_2u_1}-a_{j_1j_2u_2})^{n_{j_1u_2}}>0$ and $a_{j_2j_20}\prod_{v_2=1,v_2\neq v_1}^{p_2}(a_{j_2j_1v_1}-a_{j_2j_1v_2})^{m_{j_2v_2}}>0$,

$$(a_{j_1j_2u_1}, a_{j_2j_1v_1}) = \underbrace{([(2m_{j_1r_1}1+1),(2n_{j_1u_1}1)]^{\text{th}}\text{II}_-\text{:II}_+,[2(m_{j_2v_1}1-m_{j_1r_1}1)]^{\text{th}}{}_{\text{II:II}}\text{SO})}_{([(2m_{j_1r_1}1+1),(2n_{j_1u_1}1)],[2(m_{j_2v_1}1-m_{j_1r_1}1)])\text{-increasing-inflection source}}.$$

The equilibrium is a $([(2m_{j_1r_1}1+1),(2n_{j_1u_1}1)]^{\text{th}}\text{II}_-\text{:II}_+,[2(m_{j_2v_1}1-m_{j_1r_1}1)]^{\text{th}}{}_{\text{II:II}}\text{SO})$-increasing-inflection source at $(x_{j_2}^*, x_{j_1}^*) = (a_{j_1j_2u_1}, a_{j_2j_1v_1})$.

- For $a_{j_1j_10}\prod_{r_2=1,r_2\neq r_1}^{p_1}(a_{j_1j_1r_1}-a_{j_1j_1r_2})^{m_{j_1r_2}}\prod_{u_2=1,u_2\neq u_1}^{q_1}(a_{j_1j_2u_1}-a_{j_1j_2u_2})^{n_{j_1u_2}}<0$ and $a_{j_2j_20}\prod_{v_2=1,v_2\neq v_1}^{p_2}(a_{j_2j_1v_1}-a_{j_2j_1v_2})^{m_{j_2v_2}}>0$,

$$(a_{j_1j_2u_1},a_{j_2j_1v_1})=\underbrace{([(2m_{j_1r_11}+1),(2n_{j_1u_11})]^{\text{th}}\text{DI}_-{:}\text{DI}_+,[2(m_{j_2v_11}-m_{j_1r_11})]^{\text{th}}{}_{\text{DI:DI}}\text{SI})}_{([(2m_{j_1r_11}+1),(2n_{j_1u_11})],[2(m_{j_2v_11}-m_{j_1r_11})])\text{-decreasing-inflection sink}}.$$

The equilibrium is a $([(2m_{j_1r_11}+1),(2n_{j_1u_11})]^{\text{th}}\text{DI}_-{:}\text{DI}_+,[2(m_{j_2v_11}-m_{j_1r_11})]^{\text{th}}{}_{\text{DI:DI}}\text{SI})$-decreasing-inflection sink at $(x_{j_2}^*,x_{j_1}^*)=(a_{j_1j_2u_1},a_{j_2j_1v_1})$.

- For $a_{j_1j_10}\prod_{r_2=1,r_2\neq r_1}^{p_1}(a_{j_1j_1r_1}-a_{j_1j_1r_2})^{m_{j_1r_2}}\prod_{u_2=1,u_2\neq u_1}^{q_1}(a_{j_1j_2u_1}-a_{j_1j_2u_2})^{n_{j_1u_2}}>0$ and $a_{j_2j_20}\prod_{v_2=1,v_2\neq v_1}^{p_2}(a_{j_2j_1v_1}-a_{j_2j_1v_2})^{m_{j_2v_2}}<0$,

$$(a_{j_1j_2u_1},a_{j_2j_1v_1})=\underbrace{([(2m_{j_1r_11}+1),(2n_{j_1u_11})]^{\text{th}}\text{DI}_+{:}\text{DI}_-,[2(m_{j_2v_11}-m_{j_1r_11})]^{\text{th}}{}_{\text{DI:DI}}\text{SO})}_{([(2m_{j_1r_11}+1),(2n_{j_1u_11})],[2(m_{j_2v_11}-m_{j_1r_11})])\text{-decreasing-inflection sink}}.$$

The equilibrium is a $([(2m_{j_1r_11}+1),(2n_{j_1u_11})]^{\text{th}}\text{DI}_+{:}\text{DI}_1,[2(m_{j_2v_11}-m_{j_1r_11})]^{\text{th}}{}_{\text{DI:DI}}\text{SO})$-decreasing-inflection source at $(x_{j_2}^*,x_{j_1}^*)=(a_{j_1j_2u_1},a_{j_2j_1v_1})$.

- For $a_{j_1j_10}\prod_{r_2=1,r_2\neq r_1}^{p_1}(a_{j_1j_1r_1}-a_{j_1j_1r_2})^{m_{j_1r_2}}\prod_{u_2=1,u_2\neq u_1}^{q_1}(a_{j_1j_2u_1}-a_{j_1j_2u_2})^{n_{j_1u_2}}<0$ and $a_{j_2j_20}\prod_{v_2=1,v_2\neq v_1}^{p_2}(a_{j_2j_1v_1}-a_{j_2j_1v_2})^{m_{j_2v_2}}<0$,

$$(a_{j_1j_2u_1},a_{j_2j_1v_1})=\underbrace{([(2m_{j_1r_11}+1),(2n_{j_1u_11})]^{\text{th}}\text{II}_+{:}\text{II}_-,[2(m_{j_2v_11}-m_{j_1r_11})]^{\text{th}}{}_{\text{II:II}}\text{SI})}_{([(2m_{j_1r_11}+1),(2n_{j_1u_11})],[2(m_{j_2v_11}-m_{j_1r_11})])\text{-increasing-inflection sink}}.$$

The equilibrium is a $([(2m_{j_1r_11}+1),(2n_{j_1u_11})]^{\text{th}}\text{II}_+{:}\text{II}_-,[2(m_{j_2v_11}-m_{j_1r_11})]^{\text{th}}{}_{\text{II:II}}\text{SI})$-increasing-inflection sink at $(x_{j_2}^*,x_{j_1}^*)=(a_{j_1j_2u_1},a_{j_2j_1v_1})$.

($\text{ii}_{2\text{a}2\text{f}}$) For $m_{j_1r_1}=2m_{j_1r_11}$, $n_{j_1u_1}=2n_{j_1u_11}$, and $m_{j_2v_1}=2m_{j_1v_11}+1$, the equilibrium of $(x_{j_2}^*,x_{j_1}^*)=(a_{j_1j_2u_1},a_{j_2j_1v_1})$ on the infinite-equilibrium $x_{j_1}^*=a_{j_2j_1v_1}=a_{j_1j_1r_1}$ has the following properties as in Eqs. (1.219)–(1.222).

- For $a_{j_1j_10}\prod_{r_2=1,r_2\neq r_1}^{p_1}(a_{j_1j_1r_1}-a_{j_1j_1r_2})^{m_{j_1r_2}}\prod_{u_2=1,u_2\neq u_1}^{q_1}(a_{j_1j_2u_1}-a_{j_1j_2u_2})^{n_{j_1u_2}}>0$ and $a_{j_2j_20}\prod_{v_2=1,v_2\neq v_1}^{p_2}(a_{j_2j_1v_1}-a_{j_2j_1v_2})^{m_{j_2v_2}}>0$,

$$(a_{j_1j_2u_1},a_{j_2j_1v_1}) = \underbrace{([(2m_{j_1r_1}1),(2n_{j_1u_1}1)]^{\text{th}}\text{DI}_{-}\text{:II}_{+},[2(m_{j_2v_1}1-m_{j_1r_1}1)+1]^{\text{th}}{}_{\text{UP:UP}}\text{US})}_{([(2m_{j_1r_1}1),(2n_{j_1u_1}1)],[2(m_{j_2v_1}1-m_{j_1r_1}1)+1])\text{-up-parabola upper-saddle}}.$$

The equilibrium is a $([(2m_{j_1r_1}1),(2n_{j_1u_1}1)]^{\text{th}}\text{DI}_{-}\text{:II}_{+},[2(m_{j_2v_1}1-m_{j_1r_1}1)+1]^{\text{th}}{}_{\text{UP:UP}}\text{US})$-up-parabola upper-saddle at $(x_{j_2}^*,x_{j_1}^*)=(a_{j_1j_2u_1},a_{j_2j_1v_1})$.

- For $a_{j_1j_10}\prod_{r_2=1,r_2\neq r_1}^{p_1}(a_{j_1j_1r_1}-a_{j_1j_1r_2})^{m_{j_1r_2}}\prod_{u_2=1,u_2\neq u_1}^{q_1}(a_{j_1j_2u_1}-a_{j_1j_2u_2})^{n_{j_1u_2}}<0$ and $a_{j_2j_20}\prod_{v_2=1,v_2\neq v_1}^{p_2}(a_{j_2j_1v_1}-a_{j_2j_1v_2})^{m_{j_2v_2}}>0$,

$$(a_{j_1j_2u_1},a_{j_2j_1v_1}) = \underbrace{([(2m_{j_1r_1}1),(2n_{j_1u_1}1)]^{\text{th}}\text{II}_{-}\text{:DI}_{+},[2(m_{j_2v_1}1-m_{j_1r_1}1)+1]^{\text{th}}{}_{\text{DP:DP}}\text{LS})}_{([(2m_{j_1r_1}1),(2n_{j_1u_1}1)],[2(m_{j_2v_1}1-m_{j_1r_1}1)+1])\text{-down-parabola lower-saddle}}.$$

The equilibrium is a $([(2m_{j_1r_1}1),(2n_{j_1u_1}1)]^{\text{th}}\text{DI}_{-}\text{:II}_{+},[2(m_{j_2v_1}1-m_{j_1r_1}1)+1]^{\text{th}}{}_{\text{DP:DP}}\text{LS})$-down-parabola lower-saddle at $(x_{j_2}^*,x_{j_1}^*)=(a_{j_1j_2u_1},a_{j_2j_1v_1})$.

- For $a_{j_1j_10}\prod_{r_2=1,r_2\neq r_1}^{p_1}(a_{j_1j_1r_1}-a_{j_1j_1r_2})^{m_{j_1r_2}}\prod_{u_2=1,u_2\neq u_1}^{q_1}(a_{j_1j_2u_1}-a_{j_1j_2u_2})^{n_{j_1u_2}}>0$ and $a_{j_2j_20}\prod_{v_2=1,v_2\neq v_1}^{p_2}(a_{j_2j_1v_1}-a_{j_2j_1v_2})^{m_{j_2v_2}}<0$,

$$(a_{j_1j_2u_1},a_{j_2j_1v_1}) = \underbrace{([(2m_{j_1r_1}1),(2n_{j_1u_1}1)]^{\text{th}}\text{II}_{+}\text{:DI}_{-},[2(m_{j_2v_1}1-m_{j_1r_1}1)+1]^{\text{th}}{}_{\text{DP:DP}}\text{US})}_{([(2m_{j_1r_1}1),(2n_{j_1u_1}1)],[2(m_{j_2v_1}1-m_{j_1r_1}1)+1])\text{-down-parabola upper-saddle}}.$$

The equilibrium is a $([(2m_{j_1r_1}1),(2n_{j_1u_1}1)]^{\text{th}}\text{ DI}_{-}:\text{II}_{+},[2(m_{j_2v_1}1-m_{j_1r_1}1)+1]^{\text{th}}{}_{\text{DP:DP}}\text{US})$-down-parabola upper-saddle at $(x_{j_2}^*,x_{j_1}^*)=(a_{j_1j_2u_1},a_{j_2j_1v_1})$.

- For $a_{j_1j_10}\prod_{r_2=1,r_2\neq r_1}^{p_1}(a_{j_1j_1r_1}-a_{j_1j_1r_2})^{m_{j_1r_2}}\prod_{u_2=1,u_2\neq u_1}^{q_1}(a_{j_1j_2u_1}-a_{j_1j_2u_2})^{n_{j_1u_2}}<0$ and $a_{j_2j_20}\prod_{v_2=1,v_2\neq v_1}^{p_2}(a_{j_2j_1v_1}-a_{j_2j_1v_2})^{m_{j_2v_2}}<0$,

$$(a_{j_1j_2l_1}, a_{j_2j_1l_1}) = \underbrace{([(2m_{j_11})), (2n_{j_11}1)]^{\text{th}}\,\text{DI}_+ : \text{II}_-, [2(m_{j_21} - m_{j_11}1) + 1]^{\text{th}}\,{}_{\text{UP:UP}}\,\text{LS})}_{([(2m_{j\eta_11}),(2n_{j\eta_11}))][2(m_{j\eta_11}-m_{j\eta_11})+1])\text{-up-parabola lower-saddle}}.$$

The equilibrium is a $([(2m_{j_1r_11}), (2n_{j_1u_11})]^{\text{th}}\,\text{DI}_- : \text{II}_+, [2(m_{j_2v_11}-m_{j_1r_11})+1]^{\text{th}}\,{}_{\text{UP:UP}}\text{US})$-up-parabola lower-saddle at $(x^*_{j_2}, x^*_{j_1})=(a_{j_1j_2l_1}, a_{j_2j_1v_1})$.

(ii_{2a2g}) For $m_{j_1r_1} = 2m_{j_1r_11} + 1$, $n_{j_1u_1} = 2n_{j_1u_11}$ and $m_{j_2v_1} = 2m_{j_1v_11}$, the equilibrium of $(x^*_{j_2}, x^*_{j_1}) = (a_{j_1j_2u_1}, a_{j_2j_1v_1})$ on the infinite-equilibrium $x^*_{j_1} = a_{j_2j_1v_1} = a_{j_1j_1r_1}$ has the following properties as in Eqs. (1.223)–(1.226).

- For $a_{j_1j_10} \prod_{r_2=1,r_2\neq r_1}^{p_1} (a_{j_1j_1r_1} - a_{j_1j_1r_2})^{m_{j_1r_2}} \prod_{u_2=1,u_2\neq u_1}^{q_1} (a_{j_1j_2u_1} - a_{j_1j_2u_2})^{n_{j_1u_2}} > 0$ and $a_{j_2j_20} \prod_{v_2=1,v_2\neq v_1}^{p_2} (a_{j_2j_1v_1} - a_{j_2j_1v_2})^{m_{j_2v_2}} > 0$,

$$(a_{j_1j_2l_1}, a_{j_2j_1v_1}) = \underbrace{([(2m_{j_1r_11} + 1), (2n_{j_1l_11})]^{\text{th}}\,\text{DI}_+ : \text{II}_+, [2(m_{j_2v_11} - m_{j_1r_1}) - 1]^{\text{th}}_{\text{UP:UP}}\,\text{SO})}_{([(2m_{j_11_1}+1),(2n_{j_11_11})],[2(m_{j_21_11}-m_{j_11}1)-1])-\text{ up-parabola source}}.$$

The equilibrium is a $([(2m_{j_1r_11}+1), (2n_{j_1r_11})]^{\text{th}}\,\text{DI}_+ : \text{II}_+, [2(m_{j_2v_11}-m_{j_1r_11})-1]^{\text{th}}\,{}_{\text{UP:UP}}\text{SO})$-up-parabola source at $(x^*_{j_2}, x^*_{j_1})=(a_{j_1j_2u_1}, a_{j_2j_1v_1})$.

- For $a_{j_1j_10} \prod_{r_2=1,r_2\neq r_1}^{p_1} (a_{j_1j_1r_1} - a_{j_1j_1r_2})^{m_{j_1r_2}} \prod_{u_2=1,u_2\neq u_1}^{q_1} (a_{j_1j_2u_1} - a_{j_1j_2u_2})^{n_{j_1u_2}} < 0$ and $a_{j_2j_20} \prod_{v_2=1,v_2\neq v_1}^{p_2} (a_{j_2j_1v_1} - a_{j_2j_1v_2})^{m_{j_2v_2}} > 0$,

$$(a_{j_1j_2u_1}, a_{j_2j_1v_1}) = \underbrace{([(2m_{j_1r_11} + 1), (2n_{j_1u_11})]^{\text{th}}\text{II}_+{:}\text{DI}_+,[2(m_{j_2v_11} - m_{j_1r_11}) - 1]^{\text{th}}{}_{\text{DP:DP}}\text{SI})}_{([(2m_{j_1r_11}+1),(2n_{j_1u_11})],[2(m_{j_2v_11}-m_{j_1r_11})-1])-\textit{down}\text{-parabola sink}}.$$

The equilibrium is a $([(2m_{j_1r_11}+1), (2n_{j_1u_11})]^{\text{th}}\text{II}_+{:}\text{DI}_+,[2(m_{j_2v_11}-m_{j_1r_11})-1]^{\text{th}}\,{}_{\text{DP:DP}}\text{SI})$-down-parabola sink at $(x^*_{j_2}, x^*_{j_1}) = (a_{j_1j_2u_1}, a_{j_2j_1v_1})$.

- For $a_{j_1j_10} \prod_{r_2=1,r_2\neq r_1}^{p_1} (a_{j_1j_1r_1} - a_{j_1j_1r_2})^{m_{j_1r_2}} \prod_{u_2=1,u_2\neq u_1}^{q_1} (a_{j_1j_2u_1} - a_{j_1j_2u_2})^{n_{j_1u_2}} > 0$ and $a_{j_2j_20} \prod_{v_2=1,v_2\neq v_1}^{p_2} (a_{j_2j_1v_1} - a_{j_2j_1v_2})^{m_{j_2v_2}} < 0$,

$$(a_{j_1j_2l_1}, a_{j_2j_1v_1}) = \underbrace{([(2m_{j_1r_11} + 1), (2n_{j_1u_11})]^{\text{th}}\text{II}_- : \text{DI}_-, [2(m_{j_2v_11} - m_{j_1r_11}) - 1]^{\text{th}}{}_{\text{DP:DP}}\,\text{SO})}_{([(2m_{j\eta 1_1}+1),(2n_{j\eta_11})],[2(m_{j\eta_11}-m_{j\eta_11})-1])\text{-down-parabola source}}.$$

The equilibrium is a $([(2m_{j_1r_1}1{+}1), (2n_{j_1u_1})]^{\text{th}}\text{II}_- : \text{DI}_-, [2(m_{j_2v_1}1 - m_{j_1r_1}1) - 1]^{\text{th}}{}_{\text{DP:DP}}\text{SO})$-down-parabola source at $(x^*_{j_2}, x^*_{j_1}) = (a_{j_1j_2l_1}, a_{j_2j_1v_1})$.

- For $a_{j_1j_10} \prod_{r_2=1,r_2\neq r_1}^{p_1} (a_{j_1j_1r_1} - a_{j_1j_1r_2})^{m_{j_1r_2}} \prod_{u_2=1,u_2\neq u_1}^{q_1} (a_{j_1j_2u_1} - a_{j_1j_2u_2})^{n_{j_1u_2}} < 0$ and $a_{j_2j_20} \prod_{v_2=1,v_2\neq v_1}^{p_2} (a_{j_2j_1v_1} - a_{j_2j_1v_2})^{m_{j_2v_2}} < 0$,

$$
\begin{aligned}
&(a_{j_1j_2u_1}, a_{j_2j_1v_1}) \\
&= \underbrace{([(2m_{j_1r_1}1 + 1), (2n_{j_1u_1}1)]^{\text{th}}\text{DI}_-{:}\text{II}_-, [2(m_{j_2v_1}1 - m_{j_1r_1}1) - 1]^{\text{th}}{}_{\text{UP:UP}}\text{SI})}_{([(2m_{j_1r_1}1+1),(2n_{j_1u_1}1)],[2(m_{j_2v_1}1-m_{j_1r_1}1)-1])\text{-up-parabola sink}}.
\end{aligned}
$$

The equilibrium is a $([(2m_{j_1r_1}1{+}1), (2n_{j_1u_1}1)]^{\text{th}}\text{DI}_-{:}\text{II}_-, [2(m_{j_2v_1}1 - m_{j_1r_1}1) - 1]^{\text{th}}{}_{\text{UP:UP}}\text{SI})$-up-parabola source at $(x^*_{j_2}, x^*_{j_1}) = (a_{j_1j_2u_1}, a_{j_2j_1v_1})$.

($\text{ii}_{2\text{a}2\text{h}}$) For $m_{j_1r_1} = 2m_{j_1r_1}1$, $n_{j_1u_1} = 2n_{j_1u_1}1$, and $m_{j_2v_1} = 2m_{j_1v_1}1$, the equilibrium of $(x^*_{j_1}, x^*_{j_2}) = (a_{j_1j_1r_1}, a_{j_1j_2u_1})$ on the infinite-equilibrium $x^*_{j_1} = a_{j_2j_1v_1} = a_{j_1j_1r_1}$ has the following properties as in Eqs. (1.227)–(1.230).

- For $a_{j_1j_10} \prod_{r_2=1,r_2\neq r_1}^{p_1} (a_{j_1j_1r_1} - a_{j_1j_1r_2})^{m_{j_1r_2}} \prod_{u_2=1,u_2\neq u_1}^{q_1} (a_{j_1j_2u_1} - a_{j_1j_2u_2})^{n_{j_1u_2}} > 0$ and $a_{j_2j_20} \prod_{v_2=1,v_2\neq v_1}^{p_2} (a_{j_2j_1v_1} - a_{j_2j_1v_2})^{m_{j_2v_2}} > 0$,

$$
\begin{aligned}
&(a_{j_1j_2u_1}, a_{j_2j_1v_1}) \\
&= \underbrace{([(2m_{j_1r_1}1), (2n_{j_1u_1}1)]^{\text{th}}\text{II}_+{:}\text{II}_+, [2(m_{j_2v_1}1 - m_{j_1r_1}1)]^{\text{th}}{}_{\text{II:II}}\text{US})}_{([(2m_{j_1r_1}1),(2n_{j_1u_1}1)],[2(m_{j_2v_1}1-m_{j_1r_1}1)])\text{-increasing-inflection upper-saddle}}.
\end{aligned}
$$

The equilibrium is a $([(2m_{j_1r_1}1), (2n_{j_1u_1}1)]^{\text{th}}\text{II}_+{:}\text{II}_+, [2(m_{j_2v_1}1 - m_{j_1r_1}1)]^{\text{th}}{}_{\text{II:II}}\text{US})$-increasing-inflection upper-saddle at $(x^*_{j_2}, x^*_{j_1}) = (a_{j_1j_2u_1}, a_{j_2j_1v_1})$.

- For $a_{j_1j_10} \prod_{r_2=1,r_2\neq r_1}^{p_1} (a_{j_1j_1r_1} - a_{j_1j_1r_2})^{m_{j_1r_2}} \prod_{u_2=1,u_2\neq u_1}^{q_1} (a_{j_1j_2u_1} - a_{j_1j_2u_2})^{n_{j_1u_2}} < 0$ and $a_{j_2j_20} \prod_{v_2=1,v_2\neq v_1}^{p_2} (a_{j_2j_1v_1} - a_{j_2j_1v_2})^{m_{j_2v_2}} > 0$,

$$
\begin{aligned}
&(a_{j_1j_2u_1}, a_{j_2j_1v_1}) \\
&= \underbrace{([(2m_{j_1r_1}1), (2n_{j_1u_1}1)]^{\text{th}}\text{DI}_+{:}\text{DI}_+, [2(m_{j_2v_1}1 - m_{j_1r_1}1)]^{\text{th}}{}_{\text{DI:DI}}\text{LS})}_{([(2m_{j_1r_1}1),(2n_{j_1u_1}1)],[2(m_{j_2v_1}1-m_{j_1r_1}1)])\text{-decreasing-inflection lower-saddle}}.
\end{aligned}
$$

The equilibrium is a $([(2m_{j_1r_1}1), (2n_{j_1u_1}1)]^{\text{th}}\text{DI}_+{:}\text{DI}_+, [2(m_{j_2v_1}1 - m_{j_1r_1}1)]^{\text{th}}{}_{\text{DI:DI}}\text{US})$-decreasing-inflection lower-saddle at $(x^*_{j_2}, x^*_{j_1}) = (a_{j_1j_2u_1}, a_{j_2j_1v_1})$.

- For $a_{j_1j_10}\prod_{r_2=1,r_2\neq r_1}^{p_1}(a_{j_1j_1r_1}-a_{j_1j_1r_2})^{m_{j_1r_2}}\prod_{u_2=1,u_2\neq u_1}^{q_1}(a_{j_1j_2u_1}-a_{j_1j_2u_2})^{n_{j_1u_2}}>0$ and $a_{j_2j_20}\prod_{v_2=1,v_2\neq v_1}^{p_2}(a_{j_2j_1v_1}-a_{j_2j_1v_2})^{m_{j_2v_2}}<0$,

$$
\begin{aligned}
&(a_{j_1j_2u_1},a_{j_2j_1v_1})\\
&=\underbrace{([(2m_{j_1r_1}1),(2n_{j_1u_1}1)]^{\text{th}}\text{DI}_-\text{:DI}_-,[2(m_{j_2v_1}1-m_{j_1r_1}1)]^{\text{th}}{}_{\text{DI:DI}}\text{US})}_{([(2m_{j_1r_1}1),(2n_{j_1u_1}1)],[2(m_{j_2v_1}1-m_{j_1r_1}1)])\text{-decreasing-inflection upper-saddle}}.
\end{aligned}
$$

The equilibrium is a $([(2m_{j_1r_1}1),(2n_{j_1u_1}1)]^{\text{th}}\text{DI}_-\text{:DI}_-,[2(m_{j_2v_1}1-m_{j_1r_1}1)]^{\text{th}}{}_{\text{DI:DI}}\text{US})$-decreasing-inflection upper-saddle at $(x^*_{j_2},x^*_{j_1})=(a_{j_1j_2u_1},a_{j_2j_1v_1})$.

- For $a_{j_1j_10}\prod_{r_2=1,r_2\neq r_1}^{p_1}(a_{j_1j_1r_1}-a_{j_1j_1r_2})^{m_{j_1r_2}}\prod_{u_2=1,u_2\neq u_1}^{q_1}(a_{j_1j_2u_1}-a_{j_1j_2u_2})^{n_{j_1u_2}}<0$ and $a_{j_2j_20}\prod_{v_2=1,v_2\neq v_1}^{p_2}(a_{j_2j_1v_1}-a_{j_2j_1v_2})^{m_{j_2v_2}}<0$,

$$
\begin{aligned}
&(a_{j_1j_2u_1},a_{j_2j_1v_1})\\
&=\underbrace{([(2m_{j_1r_1}1),(2n_{j_1u_1}1)]^{\text{th}}\text{II}_-\text{:II}_-,[2(m_{j_2v_1}1-m_{j_1r_1}1)]^{\text{th}}{}_{\text{II:II}}\text{LS})}_{([(2m_{j_1r_1}1),(2n_{j_1u_1}1)],[2(m_{j_2v_1}1-m_{j_1r_1}1)])\text{-increasing-inflection lower-saddle}}.
\end{aligned}
$$

The equilibrium is a $([(2m_{j_1r_1}1),(2n_{j_1u_1}1)]^{\text{th}}\text{II}_+\text{:II}_+,[2(m_{j_2v_1}1-m_{j_1r_1}1)]^{\text{th}}{}_{\text{II:II}}\text{LS})$-increasing-inflection lower-saddle at $(x^*_{j_2},x^*_{j_1})=(a_{j_1j_2u_1},a_{j_2j_1v_1})$.

(ii_{2b}) For $m_{j_1r_1}>m_{j_2v_1}$, at $x^*_{j_1}=a_{j_1j_1r_1}=a_{j_2j_1v_1}$ and $\overline{x}_{j_2}\neq a_{j_1j_2u}$, if

$$
\begin{aligned}
&\left.\frac{dx_{j_2}}{dx_{j_1}}\right|_{x^*_{j_1}=a_{j_1j_1r_1}\pm\varepsilon}\\
&=\lim_{\varepsilon\to0}\frac{a_{j_2j_20}}{a_{j_1j_10}}\frac{\prod_{v_2=1,v_2\neq v_1}^{p_2}(a_{j_2j_1v_1}-a_{j_2j_1v_2})^{m_{j_2v_2}}}{(\pm\varepsilon)^{m_{j_1r_1}-m_{j_1v_1}}\prod_{r_2=1,r_2\neq r_1}^{p_1}(a_{j_1j_1r_1}-a_{j_1j_1r_2})^{m_{j_1r_1}}\prod_{u=1}^{q_1}(\overline{x}_{j_2}-a_{j_1j_2u})^{n_{j_1u}}}\\
&=+\infty,
\end{aligned}
$$

the flow at $x^*_{j_1}=a_{j_1j_1r_1}=a_{j_2j_1v_1}$ are a $[2(m_{j_1r_1}1-m_{j_2v_1}1)]^{\text{th}}$-order increasing-concave flows for $m_{j_1r_1}=2m_{j_1r_1}1+1$, $m_{j_2v_1}=2m_{j_2v_1}1+1$ and $m_{j_1r_1}=2m_{j_1r_1}1$, $m_{j_2v_1}=2m_{j_2v_1}1$). If

$$\left.\frac{dx_{j_2}}{dx_{j_1}}\right|_{x_{j_1}^*=a_{j_1j_1r_1}\pm\varepsilon}$$

$$=\lim_{\varepsilon\to 0}\frac{a_{j_2j_20}}{a_{j_1j_10}}\frac{\prod_{v_2=1,v_2\neq v_1}^{p_2}(a_{j_2j_1v_1}-a_{j_2j_1v_2})^{m_{j_2v_2}}}{(\pm\varepsilon)^{m_{j_1r_1}-m_{j_1v_1}}\prod_{r_2=1,r_2\neq r_1}^{p_1}(a_{j_1j_1r_1}-a_{j_1j_1r_2})^{m_{j_1r_1}}\prod_{u=1}^{q_1}(\overline{x}_{j_2}-a_{j_1j_2u})^{n_{j_1u}}}$$

$$=-\infty,$$

the flow at $x_{j_1}^*=a_{j_1j_1r_1}=a_{j_2j_1v_1}$ are a $[2(m_{j_1r_11}-m_{j_2v_11})]^{\text{th}}$-order decreasing-concave flows for $m_{j_1r_1}=2m_{j_1r_11}+1$, $m_{j_2v_1}=2m_{j_2v_11}+1$ and $m_{j_1r_1}=2m_{j_1r_11}$, $m_{j_2v_1}=2m_{j_2v_11}$).

For $x_{j_2}^*=a_{j_2j_21}=a_{j_1j_21}$ and $\overline{x}_{j_1}\neq a_{j_1j_11}$, if

$$\left.\frac{dx_{j_2}}{dx_{j_1}}\right|_{x_{j_1}^*=a_{j_1j_1r_1}\pm\varepsilon}$$

$$=\lim_{\varepsilon\to 0}\frac{a_{j_2j_20}}{a_{j_1j_10}}\frac{\prod_{v_2=1,v_2\neq v_1}^{p_2}(a_{j_2j_1v_1}-a_{j_2j_1v_2})^{m_{j_2v_2}}}{(\pm\varepsilon)^{m_{j_1r_1}-m_{j_1v_1}}\prod_{r_2=1,r_2\neq r_1}^{p_1}(a_{j_1j_1r_1}-a_{j_1j_1r_2})^{m_{j_1r_2}}\prod_{u=1}^{q_1}(\overline{x}_{j_2}-a_{j_1j_2u})^{n_{j_1u}}}$$

$$=\pm\infty,$$

the flow at $x_{j_1}^*=a_{j_1j_1r_1}=a_{j_2j_1v_1}$ are a $[2(m_{j_1r_11}-m_{j_2v_11})\pm 1]^{\text{th}}$-order up-asymptote flows for $m_{j_1r_1}=2m_{j_1r_11}$, $m_{j_2v_1}=2m_{j_2v_11}+1$ and $m_{j_1r_1}=2m_{j_1r_11}+1$, $m_{j_2v_1}=2m_{j_2v_11}$). If

$$\left.\frac{dx_{j_2}}{dx_{j_1}}\right|_{x_{j_1}^*=a_{j_1j_1r_1}\pm\varepsilon}$$

$$=\lim_{\varepsilon\to 0}\frac{a_{j_2j_20}}{a_{j_1j_10}}\frac{\prod_{v_2=1,v_2\neq v_1}^{q_2}(a_{j_2j_1v_1}-a_{j_2j_1v_2})^{m_{j_2s_2}}}{(\pm\varepsilon)^{m_{j_1r_1}-m_{j_1v_1}}\prod_{r_2=1,r_2\neq r_1}^{p_1}(a_{j_1j_1r_1}-a_{j_1j_1r_2})^{m_{j_1r_2}}\prod_{u=1}^{q_1}(\overline{x}_{j_2}-a_{j_1j_2u})^{n_{j_1u}}}$$

$$=\mp\infty,$$

the flow at $x_{j_1}^*=a_{j_1j_1r_1}=a_{j_2j_1v_1}$ is a $[2(m_{j_1r_11}-m_{j_2v_11})\pm 1]^{\text{th}}$-order down-asymptote flow for $m_{j_1r_1}=2m_{j_1r_11}$, $m_{j_2v_1}=2m_{j_2v_11}+1$ and $m_{j_1r_1}=2m_{j_1r_11}+1$, $m_{j_2v_1}=2m_{j_2v_11}$).

For $\overline{x}_{j_2}\neq a_{j_1j_2u}(u=1,2,\ldots,q_1)$, the variational equation at $x_{j_1}^*=a_{j_1j_1r_1}=a_{j_2j_1v_1}$ is

$$\Delta\dot{x}_{j_1}=a_{j_1j_10}\prod_{r_2=1,r_2\neq r_1}^{p_1}(a_{j_1j_1r_1}-a_{j_1j_1r_2})^{m_{j_1r_2}}\prod_{u=1}^{q_1}(\overline{x}_{j_2}-a_{j_1j_2u})^{n_{j_1u}}(\Delta x_{j_1})^{m_{j_1r_1}}.$$

For

$$a_{j_1j_10}\prod_{r_2=1,r_2\neq r_1}^{p_1}(a_{j_1j_1r_1}-a_{j_1j_1r_2})^{m_{j_1r_2}}\prod_{u=1}^{q_1}(\overline{x}_{j_2}-a_{j_1j_2u})^{n_{j_1u}}>0,\ \text{and}$$
$$a_{j_1j_10}\prod_{r_2=1,r_2\neq r_1}^{p_1}(a_{j_1j_1r_1}-a_{j_1j_1r_2})^{m_{j_1r_2}}\prod_{u=1}^{q_1}(\overline{x}_{j_2}-a_{j_1j_2u})^{n_{j_1u}}<0,$$

the flows at $x_{j_1}^*=a_{j_1j_1r_1}=a_{j_2j_1v_1}$ are $(2m_{j_1r_11}+1)^{\text{th}}$-order source and sink flows for $m_{j_1r_1}=2m_{j_1r_11}+1$, and are $(2m_{j_1r_11})^{\text{th}}$-order upper-saddle and lower-saddle flows for $m_{j_1r_1}=2m_{j_1r_11}$.

($\text{ii}_{2\text{b}1}$) Based on the above analysis, the flows at $x_{j_1}^*=a_{j_1j_11}=a_{j_2j_11}$ and on both sides of $x_{j_2}^*=a_{j_1j_21}$ are $((2m_{j_1r_11}+1)^{\text{th}}\text{XX}, [\cdot]^{\text{th}}\text{YY})$-source and sink flows for

$$a_{j_1j_10}\prod_{r_2=1,r_2\neq r_1}^{p_1}(a_{j_2j_1v_1}-a_{j_1j_1r_2})^{m_{j_1r_2}}\prod_{u=1}^{q_1}(\overline{x}_{j_2}-a_{j_1j_2u})^{n_{j_1u}}>0\ \text{and}$$
$$a_{j_1j_10}\prod_{r_2=1,r_2\neq r_1}^{p_1}(a_{j_2j_1v_1}-a_{j_1j_1r_2})^{m_{j_1r_2}}\prod_{u=1}^{q_1}(\overline{x}_{j_2}-a_{j_1j_2u})^{n_{j_1u}}<0,$$

respectively. The flows at $x_{j_1}^*=a_{j_1j_11}=a_{j_2j_11}$ and on both sides of $x_{j_2}^*=a_{j_1j_21}$ are $((2m_{11})^{\text{th}}\text{XX}, [\cdot]^{\text{th}}\text{YY})$-upper and lower-saddle flows for

$$a_{j_1j_10}\prod_{r_2=1,r_2\neq r_1}^{p_1}(a_{j_2j_1v_1}-a_{j_1j_1r_2})^{m_{j_1r_2}}\prod_{u=1}^{q_1}(\overline{x}_{j_2}-a_{j_1j_2u})^{n_{j_1u}}>0\ \text{and}$$
$$a_{j_1j_10}\prod_{r_2=1,r_2\neq r_1}^{p_1}(a_{j_2j_1v_1}-a_{j_1j_1r_2})^{m_{j_1r_2}}\prod_{u=1}^{q_1}(\overline{x}_{j_2}-a_{j_1j_2u})^{n_{j_1u}}<0,$$

respectively. $[2(m_{j_1r_11}-m_{j_2v_11})]^{\text{th}}\text{YY}$ are the $[2(m_{j_1r_11}-m_{j_2v_11})]^{\text{th}}$-order increasing and decreasing concave flows, and $[2(m_{j_1r_11}-m_{j_2v_11})\pm 1]^{\text{th}}\text{YY}$ are the $[2(m_{j_1r_11}-m_{j_2v_11})\pm 1]^{\text{th}}$-order down and up-asymptote flows.

Therefore, for $m_{j_1r_1}>m_{j_2v_1}$, there are 16 types of infinite-equilibriums at $x_{j_1}^*=a_{j_1j_11}=a_{j_2j_11}$.

($\text{ii}_{2\text{b}1\text{a}}$) For $m_{j_1r_1}=2m_{j_1r_11}+1$ and $m_{j_2v_1}=2m_{j_1v_11}+1$, the infinite-equilibrium of $x_{j_1}^*=a_{j_2j_1v_1}=a_{j_1j_1r_1}$ with $\overline{x}_{j_2}\neq a_{j_1j_2u_1}$ has the following properties as in Eqs. (1.231)–(1.234).

- For $a_{j_1j_10}\prod_{r_2=1,r_2\neq r_1}^{p_1}(a_{j_1j_1r_1}-a_{j_1j_1r_2})^{m_{j_1r_2}}\prod_{u=1}^{q_1}(\overline{x}_{j_2}-a_{j_1j_2u})^{n_{j_1u}}>0$ and

  $a_{j_2j_20}\prod_{v_2=1,v_2\neq v_1}^{p_2}(a_{j_2j_1v_1}-a_{j_2j_1v_2})^{m_{j_2v_2}}>0,$

$$(a_{j_1 j_1 r_1}, \overline{x}_{j_2}) = \underbrace{((2m_{j_1 r_1 1}+1)^{\text{th}}\text{SO},[2(m_{j_1 r_1 1}-m_{j_2 v_1 1})]^{\text{th}}\text{IC})}_{((2m_{j_1 r_1 1}+1),[2(m_{j_1 r_1 1}-m_{j_2 v_1 1})])\text{-increasing-concave source}}.$$

The infinite-equilibrium is a $((2m_{j_1 r_1 1}+1)^{\text{th}}\text{SO},[2(m_{j_1 r_1 1}-m_{j_2 v_1 1})]^{\text{th}}\text{IC})$-increasing-concave source at $x_{j_1}^* = a_{j_2 j_1 v_1} = a_{j_1 j_1 r_1}$.

- For $a_{j_1 j_1 0} \prod_{r_2=1, r_2\neq r_1}^{p_1} (a_{j_1 j_1 r_1} - a_{j_1 j_1 r_2})^{m_{j_1 r_2}} \prod_{u=1}^{q_1} (\overline{x}_{j_2} - a_{j_1 j_2 u})^{n_{j_1 u}} < 0$ and $a_{j_2 j_2 0} \prod_{v_2=1, v_2\neq v_1}^{p_2} (a_{j_2 j_1 v_1} - a_{j_2 j_1 v_2})^{m_{j_2 v_2}} > 0$,

$$(a_{j_1 j_1 r_1}, \overline{x}_{j_2}) = \underbrace{((2m_{j_1 r_1 1}+1)^{\text{th}}\text{SI},[2(m_{j_1 r_1 1}-m_{j_2 v_1 1})]^{\text{th}}\text{DC})}_{((2m_{j_1 r_1 1}+1),[2(m_{j_1 r_1 1}-m_{j_2 v_1 1})])\text{-decreasing-concave sink}}.$$

The infinite-equilibrium is a $((2m_{j_1 r_1 1}+1)^{\text{th}}\text{SI},[2(m_{j_1 r_1 1}-m_{j_2 v_1 1})]^{\text{th}}\text{DC})$-decreasing-concave sink at $x_{j_1}^* = a_{j_2 j_1 v_1} = a_{j_1 j_1 r_1}$.

- For $a_{j_1 j_1 0} \prod_{r_2=1, r_2\neq r_1}^{p_1} (a_{j_1 j_1 r_1} - a_{j_1 j_1 r_2})^{m_{j_1 r_2}} \prod_{u=1}^{q_1} (\overline{x}_{j_2} - a_{j_1 j_2 u})^{n_{j_1 u}} > 0$ and $a_{j_2 j_2 0} \prod_{v_2=1, v_2\neq v_1}^{p_2} (a_{j_2 j_1 v_1} - a_{j_2 j_1 v_2})^{m_{j_2 v_2}} < 0$,

$$(a_{j_1 j_1 r_1}, \overline{x}_{j_2}) = \underbrace{((2m_{j_1 r_1 1}+1)^{\text{th}}\text{SO},[2(m_{j_1 r_1 1}-m_{j_2 v_1 1})]^{\text{th}}\text{DC})}_{((2m_{j_1 r_1 1}+1),[2(m_{j_1 r_1 1}-m_{j_2 v_1 1})])\text{-decreasing-concave source}}.$$

The infinite-equilibrium is a $((2m_{j_1 r_1 1}+1)^{\text{th}}\text{SO},[2(m_{j_1 r_1 1}-m_{j_2 v_1 1})]^{\text{th}}\text{DC})$-decreasing-concave sink at $x_{j_1}^* = a_{j_2 j_1 v_1} = a_{j_1 j_1 r_1}$.

- For $a_{j_1 j_1 0} \prod_{r_2=1, r_2\neq r_1}^{p_1} (a_{j_1 j_1 r_1} - a_{j_1 j_1 r_2})^{m_{j_1 r_2}} \prod_{u=1}^{q_1} (\overline{x}_{j_2} - a_{j_1 j_2 u})^{n_{j_1 u}} < 0$ and $a_{j_2 j_2 0} \prod_{v_2=1, v_2\neq v_1}^{p_2} (a_{j_2 j_1 v_1} - a_{j_2 j_1 v_2})^{m_{j_2 v_2}} < 0$,

$$(a_{j_1 j_1 r_1}, \overline{x}_{j_2}) = \underbrace{((2m_{j_1 r_1 1}+1)^{\text{th}}\text{SI},[2(m_{j_1 r_1 1}-m_{j_2 v_1 1})]^{\text{th}}\text{IC})}_{((2m_{j_1 r_1 1}+1),[2(m_{j_1 r_1 1}-m_{j_2 v_1 1})])\text{-increasing-concave sink}}.$$

The infinite-equilibrium is a $((2m_{j_1 r_1 1}+1)^{\text{th}}\text{SI},[2(m_{j_1 r_1 1}-m_{j_2 v_1 1})]^{\text{th}}\text{IC})$-increasing-concave sink at $x_{j_1}^* = a_{j_2 j_1 v_1} = a_{j_1 j_1 r_1}$.

(ii_{2b1b}) For $m_{j_1 r_1} = 2m_{j_1 r_1 1}$ and $m_{j_2 v_1} = 2m_{j_1 v_1 1}+1$, the infinite-equilibrium of $x_{j_1}^* = a_{j_2 j_1 v_1} = a_{j_1 j_1 r_1}$ with $\overline{x}_{j_2} \neq a_{j_1 j_2 u_1}$ has the following properties as in Eqs. (1.235)–(1.238).

- For $a_{j_1j_10}\prod_{r_2=1,r_2\neq r_1}^{p_1}(a_{j_1j_1r_1}-a_{j_1j_1r_2})^{m_{j_1r_2}}\prod_{u=1}^{q_1}(\overline{x}_{j_2}-a_{j_1j_2u})^{n_{j_1u}}>0$ and $a_{j_2j_20}\prod_{v_2=1,v_2\neq v_1}^{p_2}(a_{j_2j_1v_1}-a_{j_2j_1v_2})^{m_{j_2v_2}}>0$,

$$(a_{j_1j_1r_1},\overline{x}_{j_2})=\underbrace{((2m_{j_1r_1}1)^{\text{th}}\text{US},[2(m_{j_1r_1}1-m_{j_2v_1}1)-1]^{\text{th}}\text{DA})}_{((2m_{j_1r_1}1),[2(m_{j_1r_1}1-m_{j_2v_1}1)-1])\text{-down-asymptote upper-saddle}}.$$

The infinite-equilibrium is a $((2m_{j_1r_1}1)^{\text{th}}\text{US},[2(m_{j_1r_1}1-m_{j_2v_1}1)-1]^{\text{th}}\text{DA})$-down-asymptote upper-saddle at $x^*_{j_1}=a_{j_2j_1v_1}=a_{j_1j_1r_1}$.

- For $a_{j_1j_10}\prod_{r_2=1,r_2\neq r_1}^{p_1}(a_{j_1j_1r_1}-a_{j_1j_1r_2})^{m_{j_1r_2}}\prod_{u=1}^{q_1}(\overline{x}_{j_2}-a_{j_1j_2u})^{n_{j_1u}}<0$ and $a_{j_2j_20}\prod_{v_2=1,v_2\neq v_1}^{p_2}(a_{j_2j_1v_1}-a_{j_2j_1v_2})^{m_{j_2v_2}}>0$,

$$(a_{j_1j_1r_1},\overline{x}_{j_2})=\underbrace{((2m_{j_1r_1}1)^{\text{th}}\text{LS},[2(m_{j_1r_1}1-m_{j_2v_1}1)-1]^{\text{th}}\text{UA})}_{((2m_{j_1r_1}1),[2(m_{j_1r_1}1-m_{j_2v_1}1)-1])\text{-up-asymptote lower-saddle}}.$$

The infinite-equilibrium is a $((2m_{j_1r_1}1)^{\text{th}}\text{LS},[2(m_{j_1r_1}1-m_{j_2v_1}1)-1]^{\text{th}}\text{UA})$-up-asymptote lower-saddle at $x^*_{j_1}=a_{j_2j_1v_1}=a_{j_1j_1r_1}$.

- For $a_{j_1j_10}\prod_{r_2=1,r_2\neq r_1}^{p_1}(a_{j_1j_1r_1}-a_{j_1j_1r_2})^{m_{j_1r_2}}\prod_{u=1}^{q_1}(\overline{x}_{j_2}-a_{j_1j_2u})^{n_{j_1u}}>0$ and $a_{j_2j_20}\prod_{v_2=1,v_2\neq v_1}^{p_2}(a_{j_2j_1v_1}-a_{j_2j_1v_2})^{m_{j_2v_2}}<0$,

$$(a_{j_1j_1r_1},\overline{x}_{j_2})=\underbrace{((2m_{j_1r_1}1)^{\text{th}}\text{US},[2(m_{j_1r_1}1-m_{j_2v_1}1)-1]^{\text{th}}\text{UA})}_{((2m_{j_1r_1}1),[2(m_{j_1r_1}1-m_{j_2v_1}1)-1])\text{-up-asymptote upper-saddle}}.$$

The infinite-equilibrium is a $((2m_{j_1r_1}1)^{\text{th}}\text{US},[2(m_{j_1r_1}1-m_{j_2v_1}1)-1]^{\text{th}}\text{UA})$-up-asymptote upper-saddle at $x^*_{j_1}=a_{j_2j_1v_1}=a_{j_1j_1r_1}$.

- For $a_{j_1j_10}\prod_{r_2=1,r_2\neq r_1}^{p_1}(a_{j_1j_1r_1}-a_{j_1j_1r_2})^{m_{j_1r_2}}\prod_{u=1}^{q_1}(\overline{x}_{j_2}-a_{j_1j_2u})^{n_{j_1u}}<0$ and $a_{j_2j_20}\prod_{v_2=1,v_2\neq v_1}^{p_2}(a_{j_2j_1v_1}-a_{j_2j_1v_2})^{m_{j_2v_2}}<0$,

$$(a_{j_1j_1r_1},\overline{x}_{j_2})=\underbrace{((2m_{j_1r_1}1)^{\text{th}}\text{LS},[2(m_{j_1r_1}1-m_{j_2v_1}1)-1]^{\text{th}}\text{DA})}_{((2m_{j_1r_1}1),[2(m_{j_1r_1}1-m_{j_2v_1}1)-1])\text{-down-asymptote lower-saddle}}.$$

The infinite-equilibrium is a $((2m_{j_1r_1 1})^{\text{th}}\text{LS},[2(m_{j_1r_1}1 - m_{j_2v_1}1)-1]^{\text{th}}\text{DA})$-down-asymptote lower-saddle at $x_{j_1}^* = a_{j_2j_1v_1} = a_{j_1j_1r_1}$.

($\text{ii}_{2\text{b}1\text{c}}$) For $m_{j_1r_1} = 2m_{j_1r_1}1 + 1$ and $m_{j_2v_1} = 2m_{j_1v_1}1$, the infinite-equilibrium of $x_{j_1}^* = a_{j_2j_1v_1} = a_{j_1j_1r_1}$ with $\overline{x}_{j_2} \neq a_{j_1j_2u_1}$ has the following properties as in Eqs. (1.239)–(1.242).

- For $a_{j_1j_10} \prod_{r_2=1,r_2\neq r_1}^{p_1} (a_{j_1j_1r_1} - a_{j_1j_1r_2})^{m_{j_1r_2}} \prod_{u=1}^{q_1} (\overline{x}_{j_2} - a_{j_1j_2u})^{n_{j_1u}} > 0$ and $a_{j_2j_20} \prod_{v_2=1,v_2\neq v_1}^{p_2} (a_{j_2j_1v_1} - a_{j_2j_1v_2})^{m_{j_2v_2}} > 0$,

$$(a_{j_1j_1r_1}, \overline{x}_{j_2}) = \underbrace{((2m_{j_1r_1}1 + 1)^{\text{th}}\text{SO},[2(m_{j_1r_1}1 - m_{j_2v_1}1) + 1]^{\text{th}}\text{DA})}_{((2m_{j_1r_1}1+1),[2(m_{j_1r_1}1-m_{j_2v_1}1)+1])\text{-down-asymptote source}}.$$

The infinite-equilibrium is a $((2m_{j_1r_1}1 + 1)^{\text{th}}\text{SO},[2(m_{j_1r_1}1 - m_{j_2v_1}1)+1]^{\text{th}}$ DA)-up-parabola source at $x_{j_1}^* = a_{j_2j_1v_1} = a_{j_1j_1r_1}$.

- For $a_{j_1j_10} \prod_{r_2=1,r_2\neq r_1}^{p_1} (a_{j_1j_1r_1} - a_{j_1j_1r_2})^{m_{j_1r_2}} \prod_{u=1}^{q_1} (\overline{x}_{j_2} - a_{j_1j_2u})^{n_{j_1u}} < 0$ and $a_{j_2j_20} \prod_{v_2=1,v_2\neq v_1}^{p_2} (a_{j_2j_1v_1} - a_{j_2j_1v_2})^{m_{j_2v_2}} > 0$,

$$(a_{j_1j_1r_1}, \overline{x}_{j_2}) = \underbrace{((2m_{j_1r_1}1 + 1)^{\text{th}}\text{SI},[2(m_{j_1r_1}1 - m_{j_2v_1}1) + 1]^{\text{th}}\text{UA})}_{((2m_{j_1r_1}1+1),[2(m_{j_1r_1}1-m_{j_2v_1}1)+1])\text{-up-asymptote sink}}.$$

The infinite-equilibrium is a $((2m_{j_1r_1}1 + 1)^{\text{th}}\text{SI},[2(m_{j_1r_1}1 - m_{j_2v_1}1)+1]^{\text{th}}$ UA)-up-asymptote sink at $x_{j_1}^* = a_{j_2j_1v_1} = a_{j_1j_1r_1}$.

- For $a_{j_1j_10} \prod_{r_2=1,r_2\neq r_1}^{p_1} (a_{j_1j_1r_1} - a_{j_1j_1r_2})^{m_{j_1r_2}} \prod_{u=1}^{q_1} (\overline{x}_{j_2} - a_{j_1j_2u})^{n_{j_1u}} > 0$ and $a_{j_2j_20} \prod_{v_2=1,v_2\neq v_1}^{p_2} (a_{j_2j_1v_1} - a_{j_2j_1v_2})^{m_{j_2v_2}} < 0$,

$$(a_{j_1j_1r_1}, \overline{x}_{j_2}) = \underbrace{((2m_{j_1r_1}1 + 1)^{\text{th}}\text{SO},[2(m_{j_1r_1}1 - m_{j_2v_1}1) + 1]^{\text{th}}\text{UA})}_{((2m_{j_1r_1}1+1),[2(m_{j_1r_1}1-m_{j_2v_1}1)+1])\text{-up-asymptote source}}.$$

The infinite-equilibrium is a $((2m_{j_1r_1}1 + 1)^{\text{th}}\text{SO},[2(m_{j_1r_1}1 - m_{j_2v_1}1)+1]^{\text{th}}$ UA)- up-asymptote source at $x_{j_1}^* = a_{j_2j_1v_1} = a_{j_1j_1r_1}$.

- For $a_{j_1j_10} \prod_{r_2=1,r_2\neq r_1}^{p_1} (a_{j_1j_1r_1} - a_{j_1j_1r_2})^{m_{j_1r_2}} \prod_{u=1}^{q_1} (\overline{x}_{j_2} - a_{j_1j_2u})^{n_{j_1u}} < 0$ and $a_{j_2j_20} \prod_{v_2=1,v_2\neq v_1}^{p_2} (a_{j_2j_1v_1} - a_{j_2j_1v_2})^{m_{j_2v_2}} < 0$,

$$(a_{j_1j_1r_1}, \bar{x}_{j_2}) = \underbrace{((2m_{j_1r_1}1+1)^{\text{th}}\text{SI},[2(m_{j_1r_1}1-m_{j_2v_1}1)+1]^{\text{th}}\text{DA})}_{((2m_{j_1r_1}1+1),[2(m_{j_1r_1}1-m_{j_2v_1}1)+1])\text{-down-asymptote sink}}.$$

The infinite-equilibrium is a $((2m_{j_1r_1}1+1)^{\text{th}}\text{SI},[2(m_{j_1r_1}1-m_{j_2v_1}1)+1]^{\text{th}}$ DA)-up-parabola sink at $x_{j_1}^* = a_{j_2j_1v_1} = a_{j_1j_1r_1}$.

($\text{ii}_{2\text{b}1\text{d}}$) For $m_{j_1r_1} = 2m_{j_1r_1}1$ and $m_{j_2v_1} = 2m_{j_1v_1}1$, the infinite-equilibrium of $x_{j_1}^* = a_{j_2j_1v_1} = a_{j_1j_1r_1}$ with $\bar{x}_{j_2} \neq a_{j_1j_2u_1}$ has the following properties as in Eqs. (1.243)–(1.246).

- For $a_{j_1j_10}\prod_{r_2=1,r_2\neq r_1}^{p_1}(a_{j_1j_1r_1}-a_{j_1j_1r_2})^{m_{j_1r_2}}\prod_{u=1}^{q_1}(\bar{x}_{j_2}-a_{j_1j_2u})^{n_{j_1u}} > 0$ and $a_{j_2j_20}\prod_{v_2=1,v_2\neq v_1}^{p_2}(a_{j_2j_1v_1}-a_{j_2j_1v_2})^{m_{j_2v_2}} > 0$,

$$(a_{j_1j_1r_1}, \bar{x}_{j_2}) = \underbrace{((2m_{j_1r_1}1)^{\text{th}}\text{US},[2(m_{j_1r_1}1-m_{j_2v_1}1)]^{\text{th}}\text{IC})}_{((2m_{j_1r_1}1),[2(m_{j_1r_1}1-m_{j_2v_1}1)])\text{-increasing-concave upper-saddle}}.$$

The infinite-equilibrium is a $((2m_{j_1r_1}1)^{\text{th}}\text{US},[2(m_{j_1r_1}1-m_{j_2v_1}1)]^{\text{th}}\text{IC})$-increasing-concave upper-saddle at $x_{j_1}^* = a_{j_2j_1v_1} = a_{j_1j_1r_1}$.

- For $a_{j_1j_10}\prod_{r_2=1,r_2\neq r_1}^{p_1}(a_{j_1j_1r_1}-a_{j_1j_1r_2})^{m_{j_1r_2}}\prod_{u=1}^{q_1}(\bar{x}_{j_2}-a_{j_1j_2u})^{n_{j_1u}} < 0$ and $a_{j_2j_20}\prod_{v_2=1,v_2\neq v_1}^{p_2}(a_{j_2j_1v_1}-a_{j_2j_1v_2})^{m_{j_2v_2}} > 0$,

$$(a_{j_1j_1r_1}, \bar{x}_{j_2}) = \underbrace{((2m_{j_1r_1}1)^{\text{th}}\text{LS},[2(m_{j_1r_1}1-m_{j_2v_1}1)]^{\text{th}}\text{DC})}_{((2m_{j_1r_1}1),[2(m_{j_1r_1}1-m_{j_2v_1}1)])-\textit{decreasing}\text{-concave lower-saddle}}.$$

The infinite-equilibrium is a $((2m_{j_1r_1}1)^{\text{th}}\text{LS},[2(m_{j_1r_1}1-m_{j_2v_1}1)]^{\text{th}}\text{DC})$-decreasing-concave lower-saddle at $x_{j_1}^* = a_{j_2j_1v_1} = a_{j_1j_1r_1}$.

- For $a_{j_1j_10}\prod_{r_2=1,r_2\neq r_1}^{p_1}(a_{j_1j_1r_1}-a_{j_1j_1r_2})^{m_{j_1r_2}}\prod_{u=1}^{q_1}(\bar{x}_{j_2}-a_{j_1j_2u})^{n_{j_1u}} > 0$ and $a_{j_2j_20}\prod_{v_2=1,v_2\neq v_1}^{p_2}(a_{j_2j_1v_1}-a_{j_2j_1v_2})^{m_{j_2v_2}} < 0$,

$$(a_{j_1j_1r_1}, \bar{x}_{j_2}) = \underbrace{((2m_{j_1r_1}1)^{\text{th}}\text{US},[2(m_{j_1r_1}1-m_{j_2v_1}1)]^{\text{th}}\text{DC})}_{((2m_{j_1r_1}1),[2(m_{j_1r_1}1-m_{j_2v_1}1)])-\textit{decreasing}\text{-concave upper-saddle}}.$$

The infinite-equilibrium is a $((2m_{j_1r_1}1)^{\text{th}}\text{US},[2(m_{j_1r_1}1-m_{j_2v_1}1)]^{\text{th}}\text{DC})$-decreasing-concave upper-saddle at $x_{j_1}^* = a_{j_2j_1v_1} = a_{j_1j_1r_1}$.

- For $a_{j_1j_10}\prod_{r_2=1,r_2\neq r_1}^{p_1}(a_{j_1j_1r_1}-a_{j_1j_1r_2})^{m_{j_1r_2}}\prod_{u=1}^{q_1}(\overline{x}_{j_2}-a_{j_1j_2u})^{n_{j_1u}}<0$ and $a_{j_2j_20}\prod_{v_2=1,v_2\neq v_1}^{p_2}(a_{j_2j_1v_1}-a_{j_2j_1v_2})^{m_{j_2v_2}}<0$,

$$(a_{j_1j_1r_1},\overline{x}_{j_2})=\underbrace{((2m_{j_1r_11})^{\text{th}}\text{LS},[2(m_{j_1r_11}-m_{j_2v_11})]^{\text{th}}\text{IC})}_{((2m_{j_1r_11}),[2(m_{j_1r_11}-m_{j_2v_11})])\text{-increasing-concave lower-saddle}}.$$

The infinite-equilibrium is a $((2m_{j_1r_11})^{\text{th}}\text{LS},[2(m_{j_1r_11}-m_{j_2v_11})]^{\text{th}}\text{IC})$-increasing-concave lower-saddle at $x_{j_1}^*=a_{j_2j_1v_1}=a_{j_1j_1r_1}$.

(ii$_{2b2}$) At $x_{j_2}^*=a_{j_1j_2u_1}$ and $\overline{x}_{j_1}\neq a_{j_1j_1r},a_{j_2j_1v}(r=1,2,\ldots,p_1;v=1,2,\ldots,p_2)$, in phase space,

$$\frac{dx_{j_1}}{dx_{j_2}}\bigg|_{x_{j_2}^*=a_{j_1j_2u_1}}=\frac{a_{j_1j_10}}{a_{j_2j_20}}\frac{\prod_{r=1}^{p_1}(\overline{x}_{j_1}-a_{j_1j_1r})^{m_{j_1r}}\prod_{u_2=1,u_2\neq u_1}^{q_1}(x_{j_2}-a_{j_1j_2u_2})^{n_{j_1u_2}}}{\prod_{v=1}^{p_2}(\overline{x}_{j_1}-a_{j_2j_1v})^{m_{j_2v}}}\times(x_{j_2}-a_{j_1j_2u_1})^{n_{j_1u_1}}\bigg|_{x_{j_2}^*=a_{j_1j_2u_1}}=0,$$

$$\vdots$$

$$\frac{d^{n_{j_1u_1}}x_{j_1}}{dx_{j_2}^{n_{j_1u_1}}}\bigg|_{x_{j_2}^*=a_{j_1j_2u_1}}=0.$$

If

$$\frac{d^{n_{j_1u_1}+1}x_{j_1}}{dx_{j_2}^{n_{j_1u_1}+!}}\bigg|_{x_{j_2}^*=a_{j_1j_2u_1}}=\frac{a_{j_1j_10}}{a_{j_2j_20}}\frac{\prod_{r=1}^{p_1}(\overline{x}_{j_1}-a_{j_1j_1r})^{m_{j_1r}}\prod_{u_2=1,u_2\neq u_1}^{q_1}(a_{j_1j_2u_1}-a_{j_1j_2u_2})^{n_{j_1u_2}}n_{j_1u_1}!}{\prod_{v=1}^{p_2}(\overline{x}_{j_1}-a_{j_2j_1v})^{m_{j_2v}}}>0,$$

the $(2n_{j_1u_11}+1)^{\text{th}}$-order up-parabola flows exists for $n_{j_1u_1}=2n_{j_1u_11}+1$, and the $(2n_{j_1u_11})^{\text{th}}$-order increasing-inflection flow exists for $n_{j_1u_1}=2n_{j_1u_11}$. If

$$\left.\frac{d^{n_{j_1u_1}+1}x_{j_1}}{dx_{j_2}^{n_{j_1u_1}+1}}\right|_{x_{j_2}^*=a_{j_1j_2u_1}} = \frac{a_{j_1j_10}}{a_{j_2j_20}}\frac{\prod_{r=1}^{p_1}(\overline{x}_{j_1}-a_{j_1j_1r})^{m_{j_1r}}\prod_{u_2=1,u_2\neq u_1}^{q_1}(a_{j_1j_2u_1}-a_{j_1j_2u_2})^{n_{j_1u_2}}n_{j_1u_1}!}{\prod_{v=1}^{p_2}(\overline{x}_{j_1}-a_{j_2j_1v})^{m_{j_2v}}} < 0,$$

the $(2n_{j_1u_1 1}+1)^{\text{th}}$-order down-parabola flows exists for $n_{j_1u_1}=2n_{j_1u_1 1}+1$, and the $(2n_{j_1u_1 1})^{\text{th}}$-order decreasing-inflection flow exists for $n_{j_1u_1}=2n_{j_1u_1 1}$.

At $\overline{x}_{j_1}\neq a_{j_1j_1r}, a_{j_2j_1v} (r=1,2,\ldots,p_1; v=1,2,\ldots,p_2)$, let

$$\dot{x}_{j_2}=a_{j_2j_20}\prod_{v=1}^{p_2}(\overline{x}_{j_1}-a_{j_2j_1v})^{m_{j_2v}}.$$

If $a_{j_2j_20}\prod_{v=1}^{p_2}(\overline{x}_{j_1}-a_{j_2j_1v})^{m_{j_2v}}>0$ and $a_{j_2j_20}\prod_{v=1}^{p_2}(\overline{x}_{j_1}-a_{j_2j_1v})^{m_{j_2v}}<0$, the parabola and inflection flows are positive and negative in the x_{j_2}-direction, respectively.

Therefore, from (ii$_{2b1}$), there are 32 types of equilibrium $(x_{j_2}^*, x_{j_1}^*)=(a_{j_1j_2u_1}, a_{j_2j_1v_1})$ on the infinite-equilibrium of $x_{j_1}^*=a_{j_2j_1v_1}=a_{j_1j_1r_1}$.

(ii$_{2b2a}$) For $m_{j_1r_1}=2m_{j_1r_1 1}+1, n_{j_1u_1}=2n_{j_1u_1 1}+1$, and $m_{j_2v_1}=2m_{j_1v_1 1}+1$, the equilibrium $(x_{j_2}^*, x_{j_1}^*)=(a_{j_1j_2u_1}, a_{j_2j_1v_1})$ on the infinite-equilibrium of $x_{j_1}^*=a_{j_2j_1v_1}=a_{j_1j_1r_1}$ has the following properties as in Eqs. (1.247)–(1.250).

- For $a_{j_1j_10}\prod_{r_2=1,r_2\neq r_1}^{p_1}(a_{j_1j_1r_1}-a_{j_1j_1r_2})^{m_{j_1r_2}}\prod_{u_2=1,u_2\neq u_1}^{q_1}(a_{j_1j_2u_1}-a_{j_1j_2u_2})^{n_{j_1u_2}}>0$ and $a_{j_2j_20}\prod_{v_2=1,v_2\neq v_1}^{p_2}(a_{j_2j_1v_1}-a_{j_2j_1v_2})^{m_{j_2v_2}}>0,$

$$(a_{j_1j_2u_1}, a_{j_2j_1v_1}) = \underbrace{([(2m_{j_1r_1 1}+1),(2n_{j_1u_1 1}+1)]^{\text{th}}\text{UP}_-{:}\text{UP}_+,[2(m_{j_1r_1 1}-m_{j_2v_1 1})]^{\text{th}}{}_{\text{DC:IC}}\text{US})}_{([(2m_{j_1r_1 1}+1),(2n_{j_1u_1 1}+1)],[2(m_{j_1r_1 1}-m_{j_2v_1 1})])\text{-concave up-parabola upper-saddle}}.$$

  The equilibrium is a $([(2m_{j_1r_1 1}+1),(2n_{j_1u_1 1}+1)]^{\text{th}}\text{UP}_-{:}\text{UP}_+,[2(m_{j_1r_1 1}-m_{j_2v_1 1})]^{\text{th}}{}_{\text{DC:IC}}\text{US})$-concave up-parabola upper-saddle at $(x_{j_2}^*, x_{j_1}^*)=(a_{j_1j_2u_1}, a_{j_2j_1v_1})$.

- For $a_{j_1j_10}\prod_{r_2=1,r_2\neq r_1}^{p_1}(a_{j_1j_1r_1}-a_{j_1j_1r_2})^{m_{j_1r_2}}\prod_{u_2=1,u_2\neq u_1}^{q_1}(a_{j_1j_2u_1}-a_{j_1j_2u_2})^{n_{j_1u_2}}<0$ and $a_{j_2j_20}\prod_{v_2=1,v_2\neq v_1}^{p_2}(a_{j_2j_1v_1}-a_{j_2j_1v_2})^{m_{j_2v_2}}>0,$

$$(a_{j_1j_2u_1}, a_{j_2j_1v_1}) = \underbrace{([(2m_{j_1r_1 1}+1),(2n_{j_1u_1 1}+1)]^{\text{th}}\text{DP}_-{:}\text{DP}_+,[2(m_{j_1r_1 1}-m_{j_2v_1 1})]^{\text{th}}{}_{\text{IC:DC}}\text{LS})}_{([(2m_{j_1r_1 1}+1),(2n_{j_1u_1 1}+1)],[2(m_{j_1r_1 1}-m_{j_2v_1 1})])\text{-concave down-parabola lower-saddle}}.$$

The equilibrium is a $([(2m_{j_1r_1 1}+1),(2n_{j_1u_1 1}+1)]^{\text{th}}\text{UP}_{-}{:}\text{UP}_{+},[2(m_{j_1r_1 1}-m_{j_2v_1 1})]^{\text{th}}{}_{\text{IC:DC}}\text{LS})$-concave down-parabola lower-saddle at $(x_{j_2}^*,x_{j_1}^*)=(a_{j_1j_2u_1},a_{j_2j_1v_1})$.

- For $a_{j_1j_10}\prod_{r_2=1,r_2\neq r_1}^{p_1}(a_{j_1j_1r_1}-a_{j_1j_1r_2})^{m_{j_1r_2}}\prod_{u_2=1,u_2\neq u_1}^{q_1}(a_{j_1j_2u_1}-a_{j_1j_2u_2})^{n_{j_1u_2}}>0$ and $a_{j_2j_20}\prod_{v_2=1,v_2\neq v_1}^{p_2}(a_{j_2j_1v_1}-a_{j_2j_1v_2})^{m_{j_2v_2}}<0$,

$$(a_{j_1j_2u_1},a_{j_2j_1v_1})=\underbrace{([(2m_{j_1r_1 1}+1),(2n_{j_1u_1 1}+1)]^{\text{th}}\text{DP}_{+}{:}\text{DP}_{-},[2(m_{j_1r_1 1}-m_{j_2v_1 1})]^{\text{th}}{}_{\text{IC:DC}}\text{US})}_{([(2m_{j_1r_1 1}+1),(2n_{j_1u_1 1}+1)],[2(m_{j_1r_1 1}-m_{j_2v_1 1})])\text{-concave down-parabola upper-saddle}}.$$

The equilibrium is a $([(2m_{j_1r_1 1}+1),(2n_{j_1u_1 1}+1)]^{\text{th}}\text{DP}_{+}{:}\text{DP}_{-},[2(m_{j_1r_1 1}-m_{j_2v_1 1})]^{\text{th}}{}_{\text{IC:DC}}\text{US})$-concave down-parabola upper-saddle at $(x_{j_2}^*,x_{j_1}^*)=(a_{j_1j_2u_1},a_{j_2j_1v_1})$.

- For $a_{j_1j_10}\prod_{r_2=1,r_2\neq r_1}^{p_1}(a_{j_1j_1r_1}-a_{j_1j_1r_2})^{m_{j_1r_2}}\prod_{u_2=1,u_2\neq u_1}^{q_1}(a_{j_1j_2u_1}-a_{j_1j_2u_2})^{n_{j_1u_2}}<0$ and $a_{j_2j_20}\prod_{v_2=1,v_2\neq v_1}^{p_2}(a_{j_2j_1v_1}-a_{j_2j_1v_2})^{m_{j_2v_2}}<0$,

$$(a_{j_1j_2u_1},a_{j_2j_1v_1})=\underbrace{([(2m_{j_1r_1 1}+1),(2n_{j_1u_1 1}+1)]^{\text{th}}\text{UP}_{+}{:}\text{UP}_{-},[2(m_{j_1r_1 1}-m_{j_2v_1 1})]^{\text{th}}{}_{\text{DC:IC}}\text{LS})}_{([(2m_{j_1r_1 1}+1),(2n_{j_1u_1 1}+1)],[2(m_{j_1r_1 1}-m_{j_2v_1 1})])\text{-concave up-paraobla upper-saddle}}.$$

The equilibrium is a $([(2m_{j_1r_1 1}+1),(2n_{j_1u_1 1}+1)]^{\text{th}}\text{UP}_{+}{:}\text{UP}_{-},[2(m_{j_1r_1 1}-m_{j_2v_1 1})]^{\text{th}}{}_{\text{DC:IC}}\text{LS})$-concave up-parabola lower-saddle at $(x_{j_1}^*,x_{j_2}^*)=(a_{j_1j_1r_1},a_{j_2j_2s_1})$.

(ii_{2b2b}) For $m_{j_1r_1}=2m_{j_1r_1 1}$, $n_{j_1u_1}=2n_{j_1u_1 1}+1$, and $m_{j_2v_1}=2m_{j_1v_1 1}+1$, the equilibrium $(x_{j_1}^*,x_{j_2}^*)=(a_{j_1j_1r_1},a_{j_1j_2u_1})$ on the infinite-equilibrium of $x_{j_1}^*=a_{j_2j_1v_1}=a_{j_1j_1r_1}$ has the following properties as in Eqs. (1.251)–(1.254).

- For $a_{j_1j_10}\prod_{r_2=1,r_2\neq r_1}^{p_1}(a_{j_1j_1r_1}-a_{j_1j_1r_2})^{m_{j_1r_2}}\prod_{u_2=1,u_2\neq u_1}^{q_1}(a_{j_1j_2u_1}-a_{j_1j_2u_2})^{n_{j_1u_2}}>0$ and $a_{j_2j_20}\prod_{v_2=1,v_2\neq v_1}^{p_2}(a_{j_2j_1v_1}-a_{j_2j_1v_2})^{m_{j_2v_2}}>0$,

$$(a_{j_1j_1r_1},a_{j_1j_2u_1})=\underbrace{([(2m_{j_1r_1 1}),(2n_{j_1u_1 1}+1)]^{\text{th}}\text{DP}_{-}{:}\text{UP}_{+},[2(m_{j_1r_1 1}-m_{j_2v_1 1})-1]^{\text{th}}{}_{\text{UA}}\text{LS}{:}{}_{\text{DA}}\text{US})}_{([(2m_{j_1r_1 1}),(2n_{j_1u_1 1}+1)],[2(m_{j_1r_1 1}-m_{j_2v_1 1})]-1)-\textit{hyperbolic}\text{ asymptote lower-to-upper saddle}}.$$

The equilibrium is a $([(2m_{j_1r_1 1}),(2n_{j_1u_1 1}+1)]^{\text{th}}\text{DP}_{-}{:}\text{UP}_{+},[2(m_{j_1r_1 1}-m_{j_2v_1 1})-1]^{\text{th}}{}_{\text{UA}}\text{LS}{:}{}_{\text{DA}}\text{US})$-hyperbolic asymptote lower-to-upper saddle at $(x_{j_1}^*,x_{j_2}^*)=(a_{j_1j_1r_1},a_{j_1j_2u_1})$.

- For $a_{j_1j_10}\prod_{r_2=1,r_2\neq r_1}^{p_1}(a_{j_1j_1r_1}-a_{j_1j_1r_2})^{m_{j_1r_2}}\prod_{u_2=1,u_2\neq u_1}^{q_1}(a_{j_1j_2u_1}-a_{j_1j_2u_2})^{n_{j_1u_2}}<0$ and $a_{j_2j_20}\prod_{v_2=1,v_2\neq v_1}^{p_2}(a_{j_2j_1v_1}-a_{j_2j_1v_2})^{m_{j_2v_2}}>0$,

$$(a_{j_1j_1r_1},a_{j_1j_2u_1}) = \underbrace{([(2m_{j_1r_1}1),(2n_{j_1u_1}1+1)]^{\text{th}}\text{UP}_-\text{:DP}_+,[2(m_{j_1r_1}1-m_{j_2v_1}1)-1]^{\text{th}}{}_{\text{DA}}\text{US:}_{\text{UA}}\text{LS})}_{([(2m_{j_1r_1}1),(2n_{j_1u_1}1+1)],[2(m_{j_1r_1}1-m_{j_2v_1}1)]-1)\text{-hyperbolic-secant asymptote upper-to-lower saddle}}.$$

The equilibrium is a $([(2m_{j_1r_1}1),(2n_{j_1u_1}1+1)]^{\text{th}}\text{UP}_-\text{:DP}_+,[2(m_{j_1r_1}1-m_{j_2v_1}1)-1]^{\text{th}}{}_{\text{DA}}\text{US:}_{\text{UA}}\text{LS})$-hyperbolic-secant asymptote upper-to-lower saddle at $(x_{j_1}^*,x_{j_2}^*)=(a_{j_1j_1r_1},a_{j_1j_2u_1})$.

- For $a_{j_1j_10}\prod_{r_2=1,r_2\neq r_1}^{p_1}(a_{j_1j_1r_1}-a_{j_1j_1r_2})^{m_{j_1r_2}}\prod_{u_2=1,u_2\neq u_1}^{q_1}(a_{j_1j_2u_1}-a_{j_1j_2u_2})^{n_{j_1u_2}}>0$ and $a_{j_2j_20}\prod_{v_2=1,v_2\neq v_1}^{p_2}(a_{j_2j_1v_1}-a_{j_2j_1v_2})^{m_{j_2v_2}}<0$,

$$(a_{j_1j_1r_1},a_{j_1j_2u_1}) = \underbrace{([(2m_{j_1r_1}1),(2n_{j_1u_1}1+1)]^{\text{th}}\text{UP}_+\text{:DP}_-,[2(m_{j_1r_1}1-m_{j_2v_1}1)-1]^{\text{th}}{}_{\text{DA}}\text{LS:}_{\text{UA}}\text{US})}_{([(2m_{j_1r_1}1),(2n_{j_1u_1}1+1)],[2(m_{j_1r_1}1-m_{j_2v_1}1)]-1)\text{-hyperbolic-secant asymptote lower-to-upper saddle}}.$$

The equilibrium is a $([(2m_{j_1r_1}1),(2n_{j_1u_1}1+1)]^{\text{th}}\text{UP}_+\text{:DP}_-,[2(m_{j_1r_1}1-m_{j_2v_1}1)-1]^{\text{th}}{}_{\text{DA}}\text{LS:}_{\text{UA}}\text{US})$-hyperbolic-secant asymptote upper-to-lower saddle at $(x_{j_1}^*,x_{j_2}^*)=(a_{j_1j_1r_1},a_{j_1j_2u_1})$.

- For $a_{j_1j_10}\prod_{r_2=1,r_2\neq l_1}^{p_1}(a_{j_1j_1r_1}-a_{j_1j_1r_2})^{m_{j_1r_2}}\prod_{u_2=1,u_2\neq u_1}^{q_1}(a_{j_1j_2u_1}-a_{j_1j_2u_2})^{n_{j_1u_2}}<0$ and $a_{j_2j_20}\prod_{v_2=1,v_2\neq v_1}^{p_2}(a_{j_2j_1v_1}-a_{j_2j_1v_2})^{m_{j_2v_2}}<0$,

$$(a_{j_1j_1r_1},a_{j_1j_2u_1}) = \underbrace{([(2m_{j_1r_1}1),(2n_{j_1u_1}1+1)]^{\text{th}}\text{DP}_+:\text{UP}_-,[2(m_{j_1r_1}1-m_{j_2v_1}1)-1]^{\text{th}}{}_{\text{UA}}\text{US}:_{\text{DA}}\text{LS})}_{([(2m_{j_1r_1}1),(2n_{j_1u_1}1+1)],[2(m_{j_1r_1}1-m_{j_2v_1}1)]-1)-\text{hyperbolic asymptote upper-to-lower saddle}}.$$

The equilibrium is a $([(2m_{j_1r_1}1),(2n_{j_1u_1}1+1)]^{\text{th}}\text{DP}_+\text{:UP}_-,[2(m_{j_1r_1}1-m_{j_2v_1}1)-1]^{\text{th}}{}_{\text{UA}}\text{US:}_{\text{DA}}\text{LS})$-hyperbolic-secant asymptote upper-to-lower saddle at $(x_{j_1}^*,x_{j_2}^*)=(a_{j_1j_1r_1},a_{j_1j_2u_1})$.

(ii_{2b2c}) For $m_{j_1r_1}=2m_{j_1r_1}1+1$, $n_{j_1u_1}=2n_{j_1u_1}1+1$, and $m_{j_2v_1}=2m_{j_1v_1}1$, the equilibrium $(x_{j_1}^*,x_{j_2}^*)=(a_{j_1j_1r_1},a_{j_1j_2u_1})$ on the infinite-equilibrium of $x_{j_1}^*=a_{j_2j_1v_1}=a_{j_1j_1r_1}$ has the following properties as in Eqs. (1.255)–(1.258).

- For $a_{j_1j_10}\prod_{r_2=1,r_2\neq r_1}^{p_1}(a_{j_1j_1r_1}-a_{j_1j_1r_2})^{m_{j_1r_2}}\prod_{u_2=1,u_2\neq u_1}^{q_1}(a_{j_1j_2u_1}-a_{j_1j_2u_2})^{n_{j_1u_2}}>0$ and $a_{j_2j_20}\prod_{v_2=1,v_2\neq v_1}^{p_2}(a_{j_2j_1v_1}-a_{j_2j_1v_2})^{m_{j_2v_2}}>0$,

$$
\begin{aligned}
&(a_{j_1j_1r_1},a_{j_1j_2u_1})\\
&=\underbrace{([(2m_{j_1r_1}1+1),(2n_{j_1u_1}1+1)]^{\text{th}}\text{DP}_+\text{:UP}_+,[2(m_{j_1r_1}1-m_{j_2v_1}1)+1]^{\text{th}}{}_{\text{UA}}\text{SI:}_{\text{DA}}\text{SO})}_{([(2m_{j_1r_1}1+1),(2n_{j_1u_1}1+1)],[2(m_{j_1r_1}1-m_{j_2v_1}1)]+1)\text{-hyperbolic asymptote sink-to-source}}.
\end{aligned}
$$

The equilibrium is a $([(2m_{j_1r_1}1+1),(2n_{j_1u_1}1+1)]^{\text{th}}\text{DP}_+\text{:UP}_+,[2(m_{j_1r_1}1-m_{j_2v_1}1)+1]^{\text{th}}{}_{\text{UA}}\text{SI:}_{\text{DA}}\text{SO})$ -hyperbolic asymptote sink-to-source at $(x_{j_1}^*,x_{j_2}^*)=(a_{j_1j_1r_1},a_{j_1j_2u_1})$.

- For $a_{j_1j_10}\prod_{r_2=1,r_2\neq r_1}^{p_1}\left(a_{j_1j_1r_1}-a_{j_1j_1r_2}\right)^{m_{j_1r_2}}\prod_{u_2=1,u_2\neq u_1}^{q_1}\left(a_{j_1j_2u_1}-a_{j_1j_2u_2}\right)^{n_{j_1u_2}}<0$ and $a_{j_2j_20}\prod_{v_2=1,v_2\neq v_1}^{p_2}(a_{j_2j_1v_1}-a_{j_2j_1v_2})^{m_{j_2v_2}}>0$,

$$
\begin{aligned}
&(a_{j_1j_1r_1},a_{j_1j_2u_1})\\
&=\underbrace{([(2m_{j_1r_1}1+1),(2n_{j_1u_1}1+1)]^{\text{th}}\text{UP}_+\text{:DP}_+,[2(m_{j_1r_1}1-m_{j_2v_1}1)+1]^{\text{th}}{}_{\text{DA}}\text{SO:}_{\text{DA}}\text{SI})}_{([(2m_{j_1r_1}1+1),(2n_{j_1u_1}1+1)],[2(m_{j_1r_1}1-m_{j_2v_1}1)]+1)\text{-hyperbolic-secant asymptote source-to-sink}}.
\end{aligned}
$$

The equilibrium is a $([(2m_{j_1r_1}1+1),(2n_{j_1u_1}1+1)]^{\text{th}}\text{UP}_+\text{:DP}_+,[2(m_{j_1r_1}1-m_{j_2v_1}1)+1]^{\text{th}}{}_{\text{DA}}\text{SO:}_{\text{UA}}\text{SI})$ -hyperbolic-secant asymptote source-to-sink at $(x_{j_1}^*,x_{j_2}^*)=(a_{j_1j_1r_1},a_{j_1j_2u_1})$.

- For $a_{j_1j_10}\prod_{r_2=1,r_2\neq r_1}^{p_1}(a_{j_1j_1r_1}-a_{j_1j_1r_2})^{m_{j_1r_2}}\prod_{u_2=1,u_2\neq u_1}^{q_1}(a_{j_1j_2u_1}-a_{j_1j_2u_2})^{n_{j_1u_2}}>0$ and $a_{j_2j_20}\prod_{v_2=1,v_2\neq v_1}^{p_2}(a_{j_2j_1v_1}-a_{j_2j_1v_2})^{m_{j_2v_2}}<0$,

$$
\begin{aligned}
&(a_{j_1j_1r_1},a_{j_1j_2u_1})\\
&=\underbrace{([(2m_{j_1r_1}1+1),(2n_{j_1u_1}1+1)]^{\text{th}}\text{UP}_-\text{:DP}_-,[2(m_{j_1r_1}1-m_{j_2v_1}1)+1]^{\text{th}}{}_{\text{DA}}\text{SI:}_{\text{DA}}\text{SO})}_{([(2m_{j_1r_1}1+1),(2n_{j_1u_1}1+1)],[2(m_{j_1r_1}1-m_{j_2v_1}1)]+11)\text{-hyperbolic-secant asymptote sink-to-source}}.
\end{aligned}
$$

The equilibrium is a $([(2m_{j_1r_1}1+1),(2n_{j_1u_1}1+1)]^{\text{th}}\text{UP}_-\text{:DP}_-,[2(m_{j_1r_1}1-m_{j_2v_1}1)+1]^{\text{th}}{}_{\text{DA}}\text{SI}_{\text{UA}}\text{SO})$ -hyperbolic-secant asymptote sink-to-source at $(x_{j_1}^*,x_{j_2}^*)=(a_{j_1j_1r_1},a_{j_1j_2u_1})$.

- For $a_{j_1j_10}\prod_{r_2=1,r_2\neq r_1}^{p_1}(a_{j_1j_1r_1}-a_{j_1j_1r_2})^{m_{j_1r_2}}\prod_{u_2=1,u_2\neq u_1}^{q_1}(a_{j_1j_2u_1}-a_{j_1j_2u_2})^{n_{j_1u_2}}<0$ and $a_{j_2j_20}\prod_{v_2=1,v_2\neq v_1}^{p_2}(a_{j_2j_1v_1}-a_{j_2j_1v_2})^{m_{j_2v_2}}<0$,

$$(a_{j_1j_1r_1},a_{j_1j_2u_1})=\underbrace{([(2m_{j_1r_1}1+1),(2n_{j_1u_1}1+1)]^{\text{th}}\text{DP}_+\text{:UP}_+,[2(m_{j_1r_1}1-m_{j_2v_1}1)+1]^{\text{th}}{}_{\text{UA}}\text{SO:}_{\text{DA}}\text{SI})}_{([(2m_{j_1r_1}1+1),(2n_{j_1u_1}1+1)],[2(m_{j_1r_1}1-m_{j_2v_1}1)]+11)\text{-hyperbolic asymptote source-to-sink}}.$$

The equilibrium is a $([(2m_{j_1r_1}1+1),(2n_{j_1u_1}1+1)]^{\text{th}}\text{DP}_-\text{:UP}_-,[2(m_{j_1r_1}1-m_{j_2v_1}1)+1]^{\text{th}}{}_{\text{UA}}\text{SI:}_{\text{DA}}\text{SO})$-hyperbolic asymptote sink-to-source at $(x^*_{j_1},x^*_{j_2})=(a_{j_1j_1r_1},a_{j_1j_2u_1})$.

($\text{ii}_{2\text{b}2\text{d}}$) For $m_{j_1r_1}=2m_{j_1r_1}1, n_{j_1u_1}=2n_{j_1u_1}1+1$, and $m_{j_2v_1}=2m_{j_1v_1}1$, the equilibrium $(x^*_{j_1},x^*_{j_2})=(a_{j_1j_1r_1},a_{j_1j_2u_1})$ on the infinite-equilibrium of $x^*_{j_1}=a_{j_2j_1v_1}=a_{j_1j_1r_1}$ has the following properties as in Eqs. (1.259)–(1.262).

- For $a_{j_1j_10}\prod_{r_2=1,r_2\neq r_1}^{p_1}(a_{j_1j_1r_1}-a_{j_1j_1r_2})^{m_{j_1r_2}}\prod_{u_2=1,u_2\neq u_1}^{q_1}(a_{j_1j_2u_1}-a_{j_1j_2u_2})^{n_{j_1u_2}}>0$ and $a_{j_2j_20}\prod_{v_2=1,v_2\neq v_1}^{p_2}(a_{j_2j_1v_1}-a_{j_2j_1v_2})^{m_{j_2v_2}}>0$,

$$(a_{j_1j_2u_1},a_{j_2j_1v_1})=\underbrace{([(2m_{j_1r_1}1),(2n_{j_1u_1}1+1)]^{\text{th}}\text{UP}_+\text{:UP}_+,[2(m_{j_1r_1}1-m_{j_2v_1}1)]^{\text{th}}{}_{\text{DC:IC}}\text{US})}_{([(2m_{j_1r_1}1),(2n_{j_1u_1}1+1)],[2(m_{j_1r_1}1-m_{j_2v_1}1)])\text{-concave up-parabola upper-saddle}}.$$

The equilibrium is a $([(2m_{j_1r_1}1),(2n_{j_1u_1}1+1)]^{\text{th}}\text{UP}_+\text{:UP}_+,[2(m_{j_1r_1}1-m_{j_2v_1}1)]^{\text{th}}$-${}_{\text{DC:IC}}\text{US}$)-concave up-parabola upper-saddle at $(x^*_{j_1},x^*_{j_2})=(a_{j_1j_1r_1},a_{j_2j_2s_1})$.

- For $a_{j_1j_10}\prod_{r_2=1,r_2\neq r_1}^{p_1}(a_{j_1j_1r_1}-a_{j_1j_1r_2})^{m_{j_1r_2}}\prod_{u_2=1,u_2\neq u_1}^{q_1}(a_{j_1j_2u_1}-a_{j_1j_2u_2})^{n_{j_1u_2}}<0$ and $a_{j_2j_20}\prod_{v_2=1,v_2\neq v_1}^{p_2}(a_{j_1j_1r_1}-a_{j_2j_1v_2})^{m_{j_2v_2}}>0$,

$$(a_{j_1j_2u_1},a_{j_2j_1v_1})=\underbrace{([(2m_{j_1r_1}1),(2n_{j_1u_1}1+1)]^{\text{th}}\text{DP}_+\text{:DP}_+,[2(m_{j_1r_1}1-m_{j_2v_1}1)]^{\text{th}}{}_{\text{IC:DC}}\text{US})}_{([(2m_{j_1r_1}1),(2n_{j_1u_1}1+1)],[2(m_{j_1r_1}1-m_{j_2v_1}1)])\text{-concave down-parabola upper-saddle}}.$$

The equilibrium is a $([(2m_{j_1r_1}1+1),(2n_{j_1u_1}1+1)]^{\text{th}}\text{DP}_+\text{:DP}_-,[2(m_{j_1r_1}1-m_{j_2v_1}1)]^{\text{th}}$ ${}_{\text{IC:DC}}\text{US}$)-concave down-parabola upper-saddle at $(x^*_{j_1},x^*_{j_2})=(a_{j_1j_1r_1},a_{j_2j_2s_1})$.

- For $a_{j_1j_10}\prod_{r_2=1,r_2\neq r_1}^{p_1}(a_{j_1j_1r_1}-a_{j_1j_1r_2})^{m_{j_1r_2}}\prod_{u_2=1,u_2\neq u_1}^{q_1}(a_{j_1j_2u_1}-a_{j_1j_2u_2})^{n_{j_1u_2}}>0$ and $a_{j_2j_20}\prod_{v_2=1,v_2\neq v_1}^{p_2}(a_{j_2j_1v_1}-a_{j_2j_1v_2})^{m_{j_2v_2}}<0$,

$$(a_{j_1j_2u_1},a_{j_2j_1v_1})=\underbrace{([(2m_{j_1r_11}),(2n_{j_1u_11}+1)]^{\text{th}}\text{DP}_{-}\text{:DP}_{-},[2(m_{j_1r_11}-m_{j_2v_11})]^{\text{th}}{}_{\text{IC:DC}}\text{LS})}_{([(2m_{j_1r_11}),(2n_{j_1u_11}+1)],[2(m_{j_1r_11}-m_{j_2v_11})])\text{-concave down-parabola lower-saddle}}.$$

The equilibrium is a $([(2m_{j_1r_11}),(2n_{j_1u_11}+1)]^{\text{th}}\text{DP}_{-}\text{:DP}_{-},[2(m_{j_1r_11}-m_{j_2v_11})]^{\text{th}}{}_{\text{IC:DC}}\text{LS})$- concave down-parabola lower-saddle at $(x_{j_1}^*,x_{j_2}^*)=(a_{j_1j_1r_1},a_{j_2j_2s_1})$.

- For $a_{j_1j_10}\prod_{r_2=1,r_2\neq r_1}^{p_1}(a_{j_1j_1r_1}-a_{j_1j_1r_2})^{m_{j_1r_2}}\prod_{u_2=1,u_2\neq u_1}^{q_1}(a_{j_1j_2u_1}-a_{j_1j_2u_2})^{n_{j_1u_2}}<0$ and $a_{j_2j_20}\prod_{v_2=1,v_2\neq v_1}^{p_2}(a_{j_2j_1v_1}-a_{j_2j_1v_2})^{m_{j_2v_2}}<0$,

$$(a_{j_1j_2u_1},a_{j_2j_1v_1})=\underbrace{([(2m_{j_1r_11}),(2n_{j_1u_11}+1)]^{\text{th}}\text{UP}_{-}\text{:UP}_{-},[2(m_{j_1r_11}-m_{j_2v_11})]^{\text{th}}{}_{\text{DC:IC}}\text{LS})}_{([(2m_{j_1r_11}),(2n_{j_1u_11}+1)],[2(m_{j_1r_11}-m_{j_2v_11})])\text{-concave up-parabola lower-saddle}}.$$

The equilibrium is a $([(2m_{j_1r_11}),(2n_{j_1u_11}+1)]^{\text{th}}\text{UP}_{-}\text{:UP}_{-},[2(m_{j_1r_11}-m_{j_2v_11})]^{\text{th}}{}_{\text{DC:IC}}\text{LS})$ concave up-parabola lower-saddle at $(x_{j_1}^*,x_{j_2}^*)=(a_{j_1j_1r_1},a_{j_2j_2s_1})$.

(ii_{2b2e}) For $m_{j_1r_1}=2m_{j_1r_11}+1$, $n_{j_1u_1}=2n_{j_1u_11}$, and $m_{j_2v_1}=2m_{j_1v_11}+1$, the equilibrium $(x_{j_2}^*,x_{j_1}^*)=(a_{j_1j_2u_1},a_{j_2j_1v_1})$ on the infinite-equilibrium of $x_{j_1}^*=a_{j_2j_1v_1}=a_{j_1j_1r_1}$ has the following properties as in Eqs. (1.263)–(1.266).

- For $a_{j_1j_10}\prod_{r_2=1,r_2\neq r_1}^{p_1}(a_{j_1j_1r_1}-a_{j_1j_1r_2})^{m_{j_1r_2}}\prod_{u_2=1,u_2\neq u_1}^{q_1}(a_{j_1j_2u_1}-a_{j_1j_2u_2})^{n_{j_1u_2}}>0$ and $a_{j_2j_20}\prod_{v_2=1,v_2\neq v_1}^{p_2}(a_{j_2j_1v_1}-a_{j_2j_1v_2})^{m_{j_2v_2}}>0$,

$$(a_{j_1j_2u_1},a_{j_2j_1v_1})=\underbrace{([(2m_{j_1r_11}+1),(2n_{j_1u_11})]^{\text{th}}\text{II}_{-}\text{:II}_{+},[2(m_{j_1r_11}-m_{j_2v_11})]^{\text{th}}{}_{\text{IC:IC}}\text{SO})}_{([(2m_{j_1r_11}+1),(2n_{j_1u_11})],[2(m_{j_1r_11}-m_{j_2v_11})])\text{-increasing-concave inflection-source}}.$$

The equilibrium is a $([(2m_{j_1r_11}+1),(2n_{j_1u_11})]^{\text{th}}\text{II}_{-}\text{:II}_{+},[2(m_{j_1r_11}-m_{j_2v_11})]^{\text{th}}{}_{\text{IC:IC}}\text{SO})$- increasing-concave inflection-source at $(x_{j_2}^*,x_{j_1}^*)=(a_{j_1j_2u_1},a_{j_2j_1v_1})$.

- For $a_{j_1j_10}\prod_{r_2=1,r_2\neq r_1}^{p_1}(a_{j_1j_1r_1}-a_{j_1j_1r_2})^{m_{j_1r_2}}\prod_{u_2=1,u_2\neq u_1}^{q_1}(a_{j_1j_2u_1}-a_{j_1j_2u_2})^{n_{j_1u_2}}<0$ and $a_{j_2j_20}\prod_{v_2=1,v_2\neq v_1}^{p_2}(a_{j_2j_1v_1}-a_{j_2j_1v_2})^{m_{j_2v_2}}>0$,

$$\begin{aligned}&(a_{j_1j_2u_1},a_{j_2j_1v_1})\\&=\underbrace{([(2m_{j_1r_11}+1),(2n_{j_1u_11})]^{\text{th}}\text{DI}_-\text{:DI}_+,[2(m_{j_1r_11}-m_{j_2v_11})]^{\text{th}}{}_{\text{DC:DC}}\text{SI})}_{([(2m_{j_1r_11}+1),(2n_{j_1u_11})],[2(m_{j_1r_11}-m_{j_2v_11})])\text{-decreasing-concave inflection-sink}}.\end{aligned}$$

The equilibrium is a $([(2m_{j_1r_11}+1),(2n_{j_1u_11})]^{\text{th}}\text{DI}_-\text{:DI}_+,[2(m_{j_1r_11}-m_{j_2v_11})]^{\text{th}}{}_{\text{DC:DC}}\text{SI})$-decreasing-concave inflection-sink at $(x_{j_2}^*,x_{j_1}^*)=(a_{j_1j_2u_1},a_{j_2j_1v_1})$.

- For $a_{j_1j_10}\prod_{r_2=1,r_2\neq r_1}^{p_1}(a_{j_1j_1r_1}-a_{j_1j_1r_2})^{m_{j_1r_2}}\prod_{u_2=1,u_2\neq u_1}^{q_1}(a_{j_1j_2u_1}-a_{j_1j_2u_2})^{n_{j_1u_2}}>0$ and $a_{j_2j_20}\prod_{v_2=1,v_2\neq v_1}^{p_2}(a_{j_2j_1v_1}-a_{j_2j_1v_2})^{m_{j_2v_2}}>0$,

$$\begin{aligned}&(a_{j_1j_2u_1},a_{j_2j_1v_1})\\&=\underbrace{([(2m_{j_1r_11}+1),(2n_{j_1u_11})]^{\text{th}}\text{II}_-\text{:II}_+,[2(m_{j_1r_11}-m_{j_2v_11})]^{\text{th}}{}_{\text{IC:IC}}\text{SO})}_{([(2m_{j_1r_11}+1),(2n_{j_1u_11})],[2(m_{j_1r_11}-m_{j_2v_11})])\text{-increasing-concave inflection-source}}.\end{aligned}$$

The equilibrium is a $([(2m_{j_1r_11}+1),(2n_{j_1u_11})]^{\text{th}}\text{II}_-\text{:II}_+,[2(m_{j_1r_11}-m_{j_2v_11})]^{\text{th}}{}_{\text{IC:IC}}\text{SO})$-increasing-concave inflection-source at $(x_{j_2}^*,x_{j_1}^*)=(a_{j_1j_2u_1},a_{j_2j_1v_1})$.

- For $a_{j_1j_10}\prod_{r_2=1,r_2\neq r_1}^{p_1}(a_{j_1j_1r_1}-a_{j_1j_1r_2})^{m_{j_1r_2}}\prod_{u_2=1,u_2\neq u_1}^{q_1}(a_{j_1j_2u_1}-a_{j_1j_2u_2})^{n_{j_1u_2}}<0$ and $a_{j_2j_20}\prod_{v_2=1,v_2\neq v_1}^{p_2}(a_{j_2j_1v_1}-a_{j_2j_1v_2})^{m_{j_2v_2}}>0$,

$$\begin{aligned}&(a_{j_1j_2u_1},a_{j_2j_1v_1})\\&=\underbrace{([(2m_{j_1r_11}+1),(2n_{j_1u_11})]^{\text{th}}\text{DI}_-\text{:DI}_+,[2(m_{j_1r_11}-m_{j_2v_11})]^{\text{th}}{}_{\text{DC:DC}}\text{SI})}_{([(2m_{j_1r_11}+1),(2n_{j_1u_11})],[2(m_{j_1r_11}-m_{j_2v_11})])\text{-decreasing-concave inflection-sink}}.\end{aligned}$$

The equilibrium is a $([(2m_{j_1r_11}+1),(2n_{j_1u_11})]^{\text{th}}\text{DI}_-\text{:DI}_+,[2(m_{j_1r_11}-m_{j_2v_11})]^{\text{th}}{}_{\text{DC:DC}}\text{SI})$-decreasing-concave inflection-sink at $(x_{j_2}^*,x_{j_1}^*)=(a_{j_1j_2u_1},a_{j_2j_1v_1})$.

(ii_{2b2f}) For $m_{j_1r_1}=2m_{j_1r_11}$, $n_{j_1u_1}=2n_{j_1u_11}$ and $m_{j_2v_1}=2m_{j_1v_11}+1$, the equilibrium $(x_{j_2}^*,x_{j_1}^*)=(a_{j_1j_2u_1},a_{j_2j_1v_1})$ on the infinite-equilibrium of $x_{j_1}^*=a_{j_2j_1v_1}=a_{j_1j_1r_1}$ has the following properties as in Eqs. (1.267)–(1.270).

- For $a_{j_1j_10}\prod_{r_2=1,r_2\neq r_1}^{p_1}(a_{j_1j_1r_1}-a_{j_1j_1r_2})^{m_{j_1r_2}}\prod_{u_2=1,u_2\neq u_1}^{q_1}(a_{j_1j_2u_1}-a_{j_1j_2u_2})^{n_{j_1u_2}}>0$ and $a_{j_2j_20}\prod_{v_2=1,v_2\neq v_1}^{p_2}(a_{j_2j_1v_1}-a_{j_2j_1v_2})^{m_{j_2v_2}}>0$,

$$(a_{j_1j_2u_1},a_{j_2j_1v_1}) = \underbrace{([(2m_{j_1r_1}1),(2n_{j_1u_1}1)]^{\text{th}}\text{DI}_-\text{:II}_+,[2(m_{j_1r_1}1-m_{j_2v_1}1)-1]^{\text{th}}{}_{\text{DA:DA}}\text{US})}_{([(2m_{j_1r_1}1),(2n_{j_1u_1}1)],[2(m_{j_1r_1}1-m_{j_2v_1}1)-1])\text{-inflection down-asymptote upper-saddle}}.$$

The equilibrium is a $([(2m_{j_1r_1}1),(2n_{j_1u_1}1)]^{\text{th}}\text{DI}_-\text{:II}_+,[2(m_{j_1r_1}1-m_{j_2v_1}1)-1]^{\text{th}}{}_{\text{DA:DA}}\text{US})$-inflection down-asymptote upper-saddle at $(x_{j_2}^*,x_{j_1}^*)=(a_{j_1j_2u_1},a_{j_2j_1v_1})$.

- For $a_{j_1j_10}\prod_{r_2=1,r_2\neq r_1}^{p_1}(a_{j_1j_1r_1}-a_{j_1j_1r_2})^{m_{j_1r_2}}\prod_{u_2=1,u_2\neq u_1}^{q_1}(a_{j_1j_2u_1}-a_{j_1j_2u_2})^{n_{j_1u_2}}<0$ and $a_{j_2j_20}\prod_{v_2=1,v_2\neq v_1}^{p_2}(a_{j_2j_1v_1}-a_{j_2j_1v_2})^{m_{j_2v_2}}>0$,

$$(a_{j_1j_2u_1},a_{j_2j_1v_1}) = \underbrace{([(2m_{j_1r_1}1),(2n_{j_1u_1}1)]^{\text{th}}\text{II}_-\text{:DI}_+,[2(m_{j_1r_1}1-m_{j_2v_1}1)-1]^{\text{th}}{}_{\text{UA:UA}}\text{LS})}_{([(2m_{j_1r_1}1),(2n_{j_1u_1}1)],[2(m_{j_1r_1}1-m_{j_2v_1}1)-1])\text{-inflection up-asymptote lower-saddle}}.$$

The equilibrium is a $([(2m_{j_1r_1}1),(2n_{j_1u_1}1)]^{\text{th}}\text{II}_-\text{:DI}_+,[2(m_{j_1r_1}1-m_{j_2v_1}1)-1]^{\text{th}}{}_{\text{UA:UA}}\text{LS})$-inflection up-asymptote lower-saddle at $(x_{j_2}^*,x_{j_1}^*)=(a_{j_1j_2u_1},a_{j_2j_1v_1})$.

- For $a_{j_1j_10}\prod_{r_2=1,r_2\neq r_1}^{p_1}(a_{j_1j_1r_1}-a_{j_1j_1r_2})^{m_{j_1r_2}}\prod_{u_2=1,u_2\neq u_1}^{q_1}(a_{j_1j_2u_1}-a_{j_1j_2u_2})^{n_{j_1u_2}}>0$ and $a_{j_2j_20}\prod_{v_2=1,v_2\neq v_1}^{p_2}(a_{j_2j_1v_1}-a_{j_2j_1v_2})^{m_{j_2v_2}}<0$,

$$(a_{j_1j_2u_1},a_{j_2j_1v_1}) = \underbrace{([(2m_{j_1r_1}1),(2n_{j_1u_1}1)]^{\text{th}}\text{II}_+\text{:DI}_-,[2(m_{j_1r_1}1-m_{j_2v_1}1)-1]^{\text{th}}{}_{\text{UA:UA}}\text{US})}_{([(2m_{j_1r_1}1),(2n_{j_1u_1}1)],[2(m_{j_1r_1}1-m_{j_2v_1}1)-1])\text{-inflection up-asymptote upper-saddle}}.$$

The equilibrium is a $([(2m_{j_1r_1}1),(2n_{j_1u_1}1)]^{\text{th}}\text{II}_+\text{:DI}_-,[2(m_{j_1r_1}1-m_{j_2v_1}1)-1]^{\text{th}}{}_{\text{UA:UA}}\text{US})$-inflection up-asymptote upper-saddle at $(x_{j_2}^*,x_{j_1}^*)=(a_{j_1j_2u_1},a_{j_2j_1v_1})$.

- For $a_{j_1j_10}\prod_{r_2=1,r_2\neq r_1}^{p_1}(a_{j_1j_1r_1}-a_{j_1j_1r_2})^{m_{j_1r_2}}\prod_{u_2=1,u_2\neq u_1}^{q_1}(a_{j_1j_2u_1}-a_{j_1j_2u_2})^{n_{j_1u_2}}<0$ and $a_{j_2j_20}\prod_{v_2=1,v_2\neq v_1}^{p_2}(a_{j_2j_1v_1}-a_{j_2j_1v_2})^{m_{j_2v_2}}<0$,

$$
\begin{aligned}
&(a_{j_1j_2u_1}, a_{j_2j_1v_1}) \\
&= \underbrace{([(2m_{j_1r_1}1), (2n_{j_1u_1}1)]^{\text{th}}\text{DI}_+\text{:II}_-,[2(m_{j_1r_1}1 - m_{j_2v_1}1) - 1]^{\text{th}}{}_{\text{DA:DA}}\text{LS})}_{([(2m_{j_1r_1}1),(2n_{j_1u_1}1)],[2(m_{j_1r_1}1-m_{j_2v_1}1)-1])\text{-inflection down-asymptote lower-saddle}}.
\end{aligned}
$$

The equilibrium is a $([(2m_{j_1r_1}1), (2n_{j_1u_1}1)]^{\text{th}}\text{DI}_+\text{:II}_-,[2(m_{j_1r_1}1-m_{j_2v_1}1)-1]^{\text{th}}{}_{\text{DA:DA}}\text{LS})$-inflection down-asymptote lower-saddle at$(x^*_{j_2}, x^*_{j_1})=(a_{j_1j_2u_1}, a_{j_2j_1v_1})$.

(ii_{2b2g}) For $m_{j_1r_1} = 2m_{j_1r_1}1 + 1, n_{j_1u_1} = 2n_{j_1u_1}1$, and $m_{j_2v_1} = 2m_{j_1v_1}1$, the equilibrium $(x^*_{j_2}, x^*_{j_1}) = (a_{j_1j_2u_1}, a_{j_2j_1v_1})$ on the infinite-equilibrium of $x^*_{j_1} = a_{j_2j_1v_1} = a_{j_1j_1r_1}$ has the following properties as in Eqs. (1.271)–(1.274).

- For $a_{j_1j_10}\prod_{r_2=1,r_2\neq r_1}^{p_1}(a_{j_1j_1r_1} - a_{j_1j_1r_2})^{m_{j_1r_2}}\prod_{u_2=1,u_2\neq u_1}^{q_1}(a_{j_1j_2u_1} - a_{j_1j_2u_2})^{n_{j_1u_2}} > 0$ and $a_{j_2j_20}\prod_{v_2=1,v_2\neq v_1}^{p_2}(a_{j_2j_1v_1} - a_{j_2j_1v_2})^{m_{j_2v_2}} > 0$,

$$
\begin{aligned}
&(a_{j_1j_2u_1}, a_{j_2j_1v_1}) \\
&= \underbrace{([(2m_{j_1r_1}1 + 1), (2n_{j_1u_1}1)]^{\text{th}}\text{DI}_+\text{:II}_+,[2(m_{j_1r_1}1 - m_{j_2v_1}1) + 1]^{\text{th}}{}_{\text{DA:DA}}\text{SO})}_{([(2m_{j_1r_1}1+1),(2n_{j_1u_1}1)],[2(m_{j_1r_1}1-m_{j_2v_1}1)]+1)\text{-inflection down-asymptote source}}.
\end{aligned}
$$

The equilibrium is a $([(2m_{j_1r_1}1+1), (2n_{j_1u_1}1)]^{\text{th}}\text{DI}_+\text{:II}_+,[2(m_{j_1r_1}1-m_{j_2v_1}1)+1]^{\text{th}}{}_{\text{DA:DA}}\text{SO})$-inflection down-asymptote source at $(x^*_{j_2}, x^*_{j_1}) = (a_{j_1j_2u_1}, a_{j_2j_1v_1})$.

- For $a_{j_1j_10}\prod_{r_2=1,r_2\neq r_1}^{p_1}(a_{j_1j_1r_1} - a_{j_1j_1r_2})^{m_{j_1r_2}}\prod_{u_2=1,u_2\neq u_1}^{q_1}(a_{j_1j_2u_1} - a_{j_1j_2u_2})^{n_{j_1u_2}} < 0$ and $a_{j_2j_20}\prod_{v_2=1,v_2\neq v_1}^{p_2}(a_{j_2j_1v_1} - a_{j_2j_1v_2})^{m_{j_2v_2}} > 0$,

$$
\begin{aligned}
&(a_{j_1j_2u_1}, a_{j_2j_1v_1}) \\
&= \underbrace{([(2m_{j_1r_1}1 + 1), (2n_{j_1u_1}1)]^{\text{th}}\text{II}_+\text{:DI}_+,[2(m_{j_1r_1}1 - m_{j_2v_1}1) + 1]^{\text{th}}{}_{\text{UA:UA}}\text{SI})}_{([(2m_{j_1r_1}1+1),(2n_{j_1u_1}1)],[2(m_{j_1r_1}1-m_{j_2v_1}1)]+1)\text{-inflection up-asymptote sink}}.
\end{aligned}
$$

The equilibrium is a $([(2m_{j_1r_1}1+1), (2n_{j_1u_1}1)]^{\text{th}}\text{II}_+\text{:DI}_+,[2(m_{j_1r_1}1-m_{j_2v_1}1)+1]^{\text{th}}{}_{\text{UA:UA}}\text{SI})$-inflection up-asymptote sink at $(x^*_{j_1}, x^*_{j_2}) = (a_{j_1j_1r_1}, a_{j_2j_2s_1})$.

- For $a_{j_1j_10}\prod_{r_2=1,r_2\neq r_1}^{p_1}(a_{j_1j_1r_1} - a_{j_1j_1r_2})^{m_{j_1r_2}}\prod_{u_2=1,u_2\neq u_1}^{q_1}(a_{j_1j_2u_1} - a_{j_1j_2u_2})^{n_{j_1u_2}} > 0$ and $a_{j_2j_20}\prod_{v_2=1,v_2\neq v_1}^{p_2}(a_{j_2j_1v_1} - a_{j_2j_1v_2})^{m_{j_2v_2}} < 0$,

$$
\begin{aligned}
&(a_{j_1j_2u_1}, a_{j_2j_1v_1}) \\
&= \underbrace{([(2m_{j_1r_1}1 + 1), (2n_{j_1u_1}1)]^{\text{th}}\text{II}_-\text{:DI}_-,[2(m_{j_1r_1}1 - m_{j_2v_1}1) + 1]^{\text{th}}{}_{\text{UA:UA}}\text{SO})}_{([(2m_{j_1r_1}1+1),(2n_{j_1u_1}1)],[2(m_{j_1r_1}1-m_{j_2v_1}1)]+1)\text{-inflection up-asymptote source}}.
\end{aligned}
$$

The equilibrium is a $([(2m_{j_1r_1}1+1),(2n_{j_1l_1}1\,1)]^{\text{th}}\,\text{II}_-{:}\text{DI}_-,[2(m_{j_1r_1}1-m_{j_2k_1}1)+1]^{\text{th}}{}_{\text{UA:UA}}\text{SO})$-inflection up-asymptote source at $(x_{j_2}^*,x_{j_1}^*)=(a_{j_1j_2u_1},a_{j_2j_1v_1})$.

- For $a_{j_1j_10}\prod_{r_2=1,r_2\neq r_1}^{p_1}(a_{j_1j_1r_1}-a_{j_1j_1r_2})^{m_{j_1r_2}}\prod_{u_2=1,u_2\neq u_1}^{q_1}(a_{j_1j_2u_1}-a_{j_1j_2u_2})^{n_{j_1u_2}}<0$ and $a_{j_2j_20}\prod_{v_2=1,v_2\neq v_1}^{p_2}(a_{j_2j_1v_1}-a_{j_2j_1v_2})^{m_{j_2v_2}}<0$,

$$(a_{j_1j_2u_1},a_{j_2j_1v_1})=\underbrace{([(2m_{j_1r_1}1+1),(2n_{j_1u_1}1)]^{\text{th}}\text{DI}_-{:}\text{II}_-,[2(m_{j_1r_1}1-m_{j_2v_1}1)+1]^{\text{th}}{}_{\text{DA:DA}}\text{SI})}_{([(2m_{j_1r_1}1+1),(2n_{j_1u_1}1)],[2(m_{j_1r_1}1-m_{j_2v_1}1)]+1)\text{-inflection down-asymptote sink}}.$$

The equilibrium is a $([(2m_{j_1r_1}1+1),(2n_{j_1u_1}1)]^{\text{th}}\text{DI}_-{:}\text{II}_-,[2(m_{j_1r_1}1-m_{j_2v_1}1)+1]^{\text{th}}{}_{\text{DA:DA}}\text{SI})$-inflection down-asymptote sink at $(x_{j_2}^*,x_{j_1}^*)=(a_{j_1j_2u_1},a_{j_2j_1v_1})$.

(ii_{2b2h}) For $m_{j_1r_1}=2m_{j_1r_1}1, n_{j_1u_1}=2n_{j_1u_1}1$, and $m_{j_2v_1}=2m_{j_1v_1}1$, the equilibrium of $(x_{j_1}^*,x_{j_2}^*)=(a_{j_1j_1r_1},a_{j_1j_2u_1})$ on the infinite-equilibrium of $x_{j_1}^*=a_{j_2j_1v_1}=a_{j_1j_1r_1}$ has the following properties as in Eqs. (1.275)–(1.278).

- For $a_{j_1j_10}\prod_{r_2=1,r_2\neq r_1}^{p_1}(a_{j_1j_1r_1}-a_{j_1j_1r_2})^{m_{j_1r_2}}\prod_{u_2=1,u_2\neq u_1}^{q_1}(a_{j_1j_2u_1}-a_{j_1j_2u_2})^{n_{j_1u_2}}>0$ and $a_{j_2j_20}\prod_{v_2=1,v_2\neq v_1}^{p_2}(a_{j_2j_1v_1}-a_{j_2j_1v_2})^{m_{j_2v_2}}>0$,

$$(a_{j_1j_2u_1},a_{j_2j_1v_1})=\underbrace{([(2m_{j_1r_1}1),(2n_{j_1u_1}1)]^{\text{th}}\text{II}_+{:}\text{II}_+,[2(m_{j_1r_1}1-m_{j_2v_1}1)]^{\text{th}}{}_{\text{IC:IC}}\text{US})}_{([(2m_{j_1r_1}1),(2n_{j_1u_1}1)],[2(m_{j_1r_1}1-m_{j_2v_1}1)])\text{-increasing-concave inflection upper-saddle}}.$$

The equilibrium is a $([(2m_{j_1r_1}1),(2n_{j_1u_1}1)]^{\text{th}}\text{II}_+{:}\text{II}_+,[2(m_{j_1r_1}1-m_{j_2v_1}1)]^{\text{th}}{}_{\text{IC:IC}}\text{US})$-increasing-concave inflection upper-saddle at $(x_{j_2}^*,x_{j_1}^*)=(a_{j_1j_2u_1},a_{j_2j_1v_1})$.

- For $a_{j_1j_10}\prod_{r_2=1,r_2\neq r_1}^{p_1}(a_{j_1j_1r_1}-a_{j_1j_1r_2})^{m_{j_1r_2}}\prod_{u_2=1,u_2\neq u_1}^{q_1}(a_{j_1j_2u_1}-a_{j_1j_2u_2})^{n_{j_1u_2}}<0$ and $a_{j_2j_20}\prod_{v_2=1,v_2\neq v_1}^{p_2}(a_{j_2j_1v_1}-a_{j_2j_1v_2})^{m_{j_2v_2}}>0$,

$$(a_{j_1j_2u_1},a_{j_2j_1v_1})=\underbrace{([(2m_{j_1r_1}1),(2n_{j_1u_1}1)]^{\text{th}}\text{DI}_+{:}\text{DI}_+,[2(m_{j_1r_1}1-m_{j_2v_1}1)]^{\text{th}}{}_{\text{DC:DC}}\text{LS})}_{([(2m_{j_1r_1}1),(2n_{j_1u_1}1)],[2(m_{j_1r_1}1-m_{j_2v_1}1)])\text{-decreasing-concave inflection lower-saddle}}.$$

The equilibrium is a $([(2m_{j_1r_1}1),(2n_{j_1u_1}1)]^{\text{th}}\text{DI}_+{:}\text{DI}_+,[2(m_{j_1r_1}1-m_{j_2v_1}1)]^{\text{th}}{}_{\text{DC:DC}}\text{LS})$-decreasing-concave inflection lower-saddle at $(x_{j_2}^*,x_{j_1}^*)=(a_{j_1j_2u_1},a_{j_2j_1v_1})$.

- For $a_{j_1j_10} \prod_{r_2=1,r_2\neq r_1}^{p_1}(a_{j_1j_1r_1}-a_{j_1j_1r_2})^{m_{j_1r_2}} \prod_{u_2=1,u_2\neq u_1}^{q_1}(a_{j_1j_2u_1}-a_{j_1j_2u_2})^{n_{j_1u_2}} > 0$ and $a_{j_2j_20} \prod_{v_2=1,v_2\neq v_1}^{p_2}(a_{j_2j_1v_1}-a_{j_2j_1v_2})^{m_{j_2v_2}} < 0$,

$$(a_{j_1j_2u_1}, a_{j_2j_1v_1}) = \underbrace{([(2m_{j_1r_1}1+1), (2n_{j_1u_1}1)]^{\text{th}}\text{DI}_{+}\text{:DI}_{+},[2(m_{j_1r_1}1-m_{j_2v_1}1)]^{\text{th}}{}_{\text{DC:DC}}\text{US})}_{([(2m_{j_1r_1}1),(2n_{j_1u_1}1)],[2(m_{j_1r_1}1-m_{j_2v_1}1)])\text{-decreasing-concave inflection upper-saddle}}.$$

The equilibrium is a $([(2m_{j_1r_1}1), (2n_{j_1u_1}1)]^{\text{th}}\text{DI}_{+}\text{:DI}_{+},[2(m_{j_1r_1}1-m_{j_2v_1}1)]^{\text{th}}{}_{\text{DC:DC}}\text{US})$-decreasing-concave inflection lower-saddle at$(x^{*}_{j_2}, x^{*}_{j_1})=(a_{j_1j_2u_1}, a_{j_2j_1v_1})$.

- For $a_{j_1j_10} \prod_{r_2=1,r_2\neq r_1}^{p_1}(a_{j_1j_1r_1}-a_{j_1j_1r_2})^{m_{j_1r_2}} \prod_{u_2=1,u_2\neq u_1}^{q_1}(a_{j_1j_2u_1}-a_{j_1j_2u_2})^{n_{j_1u_2}} < 0$ and $a_{j_2j_20} \prod_{v_2=1,v_2\neq v_1}^{p_2}(a_{j_2j_1v_1}-a_{j_2j_1v_2})^{m_{j_2v_2}} < 0$,

$$(a_{j_1j_2u_1}, a_{j_2j_1v_1}) = \underbrace{([(2m_{j_1r_1}1), (2n_{j_1u_1}1)]^{\text{th}}\text{II}_{-}\text{:II}_{-},[2(m_{j_1r_1}1-m_{j_2v_1}1)]^{\text{th}}{}_{\text{IC:IC}}\text{LS})}_{([(2m_{j_1r_1}1),(2n_{j_1u_1}1)],[2(m_{j_1r_1}1-m_{j_2v_1}1)])\text{-increasing-concave inflection lower-saddle}}.$$

The equilibrium is a $([(2m_{j_1r_1}1), (2n_{j_1u_1}1)]^{\text{th}}\text{II}_{-}\text{:II}_{-},[2(m_{j_1r_1}1-m_{j_2v_1}1)]^{\text{th}}{}_{\text{IC:IC}}\text{LS})$-increasing-concave inflection lower-saddle at$(x^{*}_{j_2}, x^{*}_{j_1})=(a_{j_1j_2u_1}, a_{j_2j_1v_1})$.

2.3 Hybrid Networks of Simple Equilibriums and Flows

In this section, the theorem about the hybrid networks of simple equilibriums and 1-dimensional flows will be proved, and the infinite-equilibriums will be discussed.

(iii) Consider

$$\begin{aligned} \dot{x}_{j_2} &= a_{j_2j_20}\prod_{v=1}^{m_2}(x_{j_2}-a_{j_2j_1v}) \\ \dot{x}_{j_1} &= a_{j_1j_10}\prod_{r=1}^{m_1}(x_{j_1}-a_{j_1j_1r})\prod_{u=1}^{n_1}(x_{j_2}-a_{j_1j_2u}). \end{aligned}$$

For $a_{j_1j_2u} \neq a_{j_2j_2s}$ $(u=1,2,\ldots,n_1; s=1,2,\ldots,n_2)$, using Eqs. (1.136)-(1.140),

$$\begin{aligned} q_1 &= n_1, p_2 = m_2, p_1 = m_1; \\ m_{j_1r_i} &= 1 (i = 1,2,\ldots,m_1), \end{aligned}$$

$$n_{j_2u_j} = 1(j = l, 2, \ldots, n_1),$$
$$m_{j_2v_l} = 1(l = l, 2, \ldots, m_2);$$

the first integral manifold is obtained as

$$\sum_{\nu_0=m_1}^{m_2} b_{j_2j_1\nu_0} \sum_{\substack{\mu_1+\nu_1=\nu_0-1\\ \alpha_1\in[0,\beta_0-m_1],\\ \nu_1\in[m_1-1,\nu_0-1]}} \sum_{\substack{\mu_2+\nu_2=\nu_1-2\\ \mu_2\in[0,\nu_1-m_1]\\ \nu_2\in[m_1-2,\nu_1-2]}} \cdots \sum_{\substack{\mu_{m_1-1}+\nu_{m_1-1}=\nu_{m_1-2}-m_1+1\\ \mu_{m_1-1}\in[0,\nu_{m_1-2}-m_1]\\ \nu_{m_1-1}\in[1,\nu_{m_1-2}-m_1+1]}} \prod_{j=1}^{m_1-1} \frac{(\nu_{j-1}-j)!}{\mu_j!\nu_j!}$$

$$\times (a_{j_1j_1r_{j+1}} - a_{j_1j_1r_j})^{\mu_j} \frac{1}{\nu_{m_1-1}} [(x_{j_1} - a_{j_1j_1m_1})^{\nu_{m_1-1}} - (x_{j_10} - a_{j_1j_1m_1})^{\nu_{m_1-1}}]$$

$$- \sum_{\nu_0=m_1}^{m_2} b_{j_2j_1\nu_0} \sum_{k=2}^{m_1-1} \left[\sum_{\substack{\mu_1+\nu_1=\nu_0-1\\ \mu_1\in[0,\nu_0-m_1),\\ \nu_1\in(m_1-1,\nu_0-1]}} \sum_{\substack{\mu_2+\nu_2=\nu_1-2\\ \mu_2\in[0,\nu_1-m_1]\\ \nu_2\in(m_1-2,\nu_1-2)}} \cdots \right.$$

$$\times \left(\sum_{\substack{\mu_{k-1}+\nu_{k-1}=\nu_{k-2}-k+1\\ \mu_{k-1}\in(\nu_{k-2}-m_1+1,\nu_{k-2}-k+1]\\ \nu_{k-1}\in[0,m_1-k)}} \prod_{j=1}^{k-1} \frac{(\nu_{j-1}-j)!}{\mu_j!\nu_j!} (a_{j_1j_1r_{j+1}} - a_{j_1j_1r_j})^{\mu_j} \right.$$

$$\times \sum_{\rho=k}^{m_1} A_{\rho 1}^{(r_k)} \ln \frac{|x_{j_1} - a_{j_1j_1r_\rho}|}{|x_{j_10} - a_{j_1j_1r_\rho}|}$$

$$+ \sum_{\substack{\mu_{k-1}+\nu_{k-1}=\nu_{k-2}-k+1\\ \mu_{k-1}\in(\nu_{k-2}-m_1,\nu_{k-2}-m_1+1]\\ \nu_{k-1}\in[m_1-k,m_1-k+1)}} \prod_{j=1}^{k-1} \frac{(\nu_{j-1}-j)!}{\mu_j!\nu_j!} (a_{j_1j_1r_{j+1}} - a_{j_1j_1r_j})^{\mu_j}$$

$$\left.\left. \times \sum_{\rho=k+1}^{m_1} B_{\rho 1}^{(r_k)} \ln \frac{|x_{j_1} - a_{j_1j_1s_\rho}|}{|x_{j_10} - a_{j_1j_1s_\rho}|} \right) \right] + b_{j_2j_10} \sum_{\rho=1}^{m_2} A_{\rho 1}^{(r_1)} \ln \frac{|x_{j_1} - a_{j_1j_1r_\rho}|}{|x_{j_10} - a_{j_1j_1r_\rho}|}$$

$$+ \sum_{\nu_0=1}^{m_2-1} b_{j_2j_1\nu_0} \sum_{\rho=2}^{m_2} B_{\rho 1}^{(r_1)} \ln \frac{|x_{j_1} - a_{j_1j_1r_\rho}|}{|x_{j_10} - a_{j_1j_1r_\rho}|}$$

$$= \frac{a_{j_1j_10}}{a_{j_2j_20}} \{ \frac{1}{n_1+1} [(x_{j_2} - a_{j_1j_2u_1})^{n_1+1} - (x_{j_20} - a_{j_1j_2u_1})^{n_1+1}]$$

$$+ \sum_{\alpha=1}^{n_1-1} \frac{1}{n_1-\alpha+1} b_{j_1j_2\alpha} [(x_{j_1} - a_{j_1j_2u_1})^{n_1-\alpha+1} - (x_{j_10} - a_{j_1j_2u_1})^{n_1-\alpha+1}]\},$$

where

$$b_{j_1j_2\alpha} = \sum_{l_{n_1},t_{n_1}=0}^{l_{n_1}+t_{n_1}=1} \cdots \sum_{l_2,t_2=0}^{l_2+t_2=1} [\prod_{k=2}^{n_1} (a_{j_1j_2u_1} - a_{j_1j_2u_k})^{l_k}] \delta_{t_2+\cdots+t_{n_1}}^{n_1-\alpha};$$

$$b_{j_1j_20} = 1.$$

$$b_{j_2j_1v_0} = \sum_{t_1,l_1=0}^{t_1+l_1=1} \cdots \sum_{t_{m_2},l_{m_2}=0}^{t_{m_2}+l_{m_2}=1} \prod_{k=1}^{m_2} (a_{j_1j_1r_1} - a_{j_2j_1v_k})^{t_k} \delta_{v_0}^{\Sigma_{r=1}^{m_2} l_r};$$

$$\lim_{v_{m_1-1}\to 0} \frac{1}{v_{m_1-1}}[(x_{j_1} - a_{j_1j_1r_{m_1}})^{v_{m_1-1}} - (x_{j_10} - a_{j_1j_1r_{m_1}})^{v_{m_1-1}}] = \ln \frac{|x_{j_1} - a_{j_1j_1r_{m_1}}|}{|x_{j_10} - a_{j_1j_1r_{m_1}}|};$$

$$A_{\rho 1}^{(r_k)} = \frac{1}{\prod_{\sigma=k,\sigma\neq\rho}^{m_1} (a_{j_1j_1r_\rho} - a_{j_1j_1r_\sigma})}, \text{ for } \rho \in \{k, k+1, \ldots, m_1\},$$

$$B_{\rho 1}^{(r_k)} = \frac{(a_{j_2j_2r_\rho} - a_{j_2j_2r_k})^{\beta_{k-1}-k}}{\prod_{\sigma=k+1,\sigma\neq\rho}^{m_1} (a_{j_1j_1r_\rho} - a_{j_1j_1r_\sigma})}, \text{ for} \rho \in \{k+1, k+2, \ldots, m_1\};$$

$k \in \{1, 2, \ldots, m_1 - 1\}$.

(iii$_1$) For $a_{j_2j_1v_1} \neq a_{j_1j_1r_1} (r = 1, 2, \ldots, m_1; v = 1, 2, \ldots, m_2)$, the simple 1-dimensional flows and simple equilibriums exist.

(iii$_{1a}$) At the equilibrium of $x_{j_2}^* = a_{j_1j_2u_1}$ and $\overline{x}_{j_1} \neq a_{j_1j_1r}$, in phase space,

$$\begin{aligned} \frac{dx_{j_1}}{dx_{j_2}}\Big|_{x_{j_2}^*=a_{j_1j_2s_1}} &= \frac{a_{j_1j_10}(x_{j_2} - a_{j_1j_2u_1}) \prod_{r=1}^{m_1} (\overline{x}_{j_1} - a_{j_1j_1r}) \prod_{u_2=1,u_2\neq u_1}^{n_1} (x_{j_2} - a_{j_1j_2u_2})}{a_{j_2j_20} \prod_{v=1}^{m_2} (\overline{x}_{j_1} - a_{j_2j_1v})}\Bigg|_{x_{j_2}^*=a_{j_1j_2u_1}} \\ &= 0, \end{aligned}$$

If

$$\frac{d^2x_{j_1}}{dx_{j_2}^2}\Big|_{x_{j_2}^*=a_{j_1j_2u_1}} = \frac{a_{j_1j_10} \prod_{r=1}^{m_1} (\overline{x}_{j_1} - a_{j_1j_1r}) \prod_{u_2=1,u_2\neq u_1}^{n_1} (a_{j_1j_2u_1} - a_{j_1j_2u_2})}{a_{j_2j_20} \prod_{v=1}^{m_2} (\overline{x}_{j_1} - a_{j_2j_1v})} > 0,$$

the up-parabola flow at $x_{j_2}^* = a_{j_1j_2u_1}$ exists. If

$$\frac{d^2x_{j_1}}{dx_{j_2}^2}\Big|_{x_{j_2}^*=a_{j_1j_2u_1}} = \frac{a_{j_1j_10} \prod_{r=1}^{m_1} (\overline{x}_{j_1} - a_{j_1j_1r}) \prod_{u_2=1,u_2\neq u_1}^{n_1} (a_{j_1j_2u_1} - a_{j_1j_2u_2})}{a_{j_2j_20} \prod_{v=1}^{m_2} (\overline{x}_{j_1} - a_{j_2j_1v})} < 0,$$

the down-parabola flow at $x_{j_2}^* = a_{j_1j_2u_2}$ exists.

Let

$$\dot{x}_{j_2} = a_{j_2j_20}\prod_{v=1}^{m_2}(\overline{x}_{j_1} - a_{j_2j_1v}).$$

The parabola flows at $x^*_{j_2} = a_{j_1j_2u_2}$ are positive and negative for $\dot{x}_{j_2} > 0$ and $\dot{x}_{j_2} < 0$, respectively.

At the equilibrium of $x^*_{j_1} = a_{j_1j_1r_1} \neq a_{j_2j_1v}$ and $\overline{x}_{j_2} \neq a_{j_1j_2u}$, the variational equation is

$$\begin{aligned}
\dot{x}_{j_2} &= a_{j_2j_20}\prod_{v=1}^{m_2}(a_{j_1j_1r_1} - a_{j_2j_1v}),\\
\Delta\dot{x}_{j_1} &= a_{j_1j_10}\prod_{r_1=1,r_2\neq r_1}^{m_1}(a_{j_1j_1r_1} - a_{j_1j_1r_2})\prod_{u=1}^{n_1}(\overline{x}_{j_2} - a_{j_1j_2u})(\Delta x_{j_1}).
\end{aligned}$$

For

$$\begin{aligned}
&a_{j_1j_10}\prod_{r_1=1,r_2\neq r_1}^{m_1}(a_{j_1j_1r_1} - a_{j_1j_1r_2})\prod_{u=1}^{n_1}(\overline{x}_{j_2} - a_{j_1j_2u}) > 0, \text{ and}\\
&a_{j_1j_10}\prod_{r_1=1,r_2\neq r_1}^{m_1}(a_{j_1j_1r_1} - a_{j_1j_1r_2})\prod_{u=1}^{n_1}(\overline{x}_{j_2} - a_{j_1j_2u}) < 0,
\end{aligned}$$

the flows at $x^*_{j_1} = a_{j_1j_1r_1} \neq a_{j_2j_1v}$ and $\overline{x}_{j_2} \neq a_{j_1j_2u}$ are source and sink, respectively.

Let $\overline{x}_{j_1} = a_{j_1j_1r_1}$ and $\overline{x}_{j_2} = a_{j_1j_2u_1} \neq a_{j_2j_2s}$. Therefore, the equilibrium of $(x^*_{j_1}, x^*_{j_2}) = (a_{j_1j_1r_1}, a_{j_1j_2u_1})(r_1 = 1, 2, \ldots, m_1; u_1 = 1, 2, \ldots, n_1)$ for the x_{j_1}-direction has the following properties as in Eqs. (1.282)–(1.285).

- For $a_{j_1j_10}\prod_{r_2=1,r_2\neq r_1}^{m_1}(a_{j_1j_1r_1} - a_{j_1j_1r_2})\prod_{u_2=1,u_2\neq u_1}^{n_1}(a_{j_1j_2u_2} - a_{j_1j_2u_2}) > 0$ and $a_{j_2j_20}\prod_{v=1}^{m_2}(a_{j_1j_1r_1} - a_{j_2j_1v}) > 0$,

$$(a_{j_1j_1r_1}, a_{j_1j_2u_1}) = \underbrace{(\text{DP:UP,pF})}_{\text{hyperbolic flow }(+)}.$$

- For $a_{j_1j_10}\prod_{r_2=1,r_2\neq r_1}^{m_1}(a_{j_1j_1r_1} - a_{j_1j_1r_2})\prod_{u_2=1,u_2\neq u_1}^{n_1}(a_{j_1j_2u_2} - a_{j_1j_2u_2}) < 0$ and $a_{j_2j_20}\prod_{v=1}^{m_2}(a_{j_1j_1r_1} - a_{j_2j_1v}) > 0$,

$$(a_{j_1j_1r_1}, a_{j_1j_2u_1}) = \underbrace{\text{(UP:DP,pF)}}_{\text{hyperbolic-secant flow }(+)}.$$

- For $a_{j_1j_10}\prod_{r_2=1,r_2\neq r_1}^{m_1}(a_{j_1j_1r_1}-a_{j_1j_1r_2})\prod_{u_2=1,u_2\neq u_1}^{n_1}(a_{j_1j_2u_2}-a_{j_1j_2u_2})>0$ and $a_{j_2j_20}\prod_{v=1}^{m_2}(a_{j_1j_1r_1}-a_{j_2j_1v})<0$,

$$(a_{j_1j_1r_1}, a_{j_1j_2u_1}) = \underbrace{\text{(UP:DP,nF)}}_{\text{hyperbolic-secant flow }(-)}.$$

- For $a_{j_1j_10}\prod_{r_2=1,r_2\neq r_1}^{m_1}(a_{j_1j_1r_1}-a_{j_1j_1r_2})\prod_{u_2=1,u_2\neq u_1}^{n_1}(a_{j_1j_2u_2}-a_{j_1j_2u_2})<0$ and $a_{j_2j_20}\prod_{v=1}^{m_2}(a_{j_1j_1r_1}-a_{j_2j_1v})<0$,

$$(a_{j_1j_1r_1}, a_{j_1j_2u_1}) = \underbrace{\text{(DP:UP,nF)}}_{\text{hyperbolic flow }(-)}.$$

(iii_{1b}) At the equilibrium of $x_{j_1}^* = a_{j_2j_1v_1}$ and $\overline{x}_{j_2}\neq a_{j_1j_2v}$, in phase space,

$$\begin{aligned}
&\left.\frac{dx_{j_2}}{dx_{j_1}}\right|_{x_{j_1}^*=a_{j_2j_1v_1}}\\
&=\left.\frac{a_{j_2j_20}(x_{j_1}-a_{j_2j_1v_1})\prod_{v_2=1,v_2\neq v_1}^{m_2}(x_{j_1}-a_{j_2j_1v_2})}{a_{j_1j_10}\prod_{r=1}^{m_1}(x_{j_1}-a_{j_1j_1r})\prod_{u=1}^{n_1}(\overline{x}_{j_2}-a_{j_1j_2u})}\right|_{x_{j_1}^*=a_{j_2j_1v_1}}\\
&=0,
\end{aligned}$$

If

$$\left.\frac{d^2x_{j_2}}{dx_{j_1}^2}\right|_{x_{j_1}^*=a_{j_2j_1v_1}}=\frac{a_{j_2j_20}\prod_{v_2=1,v_2\neq v_1}^{m_2}(a_{j_2j_1v_1}-a_{j_2j_1v_2})}{a_{j_1j_10}\prod_{r=1}^{m_1}(a_{j_2j_1v_1}-a_{j_1j_1r})\prod_{u=1}^{n_1}(\overline{x}_{j_2}-a_{j_1j_2u})}>0,$$

the up-parabola flow at $x_{j_1}^* = a_{j_2j_1v_1}$ exists. If

$$\left.\frac{d^2x_{j_2}}{dx_{j_1}^2}\right|_{x_{j_1}^*=a_{j_2j_1v_1}} = \frac{a_{j_2j_20}\prod_{v_2=1,v_2\neq v_1}^{m_2}(a_{j_2j_1v_1}-a_{j_2j_1v_2})}{a_{j_1j_10}\prod_{r=1}^{m_1}(a_{j_2j_1v_1}-a_{j_1j_1r})\prod_{u=1}^{n_1}(\overline{x}_{j_2}-a_{j_1j_2u})} < 0,$$

the down-parabola flow at $x_{j_1}^* = a_{j_2j_1v_1}$ exists.

Let

$$\dot{x}_{j_1} = a_{j_1j_10}\prod_{r=1}^{m_1}(a_{j_2j_1v_1}-a_{j_1j_1r})\prod_{u=1}^{n_1}(\overline{x}_{j_2}-a_{j_1j_2u}).$$

The parabola flows at $x_{j_1}^* = a_{j_2j_1v_1}$ are positive and negative for $\dot{x}_{j_2} > 0$ and $\dot{x}_{j_2} < 0$, respectively.

Therefore, from the above analysis and the proof of case (iii$_{1a}$), the equilibrium of $(x_{j_2}^*, x_{j_1}^*) = (a_{j_1j_2u_1}, a_{j_2j_1v_1})$ $(v_1 = 1, 2, \ldots, n_1;\ v_1 = 1, 2, \ldots, m_2)$ has the following properties as in Eqs. (1.286)–(1.289).

- For $a_{j_1j_10}\prod_{r=1}^{m_1}(a_{j_2j_1v_1}-a_{j_1j_1r})\prod_{u_2=1,u_2\neq u_1}^{n_1}(a_{j_1j_2u_1}-a_{j_1j_2u_2}) > 0$ and $a_{j_2j_20}\prod_{v_2=1,v_2\neq v_1}^{m_2}(a_{j_2j_1v_1}-a_{j_2j_1v_2}) > 0$,

$$(a_{j_2j_1u_1}, a_{j_2j_1v_1}) = \underbrace{(\text{UP}_+,\text{UP}_+)}_{\text{positve saddle}}.$$

- For $a_{j_1j_10}\prod_{r=1}^{m_1}(a_{j_2j_1v_1}-a_{j_1j_1r})\prod_{u_2=1,u_2\neq u_1}^{n_1}(a_{j_1j_2u_1}-a_{j_1j_2u_2}) < 0$ and $a_{j_2j_20}\prod_{v_2=1,v_2\neq v_1}^{m_2}(a_{j_2j_1v_1}-a_{j_2j_1v_2}) > 0$,

$$(a_{j_2j_1u_1}, a_{j_2j_1v_1}) = \underbrace{(\text{DP}_+,\text{DP}_-)}_{\text{CCW center}}.$$

- For $a_{j_1j_10}\prod_{r=1}^{m_1}(a_{j_2j_1v_1}-a_{j_1j_1r})\prod_{u_2=1,u_2\neq u_1}^{n_1}(a_{j_1j_2u_1}-a_{j_1j_2u_2}) > 0$ and $a_{j_2j_20}\prod_{v_2=1,v_2\neq v_1}^{m_2}(a_{j_2j_1v_1}-a_{j_2j_1v_2}) < 0$,

$$(a_{j_2j_1u_1}, a_{j_2j_1v_1}) = \underbrace{(\text{DP}_-,\text{DP}_+)}_{\text{CW center}}.$$

- For $a_{j_1j_1 0}\prod_{r=1}^{m_1}(a_{j_2j_1v_1}-a_{j_1j_1r})\prod_{u_2=1,u_2\neq u_1}^{n_1}(a_{j_1j_2u_1}-a_{j_1j_2u_2})>0$ and $a_{j_2j_2 0}\prod_{v_2=1,v_2\neq v_1}^{m_2}(a_{j_2j_1v_1}-a_{j_2j_1v_2})<0$,

$$(a_{j_2j_1u_1},a_{j_2j_1v_1})=\underbrace{(\mathrm{UP}_-,\mathrm{UP}_-)}_{\text{negative saddle}}.$$

Define

$$\begin{aligned}
&\bigcup_{l=1}^{m_1+m_2}\bigcup_{u=1}^{n_1}(c_{j_1j_1l},a_{j_1j_2u})\\
&=\bigcup_{r=1}^{m_1}\bigcup_{u=1}^{n_1}\underbrace{(a_{j_1j_1r},a_{j_1j_2u})}_{(\mathrm{XX,xF})}\oplus\bigcup_{v=1}^{m_2}\bigcup_{u=1}^{n_1}\underbrace{(a_{j_1j_2u},a_{j_2j_1v})}_{\mathrm{YY}},\\
&\{c_{j_1j_1l}|l=1,2,\ldots,m_1+m_2\}\\
&=\mathrm{sort}\{a_{j_1j_1r},a_{j_2j_1v}|r=1,2,\ldots,m_1;v=1,2,\ldots,m_2\},\\
&\text{for } c_{j_1j_1l_1}<c_{j_1j_1l_2} \text{ for } l_1<l_2;
\end{aligned}$$

where

$$\bigcup_{r=1}^{m_1}\bigcup_{u=1}^{n_1}\underbrace{(a_{j_1j_1r},a_{j_1j_2u})}_{\mathrm{XX}}
=\begin{bmatrix}
(a_{j_1j_11},a_{j_1j_2n_1}) & \cdots & (a_{j_1j_1(m_1-1)},a_{j_1j_2n_1}) & (a_{j_1j_1m_1},a_{j_1j_2n_1})\\
(a_{j_1j_11},a_{j_1j_2(n_1-1)}) & \cdots & (a_{j_1j_1(m_1-1)},a_{j_1j_2(n_1-1)}) & (a_{j_1j_1m_1},a_{j_1j_2(n_1-1)})\\
\vdots & \vdots & \vdots & \vdots\\
(a_{j_1j_11},a_{j_1j_21}) & \cdots & (a_{j_1j_1(m_1-1)},a_{j_1j_21}) & (a_{j_1j_1m_1},a_{j_1j_21})
\end{bmatrix}_{n_1\times m_1}.$$

and

$$\bigcup_{v=1}^{m_2}\bigcup_{u=1}^{n_1}\underbrace{(a_{j_1j_2u},a_{j_2j_1v})}_{\mathrm{YY}}
=\begin{bmatrix}
(a_{j_1j_2n_1},a_{j_2j_11}) & \cdots & (a_{j_1j_2(n_1-1)},a_{j_2j_1m_2}) & (a_{j_1j_2n_1},a_{j_2j_1m_2})\\
(a_{j_1j_2(n_1-1)},a_{j_2j_11}) & \cdots & (a_{j_1j_2(n_1-1)},a_{j_2j_1(m_2-1)}) & (a_{j_1j_2(n_1-1)},a_{j_2j_1m_2})\\
\vdots & \vdots & \vdots & \vdots\\
(a_{j_1j_21},a_{j_2j_11}) & \cdots & (a_{j_1j_21},a_{j_2j_1(m_2-1)}) & (a_{j_1j_21},a_{j_2j_1m_2})
\end{bmatrix}_{n_1\times m_2}.$$

(iii$_2$) For $a_{j_1j_1r_1}=a_{j_2j_1v_1}$, the dynamical system becomes

$$\dot{x}_{j_1} = a_{j_1j_10}(x_{j_1} - a_{j_1j_1r_1}) \prod_{r_2=1,r_2\neq r_1}^{m_1} (x_{j_1} - a_{j_1j_1r_2}) \prod_{u=1,}^{n_1} (x_{j_2} - a_{j_1j_2u}),$$
$$\dot{x}_{j_2} = a_{j_2j_20}(x_{j_1} - a_{j_2j_1r_1}) \prod_{v_2=1,v_2\neq v_1}^{m_2} (x_{j_1} - a_{j_2j_1v_2}).$$

The first integral manifolds can be determined from Eqs. (1.280) through the equivalent dynamical system as

$$\begin{aligned}
\dot{x}_{j_1} &= a_{j_1j_10} \prod_{r_2=1,r_2\neq r_1}^{m_1} (x_{j_1} - a_{j_1j_1r_2}) \prod_{u=1,}^{n_1} (x_{j_2} - a_{j_1j_2u}),\\
\dot{x}_{j_2} &= a_{j_2j_20} \prod_{v_2=1,v_2\neq v_1}^{m_2} (x_{j_1} - a_{j_2j_1v_2}).\\
m_1 &\Rightarrow m_1 - 1, r_{j+1} \Rightarrow r_j\ (j = 1, 2, \ldots, m_1 - 1);\\
m_2 &\Rightarrow m_2 - 1, v_{i+1} \Rightarrow v_i\ (i = 1, 2, \ldots, m_2 - 1).
\end{aligned}$$

Define a network matrix of 1-dimensional flows and equilibrium with an infinite-equilibrium as

$$\begin{aligned}
NM = &\underbrace{(a_{j_1j_2u_1}, a_{j_2j_1v_1})}_{\text{(XX,YY)}} \oplus \bigcup_{u_2=1,u_2\neq u_1}^{n_1} \underbrace{(a_{j_1j_2u_1}, a_{j_2j_1v_1})}_{\text{(XX,YY)}}\\
&\oplus \bigcup_{u=1}^{n_1} \bigcup_{r_2=1,r_2\neq r_1}^{m_1} \underbrace{(a_{j_1j_2r_2}, a_{j_2j_1u})}_{\text{(XX,xF)}}\\
&\oplus \bigcup_{u=1}^{n_1} \bigcup_{v_2=1,v_2\neq v_1}^{m_2} \underbrace{(a_{j_1j_2u}, a_{j_2j_1v_2})}_{\text{ZZ}}
\end{aligned}$$

(iii$_{2a}$) For $x_{j_1}^* = a_{j_1j_1r_1} = a_{j_2j_1v_1}$ and $\overline{x}_{j_2} \neq a_{j_1j_2u}$, if

$$\left.\frac{dx_{j_2}}{dx_{j_1}}\right|_{x_{j_1}^*=a_{j_2j_1v_1}} = \frac{a_{j_2j_20} \prod_{v_2=1,v_2\neq v_1}^{m_2} (a_{j_2j_1v_1} - a_{j_2j_1v_2})}{a_{j_1j_10} \prod_{r_2=1,r_2\neq r_1}^{m_1} (a_{j_1j_1r_1} - a_{j_1j_1r_2}) \prod_{u=1}^{n_1} (\overline{x}_{j_2} - a_{j_1j_2u})} > 0,$$

the flow at $x_{j_1}^* = a_{j_1j_1r_1} = a_{j_2j_1v_1}$ is an increasing-inflection flow, and if

$$\frac{dx_{j_2}}{dx_{j_1}}\Big|_{x^*_{j_1}=a_{j_2j_1v_1}} = \frac{a_{j_2j_20}\prod_{v_2=1,v_2\neq v_1}^{m_2}(a_{j_2j_1v_1}-a_{j_2j_1v_2})}{a_{j_1j_10}\prod_{r_2=1,r_2\neq r_1}^{m_1}(a_{j_1j_1r_1}-a_{j_1j_1r_2})\prod_{u=1}^{n_1}(\overline{x}_{j_2}-a_{j_1j_2u})} < 0,$$

the flow at $x^*_{j_1} = a_{j_1j_1r_1} = a_{j_2j_1v_1}$ is a decreasing-inflection flow.

The variational equation at $x^*_{j_1} = a_{j_1j_1r_1} = a_{j_2j_1v_1}$ is

$$\Delta\dot{x}_{j_1} = a_{j_1j_10}\prod_{r_2=1,r_2\neq r_1}^{m_1}(a_{j_1j_1r_1}-a_{j_1j_1r_2})\prod_{u=1}^{n_1}(\overline{x}_{j_2}-a_{j_1j_2u})(\Delta x_{j_1}).$$

For

$$a_{j_1j_10}\prod_{r_2=1,r_2\neq r_1}^{m_1}(a_{j_1j_1r_1}-a_{j_1j_1r_2})\prod_{u=1}^{n_1}(\overline{x}_{j_2}-a_{j_1j_2u}) > 0 \text{ and}$$
$$a_{j_1j_10}\prod_{r_2=1,r_2\neq r_1}^{m_1}(a_{j_1j_1r_1}-a_{j_1j_1r_2})\prod_{u=1}^{n_1}(\overline{x}_{j_2}-a_{j_1j_2u}) < 0,$$

the flows at $x^*_{j_1} = a_{j_1j_1r_1} = a_{j_2j_1v_1}$ are source and sink, respectively.

(iii$_{2a}$) The infinite-equilibrium of $x^*_{j_1} = a_{j_2j_1v_1} = a_{j_1j_1r_1}$ with $\overline{x}_{j_2} \neq a_{j_1j_2u_1}$ has the following properties as in Eqs. (1.296)–(1.299).

- For $a_{j_1j_10}\prod_{r_2=1,r_2\neq r_1}^{m_1}(a_{j_1j_1r_1}-a_{j_1j_1r_2})\prod_{u=1}^{n_1}(\overline{x}_{j_2}-a_{j_1j_2u}) > 0$ and $a_{j_2j_20}\prod_{v_2=1,v_2\neq v_1}^{m_2}(a_{j_2j_1v_1}-a_{j_2j_1v_2}) > 0$,

$$(a_{j_1j_1r_1},\overline{x}_{j_2}) = \underbrace{(\text{SO,II})}_{\text{increasing-inflection source}}.$$

The infinite-equilibrium is a (SO,II) -increasing-inflection source at $x^*_{j_1}=a_{j_2j_1v_1} = a_{j_1j_1r_1}$.

- For $a_{j_1j_10}\prod_{r_2=1,r_2\neq r_1}^{m_1}(a_{j_1j_1r_1}-a_{j_1j_1r_2})\prod_{u=1}^{n_1}(\overline{x}_{j_2}-a_{j_1j_2u}) < 0$ and $a_{j_2j_20}\prod_{v_2=1,v_2\neq v_1}^{m_2}(a_{j_2j_1v_1}-a_{j_2j_1v_2}) > 0$,

$$(a_{j_1j_1r_1},\overline{x}_{j_2}) = \underbrace{(\text{SI,DI})}_{\text{decreasing-inflection sink}}.$$

The infinite-equilibrium is a (SI,DI) at -decreasing-inflection sink $x_{j_1}^{*}=a_{j_2j_1v_1} = a_{j_1j_1r_1}$.

- For $a_{j_1j_10}\prod_{r_2=1,r_2\neq r_1}^{m_1}(a_{j_1j_1r_1}-a_{j_1j_1r_2})\prod_{u=1}^{n_1}(\overline{x}_{j_2}-a_{j_1j_2u})>0$ and $a_{j_2j_20}\prod_{v_2=1,v_2\neq v_1}^{m_2}(a_{j_2j_1v_1}-a_{j_2j_1v_2})<0$,

$$(a_{j_1j_1r_1},\overline{x}_{j_2})=\underbrace{\text{(SO,DI)}}_{\text{decreasing-inflection source}}.$$

The infinite-equilibrium is a (SO,DI) at -decreasing-inflection sink $x_{j_1}^{*}=a_{j_2j_1v_1} = a_{j_1j_1r_1}$.

- For $a_{j_1j_10}\prod_{r_2=1,r_2\neq r_1}^{m_1}(a_{j_1j_1r_1}-a_{j_1j_1r_2})\prod_{u=1}^{n_1}(\overline{x}_{j_2}-a_{j_1j_2u})<0$ and $a_{j_2j_20}\prod_{v_2=1,v_2\neq v_1}^{m_2}(a_{j_2j_1v_1}-a_{j_2j_1v_2})<0$,

$$(a_{j_1j_1r_1},\overline{x}_{j_2})=\underbrace{\text{(SI,II)}}_{\text{increasing-inflection sink}}.$$

The infinite-equilibrium is a (SI,II) at -increasing-inflection sink $x_{j_1}^{*}=a_{j_2j_1v_1} = a_{j_1j_1r_1}$.

(iii_{2b}) At the equilibrium of $x_{j_2}^{*}=a_{j_1j_2u_1}$ and $\overline{x}_{j_1}\neq a_{j_1j_1r}$, in phase space,

$$\begin{aligned}\frac{dx_{j_1}}{dx_{j_2}}\Big|_{x_{j_2}^{*}=a_{j_1j_2u_1}}&=\frac{a_{j_1j_10}(x_{j_2}-a_{j_1j_2u_1})\prod_{r_2=1,r_2\neq r_1}^{m_1}(\overline{x}_{j_1}-a_{j_1j_1r_2})\prod_{u_2=1,u_2\neq u_1}^{n_1}(x_{j_2}-a_{j_1j_2u_2})}{a_{j_2j_20}\prod_{v_2=1,v_2\neq v_1}^{m_2}(\overline{x}_{j_1}-a_{j_2j_1v_1})}\Bigg|_{x_{j_2}^{*}=a_{j_1j_2u_1}}\\&=0,\end{aligned}$$

If

$$\frac{d^2x_{j_1}}{dx_{j_2}^2}\Big|_{x_{j_2}^{*}=a_{j_1j_2u_1}}=\frac{a_{j_1j_10}\prod_{r_2=1,r_2\neq r_1}^{m_1}(\overline{x}_{j_1}-a_{j_1j_1r_2})\prod_{u_2=1,u_2\neq u_1}^{n_1}(a_{j_1j_2u_1}-a_{j_1j_2u_2})}{a_{j_2j_20}\prod_{v_2=1,v_2\neq v_1}^{m_2}(\overline{x}_{j_1}-a_{j_2j_1v_2})}$$

$$= \frac{a_{j_1j_10}(\overline{x}_{j_1} - a_{j_1j_1r_1}) \prod_{r_2=1, r_2\neq r_1}^{m_1} (\overline{x}_{j_1} - a_{j_1j_1r_2})}{a_{j_2j_20}(\overline{x}_{j_1} - a_{j_2j_1v_1}) \prod_{v_2=1, v_2\neq v_1}^{m_2} (\overline{x}_{j_1} - a_{j_2j_1v_2})} \prod_{u_2=1, u_2\neq u_1}^{n_1} (a_{j_1j_2u_1} - a_{j_1j_2u_2})$$

$$= \frac{a_{j_1j_10} \prod_{r=1}^{m_1} (\overline{x}_{j_1} - a_{j_1j_1r}) \prod_{u_2=1, u_2\neq u_1}^{n_1} (a_{j_1j_2u_1} - a_{j_1j_2u_2})}{a_{j_2j_20} \prod_{v=1}^{m_2} (\overline{x}_{j_1} - a_{j_2j_1v})} > 0,$$

the up-parabola flow at $x_{j_2}^* = a_{j_1j_2u_1}$ exists. If

$$\left.\frac{d^2x_{j_1}}{dx_{j_2}^2}\right|_{x_{j_2}^*=a_{j_1j_2u_1}} = \frac{a_{j_1j_10} \prod_{r=1}^{m_1} (\overline{x}_{j_1} - a_{j_1j_1r}) \prod_{u_2=1, u_2\neq u_1}^{n_1} (a_{j_1j_2u_1} - a_{j_1j_2u_2})}{a_{j_2j_20} \prod_{v=1}^{m_2} (\overline{x}_{j_1} - a_{j_2j_1v})} < 0,$$

the down-parabola flow at $x_{j_2}^* = a_{j_1j_2u_2}$ exists.

Let

$$\dot{x}_{j_2} = a_{j_2j_20} \prod_{v=1}^{m_2} (\overline{x}_{j_1} - a_{j_2j_1v}).$$

The parabola flows at $x_{j_2}^* = a_{j_1j_2u_2}$ are positive and negative for $\dot{x}_{j_2} > 0$ and $\dot{x}_{j_2} < 0$, respectively. Thus, based on the above analysis, for

$$a_{j_2j_20} \prod_{v=1}^{m_2} (\overline{x}_{j_1} - a_{j_2j_1v}) > 0, \text{ and}$$
$$a_{j_2j_20} \prod_{v=1}^{m_2} (\overline{x}_{j_1} - a_{j_2j_1v}) < 0,$$

the parabola flows at $x_{j_1}^* = a_{j_1j_1r_1} = a_{j_2j_1v_1}$ are positive and negative, respectively.

On both sides of $\overline{x}_{j_2} = a_{j_1j_2v_1}$, the inflection flows will be switched. Therefore, the equilibrium of $(x_{j_2}^*, x_{j_1}^*) =$ on the infinite-equilibrium of $x_{j_1}^* = a_{j_1j_1r_1} = a_{j_2j_1v_1}$ has the following properties as in Eqs. (1.299)–(1.303).

- For $a_{j_1j_10} \prod_{r_2=1, r_2\neq r_1}^{m_1} (a_{j_1j_1r_1} - a_{j_1j_1r_2}) \prod_{u_2=1, u_2\neq u_1}^{n_2} (a_{j_1j_2u_1} - a_{j_1j_2u_2}) > 0$ and $a_{j_2j_20} \prod_{v_2=1, v_2\neq v_1}^{m_2} (a_{j_2j_1v_1} - a_{j_2j_1v_2}) > 0,$

$$(a_{j_1j_2u_1}, a_{j_2j_1v_1}) = \underbrace{(\mathrm{UP}_- : \mathrm{UP}_{+,\mathrm{DI:II}}\mathrm{US})}_{\text{inflection up-parabola upper-saddle}}.$$

The equilibrium is a $(\mathrm{UP}_{-}\mathrm{:UP}_{+},_{\mathrm{DI:II}}\mathrm{US})$-inflection up-parabola upper-saddle at $(x_{j_2}^*, x_{j_1}^*) = (a_{j_1 j_2 u_1}, a_{j_2 j_1 v_1})$.

- For $a_{j_1 j_1 0} \prod_{r_2=1, r_2 \neq r_1}^{m_1} (a_{j_1 j_1 r_1} - a_{j_1 j_1 r_2}) \prod_{u_2=1, u_2 \neq u_1}^{n_2} (a_{j_1 j_2 u_1} - a_{j_1 j_2 u_2}) < 0$ and $a_{j_2 j_2 0} \prod_{v_2=1, v_2 \neq v_1}^{m_2} (a_{j_2 j_1 v_1} - a_{j_2 j_1 v_2}) > 0$,

$$(a_{j_1 j_2 u_1}, a_{j_2 j_1 v_1}) = \underbrace{(\mathrm{DP}_{-} : \mathrm{DP}_{+},_{\mathrm{II:DI}}\mathrm{LS})}_{\text{inflection down-parabola lower-saddle}}.$$

The equilibrium is a $(\mathrm{DP}_{-}\mathrm{:DP}_{+},_{\mathrm{II:DI}}\mathrm{LS})$-inflection down-parabola lower-saddle at $(x_{j_2}^*, x_{j_1}^*) = (a_{j_1 j_2 u_1}, a_{j_2 j_1 v_1})$.

- For $a_{j_1 j_1 0} \prod_{r_2=1, r_2 \neq r_1}^{m_1} (a_{j_1 j_1 r_1} - a_{j_1 j_1 r_2}) \prod_{u_2=1, u_2 \neq u_1}^{n_2} (a_{j_1 j_2 u_1} - a_{j_1 j_2 u_2}) > 0$ and $a_{j_2 j_2 0} \prod_{v_2=1, v_2 \neq v_1}^{m_2} (a_{j_2 j_1 v_1} - a_{j_2 j_1 v_2}) < 0$,

$$(a_{j_1 j_2 u_1}, a_{j_2 j_1 v_1}) = \underbrace{(\mathrm{DP}_{+} : \mathrm{DP}_{-},_{\mathrm{II:DI}}\mathrm{US})}_{\text{inflection down-parabola upper-saddle}}.$$

The equilibrium is a $(\mathrm{DP}_{+}\mathrm{:DP}_{-},_{\mathrm{II:DI}}\mathrm{US})$-inflection down-parabola upper-saddle at $(x_{j_2}^*, x_{j_1}^*) = (a_{j_1 j_2 u_1}, a_{j_2 j_1 v_1})$.

- For $a_{j_1 j_1 0} \prod_{r_2=1, r_2 \neq r_1}^{m_1} (a_{j_1 j_1 r_1} - a_{j_1 j_1 r_2}) \prod_{u_2=1, u_2 \neq u_1}^{n_2} (a_{j_1 j_2 u_1} - a_{j_1 j_2 u_2}) < 0$ and $a_{j_2 j_2 0} \prod_{v_2=1, v_2 \neq v_1}^{m_2} (a_{j_2 j_1 v_1} - a_{j_2 j_1 v_2}) < 0$,

$$(a_{j_1 j_2 u_1}, a_{j_2 j_1 v_1}) = \underbrace{(\mathrm{UP}_{+} : \mathrm{UP}_{-},_{\mathrm{DI:II}}\mathrm{LS})}_{\text{inflection up-parabola lower-saddle}}.$$

The equilibrium is a $(\mathrm{UP}_{+}\mathrm{:UP}_{-},_{\mathrm{DI:II}}\mathrm{LS})$-inflection up-parabola lower-saddle at $(x_{j_2}^*, x_{j_1}^*) = (a_{j_1 j_2 u_1}, a_{j_2 j_1 v_1})$.

2.4 Appearing and Switching for Equilibriums and Flows

In this section, the theorem on appearing and switching bifurcations of singular equilibriums and 1-dimensional flows in the product polynomial systems will be proved.

2.4.1 Appearing and Switching for 1-Dimensional Flows

(iv) The 1-dimensional flow at $(x_{j_1}^*, x_{j_2}^*) = (a_{j_1 j_1 1}, a_{j_1 j_2 1})$ with $a_{j_2 j_2 1} \neq a_{j_1 j_2 1}$ has four classes of 1-dimensional flow bifurcations with $\dot{x}_{j_2}^* = a_{j_2 j_2 0}(a_{j_1 j_2 1} - a_{j_2 j_2 1})^{n_2}$.

(iv$_1$) For $m_1 = 2m_{11}+1$ and $n_1 = 2n_{11}+1$ the equilibrium of $(x_{j_1}^*, x_{j_2}^*) = (a_{j_1 j_1 1}, a_{j_1 j_2 1})$ for the x_{j_1}-direction has the following bifurcation properties with

$$2m_{11} + 1 - \sum_{r_1=1}^{p_1} m_{j_1 r_1} = 2\sum_{\rho_1=1}^{l_1} m_{\rho_1}, 2n_{11} + 1 - \sum_{u_1=1}^{q_1} n_{j_1 u_1} = 2\sum_{\sigma_1=1}^{l_2} n_{\sigma_1};$$

$$\sum_{r_2=1}^{\bar{p}_1} m_{j_1 r_2} = 2m_{11} + 1, \sum_{u_2=1}^{\bar{q}_1} n_{j_1 u_2} = 2n_{11} + 1;$$

$$\sum_{r_i=1}^{p_1} m_{j_1 r_i} = 2m_{11} + 1, \sum_{u_i=1}^{q_1} n_{j_1 u_i} = 2n_{11} + 1, \mathit{for}\ i = 1, 2;$$

l_1, l_2 quadratic polynomials are without real roots and, ρ_1^{th} and σ_1^{th} quadratic polynomials are with powers m_{ρ_1} and n_{σ_1},respectively.

The bifurcation process is discussed through differential equations as follows. There are four bifurcations (I$_1$)–(I$_3$).

(I$_1$) Consider a dynamical system having a single equilibrium (A$_1$) as

$$\begin{aligned}
\dot{x}_{j_1} &= a_{j_1 j_1 0}(x_{j_1} - b_{j_1 j_1 1}) \prod_{r=1}^{m_{11}} [(x_{j_1} - a_{j_1 j_1}^{(r)})^2 + \Delta_{j_1 j_1}^{(r)}] \\
&\quad \times (x_{j_2} - b_{j_1 j_2 1}) \prod_{u=1}^{n_{11}} [(x_{j_2} - a_{j_1 j_2}^{(u)})^2 + \Delta_{j_1 j_2}^{(u)}], \\
\dot{x}_{j_2} &= a_{j_2 j_2 0}(x_{j_1} - a_{j_2 j_1 1})^{m_2};
\end{aligned}$$

where $a_{j_1 j_1 1} = b_{j_1 j_1 1} \neq a_{j_2 j_1 1}$ and $a_{j_1 j_2 1} = b_{j_1 j_2 1}$. Once

$$\begin{aligned}
\Delta_{j_1 j_1}^{(r)} &= \Delta_{j_1 j_1} = 0\,(r = 1, 2, \ldots, m_{11}), \\
\Delta_{j_1 j_2}^{(u)} &= \Delta_{j_1 j_2} = 0\,(u = 1, 2, \ldots, n_{11}),
\end{aligned}$$

if

$$\begin{aligned}
a_{j_1 j_1 1} &= b_{j_1 j_1 1} = a_{j_1 j_1}^{(r)}\ (r = 1, 2, \ldots, m_{11}), \\
a_{j_1 j_2 1} &= b_{j_1 j_2 1} = a_{j_1 j_2}^{(u)}\ (u = 1, 2, \ldots, n_{11}),
\end{aligned}$$

the foregoing differential equations become a dynamical system with a singular equilibrium (B_1) as

$$\dot{x}_{j_1} = a_{j_1 j_1 0}(x_{j_1} - a_{j_1 j_1 1})^{2m_{11}+1}(x_{j_2} - a_{j_1 j_2 1})^{2n_{11}+1},$$
$$\dot{x}_{j_2} = a_{j_2 j_2 0}(x_{j_1} - a_{j_2 j_1 1})^{2m_{21}}.$$

If

$$\Delta_{j_1 j_1}^{(r)} = \Delta_{j_1 j_1} + \delta_r \ (r = 1, 2, \ldots, m_{11}),\ \delta_r > 0;$$
$$\Delta_{j_1 j_2}^{(u)} = \Delta_{j_1 j_2} + \delta_u \ (u = 1, 2, \ldots, n_{11}),\ \delta_u > 0;$$

then

$$a_{j_1 j_1 1}^{(r)}, a_{j_1 j_1 2}^{(r)} = a_{j_1 j_1 1} \pm \varepsilon_r,\ (r = 1, 2, \ldots, m_{11}),$$
$$\{a_{j_1 j_1 1}, a_{j_1 j_1 2}, \ldots, a_{j_1 j_1 (2m_{11}+1)}\} = \mathrm{sort}\{b_{j_1 j_1 1}, a_{j_1 j_1 1}^{(r)}, a_{j_1 j_1 2}^{(r)} | r = 1, 2, \ldots, m_{11}\};$$
$$a_{j_1 j_2 1}^{(u)}, a_{j_1 j_2 2}^{(u)} = a_{j_1 j_2 1} \pm \varepsilon_u,\ (u = 1, 2, \ldots, n_{11}),$$
$$\{a_{j_1 j_2 1}, a_{j_1 j_2 2}, \ldots, a_{j_1 j_2 (2n_{11}+1)}\} = \mathrm{sort}\{b_{j_1 j_2 1}, a_{j_1 j_2 1}^{(u)}, a_{j_1 j_2 2}^{(u)} | u = 1, 2, \ldots, n_{11}\}.$$

Thus, the differential equation becomes a dynamical system with a non-singular equilibrium network (C_1) as

$$\dot{x}_{j_1} = a_{j_1 j_1 0} \prod_{r=1}^{2m_{11}+1} (x_{j_1} - a_{j_1 j_1 r}) \prod_{u=1}^{2n_{11}+1} (x_{j_2} - a_{j_1 j_2 u}),$$
$$\dot{x}_{j_2} = a_{j_2 j_2 0}(x_{j_1} - a_{j_2 j_1 1})^{m_2}.$$

- From the above bifurcation process analysis, at least, only $(m_{11}+1)$-parameter variations in the x_{j_1}-direction are engaged in such a bifurcation from a 1-dimensional flow to the 1-dimensional flow array of $1 \times (2m_{11}+1)$ through the $(2m_{11}+1)^{\text{th}}$-order source and sink bifurcations in the x_{j_1}-direction. The higher-order source and sink flow bifurcations are developed.

  Thus, the 1-dimensional flow appearing or vanishing bifurcation route is as follows.

$$\underbrace{(a_{j_1 j_1 1}, \dot{x}_{j_2})}_{\text{XX}} \rightleftharpoons \underbrace{(a_{j_1 j_1 1}, \dot{x}_{j_2})}_{(2m_{11}+1)^{\text{th}}\text{order XX}} \rightleftharpoons \bigcup_{r=1}^{2m_{11}+1} \underbrace{(a_{j_1 j_1 r}, \dot{x}_{j_2})}_{\text{XX}}.$$

- From the above bifurcation process analysis, at least, only $(n_{11}+1)$-parameter variations in the x_{j_1}-direction are engaged in such a bifurcation from a 1-dimensional flow to the 1-dimensional flow array of $(2n_{11}+1) \times 1$ through the $(2n_{11}+1)^{\text{th}}$-order up-parabola and down-parabola bifurcations in the x_{j_1}-direction. The higher-order up-parabola and down-parabola flow bifurcations are developed.

Thus, the 1-dimensional flow appearing or vanishing bifurcation route is as follows.

$$\underbrace{(a_{j_1j_2 1}, \dot{x}_{j_2})}_{\text{YY}} \rightleftharpoons \underbrace{(a_{j_1j_2 1}, \dot{x}_{j_2})}_{(2n_{11}+1)^{\text{th}}\text{order YY}} \rightleftharpoons \bigcup_{u=1}^{2n_{11}+1} \underbrace{(a_{j_1j_2 u}, \dot{x}_{j_2})}_{\text{YY}}.$$

- From the above bifurcation process analysis, $(m_{11}+n_{11}+2)$-parameter variations in the two directions are engaged in such a bifurcation from a parabola-flow to a $(2m_{11}+1)\times(2n_{11}+1)$ network of hyperbolic and hyperbolic-secant flows through the $(2m_{11}+1)$-source and sink bifurcations in the x_{j_1}-direction, and the $(2n_{11}+1)^{\text{th}}$-order up-parabola and down-parabola flows bifurcations in the x_{j_1}-direction. The higher-order hyperbolic and hyperbolic-recant flow bifurcations are developed.

  Thus, the 1-dimensional flow appearing or vanishing bifurcation route is as follows.

$$\begin{aligned}\underbrace{(a_{j_1j_1 1}, a_{j_1j_2 1})}_{\text{(XX,xF)-flow}} &\rightleftharpoons \underbrace{([(2m_{11}+1), (2n_{11}+1)]\text{XX,xF})}_{[(2m_{11}+1),(2n_{11}+1)]\text{-XX}}\\ &\rightleftharpoons \bigcup_{r=1}^{2m_{11}+1}\bigcup_{u=1}^{2n_{11}+1} \underbrace{(a_{j_1j_1 r}, a_{j_1j_2 u})}_{\text{(YY,xF)-flow}},\end{aligned}$$

  which are the higher-order hyperbolic and hyperbolic-secant flows bifurcations.

(I_2) From Case (I_1), consider a dynamical system having singular equilibriums (A_2) as

$$\begin{aligned}\dot{x}_{j_1} &= a_{j_1j_1 0}\prod_{r_1=1}^{p_1}(x_{j_1}-b_{j_1j_1 r_1})^{m_{j_1 r_1}}\prod_{r=1}^{l_1}[(x_{j_1}-a^{(r)}_{j_1j_1})^2+\Delta^{(r)}_{j_1j_1}]^{m_{j_1 r}}\\ &\quad\times\prod_{u_1=1}^{q_1}(x_{j_2}-b_{j_1j_2 u_1})^{n_{j_1 u_1}}\prod_{u=1}^{l_2}[(x_{j_2}-a^{(u)}_{j_1j_2})^2+\Delta^{(u)}_{j_1j_2}]^{n_{j_1 u}},\\ \dot{x}_{j_2} &= a_{j_2j_2 0}(x_{j_1}-a_{j_2j_1 1})^{m_2};\end{aligned}$$

where

$$2m_{11}+1-\sum_{r_1=1}^{p_1} m_{j_1 r_1} = 2\sum_{\rho_1=1}^{l_1} m_{\rho_1},\; 2n_{11}+1-\sum_{u_1=1}^{q_1} n_{j_1 u_1} = 2\sum_{\sigma_1=1}^{l_2} n_{\sigma_1};$$

$$\sum_{r_2=1}^{\bar{p}_1} m_{j_1r_2} = 2m_{11} + 1, \sum_{u_2=1}^{\bar{q}_1} n_{j_1u_2} = 2n_{11} + 1;$$

l_1, l_2-quadratic polynomials are without real roots and ρ_1^{th} and σ_1^{th}-quadratic polynomials are with powers m_{ρ_1} and n_{σ_1},respectively.

Once

$$\Delta_{j_1j_1}^{(r)} = 0\,(r = 1, 2, \ldots, l_1),$$
$$\Delta_{j_1j_2}^{(u)} = 0\,(u = 1, 2, \ldots, l_2);$$

if

$$a_{j_1j_11} \equiv a_{j_1j_1r_1} = a_{j_1j_1}^{(r)}\,(r_1 = 1, 2, \ldots, p_1; r = 1, 2, \ldots, l_1),$$
$$a_{j_1j_21} \equiv a_{j_1j_2u_1} = a_{j_1j_2}^{(u)}\,(u_1 = 1, 2, \ldots, q_1; u = 1, 2, \ldots, l_2),$$

the foregoing differential equations become a dynamical system with a 1-dimensional flow (B_2) as

$$\dot{x}_{j_1} = a_{j_1j_10}(x_{j_1} - a_{j_1j_11})^{2m_{11}+1}(x_{j_2} - a_{j_1j_21})^{2n_{11}+1},$$
$$\dot{x}_{j_2} = a_{j_2j_20}(x_{j_1} - a_{j_2j_11})^{m_2}.$$

If

$$\Delta_{j_1j_1}^{(r)} = \Delta_{j_1j_1} + \delta_r\,(r = 1, 2, \ldots, l_1),\ \delta_r > 0;$$
$$\Delta_{j_1j_2}^{(u)} = \Delta_{j_1j_2} + \delta_u\,(u = 1, 2, \ldots, l_2),\ \delta_s > 0;$$

then

$$a_{j_1j_21}^{(u)}, a_{j_1j_22}^{(u)} = a_{j_1j_21} \pm \varepsilon_u, (u = 1, 2, \ldots, l_1),$$
$$\{a_{j_1j_21}, a_{j_1j_22}, \ldots, a_{j_1j_2\bar{q}_1}\}$$
$$= \text{sort}\{a_{j_1j_21}, a_{j_1j_22}, \ldots, a_{j_1j_2q_1}; a_{j_1j_21}^{(u)}, a_{j_1j_22}^{(u)} | u = 1, 2, \ldots, l_1\};$$
$$a_{j_2j_11}^{(v)}, a_{j_2j_12}^{(v)} = a_{j_2j_11} \pm \varepsilon_v, (v = 1, 2, \ldots, l_2),$$
$$\{a_{j_2j_11}, a_{j_2j_12}, \ldots, a_{j_2j_1p_2}\}$$
$$= \text{sort}\{a_{j_2j_11}, a_{j_2j_12}, \ldots, a_{j_2j_1p_2}; a_{j_2j_11}^{(v)}, a_{j_2j_21}^{(v)} | v = 1, 2, \ldots, l_2\}.$$

Thus, the differential equation becomes a dynamical system with singular 1-dimensional flows (C_2) as

$$\begin{aligned}\dot{x}_{j_1} &= a_{j_1 j_1 0} \prod_{r_2=1}^{\overline{p}_1} (x_{j_1} - a_{j_1 j_1 r_2})^{m_{j_1 r_2}} \prod_{u_2=1}^{\overline{q}_1} (x_{j_2} - a_{j_1 j_1 u_2})^{n_{j_1 u_2}},\\ \dot{x}_{j_2} &= a_{j_2 j_2 0}(x_{j_1} - a_{j_2 j_1 1})^{m_2}\end{aligned}$$

where

$$\Sigma_{r_2=1}^{\overline{p}_1} m_{j_1 r_2} = 2m_{11} + 1,\ \Sigma_{u_2=1}^{\overline{q}_1} n_{j_1 u_2} = 2n_{11} + 1.$$

- From the above bifurcation process analysis, only $\overline{p}_1$-parameter variations in the x_{j_1}-direction from a p_1 to $\overline{p}_1$ 1-dimensional flow array through the $(2m_{11} + 1)^{\text{th}}$-order source and sink flow bifurcations in the x_{j_1}-direction.

  Thus, the 1-dimensional flow appearing or vanishing bifurcation route from a p_1 to $\overline{p}_1$ 1-dimensional flow array is expressed as

$$\bigcup_{r_1=1}^{p_1} \underbrace{(a_{j_1 j_1 r_1}, \dot{x}_{j_2})}_{(\mathrm{m}_{j_1 r_1})^{\text{th}}\text{order XX}} \rightleftharpoons \underbrace{(a_{j_1 j_1 1}, \dot{x}_{j_2})}_{(2m_{11}+1)^{\text{th}}\text{ order XX}} \rightleftharpoons \bigcup_{r_2=1}^{\overline{p}_1} \underbrace{(a_{j_1 j_1 r_2}, \dot{x}_{j_2})}_{(\mathrm{m}_{j_1 r_2})^{\text{th}}\text{ order XX}}.$$

- From the above bifurcation process analysis, only $\overline{q}_1$-parameter variations in the x_{j_1}-direction from a q_1 to $\overline{q}_1$ 1-dimensional flow array through the $(2n_{11} + 1)^{\text{th}}$-order up-parabola and down-parabola flow bifurcations in the x_{j_1}-direction.

  Thus, the 1-dimensional flow appearing or vanishing bifurcation route from a q_1 to $\overline{q}_1$ 1-dimensional flow array is expressed as

$$\bigcup_{u_1=1}^{q_1} \underbrace{(a_{j_1 j_2 u_1}, \dot{x}_{j_2})}_{(n_{j_1 u_1})^{\text{th}}\text{order YY}} \rightleftharpoons \underbrace{(a_{j_1 j_2 1}, \dot{x}_{j_2})}_{(2n_{11}+1)^{\text{th}}\text{ order YY}} \rightleftharpoons \bigcup_{u_2=1}^{\overline{q}_1} \underbrace{(a_{j_1 j_2 u_2}, \dot{x}_{j_2})}_{(n_{j_1 u_2})^{\text{th}}\text{ order YY}}.$$

- From the above bifurcation process analysis, only $\overline{q}_1 + \overline{p}_1$-parameter variations in the x_{j_1}-direction from a $q_1 \times p_1$ to $\overline{q}_1 \times \overline{p}_1$ network of 1-dimensional flows through the $(2m_{11} + 1)^{\text{th}}$-order source and sink flows bifurcations in the x_{j_1}-direction, and the $(2n_{11} + 1)^{\text{th}}$-order up-parabola and down-parabola flows bifurcations in the x_{j_1}-direction.

  Thus, the 1-dimensional flow appearing or vanishing bifurcation route is as follows:

$$\begin{aligned}\bigcup_{r_1=1}^{p_1} \bigcup_{u_1=1}^{q_1} \underbrace{(a_{j_1 j_1 r_1}, a_{j_1 j_2 u_1})}_{([(m_{j_1 r_1}),(n_{j_1 u_1})]\text{XX,xF)-flow}} &\rightleftharpoons \underbrace{([(2m_{11}+1), (2n_{11}+1)]\text{XX, xF})}_{[(2m_{11}+1),(2n_{11}+1)]\text{-XX}}\\ &\rightleftharpoons \bigcup_{r_2=1}^{\overline{p}_1} \bigcup_{u_2=1}^{\overline{q}_1} \underbrace{(a_{j_1 j_1 r_2}, a_{j_1 j_2 u_2})}_{([(m_{j_1 r_2}),(n_{j_1 u_1})]\text{XX,xF)-flow}},\end{aligned}$$

which are the higher-order hyperbolic and hyperbolic-secant flows bifurcations.

(I$_3$) Consider two dynamical systems with the same 1-dimensional flows with locations switched (A$_3$,C$_3$) for $i = 1, 2$ as

$$\dot{x}_{j_1} = a_{j_1 j_1 0} \prod_{r_i=1}^{p_1} (x_{j_1} - a_{j_1 j_1 r_i})^{m_{j_1 r_i}} \prod_{u_i=1}^{q_1} (x_{j_2} - a_{j_1 j_2 s_i})^{n_{j_1 u_i}},$$
$$\dot{x}_{j_2} = a_{j_2 j_2 0}(x_{j_1} - a_{j_2 j_1 1})^{m_2};$$

where

$$\sum_{r_i=1}^{p_1} m_{j_1 r_i} = 2m_{11} + 1, \sum_{u_i=1}^{q_1} n_{j_1 u_i} = 2n_{11} + 1, \text{ for } i = 1, 2.$$

Consider a dynamical system as a singular 1-dimensional flow (B$_3$) as

$$\dot{x}_{j_1} = a_{j_1 j_1 0}(x_{j_1} - a_{j_1 j_1 1})^{2m_{11}+1}(x_{j_2} - a_{j_1 j_2 1})^{2n_{11}+1},$$
$$\dot{x}_{j_2} = a_{j_2 j_2 0}(x_{j_1} - a_{j_2 j_1 1})^{m_2}.$$

For $r_i = 1, 2, \ldots, p_1$ and $u_i = 1, 2, \ldots, q_1$ $(i = 1, 2)$, define a function as

$$\Delta_{j_1 j_1}^{(r_1 r_2)} = (a_{j_1 j_1 r_1} - a_{j_1 j_1 r_2})^2 \; (r_1, r_2 \in \{1, 2, \ldots, p_1\}, r_1 \neq r_2),$$
$$\Delta_{j_1 j_2}^{(u_1 u_2)} = (a_{j_1 j_2 u_1} - a_{j_1 j_2 u_2})^2 \; (u_1, u_2 \in \{1, 2, \ldots, q_1\}, u_1 \neq u_2).$$

For $r_i = 1, 2, \ldots, p$ and $s_i = 1, 2, \ldots, q_1$ $(i = 1, 2)$, if

$$\Delta_{j_1 j_1}^{(r_1 r_2)} = (a_{j_1 j_1 r_1} - a_{j_1 j_1 r_2})^2 = 0 \text{ and } \Delta_{j_1 j_2}^{(u_1 u_2)} = (a_{j_1 j_2 u_1} - a_{j_1 j_2 u_2})^2 = 0,$$

there are three bifurcations as follows.

- Two 1-dimensional flows of $(a_{j_1 j_1 r_1}, \dot{x}_{j_2})$ and $(a_{j_1 j_1 r_2}, \dot{x}_{j_2})$ switching at point $(a_{j_1 j_1 1}, \dot{x}_{j_2})$ with the same order singularity are given through

$$\bigcup_{r_1=1}^{p_1} \underbrace{(a_{j_1 j_1 r_1}, \dot{x}_{j_2})}_{(m_{j_1 r_1})^{\text{th}} \text{order XX}} \rightleftharpoons \underbrace{(a_{j_1 j_1 1}, \dot{x}_{j_2})}_{(2m_{11}+1)^{\text{th}} \text{ order XX}} \rightleftharpoons \bigcup_{r_2=1}^{p_1} \underbrace{(a_{j_1 j_1 r_2}, \dot{x}_{j_2})}_{(m_{j_1 r_2})^{\text{th}} \text{order XX}}$$
$$\{a_{j_1 j_1 r_1} \rightleftharpoons a_{j_1 j_1 r_2} | m_{j_1 r_1} = m_{j_1 r_2}; r_1, r_2 \in \{1, 2, \ldots, p_1\}, r_1 \neq r_2\};$$

and

$$a_{j_1 j_1 r_1} = a_{j_1 j_1 1} - \varepsilon_{r_1}, a_{j_1 j_1 r_2} = a_{j_1 j_1 1} + \varepsilon_{r_2},$$
$$(a_{j_1 j_1 r_1} = a_{j_1 j_1 1} + \varepsilon_{r_1}, a_{j_1 j_1 r_2} = a_{j_1 j_1 1} - \varepsilon_{r_2});$$

$$\varepsilon_{r_1}, \varepsilon_{r_2} > 0;\ r_i = 1, 2, \ldots, p_1\ (i = 1, 2).$$

- Two 1-dimensional flows of $(a_{j_1 j_2 u_1}, \dot{x}_{j_2})$ and $(a_{j_1 j_2 u_2}, \dot{x}_{j_2})$ switching at point $(a_{j_1 j_2 1}, \dot{x}_{j_2})$ with the same order singularity are given through

$$\bigcup_{u_1=1}^{q_1} \underbrace{(a_{j_1 j_2 u_1}, \dot{x}_{j_2})}_{(n_{j_1 u_1})^{\text{th}} \text{order YY}} \rightleftharpoons \underbrace{(a_{j_1 j_2 1}, \dot{x}_{j_2})}_{(2n_{11}+1)^{\text{th}} \text{ order YY}} \rightleftharpoons \bigcup_{u_2=1}^{q_1} \underbrace{(a_{j_1 j_2 u_2}, \dot{x}_{j_2})}_{(n_{j_1 u_2})^{\text{th}} \text{order YY}}$$
$$\{a_{j_1 j_2 u_1} \rightleftharpoons a_{j_1 j_2 u_2} | n_{j_1 u_1} = n_{j_1 u_2};\ u_1, u_2 \in \{1, 2, \ldots, q\}, u_1 \neq u_2\};$$

and

$$\begin{aligned}
&a_{j_1 j_2 u_1} = a_{j_1 j_2 1} - \varepsilon_{u_1}, a_{j_1 j_2 u_2} = a_{j_1 j_2 1} + \varepsilon_{u_2},\\
&(a_{j_1 j_2 u_1} = a_{j_1 j_2 1} + \varepsilon_{u_1}, a_{j_1 j_2 u_2} = a_{j_1 j_2 1} - \varepsilon_{u_2});\\
&\varepsilon_{u_1}, \varepsilon_{u_2} > 0;\ u_i = 1, 2, \ldots, q_1\ (i = 1, 2).
\end{aligned}$$

- Two 1-dimensional flows of $(a_{j_1 j_1 r_1}, a_{j_1 j_2 u_1})$ and $(a_{j_1 j_1 r_2}, a_{j_1 j_2 u_2})$ switching at point $(a_{j_1 j_1 1}, a_{j_1 j_2 1})$ with the same order singularity are given through

$$\begin{aligned}
\bigcup_{r_1=1}^{p_1} \bigcup_{u_1=1}^{q_1} \underbrace{(a_{j_1 j_1 r_1}, a_{j_1 j_2 u_1})}_{([(m_{j_1 r_1}),(n_{j_1 u_1})]\text{XX,xF})\text{-flow}} &\rightleftharpoons \underbrace{([(2m_{11} + 1),(2n_{11}+1)]\text{XX, xF})}_{[(2m_{11} + 1),(2n_{11}+1)]\text{ -XX}}\\
&\rightleftharpoons \bigcup_{r_2=1}^{p_1} \bigcup_{u_2=1}^{q_1} \underbrace{(a_{j_1 j_1 r_2}, a_{j_1 j_2 u_2})}_{([(m_{j_1 r_2}),(n_{j_1 r_2})]\text{XX,xF})\text{-flow}}
\end{aligned}$$

with

$$\begin{aligned}
&a_{j_1 j_1 r_1} = a_{j_1 j_1 1} - \varepsilon_{r_1}, a_{j_1 j_1 r_2} = a_{j_1 j_1 1} + \varepsilon_{r_2},\\
&(a_{j_1 j_1 r_1} = a_{j_1 j_1 1} + \varepsilon_{r_1}, a_{j_1 j_1 r_2} = a_{j_1 j_1 1} - \varepsilon_{r_2}),\\
&\varepsilon_{r_1}, \varepsilon_{r_2} > 0,\ r_i = 1, 2, \ldots, p_1\ (i = 1, 2);\\
&a_{j_1 j_2 u_1} = a_{j_1 j_2 1} - \varepsilon_{u_1}, a_{j_1 j_2 u_2} = a_{j_1 j_2 1} + \varepsilon_{u_2},\\
&(a_{j_1 j_2 u_1} = a_{j_1 j_2 1} + \varepsilon_{u_1}, a_{j_1 j_2 u_2} = a_{j_1 j_2 1} - \varepsilon_{u_2}),\\
&\varepsilon_{u_1}, \varepsilon_{u_2} > 0,\ u_i = 1, 2, \ldots, q_1\ (i = 1, 2).
\end{aligned}$$

The switching bifurcations are the higher-order hyperbolic and hyperbolic-secant flows bifurcations .

In summary, from the analysis in case (I_1)–(I_3), the 1-dimensional flow of $(x_{j_1}^*, x_{j_2}^*) = (a_{j_1 j_1 1}, a_{j_1 j_2 1})$ for $m_1 = 2m_{11} + 1$ and $n_1 = 2n_{11} + 1$ has the bifurcation properties as stated in (iv_{1a})–(iv_{1d}) through Eqs. (1.305)–(1.324).

(iv$_2$) For $m_1 = 2m_{11}$ and $n_1 = 2n_{11}+1$ the equilibrium of $(x_{j_1}^*, x_{j_2}^*) = (a_{j_1j_11}, a_{j_1j_21})$ for the x_{j_1}-direction has the following bifurcation properties with

$$2m_{11} - \sum_{r_1=1}^{p_1} m_{j_1r_1} = 2\sum_{\rho_1=1}^{l_1} m_{\rho_1},\ 2n_{11}+1 - \sum_{u_1=1}^{q_1} n_{j_1u_1} = 2\sum_{\sigma_1=1}^{l_2} n_{\sigma_1};$$

$$\sum_{r_2=1}^{\overline{p}_1} m_{j_1r_2} = 2m_{11},\ \sum_{u_2=1}^{\overline{q}_1} n_{j_1u_2} = 2n_{11}+1;$$

$$\sum_{r_i=1}^{p_1} m_{j_1r_i} = 2m_{11},\ \sum_{u_i=1}^{q_1} n_{j_1u_i} = 2n_{11}+1,\ \text{for } i=1,2;$$

l_1, l_2-quadratic polynomials are without real roots and ρ_1^{th} and σ_1^{th}-quadratic polynomials are with powers m_{ρ_1} and n_{σ_1}, respectively.

The bifurcation process is discussed through differential equations as follows. There are four bifurcations (I$_1$)–(I$_3$).

(I$_1$) Consider a dynamical system having a single equilibrium (A$_1$) as

$$\begin{aligned}\dot{x}_{j_1} &= a_{j_1j_10}\prod_{r=1}^{m_{11}}[(x_{j_1}-a_{j_1j_1}^{(r)})^2+\Delta_{j_1j_1}^{(r)}]\\ &\quad\times(x_{j_2}-b_{j_1j_21})\prod_{u=1}^{n_{11}}[(x_{j_2}-a_{j_1j_2}^{(u)})^2+\Delta_{j_1j_2}^{(u)}],\\ \dot{x}_{j_2} &= a_{j_2j_20}(x_{j_1}-a_{j_2j_11})^{m_2};\end{aligned}$$

where $a_{j_1j_21} = b_{j_1j_21}$. Once

$$\begin{aligned}\Delta_{j_1j_1}^{(r)} &= \Delta_{j_1j_1} = 0\,(r=1,2,\ldots,m_{11}),\\ \Delta_{j_1j_2}^{(u)} &= \Delta_{j_1j_2} = 0\,(u=1,2,\ldots,n_{11}),\end{aligned}$$

if

$$\begin{aligned}a_{j_1j_11} &= b_{j_1j_11} = a_{j_1j_1}^{(r)}\ (r=1,2,\ldots,m_{11}),\\ a_{j_1j_21} &= b_{j_1j_21} = a_{j_1j_2}^{(u)}\ (u=1,2,\ldots,n_{11}),\end{aligned}$$

the foregoing differential equations become a dynamical system with a singular equilibrium (B$_1$) as

$$\begin{aligned}\dot{x}_{j_1} &= a_{j_1j_10}(x_{j_1}-a_{j_1j_11})^{2m_{11}}(x_{j_2}-a_{j_1j_21})^{2n_{11}+1},\\ \dot{x}_{j_2} &= a_{j_2j_20}(x_{j_1}-a_{j_2j_11})^{m_2}.\end{aligned}$$

If

$$\Delta_{j_1j_1}^{(r)} = \Delta_{j_1j_1} + \delta_r \, (r = 1, 2, \ldots, m_{11}), \; \delta_r > 0;$$
$$\Delta_{j_1j_2}^{(u)} = \Delta_{j_1j_2} + \delta_u \, (u = 1, 2, \ldots, n_{11}), \; \delta_s > 0;$$

then

$$\begin{aligned}
&a_{j_1j_11}^{(r)}, a_{j_1j_12}^{(r)} = a_{j_1j_11} \pm \varepsilon_r, \; (r = 1, 2, \ldots, m_{11}),\\
&\{a_{j_1j_11}, a_{j_1j_12}, \ldots, a_{j_1j_1(2m_{11}+1)}\} = \mathrm{sort}\{b_{j_1j_11}, a_{j_1j_11}^{(r)}, a_{j_1j_12}^{(r)} | r = 1, 2, \ldots, m_{11}\};\\
&a_{j_1j_21}^{(u)}, a_{j_1j_22}^{(u)} = a_{j_1j_21} \pm \varepsilon_u, \; (u = 1, 2, \ldots, n_1),\\
&\{a_{j_1j_21}, a_{j_1j_22}, \ldots, a_{j_1j_2(2n_{11}+1)}\} = \mathrm{sort}\{b_{j_1j_21}, a_{j_1j_21}^{(u)}, a_{j_1j_22}^{(u)} | u = 1, 2, \ldots, n_{11}\}.
\end{aligned}$$

Thus, the differential equation becomes a dynamical system with a non-singular equilibrium network (C_1) as

$$\begin{aligned}
\dot{x}_{j_1} &= a_{j_1j_10} \prod_{r=1}^{2m_{11}} (x_{j_1} - a_{j_1j_1r}) \prod_{u=1}^{2n_{11}+1} (x_{j_2} - a_{j_1j_2u}),\\
\dot{x}_{j_2} &= a_{j_2j_20}(x_{j_1} - a_{j_2j_11})^{m_2}.
\end{aligned}$$

- From the above bifurcation process analysis, at least, only m_{11}-parameter variations in the x_{j_1}-direction are engaged in such a bifurcation from a 1-dimensional flow to the 1-dimensional flow array of $1 \times (2m_{11})$ through the $(2m_{11})^{\text{th}}$-order source and sink bifurcations in the x_{j_1}-direction. The higher-order source and sink flow bifurcations are developed.

  Thus, the 1-dimensional flow appearing or vanishing bifurcation route is as follows,

$$\underbrace{(a_{j_1j_11}, \dot{x}_{j_2})}_{\text{XX}} \rightleftharpoons \underbrace{(a_{j_1j_11}, \dot{x}_{j_2})}_{(2m_{11})^{\text{th}}\text{order XX}} \rightleftharpoons \bigcup_{r=1}^{2m_{11}} \underbrace{(a_{j_1j_1r}, \dot{x}_{j_2})}_{\text{XX}}.$$

- From the above bifurcation process analysis, at least, only $(n_{11} + 1)$-parameter variations in the x_{j_1}-direction are engaged in such a bifurcation from a 1-dimensional flow to the 1-dimensional flow array of $(2n_{11} + 1) \times 1$ through the $(2n_{11} + 1)^{\text{th}}$-order up-parabola and down-parabola bifurcations in the x_{j_1}-direction. The higher-order up-parabola and down-parabola flow bifurcations are developed.

  Thus, the 1-dimensional flow appearing or vanishing bifurcation route is as follows,

$$\underbrace{(a_{j_1j_2 1}, \dot{x}_{j_2})}_{\text{YY}} \rightleftharpoons \underbrace{(a_{j_1j_2 1}, \dot{x}_{j_2})}_{(2n_{11}+1)^{\text{th}}\text{order YY}} \rightleftharpoons \bigcup_{r=1}^{2n_{11}+1} \underbrace{(a_{j_1j_2 u}, \dot{x}_{j_2})}_{\text{YY}}.$$

- From the above bifurcation process analysis, $(m_{11} + n_{11} + 1)$-parameter variations in the two directions are engaged in such a bifurcation from a parabola-flow to a $(2m_{11}) \times (2n_{11}+1)$ network of hyperbolic and hyperbolic-secant flows through the $(2m_{11})^{\text{th}}$-order upper-saddle and lower-saddle bifurcations in the x_{j1}-direction, and the $(2n_{11} + 1)^{\text{th}}$-order up-parabola and down-parabola flows bifurcations in the x_{j1}-direction. The higher-order hyperbolic-to-hyperbolic-secant flow bifurcations are developed.

  Thus, the 1-dimensional flow appearing or vanishing bifurcation route is as follows,

$$\begin{aligned}\underbrace{(a_{j_1j_2 1}, \dot{x}_{j_2})}_{\text{(XX,xF)-flow}} &\rightleftharpoons \underbrace{([(2m_{11}), (2n_{11} + 1)]\text{XX,xF})}_{[(2m_{11}),(2n_{11}+1)]\text{-XX}}\\ &\rightleftharpoons \bigcup_{r=1}^{2m_{11}} \bigcup_{u=1}^{2n_{11}+1} \underbrace{(a_{j_1j_1 r}, a_{j_1j_2 u})}_{\text{(YY,xF)-flow}}.\end{aligned}$$

  which are the higher-order hyperbolic-to-hyperbolic-secant flows bifurcations.

(I_2) From Case (I_1), consider a dynamical system having singular equilibriums (A_2) as

$$\begin{aligned}\dot{x}_{j_1} &= a_{j_1j_1 0} \prod_{r_1=1}^{p_1} (x_{j_1} - b_{j_1j_1 r_1})^{m_{j_1 r_1}} \prod_{r=1}^{l_1} [(x_{j_1} - a_{j_1j_1}^{(r)})^2 + \Delta_{j_1j_1}^{(r)}]^{m_{j_1 r}}\\ &\quad \times \prod_{u_1=1}^{q_1} (x_{j_2} - b_{j_1j_2 u_1})^{n_{j_1 u_1}} \prod_{u=1}^{l_2} [(x_{j_2} - a_{j_1j_2}^{(u)})^2 + \Delta_{j_1j_2}^{(u)}]^{n_{j_1 u}},\\ \dot{x}_{j_2} &= a_{j_2j_2 0}(x_{j_1} - a_{j_2j_1 1})^{m_2};\end{aligned}$$

where

$$\begin{aligned}&2m_{11} - \sum_{r_1=1}^{p_1} m_{j_1 r_1} = 2\sum_{\rho_1=1}^{l_1} m_{\rho_1},\ 2n_{11} + 1 - \sum_{u_1=1}^{q_1} n_{j_1 u_1} = 2\sum_{\sigma_1=1}^{l_2} n_{\sigma_1};\\ &\sum_{r_2=1}^{\bar{p}_1} m_{j_1 r_2} = 2m_{11},\ \sum_{u_2=1}^{\bar{q}_1} n_{j_1 u_2} = 2n_{11} + 1;\end{aligned}$$

l_1, l_2-quadratic polynomials are without real roots and ρ_1^{th} and σ_1^{th}-quadratic polynomials are with powers m_{ρ_1} and n_{σ_1},respectively.

Once

$$\begin{aligned}\Delta_{j_1j_1}^{(r)} &= 0\,(r = 1, 2, \ldots, l_1),\\ \Delta_{j_1j_2}^{(u)} &= 0\,(u = 1, 2, \ldots, l_2);\end{aligned}$$

if

$$\begin{aligned}a_{j_1j_11} \equiv a_{j_1j_1r_1} &= a_{j_1j_1}^{(r)}\,(r_1 = 1, 2, \ldots, p_1;\, r = 1, 2, \ldots, l_1),\\ a_{j_1j_21} \equiv a_{j_1j_2u_1} &= a_{j_1j_2}^{(u)}\,(u_1 = 1, 2, \ldots, q_1;\, u = 1, 2, \ldots, l_2),\end{aligned}$$

the foregoing differential equations become a dynamical system with a 1-dimensional flow (B_2) as

$$\begin{aligned}\dot{x}_{j_1} &= a_{j_1j_10}(x_{j_1} - a_{j_1j_11})^{2m_{11}}(x_{j_2} - a_{j_1j_21})^{2n_{11}+1},\\ \dot{x}_{j_2} &= a_{j_2j_20}(x_{j_1} - a_{j_2j_11})^{m_2}.\end{aligned}$$

If

$$\begin{aligned}\Delta_{j_1j_1}^{(r)} &= \Delta_{j_1j_1} + \delta_r\,(r = 1, 2, \ldots, l_1),\ \delta_r > 0;\\ \Delta_{j_1j_2}^{(u)} &= \Delta_{j_1j_2} + \delta_s\,(u = 1, 2, \ldots, l_2),\ \delta_u > 0;\end{aligned}$$

then

$$\begin{aligned}&a_{j_1j_11}^{(r)}, a_{j_1j_12}^{(r)} = a_{j_1j_11} \pm \varepsilon_r,\ (r = 1, 2, \ldots, l_1),\\ &\{a_{j_1j_11}, a_{j_1j_12}, \ldots, a_{j_1j_1\bar{p}_1}\}\\ &= \mathrm{sort}\{a_{j_1j_11}, a_{j_1j_12}, \ldots, a_{j_1j_1p_1};\, a_{j_1j_11}^{(r)}, a_{j_1j_12}^{(r)} | r = 1, 2, \ldots, l_1\};\\ &a_{j_1j_21}^{(u)}, a_{j_1j_22}^{(u)} = a_{j_1j_21} \pm \varepsilon_u,\ (u = 1, 2, \ldots, l_2),\\ &\{a_{j_1j_21}, a_{j_1j_22}, \ldots, a_{j_1j_2\bar{q}_1}\}\\ &= \mathrm{sort}\{a_{j_1j_21}, a_{j_1j_22}, \ldots, a_{j_1j_2q_1};\, a_{j_2j_11}^{(u)}, a_{j_2j_12}^{(u)} | u = 1, 2, \ldots, l_2\}.\end{aligned}$$

Thus, the differential equation becomes a dynamical system with singular 1-dimensional flows (C_2) as

$$\begin{aligned}\dot{x}_{j_1} &= a_{j_1j_10}\prod_{r_2=1}^{\bar{p}_1}(x_{j_1} - a_{j_1j_1r_2})^{m_{j_1r_2}}\prod_{u_2=1}^{\bar{q}_1}(x_{j_2} - a_{j_1j_1u_2})^{n_{j_1u_2}},\\ \dot{x}_{j_2} &= a_{j_2j_20}(x_{j_1} - a_{j_2j_11})^{m_2}\end{aligned}$$

where

$$\Sigma_{r_2=1}^{\bar{p}_1} m_{j_1r_2} = 2m_{11},\ \Sigma_{u_2=1}^{\bar{q}_1} n_{j_1u_2} = 2n_{11} + 1.$$

- From the above bifurcation process analysis, only $\overline{p}_1$-parameter variations in the x_{j_1}-direction from a p_1 to $\overline{p}_1$ 1-dimensional flow array through the $(2m_{11})^{\text{th}}$-order source and sink flow bifurcations in the x_{j_1}-direction.

  Thus, the 1-dimensional flow appearing or vanishing bifurcation route from a p_1 to p_2 1-dimensional flow array is expressed as

$$\bigcup_{r_1=1}^{p_1} \underbrace{(a_{j_1j_1r_1}, \dot{x}_{j_2})}_{(m_{j_1r_1})^{\text{th}}\text{order XX}} \rightleftharpoons \underbrace{(a_{j_1j_11}, \dot{x}_{j_2})}_{(2m_{11})^{\text{th}}\text{ order XX}} \rightleftharpoons \bigcup_{r_2=1}^{\overline{p}_1} \underbrace{(a_{j_1j_1r_2}, \dot{x}_{j_2})}_{(m_{j_1r_2})^{\text{th}}\text{ order XX}}.$$

- From the above bifurcation process analysis, only $\overline{q}_1$-parameter variations in the x_{j_1}-direction from a q_1 to $\overline{q}_1$ 1-dimensional flow array through the $(2n_{11}+1)^{\text{th}}$ order up-parabola and down-parabola flow bifurcations in the x_{j_1}-direction.

  Thus, the 1-dimensional flow appearing or vanishing bifurcation route from a q_1 to $\overline{q}_1$ 1-dimensional flow array is expressed as

$$\bigcup_{u_1=1}^{q_1} \underbrace{(a_{j_1j_2u_1}, \dot{x}_{j_2})}_{(n_{j_1u_1})^{\text{th}}\text{order YY}} \rightleftharpoons \underbrace{(a_{j_1j_21}, \dot{x}_{j_2})}_{(2n_1+1)^{\text{th}}\text{ order YY}} \rightleftharpoons \bigcup_{u_2=1}^{\overline{q}_1} \underbrace{(a_{j_1j_2u_2}, \dot{x}_{j_2})}_{(n_{j_1u_2})^{\text{th}}\text{order YY}}.$$

- From the above bifurcation process analysis, only $\overline{q}_1+\overline{p}_1$-parameter variations in the x_{j_1}-direction from a $q_1 \times p_1$ to $\overline{q}_1 \times \overline{p}_1$ network of 1-dimensional flows through the $(2m_{11})^{\text{th}}$-order upper and lower-saddle flows bifurcations in the x_{j_1}-direction, and the $(2n_{11}+1)^{\text{th}}$-order up-parabola and down-parabola flows bifurcations in the x_{j_1}-direction.

  Thus, the 1-dimensional flow appearing or vanishing bifurcation route is as follows:

$$\begin{aligned}\bigcup_{r_1=1}^{p_1}\bigcup_{u_1=1}^{q_1} \underbrace{(a_{j_1j_1r_1}, a_{j_1j_2u_1})}_{([(m_{j_1r_1}),(n_{j_1u_1})]\text{XX,xF})\text{-flow}} &\rightleftharpoons \underbrace{([(2m_{11}), (2n_{11}+1)]\text{XX, xF})}_{[(2m_{11}),(2n_{11}+1)]\text{-XX}} \\ &\rightleftharpoons \bigcup_{r_2=1}^{\overline{p}_1}\bigcup_{u_2=1}^{\overline{q}_1} \underbrace{(a_{j_1j_1r_2}, a_{j_1j_2u_2})}_{([(m_{j_1r_2}),(n_{j_1u_1})]\text{XX,xF})\text{-flow}}.\end{aligned}$$

  which are the higher-order hyperbolic-to-hyperbolic-secant flows bifurcations.

(I_3) Consider two dynamical systems with the same 1-dimensional flows with locations switched (A_3, C_3) for $i=1,2$ as

$$\dot{x}_{j_1} = a_{j_1j_10}\prod_{r_i=1}^{p_1}(x_{j_1}-a_{j_1j_1r_i})^{m_{j_1r_i}}\prod_{u_i=1}^{q_1}(x_{j_1}-a_{j_2j_1u_i})^{n_{j_1u_i}},$$

$$\dot{x}_{j_2} = a_{j_2j_20}(x_{j_1} - a_{j_2j_11})^{m_2};$$

where

$$\sum_{r_i=1}^{p_1} m_{j_1r_i} = 2m_{11}, \sum_{u_i=1}^{q_1} n_{j_1u_i} = 2n_{11} + 1, \text{ for } i = 1, 2.$$

Consider a dynamical system as a singular 1-dimensional flow (B_3) as

$$\begin{aligned}\dot{x}_{j_1} &= a_{j_1j_10}(x_{j_1} - a_{j_1j_11})^{2m_{11}}(x_{j_2} - a_{j_1j_21})^{2n_{11}+1},\\ \dot{x}_{j_2} &= a_{j_2j_10}(x_{j_1} - a_{j_2j_11})^{m_2}.\end{aligned}$$

For $r_i = 1, 2, \ldots, p_1$ and $s_i = 1, 2, \ldots, q_1$ $(i = 1, 2)$, define a function as

$$\begin{aligned}\Delta_{j_1j_1}^{(r_1r_2)} &= (a_{j_1j_1r_1} - a_{j_1j_1r_2})^2\ (r_1, r_2 \in \{1, 2, \ldots, p_1\}, r_1 \neq r_2),\\ \Delta_{j_1j_2}^{(u_1u_2)} &= (a_{j_1j_2u_1} - a_{j_1j_2u_2})^2\ (u_1, u_2 \in \{1, 2, \ldots, q_1\}, u_1 \neq u_2).\end{aligned}$$

For $r_i = 1, 2, \ldots, p_1$ and $s_i = 1, 2, \ldots, q_1$ $(i = 1, 2)$, if

$$\Delta_{j_1j_1}^{(r_1r_2)} = (a_{j_1j_1r_1} - a_{j_1j_1r_2})^2 = 0 \text{ and } \Delta_{j_1j_2}^{(u_1u_2)} = (a_{j_1j_2u_1} - a_{j_1j_2u_2})^2 = 0,$$

there are three bifurcations as follows.

- Two 1-dimensional flows of $(a_{j_1j_1r_1}, \dot{x}_{j_2})$ and $(a_{j_1j_1r_2}, \dot{x}_{j_2})$ switching at point $(a_{j_1j_11}, \dot{x}_{j_2})$ with the same order singularity are given through

$$\bigcup_{r_1=1}^{p_1} \underbrace{(a_{j_1j_1r_1}, \dot{x}_{j_2})}_{(m_{j_1r_1})^{\text{th}}\text{order } XX} \rightleftharpoons \underbrace{(a_{j_1j_11}, \dot{x}_{j_2})}_{(2m_{11})^{\text{th}} \text{ order XX}} \rightleftharpoons \bigcup_{r_2=1}^{p_1} \underbrace{(a_{j_1j_1r_2}, \dot{x}_{j_2})}_{(m_{j_1r_2})^{\text{th}}\text{order } XX}$$
$$\left\{a_{j_1j_1r_1} \rightleftharpoons a_{j_1j_1r_2} | m_{j_1r_1} = m_{j_1r_2};\ r_1, r_2 \in \{1, 2, \ldots, p_1\}, r_1 \neq r_2\right\};$$

  and

$$\begin{aligned}&a_{j_1j_1r_1} = a_{j_1j_11} - \varepsilon_{r_1}, a_{j_1j_1r_2} = a_{j_1j_11} + \varepsilon_{r_2},\\ &(a_{j_1j_1r_1} = a_{j_1j_11} + \varepsilon_{r_1}, a_{j_1j_1r_2} = a_{j_1j_11} - \varepsilon_{r_2});\\ &\varepsilon_{r_1}, \varepsilon_{r_2} > 0;\ r_i = 1, 2, \ldots, p_1\ (i = 1, 2).\end{aligned}$$

- Two 1-dimensional flows of $(a_{j_1j_2u_1}, \dot{x}_{j_2})$ and $(a_{j_1j_2u_1}, \dot{x}_{j_2})$ switching at point $(a_{j_1j_21}, \dot{x}_{j_2})$ with the same order singularity are given through

$$\bigcup_{u_1=1}^{q_1} \underbrace{(a_{j_1j_2u_1}, \dot{x}_{j_2})}_{(n_{j_1u_1})^{\text{th}}\text{order YY}} \rightleftharpoons \underbrace{(a_{j_1j_21}, \dot{x}_{j_2})}_{(2n_{11}+1)^{\text{th}} \text{ order YY}} \rightleftharpoons \bigcup_{u_2=1}^{q_1} \underbrace{(a_{j_1j_2u_2}, \dot{x}_{j_2})}_{(n_{j_1u_2})^{\text{th}}\text{order YY}}$$

$$\{a_{j_1j_2u_1} \rightleftharpoons a_{j_1j_2u_2} | n_{j_1u_1} = n_{j_1u_2}; u_1, u_2 \in \{1, 2, \ldots, q\}, u_1 \neq u_2\};$$

and

$$\begin{aligned}
&a_{j_1j_2u_1} = a_{j_1j_2 1} - \varepsilon_{u_1}, a_{j_1j_2u_2} = a_{j_1j_2 1} + \varepsilon_{u_2},\\
&(a_{j_1j_2u_1} = a_{j_1j_2 1} + \varepsilon_{u_1}, a_{j_1j_2u_2} = a_{j_1j_2 1} - \varepsilon_{u_2});\\
&\varepsilon_{u_1}, \varepsilon_{u_2} > 0;\ u_i = 1, 2, \ldots, q_1\ (i = 1, 2).
\end{aligned}$$

- Two 1-dimensional flows of $(a_{j_1j_1r_1}, a_{j_1j_2u_1})$ and $(a_{j_1j_1r_2}, a_{j_1j_2u_2})$ switching at point $(a_{j_1j_1 1}, a_{j_1j_2 1})$ with the same order singularity are given through

$$\begin{aligned}
\bigcup_{r_1=1}^{p_1}\bigcup_{u_1=1}^{q_1} \underbrace{(a_{j_1j_1r_1}, a_{j_1j_2u_1})}_{([(m_{j_1r_1}),(n_{j_1u_1})]\text{XX,xF})\text{-flow}} &\rightleftharpoons \underbrace{([(2m_{11}),(2n_{11}+1)]\text{XX, xF})}_{[(2m_{11}),(2n_{11}+1)]\text{ -XX}}\\
&\rightleftharpoons \bigcup_{r_2=1}^{\bar{p}_1}\bigcup_{u_2=1}^{\bar{q}_1} \underbrace{(a_{j_1j_1r_2}, a_{j_1j_2u_2})}_{([(m_{j_1r_2}),(n_{j_1u_2})]\text{XX,xF})\text{-flow}}
\end{aligned}$$

with

$$\begin{aligned}
&a_{j_1j_1r_1} = a_{j_1j_1 1} - \varepsilon_{r_1}, a_{j_1j_1r_2} = a_{j_1j_1 1} + \varepsilon_{r_2},\\
&(a_{j_1j_1r_1} = a_{j_1j_1 1} + \varepsilon_{r_1}, a_{j_1j_1r_2} = a_{j_1j_1 1} - \varepsilon_{r_2}),\\
&\varepsilon_{r_1}, \varepsilon_{r_2} > 0,\ r_i = 1, 2, \ldots, p_1\ (i = 1, 2);\\
&a_{j_1j_2u_1} = a_{j_1j_2 1} - \varepsilon_{u_1}, a_{j_1j_2u_2} = a_{j_1j_2 1} + \varepsilon_{u_2},\\
&(a_{j_1j_2u_1} = a_{j_1j_2 1} + \varepsilon_{u_1}, a_{j_1j_2u_2} = a_{j_1j_2 1} - \varepsilon_{u_2}),\\
&\varepsilon_{u_1}, \varepsilon_{u_2} > 0,\ u_i = 1, 2, \ldots, q_1\ (i = 1, 2).
\end{aligned}$$

The switching bifurcations are the higher-order hyperbolic-to-hyperbolic-secant flows bifurcations.

In summary, from the analysis in case (I$_1$)–(I$_3$), the 1-dimensional flow of $(x^*_{j_1}, x^*_{j_2}) = (a_{j_1j_1 1}, a_{j_1j_2 1})$ for $m_1 = 2m_{11}$ and $n_1 = 2n_{11} + 1$ has the bifurcation properties as stated in (iv$_{2a}$)–(iv$_{2d}$) through Eqs. (1.326)–(1.345).

(iv$_3$) For $m_1 = 2m_{11} + 1$ and $n_1 = 2n_{11}$ the equilibrium of $(x^*_{j_1}, x^*_{j_2}) = (a_{j_1j_1 1}, a_{j_1j_2 1})$ for the x_{j_1}-direction has the following bifurcation properties with

$$\begin{aligned}
&2m_{11} + 1 - \sum_{r_1=1}^{p_1} m_{j_1r_1} = 2\sum_{\rho_1=1}^{l_1} m_{\rho_1}, 2n_{11} - \sum_{u_1=1}^{q_1} n_{j_1u_1} = 2\sum_{\sigma_1=1}^{l_2} n_{\sigma_1};\\
&\sum_{r_2=1}^{\bar{p}_1} m_{j_1r_2} = 2m_{11} + 1, \sum_{u_2=1}^{\bar{q}_1} n_{j_1u_2} = 2n_{11};
\end{aligned}$$

$$\sum_{r_i=1}^{p_1} m_{j_1 r_i} = 2m_{11} + 1, \sum_{u_i=1}^{q_1} n_{j_1 u_i} = 2n_{11}, \, for \; i = 1, 2;$$

l_1, l_2-quadratic polynomials are without real roots and ρ_1^{th} and σ_1^{th}-quadratic polynomials are with powers m_{ρ_1} and n_{σ_1}, respectively.

The bifurcation process is discussed through differential equations as follows. There are four bifurcations (I_1)–(I_3).

(I_1) Consider a dynamical system having a single equilibrium (A_1) as

$$\begin{aligned}
\dot{x}_{j_1} &= a_{j_1 j_1 0}(x_{j_1} - b_{j_1 j_1 1}) \prod_{r=1}^{m_{11}} [(x_{j_1} - a_{j_1 j_1}^{(r)})^2 + \Delta_{j_1 j_1}^{(r)}] \\
&\quad \times \prod_{u=1}^{n_{11}} [(x_{j_2} - a_{j_1 j_2}^{(u)})^2 + \Delta_{j_1 j_2}^{(u)}], \\
\dot{x}_{j_2} &= a_{j_2 j_2 0}(x_{j_1} - a_{j_2 j_1 1})^{m_2};
\end{aligned}$$

where $a_{j_1 j_1 1} = b_{j_1 j_1 1} \neq a_{j_2 j_1 1}$. Once

$$\begin{aligned}
\Delta_{j_1 j_1}^{(r)} &= \Delta_{j_1 j_1} = 0 \, (r = 1, 2, \ldots, m_{11}), \\
\Delta_{j_1 j_2}^{(u)} &= \Delta_{j_1 j_2} = 0 \, (u = 1, 2, \ldots, n_{11}),
\end{aligned}$$

if

$$\begin{aligned}
a_{j_1 j_1 1} &= b_{j_1 j_1 1} = a_{j_1 j_1}^{(r)} \, (r = 1, 2, \ldots, m_{11}), \\
a_{j_1 j_2 1} &= b_{j_1 j_2 1} = a_{j_1 j_2}^{(u)} \, (u = 1, 2, \ldots, n_{11}),
\end{aligned}$$

the foregoing differential equations become a dynamical system with a singular equilibrium (B_1) as

$$\begin{aligned}
\dot{x}_{j_1} &= a_{j_1 j_1 0}(x_{j_1} - a_{j_1 j_1 1})^{2m_{11}+1}(x_{j_2} - a_{j_1 j_2 1})^{2n_{11}}, \\
\dot{x}_{j_2} &= a_{j_2 j_2 0}(x_{j_1} - a_{j_2 j_1 1})^{n_2}.
\end{aligned}$$

If

$$\begin{aligned}
\Delta_{j_1 j_1}^{(r)} &= \Delta_{j_1 j_1} + \delta_r \, (r = 1, 2, \ldots, m_{11}), \; \delta_r > 0; \\
\Delta_{j_1 j_2}^{(u)} &= \Delta_{j_1 j_2} + \delta_s \, (u = 1, 2, \ldots, n_{11}), \; \delta_u > 0;
\end{aligned}$$

then

$$a_{j_1 j_1 1}^{(r)}, a_{j_1 j_1 2}^{(r)} = a_{j_1 j_1 1} \pm \varepsilon_r, \, (r = 1, 2, \ldots, m_{11}),$$

$$\{a_{j_1j_1 1}, a_{j_1j_1 2}, \ldots, a_{j_1j_1(2m_{11})}\} = \text{sort}\{a_{j_1j_1 1}^{(r)}, a_{j_1j_1 2}^{(r)} | r = 1, 2, \ldots, m_{11}\};$$
$$a_{j_1j_2 1}^{(u)}, a_{j_1j_2 2}^{(u)} = a_{j_1j_2 1} \pm \varepsilon_u, \ (\text{u} = 1, 2, \ldots, n_1),$$
$$\{a_{j_1j_2 1}, a_{j_1j_2 2}, \ldots, a_{j_1j_2(2n_{11}+1)}\} = \text{sort}\{b_{j_1j_2 1}, a_{j_1j_2 1}^{(u)}, a_{j_1j_2 2}^{(u)} | u = 1, 2, \ldots, n_{11}\}.$$

Thus, the differential equation becomes a dynamical system with a non-singular equilibrium network (C_1) as

$$\dot{x}_{j_1} = a_{j_1j_1 0} \prod_{r=1}^{2m_{11}+1} (x_{j_1} - a_{j_1j_1 r}) \prod_{u=1}^{2n_{11}} (x_{j_2} - a_{j_1j_2 u}),$$
$$\dot{x}_{j_2} = a_{j_2j_2 0}(x_{j_1} - a_{j_2j_1 1})^{n_2}.$$

- From the above bifurcation process analysis, at least, only $(m_{11} + 1)$-parameter variations in the x_{j_1}-direction are engaged in such a bifurcation from a 1-dimensional flow to the 1-dimensional flow array of $1 \times (2m_{11} + 1)$ through the $(2m_{11} + 1)^{\text{th}}$-order source and sink bifurcations in the x_{j_1}-direction. The higher-order source and sink flow bifurcations are developed.

  Thus, the 1-dimensional flow appearing or vanishing bifurcation route is as follows.

$$\underbrace{(a_{j_1j_1 1}, \dot{x}_{j_2})}_{\text{XX}} \rightleftharpoons \underbrace{(a_{j_1j_1 1}, \dot{x}_{j_2})}_{(2m_{11}+1)^{\text{th}}\text{order XX}} \rightleftharpoons \bigcup_{r=1}^{2m_{11}+1} \underbrace{(a_{j_1j_1 r}, \dot{x}_{j_2})}_{\text{XX}}.$$

- From the above bifurcation process analysis, at least, only (n_{11})-parameter variations in the x_{j_1}-direction are engaged in such a bifurcation from a 1-dimensional flow to the 1-dimensional flow array of $(2n_{11}) \times 1$ through the $(2n_{11})^{\text{th}}$-order increasing and decreasing-inflection flows bifurcations in the x_{j_1}-direction. The higher-order increasing-inflection and decreasing-inflection flow bifurcations are developed.

  Thus, the 1-dimensional flow appearing or vanishing bifurcation route is as follows.

$$\underbrace{(a_{j_1j_2 1}, \dot{x}_{j_2})}_{\text{YY}} \rightleftharpoons \underbrace{(a_{j_1j_2 1}, \dot{x}_{j_2})}_{(2n_{11})^{\text{th}}\text{order YY}} \rightleftharpoons \bigcup_{u=1}^{2n_{11}} \underbrace{(a_{j_1j_2 u}, \dot{x}_{j_2})}_{\text{YY}}.$$

- From the above bifurcation process analysis, $(m_{11} + n_{11} + 1)$-parameter variations in the two directions are engaged in such a bifurcation from a parabola-flow to a $(2m_{11} + 1) \times (2n_{11})$ network of inflection flows through the $(2m_{11} + 1)^{\text{th}}$-order source and sink bifurcations in the x_{j_1}-direction, and the $(2n_{11})^{\text{th}}$-order increasing and decreasing-inflection flows bifurcations in the x_{j_1}-direction. The higher-order increasing and decreasing-inflection flow bifurcations are developed.

Thus, the 1-dimensional flow appearing or vanishing bifurcation route is as follows.

$$\underbrace{(a_{j_1j_1 1}, \dot{x}_{j_2})}_{\text{(XX,xF)-flow}} \rightleftharpoons \underbrace{([(2m_{11}+1), (2n_{11})]\text{XX, xF})}_{[(2m_{11}+1),(2n_{11})]\text{-XX}} \rightleftharpoons \bigcup_{r=1}^{2m_{11}+1}\bigcup_{u=1}^{2n_{11}} \underbrace{(a_{j_1j_1 r}, a_{j_1j_2 u})}_{\text{(YY,xF)-flow}},$$

which are the higher-order inflection source and sink flows bifurcations.

(I_2) From Case (I_1), consider a dynamical system having singular equilibriums (A_2) as

$$\begin{aligned}
\dot{x}_{j_1} &= a_{j_1j_1 0}\prod_{r_1=1}^{p_1}(x_{j_1} - b_{j_1j_1r_1})^{m_{j_1r_1}}\prod_{r=1}^{l_1}[(x_{j_1} - a^{(r)}_{j_1j_1})^2 + \Delta^{(r)}_{j_1j_1}]^{m_{j_1r}}\\
&\quad\times\prod_{u_1=1}^{q_1}(x_{j_2} - b_{j_1j_2u_1})^{m_{j_1u_1}}\prod_{u=1}^{l_2}[(x_{j_2} - a^{(u)}_{j_1j_2})^2 + \Delta^{(u)}_{j_1j_2}]^{m_{j_1u}},\\
\dot{x}_{j_2} &= a_{j_2j_2 0}(x_{j_1} - a_{j_2j_1 1})^{m_2};
\end{aligned}$$

where

$$\begin{aligned}
&2m_{11} + 1 - \sum_{r_1=1}^{p_1} m_{j_1r_1} = 2\sum_{\rho_1=1}^{l_1} m_{\rho_1},\ 2n_{11} - \sum_{u_1=1}^{q_1} n_{j_1u_1} = 2\sum_{\sigma_1=1}^{l_2} n_{\sigma_1};\\
&\sum_{r_2=1}^{\bar{p}_1} m_{j_1r_2} = 2m_{11} + 1,\ \sum_{u_2=1}^{\bar{q}_1} n_{j_1u_2} = 2n_{11};
\end{aligned}$$

l_1, l_2-quadratic polynomials are without real roots and ρ_1^{th} and σ_1^{th}-quadratic polynomials are with powers m_{ρ_1} and n_{σ_1},respectively.

Once

$$\begin{aligned}
\Delta^{(r)}_{j_1j_1} &= 0\,(r = 1, 2, \ldots, l_1),\\
\Delta^{(u)}_{j_1j_2} &= 0\,(u = 1, 2, \ldots, l_2);
\end{aligned}$$

if

$$\begin{aligned}
a_{j_1j_1 1} &\equiv a_{j_1j_1r_1} = a^{(r)}_{j_1j_1}\ (r_1 = 1, 2, \ldots, p_1;\ r = 1, 2, \ldots, l_1),\\
a_{j_1j_2 1} &\equiv a_{j_1j_2u_1} = a^{(u)}_{j_1j_2}\ (u_1 = 1, 2, \ldots, q_1;\ u = 1, 2, \ldots, l_2),
\end{aligned}$$

the foregoing differential equations become a dynamical system with a 1-dimensional flow (B_2) as

$$\dot{x}_{j_1} = a_{j_1 j_1 0}(x_{j_1} - a_{j_1 j_1 1})^{2m_{11}+1}(x_{j_2} - a_{j_1 j_2 1})^{2n_{11}},$$
$$\dot{x}_{j_2} = a_{j_2 j_2 0}(x_{j_1} - a_{j_2 j_1 1})^{n_2}.$$

If

$$\Delta^{(r)}_{j_1 j_1} = \Delta_{j_1 j_1} + \delta_r\ (r = 1, 2, \ldots, l_1),\ \delta_r > 0;$$
$$\Delta^{(u)}_{j_1 j_2} = \Delta_{j_1 j_2} + \delta_u\ (u = 1, 2, \ldots, l_2),\ \delta_u > 0;$$

then

$$\begin{aligned}
&a^{(r)}_{j_1 j_1 1}, a^{(r)}_{j_1 j_1 2} = a_{j_1 j_1 1} \pm \varepsilon_r, (r = 1, 2, \ldots, l_1),\\
&\{a_{j_1 j_1 1}, a_{j_1 j_1 2}, \ldots, a_{j_1 j_1 \overline{p}_1}\}\\
&= \text{sort}\{a_{j_1 j_1 1}, a_{j_1 j_1 2}, \ldots, a_{j_1 j_1 p_1}; a^{(r)}_{j_1 j_1 1}, a^{(r)}_{j_1 j_1 2} | r = 1, 2, \ldots, l_1\};\\
&a^{(u)}_{j_1 j_2 1}, a^{(u)}_{j_1 j_2 2} = a_{j_1 j_2 1} \pm \varepsilon_u, (u = 1, 2, \ldots, l_2),\\
&\{a_{j_1 j_2 1}, a_{j_1 j_2 2}, \ldots, a_{j_1 j_2 \overline{q}_1}\}\\
&= \text{sort}\{a_{j_1 j_2 1}, a_{j_1 j_2 2}, \ldots, a_{j_1 j_2 q_1}; a^{(u)}_{j_1 j_2 1}, a^{(u)}_{j_1 j_2 2} | u = 1, 2, \ldots, l_2\}.
\end{aligned}$$

Thus, the differential equation becomes a dynamical system with singular 1-dimensional flows (C_2) as

$$\dot{x}_{j_1} = a_{j_1 j_1 0} \prod_{r_2=1}^{\overline{p}_1} (x_{j_1} - a_{j_1 j_1 r_2})^{m_{j_1 r_2}} \prod_{u_2=1}^{\overline{q}_1} (x_{j_2} - a_{j_1 j_1 u_2})^{n_{j_1 u_2}},$$
$$\dot{x}_{j_2} = a_{j_2 j_2 0}(x_{j_1} - a_{j_2 j_1 1})^{m_2}$$

where

$$\Sigma^{\overline{p}_1}_{r_2=1} m_{j_1 r_2} = 2m_{11} + 1,\ \Sigma^{\overline{q}_1}_{u_2=1} n_{j_1 u_2} = 2n_{11}.$$

- From the above bifurcation process analysis, only $\overline{p}_1$-parameter variations in the x_{j_1}-direction from a p_1 to $\overline{p}_1$ 1-dimensional flow array through the $(2m_{11} + 1)^{\text{th}}$-order source and sink flow bifurcations in the x_{j_1}-direction.

  Thus, the 1-dimensional flow appearing or vanishing bifurcation route from a p_1 to p_2 1-dimensional flow array is expressed as

$$\bigcup_{r_1=1}^{p_1} \underbrace{(a_{j_1 j_1 r_1}, \dot{x}_{j_2})}_{(\mathrm{m}_{j_1 r_1})^{\text{th}}\text{order XX}} \rightleftharpoons \underbrace{(a_{j_1 j_1 1}, \dot{x}_{j_2})}_{(2m_{11}+1)^{\text{th}}\text{ order XX}} \rightleftharpoons \bigcup_{r_2=1}^{\overline{p}_1} \underbrace{(a_{j_1 j_1 r_2}, \dot{x}_{j_2})}_{(\mathrm{m}_{j_1 r_2})^{\text{th}}\text{order XX}}.$$

- From the above bifurcation process analysis, only $\overline{q}_1$-parameter variations in the x_{j_1}-direction from a q_1 to $\overline{q}_1$ 1-dimensional flow array through the $(2n_{11})^{\text{th}}$-order increasing and decreasing inflection flow bifurcations in the x_{j_1}-direction.

  Thus, the 1-dimensional flow appearing or vanishing bifurcation route from a q_1 to $\overline{q}_1$ 1-dimensional flow array is expressed as

$$\bigcup_{u_1=1}^{q_1} \underbrace{(a_{j_1j_2u_1}, \dot{x}_{j_2})}_{(n_{j_1u_1})^{\text{th}}\text{order YY}} \rightleftharpoons \underbrace{(a_{j_1j_21}, \dot{x}_{j_2})}_{(2n_{11})^{\text{th}}\text{ order YY}} \rightleftharpoons \bigcup_{u_2=1}^{\overline{q}_1} \underbrace{(a_{j_1j_2u_2}, \dot{x}_{j_2})}_{(n_{j_1u_2})^{\text{th}}\text{order YY}}.$$

- From the above bifurcation process analysis, only $\overline{q}_1+\overline{p}_1$-parameter variations in the x_{j_1}-direction from a $q_1 \times p_1$ to $\overline{q}_1 \times \overline{p}_1$ network of 1-dimensional flows through the $(2m_{11}+1)^{\text{th}}$-order source and sink flows bifurcations in the x_{j_1}-direction, and the $(2n_{11})^{\text{th}}$-order increasing and decreasing-inflection flows bifurcations in the x_{j_1}-direction.

  Thus, the 1-dimensional flow appearing or vanishing bifurcation route is as follows:

$$\begin{aligned}\bigcup_{r_1=1}^{p_1}\bigcup_{u_1=1}^{q_1} \underbrace{(a_{j_1j_1r_1}, a_{j_1j_2u_1})}_{([(m_{j_1r_1}),(n_{j_1u_1})]\text{XX,xF)-flow}} &\rightleftharpoons \underbrace{([(2m_{11}+1),(2n_{11})]\text{XX, xF})}_{[(2m_{11}+1),(2n_{11})]\text{-XX}}\\ &\rightleftharpoons \bigcup_{r_2=1}^{\overline{p}_1}\bigcup_{u_2=1}^{\overline{q}_1} \underbrace{(a_{j_1j_1r_2}, a_{j_1j_2u_2})}_{([(m_{j_1r_2}),(n_{j_1u_1})]\text{XX,xF)-flow}},\end{aligned}$$

  which are the higher-order inflection source and sink flows bifurcations.

(I_3) Consider two dynamical systems with the same 1-dimensional flows with locations switched (A_3, C_3) for $i = 1, 2$ as

$$\begin{aligned}\dot{x}_{j_1} &= a_{j_1j_10}\prod_{r_i=1}^{p_1}(x_{j_1} - a_{j_1j_1r_i})^{m_{j_1r_i}}\prod_{u_i=1}^{q_1}(x_{j_2} - a_{j_1j_2u_i})^{n_{j_1u_i}},\\ \dot{x}_{j_2} &= a_{j_2j_20}(x_{j_2} - a_{j_2j_11})^{m_2};\end{aligned}$$

where

$$\sum_{r_i=1}^{p_1} m_{j_1r_i} = 2m_{11}+1, \sum_{u_i=1}^{q_1} n_{j_1u_i} = 2n_{11}, \text{for } i = 1, 2.$$

Consider a dynamical system as a singular 1-dimensional flow (B_3) as

$$\begin{aligned}\dot{x}_{j_1} &= a_{j_1j_10}(x_{j_1} - a_{j_1j_11})^{2m_{11}+1}(x_{j_2} - a_{j_1j_21})^{2n_{11}},\\ \dot{x}_{j_2} &= a_{j_2j_20}(x_{j_1} - a_{j_2j_11})^{m_2}.\end{aligned}$$

For $r_i = 1, 2, \ldots, p$ and $s_i = 1, 2, \ldots, q$ $(i = 1, 2)$, define a function as

$$\Delta_{j_1 j_1}^{(r_1 r_2)} = (a_{j_1 j_1 r_1} - a_{j_1 j_1 r_2})^2 \; (r_1, r_2 \in \{1, 2, \ldots, p_1\}, r_1 \neq r_2),$$
$$\Delta_{j_1 j_2}^{(u_1 u_2)} = (a_{j_1 j_2 u_1} - a_{j_1 j_2 u_2})^2 \; (u_1, u_2 \in \{1, 2, \ldots, q_1\}, u_1 \neq u_2).$$

For $r_i = 1, 2, \ldots, p$ and $s_i = 1, 2, \ldots, q$ $(i = 1, 2)$, if

$$\Delta_{j_1 j_1}^{(r_1 r_2)} = (a_{j_1 j_1 r_1} - a_{j_1 j_1 r_2})^2 = 0 \text{ and } \Delta_{j_1 j_2}^{(u_1 u_2)} = (a_{j_1 j_2 u_1} - a_{j_1 j_2 u_2})^2 = 0,$$

there are three bifurcations as follows.

- Two 1-dimensional flows of $(a_{j_1 j_1 r_1}, \dot{x}_{j_2})$ and $(a_{j_1 j_1 r_2}, \dot{x}_{j_2})$ switching at point $(a_{j_1 j_1 1}, \dot{x}_{j_2})$ with the same order singularity are given through

$$\bigcup_{r_1=1}^{p_1} \underbrace{(a_{j_1 j_1 r_1}, \dot{x}_{j_2})}_{(m_{j_1 r_1})^{\text{th}} \text{order XX}} \rightleftharpoons \underbrace{(a_{j_1 j_1 1}, \dot{x}_{j_2})}_{(2m_{11}+1)^{\text{th}} \text{ order XX}} \rightleftharpoons \bigcup_{r_2=1}^{p_1} \underbrace{(a_{j_1 j_1 r_2}, \dot{x}_{j_2})}_{(m_{j_1 r_2})^{\text{th}} \text{order XX}}$$
$$\left\{a_{j_1 j_1 r_1} \rightleftharpoons a_{j_1 j_1 r_2} | m_{j_1 r_1} = m_{j_1 r_2}; r_1, r_2 \in \{1, 2, \ldots, p_1\}, r_1 \neq r_2\right\};$$

  and

$$a_{j_1 j_1 r_1} = a_{j_1 j_1 1} - \varepsilon_{r_1}, a_{j_1 j_1 r_2} = a_{j_1 j_1 1} + \varepsilon_{r_2},$$
$$(a_{j_1 j_1 r_1} = a_{j_1 j_1 1} + \varepsilon_{r_1}, a_{j_1 j_1 r_2} = a_{j_1 j_1 1} - \varepsilon_{r_2});$$
$$\varepsilon_{r_1}, \varepsilon_{r_2} > 0; \; r_i = 1, 2, \ldots, p_1 \; (i = 1, 2).$$

- Two 1-dimensional flows of $(a_{j_1 j_2 u_1}, \dot{x}_{j_2})$ and $(a_{j_1 j_2 u_1}, \dot{x}_{j_2})$ switching at point $(a_{j_1 j_2 1}, \dot{x}_{j_2})$ with the same order singularity are given through

$$\bigcup_{u_1=1}^{q_1} \underbrace{(a_{j_1 j_2 u_1}, \dot{x}_{j_2})}_{(n_{j_1 u_1})^{\text{th}} \text{order YY}} \rightleftharpoons \underbrace{(a_{j_1 j_2 1}, \dot{x}_{j_2})}_{(2n_1)^{\text{th}} \text{ order YY}} \rightleftharpoons \bigcup_{u_2=1}^{q_1} \underbrace{(a_{j_1 j_2 u_2}, \dot{x}_{j_2})}_{(n_{j_1 u_2})^{\text{th}} \text{order YY}}$$
$$\left\{a_{j_1 j_2 u_1} \rightleftharpoons a_{j_1 j_2 u_2} | n_{j_1 u_1} = n_{j_1 u_2}; u_1, u_2 \in \{1, 2, \ldots, q\}, u_1 \neq u_2\right\};$$

  and

$$a_{j_1 j_2 u_1} = a_{j_1 j_2 1} - \varepsilon_{u_1}, a_{j_1 j_2 u_2} = a_{j_1 j_2 1} + \varepsilon_{u_2},$$
$$(a_{j_1 j_2 u_1} = a_{j_1 j_2 1} + \varepsilon_{u_1}, a_{j_1 j_2 u_2} = a_{j_1 j_2 1} - \varepsilon_{u_2});$$
$$\varepsilon_{u_1}, \varepsilon_{u_2} > 0; \; u_i = 1, 2, \ldots, q_1 \; (i = 1, 2).$$

- Two 1-dimensional flows of $(a_{j_1 j_1 r_1}, a_{j_1 j_2 u_1})$ and $(a_{j_1 j_1 r_2}, a_{j_1 j_2 u_2})$ switching at point $(a_{j_1 j_1 1}, a_{j_1 j_2 1})$ with the same order singularity are given through

$$\bigcup_{r_1=1}^{p_1}\bigcup_{u_1=1}^{q_1}\underbrace{(a_{j_1j_1r_1},a_{j_1j_2u_1})}_{([(m_{j_1r_1}),(n_{j_1u_1})]\text{XX,xF})\text{-flow}} \rightleftharpoons \underbrace{([(2m_{11}+1),(2n_{11})]\text{XX, xF})}_{[(2m_{11}+1),(2n_{21})]\text{ -XX}}$$
$$\rightleftharpoons \bigcup_{r_2=1}^{p_1}\bigcup_{u_2=1}^{q_1}\underbrace{(a_{j_1j_1r_2},a_{j_1j_2u_2})}_{([(m_{j_1r_2}),(n_{j_1u_2})]\text{XX,xF})\text{-flow}}$$

with

$$\begin{aligned}
&a_{j_1j_1r_1}=a_{j_1j_11}-\varepsilon_{r_1},\, a_{j_1j_1r_2}=a_{j_1j_11}+\varepsilon_{r_2},\\
&(a_{j_1j_1r_1}=a_{j_1j_11}+\varepsilon_{r_1},\, a_{j_1j_1r_2}=a_{j_1j_11}-\varepsilon_{r_2}),\\
&\varepsilon_{r_1},\varepsilon_{r_2}>0,\; r_i=1,2,\ldots,p_1\,(i=1,2);\\
&a_{j_1j_2u_1}=a_{j_1j_21}-\varepsilon_{u_1},\, a_{j_1j_2u_2}=a_{j_1j_21}+\varepsilon_{u_2},\\
&(a_{j_1j_2u_1}=a_{j_1j_21}+\varepsilon_{u_1},\, a_{j_1j_2u_2}=a_{j_1j_21}-\varepsilon_{u_2}),\\
&\varepsilon_{u_1},\varepsilon_{u_2}>0,\; u_i=1,2,\ldots,q_1\,(i=1,2).
\end{aligned}$$

In summary, from the analysis in case (I$_1$)–(I$_3$), the 1-dimensional flow of $(x_{j_1}^*,x_{j_2}^*)=(a_{j_1j_11},a_{j_1j_21})$ for $m_1=2m_{11}+1$ and $n_1=2n_{11}$ has the bifurcation properties as stated in (iv$_{3a}$)–(iv$_{3d}$) through Eqs. (1.347)–(1.366).

(iv$_4$) For $m_1=2m_{11}$ and $n_1=2n_{11}$ the equilibrium of $(x_{j_1}^*,x_{j_2}^*)=(a_{j_1j_11},a_{j_1j_21})$ for the x_{j_1}-direction has the following bifurcation properties with

$$\begin{aligned}
&2m_{11}-\sum_{r_1=1}^{p_1}m_{j_1r_1}=2\sum_{\rho_1=1}^{l_1}m_{\rho_1},\, 2n_{11}-\sum_{u_1=1}^{q_1}n_{j_1u_1}=2\sum_{\sigma_1=1}^{l_2}n_{\sigma_1};\\
&\sum_{r_2=1}^{\bar p_1}m_{j_1r_2}=2m_{11},\,\sum_{u_2=1}^{\bar q_1}n_{j_1u_2}=2n_{11};\\
&\sum_{r_i=1}^{p_1}m_{j_1r_i}=2m_{11},\,\sum_{u_i=1}^{q_1}n_{j_1u_i}=2n_{11},\, for\, i=1,2;
\end{aligned}$$

l_1,l_2-quadratic polynomials are without real roots and ρ_1^{th} and σ_1^{th}-quadratic polynomials are with powers m_{ρ_1} and n_{σ_1},respectively.

The bifurcation process is discussed through differential equations as follows. There are four bifurcations (I$_1$)–(I$_3$).

(I$_1$) Consider a dynamical system having a single equilibrium (A$_1$) as

$$\begin{aligned}
\dot x_{j_1}&=a_{j_1j_10}\prod_{r=1}^{m_{11}}[(x_{j_1}-a_{j_1j_1}^{(r)})^2+\Delta_{j_1j_1}^{(r)}]\prod_{u=1}^{n_{11}}[(x_{j_2}-a_{j_1j_2}^{(u)})^2+\Delta_{j_1j_2}^{(u)}],\\
\dot x_{j_2}&=a_{j_2j_20}(x_{j_1}-a_{j_2j_11})^{m_2}.
\end{aligned}$$

Once

$$\Delta_{j_1j_1}^{(r)} = \Delta_{j_1j_1} = 0\,(r = 1, 2, \ldots, m_{11}),$$
$$\Delta_{j_1j_2}^{(u)} = \Delta_{j_1j_2} = 0\,(u = 1, 2, \ldots, n_{11}),$$

if

$$a_{j_1j_11} = b_{j_1j_11} = a_{j_1j_1}^{(r)}\ (r = 1, 2, \ldots, m_{11}),$$
$$a_{j_1j_21} = b_{j_1j_21} = a_{j_1j_2}^{(u)}\ (u = 1, 2, \ldots, n_{11}),$$

the foregoing differential equations become a dynamical system with a singular equilibrium (B_1) as

$$\dot{x}_{j_1} = a_{j_1j_10}(x_{j_1} - a_{j_1j_11})^{2m_{11}}(x_{j_2} - a_{j_1j_21})^{2n_{11}},$$
$$\dot{x}_{j_2} = a_{j_2j_20}(x_{j_1} - a_{j_2j_11})^{m_2}.$$

If

$$\Delta_{j_1j_1}^{(r)} = \Delta_{j_1j_1} + \delta_r\,(r = 1, 2, \ldots, m_{11}),\ \delta_r > 0;$$
$$\Delta_{j_1j_2}^{(u)} = \Delta_{j_1j_2} + \delta_u\,(u = 1, 2, \ldots, n_{11}),\ \delta_u > 0;$$

then

$$a_{j_1j_1}^{(r)}, a_{j_1j_12}^{(r)} = a_{j_1j_11} \pm \varepsilon_r,\ (r = 1, 2, \ldots, m_{11}),$$
$$\{a_{j_1j_11}, a_{j_1j_12}, \ldots, a_{j_1j_1(2m_{11})}\} = \mathrm{sort}\{a_{j_1j_1}^{(r)}, a_{j_1j_22}^{(r)} | r = 1, 2, \ldots, m_{11}\};$$
$$a_{j_1j_21}^{(u)}, a_{j_1j_22}^{(u)} = a_{j_1j_21} \pm \varepsilon_u,\ (u = 1, 2, \ldots, n_1),$$
$$\{a_{j_1j_21}, a_{j_1j_22}, \ldots, a_{j_1j_2(2m_11)}\} = \mathrm{sort}\{a_{j_1j_21}^{(u)}, a_{j_1j_22}^{(u)} | u = 1, 2, \ldots, n_{11}\}.$$

Thus, the differential equation becomes a dynamical system with a non-singular equilibrium network (C_1) as

$$\dot{x}_{j_1} = a_{j_1j_10} \prod_{r=1}^{2m_{11}} (x_{j_1} - a_{j_1j_1r}) \prod_{u=1}^{2n_{11}} (x_{j_2} - a_{j_1j_2u}),$$
$$\dot{x}_{j_2} = a_{j_2j_20}(x_{j_1} - a_{j_2j_11})^{m_2}.$$

- From the above bifurcation process analysis, at least, only (m_{11})-parameter variations in the x_{j_1}-direction are engaged in such a bifurcation from a 1-dimensional flow to the 1-dimensional flow array of $1 \times (2m_{11})$ through the $(2m_{11})^{\text{th}}$-order increasing and decreasing-inflection flow bifurcations in the x_{j_1}-direction. The higher-order inflection flow bifurcations are developed.

Thus, the 1-dimensional flow appearing or vanishing bifurcation route is as follows.

$$\underbrace{(a_{j_1j_11}, \dot{x}_{j_2})}_{\text{XX}} \rightleftharpoons \underbrace{(a_{j_1j_11}, \dot{x}_{j_2})}_{(2m_{11})^{\text{th}}\text{order XX}} \rightleftharpoons \bigcup_{r=1}^{2m_{11}} \underbrace{(a_{j_1j_1r}, \dot{x}_{j_2})}_{\text{XX}}.$$

- From the above bifurcation process analysis, at least, only n_{11}-parameter variations in the x_{j_1}-direction are engaged in such a bifurcation from a 1-dimensional flow to the 1-dimensional flow array of $(2n_{11}) \times 1$ through the $(2n_{11})^{\text{th}}$-order increasing and decreasing-inflection bifurcations in the x_{j_1}-direction. The higher-order inflection flow bifurcations are developed.

  Thus, the 1-dimensional flow appearing or vanishing bifurcation route is as follows.

$$\underbrace{(a_{j_1j_21}, \dot{x}_{j_2})}_{\text{YY}} \rightleftharpoons \underbrace{(a_{j_1j_21}, \dot{x}_{j_2})}_{(2n_{11})^{\text{th}}\text{order YY}} \rightleftharpoons \bigcup_{u=1}^{2n_{11}} \underbrace{(a_{j_1j_2u}, \dot{x}_{j_2})}_{\text{YY}}.$$

- From the above bifurcation process analysis, $(m_{11}+n_{11})$-parameter variations in the two directions are engaged in such a bifurcation from a parabola-flow to a $(2m_{11}) \times (2n_{11})$ network of inflection flows through the $(2m_{11})^{\text{th}}$-order source and sink bifurcations in the x_{j1}-direction, and the $(2n_{11})^{\text{th}}$ order increasing and decreasing-inflection flows bifurcations in the x_{j1}-direction. The higher-order increasing and decreasing-inflection flow bifurcations are developed.

  Thus, the 1-dimensional flow appearing or vanishing bifurcation route is as follows.

$$\underbrace{(\dot{x}_{j_1}, \dot{x}_{j_2})}_{\text{(xF, xF)-flow}} \rightleftharpoons \underbrace{([(2m_{11}), (2n_{11})]\text{XX,xF})}_{[(2m_{11}),(2n_{11})]\text{-XX}} \rightleftharpoons \bigcup_{r=1}^{2m_{11}} \bigcup_{u=1}^{2n_{11}} \underbrace{(a_{j_1j_1r}, a_{j_1j_2u})}_{\text{(YY,xF)-flow}},$$

  which are the higher-order double inflection upper-saddle and lower-saddle flows bifurcations.

(I_2) From Case (I_1), consider a dynamical system having singular equilibriums (A_2) as

$$\begin{aligned}
\dot{x}_{j_1} &= a_{j_1j_10} \prod_{r_1=1}^{p_1} (x_{j_1} - b_{j_1j_1r_1})^{m_{j_1r_1}} \prod_{r=1}^{l_1} [(x_{j_1} - a_{j_1j_1}^{(r)})^2 + \Delta_{j_1j_1}^{(r)}]^{m_{j_1r}} \\
&\quad \times \prod_{u_1=1}^{q_1} (x_{j_2} - b_{j_1j_2u_1})^{m_{j_1u_1}} \prod_{u=1}^{l_2} [(x_{j_2} - a_{j_1j_2}^{(u)})^2 + \Delta_{j_1j_2}^{(u)}]^{m_{j_1u}}, \\
\dot{x}_{j_2} &= a_{j_2j_20}(x_{j_1} - a_{j_2j_11})^{m_2};
\end{aligned}$$

where

$$2m_{11} - \sum_{r_1=1}^{p_1} m_{j_1 r_1} = 2\sum_{\rho_1=1}^{l_1} m_{\rho_1}, 2n_{11} - \sum_{u_1=1}^{q_1} n_{j_1 u_1} = 2\sum_{\sigma_1=1}^{l_2} n_{\sigma_1};$$
$$\sum_{r_2=1}^{\bar{p}_1} m_{j_1 r_2} = 2m_{11}, \sum_{u_2=1}^{\bar{q}_1} n_{j_1 u_2} = 2n_{11};$$

l_1, l_2-quadratic polynomials are without real roots, ρ_1^{th} and σ_1^{th}-quadratic polynomials are with powers m_{ρ_1} and n_{σ_1},respectively.

Once

$$\begin{aligned}\Delta_{j_1 j_1}^{(r)} &= 0\,(r = 1, 2, \ldots, l_1),\\ \Delta_{j_1 j_2}^{(u)} &= 0\,(u = 1, 2, \ldots, l_2);\end{aligned}$$

if

$$\begin{aligned}a_{j_1 j_1 1} \equiv a_{j_1 j_1 r_1} &= a_{j_1 j_1}^{(r)}\,(r_1 = 1, 2, \ldots, p_1; r = 1, 2, \ldots, l_1),\\ a_{j_1 j_2 1} \equiv a_{j_1 j_2 u_1} &= a_{j_1 j_2}^{(u)}\,(u_1 = 1, 2, \ldots, q_1; u = 1, 2, \ldots, l_2),\end{aligned}$$

the foregoing differential equations become a dynamical system with a 1-dimensional flow (B_2) as

$$\begin{aligned}\dot{x}_{j_1} &= a_{j_1 j_1 0}(x_{j_1} - a_{j_1 j_1 1})^{m_1}(x_{j_2} - a_{j_1 j_2 1})^{2n_{11}},\\ \dot{x}_{j_2} &= a_{j_2 j_2 0}(x_{j_1} - a_{j_2 j_1 1})^{2m_{21}}.\end{aligned}$$

If

$$\begin{aligned}\Delta_{j_1 j_1}^{(r)} &= \Delta_{j_1 j_1} + \delta_r\,(r = 1, 2, \ldots, l_1),\ \delta_r > 0;\\ \Delta_{j_1 j_2}^{(u)} &= \Delta_{j_1 j_2} + \delta_u\,(u = 1, 2, \ldots, l_2),\ \delta_u > 0;\end{aligned}$$

then

$$\begin{aligned}& a_{j_1 j_1 1}^{(r)}, a_{j_1 j_1 2}^{(r)} = a_{j_1 j_1 1} \pm \varepsilon_r,\ (r = 1, 2, \ldots, l_1),\\ & \{a_{j_1 j_1 1}, a_{j_1 j_1 2}, \ldots, a_{j_1 j_1 \bar{p}_1}\}\\ & = \text{sort}\{a_{j_1 j_1 1}, a_{j_1 j_1 2}, \ldots, a_{j_1 j_1 p_1}; a_{j_1 j_1 1}^{(r)}, a_{j_1 j_1 2}^{(r)} | r = 1, 2, \ldots, l_1\};\\ & a_{j_1 j_2 1}^{(u)}, a_{j_1 j_2 2}^{(u)} = a_{j_1 j_2 1} \pm \varepsilon_u,\ (u = 1, 2, \ldots, l_2),\\ & \{a_{j_1 j_2 1}, a_{j_1 j_2 2}, \ldots, a_{j_1 j_2 \bar{q}_1}\}\\ & = \text{sort}\{a_{j_1 j_2 1}, a_{j_1 j_2 2}, \ldots, a_{j_1 j_2 q_1}; a_{j_1 j_2 1}^{(u)}, a_{j_1 j_2 2}^{(u)} | u = 1, 2, \ldots, l_2\}.\end{aligned}$$

Thus, the differential equation becomes a dynamical system with singular 1-dimensional flows (C_2) as

$$\begin{aligned}\dot{x}_{j_1} &= a_{j_1 j_1 0} \prod_{r_2=1}^{\overline{p}_1} (x_{j_1} - a_{j_1 j_1 r_2})^{m_{j_1 r_2}} \prod_{u_2=1}^{\overline{q}_1} (x_{j_2} - a_{j_1 j_2 u_2})^{n_{j_1 u_2}}, \\ \dot{x}_{j_2} &= a_{j_2 j_2 0}(x_{j_1} - a_{j_2 j_1 1})^{n_2}\end{aligned}$$

where

$$\Sigma_{r_2=1}^{\overline{p}_1} m_{j_1 r_2} = 2m_{11},\ \Sigma_{u_2=1}^{\overline{q}_1} n_{j_1 u_2} = 2n_{11}.$$

- From the above bifurcation process analysis, only $\overline{p}_1$-parameter variations in the x_{j_1}-direction from a p_1 to $\overline{p}_1$ 1-dimensional flow array through the $(2m_{11})^{\text{th}}$-order inflection flow bifurcations in the x_{j_1}-direction.

  Thus, the 1-dimensional flow appearing or vanishing bifurcation route from a p_1 to $\overline{p}_1$ 1-dimensional flow array is expressed as

$$\bigcup_{r_1=1}^{p_1} \underbrace{(a_{j_1 j_1 r_1}, \dot{x}_{j_2})}_{(m_{j_1 r_1})^{\text{th}}\text{order XX}} \rightleftharpoons \underbrace{(a_{j_1 j_1 1}, \dot{x}_{j_2})}_{(2m_{11})^{\text{th}}\text{ order XX}} \rightleftharpoons \bigcup_{r_2=1}^{\overline{p}_1} \underbrace{(a_{j_1 j_1 r_2}, \dot{x}_{j_2})}_{(m_{j_1 r_2})^{\text{th}}\text{order XX}}.$$

- From the above bifurcation process analysis, only $\overline{q}_1$-parameter variations in the x_{j_1}-direction from a q_1 to $\overline{q}_1$ 1-dimensional flow array through the $(2n_{11})^{\text{th}}$-order increasing and decreasing-inflection flow bifurcations in the x_{j_1}-direction.

  Thus, the 1-dimensional flow appearing or vanishing bifurcation route from a q_1 to $\overline{q}_1$ 1-dimensional flow array is expressed as

$$\bigcup_{u_1=1}^{q_1} \underbrace{(a_{j_1 j_2 u_1}, \dot{x}_{j_2})}_{(n_{j_1 u_1})^{\text{th}}\text{order YY}} \rightleftharpoons \underbrace{(a_{j_1 j_2 1}, \dot{x}_{j_2})}_{(2n_{11})^{\text{th}}\text{ order YY}} \rightleftharpoons \bigcup_{u_2=1}^{\overline{q}_1} \underbrace{(a_{j_1 j_2 u_2}, \dot{x}_{j_2})}_{(n_{j_1 u_2})^{\text{th}}\text{order YY}}.$$

- From the above bifurcation process analysis, only $\overline{q}_1 + \overline{p}_1$-parameter variations in the x_{j_1}-direction from a $q_1 \times p_1$ to $\overline{q}_1 \times \overline{p}_1$ network of 1-dimensional flows through the $(2m_{11})^{\text{th}}$-order upper-saddle and lower-saddle flows bifurcations in the x_{j_1}-direction, and the $(2n_{11})^{\text{th}}$-order up-parabola and down-parabola flows bifurcations in the x_{j_1}-direction.

  Thus, the 1-dimensional flow appearing or vanishing bifurcation route is as follows:

$$\bigcup_{r_1=1}^{p_1} \bigcup_{u_1=1}^{q_1} \underbrace{(a_{j_1 j_1 r_1}, a_{j_1 j_2 u_1})}_{([(m_{j_1 r_1}), (n_{j_1 u_1})]\text{XX,xF})\text{-flow}} \rightleftharpoons \underbrace{([(2m_{11}), (2n_{11})]\text{XX, xF})}_{[(2m_{11}), (2n_{11})]\text{-XX}}$$

$$\rightleftharpoons \bigcup_{r_2=1}^{\overline{p}_1} \bigcup_{u_2=1}^{\overline{q}_1} \underbrace{(a_{j_1j_1r_2}, a_{j_1j_2u_2})}_{([(m_{j_1r_2}),(n_{j_1u_1})]\text{XX,xF})\text{-flow}},$$

which are the higher-order inflection upper-saddle and lower-saddle bifurcations.

(I_3) Consider two dynamical systems with the same 1-dimensional flows with locations switched (A_3,C_3) for $i = 1, 2$ as

$$\begin{aligned}\dot{x}_{j_1} &= a_{j_1j_10} \prod_{r_i=1}^{p_1} (x_{j_1} - a_{j_1j_1r_i})^{m_{j_1r_i}} \prod_{u_i=1}^{q_1} (x_{j_2} - a_{j_1j_2u_i})^{n_{j_1u_i}},\\ \dot{x}_{j_2} &= a_{j_2j_20}(x_{j_1} - a_{j_2j_11})^{m_2};\end{aligned}$$

where

$$\sum_{r_i=1}^{p_1} m_{j_1r_i} = 2m_{11} + 1, \sum_{u_i=1}^{q_1} n_{j_1u_i} = 2n_{11}, \text{ for } i = 1, 2.$$

Consider a dynamical system as a singular 1-dimensional flow (B_3) as

$$\begin{aligned}\dot{x}_{j_1} &= a_{j_1j_10}(x_{j_1} - a_{j_1j_11})^{2m_{11}}(x_{j_2} - a_{j_1j_21})^{2n_{11}},\\ \dot{x}_{j_2} &= a_{j_2j_10}(x_{j_2} - a_{j_2j_21})^{n_2}.\end{aligned}$$

For $r_i = 1, 2, \ldots, p_1$ and $s_i = 1, 2, \ldots, q_1$ $(i = 1, 2)$, define a function as

$$\begin{aligned}\Delta_{j_1j_1}^{(r_1r_2)} &= (a_{j_1j_1r_1} - a_{j_1j_1r_2})^2 \; (r_1, r_2 \in \{1, 2, \ldots, p_1\}, r_1 \neq r_2),\\ \Delta_{j_1j_2}^{(u_1u_2)} &= (a_{j_1j_2u_1} - a_{j_1j_2u_2})^2 \; (u_1, u_2 \in \{1, 2, \ldots, q_1\}, u_1 \neq u_2).\end{aligned}$$

For $r_i = 1, 2, \ldots, p_1$ and $s_i = 1, 2, \ldots, q_1$ $(i = 1, 2)$, if

$$\Delta_{j_1j_1}^{(r_1r_2)} = (a_{j_1j_1r_1} - a_{j_1j_1r_2})^2 = 0 \text{ and } \Delta_{j_1j_2}^{(u_1u_2)} = (a_{j_1j_2u_1} - a_{j_1j_2u_2})^2 = 0,$$

there are three bifurcations as follows.

- Two 1-dimensional flows of $(a_{j_1j_1r_1}, \dot{x}_{j_2})$ and $(a_{j_1j_1r_2}, \dot{x}_{j_2})$ switching at point $(a_{j_1j_11}, \dot{x}_{j_2})$ with the same order singularity are given through

$$\bigcup_{r_1=1}^{p_1} \underbrace{(a_{j_1j_1r_1}, \dot{x}_{j_2})}_{(m_{j_1r_1})^{\text{th}}\text{order XX}} \rightleftharpoons \underbrace{(a_{j_1j_11}, \dot{x}_{j_2})}_{(2m_{11})^{\text{th}}\text{ order XX}} \rightleftharpoons \bigcup_{r_2=1}^{p_1} \underbrace{(a_{j_1j_1r_2}, \dot{x}_{j_2})}_{(m_{j_1r_2})^{\text{th}}\text{order XX}}$$

$$\{a_{j_1j_1r_1} \rightleftharpoons a_{j_1j_1r_2} | m_{j_1r_1} = m_{j_1r_2}; r_1, r_2 \in \{1, 2, \ldots, p_1\}, r_1 \neq r_2\};$$

and

$$\begin{aligned}
&a_{j_1j_1r_1}=a_{j_1j_1 1}-\varepsilon_{r_1},\, a_{j_1j_1r_2}=a_{j_1j_1 1}+\varepsilon_{r_2},\\
&(a_{j_1j_1r_1}=a_{j_1j_1 1}+\varepsilon_{r_1},\, a_{j_1j_1r_2}=a_{j_1j_1 1}-\varepsilon_{r_2});\\
&\varepsilon_{r_1},\varepsilon_{r_2}>0;\ r_i=1,2,\ldots,p_1\ (i=1,2).
\end{aligned}$$

- Two 1-dimensional flows of $(a_{j_1j_2s_1},\dot{x}_{j_2})$ and $(a_{j_1j_2s_1},\dot{x}_{j_2})$ switching at point $(a_{j_1j_2 1},\dot{x}_{j_2})$ with the same order singularity are given through

$$\bigcup_{u_1=1}^{q_1}\underbrace{(a_{j_1j_2u_1},\dot{x}_{j_2})}_{(n_{j_1u_1})^{\text{th}}\text{order YY}}\rightleftharpoons\underbrace{(a_{j_1j_2 1},\dot{x}_{j_2})}_{(2n_{11})^{\text{th}}\text{ order YY}}\rightleftharpoons\bigcup_{u_2=1}^{q_1}\underbrace{(a_{j_1j_2u_2},\dot{x}_{j_2})}_{(n_{j_1u_2})^{\text{th}}\text{order YY}}$$
$$\left\{a_{j_1j_2u_1}\rightleftharpoons a_{j_1j_2u_2}|n_{j_1u_1}=n_{j_1u_2};\, u_1,u_2\in\{1,2,\ldots,q\},\, u_1\neq u_2\right\};$$

and

$$\begin{aligned}
&a_{j_1j_2u_1}=a_{j_1j_2 1}-\varepsilon_{u_1},\, a_{j_1j_2u_2}=a_{j_1j_2 1}+\varepsilon_{u_2},\\
&(a_{j_1j_2u_1}=a_{j_1j_2 1}+\varepsilon_{u_1},\, a_{j_1j_2u_2}=a_{j_1j_2 1}-\varepsilon_{u_2});\\
&\varepsilon_{u_1},\varepsilon_{u_2}>0;\ u_i=1,2,\ldots,q_1\ (i=1,2).
\end{aligned}$$

- Two 1-dimensional flows of $(a_{j_1j_1r_1},a_{j_1j_2u_1})$ and $(a_{j_1j_1r_2},a_{j_1j_2u_2})$ switching at point $(a_{j_1j_1 1},a_{j_1j_2 1})$ with the same order singularity are given through

$$\begin{aligned}
\bigcup_{r_1=1}^{p}\bigcup_{u_1=1}^{q}\underbrace{(a_{j_1j_1r_1},a_{j_1j_2u_1})}_{([(m_{j_1r_1}),(n_{j_1u_1})]\text{XX,xF})\text{-flow}}
&\rightleftharpoons\underbrace{([(2m_{11}),(2n_{11})]\text{XX, xF})}_{[(2m_{11}),(2n_{11})]\text{ -XX}}\\
&\rightleftharpoons\bigcup_{r_2=1}^{p}\bigcup_{u_2=1}^{q}\underbrace{(a_{j_1j_1r_2},a_{j_1j_2u_2})}_{([(m_{j_1r_2}),(n_{j_1u_2})]\text{XX,xF})\text{-flow}}
\end{aligned}$$

with

$$\begin{aligned}
&a_{j_1j_1r_1}=a_{j_1j_1 1}-\varepsilon_{r_1},\, a_{j_1j_1r_2}=a_{j_1j_1 1}+\varepsilon_{r_2},\\
&(a_{j_1j_1r_1}=a_{j_1j_1 1}+\varepsilon_{r_1},\, a_{j_1j_1r_2}=a_{j_1j_1 1}-\varepsilon_{r_2}),\\
&\varepsilon_{r_1},\varepsilon_{r_2}>0,\ r_i=1,2,\ldots,p_1\ (i=1,2);\\
&a_{j_1j_2u_1}=a_{j_1j_2 1}-\varepsilon_{u_1},\, a_{j_1j_2u_2}=a_{j_1j_2 1}+\varepsilon_{u_2},\\
&(a_{j_1j_2u_1}=a_{j_1j_2 1}+\varepsilon_{u_1},\, a_{j_1j_2u_2}=a_{j_1j_2 1}-\varepsilon_{u_2}),\\
&\varepsilon_{u_1},\varepsilon_{u_2}>0,\ u_i=1,2,\ldots,q_1\ (i=1,2).
\end{aligned}$$

The switching bifurcations are the higher-order inflection upper-saddle and lower-saddle flows bifurcations.

In summary, from the analysis in case (I_1)–(I_3), the 1-dimensional flow of $(x_{j_1}^*, x_{j_2}^*) = (a_{j_1j_1 1}, a_{j_1j_2 1})$ for $m_1 = 2m_{11}$ and $n_1 = 2n_{11}$ has the bifurcation properties as stated in (iv_{4a})–(iv_{4d}) through Eqs. (1.368)–(1.387).

Note that the equation in the x_{j_2}-direction can be considered as follows.

$$\dot{x}_{j_2} = a_{j_2j_2 0}\prod_{v=1}^{m_2}(x_{j_1} - a_{j_2j_1 v});\ \text{or}$$

$$\dot{x}_{j_2} = a_{j_2j_2 0}\prod_{v=1}^{p_2}(x_{j_1} - a_{j_2j_1 v})^{m_{j_2 v}}.$$

2.4.2 Appearing and Switching Bifurcations for Equilibriums

(v) The equilibrium of $(x_{j_2}^*, x_{j_1}^*) = (a_{j_1j_2 1}, a_{j_2j_1 1})$ with $a_{j_1j_1 1} \neq a_{j_2j_1 1}$ has four classes of equilibrium bifurcations.

(v_1) For $n_1 = 2n_{11}+1$ and $m_2 = 2m_{21}+1$ the equilibrium of $(x_{j_2}^*, x_{j_1}^*) = (a_{j_1j_2 1}, a_{j_2j_1 1})$ has the following bifurcation properties with

$$2n_{11} + 1 - \sum_{u_1=1}^{q_1} n_{j_1 u_1} = 2\sum_{\rho_1=1}^{l_1} n_{\rho_1},\ 2m_{21} + 1 - \sum_{s_1=1}^{p_2} m_{j_2 v_1} = 2\sum_{\sigma_1=1}^{l_2} m_{\sigma_1};$$

$$\sum_{u_2=1}^{\bar{q}_1} n_{j_1 u_2} = 2n_{11} + 1,\ \sum_{v_2=1}^{\bar{p}_2} m_{j_2 v_2} = 2m_{21} + 1;$$

$$\sum_{u_i=1}^{q_1} n_{j_1 u_i} = 2n_{11} + 1,\ \sum_{v_i=1}^{q_1} m_{j_2 v_i} = 2m_{21} + 1,\ \mathit{for}\ i = 1, 2;$$

l_1, l_2-quadratic polynomials are without real roots, ρ_1^{th} and σ_1^{th}-quadratic polynomials are with powers m_{ρ_1} and n_{σ_1},respectively.

The bifurcation process is discussed through differential equations as follows. There are four bifurcations (I_1)–(I_3).

(I_1) Consider a dynamical system having a single equilibrium (A_1) as

$$\dot{x}_{j_1} = a_{j_1j_1 0}(x_{j_2} - b_{j_1j_2 1})\prod_{u=1}^{n_{11}}[(x_{j_2} - a_{j_1j_2}^{(u)})^2 + \Delta_{j_1j_2}^{(u)}](x_{j_1} - a_{j_1j_1 1})^{m_1},$$

$$\dot{x}_{j_2} = a_{j_2j_2 0}(x_{j_1} - b_{j_2j_1 1})\prod_{v=1}^{m_{21}}[(x_{j_1} - a_{j_2j_1}^{(v)})^2 + \Delta_{j_2j_1}^{(v)}];$$

where $a_{j_1j_21}=b_{j_1j_21}$ and $a_{j_2j_11}=b_{j_2j_11}\neq a_{j_1j_11}$. Once

$$\Delta^{(u)}_{j_1j_2}=\Delta_{j_1j_2}=0\,(u=1,2,\ldots,n_{11}),$$
$$\Delta^{(v)}_{j_2j_2}=\Delta_{j_2j_2}=0\,(v=1,2,\ldots,m_{21}),$$

if

$$a_{j_1j_21}=b_{j_1j_21}=a^{(u)}_{j_1j_2}\;(u=1,2,\ldots,n_{11}),$$
$$a_{j_2j_11}=b_{j_2j_11}=a^{(v)}_{j_2j_1}\;(v=1,2,\ldots,m_{21}),$$

the foregoing differential equations become a dynamical system with a singular equilibrium (B_1) as

$$\dot{x}_{j_1}=a_{j_1j_10}(x_{j_1}-a_{j_1j_11})^{m_1}(x_{j_2}-a_{j_1j_21})^{2n_{11}+1},$$
$$\dot{x}_{j_2}=a_{j_2j_20}(x_{j_1}-a_{j_2j_11})^{2m_{21}+1}.$$

If

$$\Delta^{(\mathrm{u})}_{j_1j_2}=\Delta_{j_1j_2}+\delta_u(u=1,2,\ldots,n_{11}),\,\delta_u>0;$$
$$\Delta^{(\mathrm{v})}_{j_2j_1}=\Delta_{j_2j_1}+\delta_v(v=1,2,\ldots,m_{21}),\,\delta_v>0;$$

then

$$a^{(\mathrm{u})}_{j_1j_21},a^{(\mathrm{u})}_{j_1j_22}=a_{j_1j_21}\pm\varepsilon_u,\,(u=1,2,\ldots,n_{11}),$$
$$\{a_{j_1j_21},a_{j_1j_22},\ldots,a_{j_1j_1(2n_{11}+1)}\}=\mathrm{sort}\{b_{j_1j_21},a^{(u)}_{j_1j_21},a^{(u)}_{j_1j_22}|u=1,2,\ldots,n_{11}\};$$
$$a^{(\mathrm{v})}_{j_2j_11},a^{(\mathrm{v})}_{j_2j_12}=a_{j_2j_11}\pm\varepsilon_v,\,(v=1,2,\ldots,m_{21}),$$
$$\{a_{j_2j_11},a_{j_2j_12},\ldots,a_{j_2j_1(2m_{21}+1)}\}=\mathrm{sort}\{b_{j_2j_11},a^{(v)}_{j_2j_11},a^{(v)}_{j_2j_12}|v=1,2,\ldots,m_{21}\}.$$

Thus, the differential equation becomes a dynamical system with a non-singular equilibrium network (C_1) as

$$\dot{x}_{j_1}=a_{j_1j_10}(x_{j_1}-a_{j_1j_11})^{m_1}\prod_{u=1}^{2n_{11}+1}(x_{j_2}-a_{j_1j_2u}),$$
$$\dot{x}_{j_2}=a_{j_2j_20}\prod_{v=1}^{2m_{21}+1}(x_{j_2}-a_{j_2j_1v}).$$

- From the above bifurcation process analysis, at least, only $(n_{11}+1)$-parameter variations in the x_{j_1}-direction are engaged in such a bifurcation from a 1-dimensional flow to the 1-dimensional flow array of $1\times(2n_{11}+1)$ through the $(2n_{11}+1)^{\text{th}}$-order parabola flows bifurcations in the x_{j_1}-direction. The higher-order parabola flow bifurcations are developed.

Thus, the 1-dimensional flow appearing or vanishing bifurcation route is as follows.

$$\underbrace{(a_{j_1j_21},\dot{x}_{j_2})}_{\text{XX}} \rightleftharpoons \underbrace{(a_{j_1j_21},\dot{x}_{j_2})}_{(2n_{11}+1)^{\text{th}}\text{order XX}} \rightleftharpoons \bigcup_{u=1}^{2n_{11}+1}\underbrace{(a_{j_1j_2u},\dot{x}_{j_2})}_{\text{XX}}.$$

- From the above bifurcation process analysis, at least, only $(m_{21}+1)$-parameter variations in the x_{j_1}-direction are engaged in such a bifurcation from a 1-dimensional flow to the 1-dimensional flow array of $(2m_{21}+1)\times 1$ through the $(2m_{21}+1)^{\text{th}}$-order parabola flow flows bifurcations in the x_{j_2}-direction. The higher-order parabola flow bifurcations are developed.

  Thus, the 1-dimensional flow appearing or vanishing bifurcation route is as follows.

$$\underbrace{(\dot{x}_{j_1},a_{j_1j_21})}_{\text{YY}} \rightleftharpoons \underbrace{(\dot{x}_{j_1},a_{j_2j_11})}_{(2m_{21}+1)^{\text{th}}\text{order YY}} \rightleftharpoons \bigcup_{v=1}^{2m_{21}+1}\underbrace{(\dot{x}_{j_1},a_{j_2j_1v})}_{\text{YY}}.$$

- From the above bifurcation process analysis, $(n_{11}+m_{21}+2)$-parameter variations in the two directions are engaged in such a bifurcation from one of positive and negative saddles, CCW and CW centers to a $(2n_{11}+1)\times(2m_{21}+1)$ network of positive and negative saddles, CCW and CW centers through the $(2n_{11}+1)^{\text{th}}$-order parabola bifurcations in the x_{j_1}-direction, and the $(2m_{21}+1)^{\text{th}}$-order parabola flows bifurcations in the x_{j_2}-direction. The higher-order bifurcations of positive and negative saddle, CCW and CW centers are developed.

  Thus, the equilibrium network appearing or vanishing bifurcation route is as follows.

$$\begin{aligned}\underbrace{(a_{j_1j_21},a_{j_2j_21})}_{\text{XX}} &\rightleftharpoons \underbrace{((2n_{11}+1)\text{YY},(2m_{21}+1)\text{ZZ})}_{((2n_{11}+1),(2m_{21}+1))\text{-XX}}\\ &\rightleftharpoons \bigcup_{u=1}^{2n_{11}+1}\bigcup_{v=1}^{2m_{21}+1}\underbrace{(a_{j_1j_2u},a_{j_2j_1v})}_{\text{XX}},\end{aligned}$$

  which are the higher-order bifurcations of positive and negative saddles, and CCW and CW centers.

(I_2) From Case (I_1), consider a dynamical system having singular equilibriums (A_2) as

$$\dot{x}_{j_1} = a_{j_1j_10}(x_{j_1} - a_{j_1j_11})^{m_1} \prod_{u_1=1}^{q_1} (x_{j_2} - b_{j_1j_2u_1})^{n_{j_1u_1}} \prod_{u=1}^{l_1} [(x_{j_2} - a_{j_1j_2}^{(u)})^2 + \Delta_{j_1j_2}^{(u)}]^{m_{j_1u}},$$
$$\dot{x}_{j_2} = a_{j_2j_20} \prod_{v_1=1}^{p_2} (x_{j_1} - b_{j_2j_1v_1})^{m_{j_2v_1}} \prod_{v=1}^{l_2} [(x_{j_1} - a_{j_2j_1}^{(v)})^2 + \Delta_{j_2j_1}^{(u)}]^{m_{j_2v}};$$

where

$$2n_{11} + 1 - \sum_{u_1=1}^{q_1} n_{j_1u_1} = 2\sum_{\rho_1=1}^{l_1} n_{\rho_1}, 2m_{21} + 1 - \sum_{v_1=1}^{p_2} m_{j_2v_1} = 2\sum_{\sigma_1=1}^{l_2} m_{\sigma_1};$$
$$\sum_{u_2=1}^{\overline{q}_1} n_{j_1u_2} = 2n_{11} + 1, \sum_{v_2=1}^{\overline{p}_2} m_{j_2v_2} = 2m_{21} + 1;$$

l_1, l_2-quadratic polynomials are without real roots, ρ_1^{th} and σ_1^{th}-quadratic polynomials are with powers m_{ρ_1} and n_{σ_1},respectively.

Once

$$\Delta_{j_1j_2}^{(u)} = 0\,(u = 1, 2, \ldots, l_1),$$
$$\Delta_{j_2j_1}^{(v)} = 0\,(v = 1, 2, \ldots, l_2);$$

if

$$a_{j_1j_21} \equiv a_{j_1j_2u_1} = a_{j_1j_1}^{(u)}\,(u_1 = 1, 2, \ldots, q_1;\, u = 1, 2, \ldots, l_1),$$
$$a_{j_2j_11} \equiv a_{j_2j_1v_1} = a_{j_2j_1}^{(v)}\,(v_1 = 1, 2, \ldots, p_2;\, v = 1, 2, \ldots, l_2),$$

the foregoing differential equations become a dynamical system with a 1-dimensional flow (B_2) as

$$\dot{x}_{j_1} = a_{j_1j_10}(x_{j_1} - a_{j_1j_11})^{m_1}(x_{j_2} - a_{j_1j_21})^{2n_{11}+1},$$
$$\dot{x}_{j_2} = a_{j_2j_20}(x_{j_1} - a_{j_2j_11})^{2m_{21}+1}.$$

If

$$\Delta_{j_1j_2}^{(u)} = \Delta_{j_1j_2} + \delta_u\,(u = 1, 2, \ldots, l_1),\ \delta_u > 0;$$
$$\Delta_{j_2j_1}^{(v)} = \Delta_{j_2j_1} + \delta_u\,(v = 1, 2, \ldots, l_2),\ \delta_v > 0;$$

then

$$a_{jj_21}^{(u)}, a_{jj_22}^{(u)} = a_{j_1j_21} \pm \varepsilon_u, (u = 1, 2, \ldots, l_1),$$
$$\{a_{j_1j_21}, a_{j_1j_22}, \ldots, a_{j_1j_2\overline{q}_1}\}$$

$$
\begin{aligned}
&= \mathrm{sort}\{a_{j_1j_2 1}, a_{j_1j_2 2}, \ldots, a_{j_1j_2q_1}; a^{(u)}_{j_1j_2 1}, a^{(u)}_{j_1j_2 2} | u = 1, 2, \ldots, l_1\};\\
&a^{(v)}_{j_2j_1 1}, a^{(v)}_{j_2j_1 2} = a_{j_2j_1 1} \pm \varepsilon_v, (v = 1, 2, \ldots, l_2),\\
&\{a_{j_2j_1 1}, a_{j_2j_1 2}, \ldots, a_{j_2j_1p_2}\}\\
&= \mathrm{sort}\{a_{j_2j_1 1}, a_{j_2j_1 2}, \ldots, a_{j_2j_1p_2}; a^{(v)}_{j_2j_1 1}, a^{(v)}_{j_2j_1 2} | v = 1, 2, \ldots, l_2\}.
\end{aligned}
$$

Thus, the differential equation becomes a dynamical system with singular 1-dimensional flows (C_2) as

$$
\begin{aligned}
\dot{x}_{j_1} &= a_{j_1j_1 0}(x_{j_1} - a_{j_1j_1 1})^{m_1} \prod_{u_2=1}^{\overline{q}_1} (x_{j_1} - a_{j_1j_1u_2})^{n_{j_1u_2}},\\
\dot{x}_{j_2} &= a_{j_2j_2 0} \prod_{v_2=1}^{\overline{p}_2} (x_{j_1} - a_{j_1j_2v_2})^{m_{j_1v_2}}
\end{aligned}
$$

where

$$
\Sigma_{u_2=1}^{\overline{q}_1} n_{j_1u_2} = 2n_{11} + 1,\ \Sigma_{v_2=1}^{\overline{p}_2} m_{j_1v_2} = 2m_{21} + 1.
$$

- From the above bifurcation process analysis, only $\overline{q}_1$-parameter variations in the x_{j_1}-direction from a q_1 to $\overline{q}_1$ 1-dimensional flow array through the $(2n_{11} + 1)^{\text{th}}$-order parabola flow bifurcations in the x_{j_1}-direction.

  Thus, the 1-dimensional flow appearing or vanishing bifurcation route from a q_1 to $\overline{q}_1$ 1-dimensional flow array is expressed as

$$
\bigcup_{u_1=1}^{q_1} \underbrace{(a_{j_1j_2u_1}, \dot{x}_{j_2})}_{(n_{j_1u_1})^{\text{th}}\text{order } XX} \rightleftharpoons \underbrace{(a_{j_1j_2 1}, \dot{x}_{j_2})}_{(2n_{11}+1)^{\text{th}} \text{ order XX}} \rightleftharpoons \bigcup_{u_2=1}^{\overline{q}_1} \underbrace{(a_{j_1j_2u_2}, \dot{x}_{j_2})}_{(n_{j_1u_2})^{\text{th}}\text{order XX}}.
$$

- From the above bifurcation process analysis, only $\overline{p}_2$-parameter variations in the x_{j_2}-direction from a p_2 to $\overline{p}_2$ 1-dimensional flow array through the $(2m_{21} + 1)^{\text{th}}$-order parabola flow bifurcations in the x_{j_2}-direction.

  Thus, the 1-dimensional flow appearing or vanishing bifurcation route from a p_2 to $\overline{p}_2$ 1-dimensional flow array is expressed as

$$
\bigcup_{v_1=1}^{p_2} \underbrace{(\dot{x}_{j_1}, a_{j_2j_1v_1})}_{(m_{j_2v_1})^{\text{th}}\text{order YY}} \rightleftharpoons \underbrace{(\dot{x}_{j_1}, a_{j_2j_1 1})}_{(2m_{21}+1)^{\text{th}} \text{ order YY}} \rightleftharpoons \bigcup_{v_2=1}^{\overline{p}_2} \underbrace{(\dot{x}_{j_1}, a_{j_2j_1v_2})}_{(m_{j_2v_2})^{\text{th}}\text{order YY}}.
$$

- From the above bifurcation process analysis, only $\overline{p}_2 + \overline{q}_1$-parameter variations from a $p_2 \times q_1$ to $\overline{p}_2 \times \overline{q}_1$ network of 1-dimensional flows through the $(2n_{11} +$

$1)^{\text{th}}$-order parabola flows bifurcations in the x_{j_1}-direction, and the $(2m_{21} + 1)^{\text{th}}$-order parabola flows bifurcations in the x_{j_2}-direction.

Thus, the singular-equilibrium network appearing or vanishing bifurcation route is as follows:

$$\bigcup_{u_1=1}^{q_1}\bigcup_{v_1=1}^{p_2}\underbrace{(a_{j_1j_2u_1}, a_{j_2j_1v_1})}_{((n_{j_1u_1}),(m_{j_2v_1}))\text{-XX}} \rightleftharpoons \underbrace{((2n_{11}+1)\text{YY},(2m_{21}+1)\text{ZZ})}_{((2n_{11}+1),(2m_{21}+1))\text{-XX}}$$
$$\rightleftharpoons \bigcup_{u_2=1}^{\bar{q}_1}\bigcup_{v_2=1}^{\bar{p}_2}\underbrace{(a_{j_1j_2u_2}, a_{j_2j_1v_2})}_{((n_{j_1u_2}),(m_{j_2v_2}))\text{-XX}},$$

which are the higher-order bifurcations of positive and negative saddles, CCW and CW centers .

(I_3) Consider two dynamical systems with the same 1-dimensional flows with locations switched (A_3,C_3) for $i = 1, 2$ as

$$\dot{x}_{j_1} = a_{j_1j_10}(x_{j_1} - a_{j_1j_11})^{m_1}\prod_{u_i=1}^{q_1}(x_{j_2} - a_{j_1j_2u_i})^{n_{j_1u_i}},$$
$$\dot{x}_{j_2} = a_{j_2j_20}\prod_{v_i=1}^{q_1}(x_{j_1} - a_{j_2j_1s_i})^{n_{j_2v_i}};$$

where

$$\sum_{u_i=1}^{q_1} n_{j_1u_i} = 2n_{11}+1, \sum_{v_i=1}^{p_2} m_{j_2v_i} = 2m_{21}+1, \text{ for } i = 1, 2.$$

Consider a dynamical system as a singular 1-dimensional flow (B_3) as

$$\dot{x}_{j_1} = a_{j_1j_10}(x_{j_1} - a_{j_1j_11})^{m_1}(x_{j_2} - a_{j_1j_21})^{2n_{11}+1},$$
$$\dot{x}_{j_2} = a_{j_2j_20}(x_{j_1} - a_{j_2j_11})^{2m_{21}+1}.$$

For $u_i = 1, 2, \ldots, q_1$ and $v_i = 1, 2, \ldots, p_2$ $(i = 1, 2)$, define a function as

$$\Delta_{j_1j_2}^{(u_1u_2)} = (a_{j_1j_2u_1} - a_{j_1j_2u_2})^2\ (u_1, u_2 \in \{1, 2, \ldots, q_1\}, u_1 \neq u_2),$$
$$\Delta_{j_2j_2}^{(v_1v_2)} = (a_{j_2j_1v_1} - a_{j_2j_1v_2})^2\ (v_1, v_2 \in \{1, 2, \ldots, p_2\}, v_1 \neq v_2).$$

For $u_i = 1, 2, \ldots, q_1$ and $v_i = 1, 2, \ldots, p_2$ $(i = 1, 2)$, if

$$\Delta_{j_1j_2}^{(u_1u_2)} = (a_{j_1j_2u_1} - a_{j_1j_2u_2})^2 = 0 \text{ and } \Delta_{j_2j_1}^{(v_1v_2)} = (a_{j_2j_1v_1} - a_{j_2j_1v_2})^2 = 0,$$

there are three bifurcations as follows.

- Two 1-dimensional flows of $(a_{j_1j_2u_1},\dot{x}_{j_2})$ and $(a_{j_1j_2u_2},\dot{x}_{j_2})$ switching at point $(a_{j_1j_2 1},\dot{x}_{j_2})$ with the same order singularity are given through

$$\bigcup_{u_1=1}^{q_1}\underbrace{(a_{j_1j_2u_1},\dot{x}_{j_2})}_{(n_{j_1u_1})^{\text{th}}\text{order }XX}\rightleftharpoons\underbrace{(a_{j_1j_2 1},\dot{x}_{j_2})}_{(2n_{11}+1)^{\text{th}}\text{ order XX}}\rightleftharpoons\bigcup_{u_2=1}^{q_1}\underbrace{(a_{j_1j_2u_2},\dot{x}_{j_2})}_{(n_{j_1u_2})^{\text{th}}\text{order }XX}$$
$$\left\{a_{j_1j_2u_1}\rightleftharpoons a_{j_1j_2u_2}|n_{j_1u_1}=n_{j_1u_2};\,u_1,u_2\in\{1,2,\ldots,q_1\},u_1\neq u_2\right\};$$

and

$$\begin{aligned}&a_{j_1j_2u_1}=a_{j_1j_2 1}-\varepsilon_{u_1},a_{j_1j_2u_2}=a_{j_1j_2 1}+\varepsilon_{u_2},\\&(a_{j_1j_2u_1}=a_{j_1j_2 1}+\varepsilon_{u_1},a_{j_1j_2u_2}=a_{j_1j_2 1}-\varepsilon_{u_2});\\&\varepsilon_{u_1},\varepsilon_{u_2}>0;\;u_i=1,2,\ldots,q_1\,(i=1,2).\end{aligned}$$

- Two 1-dimensional flows of $(\dot{x}_{j_1},a_{j_2j_1v_1})$ and $(\dot{x}_{j_1},a_{j_2j_1v_2})$ switching at point $(\dot{x}_{j_1},a_{j_2j_1 1})$ with the same order singularity are given through

$$\bigcup_{v_1=1}^{p_2}\underbrace{(\dot{x}_{j_1},a_{j_2j_1v_1})}_{(m_{j_2v_1})^{\text{th}}\text{order YY}}\rightleftharpoons\underbrace{(\dot{x}_{j_1},a_{j_2j_1 1})}_{(2m_{21}+1)^{\text{th}}\text{ order YY}}\rightleftharpoons\bigcup_{v_2=1}^{p_2}\underbrace{(\dot{x}_{j_1},a_{j_2j_1v_2})}_{(v_{j_2s_2})^{\text{th}}\text{order YY}}$$
$$\left\{a_{j_2j_1v_1}\rightleftharpoons a_{j_2j_1v_2}|m_{j_2v_1}=m_{j_2v_2};\,v_1,v_2\in\{1,2,\ldots,q_2\},v_1\neq v_2\right\};$$

and

$$\begin{aligned}&a_{j_2j_1v_1}=a_{j_2j_1 1}-\varepsilon_{v_1},a_{j_2j_1v_2}=a_{j_2j_1 1}+\varepsilon_{v_2},\\&(a_{j_2j_1v_1}=a_{j_2j_1 1}+\varepsilon_{v_1},a_{j_2j_1v_2}=a_{j_2j_1 1}-\varepsilon_{v_2});\\&\varepsilon_{v_1},\varepsilon_{v_2}>0;\;v_i=1,2,\ldots,p_2\,(i=1,2).\end{aligned}$$

- Two singular-equilibrium networks of $(a_{j_1j_2u_1},a_{j_2j_1v_1})$ and $(a_{j_1j_2v_2},a_{j_2j_1v_2})$ switching at point $(a_{j_1j_2 1},a_{j_2j_1 1})$ with the same order singularity are given through

$$\begin{aligned}\bigcup_{u_1=1}^{q_1}\bigcup_{v_1=1}^{p_2}\underbrace{(a_{j_1j_2u_1},a_{j_2j_1v_1})}_{((n_{j_1u_1}),(m_{j_2v_1}))\text{-XX}}&\rightleftharpoons\underbrace{((2n_{11}+1)\text{YY},(2m_{21}+1)\text{ZZ})}_{((2n_{11}+1),(2m_{21}+1))\text{ -XX}}\\&\rightleftharpoons\bigcup_{u_2=1}^{p_1}\bigcup_{v_2=1}^{p_2}\underbrace{(a_{j_1j_2u_2},a_{j_2j_1v_2})}_{((n_{j_1u_2}),(m_{j_2v_2}))\text{-XX}}\end{aligned}$$

with

$$\begin{aligned}
&a_{j_1j_2u_1}=a_{j_1j_21}-\varepsilon_{u_1},a_{j_1j_2u_2}=a_{j_1j_21}+\varepsilon_{u_2},\\
&(a_{j_1j_2u_1}=a_{j_1j_21}+\varepsilon_{u_1},a_{j_1j_2u_2}=a_{j_1j_21}-\varepsilon_{u_2}),\\
&\varepsilon_{u_1},\varepsilon_{u_2}>0,\ u_i=1,2,\ldots,q_1\ (i=1,2);\\
&a_{j_2j_1v_1}=a_{j_2j_11}-\varepsilon_{v_1},a_{j_2j_1v_2}=a_{j_2j_11}+\varepsilon_{v_2},\\
&(a_{j_2j_1v_1}=a_{j_2j_11}+\varepsilon_{v_1},a_{j_2j_1v_2}=a_{j_2j_11}-\varepsilon_{v_2}),\\
&\varepsilon_{v_1},\varepsilon_{v_2}>0,\ v_i=1,2,\ldots,p_2\ (i=1,2).
\end{aligned}$$

The switching bifurcations of two equilibrium networks are the higher-order bifurcations of positive and negative saddles, CCW and CW centeres.

In summary, from the analysis in case (I_1)–(I_3), the 1-dimensional flow of $(x^*_{j_2},x^*_{j_1})=(a_{j_1j_21},a_{j_2j_11})$ for $n_1=2n_{11}+1$ and $m_2=2m_{21}+1$ has the bifurcation properties as stated in cases (v_{1a})–(v_{1d}) through Eqs. (1.389)–(1.414).

(v_2) For $n_1=2n_{11}$and $m_2=2m_{21}+1$ the equilibrium of $(x^*_{j_2},x^*_{j_1})=(a_{j_1j_21},a_{j_2j_11})$ has the following bifurcation properties with

$$\begin{aligned}
&2n_{11}-\sum_{u_1=1}^{q_1}n_{j_1u_1}=2\sum_{\rho_1=1}^{l_1}n_{\rho_1},2m_{21}+1-\sum_{s_1=1}^{p_2}m_{j_2v_1}=2\sum_{\sigma_1=1}^{l_2}m_{\sigma_1};\\
&\sum_{u_2=1}^{\bar q_1}n_{j_1u_2}=2n_{11},\sum_{v_2=1}^{\bar p_2}m_{j_2v_2}=2m_{21}+1;\\
&\sum_{u_i=1}^{q_1}n_{j_1u_i}=2n_{11},\sum_{v_i=1}^{q_1}m_{j_2v_i}=2m_{21}+1,\ for\, i=1,2;
\end{aligned}$$

l_1,l_2-quadratic polynomials are without real roots, ρ_1^{th} and σ_1^{th}-quadratic polynomials are with powers m_{ρ_1} and n_{σ_1},respectively.

The bifurcation process is discussed through differential equations as follows. There are four bifurcations (I_1)–(I_3).

(I_1) Consider a dynamical system having a single equilibrium (A_1) as

$$\begin{aligned}
\dot x_{j_1}&=a_{j_1j_10}\prod_{u=1}^{n_{11}}[(x_{j_2}-a^{(u)}_{j_1j_2})^2+\Delta^{(u)}_{j_1j_2}](x_{j_1}-a_{j_1j_11})^{m_1},\\
\dot x_{j_2}&=a_{j_2j_20}(x_{j_1}-b_{j_2j_11})\prod_{v=1}^{m_{21}}[(x_{j_1}-a^{(v)}_{j_2j_1})^2+\Delta^{(v)}_{j_2j_1}];
\end{aligned}$$

where $a_{j_2j_11}=b_{j_2j_11}\neq a_{j_1j_11}$. Once

$$\begin{aligned}
\Delta^{(u)}_{j_1j_2}&=\Delta_{j_1j_2}=0\,(u=1,2,\ldots,n_{11}),\\
\Delta^{(v)}_{j_2j_2}&=\Delta_{j_2j_2}=0\,(v=1,2,\ldots,m_{21}),
\end{aligned}$$

if

$$a_{j_1j_21} = b_{j_1j_21} = a_{j_1j_2}^{(u)}\ (u = 1, 2, \ldots, n_{11}),$$
$$a_{j_2j_11} = b_{j_2j_11} = a_{j_2j_1}^{(v)}\ (v = 1, 2, \ldots, m_{21}),$$

the foregoing differential equations become a dynamical system with a singular equilibrium (B_1) as

$$\dot{x}_{j_1} = a_{j_1j_10}(x_{j_1} - a_{j_1j_11})^{m_1}(x_{j_2} - a_{j_1j_21})^{2n_{11}},$$
$$\dot{x}_{j_2} = a_{j_2j_20}(x_{j_1} - a_{j_2j_11})^{2m_{21}+1}.$$

If

$$\Delta_{j_1j_2}^{(u)} = \Delta_{j_1j_2} + \delta_u\ (u = 1, 2, \ldots, n_{11}),\ \delta_u > 0;$$
$$\Delta_{j_2j_1}^{(v)} = \Delta_{j_2j_1} + \delta_v\ (v = 1, 2, \ldots, m_{21}),\ \delta_v > 0;$$

then

$$a_{j_1j_2}^{(u)}, a_{j_1j_22}^{(u)} = a_{j_1j_21} \pm \varepsilon_u,\ (u = 1, 2, \ldots, n_{11}),$$
$$\{a_{j_1j_21}, a_{j_1j_22}, \ldots, a_{j_1j_1(2m_11}\} = \mathrm{sort}\{a_{j_1j_11}^{(u)}, a_{j_1j_22}^{(u)} | u = 1, 2, \ldots, n_{11}\};$$
$$a_{j_2j_11}^{(v)}, a_{j_2j_12}^{(v)} = a_{j_2j_11} \pm \varepsilon_v,\ (v = 1, 2, \ldots, m_{21}),$$
$$\{a_{j_2j_11}, a_{j_2j_12}, \ldots, a_{j_2j_1(2m_{21})}\} = \mathrm{sort}\{a_{j_2j_11}^{(v)}, a_{j_2j_12}^{(v)} | v = 1, 2, \ldots, m_{21}\}.$$

Thus, the differential equation becomes a dynamical system with a non-singular equilibrium network (C_1) as

$$\dot{x}_{j_1} = a_{j_1j_10}(x_{j_1} - a_{j_1j_11})^{m_1} \prod_{u=1}^{2n_{11}} (x_{j_2} - a_{j_1j_2u}),$$

$$\dot{x}_{j_2} = a_{j_2j_20} \prod_{v=1}^{2m_{21}+1} (x_{j_2} - a_{j_2j_1v}).$$

- From the above bifurcation process analysis, at least, only (n_{11})-parameter variations in the x_{j_1}-direction are engaged in such a bifurcation from a 2-dimensional flow to the 1-dimensional flow array of $1 \times (2n_{11})$ through the $(2n_{11})^{\text{th}}$-order inflection flows bifurcations in the x_{j_1}-direction. The higher-order inflection flow bifurcations are developed.

  Thus, the 1-dimensional flow appearing or vanishing bifurcation route is as follows.

$$\underbrace{(\dot{x}_{j_1},\dot{x}_{j_2})}_{\text{(xF,xF)-flow}} \rightleftharpoons \underbrace{(a_{j_1j_2 1},\dot{x}_{j_2})}_{(2n_{11})^{\text{th}}\text{order-XX}} \rightleftharpoons \bigcup_{u=1}^{2n_{11}} \underbrace{(a_{j_1j_2u},\dot{x}_{j_2})}_{\text{XX}}.$$

- From the above bifurcation process analysis, at least, only $(m_{21}+1)$-parameter variations in the x_{j_1}-direction are engaged in such a bifurcation from a 2-dimensional flow to the 1-dimensional flow array of $(2m_{21}+1)\times 1$ through the $(2m_{21}+1)^{\text{th}}$-order parabola flow flows bifurcations in the x_{j_2}-direction. The higher-order parabola flow bifurcations are developed.

  Thus, the 1-dimensional flow appearing or vanishing bifurcation route is as follows.

$$\underbrace{(\dot{x}_{j_1},a_{j_2j_1 1})}_{\text{YY}} \rightleftharpoons \underbrace{(\dot{x}_{j_1},a_{j_2j_1 1})}_{(2m_{21}+1)^{\text{th}}\text{order-YY}} \rightleftharpoons \bigcup_{v=1}^{2m_{21}+1} \underbrace{(\dot{x}_{j_1},a_{j_2j_1v},)}_{\text{YY}}.$$

- From the above bifurcation process analysis, $(n_{11}+m_{21}+1)$-parameter variations in the two directions are engaged in such a bifurcation from a 1-dimensional flow to a $(2n_{11})\times(2m_{21}+1)$ network of positive saddles, and CCW and CW centers through the $(2n_{11})^{\text{th}}$-order inflection bifurcations in the x_{j_1}-direction, and the $(2m_{21}+1)^{\text{th}}$-order parabola flows bifurcations in the x_{j_2}-direction. The higher-order bifurcations of parabola-saddles are developed.

  Thus, the equilibrium network appearing or vanishing bifurcation route is as follows.

$$\underbrace{(\dot{x}_{j_1},a_{j_2j_1 1})}_{\text{YY}} \rightleftharpoons \underbrace{((2n_{11})\text{YY},(2m_{21}+1)\text{ZZ})}_{((2n_{11}),(2m_{21}+1))\text{-XX}} \rightleftharpoons \bigcup_{u=1}^{2n_{11}}\bigcup_{v=1}^{2m_{21}+1} \underbrace{(a_{j_1j_2u},a_{j_2j_1v})}_{\text{XX}},$$

which are the higher-order bifurcations of parabola-saddles.

(I_2) From Case (I_1), consider a dynamical system having singular equilibriums (A_2) as

$$\dot{x}_{j_1} = a_{j_1j_1 0}(x_{j_1}-a_{j_1j_1 1})^{m_1}\prod_{u_1=1}^{q_1}(x_{j_2}-b_{j_1j_2u_1})^{n_{j_1u_1}}\prod_{u=1}^{l_1}[(x_{j_2}-a_{j_1j_2}^{(u)})^2+\Delta_{j_1j_2}^{(u)}]^{m_{j_1u}},$$

$$\dot{x}_{j_2} = a_{j_2j_2 0}\prod_{v_1=1}^{p_2}(x_{j_1}-b_{j_2j_1v_1})^{m_{j_2v_1}}\prod_{v=1}^{l_2}[(x_{j_1}-a_{j_2j_1}^{(v)})^2+\Delta_{j_2j_1}^{(u)}]^{m_{j_2v}};$$

where

$$2n_{11}-\sum_{u_1=1}^{q_1}n_{j_1u_1}=2\sum_{\rho_1=1}^{l_1}n_{\rho_1},\ 2m_{21}+1-\sum_{v_1=1}^{p_2}m_{j_2v_1}=2\sum_{\sigma_1=1}^{l_2}m_{\sigma_1};$$

$$\sum_{u_2=1}^{\overline{q}_1} n_{j_1u_2} = 2n_{11}, \sum_{v_2=1}^{\overline{p}_2} m_{j_2v_2} = 2m_{21} + 1;$$

l_1, l_2-quadratic polynomials are without real roots, ρ_1^{th} and σ_1^{th}-quadratic polynomials are with powers m_{ρ_1} and n_{σ_1},respectively.

Once

$$\begin{aligned}
\Delta_{j_1j_2}^{(u)} &= 0\,(u = 1, 2, \ldots, l_1), \\
\Delta_{j_2j_1}^{(v)} &= 0\,(v = 1, 2, \ldots, l_2);
\end{aligned}$$

if

$$\begin{aligned}
a_{j_1j_2 1} \equiv a_{j_1j_2u_1} &= a_{j_1j_1}^{(u)}\,(u_1 = 1, 2, \ldots, q_1;\, u = 1, 2, \ldots, l_1), \\
a_{j_2j_1 1} \equiv a_{j_2j_1v_1} &= a_{j_2j_1}^{(v)}\,(v_1 = 1, 2, \ldots, p_2;\, v = 1, 2, \ldots, l_2),
\end{aligned}$$

the foregoing differential equations become a dynamical system with a 1-dimensional flow (B_2) as

$$\begin{aligned}
\dot{x}_{j_1} &= a_{j_1j_1 0}(x_{j_1} - a_{j_1j_1 1})^{m_1}(x_{j_2} - a_{j_1j_2 1})^{2n_{11}}, \\
\dot{x}_{j_2} &= a_{j_2j_2 0}(x_{j_1} - a_{j_2j_1 1})^{2m_{21}+1}.
\end{aligned}$$

If

$$\begin{aligned}
\Delta_{j_1j_2}^{(u)} &= \Delta_{j_1j_2} + \delta_u\,(u = 1, 2, \ldots, l_1),\ \delta_u > 0; \\
\Delta_{j_2j_1}^{(v)} &= \Delta_{j_2j_1} + \delta_u\,(v = 1, 2, \ldots, l_2),\ \delta_v > 0;
\end{aligned}$$

then

$$\begin{aligned}
& a_{jj_2 1}^{(u)}, a_{jj_2 2}^{(u)} = a_{j_1j_2 1} \pm \varepsilon_u, (u = 1, 2, \ldots, l_1), \\
& \{a_{j_1j_2 1}, a_{j_1j_2 2}, \ldots, a_{j_1j_2\overline{q}_1}\} \\
& = \text{sort}\{a_{j_1j_2 1}, a_{j_1j_2 2}, \ldots, a_{j_1j_2q_1}; a_{j_1j_2 1}^{(u)}, a_{j_1j_2 2}^{(u)} | u = 1, 2, \ldots, l_1\}; \\
& a_{j_2j_1 1}^{(v)}, a_{j_2j_2 2}^{(v)} = a_{j_2j_1 1} \pm \varepsilon_v, (v = 1, 2, \ldots, l_2), \\
& \{a_{j_2j_1 1}, a_{j_2j_1 2}, \ldots, a_{j_2j_1p_2}\} \\
& = \text{sort}\{a_{j_2j_1 1}, a_{j_2j_1 2}, \ldots, a_{j_2j_1p_2}; a_{j_2j_1 1}^{(v)}, a_{j_2j_1 2}^{(v)} | v = 1, 2, \ldots, l_2\}.
\end{aligned}$$

Thus, the differential equation becomes a dynamical system with singular 1-dimensional flows (C_2) as

$$\dot{x}_{j_1} = a_{j_1 j_1 0}(x_{j_1} - a_{j_1 j_1 1})^{m_1} \prod_{u_2=1}^{\overline{q}_1} (x_{j_1} - a_{j_1 j_1 u_2})^{n_{j_1 u_2}},$$

$$\dot{x}_{j_2} = a_{j_2 j_2 0} \prod_{v_2=1}^{\overline{p}_2} (x_{j_1} - a_{j_1 j_2 v_2})^{m_{j_1 v_2}},$$

where

$$\Sigma_{u_2=1}^{\overline{q}_1} n_{j_1 u_2} = 2n_{11}, \ \Sigma_{v_2=1}^{\overline{p}_2} m_{j_1 v_2} = 2m_{21} + 1.$$

- From the above bifurcation process analysis, only $\overline{q}_1$-parameter variations in the x_{j_1}-direction from a q_1 to $\overline{q}_1$ 1-dimensional flow array through the $(2n_{11})^{\text{th}}$-order inflection flow bifurcations in the x_{j_1}-direction.

  Thus, the 1-dimensional flow appearing or vanishing bifurcation route from a q_1 to $\overline{q}_1$ 1-dimensional flow array is expressed as

$$\bigcup_{u_1=1}^{q_1} \underbrace{(a_{j_1 j_2 u_1}, \dot{x}_{j_2})}_{(n_{j_1 u_1})^{\text{th}}\text{order XX}} \rightleftharpoons \underbrace{(a_{j_1 j_2 1}, \dot{x}_{j_2})}_{(2n_{11})^{\text{th}} \text{ order XX}} \rightleftharpoons \bigcup_{u_2=1}^{\overline{q}_1} \underbrace{(a_{j_1 j_2 u_2}, \dot{x}_{j_2})}_{(n_{j_1 u_2})^{\text{th}}\text{order XX}}.$$

- From the above bifurcation process analysis, only $\overline{p}_2$-parameter variations in the x_{j_2}-direction from a p_2 to$\overline{p}_2$ 1-dimensional flow array through the $(2m_{21} + 1)^{\text{th}}$-order parabola flow bifurcations in the x_{j_2}-direction.

  Thus, the 1-dimensional flow appearing or vanishing bifurcation route from a p_2 to $\overline{p}_2$ 1-dimensional flow array is expressed as.

$$\bigcup_{v_1=1}^{p_2} \underbrace{(\dot{x}_{j_1}, a_{j_2 j_1 v_1})}_{(m_{j_2 v_1})^{\text{th}}\text{order YY}} \rightleftharpoons \underbrace{(\dot{x}_{j_1}, a_{j_2 j_1 1})}_{(2m_{21}+1)^{\text{th}} \text{ order YY}} \rightleftharpoons \bigcup_{s_2=1}^{\overline{p}_2} \underbrace{(\dot{x}_{j_1}, a_{j_2 j_1 v_2})}_{(m_{j_2 v_2})^{\text{th}}\text{order YY}}.$$

- From the above bifurcation process analysis, only $\overline{p}_2+\overline{q}_1$-parameter variations from a $p_2 \times q_1$ to $\overline{p}_2 \times \overline{q}_1$ network of 1-dimensional flows through the $(2n_{11})^{\text{th}}$-order inflection flows bifurcations in the x_{j_1}-direction, and the $(2m_{21} + 1)^{\text{th}}$-order parabola flows bifurcations in the x_{j_2}-direction.

  Thus, the singular-equilibrium network appearing or vanishing bifurcation route is as follows:

$$\bigcup_{u_1=1}^{q_1} \bigcup_{v_1=1}^{p_2} \underbrace{((a_{j_1 j_2 u_1}, a_{j_2 j_1 v_1})}_{((n_{j_1 u_1}),(m_{j_2 v_1}))\text{-XX}} \rightleftharpoons \underbrace{((2n_{11})\text{YY},(2m_{21} + 1)\text{ZZ})}_{((2n_{11}),(2m_{21}+1))\text{-XX}}$$

$$\rightleftharpoons \bigcup_{u_2=1}^{\bar{q}_1} \bigcup_{v_2=1}^{\bar{p}_2} \underbrace{(a_{j_1j_2u_2}, a_{j_2j_1v_2})}_{((n_{j_1u_2}),(m_{j_2v_2}))\text{-XX}},$$

which are the higher-order bifurcations of parabola-saddles.

(I_3) Consider two dynamical systems with the same 1-dimensional flows with locations switched (A_3,C_3) for $i = 1, 2$ as

$$\dot{x}_{j_1} = a_{j_1j_10}(x_{j_1} - a_{j_1j_11})^{m_1} \prod_{u_i=1}^{q_1} (x_{j_2} - a_{j_1j_2u_i})^{n_{j_1u_i}},$$

$$\dot{x}_{j_2} = a_{j_2j_20} \prod_{v_i=1}^{q_1} (x_{j_1} - a_{j_2j_1s_i})^{n_{j_2v_i}};$$

where

$$\sum_{u_i=1}^{q_1} n_{j_1u_i} = 2n_{11}, \sum_{v_i=1}^{p_2} m_{j_2v_i} = 2m_{21} + 1, \text{ for } i = 1, 2.$$

Consider a dynamical system as a singular 1-dimensional flow (B_3) as

$$\dot{x}_{j_1} = a_{j_1j_10}(x_{j_1} - a_{j_1j_11})^{m_1}(x_{j_2} - a_{j_1j_21})^{2n_{11}},$$
$$\dot{x}_{j_2} = a_{j_2j_20}(x_{j_1} - a_{j_2j_11})^{2m_{21}+1}.$$

For $u_i = 1, 2, \ldots, q_1$ and $v_i = 1, 2, \ldots, p_2$ ($i = 1, 2$), define a function as

$$\Delta_{j_1j_2}^{(u_1u_2)} = (a_{j_1j_2u_1} - a_{j_1j_2u_2})^2 \; (u_1, u_2 \in \{1, 2, \ldots, q_1\}, u_1 \neq u_2),$$
$$\Delta_{j_2j_2}^{(v_1v_2)} = (a_{j_2j_1v_1} - a_{j_2j_1v_2})^2 \; (v_1, v_2 \in \{1, 2, \ldots, p_2\}, v_1 \neq v_2).$$

For $u_i = 1, 2, \ldots, q_1$ and $v_i = 1, 2, \ldots, p_2$ ($i = 1, 2$), if

$$\Delta_{j_1j_2}^{(u_1u_2)} = (a_{j_1j_2u_1} - a_{j_1j_2u_2})^2 = 0 \text{ and } \Delta_{j_2j_1}^{(v_1v_2)} = (a_{j_2j_1v_1} - a_{j_2j_1v_2})^2 = 0,$$

there are three bifurcations as follows.

- Two 1-dimensional flows of $(a_{j_1j_2u_1}, \dot{x}_{j_2})$ and $(a_{j_1j_2u_2}, \dot{x}_{j_2})$ switching at point $(a_{j_1j_21}, \dot{x}_{j_2})$ with the same order singularity are given through

$$\bigcup_{u_1=1}^{q_1} \underbrace{(a_{j_1j_2u_1}, \dot{x}_{j_2})}_{(n_{j_1u_1})^{\text{th}}\text{order XX}} \rightleftharpoons \underbrace{(a_{j_1j_21}, \dot{x}_{j_2})}_{(2n_{11})^{\text{th}} \text{ order XX}} \rightleftharpoons \bigcup_{u_2=1}^{q_1} \underbrace{(a_{j_1j_2u_2}, \dot{x}_{j_2})}_{(n_{j_1u_2})^{\text{th}}\text{order XX}}$$
$$\left\{a_{j_1j_2u_1} \rightleftharpoons a_{j_1j_2u_2} | n_{j_1u_1} = n_{j_1u_2}; u_1, u_2 \in \{1, 2, \ldots, q_1\}, u_1 \neq u_2\right\};$$

and

$$\begin{aligned}&a_{j_1j_2u_1}=a_{j_1j_2 1}-\varepsilon_{u_1},\, a_{j_1j_2u_2}=a_{j_1j_2 1}+\varepsilon_{u_2},\\&(a_{j_1j_2u_1}=a_{j_1j_2 1}+\varepsilon_{u_1},\, a_{j_1j_2u_2}=a_{j_1j_2 1}-\varepsilon_{u_2});\\&\varepsilon_{u_1},\varepsilon_{u_2}>0;\ u_i=1,2,\ldots,q_1\,(i=1,2).\end{aligned}$$

- Two 1-dimensional flows of $(\dot{x}_{j_1}, a_{j_2j_1v_1})$ and $(\dot{x}_{j_1}, a_{j_2j_1v_2})$ switching at point $(\dot{x}_{j_1}, a_{j_2j_1 1})$ with the same order singularity are given through

$$\bigcup_{v_1=1}^{p_2}\underbrace{(\dot{x}_{j_1}, a_{j_2j_1v_1})}_{(m_{j_2v_1})^{\text{th}}\text{order YY}}\rightleftharpoons\underbrace{(\dot{x}_{j_1}, a_{j_2j_1 1})}_{(2m_{21}+1)^{\text{th}}\text{ order YY}}\rightleftharpoons\bigcup_{v_2=1}^{p_2}\underbrace{(\dot{x}_{j_1}, a_{j_2j_1v_2})}_{(v_{j_2s_2})^{\text{th}}\text{order YY}}$$
$$\{a_{j_2j_1v_1}\rightleftharpoons a_{j_2j_1v_2}\,|\,m_{j_2v_1}=m_{j_2v_2};\, v_1,v_2\in\{1,2,\ldots,q_2\},\, v_1\neq v_2\};$$

and

$$\begin{aligned}&a_{j_2j_1v_1}=a_{j_2j_1 1}-\varepsilon_{v_1},\, a_{j_2j_1v_2}=a_{j_2j_1 1}+\varepsilon_{v_2},\\&(a_{j_2j_1v_1}=a_{j_2j_1 1}+\varepsilon_{v_1},\, a_{j_2j_1v_2}=a_{j_2j_1 1}-\varepsilon_{v_2});\\&\varepsilon_{v_1},\varepsilon_{v_2}>0;\ v_i=1,2,\ldots,p_2\,(i=1,2).\end{aligned}$$

- Two singular-equilibrium networks of $(a_{j_1j_2u_1}, a_{j_2j_1v_1})$ and $(a_{j_1j_2v_2}, a_{j_2j_1v_2})$ switching at point $(a_{j_1j_2 1}, a_{j_2j_1 1})$ with the same order singularity are given through

$$\begin{aligned}\bigcup_{u_1=1}^{q_1}\bigcup_{v_1=1}^{p_2}\underbrace{(a_{j_1j_2u_1}, a_{j_2j_1v_1})}_{((n_{j_1u_1}),(m_{j_2v_1}))\text{-XX}}&\rightleftharpoons\underbrace{((2n_{11})\text{YY},(2m_{21}+1)\text{ZZ})}_{((2n_{11}),(2m_{21}+1))\text{ -XX}}\\&\rightleftharpoons\bigcup_{u_2=1}^{p_1}\bigcup_{v_2=1}^{p_2}\underbrace{(a_{j_1j_2u_2}, a_{j_2j_1v_2})}_{((n_{j_1u_2}),(m_{j_2v_2}))\text{-XX}}\end{aligned}$$

with

$$\begin{aligned}&a_{j_1j_2u_1}=a_{j_1j_2 1}-\varepsilon_{u_1},\, a_{j_1j_2u_2}=a_{j_1j_2 1}+\varepsilon_{u_2},\\&(a_{j_1j_2u_1}=a_{j_1j_2 1}+\varepsilon_{u_1},\, a_{j_1j_2u_2}=a_{j_1j_2 1}-\varepsilon_{u_2}),\\&\varepsilon_{u_1},\varepsilon_{u_2}>0,\ u_i=1,2,\ldots,q_1\,(i=1,2);\\&a_{j_2j_1v_1}=a_{j_2j_1 1}-\varepsilon_{v_1},\, a_{j_2j_1v_2}=a_{j_2j_1 1}+\varepsilon_{v_2},\\&(a_{j_2j_1v_1}=a_{j_2j_1 1}+\varepsilon_{v_1},\, a_{j_2j_1v_2}=a_{j_2j_1 1}-\varepsilon_{v_2}),\\&\varepsilon_{v_1},\varepsilon_{v_2}>0,\ v_i=1,2,\ldots,p_2\,(i=1,2).\end{aligned}$$

The switching bifurcations of two equilibrium networks are the higher-order bifurcations of parabola-saddles.

In summary, from the analysis in case (I_1)–(I_3), the 1-dimensional flow of $(x^*_{j_2}, x^*_{j_1}) = (a_{j_1j_2 1}, a_{j_2j_1 1})$ for $n_1 = 2n_{11}$ and $m_2 = 2m_{21} + 1$ has the bifurcation properties as stated in cases (v_{1a})–(v_{1d}) through Eqs. (1.415)–(1.429).

(v_3) For $n_1 = 2n_{11} + 1$ and $m_2 = 2m_{21}$ the equilibrium of $(x^*_{j_2}, x^*_{j_1}) = (a_{j_1j_2 1}, a_{j_2j_1 1})$ has the following bifurcation properties with

$$
\begin{aligned}
&2n_{11} + 1 - \sum_{u_1=1}^{q_1} n_{j_1u_1} = 2\sum_{\rho_1=1}^{l_1} n_{\rho_1}, 2m_{21} - \sum_{s_1=1}^{p_2} m_{j_2v_1} = 2\sum_{\sigma_1=1}^{l_2} m_{\sigma_1};\\
&\sum_{u_2=1}^{\overline{q}_1} n_{j_1u_2} = 2n_{11} + 1, \sum_{v_2=1}^{\overline{p}_2} m_{j_2v_2} = 2m_{21};\\
&\sum_{u_i=1}^{q_1} n_{j_1u_i} = 2n_{11} + 1, \sum_{v_i=1}^{q_1} m_{j_2v_i} = 2m_{21}, \mathit{for}\; i = 1, 2;
\end{aligned}
$$

l_1, l_2-quadratic polynomials are without real roots and ρ_1^{th} and σ_1^{th}-quadratic polynomials are with powers m_{ρ_1} and n_{σ_1}, respectively.

The bifurcation process is discussed through differential equations as follows. There are four bifurcations (I_1)–(I_3).

(I_1) Consider a dynamical system having a single equilibrium (A_1) as

$$
\begin{aligned}
\dot{x}_{j_1} &= a_{j_1j_1 0}(x_{j_2} - b_{j_1j_2 1})\prod_{u=1}^{n_{11}}[(x_{j_2} - a^{(u)}_{j_1j_2})^2 + \Delta^{(u)}_{j_1j_2}](x_{j_1} - a_{j_1j_1 1})^{m_1},\\
\dot{x}_{j_2} &= a_{j_2j_2 0}\prod_{v=1}^{m_{21}}[(x_{j_1} - a^{(v)}_{j_2j_1})^2 + \Delta^{(v)}_{j_2j_1}];
\end{aligned}
$$

where $a_{j_1j_2 1} = b_{j_1j_2 1}$. Once

$$
\begin{aligned}
\Delta^{(u)}_{j_1j_2} &= \Delta_{j_1j_2} = 0\,(u = 1, 2, \ldots, n_{11}),\\
\Delta^{(v)}_{j_2j_2} &= \Delta_{j_2j_2} = 0\,(v = 1, 2, \ldots, m_{21}),
\end{aligned}
$$

if

$$
\begin{aligned}
a_{j_1j_2 1} &= b_{j_1j_2 1} = a^{(u)}_{j_1j_2}\;(u = 1, 2, \ldots, n_{11}),\\
a_{j_2j_1 1} &= b_{j_2j_1 1} = a^{(v)}_{j_2j_1}\;(v = 1, 2, \ldots, m_{21}),
\end{aligned}
$$

the foregoing differential equations become a dynamical system with a singular equilibrium (B_1) as

$$\dot{x}_{j_1} = a_{j_1 j_1 0}(x_{j_1} - a_{j_1 j_1 1})^{m_1}(x_{j_2} - a_{j_1 j_2 1})^{2n_{11}+1},$$
$$\dot{x}_{j_2} = a_{j_2 j_2 0}(x_{j_1} - a_{j_2 j_1 1})^{2m_{21}}.$$

If

$$\Delta_{j_1 j_2}^{(u)} = \Delta_{j_1 j_2} + \delta_u \ (u = 1, 2, \ldots, n_{11}), \ \delta_u > 0;$$
$$\Delta_{j_2 j_1}^{(v)} = \Delta_{j_2 j_1} + \delta_v \ (v = 1, 2, \ldots, m_{21}), \ \delta_v > 0;$$

then

$$a_{j_1 j_2 1}^{(u)}, a_{j_1 j_2 2}^{(u)} = a_{j_1 j_2 1} \pm \varepsilon_u, \ (u = 1, 2, \ldots, n_{11}),$$
$$\{a_{j_1 j_2 1}, a_{j_1 j_2 2}, \ldots, a_{j_1 j_1 (2n_{11}+1)}\} = \text{sort}\{b_{j_1 j_2 1}, a_{j_1 j_2 1}^{(u)}, a_{j_1 j_2 2}^{(u)} | u = 1, 2, \ldots, n_{11}\};$$
$$a_{j_2 j_1 1}^{(v)}, a_{j_2 j_1 2}^{(v)} = a_{j_2 j_1 1} \pm \varepsilon_v, \ (v = 1, 2, \ldots, m_{21}),$$
$$\{a_{j_2 j_1 1}, a_{j_2 j_1 2}, \ldots, a_{j_2 j_1 (2m_{21})}\} = \text{sort}\{b_{j_2 j_1 1}, a_{j_2 j_1 1}^{(v)}, a_{j_2 j_1 2}^{(v)} | v = 1, 2, \ldots, m_{21}\}.$$

Thus, the differential equation becomes a dynamical system with a non-singular equilibrium network (C_1) as

$$\dot{x}_{j_1} = a_{j_1 j_1 0}(x_{j_1} - a_{j_1 j_1 1})^{m_1} \prod_{u=1}^{2n_{11}+1} (x_{j_2} - a_{j_1 j_2 u}),$$
$$\dot{x}_{j_2} = a_{j_2 j_2 0} \prod_{v=1}^{2m_{21}} (x_{j_2} - a_{j_2 j_1 v}).$$

- From the above bifurcation process analysis, at least, only $(n_{11} + 1)$-parameter variations in the x_{j_1}-direction are engaged in such a bifurcation from a 1-dimensional flow to the 1-dimensional flow array of $1 \times (2n_{11} + 1)$ through the $(2n_{11} + 1)^{\text{th}}$-order parabola flows bifurcations in the x_{j_1}-direction. The higher-order parabola flow bifurcations are developed.

  Thus, the 1-dimensional flow appearing or vanishing bifurcation route is as follows.

$$\underbrace{(a_{j_2 j_1 1}, \dot{x}_{j_2})}_{\text{XX}} \rightleftharpoons \underbrace{(a_{j_2 j_1 1}, \dot{x}_{j_2})}_{(2n_{11}+1)^{\text{th}} \text{order XX}} \rightleftharpoons \bigcup_{u=1}^{2n_{11}+1} \underbrace{(a_{j_1 j_2 u}, \dot{x}_{j_2})}_{\text{XX}}.$$

- From the above bifurcation process analysis, at least, only (m_{21})-parameter variations in the x_{j_1}-direction are engaged in such a bifurcation from a 1-dimensional flow to the 1-dimensional flow array of $(2m_{21}) \times 1$ through the $(2m_{21})^{\text{th}}$-order inflection flow flows bifurcations in the x_{j_2}-direction. The higher-order inflection flow bifurcations are developed.

  Thus, the 1-dimensional flow appearing or vanishing bifurcation route is as follows.

$$\underbrace{(\dot{x}_{j_1}, \dot{x}_{j_2})}_{\text{(xF,xF)-flow}} \rightleftharpoons \underbrace{(\dot{x}_{j_1}, a_{j_2j_11})}_{(2m_{21})^{\text{th}}\text{order YY}} \rightleftharpoons \bigcup_{v=1}^{2m_{21}} \underbrace{(\dot{x}_{j_1}, a_{j_2j_1v})}_{\text{YY}}.$$

- From the above bifurcation process analysis, $(n_{11} + m_{21} + 2)$-parameter variations in the two directions are engaged in such a bifurcation from a parabola flow to a $(2n_{11}) \times (2m_{21} + 1)$ network of positive and negative saddle, CCW and CW centers through the $(2n_{11} + 1)^{\text{th}}$-order parabola bifurcations in the x_{j_1}-direction, and the $(2m_{21})^{\text{th}}$-order inflection flows bifurcations in the x_{j_2}-direction. The higher-order bifurcations of parabola-saddles are developed.

  Thus, the equilibrium network appearing or vanishing bifurcation route is as follows.

$$\begin{aligned}\underbrace{(a_{j_1j_21}, a_{j_2j_21})}_{\text{XX}} &\rightleftharpoons \underbrace{((2n_{11} + 1)\text{YY},(2m_{21})\text{ZZ})}_{((2n_{11}+1),(2m_{21}))\text{-XX}}\\ &\rightleftharpoons \bigcup_{u=1}^{2n_{11}+1} \bigcup_{v=1}^{2m_{21}} \underbrace{(a_{j_1j_2u}, a_{j_2j_1v})}_{\text{XX}},\end{aligned}$$

  which are the higher-order bifurcations of parabola-saddles.

(I_2) From Case (I_1), consider a dynamical system having singular equilibriums (A_2) as

$$\begin{aligned}\dot{x}_{j_1} &= a_{j_1j_10}(x_{j_1} - a_{j_1j_11})^{m_1} \prod_{u_1=1}^{q_1} (x_{j_2} - b_{j_1j_2u_1})^{n_{j_1u_1}} \prod_{u=1}^{l_1} [(x_{j_2} - a_{j_1j_2}^{(u)})^2 + \Delta_{j_1j_2}^{(u)}]^{n_{j_1u}},\\ \dot{x}_{j_2} &= a_{j_2j_20} \prod_{v_1=1}^{p_2} (x_{j_1} - b_{j_2j_1v_1})^{m_{j_2v_1}} \prod_{v=1}^{l_2} [(x_{j_1} - a_{j_2j_1}^{(v)})^2 + \Delta_{j_2j_1}^{(u)}]^{m_{j_2v}};\end{aligned}$$

where

$$2n_{11} + 1 - \sum_{u_1=1}^{q_1} n_{j_1u_1} = 2\sum_{\rho_1=1}^{l_1} n_{\rho_1}, 2m_{21} - \sum_{v_1=1}^{p_2} m_{j_2v_1} = 2\sum_{\sigma_1=1}^{l_2} m_{\sigma_1};$$

$$\sum_{u_2=1}^{\bar{q}_1} n_{j_1u_2} = 2n_{11}+1, \sum_{v_2=1}^{\bar{p}_2} m_{j_2v_2} = 2m_{21};$$

l_1, l_2-quadratic polynomials are without real roots, ρ_1^{th} and σ_1^{th}-quadratic polynomials are with powers m_{ρ_1} and n_{σ_1}, respectively.

Once

$$\begin{aligned}\Delta_{j_1j_2}^{(u)} &= 0\,(u=1,2,\ldots,l_1),\\ \Delta_{j_2j_1}^{(v)} &= 0\,(v=1,2,\ldots,l_2);\end{aligned}$$

if

$$\begin{aligned}a_{j_1j_21} \equiv a_{j_1j_2u_1} &= a_{j_1j_1}^{(u)}\,(u_1=1,2,\ldots,q_1;\,u=1,2,\ldots,l_1),\\ a_{j_2j_11} \equiv a_{j_2j_1v_1} &= a_{j_2j_1}^{(v)}\,(v_1=1,2,\ldots,p_2;\,v=1,2,\ldots,l_2),\end{aligned}$$

the foregoing differential equations become a dynamical system with a 1-dimensional flow (B_2) as

$$\begin{aligned}\dot{x}_{j_1} &= a_{j_1j_10}(x_{j_1}-a_{j_1j_11})^{m_1}(x_{j_2}-a_{j_1j_21})^{2n_{11}+1},\\ \dot{x}_{j_2} &= a_{j_2j_20}(x_{j_1}-a_{j_2j_11})^{2m_{21}}.\end{aligned}$$

If

$$\begin{aligned}\Delta_{j_1j_2}^{(u)} &= \Delta_{j_1j_2}+\delta_u\,(u=1,2,\ldots,l_1),\ \delta_u>0;\\ \Delta_{j_2j_1}^{(v)} &= \Delta_{j_2j_1}+\delta_u\,(v=1,2,\ldots,l_2),\ \delta_v>0;\end{aligned}$$

then

$$\begin{aligned}&a_{j_1j_21}^{(u)}, a_{j_1j_22}^{(u)} = a_{j_1j_21}\pm\varepsilon_u,\ (u=1,2,\ldots,l_1),\\ &\{a_{j_1j_21}, a_{j_1j_22},\ldots,a_{j_1j_2\bar{q}_1}\}\\ &=\mathrm{sort}\{a_{j_1j_21}, a_{j_1j_22},\ldots,a_{j_1j_2q_1};\, a_{j_1j_21}^{(u)}, a_{j_1j_22}^{(u)}|u=1,2,\ldots,l_1\};\\ &a_{j_2j_11}^{(v)}, a_{j_2j_12}^{(v)} = a_{j_2j_11}\pm\varepsilon_v,\ (v=1,2,\ldots,l_2),\\ &\{a_{j_2j_11}, a_{j_2j_12},\ldots,a_{j_2j_1\bar{p}_2}\}\\ &=\mathrm{sort}\{a_{j_2j_11}, a_{j_2j_12},\ldots,a_{j_2j_1p_2};\, a_{j_2j_11}^{(v)}, a_{j_2j_12}^{(v)}|v=1,2,\ldots,l_2\}.\end{aligned}$$

Thus, the differential equation becomes a dynamical system with singular 1-dimensional flows (C_2) as

$$\dot{x}_{j_1} = a_{j_1j_10}(x_{j_1} - a_{j_1j_11})^{m_1} \prod_{u_2=1}^{\overline{q}_1} (x_{j_1} - a_{j_1j_1u_2})^{n_{j_1u_2}},$$

$$\dot{x}_{j_2} = a_{j_2j_20} \prod_{v_2=1}^{\overline{p}_2} (x_{j_1} - a_{j_1j_2v_2})^{m_{j_1v_2}},$$

where

$$\Sigma_{u_2=1}^{\overline{q}_1} n_{j_1u_2} = 2n_{11} + 1,\ \Sigma_{v_2=1}^{\overline{p}_2} m_{j_1v_2} = 2m_{21}.$$

- From the above bifurcation process analysis, only $\overline{q}_1$-parameter variations in the x_{j_1}-direction from a q_1 to $\overline{q}_1$ 1-dimensional flow array through the $(2n_{11}+1)^{\text{th}}$-order parabola flow bifurcations in the x_{j_1}-direction.

  Thus, the 1-dimensional flow appearing or vanishing bifurcation route from a q_1 to $\overline{q}_1$ 1-dimensional flow array is expressed as

$$\bigcup_{u_1=1}^{q_1} \underbrace{(a_{j_1j_2u_1}, \dot{x}_{j_2})}_{(n_{j_1u_1})^{\text{th}}\text{order XX}} \rightleftharpoons \underbrace{(a_{j_1j_21}, \dot{x}_{j_2})}_{(2n_{11}+1)^{\text{th}}\text{ order XX}} \rightleftharpoons \bigcup_{u_2=1}^{\overline{q}_1} \underbrace{(a_{j_1j_2u_2}, \dot{x}_{j_2})}_{(n_{j_1u_2})^{\text{th}}\text{ order XX}}.$$

- From the above bifurcation process analysis, only $\overline{p}_2$-parameter variations in the x_{j_2}-direction from a p_2 to $\overline{p}_2$ 1-dimensional flow array through the $(2m_{21})^{\text{th}}$-order inflection flow bifurcations in the x_{j_2}-direction.

  Thus, the 1-dimensional flow appearing or vanishing bifurcation route from a p_2 to $\overline{p}_2$ 1-dimensional flow array is expressed as

$$\bigcup_{v_1=1}^{p_2} \underbrace{(\dot{x}_{j_1}, a_{j_2j_1v_1})}_{(m_{j_2v_1})^{\text{th}}\text{order } YY} \rightleftharpoons \underbrace{(\dot{x}_{j_1}, a_{j_2j_11})}_{(2\text{m}_{21})^{\text{th}}\text{ order YY}} \rightleftharpoons \bigcup_{v_2=1}^{\overline{p}_2} \underbrace{(\dot{x}_{j_1}, a_{j_2j_1v_2})}_{(m_{j_2v_2})^{\text{th}}\text{order } YY}.$$

- From the above bifurcation process analysis, only $\overline{p}_2+\overline{q}_1$-parameter variations from a $p_2 \times q_1$ to $\overline{p}_2 \times \overline{q}_1$ network of 1-dimensional flows through the $(2n_{11}+1)^{\text{th}}$-order parabola flows bifurcations in the x_{j_1}-direction, and the $(2m_{21})^{\text{th}}$-order inflection flows bifurcations in the x_{j_2}-direction.

  Thus, the singular-equilibrium network appearing or vanishing bifurcation route is as follows:

$$\bigcup_{u_1=1}^{q_1} \bigcup_{v_1=1}^{p_2} \underbrace{(a_{j_1j_2u_1}, a_{j_2j_1v_1})}_{((n_{j_1u_1}),(m_{j_2v_1}))\text{-XX}} \rightleftharpoons \underbrace{((2n_{11}+1)\text{YY},(2m_{21})\text{ZZ})}_{((2n_{11}+1),(2m_{21}))-XX}$$

$$\rightleftharpoons \bigcup_{u_2=1}^{\bar{q}_1} \bigcup_{v_2=1}^{\bar{p}_2} \underbrace{(a_{j_1j_2u_2}, a_{j_2j_1v_2})}_{((n_{j_1u_2}),(m_{j_2v_2}))\text{-XX}},$$

which are the higher-order bifurcations of parabola-saddles.

(I_3) Consider two dynamical systems with the same 1-dimensional flows with locations switched (A_3,C_3) for $i = 1, 2$ as

$$\dot{x}_{j_1} = a_{j_1j_10}(x_{j_1} - a_{j_1j_11})^{m_1} \prod_{u_i=1}^{q_1} (x_{j_2} - a_{j_1j_2u_i})^{n_{j_1u_i}},$$
$$\dot{x}_{j_2} = a_{j_2j_20} \prod_{v_i=1}^{q_1} (x_{j_1} - a_{j_2j_1s_i})^{n_{j_2v_i}};$$

where

$$\sum_{u_i=1}^{q_1} n_{j_1u_i} = 2n_{11} + 1, \sum_{v_i=1}^{p_2} m_{j_2v_i} = 2m_{21}, \text{ for } i = 1, 2.$$

Consider a dynamical system as a singular 1-dimensional flow (B_3) as

$$\dot{x}_{j_1} = a_{j_1j_10}(x_{j_1} - a_{j_1j_11})^{m_1}(x_{j_2} - a_{j_1j_21})^{2n_{11}+1},$$
$$\dot{x}_{j_2} = a_{j_2j_20}(x_{j_1} - a_{j_2j_11})^{2m_{21}}.$$

For $u_i = 1, 2, \ldots, q_1$ and $v_i = 1, 2, \ldots, p_2$ ($i = 1, 2$), define a function as

$$\Delta_{j_1j_2}^{(u_1u_2)} = (a_{j_1j_2u_1} - a_{j_1j_2u_2})^2 \ (u_1, u_2 \in \{1, 2, \ldots, q_1\}, u_1 \neq u_2),$$
$$\Delta_{j_2j_2}^{(v_1v_2)} = (a_{j_2j_1v_1} - a_{j_2j_1v_2})^2 \ (v_1, v_2 \in \{1, 2, \ldots, p_2\}, v_1 \neq v_2).$$

For $u_i = 1, 2, \ldots, q_1$ and $v_i = 1, 2, \ldots, p_2$ ($i = 1, 2$), if

$$\Delta_{j_1j_2}^{(u_1u_2)} = (a_{j_1j_2u_1} - a_{j_1j_2u_2})^2 = 0 \text{ and } \Delta_{j_2j_1}^{(v_1v_2)} = (a_{j_2j_1v_1} - a_{j_2j_1v_2})^2 = 0,$$

there are three bifurcations as follows.

- Two 1-dimensional flows of $(a_{j_1j_2u_1}, \dot{x}_{j_2})$ and $(a_{j_1j_2u_2}, \dot{x}_{j_2})$ switching at point $(a_{j_1j_21}, \dot{x}_{j_2})$ with the same order singularity are given through

$$\bigcup_{u_1=1}^{q_1} \underbrace{(a_{j_1j_2u_1}, \dot{x}_{j_2})}_{(n_{j_1u_1})^{\text{th}}\text{order XX}} \rightleftharpoons \underbrace{(a_{j_1j_21}, \dot{x}_{j_2})}_{(2n_{11}+1)^{\text{th}}\text{ order XX}} \rightleftharpoons \bigcup_{u_2=1}^{q_1} \underbrace{(a_{j_1j_2u_2}, \dot{x}_{j_2})}_{(n_{j_1u_2})^{\text{th}}\text{order XX}}$$
$$\{a_{j_1j_2u_1} \rightleftharpoons a_{j_1j_2u_2} | n_{j_1u_1} = n_{j_1u_2}; u_1, u_2 \in \{1, 2, \ldots, q_1\}, u_1 \neq u_2\};$$

and

$$\begin{aligned}
&a_{j_1j_2u_1} = a_{j_1j_2 1} - \varepsilon_{u_1}, a_{j_1j_2u_2} = a_{j_1j_2 1} + \varepsilon_{u_2},\\
&(a_{j_1j_2u_1} = a_{j_1j_2 1} + \varepsilon_{u_1}, a_{j_1j_2u_2} = a_{j_1j_2 1} - \varepsilon_{u_2});\\
&\varepsilon_{u_1}, \varepsilon_{u_2} > 0;\ u_i = 1, 2, \ldots, q_1\ (i = 1, 2).
\end{aligned}$$

- Two 1-dimensional flows of $(\dot{x}_{j_1}, a_{j_2j_1v_1})$ and $(\dot{x}_{j_1}, a_{j_2j_1v_2})$ switching at point $(\dot{x}_{j_1}, a_{j_2j_1 1})$ with the same order singularity are given through

$$\begin{aligned}
&\bigcup_{v_1=1}^{p_2} \underbrace{(\dot{x}_{j_1}, a_{j_2j_1v_1})}_{(m_{j_2v_1})^{\text{th}}\text{order YY}} \rightleftharpoons \underbrace{(\dot{x}_{j_1}, a_{j_2j_1 1})}_{(2m_{21})^{\text{th}}\text{ order YY}} \rightleftharpoons \bigcup_{v_2=1}^{p_2} \underbrace{(\dot{x}_{j_1}, a_{j_2j_1v_2})}_{(v_{j_2s_2})^{\text{th}}\text{order YY}}\\
&\{a_{j_2j_1v_1} \rightleftharpoons a_{j_2j_1v_2} | m_{j_2v_1} = m_{j_2v_2};\ v_1, v_2 \in \{1, 2, \ldots, q_2\}, v_1 \neq v_2\};
\end{aligned}$$

and

$$\begin{aligned}
&a_{j_2j_1v_1} = a_{j_2j_1 1} - \varepsilon_{v_1}, a_{j_2j_1v_2} = a_{j_2j_1 1} + \varepsilon_{v_2},\\
&(a_{j_2j_1v_1} = a_{j_2j_1 1} + \varepsilon_{v_1}, a_{j_2j_1v_2} = a_{j_2j_1 1} - \varepsilon_{v_2});\\
&\varepsilon_{v_1}, \varepsilon_{v_2} > 0;\ v_i = 1, 2, \ldots, p_2\ (i = 1, 2).
\end{aligned}$$

- Two singular-equilibrium networks of $(a_{j_1j_2u_1}, a_{j_2j_1v_1})$ and $(a_{j_1j_2v_2}, a_{j_2j_1v_2})$ switching at point $(a_{j_1j_2 1}, a_{j_2j_1 1})$ with the same order singularity are given through

$$\begin{aligned}
\bigcup_{u_1=1}^{q_1}\bigcup_{v_1=1}^{p_2} \underbrace{(a_{j_1j_2u_1}, a_{j_2j_1v_1})}_{((n_{j_1u_1}),(m_{j_2v_1}))\text{-XX}} &\rightleftharpoons \underbrace{((2n_{11}+1)\text{YY},(2m_{21})\text{ZZ})}_{((2n_{11}+1),(2m_{21}))\text{ -XX}}\\
&\rightleftharpoons \bigcup_{u_2=1}^{p_1}\bigcup_{v_2=1}^{p_2} \underbrace{(a_{j_1j_2u_2}, a_{j_2j_1v_2})}_{((n_{j_1u_2}),(m_{j_2v_2}))\text{-XX}}
\end{aligned}$$

with

$$\begin{aligned}
&a_{j_1j_2u_1} = a_{j_1j_2 1} - \varepsilon_{u_1}, a_{j_1j_2u_2} = a_{j_1j_2 1} + \varepsilon_{u_2},\\
&(a_{j_1j_2u_1} = a_{j_1j_2 1} + \varepsilon_{u_1}, a_{j_1j_2u_2} = a_{j_1j_2 1} - \varepsilon_{u_2}),\\
&\varepsilon_{u_1}, \varepsilon_{u_2} > 0,\ u_i = 1, 2, \ldots, q_1\ (i = 1, 2);\\
&a_{j_2j_1v_1} = a_{j_2j_1 1} - \varepsilon_{v_1}, a_{j_2j_1v_2} = a_{j_2j_1 1} + \varepsilon_{v_2},\\
&(a_{j_2j_1v_1} = a_{j_2j_1 1} + \varepsilon_{v_1}, a_{j_2j_1v_2} = a_{j_2j_1 1} - \varepsilon_{v_2}),\\
&\varepsilon_{v_1}, \varepsilon_{v_2} > 0,\ v_i = 1, 2, \ldots, p_2\ (i = 1, 2).
\end{aligned}$$

The switching bifurcations of two equilibrium networks are the higher-order bifurcations of parabola-saddles.

In summary, from the analysis in case (I_1)–(I_3), the 1-dimensional flow of $(x^*_{j_2}, x^*_{j_1}) = (a_{j_1j_2 1}, a_{j_2j_1 1})$ for $n_1 = 2n_{11} + 1$ and $m_2 = 2m_{21}$ has the bifurcation properties as stated in cases (v_{3a})–(v_{3d}) through Eqs. (1.431)–(1.450).

(v_4) For $n_1 = 2n_{11}$ and $m_2 = 2m_{21}$ the equilibrium of $(x^*_{j_2}, x^*_{j_1}) = (a_{j_1j_2 1}, a_{j_2j_1 1})$ has the following bifurcation properties with

$$
\begin{aligned}
&2n_{11} - \sum_{u_1=1}^{q_1} n_{j_1u_1} = 2\sum_{\rho_1=1}^{l_1} n_{\rho_1},\, 2m_{21} - \sum_{s_1=1}^{p_2} m_{j_2v_1} = 2\sum_{\sigma_1=1}^{l_2} m_{\sigma_1};\\
&\sum_{u_2=1}^{\bar{q}_1} n_{j_1u_2} = 2n_{11},\, \sum_{v_2=1}^{\bar{p}_2} m_{j_2v_2} = 2m_{21};\\
&\sum_{u_i=1}^{q_1} n_{j_1u_i} = 2n_{11},\, \sum_{v_i=1}^{q_1} m_{j_2v_i} = 2m_{21},\, for\, i = 1, 2;
\end{aligned}
$$

l_1, l_2-quadratic polynomials are without real roots,ρ_1^{th} and σ_1^{th}-quadratic polynomials are with powers m_{ρ_1} and n_{σ_1}, respectively.

The bifurcation process is discussed through differential equations as follows. There are four bifurcations (I_1)–(I_3).

(I_1) Consider a dynamical system having a single equilibrium (A_1) as

$$
\begin{aligned}
\dot{x}_{j_1} &= a_{j_1j_1 0}\prod_{u=1}^{n_{11}}[(x_{j_2} - a^{(u)}_{j_1j_2})^2 + \Delta^{(u)}_{j_1j_2}](x_{j_1} - a_{j_1j_1 1})^{m_1},\\
\dot{x}_{j_2} &= a_{j_2j_2 0}\prod_{v=1}^{m_{21}}[(x_{j_1} - a^{(v)}_{j_2j_1})^2 + \Delta^{(v)}_{j_2j_1}].
\end{aligned}
$$

Once

$$
\begin{aligned}
\Delta^{(u)}_{j_1j_2} &= \Delta_{j_1j_2} = 0\,(u = 1, 2, \ldots, n_{11}),\\
\Delta^{(v)}_{j_2j_1} &= \Delta_{j_2j_1} = 0\,(v = 1, 2, \ldots, m_{21}),
\end{aligned}
$$

if

$$
\begin{aligned}
a_{j_1j_2 1} &= b_{j_1j_2 1} = a^{(u)}_{j_1j_2}\,(u = 1, 2, \ldots, n_{11}),\\
a_{j_2j_1 1} &= b_{j_2j_1 1} = a^{(v)}_{j_2j_1}\,(v = 1, 2, \ldots, m_{21}),
\end{aligned}
$$

the foregoing differential equations become a dynamical system with a singular equilibrium (B_1) as

$$\begin{aligned}\dot{x}_{j_1} &= a_{j_1j_10}(x_{j_1} - a_{j_1j_11})^{m_1}(x_{j_2} - a_{j_1j_21})^{2n_{11}},\\ \dot{x}_{j_2} &= a_{j_2j_20}(x_{j_1} - a_{j_2j_11})^{2m_{21}}.\end{aligned}$$

If

$$\begin{aligned}\Delta_{j_1j_2}^{(u)} &= \Delta_{j_1j_2} + \delta_u\ (u = 1, 2, \ldots, n_{11}),\ \delta_u > 0;\\ \Delta_{j_2j_1}^{(v)} &= \Delta_{j_2j_1} + \delta_v\ (v = 1, 2, \ldots, m_{21}),\ \delta_v > 0;\end{aligned}$$

then

$$\begin{aligned}&a_{j_1j_2}^{(\mathrm{u})}, a_{j_1j_2}^{(\mathrm{u})} = a_{j_1j_21} \pm \varepsilon_u, (u = 1, 2, \ldots, n_{11}),\\ &\{a_{j_1j_21}, a_{j_1j_22}, \ldots, a_{j_1j_1(2m_11}\} = \mathrm{sort}\{a_{j_1j_21}^{(\mathrm{u})}, a_{j_1j_22}^{(\mathrm{u})} | u = 1, 2, \ldots, n_{11}\};\\ &a_{j_2j_11}^{(\mathrm{v}}, a_{j_2j_12}^{(\mathrm{v})} = a_{j_2j_11} \pm \varepsilon_v, (v = 1, 2, \ldots, m_{21}),\\ &\{a_{j_2j_11}, a_{j_2j_12}, \ldots, a_{j_2j_1(2m_{21}+1)}\} = \mathrm{sort}\{b_{j_2j_11}, a_{j_2j_11}^{(\mathrm{v})}, a_{j_2j_12}^{(\mathrm{v})} | v = 1, 2, \ldots, n_{21}\}.\end{aligned}$$

Thus, the differential equation becomes a dynamical system with a non-singular equilibrium network (C_1) as

$$\begin{aligned}\dot{x}_{j_1} &= a_{j_1j_10}(x_{j_1} - a_{j_1j_11})^{m_1}\prod_{u=1}^{2n_{11}}(x_{j_2} - a_{j_1j_2u}),\\ \dot{x}_{j_2} &= a_{j_2j_20}\prod_{v=1}^{2m_{21}}(x_{j_2} - a_{j_2j_1v}).\end{aligned}$$

- From the above bifurcation process analysis, at least, only (n_{11})-parameter variations in the x_{j_1}-direction are engaged in such a bifurcation from a 2-dimensional flow to the 1-dimensional flow array of $1 \times (2n_{11})$ through the $(2n_{11})^{\text{th}}$-order inflection flows bifurcations in the x_{j_1}-direction. The higher-order inflection bifurcations are developed.

  Thus, the 1-dimensional flow appearing or vanishing bifurcation route is as follows.

$$\underbrace{(\dot{x}_{j_1}, \dot{x}_{j_2})}_{(\mathrm{xF,xF})\text{-flow}} \rightleftharpoons \underbrace{(a_{j_1j_21}, \dot{x}_{j_2})}_{(2n_{11})^{\text{th}}\text{order XX}} \rightleftharpoons \bigcup_{u=1}^{2n_{11}} \underbrace{(a_{j_1j_2u}, \dot{x}_{j_2})}_{\mathrm{XX}}.$$

- From the above bifurcation process analysis, at least, only (m_{21})-parameter variations in the x_{j_1}-direction are engaged in such a bifurcation from a 2-dimensional flow to the 1-dimensional flow array of $(2m_{21}) \times 1$ through the $(2m_{21})^{\text{th}}$-order inflection flow flows bifurcations in the x_{j_2}-direction. The higher-order inflection flow bifurcations are developed.

  Thus, the 1-dimensional flow appearing or vanishing bifurcation route is as follows.

$$\underbrace{(\dot{x}_{j_1}, \dot{x}_{j_2})}_{\text{(xF,xF)-flow}} \rightleftharpoons \underbrace{(\dot{x}_{j_1}, a_{j_2j_11})}_{(2m_{21})^{\text{th}}\text{order YY}} \rightleftharpoons \bigcup_{v=1}^{2m_{21}} \underbrace{(\dot{x}_{j_1}, a_{j_2j_1v},)}_{\text{YY}}.$$

- From the above bifurcation process analysis, $(n_{11} + m_{21})$-parameter variations in the two directions are engaged in such a bifurcation from a 2-dimensional flow to a $(2n_{11}) \times (2m_{21})$ network of positive and negative saddles and CCW and CW centers through the $(2n_{11})^{\text{th}}$-order inflection bifurcations in the x_{j_1}-direction, and the $(2m_{21})^{\text{th}}$-order inflection flows bifurcations in the x_{j_2}-direction. The higher-order bifurcations of double-inflections are developed.

  Thus, the equilibrium network appearing or vanishing bifurcation route is as follows.

$$\underbrace{(\dot{x}_{j_1}, \dot{x}_{j_2})}_{\text{(xF,xF)-flow}} \rightleftharpoons \underbrace{((2n_{11})\text{YY},(2m_{21})\text{ZZ})}_{((2n_{11}),(2m_{21}))\text{-XX}} \rightleftharpoons \bigcup_{u=1}^{2n_{11}} \bigcup_{v=1}^{2m_{21}} \underbrace{(a_{j_1j_2u}, a_{j_2j_1v})}_{\text{XX}},$$

  which are the higher-order bifurcations of double-inflections.

(I_2) From Case (I_1), consider a dynamical system having singular equilibriums (A_2) as

$$\dot{x}_{j_1} = a_{j_1j_10}(x_{j_1} - a_{j_1j_11})^{m_1} \prod_{u_1=1}^{q_1} (x_{j_2} - b_{j_1j_2u_1})^{n_{j_1u_1}} \prod_{u=1}^{l_1} [(x_{j_2} - a_{j_1j_2}^{(u)})^2 + \Delta_{j_1j_2}^{(u)}]^{m_{j_1u}},$$

$$\dot{x}_{j_2} = a_{j_2j_20} \prod_{v_1=1}^{p_2} (x_{j_1} - b_{j_2j_1v_1})^{m_{j_2v_1}} \prod_{v=1}^{l_2} [(x_{j_1} - a_{j_2j_1}^{(v)})^2 + \Delta_{j_2j_1}^{(u)}]^{m_{j_2v}};$$

where

$$2n_{11} - \sum_{u_1=1}^{q_1} n_{j_1u_1} = 2\sum_{\rho_1=1}^{l_1} n_{\rho_1}, 2m_{21} - \sum_{v_1=1}^{p_2} m_{j_2v_1} = 2\sum_{\sigma_1=1}^{l_2} m_{\sigma_1};$$

$$\sum_{u_2=1}^{\bar{q}_1} n_{j_1u_2} = 2n_{11}, \sum_{v_2=1}^{\bar{p}_2} m_{j_2v_2} = 2m_{21};$$

l_1, l_2-quadratic polynomials are without real roots, ρ_1^{th} and σ_1^{th}-quadratic polynomials are with powers m_{ρ_1} and n_{σ_1},respectively.

Once

$$\Delta_{j_1j_2}^{(u)} = 0\,(u = 1, 2, \ldots, l_1),$$
$$\Delta_{j_2j_1}^{(v)} = 0\,(v = 1, 2, \ldots, l_2);$$

if

$$a_{j_1j_2 1} \equiv a_{j_1j_2u_1} = a_{j_1j_1}^{(u)}\,(u_1 = 1, 2, \ldots, q_1;\, u = 1, 2, \ldots, l_1),$$
$$a_{j_2j_1 1} \equiv a_{j_2j_1v_1} = a_{j_2j_1}^{(v)}\,(v_1 = 1, 2, \ldots, p_2;\, v = 1, 2, \ldots, l_2),$$

the foregoing differential equations become a dynamical system with a 1-dimensional flow (B_2) as

$$\dot{x}_{j_1} = a_{j_1j_1 0}(x_{j_1} - a_{j_1j_1 1})^{m_1}(x_{j_2} - a_{j_1j_2 1})^{2n_{11}},$$
$$\dot{x}_{j_2} = a_{j_2j_2 0}(x_{j_1} - a_{j_2j_1 1})^{2m_{21}}.$$

If

$$\Delta_{j_1j_2}^{(u)} = \Delta_{j_1j_2} + \delta_u\,(u = 1, 2, \ldots, l_1),\ \delta_u > 0;$$
$$\Delta_{j_2j_1}^{(v)} = \Delta_{j_2j_1} + \delta_u\,(v = 1, 2, \ldots, l_2),\ \delta_v > 0;$$

then

$$a_{j_1j_2 1}^{(u)}, a_{j_1j_2 2}^{(u)} = a_{j_1j_2 1} \pm \varepsilon_u, (u = 1, 2, \ldots, l_1),$$
$$\{a_{j_1j_2 1}, a_{j_1j_2 2}, \ldots, a_{j_2j_1 f_2}\}$$
$$= \text{sort}\{a_{j_1j_1 1}, a_{j_1j_2 2}, \ldots, a_{j_1j_2 q_1}; a_{j_1j_2 1}^{(u)}, a_{j_1j_2 2}^{(u)} | u = 1, 2, \ldots, l_1\};$$
$$a_{j_2j_1 1}^{(v)}, a_{j_2j_2 2}^{(v)} = a_{j_2j_1 1} \pm \varepsilon_v, (v = 1, 2, \ldots, l_2),$$
$$\{a_{j_2j_1 1}, a_{j_2j_2 2}, \ldots, a_{j_2j_1 p_2}\}$$
$$= \text{sort}\{a_{j_2j_1 1}, a_{j_2j_1 2}, \ldots, a_{j_2j_1 p_2}; a_{j_2j_1 1}^{(v)}, a_{j_2j_1 2}^{(v)} | v = 1, 2, \ldots, l_2\}.$$

Thus, the differential equation becomes a dynamical system with singular 1-dimensional flows (C_2) as

$$\dot{x}_{j_1} = a_{j_1j_1 0}(x_{j_1} - a_{j_1j_1 1})^{m_1} \prod_{u_2=1}^{\bar{q}_1} (x_{j_1} - a_{j_1j_1u_2})^{n_{j_1u_2}},$$

$$\dot{x}_{j_2} = a_{j_2j_2 0} \prod_{v_2=1}^{\bar{p}_2} (x_{j_1} - a_{j_1j_2v_2})^{m_{j_1v_2}}$$

where

$$\Sigma_{u_2=1}^{\overline{q}_1} n_{j_1 u_2} = 2n_{11},\ \Sigma_{v_2=1}^{\overline{p}_2} m_{j_1 v_2} = 2m_{21}.$$

- From the above bifurcation process analysis, only $\overline{q}_1$-parameter variations in the x_{j_1}-direction from a q_1 to $\overline{q}_1$ 1-dimensional flow array through the $(2n_{11})^{\text{th}}$-order inflection flow bifurcations in the x_{j_1}-direction.

  Thus, the 1-dimensional flow appearing or vanishing bifurcation route from a q_1 to $\overline{q}_1$ 1-dimensional flow array is expressed as

$$\bigcup_{u_1=1}^{q_1} \underbrace{(a_{j_1 j_2 u_1}, \dot{x}_{j_2})}_{(n_{j_1 u_1})^{\text{th}}\text{order XX}} \rightleftharpoons \underbrace{(a_{j_1 j_2 1}, \dot{x}_{j_2})}_{(2n_{11})^{\text{th}}\text{ order XX}} \rightleftharpoons \bigcup_{u_2=1}^{\overline{q}_1} \underbrace{(a_{j_1 j_2 u_2}, \dot{x}_{j_2})}_{(n_{j_1 u_2})^{\text{th}}\text{order XX}}.$$

- From the above bifurcation process analysis, only $\overline{p}_2$-parameter variations in the x_{j_2}-direction from a p_2 to $\overline{p}_2$ 1-dimensional flow array through the $(2m_{21})^{\text{th}}$-order inflection flow bifurcations in the x_{j_2}-direction.

  Thus, the 1-dimensional flow appearing or vanishing bifurcation route from a p_2 to $\overline{p}_2$ 1-dimensional flow array is expressed as

$$\bigcup_{v_1=1}^{p_2} \underbrace{(\dot{x}_{j_1}, a_{j_2 j_1 v_1})}_{(m_{j_2 v_1})^{\text{th}}\text{order YY}} \rightleftharpoons \underbrace{(\dot{x}_{j_1}, a_{j_2 j_1 1})}_{(2\text{m}_{21})^{\text{th}}\text{ order YY}} \rightleftharpoons \bigcup_{v_2=1}^{\overline{p}_2} \underbrace{(\dot{x}_{j_1}, a_{j_2 j_1 v_2})}_{(m_{j_2 v_2})^{\text{th}}\text{order YY}}.$$

- From the above bifurcation process analysis, only $\overline{p}_2+\overline{q}_1$-parameter variations from a $p_2 \times q_1$ to $\overline{p}_2 \times \overline{q}_1$ network of 1-dimensional flows through the $(2n_{11})^{\text{th}}$-order inflection flows bifurcations in the x_{j_1}-direction, and the $(2m_{21})^{\text{th}}$-order inflection flows bifurcations in the x_{j_2}-direction.

  Thus, the singular-equilibrium network appearing or vanishing bifurcation route is as follows:

$$\begin{aligned}\bigcup_{u_1=1}^{q_1} \bigcup_{v_1=1}^{p_2} \underbrace{(a_{j_1 j_2 u_1}, a_{j_2 j_1 v_1})}_{((n_{j_1 u_1}),(m_{j_2 v_1}))\text{-XX}} &\rightleftharpoons \underbrace{((2n_{11})\text{YY},(2m_{21})\text{ZZ}))}_{((2n_{11}),(2m_{21}))\text{-XX}} \\ &\rightleftharpoons \bigcup_{u_2=1}^{\overline{q}_1} \bigcup_{v_2=1}^{\overline{p}_2} \underbrace{(a_{j_1 j_2 u_2}, a_{j_2 j_1 v_2})}_{((n_{j_1 u_2}),(m_{j_2 v_2}))\text{-XX}},\end{aligned}$$

  which are the higher-order bifurcations of double-inflections.

(I_3) Consider two dynamical systems with the same 1-dimensional flows with locations switched (A_3,C_3) for $i = 1, 2$ as

$$\dot{x}_{j_1} = a_{j_1j_10}(x_{j_1} - a_{j_1j_11})^{m_1} \prod_{u_i=1}^{q_1} (x_{j_2} - a_{j_1j_2u_i})^{n_{j_1u_i}},$$

$$\dot{x}_{j_2} = a_{j_2j_20} \prod_{v_i=1}^{q_1} (x_{j_1} - a_{j_2j_1s_i})^{n_{j_2v_i}};$$

where

$$\sum_{u_i=1}^{q_1} n_{j_1u_i} = 2n_{11}, \sum_{v_i=1}^{p_2} m_{j_2v_i} = 2m_{21}, \text{ for } i = 1, 2.$$

Consider a dynamical system as a singular 1-dimensional flow (B_3) as

$$\begin{aligned}\dot{x}_{j_1} &= a_{j_1j_10}(x_{j_1} - a_{j_1j_11})^{m_1}(x_{j_2} - a_{j_1j_21})^{2n_{11}},\\ \dot{x}_{j_2} &= a_{j_2j_20}(x_{j_1} - a_{j_2j_11})^{2m_{21}}.\end{aligned}$$

For $u_i = 1, 2, \ldots, q_1$ and $v_i = 1, 2, \ldots, p_2$ ($i = 1, 2$), define a function as

$$\begin{aligned}\Delta_{j_1j_2}^{(u_1u_2)} &= (a_{j_1j_2u_1} - a_{j_1j_2u_2})^2 \ (u_1, u_2 \in \{1, 2, \ldots, q_1\}, u_1 \neq u_2),\\ \Delta_{j_2j_2}^{(v_1v_2)} &= (a_{j_2j_1v_1} - a_{j_2j_1v_2})^2 \ (v_1, v_2 \in \{1, 2, \ldots, p_2\}, v_1 \neq v_2).\end{aligned}$$

For $u_i = 1, 2, \ldots, q_1$ and $v_i = 1, 2, \ldots, p_2$ ($i = 1, 2$), if

$$\Delta_{j_1j_2}^{(u_1u_2)} = (a_{j_1j_2u_1} - a_{j_1j_2u_2})^2 = 0 \text{ and } \Delta_{j_2j_1}^{(v_1v_2)} = (a_{j_2j_1v_1} - a_{j_2j_1v_2})^2 = 0,$$

there are three bifurcations as follows.

- Two 1-dimensional flows of $(a_{j_1j_2u_1}, \dot{x}_{j_2})$ and $(a_{j_1j_2u_2}, \dot{x}_{j_2})$ switching at point $(a_{j_1j_21}, \dot{x}_{j_2})$ with the same order singularity are given through

$$\bigcup_{u_1=1}^{q_1} \underbrace{(a_{j_1j_2u_1}, \dot{x}_{j_2})}_{(n_{j_1u_1})^{\text{th}}\text{order XX}} \rightleftharpoons \underbrace{(a_{j_1j_21}, \dot{x}_{j_2})}_{(2n_{11})^{\text{th}} \text{ order XX}} \rightleftharpoons \bigcup_{u_2=1}^{q_1} \underbrace{(a_{j_1j_2u_2}, \dot{x}_{j_2})}_{(n_{j_1u_2})^{\text{th}}\text{order XX}}$$

$$\{a_{j_1j_2u_1} \rightleftharpoons a_{j_1j_2u_2} | n_{j_1u_1} = n_{j_1u_2}; u_1, u_2 \in \{1, 2, \ldots, q_1\}, u_1 \neq u_2\};$$

and

$$\begin{aligned}&a_{j_1j_2u_1} = a_{j_1j_21} - \varepsilon_{u_1}, a_{j_1j_2u_2} = a_{j_1j_21} + \varepsilon_{u_2},\\ &(a_{j_1j_2u_1} = a_{j_1j_21} + \varepsilon_{u_1}, a_{j_1j_2u_2} = a_{j_1j_21} - \varepsilon_{u_2});\\ &\varepsilon_{u_1}, \varepsilon_{u_2} > 0; \ u_i = 1, 2, \ldots, q_1 \ (i = 1, 2).\end{aligned}$$

- Two 1-dimensional flows of $(\dot{x}_{j_1}, a_{j_2j_1v_1})$ and $(\dot{x}_{j_1}, a_{j_2j_1v_2})$ switching at point $(\dot{x}_{j_1}, a_{j_2j_11})$ with the same order singularity are given through

$$\bigcup_{v_1=1}^{p_2} \underbrace{(\dot{x}_{j_1}, a_{j_2 j_1 v_1})}_{(m_{j_2 v_1})^{\text{th}}\text{order YY}} \rightleftharpoons \underbrace{(\dot{x}_{j_1}, a_{j_2 j_1 1})}_{(2m_{21})^{\text{th}}\text{ order YY}} \rightleftharpoons \bigcup_{v_2=1}^{p_2} \underbrace{(\dot{x}_{j_1}, a_{j_2 j_1 v_2})}_{(v_{j_2 v_2})^{\text{th}}\text{order YY}}$$

$$\{a_{j_2 j_1 v_1} \rightleftharpoons a_{j_2 j_1 v_2} | m_{j_2 v_1} = m_{j_2 v_2}; v_1, v_2 \in \{1, 2, \ldots, q_2\}, v_1 \neq v_2\};$$

and

$$\begin{aligned}
&a_{j_2 j_1 v_1} = a_{j_2 j_1 1} - \varepsilon_{v_1}, a_{j_2 j_1 v_2} = a_{j_2 j_1 1} + \varepsilon_{v_2},\\
&(a_{j_2 j_1 v_1} = a_{j_2 j_1 1} + \varepsilon_{v_1}, a_{j_2 j_1 v_2} = a_{j_2 j_1 1} - \varepsilon_{v_2});\\
&\varepsilon_{v_1}, \varepsilon_{v_2} > 0;\ v_i = 1, 2, \ldots, p_2\, (i = 1, 2).
\end{aligned}$$

- Two singular-equilibrium networks of $(a_{j_1 j_2 u_1}, a_{j_2 j_1 v_1})$ and $(a_{j_1 j_2 v_2}, a_{j_2 j_1 v_2})$ switching at point $(a_{j_1 j_2 1}, a_{j_2 j_1 1})$ with the same order singularity are given through

$$\begin{aligned}
\bigcup_{u_1=1}^{q_1} \bigcup_{v_1=1}^{p_2} \underbrace{(a_{j_1 j_2 u_1}, a_{j_2 j_1 v_1})}_{((n_{j_1 u_1}), (m_{j_2 v_1}))\text{-XX}} &\rightleftharpoons \underbrace{((2n_{11})\text{YY}, (2m_{21})\text{ZZ})}_{((2n_{11}), (2m_{21}))\text{ -XX}}\\
&\rightleftharpoons \bigcup_{u_2=1}^{p_1} \bigcup_{v_2=1}^{p_2} \underbrace{(a_{j_1 j_2 u_2}, a_{j_2 j_1 v_2})}_{((n_{j_1 u_2}), (m_{j_2 v_2}))\text{-XX}}
\end{aligned}$$

with

$$\begin{aligned}
&a_{j_1 j_2 u_1} = a_{j_1 j_2 1} - \varepsilon_{u_1}, a_{j_1 j_2 u_2} = a_{j_1 j_2 1} + \varepsilon_{u_2},\\
&(a_{j_1 j_2 u_1} = a_{j_1 j_2 1} + \varepsilon_{u_1}, a_{j_1 j_2 u_2} = a_{j_1 j_2 1} - \varepsilon_{u_2}),\\
&\varepsilon_{u_1}, \varepsilon_{u_2} > 0,\ u_i = 1, 2, \ldots, q_1\ (i = 1, 2);\\
&a_{j_2 j_1 v_1} = a_{j_2 j_1 1} - \varepsilon_{v_1}, a_{j_2 j_1 v_2} = a_{j_2 j_1 1} + \varepsilon_{v_2},\\
&(a_{j_2 j_1 v_1} = a_{j_2 j_1 1} + \varepsilon_{v_1}, a_{j_2 j_1 v_2} = a_{j_2 j_1 1} - \varepsilon_{v_2}),\\
&\varepsilon_{v_1}, \varepsilon_{v_2} > 0,\ v_i = 1, 2, \ldots, p_2\ (i = 1, 2).
\end{aligned}$$

The switching bifurcations of two equilibrium networks are the higher-order bifurcations of double-inflections.

In summary, from the analysis in case (I_1)–(I_3), the 1-dimensional flow of $(x^*_{j_2}, x^*_{j_1}) = (a_{j_1 j_2 1}, a_{j_2 j_1 1})$ for $n_1 = 2n_{11}$ and $m_2 = 2m_{21}$ has the bifurcation properties as stated in cases (v_{4a})–(v_{4d}) through Eqs. (1.428)–(1.467).

Note that the equation in the x_{j_1}-direction can be considered as follows.

$$\dot{x}_{j_1} = a_{j_1 j_1 0} \prod_{r=1}^{p_1} (x_{j_1} - a_{j_1 j_1 r})^{m_{j_1 r}} \prod_{u_i=1}^{q_1} (x_{j_2} - a_{j_1 j_2 u_i})^{n_{j_1 u}},$$

$$\dot{x}_{j_1} = a_{j_1j_10}\prod_{r_i=1}^{p_1}(x_{j_1} - a_{j_1j_1r_i})^{m_{j_1r_i}}\prod_{u=1}^{n_1}(x_{j_2} - a_{j_1j_2u});$$

which can be used for discussion for case (v).

2.5 Network Switching for Equilibriums and Flows

In this section, the theorem about the switching bifurcations of hybrid networks of simple and singular equilibriums and 1-dimensional flows in product polynomial systems will be proved.

(vi) Consider

$$\begin{aligned}\dot{x}_{j_1} &= a_{j_1j_10}(x_{j_1} - a_{j_1j_11})^{m_1}(x_{j_2} - a_{j_1j_21})^{n_1},\\ \dot{x}_{j_2} &= a_{j_2j_20}(x_{j_1} - a_{j_2j_11})^{m_2}.\end{aligned}$$

For $a_{j_1j_11} = a_{j_2j_11}$, there are 64 types of singular infinite-equilibriums at $x^*_{j_1} = a_{j_1j_11} = a_{j_2j_11}$. For $m_2 \geq m_1$, 32 types of singular infinite-equilibriums are

$$(a_{j_1j_21}, a_{j_2j_11}) = \underbrace{([m_1,n_1]^{\text{th}}\text{YY},[m_1 - m_2]^{\text{th}}\text{XX})}_{([m_1,n_1],[m_2-m_1])\text{-YYXX}},$$

where

$$\begin{aligned}&\text{YY} \in \{\,\text{UP}_-{:}\text{UP}_+,\ \text{DP}_-{:}\text{DP}_+,\ \text{DP}_+{:}\text{DP}_-,\text{UP}_+{:}\text{UP}_-\,\},\\ &\text{XX} \in \{\,{}_{\text{DI:II}}\text{US},{}_{\text{DI:II}}\text{LS},{}_{\text{II:DI}}\text{US},{}_{\text{II:DI}}\text{LS}\};\end{aligned}$$

$$\begin{aligned}&\text{YY} \in \{\,\text{DP}_-{:}\text{UP}_+,\text{UP}_-{:}\text{DP}_+,\ \text{UP}_+{:}\text{DP}_-,\text{DP}_+{:}\text{UP}_-\,\},\\ &\text{XX} \in \{\,{}_{\text{DP}}\text{LS}{:}{}_{\text{UP}}\text{US},\ {}_{\text{DP}}\text{US}{:}{}_{\text{UP}}\text{LS},\ {}_{\text{UP}}\text{US}{:}{}_{\text{DP}}\text{LS},\ {}_{\text{UP}}\text{LS}{:}{}_{\text{DP}}\text{US}\};\end{aligned}$$

$$\begin{aligned}&\text{YY} \in \{\,\text{DP}_+{:}\text{UP}_+,\ \text{UP}_+{:}\text{DP}_+,\text{UP}_-{:}\text{DP}_-,\text{DP}_-{:}\text{UP}_-\,\},\\ &\text{XX} \in \{\,{}_{\text{DP}}\text{SI}{:}{}_{\text{UP}}\text{SO},\ {}_{\text{DP}}\text{SO}{:}{}_{\text{UP}}\text{SI},\ {}_{\text{UP}}\text{SI}{:}{}_{\text{DP}}\text{SO},\ {}_{\text{UP}}\text{SO}{:}{}_{\text{DP}}\text{SI}\};\end{aligned}$$

$$\begin{aligned}&\text{YY} \in \{\,\text{UP}_+{:}\text{UP}_+,\ \text{DP}_+{:}\text{DP}_+,\ \text{DP}_-{:}\text{DP}_-,\text{UP}_-{:}\text{UP}_-\,\},\\ &\text{XX} \in \{\,{}_{\text{DI:II}}\text{US},{}_{\text{DI:II}}\text{LS},{}_{\text{II:DI}}\text{US},{}_{\text{II:DI}}\text{LS}\};\end{aligned}$$

$$\begin{aligned}&\text{YY} \in \{\,\text{II}_-{:}\text{II}_+,\ \text{DI}_-{:}\text{DI}_+,\ \text{DI}_+{:}\text{DI}_-,\text{II}_+{:}\text{II}_-\,\},\\ &\text{XX} \in \{\,{}_{\text{II:II}}\text{SO},{}_{\text{DI:DI}}\text{SI}{}_{\text{DI:DI}}\text{SO},{}_{\text{II:II}}\text{SI}\};\end{aligned}$$

$$\mathrm{YY} \in \{\ \mathrm{DI}_{-}{:}\mathrm{II}_{+},\ \mathrm{II}_{-}{:}\mathrm{DI}_{+},\ \mathrm{II}_{+}{:}\mathrm{DI}_{-},\ \mathrm{DI}_{+}{:}\mathrm{II}_{-}\}\ ,$$
$$\mathrm{XX} \in \{\ {}_{\mathrm{UP:UP}}\mathrm{US},\ {}_{\mathrm{DP:DP}}\mathrm{LS},\ {}_{\mathrm{DP:DP}}\mathrm{US},\ {}_{\mathrm{UP:UP}}\mathrm{LS}\}\ ;$$

$$\mathrm{YY} \in \{\ \mathrm{DI}_{+}{:}\mathrm{II}_{+},\ \mathrm{II}_{+}{:}\mathrm{DI}_{+},\ \mathrm{II}_{-}{:}\mathrm{DI}_{-},\ \mathrm{DI}_{-}{:}\mathrm{II}_{-}\}\ ,$$
$$\mathrm{XX} \in \{\ {}_{\mathrm{UP:UP}}\mathrm{SO},\ {}_{\mathrm{DP:DP}}\mathrm{SI},\ {}_{\mathrm{DP:DP}}\mathrm{SO},\ {}_{\mathrm{UP:UP}}\mathrm{SI}\}\ ;$$

$$\mathrm{YY} \in \{\ \mathrm{II}_{+}{:}\mathrm{II}_{+},\ \mathrm{DI}_{+}{:}\mathrm{DI}_{+},\ \mathrm{DI}_{-}{:}\mathrm{DI}_{-},\mathrm{II}_{-}{:}\mathrm{II}_{-}\}\ ,$$
$$\mathrm{XX} \in \{\ {}_{\mathrm{II:II}}\mathrm{US},{}_{\mathrm{DI:DI}}\mathrm{LS}{}_{\mathrm{DI:DI}}\mathrm{US},{}_{\mathrm{II:II}}\mathrm{LS}\}\ .$$

For $m_2 < m_1$, 32 types of singular infinite-equilibriums are

$$(a_{j_1 j_2 1}, a_{j_2 j_1 1}) = \underbrace{([m_1,n_1]^{\mathrm{th}}\mathrm{YY},[m_1 - m_2]^{\mathrm{th}}\mathrm{ZZ})}_{([m_1,n_1],[m_2-m_1])\text{-YYZZ}},$$

where

$$\mathrm{YY} \in \{\ \mathrm{UP}_{-}{:}\mathrm{UP}_{+},\ \mathrm{DP}_{-}{:}\mathrm{DP}_{+},\ \mathrm{DP}_{+}{:}\mathrm{DP}_{-},\mathrm{UP}_{+}{:}\mathrm{UP}_{-}\}\ ,$$
$$\mathrm{ZZ} \in \{\ {}_{\mathrm{DC:IC}}\mathrm{US},{}_{\mathrm{IC:DC}}\mathrm{LS},{}_{\mathrm{IC:DC}}\mathrm{US},{}_{\mathrm{IC:DC}}\mathrm{LS}\}\ ;$$

$$\mathrm{YY} \in \{\ \mathrm{DP}_{-}{:}\mathrm{UP}_{+},\mathrm{UP}_{-}{:}\mathrm{DP}_{+},\ \mathrm{UP}_{+}{:}\mathrm{DP}_{-},\mathrm{DP}_{+}{:}\mathrm{UP}_{-}\}\ ,$$
$$\mathrm{ZZ} \in \{\ {}_{\mathrm{UA}}\mathrm{LS}{:}{}_{\mathrm{DA}}\mathrm{US},\ {}_{\mathrm{DA}}\mathrm{US}{:}{}_{\mathrm{UA}}\mathrm{LS},\ {}_{\mathrm{DA}}\mathrm{US}{:}{}_{\mathrm{UA}}\mathrm{LS},\ {}_{\mathrm{UA}}\mathrm{LS}{:}{}_{\mathrm{DA}}\mathrm{US}\}\ ;$$

$$\mathrm{YY} \in \{\ \mathrm{DP}_{+}{:}\mathrm{UP}_{+},\ \mathrm{UP}_{+}{:}\mathrm{DP}_{+},\mathrm{UP}_{-}{:}\mathrm{DP}_{-},\mathrm{DP}_{-}{:}\mathrm{UP}_{-}\}\ ,$$
$$\mathrm{ZZ} \in \{\ {}_{\mathrm{UA}}\mathrm{SI}{:}{}_{\mathrm{DA}}\mathrm{SO},\ {}_{\mathrm{DA}}\mathrm{SO}{:}{}_{\mathrm{UA}}\mathrm{SI},\ {}_{\mathrm{DA}}\mathrm{SI}{:}{}_{\mathrm{UA}}\mathrm{SO},\ {}_{\mathrm{UA}}\mathrm{SO}{:}{}_{\mathrm{DA}}\mathrm{SI}\}\ ;$$

$$\mathrm{YY} \in \{\ \mathrm{UP}_{+}{:}\mathrm{UP}_{+},\ \mathrm{DP}_{+}{:}\mathrm{DP}_{+},\ \mathrm{DP}_{-}{:}\mathrm{DP}_{-},\mathrm{UP}_{-}{:}\mathrm{UP}_{-}\}\ ,$$
$$\mathrm{ZZ} \in \{\ {}_{\mathrm{DC:IC}}\mathrm{US},{}_{\mathrm{IC:DC}}\mathrm{LS},{}_{\mathrm{IC:DC}}\mathrm{US},{}_{\mathrm{DC:IC}}\mathrm{LS}\}\ ;$$

$$\mathrm{YY} \in \{\ \mathrm{II}_{-}{:}\mathrm{II}_{+},\ \mathrm{DI}_{-}{:}\mathrm{DI}_{+},\ \mathrm{DI}_{+}{:}\mathrm{DI}_{-},\mathrm{II}_{+}{:}\mathrm{II}_{-}\}\ ,$$
$$\mathrm{ZZ} \in \{\ {}_{\mathrm{IC:IC}}\mathrm{SO},{}_{\mathrm{DC:DC}}\mathrm{SI}{}_{\mathrm{DC:DC}}\mathrm{SO},{}_{\mathrm{IC:IC}}\mathrm{SI}\}\ ;$$

$$\mathrm{YY} \in \{\ \mathrm{DI}_{-}{:}\mathrm{II}_{+},\ \mathrm{II}_{-}{:}\mathrm{DI}_{+},\ \mathrm{II}_{+}{:}\mathrm{DI}_{-},\ \mathrm{DI}_{+}{:}\mathrm{II}_{-}\}\ ,$$
$$\mathrm{ZZ} \in \{\ {}_{\mathrm{DA:DA}}\mathrm{US},\ {}_{\mathrm{UA:UA}}\mathrm{LS},\ {}_{\mathrm{UA:UA}}\mathrm{US},\ {}_{\mathrm{DA:DA}}\mathrm{LS}\}\ ;$$

$$\mathrm{YY} \in \{\ \mathrm{DI}_{+}{:}\mathrm{II}_{+},\ \mathrm{II}_{+}{:}\mathrm{DI}_{+},\ \mathrm{II}_{-}{:}\mathrm{DI}_{-},\ \mathrm{DI}_{-}{:}\mathrm{II}_{-}\}\ ,$$
$$\mathrm{ZZ} \in \{\ {}_{\mathrm{DA:DA}}\mathrm{SO},\ {}_{\mathrm{UA:UA}}\mathrm{SI},\ {}_{\mathrm{UA:UA}}\mathrm{SO},\ {}_{\mathrm{DA:DA}}\mathrm{SI}\}\ ;$$

$$
\begin{aligned}
&\text{YY} \in \{ \text{II}_+\text{:II}_+, \text{DI}_+\text{:DI}_+, \text{DI}_-\text{:DI}_-, \text{II}_-\text{:II}_- \} , \\
&\text{ZZ} \in \{ {}_{\text{IC:IC}}\text{US}, {}_{\text{DC:DC}}\text{LS}, {}_{\text{DC:DC}}\text{US}, {}_{\text{IC:IC}}\text{LS} \} .
\end{aligned}
$$

(vi$_1$) For $m_1 < m_2$, the singularity in the x_{j_2}-direction higher than singularity in the x_{j_1}-direction, and the following three $([m_1,n_1]^{\text{th}}\text{YY},[m_2 - m_1]^{\text{th}}\text{XX})$-YYXX switching bifurcations are given as follows.

(vi$_{1a}$) For

$$
\begin{aligned}
\dot{x}_{j_1} &= a_{j_1 j_1 0}(x_{j_1} - a_{j_1 j_1 1})^{m_1}(x_{j_2} - a_{j_1 j_2 1})^{n_1}, \\
\dot{x}_{j_2} &= a_{j_2 j_2 0}(x_{j_1} - a_{j_2 j_1 1})^{m_2}.
\end{aligned}
$$

with $a_{j_1 j_1 1} = a_{j_2 j_1 1}$, once

$$
a^{-}_{j_1 j_1 1} \rightleftharpoons a^{+}_{j_2 j_1 1}, a^{-}_{j_2 j_1 1} \rightleftharpoons a^{+}_{j_1 j_1 1}
$$

the $([m_1,n_1]^{\text{th}}\text{YY},[m_2 - m_1]^{\text{th}}\text{XX})$-YYXX switching bifurcation is from a singular equilibrium with a singular 1-dimensional flow switching to a singular 1-dimensional flow with a singular equilibrium.

$$
\left\{
\begin{array}{c}
\underbrace{(a_{j_1 j_2 1}, a_{j_2 j_1 1})}_{(n_1,m_2)\text{-YY}_1} \\
\underbrace{(a_{j_1 j_1 1}, a_{j_1 j_2 1})}_{((m_1,n_1)^{\text{th}}\text{XX}_1,\text{xF})}
\end{array}
\right\}
\rightleftharpoons
\underbrace{([m_1,n_1]^{\text{th}}\text{YY},[m_2 - m_1]^{\text{th}}\text{XX})}_{([m_1,n_1],[m_2-m_1])\text{-YYXX}}
\rightleftharpoons
\left\{
\begin{array}{c}
\underbrace{(a_{j_1 j_1 1}, a_{j_1 j_2 1})}_{((m_1,n_1)^{\text{th}}\text{XX}_2,\text{xF})} \\
\underbrace{(a_{j_1 j_2 1}, a_{j_2 j_1 1})}_{(n_1,m_2)\text{-YY}_2}
\end{array}
\right\}
$$

with

$$
\begin{aligned}
&\text{XX}_1, \text{XX}_2 \in \{ \text{DP:UP, UP:DP; UP:UP, DP:DP; } {}_{\text{II:DI}}\text{SO}, {}_{\text{DI:II}}\text{SO}, \\
&{}_{\text{II:DI}}\text{SI}, {}_{\text{DI:II}}\text{SI}; {}_{\text{II:II}}\text{US}, {}_{\text{II:II}}\text{LS}, {}_{\text{DI:DI}}\text{US}, {}_{\text{DI:DI}}\text{LS} \} ; \\
&\text{xF}_1, \text{xF}_2 \in \{ \text{pF, nF} \} ; \\
&\text{YY}_1, \text{YY}_2 \in \{ \text{positive saddle, negative saddle, CCW center,} \\
&\text{CW center, up-parabola upper-saddle (lower-saddle),} \\
&\text{down-parabola upper-saddle (lower-saddle)} \\
&\text{double inflection-saddlles} \} .
\end{aligned}
$$

(vi$_{1b}$) The $([m_1,n_1]^{\text{th}}\text{YY},[m_2 - m_1]^{\text{th}}\text{XX})$-YYXX switching bifurcation is for two $n_1 \times (m_1 + m_2)$ networks of non-singular equilibrium and hyperbolic and hyperbolic-secant flows.

$$\bigcup_{r_1=1}^{m_1}\bigcup_{u_1=1}^{n_1}\underbrace{(a_{j_1j_1r_1},a_{j_1j_2u})}_{(\mathrm{XX}_1,\mathrm{xF}_1)}\oplus\bigcup_{u=1}^{n_1}\bigcup_{v_1=1}^{m_2}\underbrace{(a_{j_1j_2u},a_{j_2j_1v_1})}_{\mathrm{YY}_1}$$
$$\rightleftharpoons\underbrace{([m_1,n_1]^{\mathrm{th}}\mathrm{YY},[m_2-m_1]^{\mathrm{th}}\mathrm{XX})}_{([m_1,n_1],[m_2-m_1])\text{-YYXX}}$$
$$\rightleftharpoons\bigcup_{r_2=1}^{m_1}\bigcup_{u=1}^{n_1}\underbrace{(a_{j_1j_1r_2},a_{j_1j_2u_2})}_{(\mathrm{XX}_2,\mathrm{xF}_2)}\oplus\bigcup_{u=1}^{n_1}\bigcup_{v_1=1}^{m_2}\underbrace{(a_{j_1j_2u},a_{j_2j_1v_2})}_{\mathrm{YY}_2},$$

with

$\mathrm{YY}_1,\mathrm{YY}_2 \in$ { positive saddle, negative saddle, CCW center, CW center} ;
xF_1, $\mathrm{xF}_2 \in$ { pF, nF} ;
XX_1, $\mathrm{XX}_2 \in$ { hyperbolic flow, hyperbolic-secant flow} .

(vi_{1c}) The $([m_1,n_1]^{\mathrm{th}}\mathrm{YY},[m_2-m_1]^{\mathrm{th}}\mathrm{XX})$-YYXX switching bifurcation is for two $p_1\times(q_1+q_2)$ networks of singular equilibriums and singular 1-dimensional flows.

$$\bigcup_{r_1=1}^{p_1}\bigcup_{u=1}^{q_1}\underbrace{(a_{j_1j_1r_1},a_{j_1j_2u})}_{([m_{r_1},n_u]\mathrm{XX}_1,\mathrm{xF}_1)}\oplus\bigcup_{u=1}^{q_1}\bigcup_{v_1=1}^{p_2}\underbrace{(a_{j_1j_2u},a_{j_2j_1v_1})}_{(n_u,m_{v_1})\text{-}\mathrm{YY}_1}$$
$$\rightleftharpoons\underbrace{([m_1,n_1]^{\mathrm{th}}\mathrm{YY},[m_2-m_1]^{\mathrm{th}}\mathrm{XX})}_{([m_1,n_1],[m_2-m_1])\text{-YYXX}}$$
$$\rightleftharpoons\bigcup_{r_2=1}^{p_1}\bigcup_{u_2=1}^{q_1}\underbrace{(a_{j_1j_1r_2},a_{j_1j_2u})}_{([m_{r_2},n_u]\mathrm{XX}_1,\mathrm{xF}_2)}\oplus\bigcup_{u=1}^{q_1}\bigcup_{v_2=1}^{p_2}\underbrace{(a_{j_1j_2u},a_{j_2j_1v_2})}_{(n_u,m_{v_2})\text{-}\mathrm{YY}_2}$$

with

XX_1, $\mathrm{XX}_2 \in$ { DP:UP, UP:DP; UP:UP, DP:DP; ${}_{\mathrm{II:DI}}\mathrm{SO}$,${}_{\mathrm{DI:II}}\mathrm{SO}$, ${}_{\mathrm{II:DI}}\mathrm{SI}$,${}_{\mathrm{DI:II}}\mathrm{SI}$; ${}_{\mathrm{II:II}}\mathrm{US}$,${}_{\mathrm{II:II}}\mathrm{LS}$,${}_{\mathrm{DI:DI}}\mathrm{US}$,${}_{\mathrm{DI:DI}}\mathrm{LS}$} ;
xF_1, $\mathrm{xF}_2 \in$ { pF, nF} ;
$\mathrm{YY}_1,\mathrm{YY}_2 \in$ { positive saddle, negative saddle, CCW center, CW center, up-parabola upper-saddle (lower-saddle), down-parabola upper-saddle (lower-saddle) double inflection-saddlles} .

and

$$m_1 = \sum_{r_i=1}^{p_1} m_{j_1 r_i}, n_1 = \sum_{u_i=1}^{q_1} n_{j_1 u_i}, m_2 = \sum_{v_i=1}^{p_1} m_{j_2 v_i};\ i = 1, 2.$$

(vi$_2$) For $m_1 < m_2$, the singularity in the x_{j_2}-direction higher than singularity in x_{j_1}-direction, and the following three $([m_1,n_1]^{\text{th}}\text{YY},[m_1 - m_2]^{\text{th}}\text{ZZ})$-YYZZ switching bifurcations are given as follows.

(vi$_{2a}$) The $([m_1,n_1]^{\text{th}}\text{YY},[m_1-m_2]^{\text{th}}\text{ZZ})$-YYZZ switching bifurcation is for a singular equilibrium with a singular 1-dimensional flow switching to a singular 1-dimensional flow with a singular equilibrium.

$$\left\{\begin{array}{c}\underbrace{(a_{j_1j_2}1, a_{j_2j_1}1)}_{(n_1,m_2)\text{-YY}_1}\\ \underbrace{(a_{j_1j_1}1, a_{j_1j_2}1)}_{((m_1,n_1)^{\text{th}}\text{XX}_1,xF)}\end{array}\right\} \rightleftharpoons \underbrace{([m_1,n_1]^{\text{th}}\text{YY},[m_1 - m_2]^{\text{th}}\text{ZZ})}_{([m_1,n_1],[m_2-m_1])\text{-YYZZ}} \rightleftharpoons \left\{\begin{array}{c}\underbrace{(a_{j_1j_1}1, a_{j_1j_2}1)}_{((m_1,n_1)^{\text{th}}\text{XX}_2,xF)}\\ \underbrace{(a_{j_1j_2}1, a_{j_2j_1}1)}_{(n_1,m_2)\text{-YY}_2}\end{array}\right\}$$

with

$$\begin{array}{l}\text{XX}_1, \text{XX}_2 \in \{\text{ DP:UP, UP:DP; UP:UP, DP:DP; }_{\text{II:DI}}\text{SO},_{\text{DI:II}}\text{SO},\\ _{\text{II:DI}}\text{SI},_{\text{DI:II}}\text{SI};\ _{\text{II:II}}\text{US},_{\text{II:II}}\text{LS},_{\text{DI:DI}}\text{US},_{\text{DI:DI}}\text{LS}\}\ ;\\ \text{xF}_1, \text{xF}_2 \in \{\text{ pF, nF}\}\ ;\\ \text{YY}_1,\text{YY}_2 \in \{\text{ positive saddle, negative saddle, CCW center,}\\ \text{CW center, up-parabola upper-saddle (lower-saddle),}\\ \text{down-parabola upper-saddle (lower-saddle)}\\ \text{double inflection-saddlles}\}\ .\end{array}$$

(vi$_{2b}$) The $([m_1,n_1]^{\text{th}}\text{YY},[m_1-m_2]^{\text{th}}\text{ZZ})$-YYZZ switching bifurcation is for two $m_1 \times (n_1+n_2)$ networks of non-singular equilibrium and hyperbolic and hyperbolic-secant flows.

$$\begin{array}{l}\bigcup_{r=1}^{m_1}\bigcup_{u_1=1}^{n_1}\underbrace{(a_{j_1j_1r}, a_{j_1j_2u_1})}_{(\text{XX}_1,\text{xF}_1)} \oplus \bigcup_{r=1}^{m_1}\bigcup_{s_1=1}^{n_2}\underbrace{(a_{j_1j_1r}, a_{j_2j_2s_1})}_{\text{YY}_1}\\ \rightleftharpoons \underbrace{([m_1,n_1]^{\text{th}}\text{YY},[m_1 - m_2]^{\text{th}}\text{ZZ})}_{([m_1,n_1],[m_2-m_1])\text{-YYZZ}}\\ \rightleftharpoons \bigcup_{r=1}^{m_1}\bigcup_{u_2=1}^{n_1}\underbrace{(a_{j_1j_1r}, a_{j_1j_2u_1})}_{(\text{XX}_2,\text{xF}_2)} \oplus \bigcup_{r=1}^{m_1}\bigcup_{s_2=1}^{n_1}\underbrace{(a_{j_1j_1r}, a_{j_2j_2s_2})}_{\text{YY}_2},\end{array}$$

with

$YY_1, YY_2 \in$ { positive and negative saddle, CCW center, CW center} ;
$xF_1, xF_2 \in$ { pF, nF} ;
$XX_1, XX_2 \in$ { hyperbolic flow, hyperbolic-secant flow} .

(vi_{2c}) The $([n_2 - n_1]^{th}ZZ,[m_1,n_2]^{th}YY)$-YYZZ switching bifurcation is for two $p_1 \times (q_1 + q_2)$ networks of singular equilibriums and singular 1-dimensional flows.

$$
\begin{aligned}
&\bigcup_{r_1=1}^{p_1}\bigcup_{u=1}^{q_1}\underbrace{(a_{j_1j_1r}, a_{j_1j_2u})}_{([m_{j_1r_1}, n_{j_1u}]XX_1, xF_1)} \oplus \bigcup_{u=1}^{q_1}\bigcup_{v_1=1}^{p_2}\underbrace{(a_{j_1j_2u}, a_{j_2j_1v_1})}_{(n_{j_1u}, m_{j_2v_1})\text{-}YY_1} \\
&\rightleftharpoons \underbrace{([m_1,n_1]^{th}YY,[m_1 - m_2]^{th}ZZ)}_{([m_1,n_1],[m_2-m_1])\text{-}YYZZ} \\
&\rightleftharpoons \bigcup_{r_2=1}^{p_1}\bigcup_{u=1}^{q_1}\underbrace{(a_{j_1j_1r}, a_{j_1j_2u})}_{([m_{j_1r_2}, n_{j_1u}]XX_1, xF_2)} \oplus \bigcup_{u=1}^{q_1}\bigcup_{v_2=1}^{p_2}\underbrace{(a_{j_1j_2u}, a_{j_2j_1v_2})}_{(n_{j_1u}, m_{j_2v_2})\text{-}YY_2}
\end{aligned}
$$

with

$XX_1, XX_2 \in$ { DP:UP, UP:DP; UP:UP, DP:DP; ${}_{II:DI}SO, {}_{DI:II}SO$, ${}_{II:DI}SI, {}_{DI:II}SI$; ${}_{II:II}US, {}_{II:II}LS, {}_{DI:DI}US, {}_{DI:DI}LS$} ;
$xF_1, xF_2 \in$ { pF, nF} ;
$YY_1, YY_2 \in$ { positive saddle, negative saddle, CCW center,
CW center, up-parabola upper-saddle (lower-saddle),
down-parabola upper-saddle (lower-saddle)
double inflection-saddlles} .

and

$$
m_1 = \sum_{r_i=1}^{p_1} m_{j_1r_i},\ n_1 = \sum_{u_i=1}^{q_1} n_{j_1u_i},\ n_2 = \sum_{v_i=1}^{p_2} n_{j_2v_i};\ i = 1, 2.
$$

In the end, the theorem is proved. □

Chapter 3
Singular Equilibria and Flows, and Simple Networks

In this chapter, the singular 1-dimensional flows and equilibriums in 2-dimensional polynomial systems are illustrated to help one understand the global bifurcations for networks of the hyperbolic and hyperbolic-secant flows with saddles and centers. The singular saddles and centers, parabola-saddles and inflection-saddles are presented, and the singular hyperbolic-flows, hyperbolic-to-hyperbolic-secant flows, inflection-source and sink flows, inflection-saddle flows are also presented. The infinite-equilibriums in the crossing and product polynomial systems are the appearing and switching bifurcations for hybrid networks of simple and singular equilibriums and flows. Such hybrid networks of simple equilibriums and flows are illustrated.

3.1 $[(2m_{11}+1), (2n_{11}+1)]$ and $((2n_{11}+1), (2m_{21}+1))$-Singular Systems

In this section, there are four global bifurcations of the 1-dimensional flows through singular hyperbolic flows, and four global bifurcations of equilibriums, through singular positive and negative saddles and CCW and CW centers bifurcations. The corresponding global bifurcations generate networks of hyperbolic and hyperbolic-secant flows with saddles and centers. The infinite-equilibriums are the appearing and switching bifurcations for the singular flows and singular equilibriums and for hybrid networks of hyperbolic flows and saddles and centers.

A. C. J. Luo, *Two-dimensional Crossing and Product Polynomial Systems*,
https://doi.org/10.1007/978-981-96-5715-5_3

3.1.1 $[(2m_{11}+1),(2n_{11}+1)]$-Flows and $((2n_{11}+1),(2m_{21}+1))$-Equilibriums

Consider a polynomial dynamical system for $a_{111} \neq a_{211}$ as

$$\begin{aligned} \dot{x}_2 &= a_{220}(x_1 - a_{211})^{2m_{21}+1}, \\ \dot{x}_1 &= a_{110}(x_1 - a_{111})^{2m_{11}+1}(x_2 - a_{121})^{2n_{11}+1}. \end{aligned} \tag{3.1}$$

The first integral manifold is

$$\begin{aligned} &\sum_{\alpha,\beta=0}^{\alpha+\beta=2m_{21}+1} \frac{(2m_{21}+1)!}{\alpha!\beta!} \frac{(a_{111}-a_{211})^{\alpha}}{\beta - 2m_{11}} [(x_1 - a_{111})^{\beta-2m_{11}} - (x_{10} - a_{111})^{\beta-2m_{11}}] \\ &\quad = \frac{a_{j_1j_10}}{a_{j_2j_20}} \frac{1}{2n_{11}+2} [(x_2 - a_{121})^{2n_{11}+2} - (x_{20} - a_{121})^{2n_{11}+2}], \\ &\quad \text{for } \beta - 2m_{11} \neq 0; \\ &\frac{1}{\beta - 2m_{11}} [(x_1 - a_{111})^{\beta-2m_{11}} - (x_{10} - a_{111})^{\beta-2m_{11}}] = \ln \frac{|x_1 - a_{111}|}{|x_{10} - a_{111}|}, \\ &\quad \text{for } \beta - 2m_{11} = 0. \end{aligned} \tag{3.2}$$

The singular hyperbolic and hyperbolic-secant flows with singular positive and negative saddles and CCW and CW centers are illustrated in Figs. 3.1 and 3.2. Through such singular flows, the corresponding global bifurcations generate from a hyperbolic (hyperbolic-secant) flow to a network of singular hyperbolic and hyperbolic-secant flows. Through such singular equilibriums, the corresponding global bifurcations generate from one of saddles and centers to a network of singular saddles and centers equilibriums.

For $(x_1^*, x_2^*) = (a_{111}, a_{121})$ with $\dot{x}_1 = 0$, the $[(2m_{11}+1),(2n_{11}+1)]$-singular flows exist with $\dot{x}_2 = a_{220}(a_{111} - a_{211})^{2m_{21}+1}$. For $(x_1^*, x_2^*) = (a_{111}, a_{121})$, the 1-dimensional flow direction is determined by $\dot{x}_2 \neq 0$. For $(x_2^*, x_1^*) = (a_{121}, a_{211})$ with $\dot{x}_1 = \dot{x}_2 = 0$, the $((2n_{11}+1),(2m_{21}+1))$-singular equilibriums exist.

(A) For $a_{111} > a_{211}$,

$$\left\{ \begin{array}{c} (a_{111}, a_{121}) \\ (a_{121}, a_{211}) \end{array} \right\} = \left\{ \begin{array}{c} \underbrace{([(2m_{11}+1),(2n_{11}+1)]\text{DP:UP,pF})}_{[(2m_{11}+1),(2n_{11}+1)]\text{-hyperbolic flow }(+)} \\ \underbrace{((2n_{11}+1)\text{DP}_{+},(2m_{21}+1)\text{DP}_{-})}_{[(2n_{11}+1),(2m_{21}+1)]\text{-CCW center}} \end{array} \right\},$$

for $a_{110} > 0, a_{220} > 0$;

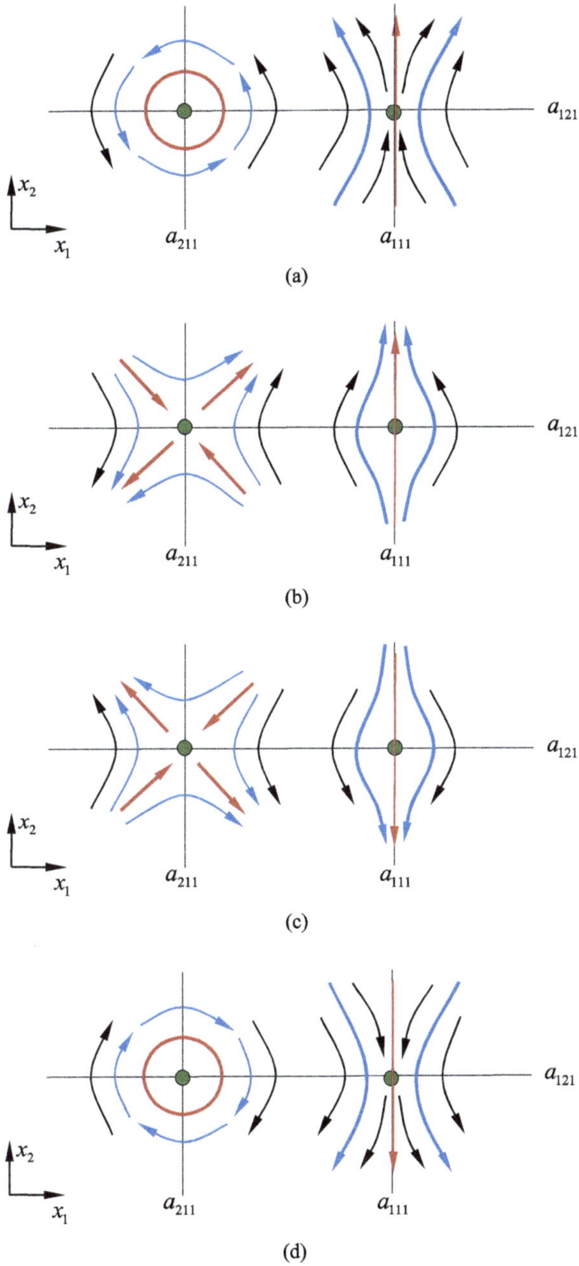

Fig. 3.1 Phase portraits ($a_{111} > a_{211}$) of the ($[(2m_{11}+1), (2n_{11}+1)]$, xF)-hyperbolic and hyperbolic-secant singular flows at $(x_1^*, x_2^*) = (a_{111}, a_{121})$ and $((2n_{11}+1), (2m_{21}+1))$-positive and negative saddles, CCW and CW centers at $(x_2^*, x_1^*) = (a_{121}, a_{211})$ for 2-dimensional polynomial systems: **a** ($a_{110} > 0, a_{220} > 0$), **b** ($a_{110} < 0, a_{220} > 0$), **c** ($a_{110} > 0, a_{220} < 0$), **d** ($a_{110} < 0, a_{220} < 0$)

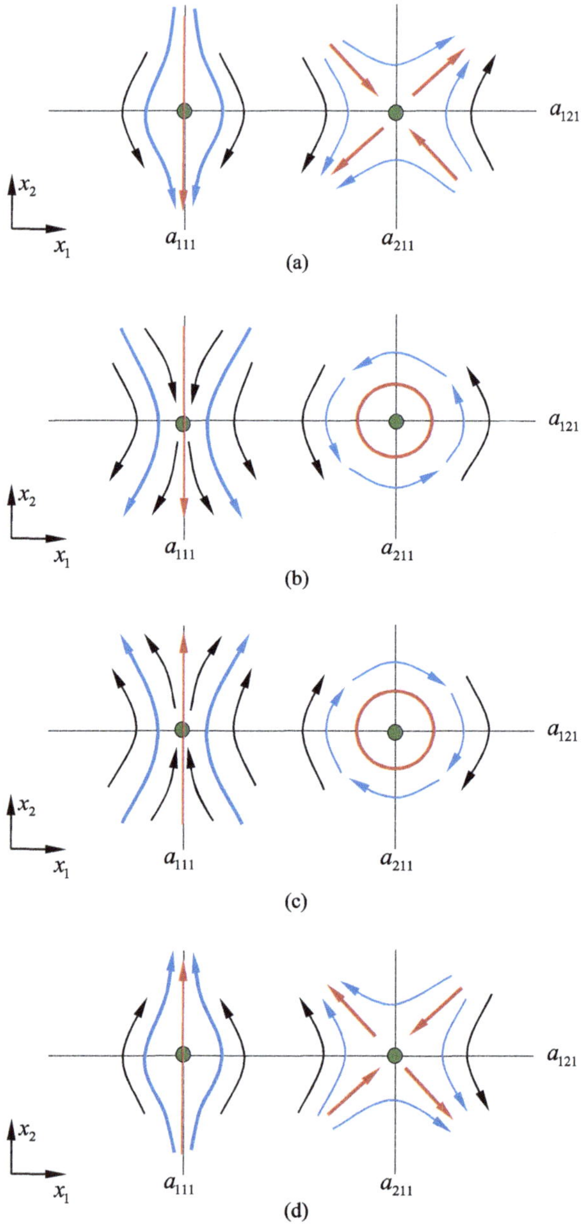

Fig. 3.2 Phase portraits ($a_{211} > a_{111}$) of the ([$(2m_{11}+1)$, $(2n_{11}+1)$], xF)-hyperbolic and hyperbolic-secant singular flows at $(x_1^*, x_2^*) = (a_{111}, a_{121})$ and ($(2n_{11}+1)$, $(2m_{21}+1)$)-positive and negative saddles, CCW and CW centers at $(x_2^*, x_1^*) = (a_{121}, a_{211})$ for 2-dimensional polynomial systems: **a** ($a_{110} > 0, a_{220} > 0$), **b** ($a_{110} < 0, a_{220} > 0$), **c** ($a_{110} > 0, a_{220} < 0$), **d** ($a_{110} < 0, a_{220} < 0$)

$$\begin{Bmatrix}(a_{111},a_{121})\\(a_{121},a_{211})\end{Bmatrix}=\begin{Bmatrix}\underbrace{([(2m_{11}+1),(2n_{11}+1)]\text{UP:DP,pF})}_{[(2m_{11}+1),(2n_{11}+1)]\text{-hyperbolic-secant flow }(+)}\\\underbrace{((2n_{11}+1)\text{UP}_{+},(2m_{21}+1)\text{UP}_{+})}_{[(2n_{11}+1),(2m_{21}+1)]\text{-postive saddle}}\end{Bmatrix},$$

for $a_{110}<0, a_{220}>0$;

$$\begin{Bmatrix}(a_{111},a_{121})\\(a_{121},a_{211})\end{Bmatrix}=\begin{Bmatrix}\underbrace{([(2m_{11}+1),(2n_{11}+1)]\text{UP:DP,nF})}_{[(2m_{11}+1),(2n_{11}+1)]\text{-hyperbolic-secant flow }(-)}\\\underbrace{((2n_{11}+1)\text{UP}_{-},(2m_{21}+1)\text{UP}_{-})}_{[(2n_{11}+1),(2m_{21}+1)]\text{-negative saddle}}\end{Bmatrix},$$

for $a_{110}>0, a_{220}<0$;

$$\begin{Bmatrix}(a_{111},a_{121})\\(a_{121},a_{211})\end{Bmatrix}=\begin{Bmatrix}\underbrace{([(2m_{11}+1),(2n_{11}+1)]\text{DP:UP,nF})}_{[(2m_{11}+1),(2n_{11}+1)]\text{-hyperbolic flow }(-)}\\\underbrace{((2n_{11}+1)\text{DP}_{-},(2m_{21}+1)\text{DP}_{+})}_{[(2n_{11}+1),(2m_{21}+1)]\text{-CW center}}\end{Bmatrix},$$

for $a_{110}<0, a_{220}<0$. (3.3)

(B) For $a_{111}<a_{211}$,

$$\begin{Bmatrix}(a_{121},a_{211})\\(a_{111},a_{121})\end{Bmatrix}=\begin{Bmatrix}\underbrace{((2n_{11}+1)\text{UP}_{+},(2m_{21}+1)\text{UP}_{+})}_{[(2n_{11}+1),(2m_{21}+1)]\text{-positive saddle}}\\\underbrace{([(2m_{11}+1),(2n_{11}+1)]\text{UP:DP,nF})}_{[(2m_{11}+1),(2n_{11}+1)]\text{-hyperbolic-secant flow }(-)}\end{Bmatrix},$$

for $a_{110}>0, a_{220}>0$;

$$\begin{Bmatrix}(a_{121},a_{211})\\(a_{111},a_{121})\end{Bmatrix}=\begin{Bmatrix}\underbrace{((2n_{11}+1)\text{DP}_{+},(2m_{21}+1)\text{DP}_{-})}_{[(2n_{11}+1),(2m_{21}+1)]\text{-CCW center}}\\\underbrace{([(2m_{11}+1),(2n_{11}+1)]\text{DP:UP,nF})}_{[(2m_{11}+1),(2n_{11}+1)]\text{-hyperbolic flow }(-)}\end{Bmatrix},$$

for $a_{110}<0, a_{220}>0$;

$$\begin{Bmatrix}(a_{121},a_{211})\\(a_{111},a_{121})\end{Bmatrix}=\begin{Bmatrix}\underbrace{((2n_{11}+1)\text{DP}_{-},(2m_{21}+1)\text{DP}_{+})}_{[(2n_{11}+1),(2m_{21}+1)]\text{-CW center}}\\\underbrace{([(2m_{11}+1),(2n_{11}+1)]\text{DP:UP,pF})}_{[(2m_{11}+1),(2n_{11}+1)]\text{-hyperbolic flow }(+)}\end{Bmatrix},$$

for $a_{110}>0, a_{220}<0$;

$$\begin{Bmatrix}(a_{121},a_{211})\\(a_{111},a_{121})\end{Bmatrix}=\begin{Bmatrix}\underbrace{((2n_{11}+1)\text{UP}_{-},(2m_{21}+1)\text{UP}_{-})}_{[(2n_{11}+1),(2m_{21}+1)]\text{-negative saddle}}\\\underbrace{([(2m_{11}+1),(2n_{11}+1)]\text{UP:DP,pF})}_{[(2m_{11}+1),(2n_{11}+1)]\text{-hyperbolic-secant flow }(+)}\end{Bmatrix},$$

$$\text{for } a_{110} < 0, a_{220} < 0. \tag{3.4}$$

- The ([$(2m_{11}+1)$, $(2n_{11}+1)$]DP:UP,pF) is for a positive singular hyperbolic flow.
- The ([$(2m_{11}+1)$, $(2n_{11}+1)$]UP:DP,pF) is for a positive singular hyperbolic-secant flow.
- The ([$(2m_{11}+1)$, $(2n_{11}+1)$]UP:DP,nF) is for a negative singular hyperbolic-secant flow.
- The ([$(2m_{11}+1)$, $(2n_{11}+1)$]DP:UP,nF) is for a negative singular hyperbolic flow.

The acronyms "UP" and "DP" are for up-parabola and down-parabola flows, respectively. "pF" and "nF" are for positive and negative flows, respectively.

- The (($2n_{11}+1$)UP$_+$,($2m_{21}+1$)UP$_+$] is for a (($2n_{11}+1$), ($2m_{21}+1$)]-singular positive saddle.
- The (($2n_{11}+1$)UP$_-$,($2m_{21}+1$)UP$_-$] is for a (($2n_{11}+1$), ($2m_{21}+1$)]-singular negative saddle.
- The (($2n_{11}+1$)DP$_+$,($2m_{21}+1$)DP$_-$] is for a (($2n_{11}+1$), ($2m_{21}+1$)]-singular CCW center.
- The (($2n_{11}+1$)DP$_-$, ($2m_{21}+1$)DP$_+$] is for a (($2n_{11}+1$), ($2m_{21}+1$)]-singular CW center.

3.1.2 $[(2m_{11}+1), (2n_{11}+1)]$ and $((2n_{11}+1), (2m_{21}+1))$-*Infinite-Equilibriums*

Consider a polynomial dynamical system for $a_{111} = a_{211}$ as

$$\begin{aligned} \dot{x}_2 &= a_{220}(x_1 - a_{211})^{2m_{21}+1}, \\ \dot{x}_1 &= a_{110}(x_1 - a_{111})^{2m_{11}+1}(x_2 - a_{121})^{2n_{11}+1}. \end{aligned} \tag{3.5}$$

The first integral manifold is

$$\begin{aligned} &\frac{1}{2(m_{21}-m_{11})+1}\left[(x_1 - a_{211})^{2(m_{21}-m_{11})+1} - (x_{10} - a_{211})^{2(m_{21}-m_{11})+1}\right] \\ &= \frac{a_{110}}{a_{220}}\frac{1}{2n_{11}+2}\left[(x_2 - a_{121})^{2n_{11}+2} - (x_{20} - a_{121})^{2n_{11}+2}\right]. \end{aligned} \tag{3.6}$$

(**A**) For $2m_{21}+1 \geq 2m_{11}+1$, the corresponding infinite-equilibriums are given.

$$(a_{121}, a_{211}) = \underbrace{([(2m_{11}+1), (2n_{11}+1)]^{\text{th}}\text{UP}_-{:}\text{UP}_+,[2(m_{21}-m_{11})]^{\text{th}}{}_{\text{DI:II}}\text{US})}_{([(2m_{11}+1),(2n_{11}+1)],[2(m_{21}-m_{11})])\text{-inflection up-parabola upper-saddle}},$$

for $a_{110} > 0, a_{220} > 0$;

$$(a_{121},a_{211}) = \underbrace{([(2m_{11}+1),(2n_{11}+1)]^{\text{th}}\text{DP}_{-}\text{:DP}_{+},[2(m_{21}-m_{11})]^{\text{th}}{}_{\text{II:DI}}\text{LS})}_{([(2m_{11}+1),(2n_{11}+1)],[2(m_{21}-m_{11})])\text{-inflection down-parabola lower-saddle}},$$
for $a_{110} < 0, a_{220} > 0$;
$$(a_{121},a_{211}) = \underbrace{([(2m_{11}+1),(2n_{11}+1)]^{\text{th}}\text{DP}_{+}\text{:DP}_{-},[2(m_{21}-m_{11})]^{\text{th}}{}_{\text{II:DI}}\text{US})}_{([(2m_{11}+1),(2n_{11}+1)],[2(m_{21}-m_{11})])\text{-inflection down-parabola upper-saddle}},$$
for $a_{110} > 0, a_{220} < 0$;
$$(a_{121},a_{211}) = \underbrace{([(2m_{11}+1),(2n_{11}+1)]^{\text{th}}\text{UP}_{+}\text{:UP}_{-},[2(m_{21}-m_{11})]^{\text{th}}{}_{\text{DI:II}}\text{LS})}_{([(2m_{11}+1),(2n_{11}+1)],[2(m_{21}-m_{11})])\text{-inflection up-parabola lower-saddle}},$$
for $a_{110} < 0, a_{220} < 0$. (3.7)

(B) For $2m_{21}+1 < 2m_{11}+1$, the infinite-equilibriums are given as follows.

$$(a_{121},a_{211}) = \underbrace{([(2m_{11}+1),(2n_{11}+1)]^{\text{th}}\text{UP}_{-}\text{:UP}_{+},[2(m_{11}-m_{21})]^{\text{th}}{}_{\text{DC:IC}}\text{US})}_{([(2m_{11}+1),(2n_{11}+1)],[2(m_{21}-m_{11})])\text{-concave up-parabola upper-saddle}},$$
for $a_{110} > 0, a_{220} > 0$;
$$(a_{121},a_{211}) = \underbrace{([(2m_{11}+1),(2n_{11}+1)]^{\text{th}}\text{DP}_{-}\text{:DP}_{+},[2(m_{11}-m_{21})]^{\text{th}}{}_{\text{IC:DC}}\text{LS})}_{([(2m_{11}+1),(2n_{11}+1)],[2(m_{21}-m_{11})])\text{-concave down-parabola lower-saddle}},$$
for $a_{110} < 0, a_{220} > 0$;
$$(a_{121},a_{211}) = \underbrace{([(2m_{11}+1),(2n_{11}+1)]^{\text{th}}\text{DP}_{+}\text{:DP}_{-},[2(m_{11}-m_{21})]^{\text{th}}{}_{\text{IC:DC}}\text{US})}_{([(2m_{11}+1),(2n_{11}+1)],[2(m_{21}-m_{11})])\text{-concave down-parabola upper-saddle}},$$
for $a_{110} > 0, a_{220} < 0$;
$$(a_{121},a_{211}) = \underbrace{([(2m_{11}+1),(2n_{11}+1)]^{\text{th}}\text{UP}_{+}\text{:UP}_{-},[2(m_{11}-m_{21})]^{\text{th}}{}_{\text{DC:IC}}\text{LS})}_{([(2m_{11}+1),(2n_{11}+1)],[2(m_{21}-m_{11})])\text{-concave up-parabola lower-saddle}},$$
for $a_{110} < 0, a_{220} < 0$. (3.8)

For $x_1^* = a_{111} = a_{211}$, the infinite-equilibrium exists. The infinite-equilibriums are illustrated in Figs. 3.3 and 3.4 for $m_{21} \geq m_{11}$ and $m_{21} < m_{11}$, respectively. For $m_{21} \geq m_{11}$, the singular inflection parabola-saddles are presented. For $m_{11} > m_{21}$, the concave parabola-saddles are presented.

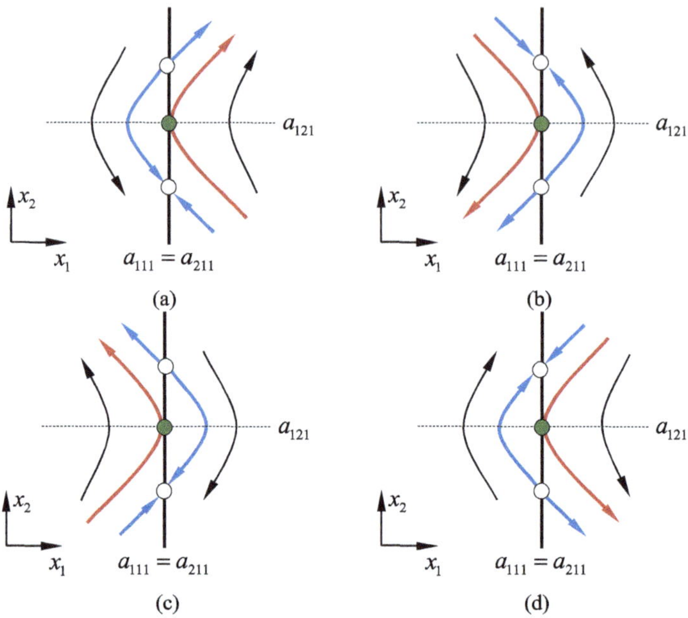

Fig. 3.3 The infinite-equilibriums ($a_{211} = a_{111}, 2m_{21} + 1 \geq 2m_{11} + 1, n_1 = 2n_{11} + 1$) for 2-dimensional polynomial systems: **a** ($a_{110} > 0, a_{220} > 0$), **b** ($a_{110} < 0, a_{220} > 0$), **c** ($a_{110} > 0, a_{220} < 0$), **d** ($a_{110} < 0, a_{220} < 0$)

3.1.3 [(2m₁₁ + 1), (2n₁₁ + 1)] and ((2n₁₁ + 1), (2m₂₁ + 1))-Hybrid Networks

Consider

$$\begin{aligned}
\dot{x}_2 &= a_{220} \prod_{v=1}^{2m_{21}+1} (x_1 - a_{21v}), \\
\dot{x}_1 &= a_{110} \prod_{r=1}^{2m_{11}+1} (x_1 - a_{11r}) \prod_{u=1}^{2n_{11}+1} (x_2 - a_{12u}).
\end{aligned} \tag{3.9}$$

The first integral manifold is

$$\sum_{\nu_0=2m_{11}+1}^{2m_{21}+1} b_{21\nu_0} \sum_{\substack{\mu_1 \in [0,\beta_0-2m_{11}-1],\\ \nu_1 \in [2m_{11},\nu_0-1]}}^{\mu_1+\nu_1=\nu_0-1} \sum_{\substack{\mu_2 \in [0,\nu_1-2m_{11}-1]\\ \nu_2 \in [2m_{11}-1,\nu_1-2]}}^{\mu_2+\nu_2=\nu_1-2} \cdots \sum_{\substack{\mu_{2m_{11}} \in [0,\nu_{2m_{11}-1}-2m_{11}-1]\\ \nu_{2m_{11}} \in [1,\nu_{2m_{11}-1}-2m_{11}]}}^{\mu_{2m_{11}}+\nu_{2m_{11}}=\nu_{2m_{11}-1}-2m_{11}}$$

$$\prod_{j=1}^{2m_{11}} \frac{(\nu_{j-1}-j)!}{\mu_j!\nu_j!} \frac{(a_{11r_{j+1}} - a_{11r_j})^{\mu_j}}{\nu_{2m_{11}}} [(x_1 - a_{11r_{2m_{11}+1}})^{\nu_{2m_{11}}} - (x_{10} - a_{11r_{2m_{11}+1}})^{\nu_{2m_{11}}}]$$

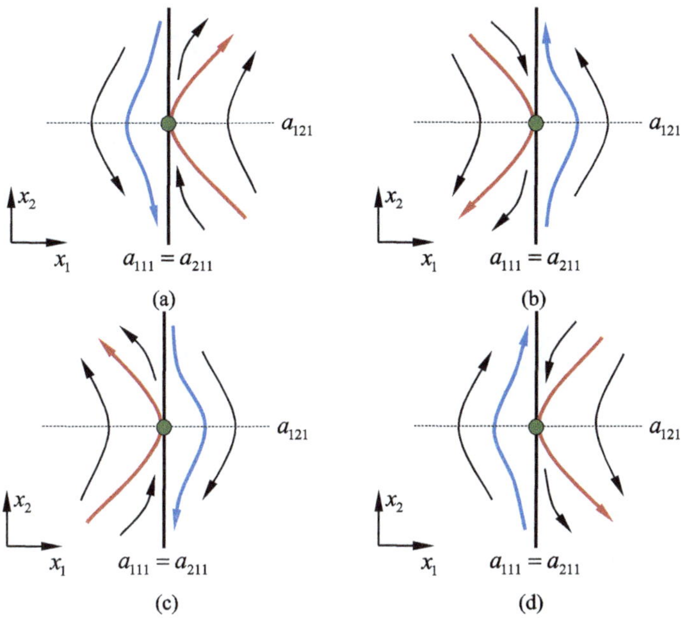

Fig. 3.4 The infinite-equilibriums ($a_{211}=a_{111}, 2m_{21}+1<2m_{11}+1, n_1=2n_{11}+1$) for 2-dimensional polynomial systems: **a** ($a_{110}>0, a_{220}>0$), **b** ($a_{110}<0, a_{220}>0$), **c** ($a_{110}>0, a_{220}<0$), **d** ($a_{110}<0, a_{220}<0$)

$$
\begin{aligned}
&-\sum_{\nu_0=2m_{11}+1}^{2m_{21}+1} b_{21\nu_0}\sum_{k=2}^{2m_{11}}\Bigg[\sum_{\substack{\mu_1\in[0,\nu_0-2m_{11}-1),\\ \nu_1\in(2m_{11},\nu_0-1]}}^{\mu_1+\nu_1=\nu_0-1}\ \sum_{\substack{\mu_2\in[0,\nu_1-2m_{11}-1)\\ \nu_2\in(2m_{11}-1,\nu_1-2)}}^{\mu_2+\nu_2=\nu_1-2}\cdots\\
&\times\Bigg(\sum_{\substack{\mu_{k-1}\in(\nu_{k-2}-2m_{11}+2,\nu_{k-2}-k+1]\\ \nu_{k-1}\in[0,2m_{11}+1-k)}}^{\mu_{k-1}+\nu_{k-1}=\nu_{k-2}-k+1}\prod_{j=1}^{k-1}\frac{(\nu_{j-1}-j)!}{\mu_j!\nu_j!}(a_{11r_{j+1}}-a_{11r_j})^{\mu_j}\\
&\times\sum_{\rho=k}^{2m_{11}+1}A_{\rho 1}^{(r_k)}\ln\frac{|x_1-a_{11r_\rho}|}{|x_{10}-a_{11r_\rho}|}\\
&+\sum_{\substack{\mu_{k-1}\in(\nu_{k-2}-2m_{11}-1,\nu_{k-2}-2m_{11}]\\ \nu_{k-1}\in[2m_{11}+1-k,2m_{11}-k+2)}}^{\mu_{k-1}+\nu_{k-1}=\nu_{k-2}-k+1}\prod_{j=1}^{k-1}\frac{(\nu_{j-1}-j)!}{\mu_j!\nu_j!}(a_{11r_{j+1}}-a_{11r_j})^{\mu_j}\\
&\times\sum_{\rho=k+1}^{2m_{11}+1}B_{\rho 1}^{(r_k)}\ln\frac{|x_1-a_{11r_\rho}|}{|x_{10}-a_{11r_\rho}|}\Bigg)\Bigg]+b_{210}\sum_{\rho=1}^{2m_{11}+1}A_{\rho 1}^{(r_1)}\ln\frac{|x_1-a_{11r_\rho}|}{|x_{10}-a_{11r_\rho}|}\\
&+\sum_{\nu_0=1}^{2m_{11}}b_{21\nu_0}\sum_{\rho=2}^{2m_{11}+1}B_{\rho 1}^{(r_1)}\ln\frac{|x_1-a_{11r_\rho}|}{|x_{10}-a_{11r_\rho}|}
\end{aligned}
$$

$$
\begin{aligned}
&= \frac{a_{110}}{a_{220}}\Bigg\{ \frac{1}{2n_{11}+2}\big[(x_2 - a_{12u_1})^{2n_{11}+2} - (x_{j_20} - a_{12u_1})^{2n_{11}+2}\big] \\
&+ \sum\nolimits_{\alpha=1}^{2n_{11}-1} \frac{1}{2n_{11}-\alpha+2} b_{12\alpha}\big[(x_2 - a_{12u_1})^{2n_{11}-\alpha+2} - (x_{20} - a_{12u_1})^{2n_{11}-\alpha+2}\big] \Bigg\},
\end{aligned} \tag{3.10}
$$

where

$$
\begin{aligned}
b_{12\alpha} &= \sum_{l_{2n_{11}+1}, t_{2n_{11}+1}=0}^{l_{2n_{11}+1}+t_{2n_{11}+1}=1} \cdots \sum_{l_2, t_2=0}^{l_2+t_2=1} [\prod_{k=2}^{2n_{11}+1} (a_{12u_1} - a_{12u_k})^{l_k}] \delta_{t_2+\cdots+t_{2n_{11}+1}}^{2n_{11}+1-\alpha};\\
b_{120} &= 1.\\
b_{21\nu_0} &= \sum_{t_1, l_1=0}^{t_1+l_1=1} \cdots \sum_{t_{2m_{21}+1}, l_{2m_{21}+1}=0}^{t_{2m_{21}+1}+l_{2m_{21}+1}=1} \prod_{k=1}^{2m_{21}+1} (a_{11r_1} - a_{21\nu_k})^{t_k} \delta_{\nu_0}^{\Sigma_{r=1}^{2m_{21}+1} l_r};\\
&\lim_{\nu_{2m_{11}}\to 0} \frac{1}{\nu_{2m_{11}}}\big[(x_1 - a_{11r_{2m_{11}+1}})^{\nu_{2m_{11}}} - (x_{j_10} - a_{11r_{2m_{11}+1}})^{\nu_{2m_{11}}}\big]\\
&= \ln \frac{|x_1 - a_{11r_{2m_{11}+1}}|}{|x_{10} - a_{11r_{2m_{11}+1}}|};
\end{aligned} \tag{3.11}
$$

$$
\begin{aligned}
A_{\rho 1}^{(r_k)} &= \frac{1}{\prod_{\sigma=k,\sigma\neq\rho}^{2m_{11}+1} (a_{11r_\rho} - a_{11r_\sigma})}, \text{ for } \rho \in \{k, k+1, \ldots, 2m_{11}+1\},\\
B_{\rho 1}^{(r_k)} &= \frac{(a_{11r_\rho} - a_{11r_k})^{\beta_{k-1}-k}}{\prod_{\sigma=k+1,\sigma\neq\rho}^{2m_{11}+1} (a_{11r_\rho} - a_{11r_\sigma})}, \text{ for } \rho \in \{k+1, k+2, \ldots, 2m_{11}+1\};\\
&k \in \{1, 2, \ldots, 2m_{11}\}.
\end{aligned} \tag{3.12}
$$

For all cases, the $(2n_{11}+1) \times (2m_{21}+1)$-simple equilibriums plus $(2m_{11}+1) \times (2n_{11}+1)$-hyperbolic flows are based on the following inflection parabola-saddle bifurcations for $2m_{21}+1 \geq 2m_{11}+1$:

- $([(2m_{11}+1), (2n_{11}+1)]^{\text{th}}\text{UP}_-\text{:UP}_+,[2(m_{21}-m_{11})]^{\text{th}}{}_{\text{DI:II}}\text{US})$-inflection up-parabola upper-saddle bifurcation for $(a_{110} > 0, a_{220} > 0)$,
- $([(2m_{11}+1), (2n_{11}+1)]^{\text{th}}\text{DP}_-\text{:DP}_+,[2(m_{21}-m_{11})]^{\text{th}}{}_{\text{II:DI}}\text{LS})$-inflection down-parabola lower-saddle bifurcation for $(a_{110} < 0, a_{220} > 0)$,
- $([(2m_{11}+1), (2n_{11}+1)]^{\text{th}}\text{DP}_+\text{:DP}_-,[2(m_{21}-m_{11})]^{\text{th}}{}_{\text{II:DI}}\text{US})$-inflection down-parabola upper-saddle bifurcation for $(a_{110} > 0, a_{220} < 0)$,
- $([(2m_{11}+1), (2n_{11}+1)]^{\text{th}}\text{UP}_+\text{:UP}_-,[2(m_{21}-m_{11})]^{\text{th}}{}_{\text{DI:II}}\text{LS})$-inflection up-parabola lower-saddle bifurcation for $(a_{110} < 0, a_{220} < 0)$;

or based on the following concave-parabola-saddle bifurcations for $2m_{21}+1 < 2m_{11}+1$:

- $([(2m_{11}+1),(2n_{11}+1)]^{\text{th}}\text{UP}_{-}\text{:UP}_{+},[2(m_{11}-m_{21})]^{\text{th}}{}_{\text{DC:IC}}\text{US})$-concave up-parabola upper-saddle bifurcation for $(a_{110} > 0, a_{220} > 0)$,
- $([(2m_{11}+1),(2n_{21}+1)]^{\text{th}}\text{DP}_{-}\text{:DP}_{+},[2(m_{11}-m_{21})]^{\text{th}}{}_{\text{IC:DC}}\text{LS})$-concave down-parabola lower-saddle bifurcation for $(a_{110} > 0, a_{220} < 0)$,
- $([(2m_{11}+1),(2n_{11}+1)]^{\text{th}}\text{DP}_{+}\text{:DP}_{-},[2(m_{11}-m_{21})]^{\text{th}}{}_{\text{IC:DC}}\text{US})$-concave down-parabola upper-saddle bifurcation for $(a_{110} > 0, a_{220} < 0)$,
- $([(2m_{11}+1),(2n_{11}+1)]^{\text{th}}\text{UP}_{+}\text{:UP}_{-},[2(m_{11}-m_{21})]^{\text{th}}{}_{\text{DC:IC}}\text{LS})$-concave up-parabola lower-saddle bifurcation for $(a_{110} < 0, a_{220} < 0)$.

(A) Consider $2m_{11}+1 = 2m_{21}+1$ for illustrations, phase portraits for the 2-dimensional systems for the hybrid networks of hyperbolic and hyperbolic-secant flows with the simple equilibriums of the saddles, sinks and sources with $x_1^* = a_{111}, a_{211}, a_{112}, \cdots, a_{11(2m_{11}+1)}, a_{21(2m_{21}+1)}$ and $x_2^* = a_{121}, a_{122}, \cdots, a_{12(2n_{11}+1)}$ are presented in Fig. 3.5a–d for $(a_{110} > 0, a_{220} > 0)$, $(a_{110} < 0, a_{220} > 0)$, $(a_{110} > 0, a_{220} < 0)$ and $(a_{110} < 0, a_{220} < 0)$.

(B) Consider $2m_{11}+1 = 2m_{21}+1$ for illustrations. Phase portraits for the 2-dimensional systems for the hybrid networks of hyperbolic and hyperbolic-secant flows with the simple equilibriums of saddles and centers with $x_1^* = a_{211}, a_{111}, \cdots, a_{11(2m_{11})}, a_{21(2m_{21}+1)}, a_{11(2m_{11}+1)}$ and $x_2^* = a_{121}, a_{122}, \cdots, a_{12(2n_{11}+1)}$ are presented in Fig. 3.6a–d for $(a_{110} > 0, a_{220} > 0)$, $(a_{110} < 0, a_{220} > 0)$, $(a_{110} > 0, a_{220} < 0)$ and $(a_{110} < 0, a_{220} < 0)$.

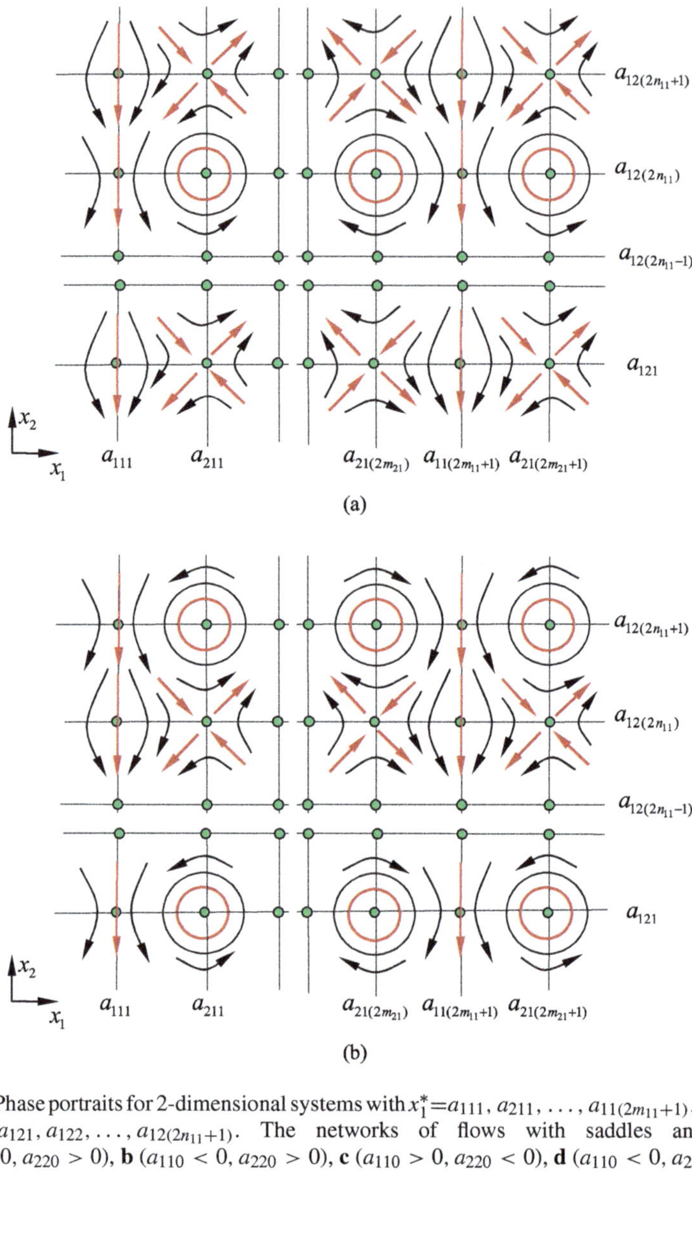

Fig. 3.5 Phase portraits for 2-dimensional systems with $x_1^*=a_{111}, a_{211}, \ldots, a_{11(2m_{11}+1)}, a_{121(2m_{21}+1)}$ and $x_2^*=a_{121}, a_{122}, \ldots, a_{12(2n_{11}+1)}$. The networks of flows with saddles and centers: **a** ($a_{110} > 0, a_{220} > 0$), **b** ($a_{110} < 0, a_{220} > 0$), **c** ($a_{110} > 0, a_{220} < 0$), **d** ($a_{110} < 0, a_{220} < 0$)

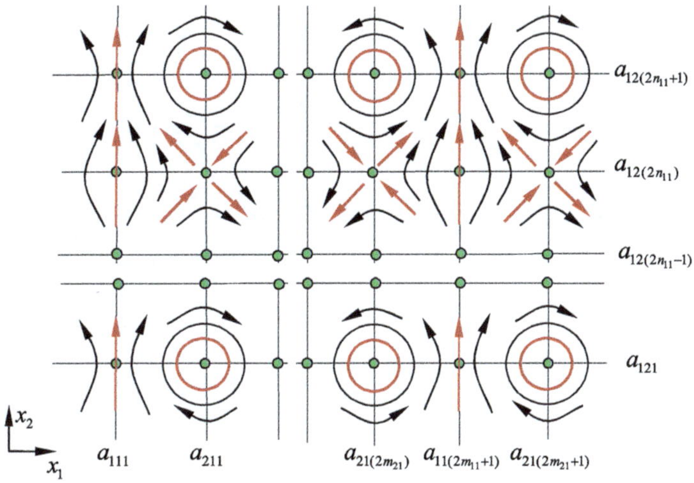

(c)

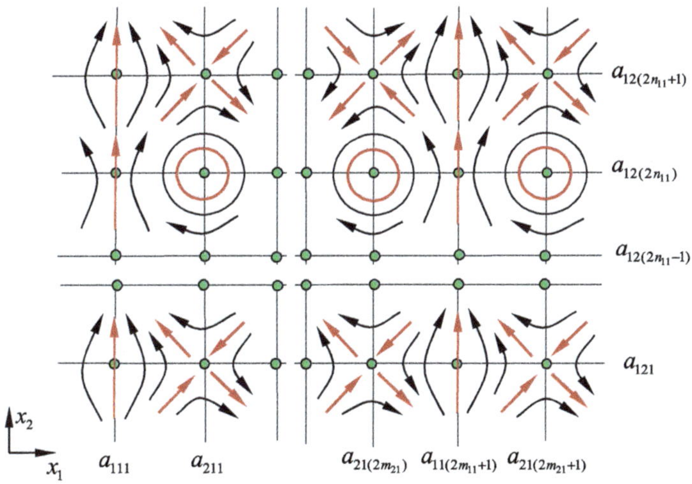

(d)

Fig. 3.5 (continued)

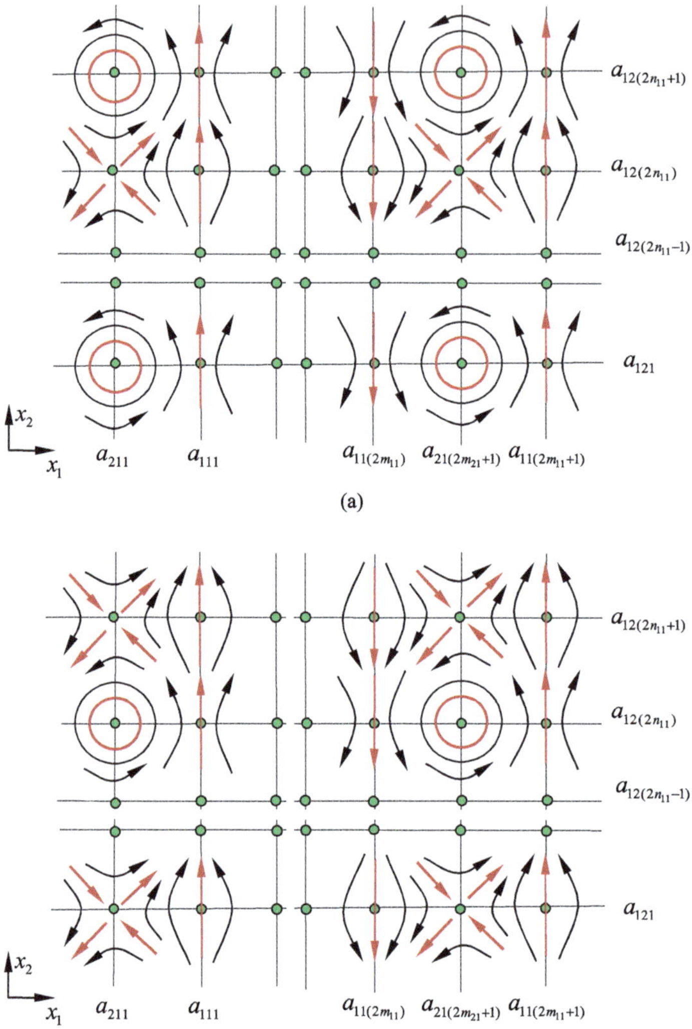

Fig. 3.6 Phase portraits for 2-dimensional systems with $x_1^*=a_{211}, a_{111}, \cdots, a_{21(2m_{21}+1)}, a_{11(2m_{11}+1)}$ and $x_2^*=a_{121}, a_{122}, \cdots, a_{12(2n_{11}+1)}$. The networks of flows with saddles and centers: **a** $(a_{110} > 0, a_{220} > 0)$, **b** $(a_{110} < 0, a_{220} > 0)$, **c** $(a_{110} > 0, a_{220} < 0)$, **d** $(a_{110} < 0, a_{220} < 0)$

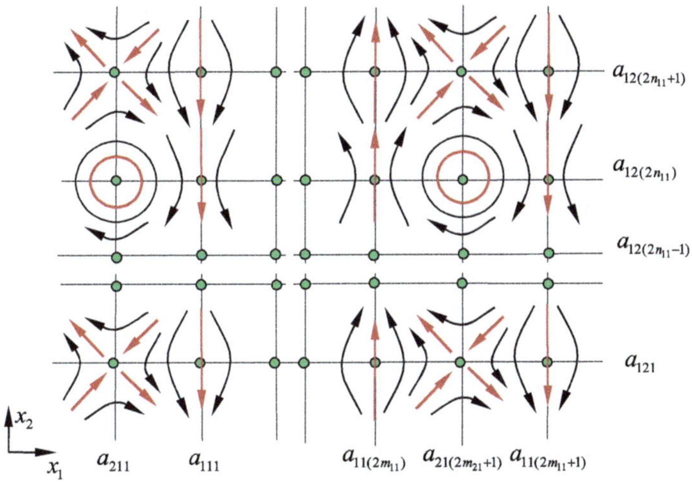

(c)

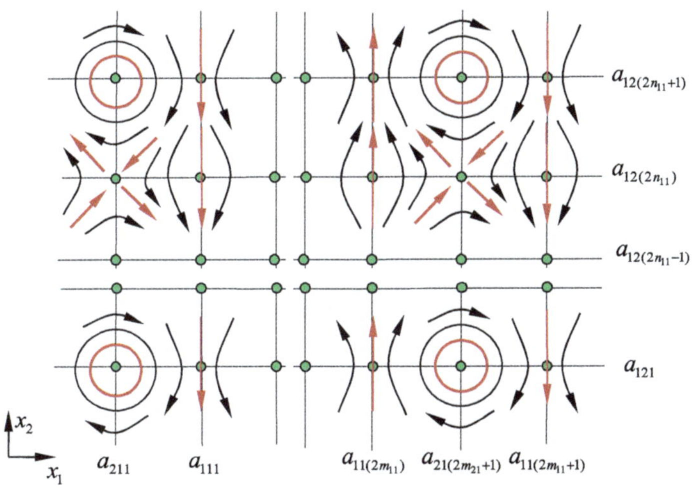

(d)

Fig. 3.6 (continued)

$$
\left\{\begin{array}{cccccc}
(a_{111},a_{12(2n_{11}+1)}) & (a_{12(2n_{11}+1)},a_{211}) & \cdots & (a_{12(2n_{11}+1)},a_{21(2m_{21})}) & (a_{11(2m_{11}+1)},a_{12(2n_{11}+1)}) & (a_{12(2n_{11}+1)},a_{21(2m_{21}+1)}) \\
(a_{111},a_{12(2n_{11})}) & (a_{12(2n_{11})},a_{211}) & \cdots & (a_{12(2n_{11})},a_{21(2m_{21})}) & (a_{11(2m_{11}+1)},a_{12(2n_{11})}) & (a_{12(2n_{11})},a_{21(2m_{21}+1)}) \\
\vdots & \vdots & \vdots & \vdots & \vdots & \vdots \\
(a_{111},a_{121}) & (a_{121},a_{211}) & \cdots & (a_{121},a_{21(2m_{21})}) & (a_{11(2m_{11}+1)},a_{121}) & (a_{121},a_{21(2m_{21}+1)})
\end{array}\right\}
$$

$$
= \left\{\begin{array}{cccccc}
\underbrace{(\mathrm{UP{:}DP,nF})}_{\text{hyperbolic-secant flow }(-)} & \underbrace{(\mathrm{UP}_{+},\mathrm{UP}_{+})}_{\text{positive saddle}} & \cdots & \underbrace{(\mathrm{UP}_{-},\mathrm{UP}_{-})}_{\text{negative saddle}} & \underbrace{(\mathrm{UP{:}DP,nF})}_{\text{hyperbolic-secant flow }(-)} & \underbrace{(\mathrm{UP}_{+},\mathrm{UP}_{+})}_{\text{positive saddle}} \\
\underbrace{(\mathrm{DP{:}UP,nF})}_{\text{hyperbolic flow }(-)} & \underbrace{(\mathrm{DP}_{+},\mathrm{DP}_{-})}_{\text{CCW center}} & \cdots & \underbrace{(\mathrm{DP}_{-},\mathrm{DP}_{+})}_{\text{CW center}} & \underbrace{(\mathrm{DP{:}UP,nF})}_{\text{hyperbolic flow }(-)} & \underbrace{(\mathrm{DP}_{+},\mathrm{DP}_{-})}_{\text{CCW center}} \\
\vdots & \vdots & \vdots & \vdots & \vdots & \vdots \\
\underbrace{(\mathrm{UP{:}DP,nF})}_{\text{hyperbolic-secant flow }(-)} & \underbrace{(\mathrm{UP}_{+},\mathrm{UP}_{+})}_{\text{positive saddle}} & \cdots & \underbrace{(\mathrm{UP}_{-},\mathrm{UP}_{-})}_{\text{negative saddle}} & \underbrace{(\mathrm{UP{:}DP,nF})}_{\text{hyperbolic-secant flow }(-)} & \underbrace{(\mathrm{UP}_{+},\mathrm{UP}_{+})}_{\text{positive saddle}}
\end{array}\right\}_{(2n_{11}+1)\times(2m_{11}+2m_{21}+2)}
$$

for $a_{110} > 0, a_{220} > 0$; (3.13)

$$
\left\{\begin{array}{cccccc}
(a_{111},a_{12(2n_{11}+1)}) & (a_{12(2n_{11}+1)},a_{211}) & \cdots & (a_{12(2n_{11}+1)},a_{21(2m_{21})}) & (a_{11(2m_{11}+1)},a_{12(2n_{11}+1)}) & (a_{12(2n_{11}+1)},a_{21(2m_{21}+1)}) \\
(a_{111},a_{12(2n_{11})}) & (a_{12(2n_{11})},a_{211}) & \cdots & (a_{12(2n_{11})},a_{21(2m_{21})}) & (a_{11(2m_{11}+1)},a_{12(2n_{11})}) & (a_{12(2n_{11})},a_{21(2m_{21}+1)}) \\
\vdots & \vdots & \vdots & \vdots & \vdots & \vdots \\
(a_{111},a_{121}) & (a_{121},a_{211}) & \cdots & (a_{121},a_{21(2m_{21})}) & (a_{11(2m_{11}+1)},a_{121}) & (a_{121},a_{21(2m_{21}+1)})
\end{array}\right\}
$$

$$
=\left\{\begin{array}{cccccc}
\underbrace{(\text{DP:UP,nF})}_{\text{hyperbolic flow }(-)} & \underbrace{(\text{DP}_+,\text{DP}_-)}_{\text{CCW center}} & \cdots & \underbrace{(\text{DP}_-,\text{DP}_+)}_{\text{CW center}} & \underbrace{(\text{DP:UP,nF})}_{\text{hyperbolic flow }(-)} & \underbrace{(\text{DP}_+,\text{DP}_-)}_{\text{CCW center}} \\
\underbrace{(\text{UP:DP,nF})}_{\text{hyperbolic-secant flow }(-)} & \underbrace{(\text{UP}_+,\text{UP}_+)}_{\text{positive saddle}} & \cdots & \underbrace{(\text{UP}_-,\text{UP}_-)}_{\text{negative saddle}} & \underbrace{(\text{UP:DP,nF})}_{\text{hyperbolic-secant flow }(-)} & \underbrace{(\text{UP}_+,\text{UP}_+)}_{\text{positive saddle}} \\
\vdots & \vdots & \vdots & \vdots & \vdots & \vdots \\
\underbrace{(\text{DP:UP,nF})}_{\text{hyperbolic flow }(-)} & \underbrace{(\text{DP}_+,\text{DP}_-)}_{\text{CCW center}} & \cdots & \underbrace{(\text{DP}_-,\text{DP}_+)}_{\text{CW center}} & \underbrace{(\text{DP:UP,nF})}_{\text{hyperbolic flow }(-)} & \underbrace{(\text{DP}_+,\text{DP}_-)}_{\text{CCW center}}
\end{array}\right\}_{(2n_{11}+1)\times(2m_{11}+2m_{21}+2)}
$$

for $a_{110} < 0, a_{220} > 0$; (3.14)

$$
\left\{\begin{array}{cccccc}
(a_{111},a_{12(2n_{11}+1)}) & (a_{12(2n_{11}+1)},a_{211}) & \cdots & (a_{12(2n_{11}+1)},a_{21(2m_{21})}) & (a_{11(2m_{11}+1)},a_{12(2n_{11}+1)}) & (a_{12(2n_{11}+1)},a_{21(2m_{21}+1)}) \\
(a_{111},a_{12(2n_{11})}) & (a_{12(2n_{11})},a_{211}) & \cdots & (a_{12(2n_{11})},a_{21(2m_{21})}) & (a_{11(2m_{11}+1)},a_{12(2n_{11})}) & (a_{12(2n_{11})},a_{21(2m_{21}+1)}) \\
\vdots & \vdots & \vdots & \vdots & \vdots & \vdots \\
(a_{111},a_{121}) & (a_{121},a_{211}) & \cdots & (a_{121},a_{21(2m_{21})}) & (a_{11(2m_{11}+1)},a_{121}) & (a_{121},a_{21(2m_{21}+1)})
\end{array}\right\}
$$

$$
=\left\{\begin{array}{cccccc}
\underbrace{(\mathrm{DP:UP,pF})}_{\text{hyperbolic flow }(+)} & \underbrace{(\mathrm{DP}_{-},\mathrm{DP}_{+})}_{\text{CW center}} & \cdots & \underbrace{(\mathrm{DP}_{+},\mathrm{DP}_{-})}_{\text{CCW center}} & \underbrace{(\mathrm{DP:UP,pF})}_{\text{hyperbolic flow }(+)} & \underbrace{(\mathrm{DP}_{-},\mathrm{DP}_{+})}_{\text{CW center}} \\
\underbrace{(\mathrm{UP:DP,pF})}_{\text{hyperbolic-secant flow }(+)} & \underbrace{(\mathrm{UP}_{-},\mathrm{UP}_{-})}_{\text{negative saddle}} & \cdots & \underbrace{(\mathrm{UP}_{+},\mathrm{UP}_{+})}_{\text{positive saddle}} & \underbrace{(\mathrm{UP:DP,pF})}_{\text{hyperbolic-secant flow }(+)} & \underbrace{(\mathrm{UP}_{-},\mathrm{UP}_{-})}_{\text{negative saddle}} \\
\vdots & \vdots & \vdots & \vdots & \vdots & \vdots \\
\underbrace{(\mathrm{DP:UP,pF})}_{\text{hyperbolic flow }(+)} & \underbrace{(\mathrm{DP}_{-},\mathrm{DP}_{+})}_{\text{CW center}} & \cdots & \underbrace{(\mathrm{DP}_{+},\mathrm{DP}_{-})}_{\text{CCW center}} & \underbrace{(\mathrm{DP:UP,pF})}_{\text{hyperbolic flow }(+)} & \underbrace{(\mathrm{DP}_{-},\mathrm{DP}_{+})}_{\text{CW center}}
\end{array}\right\}_{(2n_{11}+1)\times(2m_{11}+2m_{21}+2)}
$$

for $a_{110} < 0, a_{220} > 0;$ (3.15)

$$
\begin{Bmatrix}
(a_{111},a_{12(2n_{11}+1)}) & (a_{12(2n_{11}+1)},a_{211}) & \cdots & (a_{12(2n_{11}+1)},a_{21(2m_{21})}) & (a_{11(2m_{11}+1)},a_{12(2n_{11}+1)}) & (a_{12(2n_{11}+1)},a_{21(2m_{21}+1)}) \\
(a_{111},a_{12(2n_{11})}) & (a_{12(2n_{11})},a_{211}) & \cdots & (a_{12(2n_{11})},a_{21(2m_{21})}) & (a_{11(2m_{11}+1)},a_{12(2n_{11})}) & (a_{12(2n_{11})},a_{21(2m_{21}+1)}) \\
\vdots & \vdots & \vdots & \vdots & \vdots & \vdots \\
(a_{111},a_{121}) & (a_{121},a_{211}) & \cdots & (a_{121},a_{21(2m_{21})}) & (a_{11(2m_{11}+1)},a_{121}) & (a_{121},a_{21(2m_{21}+1)})
\end{Bmatrix}
$$
$$
= \begin{Bmatrix}
\underbrace{(\mathrm{UP{:}DP,pF})}_{\text{hyperbolic-secant flow }(+)} & \underbrace{(\mathrm{UP}_-,\mathrm{UP}_-)}_{\text{negative saddle}} & \cdots & \underbrace{(\mathrm{UP}_+,\mathrm{UP}_+)}_{\text{positive saddle}} & \underbrace{(\mathrm{UP{:}DP,pF})}_{\text{hyperbolic-secant flow }(+)} & \underbrace{(\mathrm{UP}_-,\mathrm{UP}_-)}_{\text{negative saddle}} \\
\underbrace{(\mathrm{DP{:}UP,pF})}_{\text{hyperbolic flow }(+)} & \underbrace{(\mathrm{DP}_-,\mathrm{DP}_+)}_{\text{CW center}} & \cdots & \underbrace{(\mathrm{DP}_+,\mathrm{DP}_-)}_{\text{CCW center}} & \underbrace{(\mathrm{DP{:}UP,pF})}_{\text{hyperbolic flow }(+)} & \underbrace{(\mathrm{DP}_-,\mathrm{DP}_+)}_{\text{CW center}} \\
\vdots & \vdots & \vdots & \vdots & \vdots & \vdots \\
\underbrace{(\mathrm{UP{:}DP,pF})}_{\text{hyperbolic-secant flow }(+)} & \underbrace{(\mathrm{UP}_-,\mathrm{UP}_-)}_{\text{negative saddle}} & \cdots & \underbrace{(\mathrm{UP}_+,\mathrm{UP}_+)}_{\text{positive saddle}} & \underbrace{(\mathrm{UP{:}DP,pF})}_{\text{hyperbolic-secant flow }(+)} & \underbrace{(\mathrm{UP}_-,\mathrm{UP}_-)}_{\text{negative saddle}}
\end{Bmatrix}_{(2n_{11}+1)\times(2m_{11}+2m_{21}+2)}
$$

for $a_{110} < 0, a_{220} < 0$; (3.16)

$$
\begin{Bmatrix}
(a_{12(2n_{11}+1)},a_{211}) & (a_{111},a_{12(2n_{11}+1)}) & \cdots & (a_{11(2m_{11})},a_{12(2n_{11}+1)},) & (a_{12(2n_{11}+1)},a_{21(2m_{21}+1)}) & (a_{11(2m_{11}+1)},a_{12(2n_{11}+1)}) \\
(a_{12(2n_{11})},a_{211}) & (a_{111},a_{12(2n_{11})}) & \cdots & (a_{11(2m_{11})},a_{12(2n_{11})}) & (a_{12(2n_{11})},a_{21(2m_{21}+1)}) & (a_{11(2m_{11}+1)},a_{12(2n_{11})}) \\
\vdots & \vdots & \vdots & \vdots & \vdots & \vdots \\
(a_{121},a_{211}) & (a_{111},a_{121}) & \cdots & (a_{11(2m_{11})},a_{121}) & (a_{121},a_{21(2m_{21}+1)}) & (a_{11(2m_{11}+1)},a_{121})
\end{Bmatrix}
$$

$$
= \begin{Bmatrix}
\underbrace{(\text{DP}_+,\text{DP}_-)}_{\text{CCW center}} & \underbrace{(\text{DP:UP,pF})}_{\text{hyperbolic flow }(+)} & \cdots & \underbrace{(\text{DP:UP,nF})}_{\text{hyperbolic flow }(-)} & \underbrace{(\text{DP}_+,\text{DP}_-)}_{\text{CCW center}} & \underbrace{(\text{DP:UP,pF})}_{\text{hyperbolic flow }(+)} \\
\underbrace{(\text{UP}_+,\text{UP}_+)}_{\text{positive saddle}} & \underbrace{(\text{UP:DP,pF})}_{\text{hyperbolic-secant flow }(+)} & \cdots & \underbrace{(\text{UP:DP,nF})}_{\text{hyperbolic-secant flow }(-)} & \underbrace{(\text{UP}_+,\text{UP}_+)}_{\text{positive saddle}} & \underbrace{(\text{UP:DP,pF})}_{\text{hyperbolic-secant flow }(+)} \\
\vdots & \vdots & \vdots & \vdots & \vdots & \vdots \\
\underbrace{(\text{DP}_+,\text{DP}_-)}_{\text{CCW center}} & \underbrace{(\text{DP:UP,pF})}_{\text{hyperbolic flow }(+)} & \cdots & \underbrace{(\text{DP:UP,nF})}_{\text{hyperbolic flow }(-)} & \underbrace{(\text{DP}_+,\text{DP}_-)}_{\text{CCW center}} & \underbrace{(\text{DP:UP,pF})}_{\text{hyperbolic flow }(+)}
\end{Bmatrix}_{(2n_{11}+1)\times(2m_{11}+2m_{21}+2)}
\tag{3.17}
$$

for $a_{110} > 0, a_{220} > 0$;

$$\left\{\begin{array}{cccccc} (a_{12(2n_{11}+1)},a_{211}) & (a_{111},a_{12(2n_{11}+1)}) & \cdots & (a_{11(2m_{11})},a_{12(2n_{11}+1)},) & (a_{12(2n_{11}+1)},a_{21(2m_{21}+1)}) & (a_{11(2m_{11}+1)},a_{12(2n_{11}+1)}) \\ (a_{12(2n_{11})},a_{211}) & (a_{111},a_{12(2n_{11})}) & \cdots & (a_{11(2m_{11})},a_{12(2n_{11})}) & (a_{12(2n_{11})},a_{21(2m_{21}+1)}) & (a_{11(2m_{11}+1)},a_{12(2n_{11})}) \\ \vdots & \vdots & \vdots & \vdots & \vdots & \vdots \\ (a_{121},a_{211}) & (a_{111},a_{121}) & \cdots & (a_{11(2m_{11})},a_{121}) & (a_{121},a_{21(2m_{21}+1)}) & (a_{11(2m_{11}+1)},a_{121}) \end{array}\right\}$$
$$= \left\{\begin{array}{cccccc} \underbrace{(\mathrm{UP}_+,\mathrm{UP}_+)}_{\text{positive saddle}} & \underbrace{(\mathrm{UP:DP,pF})}_{\text{hyperbolic-secant flow }(+)} & \cdots & \underbrace{(\mathrm{UP:DP,nF})}_{\text{hyperbolic-secant flow }(-)} & \underbrace{(\mathrm{UP}_+,\mathrm{UP}_+)}_{\text{positive saddle}} & \underbrace{(\mathrm{UP:DP,pF})}_{\text{hyperbolic-secant flow }(+)} \\ \underbrace{(\mathrm{DP}_+,\mathrm{DP}_-)}_{\text{CCW center}} & \underbrace{(\mathrm{DP:UP,pF})}_{\text{hyperbolic flow }(+)} & \cdots & \underbrace{(\mathrm{DP:UP,nF})}_{\text{hyperbolic flow }(-)} & \underbrace{(\mathrm{DP}_+,\mathrm{DP}_-)}_{\text{CCW center}} & \underbrace{(\mathrm{DP:UP,pF})}_{\text{hyperbolic flow }(+)} \\ \vdots & \vdots & \vdots & \vdots & \vdots & \vdots \\ \underbrace{(\mathrm{UP}_+,\mathrm{UP}_+)}_{\text{positive saddle}} & \underbrace{(\mathrm{UP:DP,pF})}_{\text{hyperbolic-secant flow }(+)} & \cdots & \underbrace{(\mathrm{UP:DP,nF})}_{\text{hyperbolic-secant flow }(-)} & \underbrace{(\mathrm{UP}_+,\mathrm{UP}_+)}_{\text{positive saddle}} & \underbrace{(\mathrm{DP:UP,pF})}_{\text{hyperbolic flow }(+)} \end{array}\right\}_{(2n_{11}+1)\times(2m_{11}+2m_{21}+2)}$$

for $a_{110} < 0, a_{220} > 0;$ (3.18)

$$
\left\{\begin{array}{cccccc}
(a_{12(2n_{11}+1)},a_{211}) & (a_{111},a_{12(2n_{11}+1)}) & \cdots & (a_{11(2m_{11})},a_{12(2n_{11}+1)},) & (a_{12(2n_{11}+1)},a_{21(2m_{21}+1)}) & (a_{11(2m_{11}+1)},a_{12(2n_{11}+1)}) \\
(a_{12(2n_{11})},a_{211}) & (a_{111},a_{12(2n_{11})}) & \cdots & (a_{11(2m_{11})},a_{12(2n_{11})}) & (a_{12(2n_{11})},a_{21(2m_{21}+1)}) & (a_{11(2m_{11}+1)},a_{12(2n_{11})}) \\
\vdots & \vdots & \vdots & \vdots & \vdots & \vdots \\
(a_{121},a_{211}) & (a_{111},a_{121}) & \cdots & (a_{11(2m_{11})},a_{121}) & (a_{121},a_{21(2m_{21}+1)}) & (a_{11(2m_{11}+1)},a_{121})
\end{array}\right\}
$$

$$
=\left\{\begin{array}{cccccc}
\underbrace{(\mathrm{UP}_-,\mathrm{UP}_-)}_{\text{negative saddle}} & \underbrace{(\mathrm{UP{:}DP,nF})}_{\text{hyperbolic-secant flow }(-)} & \cdots & \underbrace{(\mathrm{UP{:}DP,pF})}_{\text{hyperbolic-secant flow }(+)} & \underbrace{(\mathrm{UP}_-,\mathrm{UP}_-)}_{\text{negative saddle}} & \underbrace{(\mathrm{UP{:}DP,nF})}_{\text{hyperbolic-secant flow }(-)} \\
\underbrace{(\mathrm{DP}_-,\mathrm{DP}_+)}_{\text{CW center}} & \underbrace{(\mathrm{DP{:}UP,nF})}_{\text{hyperbolic flow }(-)} & \cdots & \underbrace{(\mathrm{DP{:}UP,pF})}_{\text{hyperbolic flow }(+)} & \underbrace{(\mathrm{DP}_-,\mathrm{DP}_+)}_{\text{CW center}} & \underbrace{(\mathrm{DP{:}UP,nF})}_{\text{hyperbolic flow }(-)} \\
\vdots & \vdots & \vdots & \vdots & \vdots & \vdots \\
\underbrace{(\mathrm{UP}_-,\mathrm{UP}_-)}_{\text{negative saddle}} & \underbrace{(\mathrm{UP{:}DP,nF})}_{\text{hyperbolic-secant flow }(-)} & \cdots & \underbrace{(\mathrm{UP{:}DP,pF})}_{\text{hyperbolic-secant flow }(+)} & \underbrace{(\mathrm{UP}_-,\mathrm{UP}_-)}_{\text{negative saddle}} & \underbrace{(\mathrm{UP{:}DP,nF})}_{\text{hyperbolic-secant flow }(-)}
\end{array}\right\}_{(2n_{11}+1)\times(2m_{11}+2m_{21}+2)}
$$

for $a_{110} > 0$, $a_{220} < 0$; (3.19)

$$
\left\{\begin{array}{cccccc}
(a_{12(2n_{11}+1)},a_{211}) & (a_{111},a_{12(2n_{11}+1)}) & \cdots & (a_{11(2m_{11})},a_{12(2n_{11}+1)},) & (a_{12(2n_{11}+1)},a_{21(2m_{21}+1)}) & (a_{11(2m_{11}+1)},a_{12(2n_{11}+1)}) \\
(a_{12(2n_{11})},a_{211}) & (a_{111},a_{12(2n_{11})}) & \cdots & (a_{11(2m_{11})},a_{12(2n_{11})}) & (a_{12(2n_{11})},a_{21(2m_{21}+1)}) & (a_{11(2m_{11}+1)},a_{12(2n_{11})}) \\
\vdots & \vdots & \vdots & \vdots & \vdots & \vdots \\
(a_{121},a_{211}) & (a_{111},a_{121}) & \cdots & (a_{11(2m_{11})},a_{121}) & (a_{121},a_{21(2m_{21}+1)}) & (a_{11(2m_{11}+1)},a_{121})
\end{array}\right\}
$$

$$
=\left\{\begin{array}{cccccc}
\underbrace{(\mathrm{DP}_-,\mathrm{DP}_+)}_{\text{CW center}} & \underbrace{(\mathrm{DP{:}UP,nF})}_{\text{hyperbolic flow }(-)} & \cdots & \underbrace{(\mathrm{DP{:}UP,pF})}_{\text{hyperbolic flow }(+)} & \underbrace{(\mathrm{DP}_-,\mathrm{DP}_+)}_{\text{CW center}} & \underbrace{(\mathrm{DP{:}UP,nF})}_{\text{hyperbolic flow }(-)} \\
\underbrace{(\mathrm{UP}_-,\mathrm{UP}_-)}_{\text{negative saddle}} & \underbrace{(\mathrm{UP{:}DP,nF})}_{\text{hyperbolic-secant flow }(-)} & \cdots & \underbrace{(\mathrm{UP{:}DP,pF})}_{\text{hyperbolic-secant flow }(+)} & \underbrace{(\mathrm{UP}_-,\mathrm{UP}_-)}_{\text{negative saddle}} & \underbrace{(\mathrm{UP{:}DP,nF})}_{\text{hyperbolic-secant flow }(-)} \\
\vdots & \vdots & \vdots & \vdots & \vdots & \vdots \\
\underbrace{(\mathrm{DP}_-,\mathrm{DP}_+)}_{\text{CW center}} & \underbrace{(\mathrm{DP{:}UP,pF})}_{\text{hyperbolic flow }(-)} & \cdots & \underbrace{(\mathrm{DP{:}UP,pF})}_{\text{hyperbolic flow }(+)} & \underbrace{(\mathrm{DP}_-,\mathrm{DP}_+)}_{\text{CW center}} & \underbrace{(\mathrm{DP{:}UP,pF})}_{\text{hyperbolic flow }(-)}
\end{array}\right\}_{(2n_{11}+1)\times(2m_{11}+2m_{21}+2)}
$$

for $a_{110} < 0$, $a_{220} < 0$; (3.20)

3.2 $[(2m_{11}), (2n_{11}+1)]$ and $((2n_{11}+1), (2m_{21}+1))$-Singular Systems

In this section, there are four global bifurcations of the 1-dimensional flows through singular hyperbolic-to-hyperbolic-secant flows, and four global bifurcations of equilibriums, through singular saddles and centers bifurcations. The corresponding global bifurcations generate networks of hyperbolic and hyperbolic-secant flows with saddles and centers. The infinite-equilibriums are the appearing and switching bifurcations for the singular flows and singular equilibriums and for hybrid networks of hyperbolic flows, saddles and centers.

3.2.1 $[(2m_{11}), (2n_{11}+1)]$-*Flows and* $((2n_{11}+1), (2m_{21}+1))$-*Equilibriums*

Consider a polynomial dynamical system for $a_{111} \neq a_{211}$ as

$$\begin{aligned} \dot{x}_2 &= a_{220}(x_1 - a_{211})^{2n_{21}+1}, \\ \dot{x}_1 &= a_{110}(x_1 - a_{111})^{2m_{11}}(x_2 - a_{121})^{2n_{11}+1}. \end{aligned} \tag{3.21}$$

The first integral manifold is

$$\begin{aligned} &\sum_{\alpha,\beta=0}^{\alpha+\beta=2m_{21}+1} \frac{(2m_{21}+1)!}{\alpha!\beta!}(a_{111}-a_{211})^{\alpha} \\ &\quad \times \frac{1}{\beta-2m_{11}+1}\left[(x_1-a_{111})^{\beta-2m_{11}+1} - (x_{10}-a_{111})^{\beta-2m_{11}+1}\right] \\ &= \frac{a_{j_1j_10}}{a_{j_2j_20}}\frac{1}{2n_{11}+2}\left[(x_2-a_{121})^{2n_{11}+2} - (x_{20}-a_{121})^{2n_{11}+2}\right], \\ &\quad \text{for } \beta - 2m_{11} \neq 0; \\ &\frac{1}{\beta-2m_{11}+1}\left[(x_1-a_{111})^{\beta-2m_{11}+1} - (x_{10}-a_{111})^{\beta-2m_{11}+1}\right] = \ln\frac{|x_1-a_{111}|}{|x_{10}-a_{111}|}, \\ &\text{for } \beta - 2m_{11} + 1 = 0. \end{aligned} \tag{3.22}$$

For $(x_1^*,x_2^*)=(a_{111},a_{121})$ with $\dot{x}_1=0$, the $[(2m_{11}),(2n_{11}+1)]$-singular flows exist with $\dot{x}_2=a_{220}(a_{111}-a_{211})^{2m_{21}+1}$. For $(x_1^*,x_2^*)=(a_{111},a_{121})$, the 1-dimensional flow direction is determined by $\dot{x}_2\neq 0$. For $(x_2^*,x_1^*)=(a_{121},a_{211})$ with $\dot{x}_1=\dot{x}_2=0$, the $((2n_{11}+1),(2m_{21}+1))$-singular equilibriums exist.

(A) For $a_{111}>a_{211}$,

$$\left\{\begin{array}{l}(a_{111},a_{121})\\(a_{121},a_{211})\end{array}\right\}=\left\{\begin{array}{c}\underbrace{([(2m_{11}),(2n_{11}+1)]\text{UP:UP,pF})}_{[(2m_{11}),(2n_{11}+1)]\text{-hyperbolic-secant-to-hyperbolic flow }(+)}\\\underbrace{((2n_{11}+1)\text{UP}_+,(2m_{21}+1)\text{UP}_+)}_{[(2n_{11}+1),(2m_{21}+1)]\text{-postive saddle}}\end{array}\right\},$$

for $a_{110}>0,a_{220}>0$;

$$\left\{\begin{array}{l}(a_{111},a_{121})\\(a_{121},a_{211})\end{array}\right\}=\left\{\begin{array}{c}\underbrace{([(2m_{11}),(2n_{11}+1)]\text{DP:DP,pF})}_{[(2m_{11}),(2n_{11}+1)]\text{-hyperbolic-to-hyperbolic-secant flow }(+)}\\\underbrace{((2n_{11}+1)\text{DP}_+,(2m_{21}+1)\text{DP}_-)}_{[(2n_{11}+1),(2m_{21}+1)]\text{-CCW center}}\end{array}\right\},$$

for $a_{110}<0,a_{220}>0$;

$$\left\{\begin{array}{l}(a_{111},a_{121})\\(a_{121},a_{211})\end{array}\right\}=\left\{\begin{array}{c}\underbrace{([(2m_{11}),(2n_{11}+1)]\text{DP:DP,nF})}_{[(2m_{11}),(2n_{11}+1)]\text{-hyperbolic-to-hyperbolic-secant flow }(-)}\\\underbrace{((2n_{11}+1)\text{DP}_-,(2m_{21}+1)\text{DP}_+)}_{[(2n_{11}+1),(2m_{21}+1)]\text{-CW center}}\end{array}\right\},$$

for $a_{110}>0,a_{220}<0$;

$$\left\{\begin{array}{l}(a_{111},a_{121})\\(a_{121},a_{211})\end{array}\right\}=\left\{\begin{array}{c}\underbrace{([(2m_{11}),(2n_{11}+1)]\text{UP:UP,nF})}_{[(2m_{11}),(2n_{11}+1)]\text{-hyperbolic-secant-to-hyperbolic flow }(-)}\\\underbrace{((2n_{11}+1)\text{UP}_-,(2m_{21}+1)\text{UP}_-)}_{[(2n_{11}+1),(2m_{21}+1)]\text{-negative saddle}}\end{array}\right\},$$

for $a_{110}<0,a_{220}<0$. (3.23)

(B) For $a_{211}<a_{111}$,

$$\left\{\begin{array}{l}(a_{121},a_{211})\\(a_{111},a_{121})\end{array}\right\}=\left\{\begin{array}{c}\underbrace{((2n_{11}+1)\text{UP}_+,(2m_{21}+1)\text{UP}_+)}_{[(2n_{11}+1),(2m_{21}+1)]\text{-positive saddle}}\\\underbrace{([(2m_{11}),(2n_{11}+1)]\text{DP:DP,nF})}_{[(2m_{11}),(2n_{11}+1)]\text{-hyperbolic-to-hyperbolic-secant flow }(-)}\end{array}\right\},$$

for $a_{110}>0,a_{220}>0$;

$$\left\{\begin{array}{l}(a_{121},a_{211})\\(a_{111},a_{121})\end{array}\right\}=\left\{\begin{array}{c}\underbrace{((2n_{11}+1)\text{DP}_+,(2m_{21}+1)\text{DP}_-)}_{[(2n_{11}+1),(2m_{21}+1)]\text{-CCW center}}\\\underbrace{([(2m_{11}),(2n_{11}+1)]\text{UP:UP,nF})}_{[(2m_{11}),(2n_{11}+1)]\text{-hyperbolic-secant-to-hyperbolic flow }(-)}\end{array}\right\},$$

for $a_{110} < 0, a_{220} > 0$;

$$\left\{\begin{array}{c}(a_{121}, a_{211}) \\ (a_{111}, a_{121})\end{array}\right\} = \left\{\begin{array}{c}\underbrace{((2n_{11}+1)\text{DP}_{-},(2m_{21}+1)\text{DP}_{+})}_{[(2n_{11}+1),(2m_{21}+1)]\text{-CW center}} \\ \underbrace{([(2m_{11}),(2n_{11}+1)]\text{UP:UP,pF})}_{[(2m_{11}),(2n_{11}+1)]\text{-hyperbolic-secnt-to-hyperbolic flow } (+)}\end{array}\right\},$$

for $a_{110} > 0, a_{220} < 0$;

$$\left\{\begin{array}{c}(a_{121}, a_{211}) \\ (a_{111}, a_{121})\end{array}\right\} = \left\{\begin{array}{c}\underbrace{((2n_{11}+1)\text{UP}_{-},(2m_{21}+1)\text{UP}_{-})}_{[(2n_{11}+1),(2m_{21}+1)]\text{-negative saddle}} \\ \underbrace{([(2m_{11}),(2n_{11}+1)]\text{DP:DP,pF})}_{[(2m_{11}),(2n_{11}+1)]\text{-hyperbolic-to-hyperbolic-secant flow } (+)}\end{array}\right\},$$

for $a_{110} < 0, a_{220} < 0$. (3.24)

The singular hyperbolic-to-hyperbolic-secant flows with singular saddles and centers are illustrated in Figs. 3.7 and 3.8. Through such singular flows, the corresponding global bifurcations generate from a hyperbolic (hyperbolic-secant) flow to a network of singular hyperbolic and hyperbolic-secant flows. Through such singular equilibriums, the corresponding global bifurcations generate from one of saddle and center to a network of singular saddles and centers.

- The ($[(2m_{11}), (2n_{11} + 1)]$UP:UP,pF) is for a positive singular hyperbolic-secant-to-hyperbolic flow.
- The ($[(2m_{11}), (2n_{11} + 1)]$DP:DP,pF) is for a positive singular hyperbolic-to-hyperbolic-secant flow.
- The ($[(2m_{11}), (2n_{11} + 1)]$DP:DP,nF) is for a negative singular hyperbolic-to-hyperbolic-secant flow.
- The ($[(2m_{11}), (2n_{11} + 1)]$UP:UP,nF) is for a negative singular hyperbolic-secant-to-hyperbolic flow.

UP:UP is for a hyperbolic-secant-to-hyperbolic flow, and UP:DP is for a hyperbolic-to-hyperbolic-secant flow.

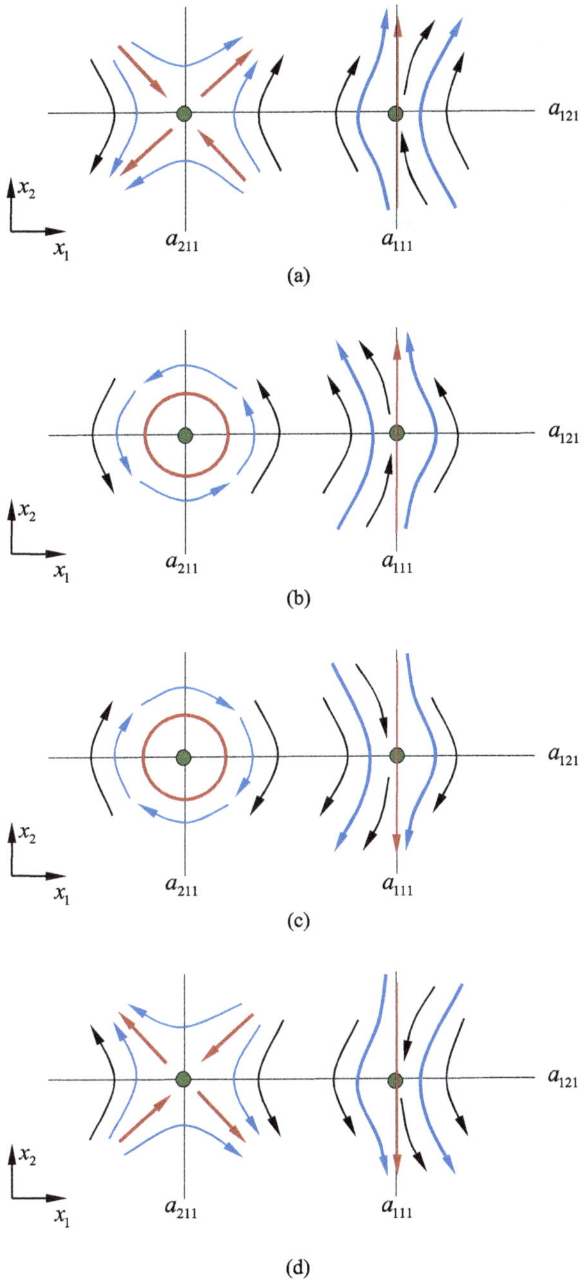

Fig. 3.7 Phase portraits ($a_{111} > a_{211}$) of the ($[(2m_{11}), (2n_{11}+1)]$, xF)-hyperbolic and hyperbolic-secant singular flows at $(x_1^*, x_2^*) = (a_{111}, a_{121})$ and $((2n_{11}+1), (2m_{21}+1))$-positive and negative saddles, CCW and CW centers at $(x_2^*, x_1^*) = (a_{121}, a_{211})$ for 2-dimensional polynomial systems: **a** ($a_{110} > 0, a_{220} > 0$), **b** ($a_{110} < 0, a_{220} > 0$), **c** ($a_{110} > 0, a_{220} < 0$), **d** ($a_{110} < 0, a_{220} < 0$)

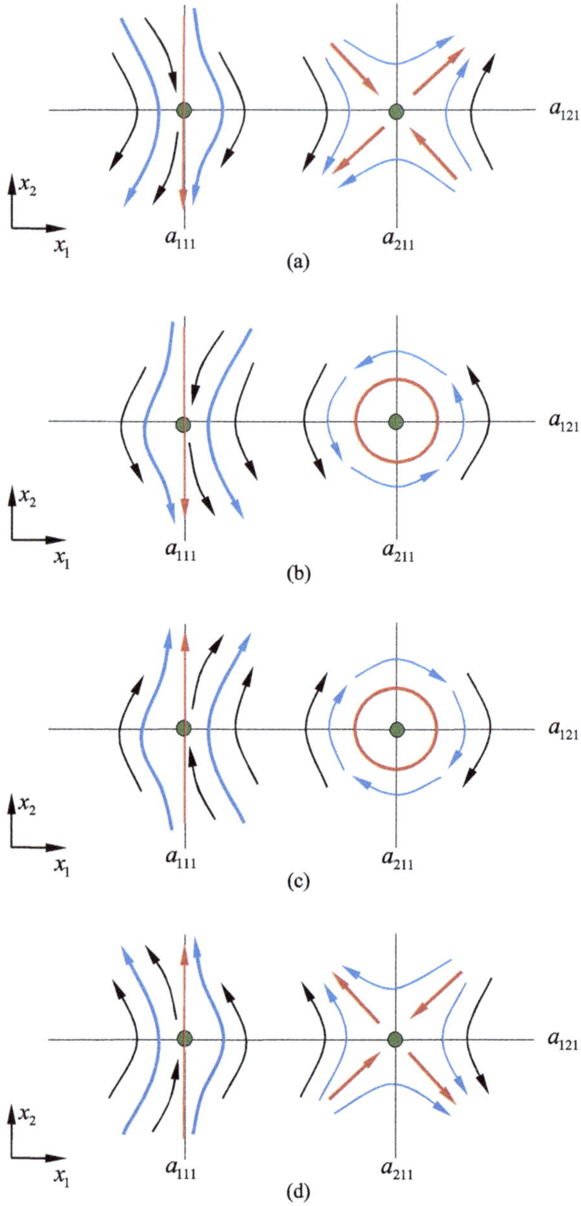

Fig. 3.8 Phase portraits ($a_{211} > a_{111}$) of the $([(2m_{11}), (2n_{11}+1)], \text{xF})$-hyperbolic and hyperbolic-secant singular flows at $(x_1^*, x_2^*) = (a_{111}, a_{121})$ and $((2n_{12}+1), (2m_{21}+1))$-positive and negative saddles, CCW and CW centers at $(x_2^*, x_1^*) = (a_{121}, a_{211})$ for 2-dimensional polynomial systems:
a ($a_{110} > 0, a_{220} > 0$),
b ($a_{110} < 0, a_{220} > 0$),
c ($a_{110} > 0, a_{220} < 0$),
d ($a_{110} < 0, a_{220} < 0$)

3.2.2 $[(2m_{11}), (2n_{11}+1)]$ and $((2n_{11}+1), (2m_{21}+1))$-Infinite-Equilibriums

Consider a polynomial dynamical system for $a_{111} = a_{211}$ as

$$\begin{aligned}\dot{x}_2 &= a_{220}(x_1 - a_{211})^{2m_{21}+1},\\ \dot{x}_1 &= a_{110}(x_1 - a_{111})^{2m_{11}}(x_2 - a_{121})^{2n_{11}+1}.\end{aligned} \tag{3.25}$$

The first integral manifold is

$$\begin{aligned}&\frac{1}{2(m_{21}-m_{11})+2}\left[(x_1 - a_{211})^{2(m_{21}-m_{11})+2} - (x_{10} - a_{211})^{2(m_{21}-m_{11})+2}\right]\\ &\quad = \frac{a_{110}}{a_{220}}\frac{1}{2n_{11}+2}\left[(x_2 - a_{121})^{2n_{11}+2} - (x_{20} - a_{121})^{2n_{11}+2}\right]\\ &\quad \text{for } 2(m_{21}-m_{11})+2 \neq 0;\\ &\frac{1}{2(m_{21}-m_{11})+2}\left[(x_1 - a_{211})^{2(m_{21}-m_{11})+2} - (x_{10} - a_{211})^{2(m_{21}-m_{11})+2}\right]\\ &\quad = \ln\frac{|x_1 - a_{111}|}{|x_{10} - a_{111}|},\ \text{for } 2(m_{21}-m_{11})+2 = 0.\end{aligned} \tag{3.26}$$

(**A**) For $2m_{21}+1 \geq 2m_{11}$, the corresponding infinite-equilibriums are given.

$$\begin{aligned}&(a_{121}, a_{211})\\ &= \underbrace{([(2m_{11}), (2n_{11}+1)]^{\text{th}}\text{DP}_{-}\text{:UP}_{+}, [2(m_{21}-m_{11})+1]^{\text{th}}{}_{\text{DP}}\text{LS:}_{\text{UP}}\text{US})}_{([(2m_{11}),(2n_{11}+1)],[2(m_{21}-m_{11})+1])\text{-positive hyperbolic lower-to-upper saddle}},\\ &\text{for } a_{110} > 0, a_{220} > 0;\\ &(a_{121}, a_{211})\\ &= \underbrace{([(2m_{11}), (2n_{11}+1)]^{\text{th}}\text{UP}_{-}\text{:DP}_{+}, [2(m_{21}-m_{11})+1]^{\text{th}}{}_{\text{UP}}\text{US:}_{\text{DP}}\text{LS})}_{([(2m_{11}),(2n_{11}+1)],[2(m_{21}-m_{11})+1])\text{-CCW circular upper-to-lower saddle}},\\ &\text{for } a_{110} < 0, a_{220} > 0;\\ &(a_{121}, a_{211})\\ &= \underbrace{([(2m_{11}), (2n_{11}+1)]^{\text{th}}\text{UP}_{+}\text{:DP}_{-}, [2(m_{21}-m_{11})+1]^{\text{th}}{}_{\text{UP}}\text{LS:}_{\text{DP}}\text{US})}_{([(2m_{11}),(2n_{11}+1)],[2(m_{21}-m_{11})+1])\text{-CW circular lower-to-upper saddle}},\\ &\text{for } a_{110} > 0, a_{220} < 0;\end{aligned}$$

$$
\begin{aligned}
&(a_{121}, a_{211}) \\
&= \underbrace{([(2m_{11}), (2n_{11}+1)]^{\text{th}}\text{DP}_{+}\text{:UP}_{-},[2(m_{21} - m_{11}) + 1]^{\text{th}}{}_{\text{DP}}\text{US:}_{\text{UP}}\text{LS})}_{([(2m_{11}),(2n_{11}+1)],[2(m_{21}-m_{11})+1])\text{-positive hyperbolic upper-to-lower saddle}}, \\
&\text{for } a_{110} < 0, a_{220} < 0.
\end{aligned} \tag{3.27}
$$

(**B**) For $2m_{21} + 1 < 2m_{11} + 1$, the infinite-equilibriums are given as follows.

$$
\begin{aligned}
&(a_{121}, a_{211}) \\
&= \underbrace{([(2m_{11}), (2n_{11}+1)]^{\text{th}}\text{DP}_{-}\text{:UP}_{+},[2(m_{11} - m_{21}) - 1]^{\text{th}}{}_{\text{UA}}\text{LS:}_{\text{DA}}\text{US})}_{([(2m_{11}),(2n_{11}+1)],[2(m_{11}-m_{21})-1])\text{-hyperbolic asymptote lower-to-upper saddle}}, \\
&\text{for } a_{110} > 0, a_{220} > 0; \\
&(a_{121}, a_{211}) \\
&= \underbrace{([(2m_{11}), (2n_{11}+1)]^{\text{th}}\text{UP}_{-}\text{:DP}_{+},[2(m_{11} - m_{21}) - 1]^{\text{th}}{}_{\text{DA}}\text{US:}_{\text{UA}}\text{LS})}_{([(2m_{11}),(2n_{11}+1)],[2(m_{11}-m_{21})-1])\text{-hyperbolic-secant asymptote upper-to-lower saddle}}, \\
&\text{for } a_{110} < 0, a_{220} > 0; \\
&(a_{121}, a_{211}) \\
&= \underbrace{([(2m_{11}), (2n_{11}+1)]^{\text{th}}\text{UP}_{+}\text{:DP}_{-},[2(m_{11} - m_{21}) - 1]^{\text{th}}{}_{\text{DA}}\text{LS:}_{\text{UA}}\text{US})}_{([(2m_{11}),(2n_{11}+1)],[2(m_{11}-m_{21})-1])\text{-hyperbolic-secant asymptote lower-to-upper saddle}}, \\
&\text{for } a_{110} > 0, a_{220} < 0; \\
&(a_{121}, a_{211}) \\
&= \underbrace{([(2m_{11}), (2n_{11}+1)]^{\text{th}}\text{DP}_{+}\text{:UP}_{-},[2(m_{11} - m_{21}) - 1]^{\text{th}}{}_{\text{UA}}\text{US:}_{\text{DA}}\text{LS})}_{([(2m_{11}),(2n_{11}+1)],[2(m_{11}-m_{21})-1])\text{-hyperbolic asymptote upper-to-lower saddle}}, \\
&\text{for } a_{110} < 0, a_{220} < 0.
\end{aligned} \tag{3.28}
$$

For $x_1^* = a_{111} = a_{211}$, the infinite-equilibrium exists. The infinite-equilibriums are illustrated in Figs. 3.9 and 3.10 for $2m_{21} + 1 \geq 2m_{11}$ and $2m_{21} + 1 < 2m_{11}$, respectively. For $2m_{21} + 1 \geq 2m_{11}$, the singular hyperbolic and circular lower-to-upper saddle are presented. For $2m_{11} > 2m_{21} + 1$, the hyperbolic and hyperbolic-secant asymptote lower-to-upper saddles are presented.

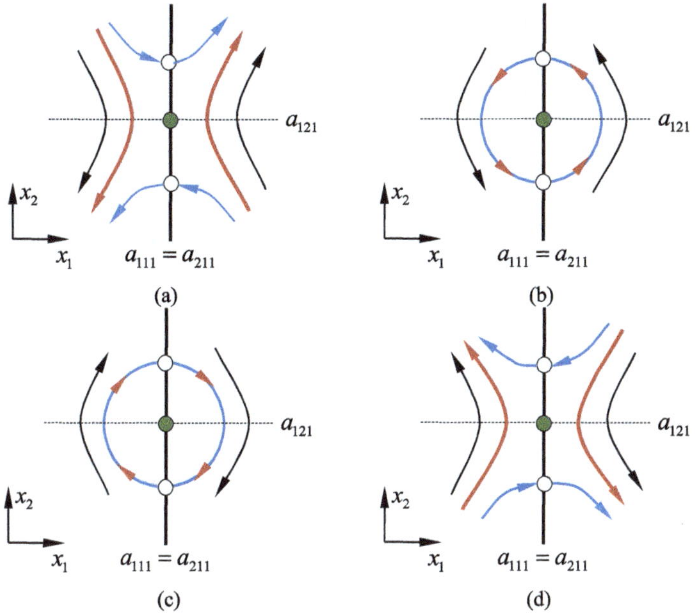

Fig. 3.9 The infinite-equilibriums ($a_{211}=a_{111}, 2m_{21}+1 \geq 2m_{11}, n_1=2n_{11}+1$) for 2-dimensional polynomial systems: **a** ($a_{110} > 0, a_{220} > 0$), **b** ($a_{110} < 0, a_{220} > 0$), **c** ($a_{110} > 0, a_{220} < 0$), **d** ($a_{110} < 0, a_{220} < 0$)

3.2.3 $[(2m_{11}), (2n_{11}+1)]$ *and* $((2n_{11}+1), (2m_{21}+1))$*-Hybrid Networks*

Consider

$$\begin{aligned} \dot{x}_2 &= a_{220} \prod_{v=1}^{2m_{21}+1} (x_1 - a_{21v}), \\ \dot{x}_1 &= a_{110} \prod_{r=1}^{2m_{11}} (x_1 - a_{11r}) \prod_{u=1}^{2n_{11}+1} (x_2 - a_{12u}). \end{aligned} \tag{3.29}$$

The first integral manifold is

$$\sum_{v_0=2m_{11}}^{2m_{21}+1} b_{21v_0} \sum_{\substack{\mu_1\in[0,\beta_0-2m_{11}-2],\\ v_1\in[2m_{11}-1,v_0-1]}}^{\mu_1+v_1=v_0-1} \sum_{\substack{\mu_2\in[0,v_1-2m_{11}]\\ v_2\in[2m_{11}-2,v_1-2]}}^{\mu_2+v_2=v_1-2} \cdots \sum_{\substack{\mu_{2m_{11}}\in[0,v_{2m_{11}-1}-2m_{11}]\\ v_{2m_{11}}\in[1,v_{2m_{11}-1}-2m_{11}+1]}}^{\mu_{2m_{11}}+v_{2m_{11}}=v_{2m_{11}-1}-2m_{11}+1} \prod_{j=1}^{2m_{11}-1}$$
$$\frac{(v_{j-1}-j)!}{\mu_j!v_j!} \times (a_{11r_{j+1}}-a_{11r_j})^{\mu_j} \frac{1}{v_{2m_{11}-1}} \Big[(x_1-a_{11r_{2m_{11}}})^{v_{2m_{11}-1}} - (x_{10}-a_{11r_{2m_{11}}})^{v_{2m_{11}-1}}\Big]$$

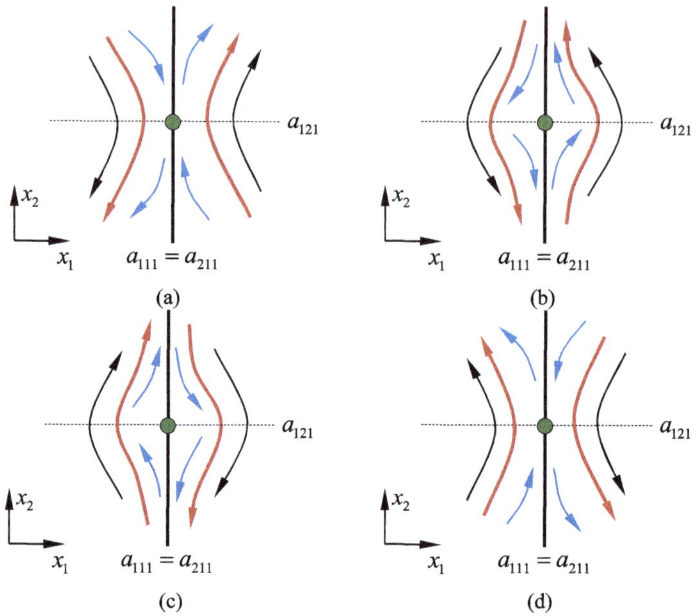

Fig. 3.10 The infinite-equilibriums ($a_{211}=a_{111}$, $2m_{21}+1 < 2m_{11}$, $n_1=2n_{11}+1$) for 2-dimensional polynomial systems: **a** ($a_{110} > 0, a_{220} > 0$), **b** ($a_{110} < 0, a_{220} > 0$), **c** ($a_{110} > 0, a_{220} < 0$), **d** ($a_{110} < 0, a_{220} < 0$)

$$
\begin{aligned}
&- \sum_{\nu_0=2m_{11}}^{2m_{21}+1} b_{21\nu_0} \sum_{k=2}^{2m_{11}-1} \Bigg[\sum_{\substack{\mu_1+\nu_1=\nu_0-1\\ \mu_1\in[0,\nu_0-2m_{11}),\\ \nu_1\in(2m_{11}-1,\nu_0-1]}} \sum_{\substack{\mu_2+\nu_2=\nu_1-2\\ \mu_2\in[0,\nu_1-2m_{11})\\ \nu_2\in(2m_{11}-2,\nu_1-2]}} \cdots \\
&\times \Bigg(\sum_{\substack{\mu_{k-1}+\nu_{k-1}=\nu_{k-2}-k+1\\ \mu_{k-1}\in(\nu_{k-2}-2m_{11},\nu_{k-2}-k+1]\\ \nu_{k-1}\in[0,2m_{11}+1-k)}} \prod_{j=1}^{k-1} \frac{(\nu_{j-1}-j)!}{\mu_j!\nu_j!}(a_{11r_{j+1}}-a_{11r_j})^{\mu_j} \\
&\times \sum_{\rho=k}^{2m_{11}} A_{\rho 1}^{(r_k)} \ln \frac{|x_1-a_{11r_\rho}|}{|x_{10}-a_{11r_\rho}|} \\
&+ \sum_{\substack{\mu_{k-1}+\nu_{k-1}=\nu_{k-2}-k+1\\ \mu_{k-1}\in(\nu_{k-2}-2m_{11}-1,\nu_{k-2}-2m_{11}+1]\\ \nu_{k-1}\in[2m_{11}+1-k,2m_{11}-k+2)}} \prod_{j=1}^{k-1} \frac{(\nu_{j-1}-j)!}{\mu_j!\nu_j!}(a_{11r_{j+1}}-a_{11r_j})^{\mu_j} \\
&\times \sum_{\rho=k+1}^{2m_{11}} B_{\rho 1}^{(r_k)} \ln \frac{|x_1-a_{11r_\rho}|}{|x_{10}-a_{11r_\rho}|} \Bigg) \Bigg] + b_{210} \sum_{\rho=1}^{2m_{11}} A_{\rho 1}^{(r_1)} \ln \frac{|x_1-a_{11r_\rho}|}{|x_{10}-a_{11r_\rho}|} \\
&+ \sum_{\nu_0=1}^{2m_{11}-1} b_{21\nu_0} \sum_{\rho=2}^{2m_{11}} B_{\rho 1}^{(r_1)} \ln \frac{|x_1-a_{11r_\rho}|}{|x_{10}-a_{11s_\rho}|}
\end{aligned}
$$

$$= \frac{a_{110}}{a_{220}}\Bigg\{ \frac{1}{2n_{11}+2}\left[(x_2 - a_{12u_1})^{2n_{11}+2} - (x_{j_2 0} - a_{12u_1})^{2n_{11}+2}\right]$$

$$+ \sum_{\alpha=1}^{2n_{11}-1} \frac{1}{2n_{11}-\alpha+2} b_{12\alpha}\left[(x_2 - a_{12u_1})^{2n_{11}-\alpha+2} - (x_{20} - a_{12u_1})^{2n_{11}-\alpha+2}\right]\Bigg\}, \tag{3.30}$$

where

$$b_{12\alpha} = \sum_{l_{2n_{11}+1},t_{2n_{11}+1}=0}^{l_{2n_{11}+1}+t_{2n_{11}+1}=1} \cdots \sum_{l_2,t_2=0}^{l_2+t_2=1} \prod_{k=2}^{2n_{11}+1} (a_{12u_1} - a_{12u_k})^{l_k} \delta_{t_2+\cdots+t_{2n_{11}+1}}^{2n_{11}+1-\alpha};$$

$$b_{120} = 1.$$

$$b_{21v_0} = \sum_{t_1,l_1=0}^{t_1+l_1=1} \cdots \sum_{t_{2m_{21}+1},l_{2m_{21}+1}=0}^{t_{2m_{21}+1}+l_{2m_{21}+1}=1} \prod_{k=1}^{2m_{21}+1} (a_{11r_1} - a_{21v_k})^{t_k} \delta_{v_0}^{\Sigma_{r=1}^{2m_{21}+1} l_r};$$

$$\lim_{v_{2m_{11}-1}\to 0} \frac{1}{v_{2m_{11}-1}}\left[(x_1 - a_{11r_{2m_{11}}})^{v_{2m_{11}-1}} - (x_{j_1 0} - a_{11r_{2m_{11}}})^{v_{2m_{11}-1}}\right] = \ln \frac{|x_1 - a_{11r_{2m_{11}}}|}{|x_{10} - a_{11r_{2m_{11}}}|}; \tag{3.31}$$

$$A_{\rho 1}^{(r_k)} = \frac{1}{\prod_{\sigma=k,\sigma\neq\rho}^{2m_{11}} (a_{11r_\rho} - a_{11r_\sigma})}, \text{ for } \rho \in \{k, k+1, \ldots, 2m_{11}\},$$

$$B_{\rho 1}^{(r_k)} = \frac{(a_{11r_\rho} - a_{11r_k})^{\beta_{k-1}-k}}{\prod_{\sigma=k+1,\sigma\neq\rho}^{2m_{11}} (a_{11r_\rho} - a_{11r_\sigma})}, \text{ for } \rho \in \{k+1, k+2, \ldots, 2m_{11}\};$$

$$k \in \{1, 2, \ldots, 2m_{11} - 1\}. \tag{3.32}$$

For all cases, the $(2n_{11}+1) \times (2m_{21}+1)$-simple equilibriums plus $(2m_{11}) \times (2n_{11}+1)$-hyperbolic flows are based on the following hyperbolic upper-to-lower/lower-to-upper saddles and CCW and CW circular upper-to-lower/lower-to-upper saddle bifurcations for $2m_{21}+1 \geq 2m_{11}$:

- ($[(2m_{11}), (2n_{11}+1)]^{\text{th}}DP_-$:UP$_+$,$[2(m_{21}-m_{11})+1]^{\text{th}}{}_{\text{DP}}$LS : $_{\text{UP}}$US)-positive hyperbolic lower-to-upper saddle bifurcation for $(a_{110} > 0, a_{220} > 0)$,
- ($[(2m_{11}), (2n_{11}+1)]^{\text{th}}UP_-$:DP$_+$,$[2(m_{21}-m_{11})+1]^{\text{th}}{}_{\text{UP}}$US:$_{\text{DP}}$LS)-CCW circular upper-to-lower saddle bifurcation for $(a_{110} < 0, a_{220} > 0)$,
- ($[(2m_{11}), (2n_{11}+1)]^{\text{th}}UP_+$:DP$_-$,$[2(m_{21}-m_{11})+1]^{\text{th}}{}_{\text{UP}}$LS:$_{\text{DP}}$US)-CW circular lower-to-upper saddle bifurcation for $(a_{110} > 0, a_{220} < 0)$,
- ($[(2m_{11}), (2n_{11}+1)]^{\text{th}}DP_+$:UP$_-$,$[2(m_{21}-m_{11})+1]^{\text{th}}{}_{\text{DP}}$US : $_{\text{UP}}$LS)-positive hyperbolic upper-to-lower saddle bifurcation for $(a_{110} < 0, a_{220} < 0)$;

or based on the following hyperbolic and hyperbolic-secant asymptote lower-to-upper/upper-to-lower saddles bifurcations for $2m_{21}+1 < 2m_{11}$:

- $([(2m_{11}), (2n_{11}+1)]^{\text{th}}\text{DP}_{-}\text{:UP}_{+},[2(m_{11}-m_{21})-1]^{\text{th}}{}_{\text{UA}}\text{LS} : {}_{\text{DA}}\text{US})$-hyperbolic asymptote lower-to-upper saddle bifurcation for $(a_{110} > 0, a_{220} > 0)$,
- $([(2m_{11}), (2n_{11}+1)]^{\text{th}}\text{UP}_{-}\text{:DP}_{+},[2(m_{11}-m_{21})-1]^{\text{th}}{}_{\text{DA}}\text{US:}{}_{\text{UA}}\text{LS})$-hyperbolic-secant asymptote lower-to-upper saddle bifurcation for $(a_{110}<0, a_{220}>0)$,
- $([(2m_{11}), (2n_{11}+1)]^{\text{th}}\text{UP}_{+}\text{:DP}_{-},[2(m_{11}-m_{21})-1]^{\text{th}}{}_{\text{DA}}\text{LS:}{}_{\text{UA}}\text{US})$-hyperbolic-secant asymptote upper-to-lower saddle bifurcation for $(a_{110}>0, a_{220}<0)$,
- $([(2m_{11}), (2n_{11}+1)]^{\text{th}}\text{DP}_{+}\text{:UP}_{-},[2(m_{11}-m_{21})-1]^{\text{th}}{}_{\text{UA}}\text{US} : {}_{\text{DA}}\text{LS})$-hyperbolic asymptote upper-to-lower saddle bifurcation for $(a_{110} < 0, a_{220} < 0)$.

(**A**) Consider $2m_{11}=2m_{21}$ for illustrations, phase portraits for the 2-dimensional systems for the hybrid networks of hyperbolic and hyperbolic-secant flows with the simple equilibriums of the saddles and centers with $x_1^*=a_{111}, a_{211}, \cdots a_{11(2m_{11})}, a_{21(2m_{21}+1)}$ and $x_2^*=a_{121}, a_{122}, \cdots, a_{12(2n_{11}+1)}$ are presented in Fig. 3.11a–d for $(a_{110}>0, a_{220}>0)$, $(a_{110}<0, a_{220}>0)$, $(a_{110}>0, a_{220}<0)$ and $(a_{110}<0, a_{220}<0)$.

(**B**) Consider $m_{11}=m_{21}$ for illustrations. Phase portraits for the 2-dimensional systems for the hybrid networks of hyperbolic and hyperbolic-secant flows with the simple equilibriums of saddles and centers with $x_1^*=a_{211}, a_{111}, \cdots, a_{11(2m_{11})}, a_{21(2m_{21}+1)}$ and $x_2^*=a_{121}, a_{122}, \cdots, a_{12(2n_{11}+1)}$ are presented in Fig. 3.12a–d for $(a_{110} > 0, a_{220} > 0)$, $(a_{110} < 0, a_{220} > 0)$, $(a_{110} > 0, a_{220} < 0)$ and $(a_{110} < 0, a_{220} < 0)$.

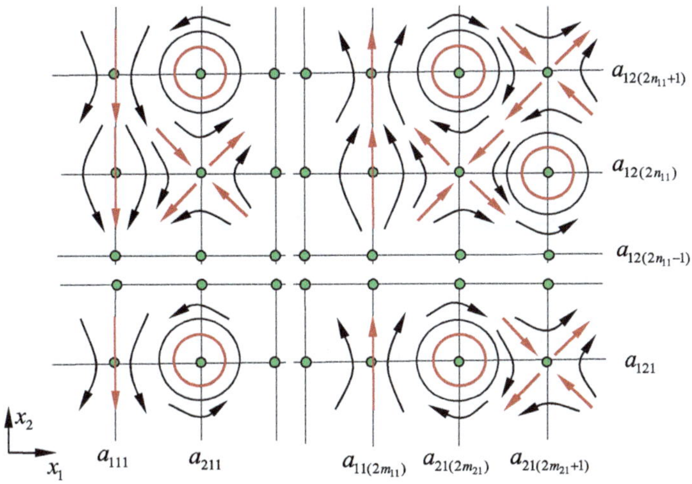

(a)

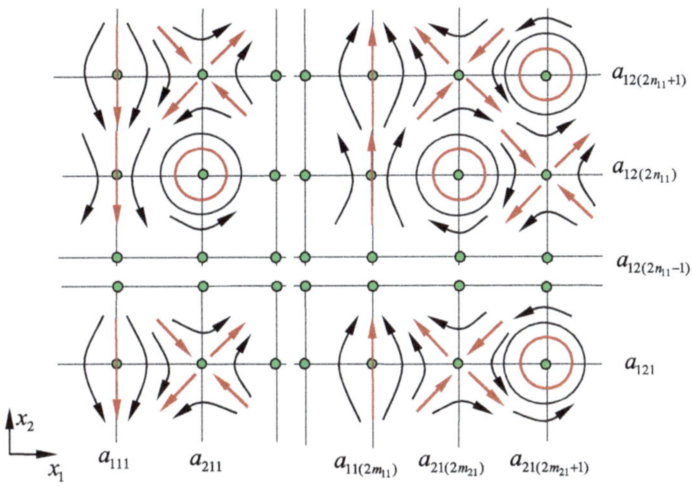

(b)

Fig. 3.11 Phase portraits for 2-dimensional systems with $x_1^* = a_{111}, a_{211}, \ldots, a_{11(2m_{11})}, a_{21(2m_{21})}, a_{21(2m_{21}+1)}$ and $x_2^* = a_{121}, \ldots, a_{12(2n_{11})}, a_{12(2n_{11}+1)}$. The networks of flows with saddles and centers: **a** ($a_{110} > 0, a_{220} > 0$), **b** ($a_{110} < 0, a_{220} > 0$), **c** ($a_{110} > 0, a_{220} < 0$), **d** ($a_{110} < 0, a_{220} < 0$)

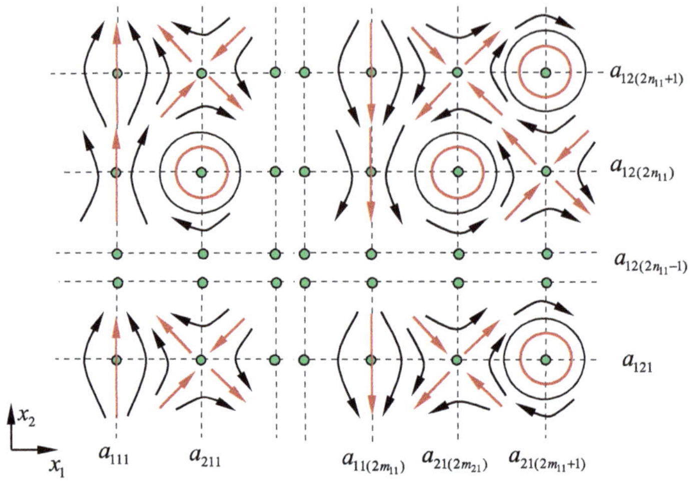

(c)

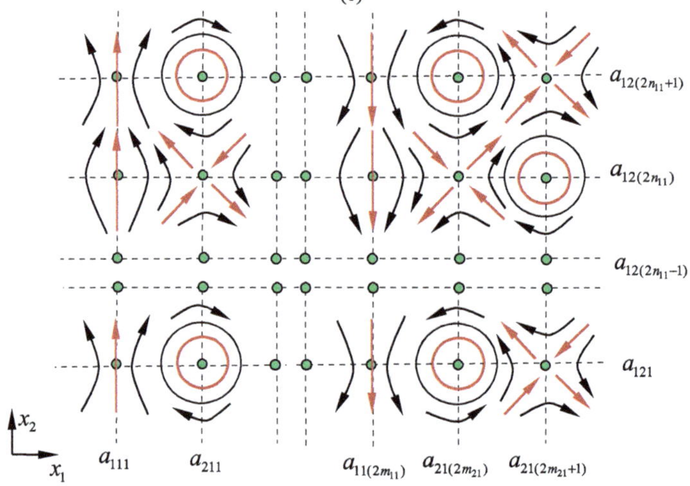

(d)

Fig. 3.11 (continued)

$$
\begin{gathered}
\left\{\begin{array}{cccccc}
(a_{111},a_{12(2n_{11}+1)}) & (a_{12(2n_{11}+1)},a_{211}) & \cdots & (a_{11(2m_{11})},a_{12(2n_{11}+1)}) & (a_{12(2n_{11}+1)},a_{21(2m_{21})}) & (a_{12(2n_{11}+1)},a_{21(2m_{21}+1)},) \\
(a_{111},a_{12(2n_{11})}) & (a_{12(2n_{11})},a_{211}) & \cdots & (a_{11(2m_{11})},a_{12(2n_{11})}) & (a_{12(2n_{11})},a_{21(2m_{21})}) & (a_{12(2n_{11})},a_{21(2m_{21}+1)}) \\
\vdots & \vdots & \vdots & \vdots & \vdots & \vdots \\
(a_{111},a_{121}) & (a_{121},a_{211}) & \cdots & (a_{11(2m_{11})},a_{121}) & (a_{121},a_{21(2m_{21})}) & (a_{121},a_{21(2m_{21}+1)})
\end{array}\right\} \\
= \left\{\begin{array}{cccccc}
\underbrace{(\text{DP:UP,nF})}_{\text{hyperbolic flow }(-)} & \underbrace{(\text{DP}_+,\text{DP}_-)}_{\text{CCW center}} & \cdots & \underbrace{(\text{DP:UP,pF})}_{\text{hyperbolic flow }(+)} & \underbrace{(\text{DP}_-,\text{DP}_+)}_{\text{CW center}} & \underbrace{(\text{UP}_+,\text{UP}_+)}_{\text{positive saddle}} \\
\underbrace{(\text{UP:DP,nF})}_{\text{hyperbolic-secant flow }(-)} & \underbrace{(\text{UP}_+,\text{UP}_+)}_{\text{positive saddle}} & \cdots & \underbrace{(\text{UP:DP,pF})}_{\text{hyperbolic-secant flow }(+)} & \underbrace{(\text{UP}_-,\text{UP}_-)}_{\text{negative saddle}} & \underbrace{(\text{DP}_+,\text{DP}_-)}_{\text{CCW center}} \\
\vdots & \vdots & \vdots & \vdots & \vdots & \vdots \\
\underbrace{(\text{DP:UP,nF})}_{\text{hyperbolic flow }(-)} & \underbrace{(\text{DP}_+,\text{DP}_-)}_{\text{CCW center}} & \cdots & \underbrace{(\text{DP:UP,pF})}_{\text{hyperbolic flow }(+)} & \underbrace{(\text{DP}_-,\text{DP}_+)}_{\text{CW center}} & \underbrace{(\text{UP}_+,\text{UP}_+)}_{\text{positive saddle}}
\end{array}\right\}_{(2n_{11}+1)\times(2m_{11}+2m_{21}+1)}
\end{gathered}
\tag{3.33}
$$

for $a_{110} > 0, a_{220} > 0$;

$$
\begin{Bmatrix}
(a_{111},a_{12(2n_{11}+1)}) & (a_{12(2n_{11}+1)},a_{211}) & \cdots & (a_{11(2m_{11})},a_{12(2n_{11}+1)}) & (a_{12(2n_{11}+1)},a_{21(2m_{21})}) & a_{21(2m_{21}+1)},(a_{12(2n_{11}+1)}) \\
(a_{111},a_{12(2n_{11})}) & (a_{12(2n_{11})},a_{211}) & \cdots & (a_{11(2m_{11})},a_{12(2n_{11})}) & (a_{12(2n_{11})},a_{21(2m_{21})}) & (a_{21(2m_{21}+1)},a_{12(2n_{11})}) \\
\vdots & \vdots & \vdots & \vdots & \vdots & \vdots \\
(a_{111},a_{121}) & (a_{121},a_{211}) & \cdots & (a_{11(2m_{11}+1)},a_{121}) & (a_{121},a_{21(2m_{21}+1)}) & (a_{21(2m_{21}+1)},a_{121})
\end{Bmatrix}
$$
$$
=\begin{Bmatrix}
\underbrace{(\mathrm{UP{:}DP,nF})}_{\text{hyperbolic-secant flow }(-)} & \underbrace{(\mathrm{UP}_+,\mathrm{UP}_+)}_{\text{positive saddle}} & \cdots & \underbrace{(\mathrm{UP{:}DP,pF})}_{\text{hyperbolic-secant flow }(+)} & \underbrace{(\mathrm{UP}_-,\mathrm{UP}_-)}_{\text{negative saddle}} & \underbrace{(\mathrm{DP}_+,\mathrm{DP}_-)}_{\text{CCW center}} \\
\underbrace{(\mathrm{DP{:}UP,nF})}_{\text{hyperbolic flow }(-)} & \underbrace{(\mathrm{DP}_+,\mathrm{DP}_-)}_{\text{CCW center}} & \cdots & \underbrace{(\mathrm{DP{:}UP,pF})}_{\text{hyperbolic flow }(+)} & \underbrace{(\mathrm{DP}_-,\mathrm{DP}_+)}_{\text{CW center}} & \underbrace{(\mathrm{UP}_+,\mathrm{UP}_+)}_{\text{positive saddle}} \\
\vdots & \vdots & \vdots & \vdots & \vdots & \vdots \\
\underbrace{(\mathrm{UP{:}DP,nF})}_{\text{hyperbolic-secant flow }(-)} & \underbrace{(\mathrm{UP}_+,\mathrm{UP}_+)}_{\text{positive saddle}} & \cdots & \underbrace{(\mathrm{UP{:}DP,pF})}_{\text{hyperbolic-secant flow }(+)} & \underbrace{(\mathrm{UP}_-,\mathrm{UP}_-)}_{\text{negative saddle}} & \underbrace{(\mathrm{DP}_+,\mathrm{DP}_-)}_{\text{CCW center}}
\end{Bmatrix}_{(2n_{11}+1)\times(2m_{11}+2m_{21}+1)}
$$

for $a_{110} < 0, a_{220} > 0$; (3.34)

$$
\left\{\begin{array}{cccccc}
(a_{111},a_{12(2n_{11}+1)}) & (a_{12(2n_{11}+1)},a_{211}) & \cdots\ (a_{11(2m_{11})},a_{12(2n_{11}+1)}) & (a_{12(2n_{11}+1)},a_{21(2m_{21})}) & a_{21(2m_{21}+1)},(a_{12(2n_{11}+1)}) \\
(a_{111},a_{12(2n_{11})}) & (a_{12(2n_{11})},a_{211}) & \cdots\ (a_{11(2m_{11})},a_{12(2n_{11})}) & (a_{12(2n_{11})},a_{21(2m_{21})}) & (a_{21(2m_{21}+1)},a_{12(2n_{11})}) \\
\vdots & \vdots & \vdots\quad \vdots & \vdots & \vdots \\
(a_{111},a_{121}) & (a_{121},a_{211}) & \cdots\ (a_{11(2m_{11}+1)},a_{121}) & (a_{121},a_{21(2m_{21}+1)}) & (a_{21(2m_{21}+1)},a_{121})
\end{array}\right\}
$$

$$
=\left\{\begin{array}{cccccc}
\underbrace{(\mathrm{UP{:}DP,pF})}_{\text{hyperbolic-secant flow }(+)} & \underbrace{(\mathrm{UP}_-,\mathrm{UP}_-)}_{\text{negative saddle}} & \cdots & \underbrace{(\mathrm{UP{:}DP,nF})}_{\text{hyperbolic-secant flow }(-)} & \underbrace{(\mathrm{UP}_+,\mathrm{UP}_+)}_{\text{negative saddle}} & \underbrace{(\mathrm{DP}_-,\mathrm{DP}_+)}_{\text{CW center}} \\
\underbrace{(\mathrm{DP{:}UP,pF})}_{\text{hyperbolic flow }(+)} & \underbrace{(\mathrm{DP}_-,\mathrm{DP}_+)}_{\text{CW center}} & \cdots & \underbrace{(\mathrm{DP{:}UP,nF})}_{\text{hyperbolic flow }(-)} & \underbrace{(\mathrm{DP}_+,\mathrm{DP}_-)}_{\text{CCW center}} & \underbrace{(\mathrm{UP}_-,\mathrm{UP}_-)}_{\text{negative saddle}} \\
\vdots & \vdots & \vdots & \vdots & \vdots & \vdots \\
\underbrace{(\mathrm{UP{:}DP,pF})}_{\text{hyperbolic-secant flow }(+)} & \underbrace{(\mathrm{UP}_-,\mathrm{UP}_-)}_{\text{negative saddle}} & \cdots & \underbrace{(\mathrm{UP{:}DP,nF})}_{\text{hyperbolic-secant flow }(-)} & \underbrace{(\mathrm{UP}_+,\mathrm{UP}_+)}_{\text{negative saddle}} & \underbrace{(\mathrm{DP}_-,\mathrm{DP}_+)}_{\text{CW center}}
\end{array}\right\}_{(2n_{11}+1)\times(2m_{11}+2m_{21}+1)}
$$

for $a_{110} > 0, a_{220} < 0$; (3.35)

$$\left\{\begin{array}{cccccc}
(a_{111},a_{12(2n_{11}+1)}) & (a_{12(2n_{11}+1)},a_{211}) & \cdots & (a_{11(2m_{11})},a_{12(2n_{11}+1)}) & (a_{12(2n_{11}+1)},a_{21(2m_{21})}) & a_{21(2m_{21}+1)},(a_{12(2n_{11}+1)}) \\
(a_{111},a_{12(2n_{11})}) & (a_{12(2n_{11})},a_{211}) & \cdots & (a_{11(2m_{11})},a_{12(2n_{11})}) & (a_{12(2n_{11})},a_{21(2m_{21})}) & (a_{21(2m_{21}+1)},a_{12(2n_{11})}) \\
\vdots & \vdots & \vdots & \vdots & \vdots & \vdots \\
(a_{111},a_{121}) & (a_{121},a_{211}) & \cdots & (a_{11(2m_{11}+1)},a_{121}) & (a_{121},a_{21(2m_{21}+1)}) & (a_{21(2m_{21}+1)},a_{121})
\end{array}\right\}$$

$$=\left\{\begin{array}{cccccc}
\underbrace{(\mathrm{DP{:}UP,pF})}_{\text{hyperbolic flow }(+)} & \underbrace{(\mathrm{DP}_-,\mathrm{DP}_+)}_{\text{CW center}} & \cdots & \underbrace{(\mathrm{DP{:}UP,nF})}_{\text{hyperbolic flow }(-)} & \underbrace{(\mathrm{DP}_+,\mathrm{DP}_-)}_{\text{CCW center}} & \underbrace{(\mathrm{UP}_-,\mathrm{UP}_-)}_{\text{negative saddle}} \\
\underbrace{(\mathrm{UP{:}DP,pF})}_{\text{hyperbolic-secant flow }(+)} & \underbrace{(\mathrm{UP}_-,\mathrm{UP}_-)}_{\text{negative saddle}} & \cdots & \underbrace{(\mathrm{UP{:}DP,nF})}_{\text{hyperbolic-secant flow }(-)} & \underbrace{(\mathrm{UP}_+,\mathrm{UP}_+)}_{\text{positive saddle}} & \underbrace{(\mathrm{DP}_-,\mathrm{DP}_+)}_{\text{CW center}} \\
\vdots & \vdots & \vdots & \vdots & \vdots & \vdots \\
\underbrace{(\mathrm{DP{:}UP,pF})}_{\text{hyperbolic flow }(+)} & \underbrace{(\mathrm{DP}_-,\mathrm{DP}_+)}_{\text{CW center}} & \cdots & \underbrace{(\mathrm{DP{:}UP,nF})}_{\text{hyperbolic flow }(-)} & \underbrace{(\mathrm{DP}_+,\mathrm{DP}_-)}_{\text{CCW center}} & \underbrace{(\mathrm{UP}_-,\mathrm{UP}_-)}_{\text{negative saddle}}
\end{array}\right\}_{(2n_{11}+1)\times(2m_{11}+2m_{21}+1)}$$

for $a_{110} < 0, a_{220} < 0.$ (3.36)

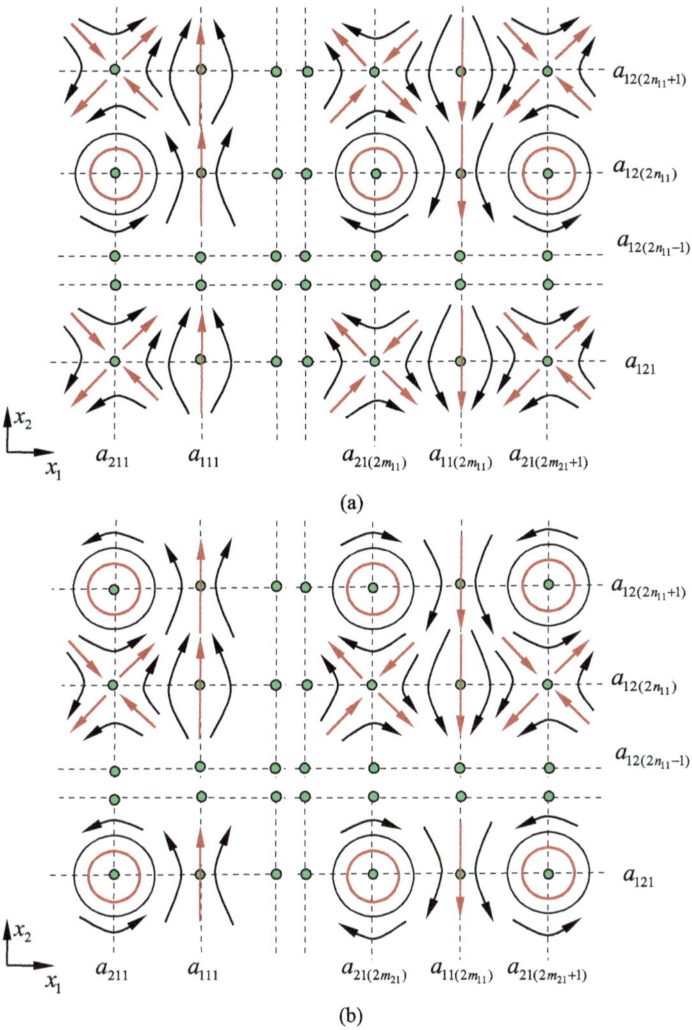

Fig. 3.12 Phase portraits for 2-dimensional systems with $x_1^*=a_{211}, a_{111}, \ldots, a_{11(2m_{11})}, a_{21(2m_{21}+1)}$ and $x_2^*=a_{121}, a_{122}, \ldots, a_{12(2n_{11}+1)}$. The networks of flows with saddles and centers: **a** $(a_{110}>0, a_{220}>0)$, **b** $(a_{110}<0, a_{220}>0)$, **c** $(a_{110}>0, a_{220}<0)$, **d** $(a_{110}<0, a_{220}<0)$

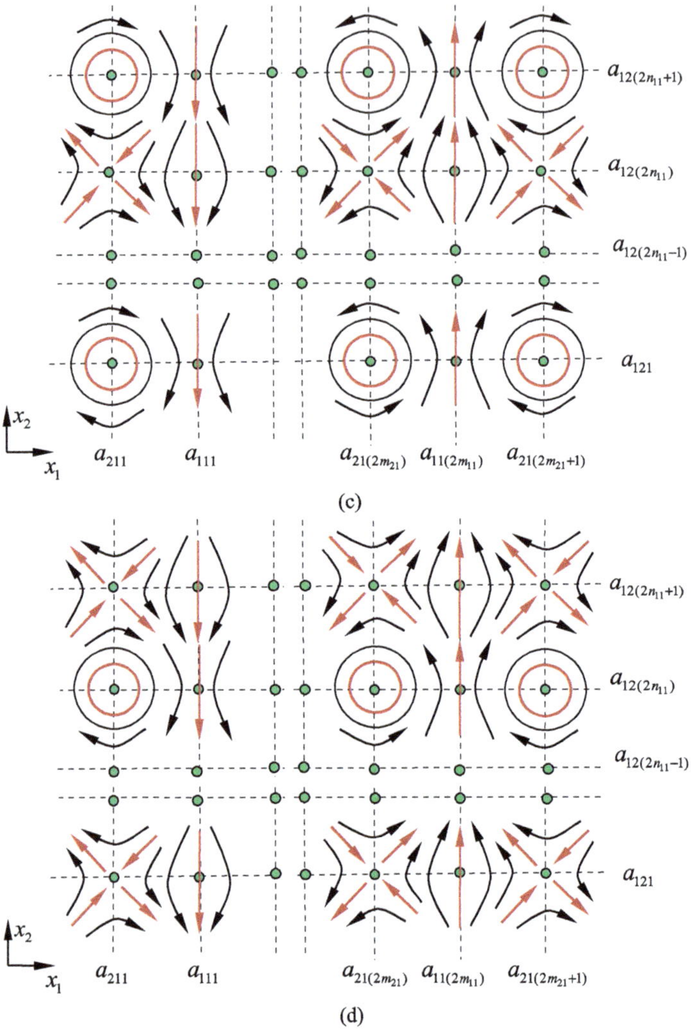

Fig. 3.12 (continued)

$$
\begin{aligned}
&\left\{\begin{array}{cccccc}
(a_{12(2n_{11}+1)},a_{211}) & (a_{111},a_{12(2n_{11}+1)}) & \cdots & (a_{12(2n_{11}+1)},a_{21(2m_{21})}) & (a_{11(2m_{11})},a_{12(2n_{11}+1)}) & (a_{12(2n_{11}+1)},a_{21(2m_{21}+1)}) \\
(a_{12(2n_{11})},a_{211}) & (a_{111},a_{12(2n_{11})}) & \cdots & (a_{12(2n_{11})},a_{21(2m_{21})},) & (a_{11(2m_{11})},a_{12(2n_{11})}) & (a_{12(2n_{11})},a_{21(2m_{21}+1)}) \\
\vdots & \vdots & \vdots & \vdots & \vdots & \vdots \\
(a_{121},a_{211}) & (a_{111},a_{121}) & \cdots & (a_{121},a_{21(2m_{21})},) & (a_{11(2m_{11})},a_{121}) & (a_{121},a_{21(2m_{21}+1)})
\end{array}\right\} \\
&= \left\{\begin{array}{cccccc}
\underbrace{(\mathrm{UP}_{+},\mathrm{UP}_{+})}_{\text{positive saddle}} & \underbrace{(\mathrm{UP{:}DP,pF})}_{\text{hyperbolic-secant flow }(+)} & \cdots & \underbrace{(\mathrm{UP}_{-},\mathrm{UP}_{-})}_{\text{negative saddle}} & \underbrace{(\mathrm{UP{:}DP,nF})}_{\text{hyperbolic-secant flow }(-)} & \underbrace{(\mathrm{UP}_{+},\mathrm{UP}_{+})}_{\text{positive saddle}} \\
\underbrace{(\mathrm{DP}_{+},\mathrm{DP}_{-})}_{\text{CCW center}} & \underbrace{(\mathrm{DP{:}UP,pF})}_{\text{hyperbolic flow }(+)} & \cdots & \underbrace{(\mathrm{DP}_{-},\mathrm{DP}_{+})}_{\text{CW center}} & \underbrace{(\mathrm{DP{:}UP,nF})}_{\text{hyperbolic flow }(-)} & \underbrace{(\mathrm{DP}_{+},\mathrm{DP}_{-})}_{\text{CCW center}} \\
\vdots & \vdots & \vdots & \vdots & \vdots & \vdots \\
\underbrace{(\mathrm{UP}_{+},\mathrm{UP}_{+})}_{\text{positive saddle}} & \underbrace{(\mathrm{UP{:}DP,pF})}_{\text{hyperbolic-secant flow }(+)} & \cdots & \underbrace{(\mathrm{UP}_{-},\mathrm{UP}_{-})}_{\text{negative saddle}} & \underbrace{(\mathrm{UP{:}DP,nF})}_{\text{hyperbolic-secant flow }(-)} & \underbrace{(\mathrm{UP}_{+},\mathrm{UP}_{+})}_{\text{positive saddle}}
\end{array}\right\}_{(2n_{21}+1)\times(2m_{11}+2m_{21}+1)}
\end{aligned}
\tag{3.37}
$$

for $a_{110} > 0, a_{220} > 0$;

$$
\left\{\begin{array}{cccccc}
(a_{12(2n_{11}+1)},a_{211}) & (a_{111},a_{12(2n_{11}+1)}) & \cdots & (a_{12(2n_{11}+1)},a_{21(2m_{21})}) & (a_{11(2m_{11})},a_{12(2n_{11}+1)}) & (a_{12(2n_{11}+1)},a_{21(2m_{21}+1)}) \\
(a_{12(2n_{11})},a_{211}) & (a_{111},a_{12(2n_{11})}) & \cdots & (a_{12(2n_{11})},a_{21(2m_{21})},) & (a_{11(2m_{11})},a_{12(2n_{11})}) & (a_{12(2n_{11})},a_{21(2m_{21}+1)}) \\
\vdots & \vdots & \vdots & \vdots & \vdots & \vdots \\
(a_{121},a_{211}) & (a_{111},a_{121}) & \cdots & (a_{121},a_{21(2m_{21})},) & (a_{11(2m_{11})},a_{121}) & (a_{121},a_{21(2m_{21}+1)})
\end{array}\right\}
$$
$$
= \left\{\begin{array}{cccccc}
\underbrace{(\mathrm{DP}_+,\mathrm{DP}_-)}_{\text{CCW center}} & \underbrace{(\mathrm{DP{:}UP,pF})}_{\text{hyperbolic flow }(+)} & \cdots & \underbrace{(\mathrm{DP}_-,\mathrm{DP}_+)}_{\text{CW center}} & \underbrace{(\mathrm{DP{:}UP,nF})}_{\text{hyperbolic flow }(-)} & \underbrace{(\mathrm{DP}_+,\mathrm{DP}_-)}_{\text{CCW center}} \\
\underbrace{(\mathrm{UP}_+,\mathrm{UP}_+)}_{\text{positive saddle}} & \underbrace{(\mathrm{UP{:}DP,pF})}_{\text{hyperbolic-secant flow }(+)} & \cdots & \underbrace{(\mathrm{UP}_-,\mathrm{UP}_-)}_{\text{negative saddle}} & \underbrace{(\mathrm{UP{:}DP,nF})}_{\text{hyperbolic-secant flow }(-)} & \underbrace{(\mathrm{UP}_+,\mathrm{UP}_+)}_{\text{positive saddle}} \\
\vdots & \vdots & \vdots & \vdots & \vdots & \vdots \\
\underbrace{(\mathrm{DP}_+,\mathrm{DP}_-)}_{\text{CCW center}} & \underbrace{(\mathrm{DP{:}UP,pF})}_{\text{hyperbolic flow }(+)} & \cdots & \underbrace{(\mathrm{DP}_-,\mathrm{DP}_+)}_{\text{CW center}} & \underbrace{(\mathrm{DP{:}UP,nF})}_{\text{hyperbolic flow }(-)} & \underbrace{(\mathrm{DP}_+,\mathrm{DP}_-)}_{\text{CCW center}}
\end{array}\right\}_{(2n_{11}+1)\times(2m_{11}+2m_{21}+1)}
\tag{3.38}
$$

for $a_{110} < 0, a_{220} > 0$;

$$
\begin{Bmatrix}
(a_{12(2n_{11}+1)},a_{211}) & (a_{111},a_{12(2n_{11}+1)}) & \cdots & (a_{12(2n_{11}+1)},a_{21(2m_{21})}) & (a_{11(2m_{11})},a_{12(2n_{11}+1)}) & (a_{12(2n_{11}+1)},a_{21(2m_{21}+1)})\\
(a_{12(2n_{11})},a_{211}) & (a_{111},a_{12(2n_{11})}) & \cdots & (a_{12(2n_{11})},a_{21(2m_{21})},) & (a_{11(2m_{11})},a_{12(2n_{11})}) & (a_{12(2n_{11})},a_{21(2m_{21}+1)})\\
\vdots & \vdots & \vdots & \vdots & \vdots & \vdots\\
(a_{121},a_{211}) & (a_{111},a_{121}) & \cdots & (a_{121},a_{21(2m_{21})},) & (a_{11(2m_{11})},a_{121}) & (a_{121},a_{21(2m_{21}+1)})
\end{Bmatrix}
$$

$$
= \begin{Bmatrix}
\underbrace{(\mathrm{DP}_-,\mathrm{DP}_+)}_{\text{CW center}} & \underbrace{(\mathrm{DP{:}UP,nF})}_{\text{hyperbolic flow }(-)} & \cdots & \underbrace{(\mathrm{DP}_+,\mathrm{DP}_-)}_{\text{CCW center}} & \underbrace{(\mathrm{DP{:}UP,pF})}_{\text{hyperbolic flow }(+)} & \underbrace{(\mathrm{DP}_-,\mathrm{DP}_+)}_{\text{CW center}}\\
\underbrace{(\mathrm{UP}_-,\mathrm{UP}_-)}_{\text{negative saddle}} & \underbrace{(\mathrm{UP{:}DP,nF})}_{\text{hyperbolic-secant flow }(-)} & \cdots & \underbrace{(\mathrm{UP}_+,\mathrm{UP}_+)}_{\text{positive saddle}} & \underbrace{(\mathrm{UP{:}DP,pF})}_{\text{hyperbolic-secant flow }(+)} & \underbrace{(\mathrm{UP}_-,\mathrm{UP}_-)}_{\text{negative saddle}}\\
\vdots & \vdots & \vdots & \vdots & \vdots & \vdots\\
\underbrace{(\mathrm{DP}_-,\mathrm{DP}_+)}_{\text{CW center}} & \underbrace{(\mathrm{DP{:}UP,nF})}_{\text{hyperbolic flow }(-)} & \cdots & \underbrace{(\mathrm{DP}_+,\mathrm{DP}_-)}_{\text{CCW center}} & \underbrace{(\mathrm{DP{:}UP,pF})}_{\text{hyperbolic flow }(+)} & \underbrace{(\mathrm{DP}_-,\mathrm{DP}_+)}_{\text{CW center}}
\end{Bmatrix}_{(2n_{11}+1)\times(2m_{11}+2m_{21}+1)}
$$

for $a_{110} > 0, a_{220} < 0;$ (3.39)

$$\left\{\begin{array}{cccccc}(a_{12(2n_{11}+1)},a_{211}) & (a_{111},a_{12(2n_{11}+1)}) & \cdots & (a_{12(2n_{11}+1)},\, a_{21(2m_{21})}) & (a_{11(2m_{11})},\, a_{12(2n_{11}+1)}) & (a_{12(2n_{11}+1)},\, a_{21(2m_{21}+1)})\\ (a_{12(2n_{11})},a_{211}) & (a_{111},a_{12(2n_{11})}) & \cdots & (a_{12(2n_{11})},\, a_{21(2m_{21})},\,) & (a_{11(2m_{11})},\, a_{12(2n_{11})}) & (a_{12(2n_{11})},\, a_{21(2m_{21}+1)})\\ \vdots & \vdots & \vdots & \vdots & \vdots & \vdots\\ (a_{121},a_{211}) & (a_{111},a_{121}) & \cdots & (a_{121},\, a_{21(2m_{21})},\,) & (a_{11(2m_{11})},\, a_{121}) & (a_{121},\, a_{21(2m_{21}+1)})\end{array}\right\}$$

$$= \left\{\begin{array}{cccccc}\underbrace{(\mathrm{UP}_-,\mathrm{UP}_-)}_{\text{negative saddle}} & \underbrace{(\mathrm{UP{:}DP,nF})}_{\text{hyperbolic-secant flow }(-)} & \cdots & \underbrace{(\mathrm{UP}_+,\mathrm{UP}_+)}_{\text{positive saddle}} & \underbrace{(\mathrm{UP{:}DP,pF})}_{\text{hyperbolic-secant flow }(+)} & \underbrace{(\mathrm{UP}_-,\mathrm{UP}_-)}_{\text{negative saddle}}\\ \underbrace{(\mathrm{DP}_-,\mathrm{DP}_+)}_{\text{CCW center}} & \underbrace{(\mathrm{DP{:}UP,nF})}_{\text{hyperbolic flow }(-)} & \cdots & \underbrace{(\mathrm{DP}_+,\mathrm{DP}_-)}_{\text{CW center}} & \underbrace{(\mathrm{DP{:}UP,pF})}_{\text{hyperbolic flow }(+)} & \underbrace{(\mathrm{DP}_-,\mathrm{DP}_+)}_{\text{CCW center}}\\ \vdots & \vdots & \vdots & \vdots & \vdots & \vdots\\ \underbrace{(\mathrm{UP}_-,\mathrm{UP}_-)}_{\text{negative saddle}} & \underbrace{(\mathrm{UP{:}DP,nF})}_{\text{hyperbolic-secant flow }(-)} & \cdots & \underbrace{(\mathrm{UP}_+,\mathrm{UP}_+)}_{\text{positive saddle}} & \underbrace{(\mathrm{UP{:}DP,pF})}_{\text{hyperbolic-secant flow }(+)} & \underbrace{(\mathrm{UP}_-,\mathrm{UP}_-)}_{\text{negative saddle}}\end{array}\right\}_{(2n_{11}+1)\times(2m_{11}+2m_{21}+1)} \tag{3.40}$$

for $a_{110} < 0$, $a_{220} < 0$;

3.3 $[(2m_{11}+1),(2n_{11}+1)]$ and $((2n_{11}+1),(2m_{21}))$-Singular Systems

In this section, there are four global bifurcations of the 1-dimensional flows through singular hyperbolic flows, and four global bifurcations of equilibriums, through singular positive and negative saddles and CCW and CW centers bifurcations. The corresponding global bifurcations generate networks of hyperbolic and hyperbolic-secant flows with saddles and centers. The infinite-equilibriums are the appearing and switching bifurcations for the singular flows and singular equilibriums and for hybrid networks of hyperbolic flows and saddles and centers.

3.3.1 $[(2m_{11}+1),(2n_{11}+1)]$-*Flows and* $((2n_{11}+1),(2m_{21}))$-*Equilibriums*

Consider a polynomial dynamical system for $a_{111}\neq a_{211}$ as

$$\begin{aligned}\dot{x}_2&=a_{220}(x_1-a_{211})^{2m_{21}},\\ \dot{x}_1&=a_{110}(x_1-a_{111})^{2m_{11}+1}(x_2-a_{121})^{2n_{11}+1}.\end{aligned}\tag{3.41}$$

The first integral manifold is

$$\begin{aligned}&\sum_{\alpha,\beta=0}^{\alpha+\beta=2m_{21}}\frac{(2m_{21})!}{\alpha!\beta!}\frac{(a_{111}-a_{211})^{\alpha}}{\beta-2m_{11}}\big[(x_1-a_{111})^{\beta-2m_{11}}-(x_{10}-a_{111})^{\beta-2m_{11}}\big]\\ &=\frac{a_{j_1j_10}}{a_{j_2j_20}}\frac{1}{2n_{11}+2}[(x_2-a_{121})^{2n_{11}+2}-(x_{20}-a_{121})^{2n_{11}+2}],\\ &\text{for } \beta-2m_{11}\neq 0;\\ &\frac{1}{\beta-2m_{11}}\big[(x_1-a_{111})^{\beta-2m_{11}}-(x_{10}-a_{111})^{\beta-2m_{11}}\big]=\ln\frac{|x_1-a_{111}|}{|x_{10}-a_{111}|},\\ &\text{for } \beta-2m_{11}=0.\end{aligned}\tag{3.42}$$

The singular hyperbolic and hyperbolic-secant flows with singular parabola upper and lower-saddle are illustrated in Figs. 3.13 and 3.14. Through such singular flows, the corresponding global bifurcations generate from a hyperbolic (hyperbolic-secant) flow to a network of singular hyperbolic and hyperbolic-secant flows. Through such singular equilibriums, the corresponding global bifurcations generate from one of 1-dimensional flows to a network of singular saddles and centers equilibriums.

For $(x_1^*,x_2^*)=(a_{111},a_{121})$ with $\dot{x}_1=0$, the $[(2m_{11}+1),(2n_{11}+1)]$-singular flows exist with $\dot{x}_2=a_{220}(a_{111}-a_{211})^{2m_{21}}$. For $(x_1^*,x_2^*)=(a_{111},a_{121})$, the 1-dimensional flow direction is determined by $\dot{x}_2\neq 0$. For $(x_2^*,x_1^*)=(a_{121},a_{211})$ with $\dot{x}_1=\dot{x}_2=0$, the $((2n_{11}+1),(2m_{21}))$-singular equilibriums exist.

Fig. 3.13 Phase portraits ($a_{111} > a_{211}$) of the ([(2m_{11} + 1), (2n_{11} + 1)], xF)-hyperbolic and hyperbolic-secant singular flows at $(x_1^*, x_2^*) = (a_{111}, a_{121})$ and ((2n_{11} + 1), (2m_{21}))-up-parabola and down-parabola upper-saddle and lower-saddle at $(x_2^*, x_1^*) = (a_{121}, a_{211})$ for 2-dimensional polynomial systems: **a** ($a_{110} > 0, a_{220} > 0$), **b** ($a_{110} < 0, a_{220} > 0$), **c** ($a_{110} > 0, a_{220} < 0$), **d** ($a_{110} < 0, a_{220} < 0$)

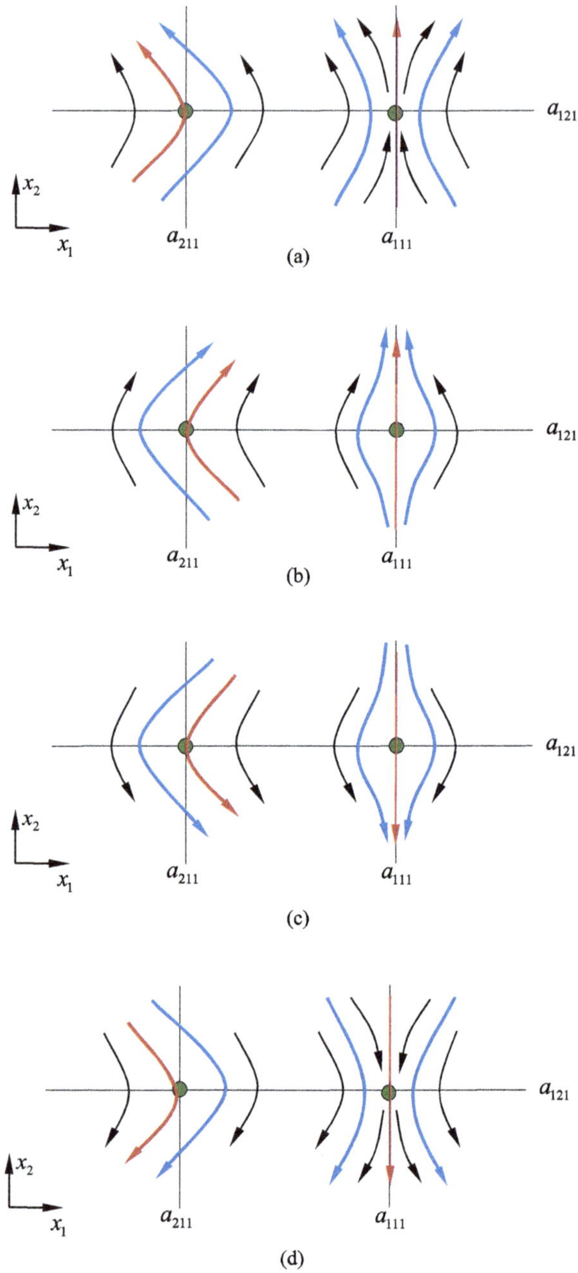

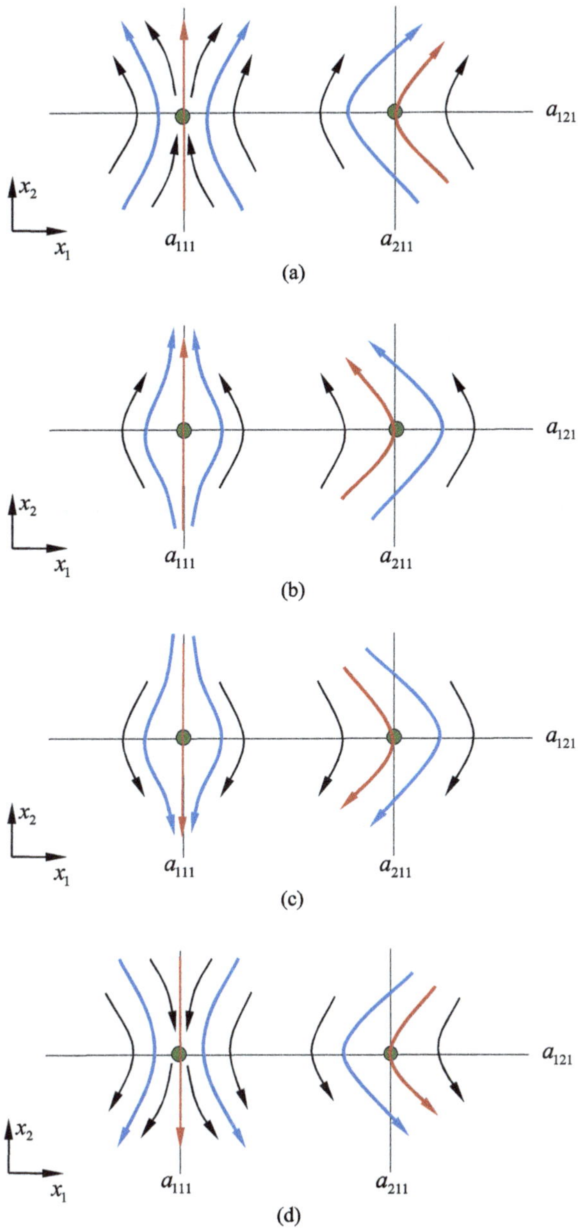

Fig. 3.14 Phase portraits ($a_{211} > a_{111}$) of the ($[(2m_{11}+1), (2n_{11}+1)]$, xF)-hyperbolic and hyperbolic-secant singular flows at $(x_1^*, x_2^*) = (a_{111}, a_{121})$ and $((2n_{12}+1), (2m_{21}))$-positive and negative saddles, CCW and CW centers at $(x_2^*, x_1^*) = (a_{121}, a_{211})$ for 2-dimensional polynomial systems: **a** ($a_{110} > 0, a_{220} > 0$), **b** ($a_{110} < 0, a_{220} > 0$), **c** ($a_{110} > 0, a_{220} < 0$), **d** ($a_{110} < 0, a_{220} < 0$)

(**A**) For $a_{111} > a_{211}$,

$$\left\{\begin{array}{c}(a_{111}, a_{121})\\(a_{121}, a_{211})\end{array}\right\} = \left\{\begin{array}{c}\underbrace{([(2m_{11}+1),(2n_{11}+1)]\text{DP:UP,pF})}_{[(2m_{11}+1),(2n_{11}+1)]\text{-hyperbolic flow }(+)}\\\underbrace{((2n_{11}+1)\text{DP},(2m_{21})\text{US})}_{[(2n_{11}+1),(2m_{21})]\text{-down-parabola upper-saddle}}\end{array}\right\},$$

for $a_{110} > 0, a_{220} > 0$;

$$\left\{\begin{array}{c}(a_{111}, a_{121})\\(a_{121}, a_{211})\end{array}\right\} = \left\{\begin{array}{c}\underbrace{([(2m_{11}+1),(2n_{11}+1)]\text{UP:DP,pF})}_{[(2m_{11}+1),(2n_{11}+1)]\text{-hyperbolic-secant flow }(+)}\\\underbrace{((2n_{11}+1)\text{UP},(2m_{21})\text{US})}_{[(2n_{11}+1),(2m_{21})]\text{-up-parabola upper-saddle}}\end{array}\right\},$$

for $a_{110} < 0, a_{220} > 0$;

$$\left\{\begin{array}{c}(a_{111}, a_{121})\\(a_{121}, a_{211})\end{array}\right\} = \left\{\begin{array}{c}\underbrace{([(2m_{11}+1),(2n_{11}+1)]\text{UP:DP,nF})}_{[(2m_{11}+1),(2n_{11}+1)]\text{-hyperbolic-secant flow }(-)}\\\underbrace{((2n_{11}+1)\text{UP},(2m_{21})\text{LS})}_{[(2n_{11}+1),(2m_{21})]\text{-up-parabola lower-saddle}}\end{array}\right\},$$

for $a_{110} > 0, a_{220} < 0$;

$$\left\{\begin{array}{c}(a_{111}, a_{121})\\(a_{121}, a_{211})\end{array}\right\} = \left\{\begin{array}{c}\underbrace{([(2m_{11}+1),(2n_{11}+1)]\text{DP:UP,nF})}_{[(2m_{11}+1),(2n_{11}+1)]\text{-hyperbolic flow }(-)}\\\underbrace{((2n_{11}+1)\text{DP},(2m_{21})\text{LS})}_{[(2n_{11}+1),(2m_{21})]\text{-down-parabola lower-saddle}}\end{array}\right\},$$

for $a_{110} < 0, a_{220} < 0$. (3.43)

(**B**) For $a_{211} < a_{111}$,

$$\left\{\begin{array}{c}(a_{121}, a_{211})\\(a_{111}, a_{121})\end{array}\right\} = \left\{\begin{array}{c}\underbrace{((2n_{11}+1)\text{UP},(2m_{21})\text{US})}_{[(2n_{11}+1),(2m_{21})]\text{-up-parabola upper-saddle}}\\\underbrace{([(2m_{11}+1),(2n_{11}+1)]\text{DP:UP,pF})}_{[(2m_{11}+1),(2n_{11}+1)]\text{-hyperbolic flow }(+)}\end{array}\right\},$$

for $a_{110} > 0, a_{220} > 0$;

$$\left\{\begin{array}{c}(a_{121}, a_{211})\\(a_{111}, a_{121})\end{array}\right\} = \left\{\begin{array}{c}\underbrace{((2n_{11}+1)\text{DP},(2m_{21})\text{US})}_{[(2n_{11}+1),(2m_{21})]\text{-down-parabola upper-saddle}}\\\underbrace{([(2m_{11}+1),(2n_{11}+1)]\text{UP:DP,pF})}_{[(2m_{11}+1),(2n_{11}+1)]\text{-hyperbolic-secant flow }(+)}\end{array}\right\},$$

for $a_{110} < 0, a_{220} > 0$;

$$\left\{\begin{array}{l}(a_{121}, a_{211})\\(a_{111}, a_{121})\end{array}\right\} = \left\{\begin{array}{c}\underbrace{((2n_{11}+1)\text{DP},(2m_{21})\text{LS})}_{[(2n_{11}+1),(2m_{21})]\text{-down-parabola lower-saddle}}\\\underbrace{([(2m_{11}+1),(2n_{11}+1)]\text{UP:DP,nF})}_{[(2m_{11}+1),(2n_{11}+1)]\text{-hyperbolic-secant flow }(-)}\end{array}\right\},$$
for $a_{110} > 0, a_{220} < 0$;
$$\left\{\begin{array}{l}(a_{121}, a_{211})\\(a_{111}, a_{121})\end{array}\right\} = \left\{\begin{array}{c}\underbrace{((2n_{11}+1)\text{UP},(2m_{21})\text{LS})}_{[(2n_{11}+1),(2m_{21})]\text{-up-parabola lower-saddle}}\\\underbrace{([(2m_{11}+1),(2n_{11}+1)]\text{DP:UP,nF})}_{[(2m_{11}+1),(2n_{11}+1)]\text{-hyperbolic flow }(-)}\end{array}\right\},$$
for $a_{110} < 0, a_{220} < 0$. (3.44)

- The $((2n_{11}+1)\text{UP},(2m_{21})\text{US}]$ is for a $((2n_{11}+1), (2m_{21})]$-singular up-parabola upper-saddle.
- The $((2n_{11}+1)\text{UP},(2m_{21})\text{LS}]$ is for a $((2n_{11}+1), (2m_{21})]$-singular up-parabola lower-saddle.
- The $((2n_{11}+1)\text{DP},(2m_{21})\text{US}]$ is for a $((2n_{11}+1), (2m_{21})]$-singular down-parabola upper-saddle.
- The $((2n_{11}+1)\text{DP},(2m_{21})\text{LS}]$ is for a $((2n_{11}+1), (2m_{21})]$-singular down-parabola lower-saddle.

3.3.2 $[(2m_{11}+1), (2n_{11}+1)]$ and $((2n_{11}+1), (2m_{21}))$-*Infinite-Equilibriums*

Consider a polynomial dynamical system for $a_{111} = a_{211}$ as

$$\begin{aligned}\dot{x}_2 &= a_{220}(x_1 - a_{211})^{2m_{21}},\\\dot{x}_1 &= a_{110}(x_1 - a_{111})^{2m_{11}+1}(x_2 - a_{121})^{2n_{11}+1}.\end{aligned}\tag{3.45}$$

The first integral manifold is

$$\begin{aligned}&\frac{1}{2(m_{21}-m_{11})}\left[(x_1 - a_{111})^{2(m_{21}-m_{11})} - (x_{10} - a_{111})^{2(m_{21}-m_{11})}\right]\\&= \frac{a_{110}}{a_{220}}\frac{1}{2n_{11}+2}\left[(x_2 - a_{121})^{2n_{11}+2} - (x_{20} - a_{121})^{2n_{11}+2}\right],\\&\text{for } (m_{21}-m_{11}) \neq 0;\\&\frac{1}{2(m_{21}-m_{11})}\left[(x_1 - a_{111})^{2(m_{21}-m_{11})} - (x_{10} - a_{111})^{2(m_{21}-m_{11})}\right]\\&= \ln\frac{|x_1 - a_{111}|}{|x_{10} - a_{111}|},\\&\text{for } 2(m_{21}-m_{11}) = 0.\end{aligned}\tag{3.46}$$

(**A**) For $2m_{21} \geq 2m_{11} + 1$, the corresponding infinite-equilibriums are given.

$$\begin{aligned}
&(a_{121}, a_{211}) \\
&= \underbrace{([(2m_{11}{+}1), (2n_{11}{+}1)]^{\text{th}}\text{DP}_{+}{:}\text{UP}_{+}, [2(m_{21} - m_{11}) - 1]^{\text{th}}{}_{\text{DP}}\text{SI}{:}_{\text{UP}}\text{SO})}_{([(2m_{11}+1),(2n_{11}+1)],[2(m_{21}-m_{11})-1])\text{-positive hyperbolic parabola sink-to-source}}, \\
&\text{for } a_{110} > 0, a_{220} > 0; \\
&(a_{121}, a_{211}) \\
&= \underbrace{([(2m_{11}{+}1), (2n_{11}{+}1)]^{\text{th}}\text{UP}_{+}{:}\text{DP}_{+}, [2(m_{21} - m_{11}) - 1]^{\text{th}}{}_{\text{UP}}\text{SO}{:}_{\text{DP}}\text{SI})}_{([(2m_{11}+1),(2n_{11}+1)],[2(m_{21}-m_{11})-1])\text{-positive circular parabola source-to-sink}}, \\
&\text{for } a_{110} < 0, a_{220} > 0; \\
&(a_{121}, a_{211}) \\
&= \underbrace{([(2m_{11}{+}1), (2n_{11}{+}1)]^{\text{th}}\text{UP}_{-}{:}\text{DP}_{-}, [2(m_{21} - m_{11}) - 1]^{\text{th}}{}_{\text{UP}}\text{SI}{:}_{\text{DP}}\text{SO})}_{([(2m_{11}+1),(2n_{11}+1)],[2(m_{21}-m_{11})-1])\text{-negative circular parabola sink-to-source}}, \\
&\text{for } a_{110} > 0, a_{220} < 0; \\
&(a_{121}, a_{211}) \\
&= \underbrace{([(2m_{11}{+}1), (2n_{11}{+}1)]^{\text{th}}\text{DP}_{-}{:}\text{UP}_{-}, [2(m_{21} - m_{11}) - 1]^{\text{th}}{}_{\text{DP}}\text{SO}{:}_{\text{UP}}\text{SI})}_{([(2m_{11}+1),(2n_{11}+1)],[2(m_{21}-m_{11})-1])\text{-negative hyperbolic parabola source-to-sink}}, \\
&\text{for } a_{110} < 0, a_{220} < 0.
\end{aligned} \tag{3.47}$$

(**B**) For $2m_{21} < 2m_{11} + 1$, the infinite-equilibriums are given as follows.

$$\begin{aligned}
&(a_{121}, a_{211}) \\
&= \underbrace{([(2m_{11}{+}1), (2n_{11}{+}1)]^{\text{th}}\text{DP}_{+}{:}\text{UP}_{+}, [2(m_{11} - m_{21}){+}1]^{\text{th}}{}_{\text{UA}}\text{SI}{:}_{\text{DA}}\text{SO})}_{([(2m_{11}+1),(2n_{11}+1)],[2(m_{11}-m_{21})+1])\text{-positive hyperbolic asymptote sink-to-source}}, \\
&\text{for } a_{110} > 0, a_{220} > 0; \\
&(a_{121}, a_{211}) \\
&= \underbrace{([(2m_{11}{+}1), (2n_{11}{+}1)]^{\text{th}}\text{UP}_{+}{:}\text{DP}_{+}, [2(m_{11} - m_{21}){+}1]^{\text{th}}{}_{\text{DA}}\text{SO}{:}_{\text{UA}}\text{SI})}_{([(2m_{11}+1),(2n_{11}+1)],[2(m_{11}-m_{21})+1])\text{-positive hyperbolic-secant asymptote source-to-sink}}, \\
&\text{for } a_{110} < 0, a_{220} > 0; \\
&(a_{121}, a_{211}) \\
&= \underbrace{([(2m_{11}{+}1), (2n_{11}{+}1)]^{\text{th}}\text{UP}_{-}{:}\text{DP}_{-}, [2(m_{11} - m_{21}){+}1]^{\text{th}}{}_{\text{DA}}\text{SI}{:}_{\text{UA}}\text{SO})}_{([(2m_{11}+1),(2n_{11}+1)],[2(m_{11}-m_{21})+1])\text{-negative hyperbolic-secant asymptote sink-to-source}}, \\
&\text{for } a_{110} > 0, a_{220} < 0; \\
&(a_{121}, a_{211}) \\
&= \underbrace{([(2m_{11}{+}1), (2n_{11}{+}1)]^{\text{th}}\text{DP}_{-}{:}\text{UP}_{-}, [2(m_{11} - m_{21}){+}1]^{\text{th}}{}_{\text{UA}}\text{SO}{:}_{\text{DA}}\text{SI})}_{([(2m_{11}+1),(2n_{11}+1)],[2(m_{11}-m_{21})+1])\text{-negative hyperbolic asymptote source-to-sink}},
\end{aligned}$$

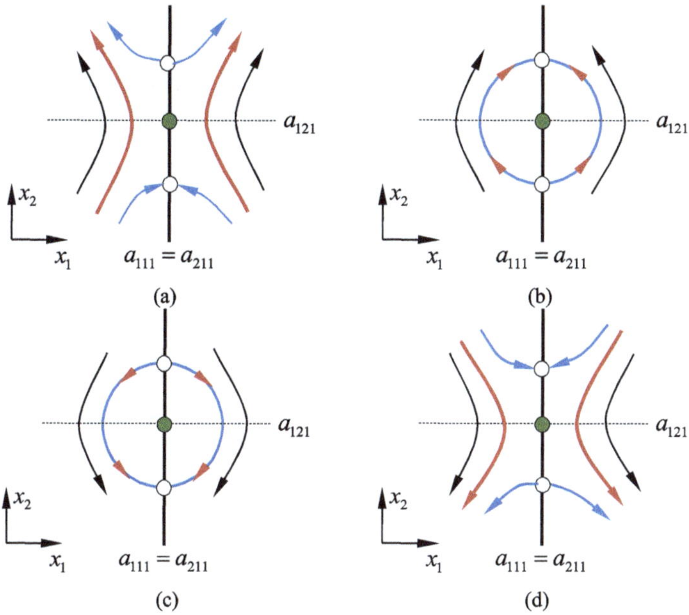

Fig. 3.15 The infinite-equilibriums ($a_{211} = a_{111}, 2m_{21} \geq 2m_{11} + 1, n_1 = 2n_{11} + 1$) for 2-dimensional polynomial systems: **a** ($a_{110} > 0, a_{220} > 0$), **b** ($a_{110} < 0, a_{220} > 0$), **c** ($a_{110} > 0, a_{220} < 0$), **d** ($a_{110} < 0, a_{220} < 0$)

$$\text{for } a_{110} < 0, a_{220} < 0. \tag{3.48}$$

For $x_1^* = a_{111} = a_{211}$, the infinite-equilibrium exists. The infinite-equilibriums are illustrated in Figs. 3.15 and 3.16 for $2m_{21} \geq 2m_{11} + 1$ and $2m_{21} < 2m_{11} + 1$, respectively. For $2m_{21} \geq 2m_{11} + 1$, the singular hyperbolic and circular sink-to-source/source-to-sink are presented. For $2m_{11} + 1 > 2m_{21}$, the hyperbolic and hyperbolic-secant sink-to-source/source-to-sink are presented.

3.3.3 $[(2m_{11}+1), (2n_{11}+1)]$ and $((2n_{11}+1), (2m_{21}))$-*Hybrid Networks*

Consider

$$\dot{x}_2 = a_{220} \prod_{v=1}^{2m_{21}} (x_1 - a_{21v}),$$
$$\dot{x}_1 = a_{110} \prod_{r=1}^{2m_{11}+1} (x_1 - a_{11r}) \prod_{u=1}^{2n_{11}+1} (x_2 - a_{12u}). \tag{3.49}$$

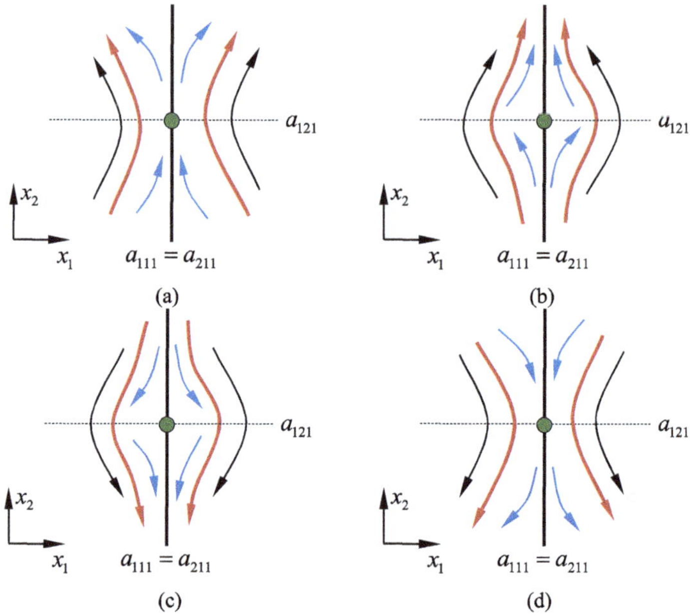

Fig. 3.16 The infinite-equilibriums ($a_{211} = a_{111}, 2m_{21} < 2m_{11} + 1, n_1 = 2n_{11} + 1$) for 2-dimensional polynomial systems: **a** ($a_{110} > 0, a_{220} > 0$), **b** ($a_{110} < 0, a_{220} > 0$), **c** ($a_{110} > 0, a_{220} < 0$), **d** ($a_{110} < 0, a_{220} < 0$)

The first integral manifold is

$$
\begin{aligned}
&\sum_{\nu_0=2m_{11}+1}^{2m_{21}} b_{21\nu_0} \sum_{\substack{\mu_1\in[0,\beta_0-2m_{11}-1],\\ \nu_1\in[2m_{11},\nu_0-1]}}^{\mu_1+\nu_1=\nu_0-1} \sum_{\substack{\mu_2\in[0,\nu_1-2m_{11}-1]\\ \nu_2\in[2m_{11}-1,\nu_1-2]}}^{\mu_2+\nu_2=\nu_1-2} \cdots \sum_{\substack{\mu_{2m_{11}}\in[0,\nu_{2m_{11}-1}-2m_{11}-1]\\ \nu_{2m_{11}}\in[1,\nu_{2m_{11}-1}-2m_{11}]}}^{\mu_{2m_{11}}+\nu_{2m_{11}}=\nu_{2m_{11}-1}-2m_{11}} \prod_{j=1}^{2m_{11}} \frac{(\nu_{j-1}-j)!}{\mu_j!\nu_j!} \\
&\quad \times (a_{11r_{j+1}} - a_{11r_j})^{\mu_j} \frac{1}{\nu_{2m_{11}}} \left[(x_1 - a_{11r_{2m_{11}+1}})^{\nu_{2m_{11}}} - (x_{10} - a_{11r_{2m_{11}+1}})^{\nu_{2m_{11}}} \right] \\
&\quad - \sum_{\nu_0=2m_{11}+1}^{2m_{21}} b_{21\nu_0} \sum_{k=2}^{2m_{11}} \left[\sum_{\substack{\mu_1\in[0,\nu_0-2m_{11}-1),\\ \nu_1\in(2m_{11},\nu_0-1]}}^{\mu_1+\nu_1=\nu_0-1} \sum_{\substack{\mu_2\in[0,\nu_1-2m_{11}-1]\\ \nu_2\in(2m_{11}-1,\nu_1-2)}}^{\mu_2+\nu_2=\nu_1-2} \cdots \right. \\
&\quad \times \left(\sum_{\substack{\mu_{k-1}\in(\nu_{k-2}-2m_{11},\nu_{k-2}-k+1]\\ \nu_{k-1}\in[0,2m_{11}+1-k)}}^{\mu_{k-1}+\nu_{k-1}=\nu_{k-2}-k+1} \prod_{j=1}^{k-1} \frac{(\nu_{j-1}-j)!}{\mu_j!\nu_j!} (a_{11r_{j+1}} - a_{11r_j})^{\mu_j} \right. \\
&\quad \times \sum_{\rho=k}^{2m_{11}+1} A_{\rho 1}^{(r_k)} \ln \frac{|x_1 - a_{11r_\rho}|}{|x_{10} - a_{11r_\rho}|}
\end{aligned}
$$

$$
\begin{aligned}
&+ \sum_{\substack{\mu_{k-1}=(\nu_{k-2}-2m_{11}-1,\nu_{k-2}-2m_{11}] \\ \nu_{k-1}\in[2m_{11}+1-k,2m_{11}-k+2)}}^{\mu_{k-1}+\nu_{k-1}=\nu_{k-2}-k+1} \prod_{j=1}^{k-1} \frac{(\nu_{j-1}-j)!}{\mu_j!\nu_j!}(a_{11r_{j+1}}-a_{11r_j})^{\mu_j} \\
&\times \sum_{\rho=k+1}^{2m_{11}+1} B_{\rho 1}^{(r_k)} \ln \frac{|x_1-a_{11r_\rho}|}{|x_{10}-a_{11r_\rho}|}\Bigg)\Bigg] + b_{210}\sum_{\rho=1}^{2m_{11}+1} A_{\rho 1}^{(r_1)} \ln \frac{|x_1-a_{11r_\rho}|}{|x_{10}-a_{11r_\rho}|} \\
&+ \sum_{\nu_0=1}^{2m_{11}} b_{21\nu_0} \sum_{\rho=2}^{2m_{11}+1} B_{\rho 1}^{(r_1)} \ln \frac{|x_1-a_{11r_\rho}|}{|x_{10}-a_{11r_\rho}|} \\
&= \frac{a_{110}}{a_{220}}\Bigg\{ \frac{1}{2n_{11}+2}\left[(x_2-a_{12u_1})^{2n_{11}+2}-(x_{j_2 0}-a_{12u_1})^{2n_{11}+2}\right] \\
&+ \sum_{\alpha=1}^{2n_{11}-1} \frac{1}{2n_{11}-\alpha+2} b_{12\alpha}\left[(x_2-a_{12u_1})^{2n_{11}-\alpha+2}-(x_{20}-a_{12u_1})^{2n_{11}-\alpha+2}\right]\Bigg\},
\end{aligned}
\tag{3.50}
$$

where

$$
\begin{aligned}
b_{12\alpha} &= \sum_{l_{2n_{11}+1},t_{2n_{11}+1}=0}^{l_{2n_{11}}+t_{2n_{11}+1}=1} \cdots \sum_{l_2,t_2=0}^{l_2+t_2=1} \Big[\prod_{k=2}^{2n_{11}+1} (a_{12u_1}-a_{12u_k})^{l_k}\Big]\delta_{t_2+\cdots+t_{2n_{11}+1}}^{2n_{11}+1-\alpha}; \\
b_{120} &= 1. \\
b_{21\nu_0} &= \sum_{t_1,l_1=0}^{t_1+l_1=1} \cdots \sum_{t_{2m_{21}},l_{2m_{21}}=0}^{t_{2m_{21}}+l_{2m_{21}}=1} \prod_{k=1}^{2m_{21}} (a_{11r_1}-a_{21\nu_k})^{t_k}\delta_{\nu_0}^{\Sigma_{r=1}^{2m_{21}} l_r}; \\
&\lim_{\nu_{2m_{11}}\to 0}\frac{1}{\nu_{2m_{11}}}\left[(x_1-a_{11r_{2m_{11}+1}})^{\nu_{2m_{11}}}-(x_{j_1 0}-a_{11r_{2m_{11}+1}})^{\nu_{2m_{11}}}\right] = \ln\frac{|x_1-a_{11r_{2m_{11}+1}}|}{|x_{10}-a_{11r_{2m_{11}+1}}|};
\end{aligned}
\tag{3.51}
$$

$$
\begin{aligned}
A_{\rho 1}^{(r_k)} &= \frac{1}{\prod_{\sigma=k,\sigma\neq\rho}^{2m_{11}+1}(a_{11r_\rho}-a_{11r_\sigma})}, \text{ for } \rho=\{k,k+1,\ldots,2m_{11}+1\}, \\
B_{\rho 1}^{(r_k)} &= \frac{(a_{11r_\rho}-a_{11r_k})^{\beta_{k-1}-k}}{\prod_{\sigma=k+1,\sigma\neq\rho}^{2m_{11}+1}(a_{11r_\rho}-a_{11r_\sigma})}, \text{ for } \rho=\{k+1,k+2,\ldots,2m_{11}+1\}; \\
k &\in \{1,2,\ldots,2m_{11}\}.
\end{aligned}
\tag{3.52}
$$

For all cases, the $(2n_{11}+1)\times(2m_{21})$-simple equilibriums plus $(2m_{11}+1)\times(2n_{11}+1)$-hyperbolic flows are based on the following hyperbolic and circular sink-to-source and source-to-sink bifurcations for $2m_{21}\geq 2m_{11}+1$:

- $([(2m_{11}+1),(2n_{11}+1)]^{\text{th}}\text{DP}_{+}\text{:UP}_{+},[2(m_{21}-m_{11})-1]^{\text{th}}{}_{\text{DP}}\text{SI:}{}_{\text{UP}}\text{SO})$-positive hyperbolic parabola sink-to-source bifurcation for $(a_{110}>0,a_{220}>0)$,
- $([(2m_{11}+1),(2n_{11}+1)]^{\text{th}}\text{UP}_{+}\text{:DP}_{+},[2(m_{21}-m_{11})-1]^{\text{th}}{}_{\text{UP}}\text{SO:}{}_{\text{DP}}\text{SI})$-positive circular parabola sink-to-source bifurcation for $(a_{110}>0,a_{220}>0)$,
- $([(2m_{11}+1),(2n_{11}+1)]^{\text{th}}\text{UP}_{-}\text{:DP}_{-},[2(m_{21}-m_{11})-1]^{\text{th}}{}_{\text{UP}}\text{SI:}{}_{\text{DP}}\text{SO})$-negative circular parabola source-to-sink bifurcation for $(a_{110}>0,a_{220}<0)$,
- $([(2m_{11}+1),(2n_{11}+1)]^{\text{th}}\text{DP}_{-}\text{:UP}_{-},[2(m_{21}-m_{11})-1]^{\text{th}}{}_{\text{DP}}\text{SO:}{}_{\text{UP}}\text{SI})$-negative hyperbolic parabola source-to-sink bifurcation for $(a_{110}<0,a_{220}<0)$;

or based on the following hyperbolic and hyperbolic-secant asymptote source-to-sink or sink-to-source bifurcations for $2m_{21}<2m_{11}+1$:

- $([(2m_{11}+1),(2n_{11}+1)]^{\text{th}}\text{DP}_{+}\text{:UP}_{+},[2(m_{11}-m_{21})+1]^{\text{th}}{}_{\text{UA}}\text{SI:}{}_{\text{DA}}\text{SO})$-positive hyperbolic asymptote sink-to-source bifurcation for $(a_{110}>0,a_{220}>0)$,
- $([(2m_{11}+1),(2n_{11}+1)]^{\text{th}}\text{UP}_{+}\text{:DP}_{+},[2(m_{11}-m_{21})+1]^{\text{th}}{}_{\text{DA}}\text{SO:}{}_{\text{UA}}\text{SI})$-positive hyperbolic-secant asymptote sink-to-source bifurcation for $(a_{110}>0,a_{220}>0)$,
- $([(2m_{11}+1),(2n_{11}+1)]^{\text{th}}\text{UP}_{-}\text{:DP}_{-},[2(m_{11}-m_{21})+1]^{\text{th}}{}_{\text{DA}}\text{SI:}{}_{\text{UA}}\text{SO})$-negative hyperbolic-secant asymptote source-to-sink bifurcation for $(a_{110}>0,a_{220}<0)$,
- $([(2m_{11}+1),(2n_{11}+1)]^{\text{th}}\text{DP}_{-}\text{:UP}_{-},[2(m_{11}-m_{21})+1]^{\text{th}}{}_{\text{UA}}\text{SO:}{}_{\text{DA}}\text{SI})$-negative hyperbolic asymptote source-to-sink bifurcation for $(a_{110}<0,a_{220}<0)$.

(**A**) Consider $2m_{11}=2m_{21}$ for illustrations, phase portraits for the 2-dimensional systems for the hybrid networks of hyperbolic and hyperbolic-secant flows with the simple equilibriums of the saddles, sinks and sources with $x_1^*=a_{111},a_{211},a_{112},\cdots,a_{21(2m_{21})},a_{11(2m_{11}+1)}$ and $x_2^*=a_{121},a_{122},\cdots,a_{12(2n_{11}+1)}$ are presented in Fig. 3.17a–d for $(a_{110}>0,a_{220}>0)$, $(a_{110}<0,a_{220}>0)$, $(a_{110}>0,a_{220}<0)$ and $(a_{110}<0,a_{220}<0)$.

(**B**) Consider $2m_{11}=2m_{21}$ for illustrations. Phase portraits for the 2-dimensional systems for the hybrid networks of hyperbolic and hyperbolic-secant flows with the simple equilibriums of saddles and centers with $x_1^*=a_{211},a_{111},\cdots,a_{21(2m_{21})},a_{11(2m_{11})},a_{11(2m_{11}+1)}$ and $x_2^*=a_{121},a_{122},\cdots,a_{12(2n_{11}+1)}$ are presented in Fig. 3.18a–d for $(a_{110}>0,a_{220}>0)$, $(a_{110}<0,a_{220}>0)$, $(a_{110}>0,a_{220}<0)$ and $(a_{110}<0,a_{220}<0)$.

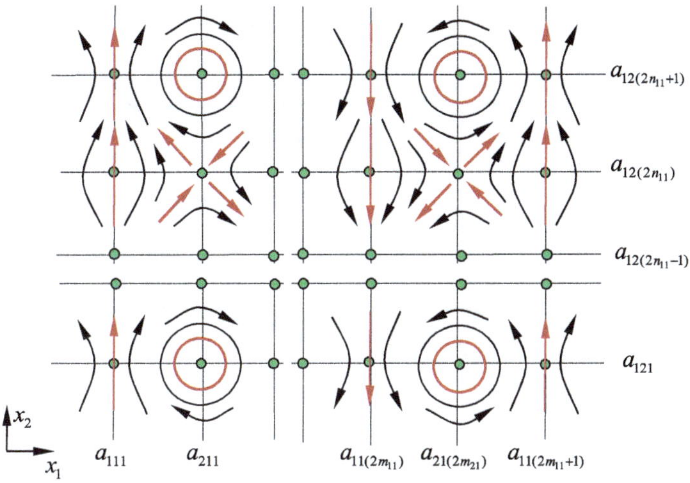

(a)

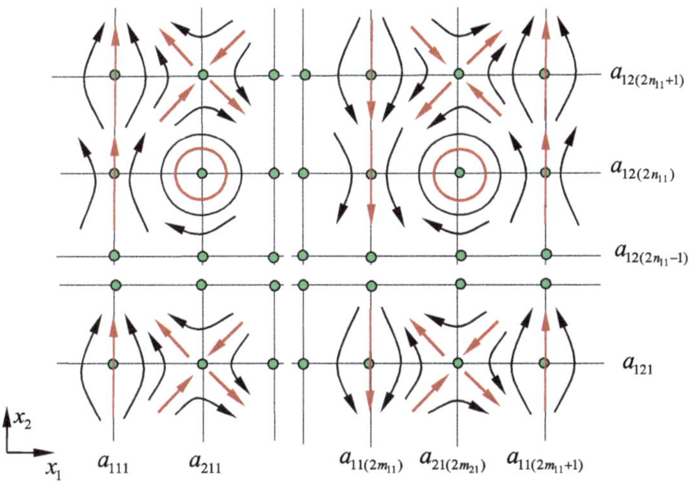

(b)

Fig. 3.17 Phase portraits for 2-dimensional systems with $x_1^* = a_{111}, a_{211}, \ldots, a_{11(2m_{11})}, a_{21(2m_{21})}, a_{11(2m_{11}+1)}$ and $x_2^* = a_{121}, a_{122}, \ldots, a_{12(2n_{11}+1)}$. The networks of flows with saddles and centers: **a** ($a_{110} > 0, a_{220} > 0$), **b** ($a_{110} < 0, a_{220} > 0$), **c** ($a_{110} > 0, a_{220} < 0$), **d** ($a_{110} < 0, a_{220} < 0$)

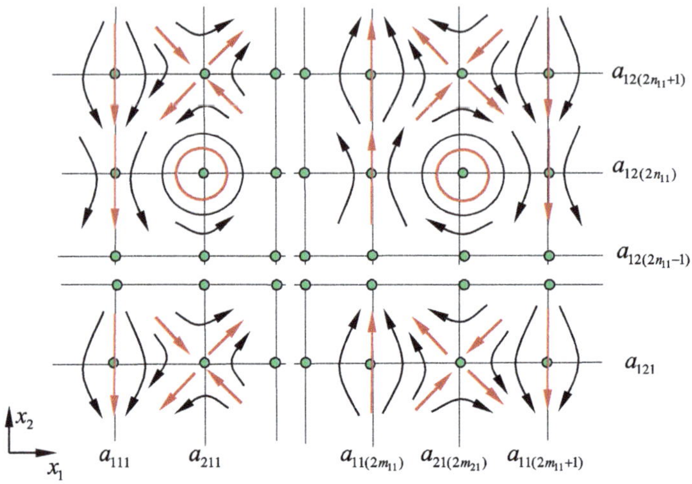

(c)

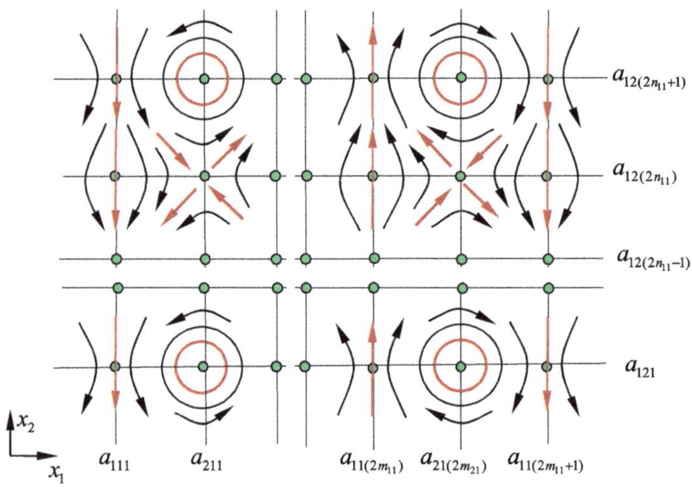

(d)

Fig. 3.17 (continued)

$$
\left\{\begin{array}{cccccc}
(a_{111},a_{12(2n_{11}+1)}) & (a_{12(2n_{11}+1)},a_{211}) & \cdots & (a_{11(2m_{11})},a_{12(2n_{11}+1)}) & (a_{12(2n_{11}+1)},a_{21(2m_{21})}) & (a_{11(2m_{11}+1)},a_{12(2n_{11}+1)}) \\
(a_{111},a_{12(2n_{11})}) & (a_{12(2n_{11})},a_{211}) & \cdots & (a_{11(2m_{11})},a_{12(2n_{11})}) & (a_{12(2n_{11})},a_{21(2m_{21})}) & (a_{11(2m_{11}+1)},a_{12(2n_{11})}) \\
\vdots & \vdots & \vdots & \vdots & \vdots & \vdots \\
(a_{111},a_{121}) & (a_{121},a_{211}) & \cdots & (a_{11(2m_{11})},a_{121}) & (a_{121},a_{21(2m_{21})}) & (a_{11(2m_{11}+1)},a_{121})
\end{array}\right\}
$$

$$
=\left\{\begin{array}{cccccc}
\underbrace{(\mathrm{DP{:}UP,pF})}_{\text{hyperbolic flow }(+)} & \underbrace{(\mathrm{DP}_{-},\mathrm{DP}_{+})}_{\text{CW center}} & \cdots & \underbrace{(\mathrm{UP{:}DP,nF})}_{\text{hyperbolic-secant flow }(-)} & \underbrace{(\mathrm{DP}_{+},\mathrm{DP}_{-})}_{\text{CCW center}} & \underbrace{(\mathrm{DP{:}UP,pF})}_{\text{hyperbolic flow }(+)} \\
\underbrace{(\mathrm{UP{:}DP,pF})}_{\text{hyperbolic-secant flow }(+)} & \underbrace{(\mathrm{UP}_{-},\mathrm{UP}_{-})}_{\text{negative saddle}} & \cdots & \underbrace{(\mathrm{DP{:}UP,nF})}_{\text{hyperbolic flow }(-)} & \underbrace{(\mathrm{UP}_{+},\mathrm{UP}_{+})}_{\text{positive saddle}} & \underbrace{(\mathrm{UP{:}DP,pF})}_{\text{hyperbolic-secant flow }(+)} \\
\vdots & \vdots & \vdots & \vdots & \vdots & \vdots \\
\underbrace{(\mathrm{DP{:}UP,pF})}_{\text{hyperbolic flow }(+)} & \underbrace{(\mathrm{DP}_{-},\mathrm{DP}_{+})}_{\text{CW center}} & \cdots & \underbrace{(\mathrm{UP{:}DP,nF})}_{\text{hyperbolic-secant flow }(-)} & \underbrace{(\mathrm{DP}_{+},\mathrm{DP}_{-})}_{\text{CCW center}} & \underbrace{(\mathrm{DP{:}UP,pF})}_{\text{hyperbolic flow }(+)}
\end{array}\right\}_{(2n_{11}+1)\times(2m_{11}+2m_{21}+1)}
$$

for $a_{110} > 0, a_{220} > 0;$ (3.53)

$$
\begin{aligned}
&\left\{\begin{array}{cccccc}
(a_{111},a_{12(2n_{11}+1)}) & (a_{12(2n_{11}+1)},a_{211}) & \cdots & (a_{11(2m_{11})},a_{12(2n_{11}+1)}) & (a_{12(2n_{11}+1)},a_{21(2m_{21})}) & (a_{11(2m_{11}+1)},a_{12(2n_{11}+1)}) \\
(a_{111},a_{12(2n_{11})}) & (a_{12(2n_{11})},a_{211}) & \cdots & (a_{11(2m_{11})},a_{12(2n_{11})}) & (a_{12(2n_{11})},a_{21(2m_{21})}) & (a_{11(2m_{11}+1)},a_{12(2n_{11})}) \\
\vdots & \vdots & \vdots & \vdots & \vdots & \vdots \\
(a_{111},a_{121}) & (a_{121},a_{211}) & \cdots & (a_{11(2m_{11})},a_{121}) & (a_{121},a_{21(2m_{21})}) & (a_{11(2m_{11}+1)},a_{121})
\end{array}\right\} \\
&=\left\{\begin{array}{cccccc}
\underbrace{(\mathrm{UP{:}DP,pF})}_{\text{hyperbolic-secant flow }(+)} & \underbrace{(\mathrm{UP}_-,\mathrm{UP}_-)}_{\text{negative saddle}} & \cdots & \underbrace{(\mathrm{UP{:}DP,nF})}_{\text{hyperbolic-secant flow }(-)} & \underbrace{(\mathrm{UP}_+,\mathrm{UP}_+)}_{\text{positive saddle}} & \underbrace{(\mathrm{UP{:}DP,pF})}_{\text{hyperbolic-secant flow }(+)} \\
\underbrace{(\mathrm{DP{:}UP,pF})}_{\text{hyperbolic flow }(+)} & \underbrace{(\mathrm{DP}_-,\mathrm{DP}_+)}_{\text{CW center}} & \cdots & \underbrace{(\mathrm{DP{:}UP,nF})}_{\text{hyperbolic flow }(-)} & \underbrace{(\mathrm{DP}_+,\mathrm{DP}_-)}_{\text{CCW center}} & \underbrace{(\mathrm{DP{:}UP,pF})}_{\text{hyperbolic flow }(+)} \\
\vdots & \vdots & \vdots & \vdots & \vdots & \vdots \\
\underbrace{(\mathrm{UP{:}DP,pF})}_{\text{hyperbolic-secant flow }(+)} & \underbrace{(\mathrm{UP}_-,\mathrm{UP}_-)}_{\text{negative saddle}} & \cdots & \underbrace{(\mathrm{DP{:}UP,nF})}_{\text{hyperbolic flow }(-)} & \underbrace{(\mathrm{UP}_+,\mathrm{UP}_+)}_{\text{positive saddle}} & \underbrace{(\mathrm{UP{:}DP,pF})}_{\text{hyperbolic-secant flow }(+)}
\end{array}\right\}_{(2n_{11}+1)\times(2m_{11}+2m_{21}+1)}
\end{aligned}
\tag{3.54}
$$

for $a_{110} < 0, a_{220} > 0$;

$$\left\{\begin{array}{cccccc}(a_{111},a_{12(2n_{11}+1)}) & (a_{12(2n_{11}+1)},a_{211}) & \cdots & (a_{11(2m_{11})},a_{12(2n_{11}+1)}) & (a_{12(2n_{11}+1)},a_{21(2m_{21})}) & (a_{11(2m_{11}+1)},a_{12(2n_{11}+1)})\\ (a_{111},a_{12(2n_{11})}) & (a_{12(2n_{11})},a_{211}) & \cdots & (a_{11(2m_{11})},a_{12(2n_{11})}) & (a_{12(2n_{11})},a_{21(2m_{21})}) & (a_{11(2m_{11}+1)},a_{12(2n_{11})})\\ \vdots & \vdots & \vdots & \vdots & \vdots & \vdots\\ (a_{111},a_{121}) & (a_{121},a_{211}) & \cdots & (a_{11(2m_{11})},a_{121}) & (a_{121},a_{21(2m_{21})}) & (a_{11(2m_{11}+1)},a_{121})\end{array}\right\}$$
$$=\left\{\begin{array}{cccccc}\underbrace{(\text{UP:DP,nF})}_{\text{hyperbolic-secant flow }(-)} & \underbrace{(\text{UP}_+,\text{UP}_+)}_{\text{positive saddle}} & \cdots & \underbrace{(\text{UP:DP,pF})}_{\text{hyperbolic-secant flow }(+)} & \underbrace{(\text{UP}_-,\text{UP}_-)}_{\text{negative saddle}} & \underbrace{(\text{UP:DP,nF})}_{\text{hyperbolic-secant flow }(-)}\\ \underbrace{(\text{DP:UP,nF})}_{\text{hyperbolic flow }(-)} & \underbrace{(\text{DP}_+,\text{DP}_-)}_{\text{CCW center}} & \cdots & \underbrace{(\text{DP:UP,pF})}_{\text{hyperbolic flow }(+)} & \underbrace{(\text{DP}_-,\text{DP}_+)}_{\text{CW center}} & \underbrace{(\text{DP:UP,nF})}_{\text{hyperbolic flow }(-)}\\ \vdots & \vdots & \vdots & \vdots & \vdots & \vdots\\ \underbrace{(\text{UP:DP,nF})}_{\text{hyperbolic-secant flow }(-)} & \underbrace{(\text{UP}_+,\text{UP}_+)}_{\text{positive saddle}} & \cdots & \underbrace{(\text{UP:DP,pF})}_{\text{hyperbolic-secant flow }(+)} & \underbrace{(\text{UP}_-,\text{UP}_-)}_{\text{negative saddle}} & \underbrace{(\text{UP:DP,nF})}_{\text{hyperbolic-secant flow }(-)}\end{array}\right\}_{(2n_{11}+1)\times(2m_{11}+2m_{21}+1)}$$

for $a_{110}>0$, $a_{220}<0$; (3.55)

$$\begin{Bmatrix} (a_{111},a_{12(2n_{11}+1)}) & (a_{12(2n_{11}+1)},a_{211}) & \cdots & (a_{11(2m_{11})},a_{12(2n_{11}+1)}) & (a_{12(2n_{11}+1)},a_{21(2m_{21})}) & (a_{11(2m_{11}+1)},a_{12(2n_{11}+1)}) \\ (a_{111},a_{12(2n_{11})}) & (a_{12(2n_{11})},a_{211}) & \cdots & (a_{11(2m_{11})},a_{12(2n_{11})}) & (a_{12(2n_{11})},a_{21(2m_{21})}) & (a_{11(2m_{11}+1)},a_{12(2n_{11})}) \\ \vdots & \vdots & \vdots & \vdots & \vdots & \vdots \\ (a_{111},a_{121}) & (a_{121},a_{211}) & \cdots & (a_{11(2m_{11})},a_{121}) & (a_{121},a_{21(2m_{21})}) & (a_{11(2m_{11}+1)},a_{121}) \end{Bmatrix}$$
$$= \begin{Bmatrix} \underbrace{(\mathrm{DP{:}UP,nF})}_{\text{hyperbolic flow }(-)} & \underbrace{(\mathrm{DP}_+,\mathrm{DP}_-)}_{\text{CCW center}} & \cdots & \underbrace{(\mathrm{DP{:}UP,pF})}_{\text{hyperbolic flow }(+)} & \underbrace{(\mathrm{DP}_-,\mathrm{DP}_+)}_{\text{CW center}} & \underbrace{(\mathrm{DP{:}UP,nF})}_{\text{hyperbolic flow }(-)} \\ \underbrace{(\mathrm{UP{:}DP,nF})}_{\text{hyperbolic-secant flow }(-)} & \underbrace{(\mathrm{UP}_+,\mathrm{UP}_+)}_{\text{positive saddle}} & \cdots & \underbrace{(\mathrm{UP{:}DP,pF})}_{\text{hyperbolic-secant flow }(+)} & \underbrace{(\mathrm{UP}_-,\mathrm{UP}_-)}_{\text{negative saddle}} & \underbrace{(\mathrm{UP{:}DP,nF})}_{\text{hyperbolic-secant flow }(-)} \\ \vdots & \vdots & \vdots & \vdots & \vdots & \vdots \\ \underbrace{(\mathrm{DP{:}UP,nF})}_{\text{hyperbolic flow }(-)} & \underbrace{(\mathrm{DP}_+,\mathrm{DP}_-)}_{\text{CCW center}} & \cdots & \underbrace{(\mathrm{DP{:}UP,pF})}_{\text{hyperbolic flow }(+)} & \underbrace{(\mathrm{DP}_-,\mathrm{DP}_+)}_{\text{CW center}} & \underbrace{(\mathrm{DP{:}UP,nF})}_{\text{hyperbolic flow }(-)} \end{Bmatrix}_{(2n_{11}+1)\times(2m_{11}+2m_{21}+1)} \tag{3.56}$$

for $a_{110} < 0, a_{220} < 0$;

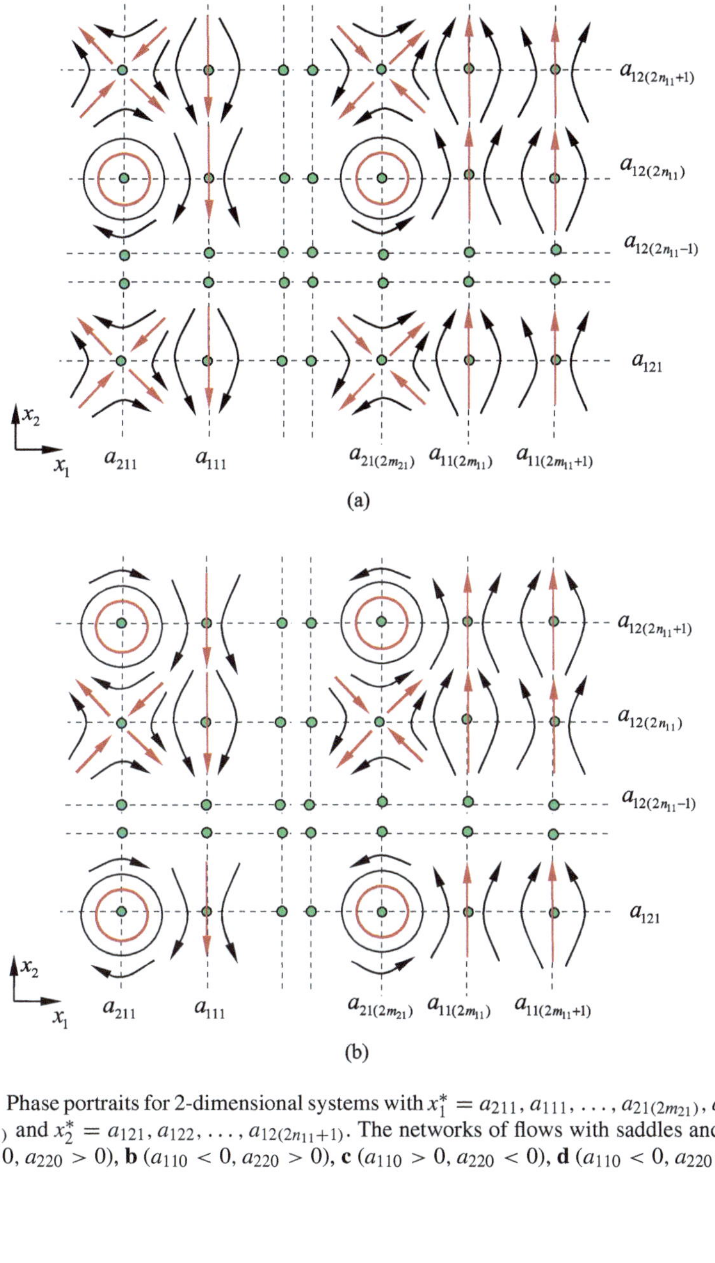

Fig. 3.18 Phase portraits for 2-dimensional systems with $x_1^* = a_{211}, a_{111}, \ldots, a_{21(2m_{21})}, a_{11(2m_{11})}, a_{11(2m_{11}+1)}$ and $x_2^* = a_{121}, a_{122}, \ldots, a_{12(2n_{11}+1)}$. The networks of flows with saddles and centers: **a** ($a_{110} > 0, a_{220} > 0$), **b** ($a_{110} < 0, a_{220} > 0$), **c** ($a_{110} > 0, a_{220} < 0$), **d** ($a_{110} < 0, a_{220} < 0$)

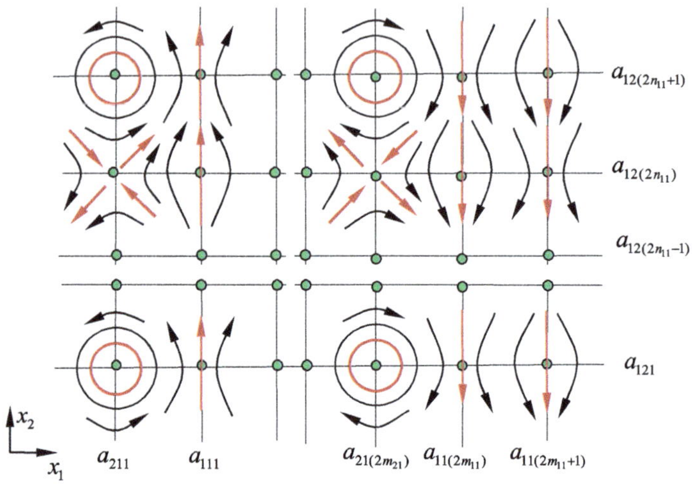

(c)

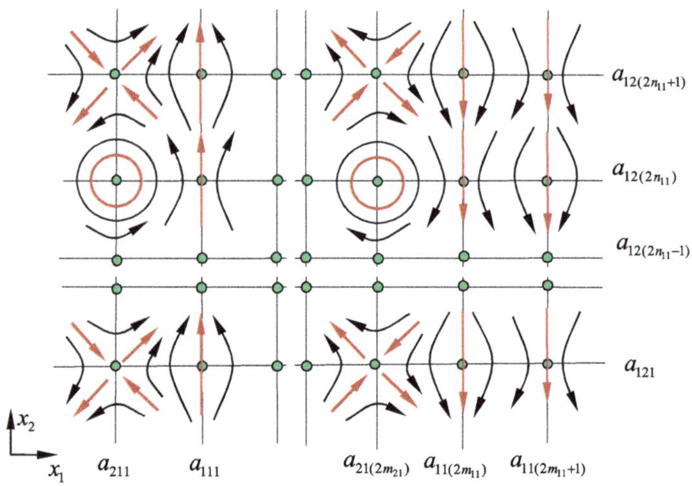

(d)

Fig. 3.18 (continued)

$$
\begin{Bmatrix}
(a_{12(2n_{11}+1)},a_{211}) & (a_{111},a_{12(2n_{11}+1)}) & \cdots & (a_{12(2n_{11}+1)},a_{21(2m_{21})}) & (a_{11(2m_{11})},a_{12(2n_{11}+1)}) & (a_{11(2m_{11}+1)},a_{12(2n_{11}+1)}) \\
(a_{12(2n_{11})},a_{211}) & (a_{111},a_{12(2n_{11})}) & \cdots & (a_{12(2n_{11})},a_{21(2m_{21})}) & (a_{11(2m_{11})},a_{12(2n_{11})}) & (a_{11(2m_{11}+1)},a_{12(2n_{11})}) \\
\vdots & \vdots & \vdots & \vdots & \vdots & \vdots \\
(a_{121},a_{211}) & (a_{111},a_{121}) & \cdots & (a_{121},a_{21(2m_{21})}) & (a_{11(2m_{11})},a_{121}) & (a_{11(2m_{11}+1)},a_{121})
\end{Bmatrix}
$$

$$
= \begin{Bmatrix}
\underbrace{(\mathrm{UP}_-,\mathrm{UP}_-)}_{\text{positive saddle}} & \underbrace{(\mathrm{UP{:}DP,nF})}_{\text{hyperbolic-secant flow }(-)} & \cdots & \underbrace{(\mathrm{UP}_+,\mathrm{UP}_+)}_{\text{positive saddle}} & \underbrace{(\mathrm{UP{:}DP,pF})}_{\text{hyperbolic-secant flow }(+)} & \underbrace{(\mathrm{DP{:}UP,pF})}_{\text{hyperbolic flow }(+)} \\
\underbrace{(\mathrm{DP}_-,\mathrm{DP}_+)}_{\text{CW center}} & \underbrace{(\mathrm{DP{:}UP,nF})}_{\text{hyperbolic flow }(-)} & \cdots & \underbrace{(\mathrm{DP}_+,\mathrm{DP}_-)}_{\text{CCW center}} & \underbrace{(\mathrm{DP{:}UP,pF})}_{\text{hyperbolic flow }(+)} & \underbrace{(\mathrm{UP{:}DP,pF})}_{\text{hyperbolic-secant flow }(+)} \\
\vdots & \vdots & \vdots & \vdots & \vdots & \vdots \\
\underbrace{(\mathrm{UP}_-,\mathrm{UP}_-)}_{\text{positive saddle}} & \underbrace{(\mathrm{UP{:}DP,nF})}_{\text{hyperbolic-secant flow }(-)} & \cdots & \underbrace{(\mathrm{UP}_+,\mathrm{UP}_+)}_{\text{positive saddle}} & \underbrace{(\mathrm{UP{:}DP,pF})}_{\text{hyperbolic-secant flow }(+)} & \underbrace{(\mathrm{DP{:}UP,pF})}_{\text{hyperbolic flow }(+)}
\end{Bmatrix}_{(2n_{11}+1)\times(2m_{11}+2m_{21}+1)}
$$

for $a_{110}>0,\ a_{220}>0$; (3.57)

$$
\left\{
\begin{array}{cccccc}
(a_{12(2n_{11}+1)},a_{211}) & (a_{111},a_{12(2n_{11}+1)}) & \cdots & (a_{12(2n_{11}+1)},a_{21(2m_{21})}) & (a_{11(2m_{11})},a_{12(2n_{11}+1)}) & (a_{11(2m_{11}+1)},a_{12(2n_{11}+1)}) \\
(a_{12(2n_{11})},a_{211}) & (a_{111},a_{12(2n_{11})}) & \cdots & (a_{12(2n_{11})},a_{21(2m_{21})}) & (a_{11(2m_{11})},a_{12(2n_{11})}) & (a_{11(2m_{11}+1)},a_{12(2n_{11})}) \\
\vdots & \vdots & \vdots & \vdots & \vdots & \vdots \\
(a_{121},a_{211}) & (a_{111},a_{121}) & \cdots & (a_{121},a_{21(2m_{21})}) & (a_{11(2m_{11})},a_{121}) & (a_{11(2m_{11}+1)},a_{121})
\end{array}
\right\}
$$
$$
= \left\{
\begin{array}{cccccc}
\underbrace{(\mathrm{DP}_-,\mathrm{DP}_+)}_{\text{CW center}} & \underbrace{(\mathrm{DP{:}UP,nF})}_{\text{hyperbolic flow }(-)} & \cdots & \underbrace{(\mathrm{DP}_+,\mathrm{DP}_-)}_{\text{CCW center}} & \underbrace{(\mathrm{DP{:}UP,pF})}_{\text{hyperbolic flow }(+)} & \underbrace{(\mathrm{UP{:}DP,pF})}_{\text{hyperbolic-secant flow }(+)} \\
\underbrace{(\mathrm{UP}_-,\mathrm{UP}_-)}_{\text{negative saddle}} & \underbrace{(\mathrm{UP{:}DP,nF})}_{\text{hyperbolic-secant flow }(-)} & \cdots & \underbrace{(\mathrm{UP}_+,\mathrm{UP}_+)}_{\text{positive saddle}} & \underbrace{(\mathrm{UP{:}DP,pF})}_{\text{hyperbolic-secant flow }(+)} & \underbrace{(\mathrm{DP{:}UP,pF})}_{\text{hyperbolic flow }(+)} \\
\vdots & \vdots & \vdots & \vdots & \vdots & \vdots \\
\underbrace{(\mathrm{DP}_-,\mathrm{DP}_+)}_{\text{CW center}} & \underbrace{(\mathrm{DP{:}UP,nF})}_{\text{hyperbolic flow }(-)} & \cdots & \underbrace{(\mathrm{DP}_+,\mathrm{DP}_-)}_{\text{CCW center}} & \underbrace{(\mathrm{DP{:}UP,pF})}_{\text{hyperbolic flow }(+)} & \underbrace{(\mathrm{DP{:}UP,pF})}_{\text{hyperbolic-secant flow }(+)}
\end{array}
\right\}_{(2n_{11}+1)\times(2m_{11}+2m_{21}+1)}
$$

for $a_{110} < 0, a_{220} > 0$; (3.58)

$$
\left\{\begin{array}{cccccc}
(a_{12(2n_{11}+1)},a_{211}) & (a_{111},a_{12(2n_{11}+1)}) & \cdots & (a_{12(2n_{11}+1)},a_{21(2m_{21})}) & (a_{11(2m_{11})},a_{12(2n_{11}+1)}) & (a_{11(2m_{11}+1)},a_{12(2n_{11}+1)}) \\
(a_{12(2n_{11})},a_{211}) & (a_{111},a_{12(2n_{11})}) & \cdots & (a_{12(2n_{11})},a_{21(2m_{21})}) & (a_{11(2m_{11})},a_{12(2n_{11})}) & (a_{11(2m_{11}+1)},a_{12(2n_{11})}) \\
\vdots & \vdots & \vdots & \vdots & \vdots & \vdots \\
(a_{121},a_{211}) & (a_{111},a_{121}) & \cdots & (a_{121},a_{21(2m_{21})}) & (a_{11(2m_{11})},a_{121}) & (a_{11(2m_{11}+1)},a_{121})
\end{array}\right\}
$$

$$
=\left\{\begin{array}{cccccc}
\underbrace{(\mathrm{UP}_+,\mathrm{UP}_+)}_{\text{positive saddle}} & \underbrace{(\mathrm{DP:UP,pF})}_{\text{hyperbolic flow }(+)} & \cdots & \underbrace{(\mathrm{UP}_-,\mathrm{UP}_-)}_{\text{negative saddle}} & \underbrace{(\mathrm{DP:UP,nF})}_{\text{hyperbolic flow }(-)} & \underbrace{(\mathrm{UP:DP,nF})}_{\text{hyperbolic-secant flow }(-)} \\
\underbrace{(\mathrm{DP}_+,\mathrm{DP}_-)}_{\text{CCW center}} & \underbrace{(\mathrm{UP:DP,pF})}_{\text{hyperbolic-secant flow }(+)} & \cdots & \underbrace{(\mathrm{DP}_-,\mathrm{DP}_+)}_{\text{CW center}} & \underbrace{(\mathrm{UP:DP,nF})}_{\text{hyperbolic-secant flow }(-)} & \underbrace{(\mathrm{DP:UP,nF})}_{\text{hyperbolic flow }(-)} \\
\vdots & \vdots & \vdots & \vdots & \vdots & \vdots \\
\underbrace{(\mathrm{UP}_+,\mathrm{UP}_+)}_{\text{positive saddle}} & \underbrace{(\mathrm{DP:UP,pF})}_{\text{hyperbolic flow }(+)} & \cdots & \underbrace{(\mathrm{UP}_-,\mathrm{UP}_-)}_{\text{negative saddle}} & \underbrace{(\mathrm{DP:UP,nF})}_{\text{hyperbolic flow }(-)} & \underbrace{(\mathrm{UP:DP,nF})}_{\text{hyperbolic-secant flow }(-)}
\end{array}\right\}_{(2n_{11}+1)\times(2m_{11}+2m_{21}+1)}
$$

for $a_{110}>0,\ a_{220}<0;$ (3.59)

$$
\left\{\begin{array}{cccccc}
(a_{12(2n_{11}+1)},a_{211}) & (a_{111},a_{12(2n_{11}+1)}) & \cdots & (a_{12(2n_{11}+1)},a_{21(2m_{21})}) & (a_{11(2m_{11})},a_{12(2n_{11}+1)}) & (a_{11(2m_{11}+1)},a_{12(2n_{11}+1)}) \\
(a_{12(2n_{11})},a_{211}) & (a_{111},a_{12(2n_{11})}) & \cdots & (a_{12(2n_{11})},a_{21(2m_{21})}) & (a_{11(2m_{11})},a_{12(2n_{11})}) & (a_{11(2m_{11}+1)},a_{12(2n_{11})}) \\
\vdots & \vdots & \vdots & \vdots & \vdots & \vdots \\
(a_{121},a_{211}) & (a_{111},a_{121}) & \cdots & (a_{121},a_{21(2m_{21})}) & (a_{11(2m_{11})},a_{121}) & (a_{11(2m_{11}+1)},a_{121})
\end{array}\right\}
$$

$$
= \left\{\begin{array}{cccccc}
\underbrace{(\mathrm{UP}_+,\mathrm{UP}_+)}_{\text{positive saddle}} & \underbrace{(\mathrm{UP{:}DP,pF})}_{\text{hyperbolic-secant flow }(+)} & \cdots & \underbrace{(\mathrm{UP}_-,\mathrm{UP}_-)}_{\text{negative saddle}} & \underbrace{(\mathrm{UP{:}DP,nF})}_{\text{hyperbolic-secant flow }(-)} & \underbrace{(\mathrm{DP{:}UP,nF})}_{\text{hyperbolic flow }(-)} \\
\underbrace{(\mathrm{DP}_+,\mathrm{DP}_-)}_{\text{CCW center}} & \underbrace{(\mathrm{DP{:}UP,pF})}_{\text{hyperbolic flow }(+)} & \cdots & \underbrace{(\mathrm{DP}_-,\mathrm{DP}_+)}_{\text{CCW center}} & \underbrace{(\mathrm{DP{:}UP,nF})}_{\text{hyperbolic flow }(-)} & \underbrace{(\mathrm{UP{:}DP,nF})}_{\text{hyperbolic-secant flow }(-)} \\
\vdots & \vdots & \vdots & \vdots & \vdots & \vdots \\
\underbrace{(\mathrm{UP}_+,\mathrm{UP}_+)}_{\text{positive saddle}} & \underbrace{(\mathrm{UP{:}DP,pF})}_{\text{hyperbolic-secant flow }(+)} & \cdots & \underbrace{(\mathrm{UP}_-,\mathrm{UP}_-)}_{\text{negative saddle}} & \underbrace{(\mathrm{UP{:}DP,nF})}_{\text{hyperbolic-secant flow }(-)} & \underbrace{(\mathrm{DP{:}UP,nF})}_{\text{hyperbolic flow }(-)}
\end{array}\right\}_{(2n_{11}+1)\times(2m_{11}+2m_{21}+1)}
$$

for $a_{110} < 0,\ a_{220} < 0$; (3.60)

3.4 $[(2m_{11}), (2n_{11}+1)]$ and $((2n_{11}+1), (2m_{21}))$-Singular Systems

In this section, there are four global bifurcations of the 1-dimensional flows through singular hyperbolic-to-hyperbolic-secant flows, and four global bifurcations of equilibriums through singular parabola-saddle bifurcations. The corresponding global bifurcations generate networks of hyperbolic and hyperbolic-secant flows with saddles and centers. The infinite-equilibriums are the appearing and switching bifurcations for the singular flows and singular equilibriums and for hybrid networks of hyperbolic flows and saddles and centers.

3.4.1 $[(2m_{11}), (2n_{11}+1)]$*-Flows and* $((2n_{11}+1), (2m_{21}))$*-Equilibriums*

Consider a polynomial dynamical system for $a_{111} \neq a_{211}$ as

$$\begin{aligned} \dot{x}_2 &= a_{220}(x_1 - a_{211})^{2m_{21}}, \\ \dot{x}_1 &= a_{110}(x_1 - a_{111})^{2m_{11}}(x_2 - a_{121})^{2n_{11}+1}. \end{aligned} \tag{3.61}$$

The first integral manifold is

$$\begin{aligned} &\sum_{\alpha,\beta=0}^{\alpha+\beta=2m_{21}} \frac{(2m_{21})!}{\alpha!\beta!}\frac{(a_{111}-a_{211})^{\alpha}}{\beta-2m_{11}+1}\left[(x_1-a_{111})^{\beta-2m_{11}+1}-(x_{10}-a_{111})^{\beta-2m_{11}+1}\right] \\ &= \frac{a_{j_1j_10}}{a_{j_2j_20}}\frac{1}{2n_{11}+2}\left[(x_2-a_{121})^{2n_{11}+2}-(x_{20}-a_{121})^{2n_{11}+2}\right], \\ &\text{for } \beta-2m_{11}+1\neq 0; \\ &\frac{1}{\beta-2m_{11}+1}\left[(x_1-a_{111})^{\beta-2m_{11}+1}-(x_{10}-a_{111})^{\beta-2m_{11}+1}\right] = \ln\frac{|x_1-a_{111}|}{|x_{10}-a_{111}|}, \\ &\text{for } \beta-2m_{11}+1=0. \end{aligned} \tag{3.62}$$

The singular hyperbolic-to-hyperbolic-secant flows with singular parabola-saddle are illustrated in Figs. 3.19 and 3.20. Through such singular flows, the corresponding global bifurcations generate from a hyperbolic (hyperbolic-secant) flow to a network of singular hyperbolic and hyperbolic-secant flows. Through such singular equilibriums, the corresponding global bifurcations generate from one of a flow to a network of singular saddles and centers equilibriums.

For $(x_1^*, x_2^*) = (a_{111}, a_{121})$ with $\dot{x}_1 = 0$, the $[(2m_{11}), (2n_{11}+1)]$-singular flows exist with $\dot{x}_2 = a_{220}(a_{111} - a_{211})^{2m_{21}}$. For $(x_1^*, x_2^*) = (a_{111}, a_{121})$, the 1-dimensional flow direction is determined by $\dot{x}_2 \neq 0$. For $(x_2^*, x_1^*) = (a_{121}, a_{211})$ with $\dot{x}_1 = \dot{x}_2 = 0$, the $((2n_{11}+1), (2m_{21}))$-singular equilibriums exist.

Fig. 3.19 Phase portraits ($a_{111} > a_{211}$) of the ([$(2m_{11})$, $(2n_{11}+1)$], xF)-hyperbolic-secant-to-hyperbolic singular flows at $(x_1^*, x_2^*) = (a_{111}, a_{121})$ and $((2n_{11}+1), (2m_{21}))$-parabola-saddle at $(x_2^*, x_1^*) = (a_{121}, a_{211})$ for 2-dimensional polynomial systems: **a** ($a_{110} > 0, a_{220} > 0$), **b** ($a_{110} < 0, a_{220} > 0$), **c** ($a_{110} > 0, a_{220} < 0$), **d** ($a_{110} < 0, a_{220} < 0$)

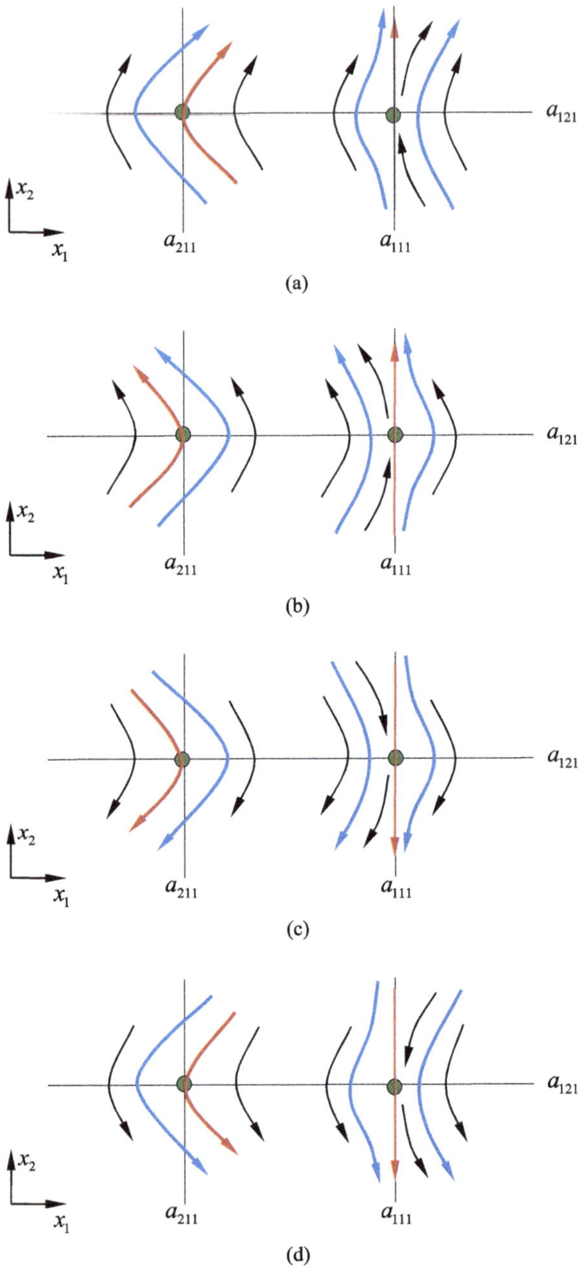

Fig. 3.20 Phase portraits ($a_{211} > a_{111}$) of the ([$(2m_{11})$, $(2n_{11}+1)$], xF)-hyperbolic and hyperbolic-secant singular flows at $(x_1^*, x_2^*) = (a_{111}, a_{121})$ and ($(2n_{11}+1)$, $(2m_{21})$)-positive and negative saddles, CCW and CW centers at $(x_2^*, x_1^*) = (a_{121}, a_{211})$ for 2-dimensional polynomial systems: **a** ($a_{110} > 0, a_{220} > 0$), **b** ($a_{110} < 0, a_{220} > 0$), **c** ($a_{110} > 0, a_{220} < 0$), **d** ($a_{110} < 0, a_{220} < 0$)

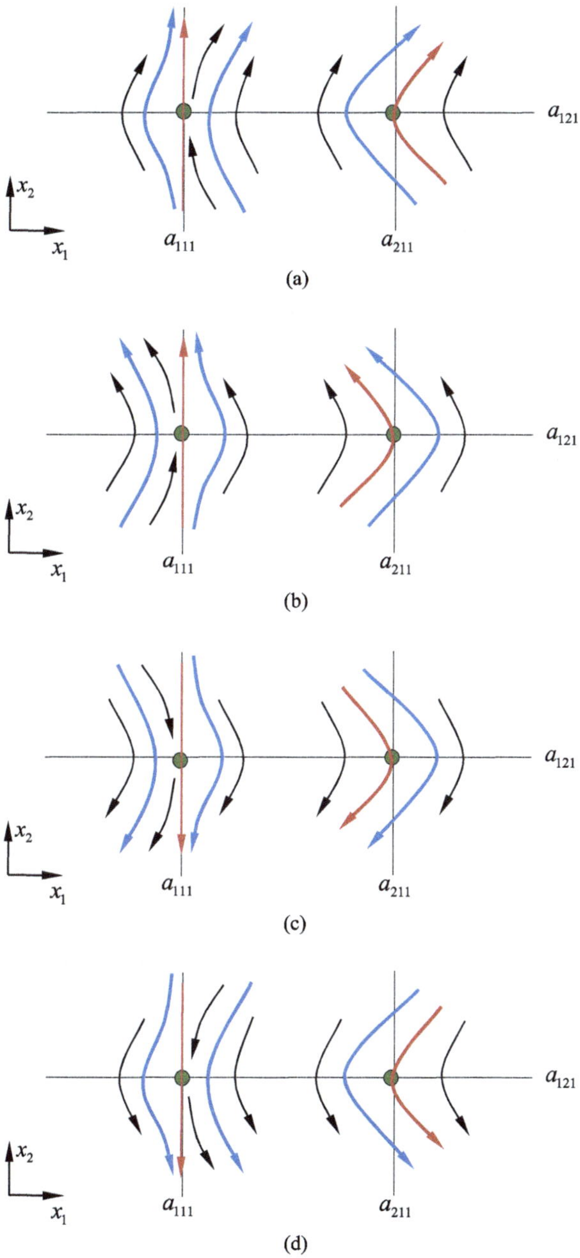

(**A**) For $a_{111} > a_{211}$,

$$\left\{\begin{array}{c}(a_{111}, a_{121})\\(a_{121}, a_{211})\end{array}\right\} = \left\{\begin{array}{c}\underbrace{([(2m_{11}),(2n_{11}+1)]\text{UP:UP,pF})}_{[(2m_{11}),(2n_{11}+1)]\text{-hyperbolic-secant-to-hyperbolic flow }(+)}\\ \underbrace{((2n_{11}+1)\text{UP},(2m_{21})\text{US})}_{[(2n_{11}+1),(2m_{21})]\text{-up-parabola upper-saddle}}\end{array}\right\},$$

for $a_{110} > 0, a_{220} > 0$;

$$\left\{\begin{array}{c}(a_{111}, a_{121})\\(a_{121}, a_{211})\end{array}\right\} = \left\{\begin{array}{c}\underbrace{([(2m_{11}),(2n_{11}+1)]\text{DP:DP,pF})}_{[(2m_{11}),(2n_{11}+1)]\text{-hyperbolic-to-hyperbolic-secant flow }(+)}\\ \underbrace{((2n_{11}+1)\text{DP},(2m_{21})\text{US})}_{[(2n_{11}+1),(2m_{21})]\text{-down-parabola lower-saddle}}\end{array}\right\},$$

for $a_{110} < 0, a_{220} > 0$;

$$\left\{\begin{array}{c}(a_{111}, a_{121})\\(a_{121}, a_{211})\end{array}\right\} = \left\{\begin{array}{c}\underbrace{([(2m_{11}),(2n_{11}+1)]\text{DP:DP,nF})}_{[(2m_{11}),(2n_{11}+1)]\text{-hyperbolic-to-hyperbolic-secant flow }(-)}\\ \underbrace{((2n_{11}+1)\text{DP},(2m_{21})\text{LS})}_{[(2n_{11}+1),(2m_{21})]\text{-down-parabola lower-saddle}}\end{array}\right\},$$

for $a_{110} > 0, a_{220} < 0$;

$$\left\{\begin{array}{c}(a_{111}, a_{121})\\(a_{121}, a_{211})\end{array}\right\} = \left\{\begin{array}{c}\underbrace{([(2m_{11}),(2n_{11}+1)]\text{UP:UP,nF})}_{[(2m_{11}),(2n_{11}+1)]\text{-hyperbolic-secant-to-hyperbolic flow }(-)}\\ \underbrace{((2n_{11}+1)\text{UP},(2m_{21})\text{LS})}_{[(2n_{11}+1),(2m_{21})]\text{-up-parabola lower-saddle}}\end{array}\right\},$$

for $a_{110} < 0, a_{220} < 0$. (3.63)

(**B**) For $a_{211} > a_{111}$,

$$\left\{\begin{array}{c}(a_{121}, a_{211})\\(a_{111}, a_{121})\end{array}\right\} = \left\{\begin{array}{c}\underbrace{((2n_{11}+1)\text{UP},(2m_{21})\text{US})}_{[(2n_{11}+1),(2m_{21})]\text{-up-parabola upper-saddle}}\\ \underbrace{([(2m_{11}),(2n_{11}+1)]\text{UP:UP,pF})}_{[(2m_{11}),(2n_{11}+1)]\text{-hyperbolic-secant-to-hyperbolic flow }(+)}\end{array}\right\},$$

for $a_{110} > 0, a_{220} > 0$;

$$\left\{\begin{array}{c}(a_{121}, a_{211})\\(a_{111}, a_{121})\end{array}\right\} = \left\{\begin{array}{c}\underbrace{((2n_{11}+1)\text{DP},(2m_{21})\text{US})}_{[(2n_{11}+1),(2m_{21})]\text{-down-parabola lower-saddle}}\\ \underbrace{([(2m_{11}),(2n_{11}+1)]\text{DP:DP,pF})}_{[(2m_{11}),(2n_{11}+1)]\text{-hyperbolic-to-hyperbolic-secant flow }(+)}\end{array}\right\},$$

for $a_{110} < 0, a_{220} > 0$;

$$\begin{Bmatrix} (a_{121}, a_{211}) \\ (a_{111}, a_{121}) \end{Bmatrix} = \left\{ \begin{array}{c} \underbrace{((2n_{11}+1)\text{DP},(2m_{21})\text{LS})}_{[(2n_{11}+1),(2m_{21})]\text{-down-parabola lower-saddle}} \\ \underbrace{([(2m_{11}),(2n_{11}+1)]\text{DP:DP,nF})}_{[(2m_{11}),(2n_{11}+1)]\text{-hyperbolic-to-hyperbolic-secant flow }(-)} \end{array} \right\},$$

for $a_{110} > 0, a_{220} < 0$;

$$\begin{Bmatrix} (a_{121}, a_{211}) \\ (a_{111}, a_{121}) \end{Bmatrix} = \left\{ \begin{array}{c} \underbrace{((2n_{11}+1)\text{UP},(2m_{21})\text{LS})}_{[(2n_{11}+1),(2m_{21})]\text{-up-parabola lower-saddle}} \\ \underbrace{([(2m_{11}),(2n_{11}+1)]\text{UP:UP,nF})}_{[(2m_{11}),(2n_{11}+1)]\text{-hyperbolic-secant-to-hyperbolic flow }(-)} \end{array} \right\},$$

for $a_{110} < 0, a_{220} < 0$. (3.64)

3.4.2 $[(2m_{11}), (2n_{11}+1)]$ and $((2n_{11}+1), (2m_{21}))$-*Infinite-Equilibriums*

Consider a polynomial dynamical system for $a_{111} = a_{211}$ as

$$\begin{aligned} \dot{x}_2 &= a_{220}(x_1 - a_{211})^{2m_{21}}, \\ \dot{x}_1 &= a_{110}(x_1 - a_{111})^{2m_{11}}(x_2 - a_{121})^{2n_{11}+1}. \end{aligned} \tag{3.65}$$

The first integral manifold is

$$\begin{aligned} &\frac{1}{2(m_{21}-m_{11})+1}\left[(x_1 - a_{211})^{2(m_{21}-m_{11})+1} - (x_{10} - a_{211})^{2(m_{21}-m_{11})+1}\right] \\ &= \frac{a_{110}}{a_{220}}\frac{1}{2n_{11}+2}\left[(x_2 - a_{121})^{2n_{11}+2} - (x_{20} - a_{121})^{2n_{11}+2}\right]. \end{aligned} \tag{3.66}$$

(**A**) For $2m_{21} \geq 2m_{11}$, the corresponding infinite-equilibriums are given.

$$(a_{121}, a_{211}) = \underbrace{([(2m_{11}), (2n_{11}+1)]^{\text{th}}\text{UP}_{+}\text{:UP}_{+},[2(m_{21}-m_{11})]^{\text{th}}{}_{\text{DI:II}}\text{US})}_{([(2m_{11}),(2n_{11}+1)],[2(m_{21}-m_{11})])\text{-inflection up-parabola upper-saddle}},$$

for $a_{110} > 0, a_{220} > 0$;

$$(a_{121}, a_{211}) = \underbrace{([(2m_{11}), (2n_{11}+1)]^{\text{th}}\text{DP}_{-}\text{:DP}_{-},[2(m_{21}-m_{11})]^{\text{th}}{}_{\text{II:DI}}\text{US})}_{([(2m_{11}),(2n_{11}+1)],[2(m_{21}-m_{11})])\text{-inflection down-parabola upper-saddle}},$$

for $a_{110} < 0, a_{220} > 0$;

$$(a_{121}, a_{211}) = \underbrace{([(2m_{11}), (2n_{11}+1)]^{\text{th}}\text{DP}_{+}\text{:DP}_{+},[2(m_{21}-m_{11})]^{\text{th}}{}_{\text{II:DI}}\text{LS})}_{([(2m_{11}+1),(2n_{11}+1)],[2(m_{21}-m_{11})])\text{-inflection down-parabola lower-saddle}},$$

for $a_{110} > 0, a_{220} < 0$;

$$(a_{121}, a_{211}) = \underbrace{([(2m_{11}), (2n_{11}+1)]^{\text{th}}\text{UP}_{-}\text{:UP}_{-},[2(m_{21} - m_{11})]^{\text{th}}{}_{\text{DI:II}}\text{LS})}_{([(2m_{11}+1),(2n_{11}+1)],[2(m_{21}-m_{11})])\text{-inflection up-parabola lower-saddle}},$$
$$\text{for } a_{110} < 0, a_{220} < 0. \tag{3.67}$$

(**B**) For $2m_{21} < 2m_{11}$, the infinite-equilibriums are given as follows.

$$(a_{121}, a_{211}) = \underbrace{([(2m_{11}), (2n_{11}+1)]^{\text{th}}\text{UP}_{+}\text{:UP}_{+},[2(m_{11} - m_{21})]^{\text{th}}{}_{\text{DC:IC}}\text{US})}_{([(2m_{11}),(2n_{11}+1)],[2(m_{11}-m_{21})])\text{-concave up-parabola upper-saddle}},$$
$$\text{for } a_{110} > 0, a_{220} > 0;$$
$$(a_{121}, a_{211}) = \underbrace{([(2m_{11}), (2n_{11}+1)]^{\text{th}}\text{DP}_{-}\text{:DP}_{-},[2(m_{11} - m_{21})]^{\text{th}}{}_{\text{IC:DC}}\text{US})}_{([(2m_{11}),(2n_{11}+1)],[2(m_{11}-m_{21})])\text{-concave down-parabola upper-saddle}},$$
$$\text{for } a_{110} < 0, a_{220} > 0;$$
$$(a_{121}, a_{211}) = \underbrace{([(2m_{11}), (2n_{11}+1)]^{\text{th}}\text{DP}_{+}\text{:DP}_{+},[2(m_{11} - m_{21})]^{\text{th}}{}_{\text{IC:DC}}\text{LS})}_{([(2m_{11}),(2n_{11}+1)],[2(m_{11}-m_{21})])\text{-concave down-parabola lower-saddle}},$$
$$\text{for } a_{110} > 0, a_{220} < 0;$$
$$(a_{121}, a_{211}) = \underbrace{([(2m_{11}), (2n_{11}+1)]^{\text{th}}\text{UP}_{-}\text{:UP}_{-},[2(m_{11} - m_{21})]^{\text{th}}{}_{\text{DC:IC}}\text{LS})}_{([(2m_{11}),(2n_{11}+1)],[2(m_{11}-m_{21})])\text{-concave up-parabola lower-saddle}},$$
$$\text{for } a_{110} < 0, a_{220} < 0. \tag{3.68}$$

For $x_1^* = a_{111} = a_{211}$, the infinite-equilibrium exists. The infinite-equilibriums are illustrated in Figs. 3.21 and 3.22 for $m_{21} \geq m_{11}$ and $m_{21} < m_{11}$, respectively. For $m_{21} \geq m_{11}$, the singular inflection parabola-saddles are presented. For $m_{11} > m_{21}$, the concave parabola-saddles are presented.

3.4.3 $[(2m_{11}), (2n_{11}+1)]$ *and* $((2n_{11}+1), (2m_{21}))$*-Hybrid Networks*

Consider

$$\dot{x}_2 = a_{220} \prod_{v=1}^{2m_{21}} (x_1 - a_{21v}),$$
$$\dot{x}_1 = a_{110} \prod_{r=1}^{2m_{11}} (x_1 - a_{11r}) \prod_{u=1}^{2n_{11}+1} (x_2 - a_{12u}). \tag{3.69}$$

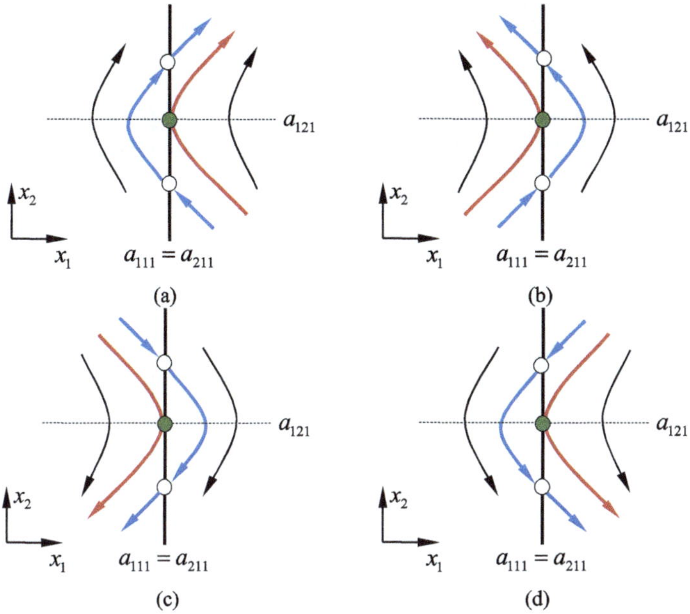

Fig. 3.21 The infinite-equilibriums $(a_{211} = a_{111}, 2m_{21} \geq 2m_{11}, n_1 = 2n_{11}+1)$ for 2-dimensional polynomial systems: **a** $(a_{110} > 0, a_{220} > 0)$, **b** $(a_{110} < 0, a_{220} > 0)$, **c** $(a_{110} > 0, a_{220} < 0)$, **d** $(a_{110} < 0, a_{220} < 0)$

The first integral manifold is

$$
\begin{aligned}
&\sum_{\nu_0=2m_{11}+1}^{2m_{21}} b_{21\nu_0} \sum_{\substack{\mu_1\in[0,\beta_0-2m_{11}],\\ \nu_1\in[2m_{11}-1,\nu_0-1]}}^{\mu_1+\nu_1=\nu_0-1} \sum_{\substack{\mu_2\in[0,\nu_1-2m_{11}]\\ \nu_2\in[2m_{11}-2,\nu_1-2]}}^{\mu_2+\nu_2=\nu_1-2} \cdots \sum_{\substack{\mu_{2m_{11}}\in[0,\nu_{2m_{11}-1}-2m_{11}]\\ \nu_{2m_{11}}\in[1,\nu_{2m_{11}-1}-2m_{11}+1]}}^{\mu_{2m_{11}}+\nu_{2m_{11}}=\nu_{2m_{11}-1}-2m_{11}+1} \\
&\prod_{j=1}^{2m_{11}-1} \frac{(\nu_{j-1}-j)!}{\mu_j!\nu_j!} \times (a_{11r_{j+1}}-a_{11r_j})^{\mu_j} \frac{1}{\nu_{2m_{11}-1}} \Big[(x_1-a_{11r_{2m_{11}}})^{\nu_{2m_{11}-1}}-(x_{10}-a_{11r_{2m_{11}}})^{\nu_{2m_{11}-1}}\Big] \\
&\quad - \sum_{\nu_0=2m_{11}+1}^{2m_{21}} b_{21\nu_0} \sum_{k=2}^{2m_{11}-1} \Bigg[\sum_{\substack{\mu_1\in[0,\nu_0-2m_{11}-1),\\ \nu_1\in(2m_{11},\nu_0-1]}}^{\mu_1+\nu_1=\nu_0-1} \sum_{\substack{\mu_2\in[0,\nu_1-2m_{11}-1)\\ \nu_2\in(2m_{11}-1,\nu_1-2]}}^{\mu_2+\nu_2=\nu_1-2} \cdots \\
&\quad \times \Bigg(\sum_{\substack{\mu_{k-1}\in(\nu_{k-2}-2m_{11}+1,\nu_{k-2}-k+1]\\ \nu_{k-1}\in[0,2m_{11}-k)}}^{\mu_{k-1}+\nu_{k-1}=\nu_{k-2}-k+1} \prod_{j=1}^{k-1} \frac{(\nu_{j-1}-j)!}{\mu_j!\nu_j!} (a_{11r_{j+1}}-a_{11r_j})^{\mu_j} \\
&\quad \times \sum_{\rho=k}^{2m_{11}} A_{\rho 1}^{(s_k)} \ln \frac{|x_1-a_{11s_\rho}|}{|x_{10}-a_{11s_\rho}|}
\end{aligned}
$$

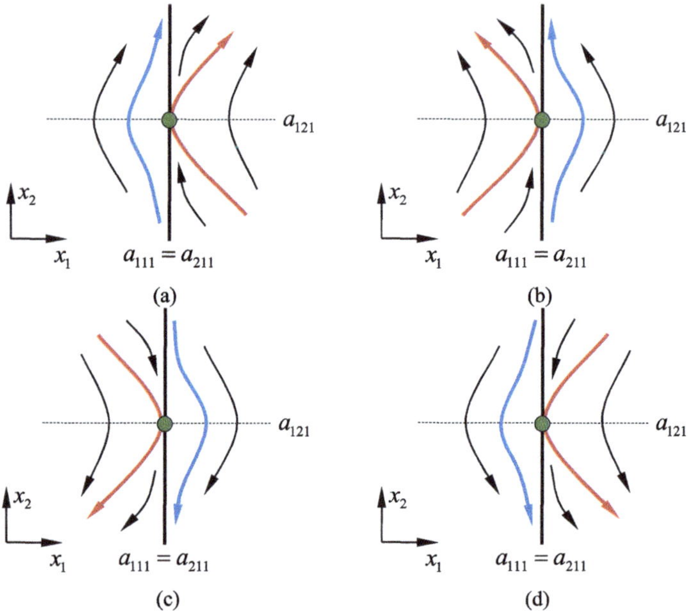

Fig. 3.22 The infinite-equilibriums ($a_{211} = a_{111}, 2m_{21} < 2m_{11}, n_1 = 2n_{11} + 1$) for 2-dimensional polynomial systems: **a** ($a_{110} > 0, a_{220} > 0$), **b** ($a_{110} < 0, a_{220} > 0$), **c** ($a_{110} > 0, a_{220} < 0$), **d** ($a_{110} < 0, a_{220} < 0$)

$$
\begin{aligned}
&+ \sum_{\substack{\mu_{k-1}\in(\nu_{k-2}-2m_{11},\nu_{k-2}-2m_{11}+1]\\ \nu_{k-1}\in[2m_{11}-k,2m_{11}-k+1)}}^{\mu_{k-1}+\nu_{k-1}=\nu_{k-2}-k+1} \prod_{j=1}^{k-1} \frac{(\nu_{j-1}-j)!}{\mu_j!\nu_j!}(a_{11r_{j+1}}-a_{11r_j})^{\mu_j}\\
&\times \sum_{\rho=k+1}^{2m_{11}} B_{\rho 1}^{(r_k)} \ln\frac{|x_1-a_{11r_\rho}|}{|x_{10}-a_{11r_\rho}|}\Bigg)\Bigg] + b_{210}\sum_{\rho=1}^{2m_{11}} A_{\rho 1}^{(r_1)} \ln\frac{|x_1-a_{11r_\rho}|}{|x_{10}-a_{11r_\rho}|}\\
&+ \sum_{\nu_0=1}^{2m_{11}-1} b_{21\nu_0}\sum_{\rho=2}^{2m_{11}} B_{\rho 1}^{(r_1)} \ln\frac{|x_1-a_{11r_\rho}|}{|x_{10}-a_{11r_\rho}|}\\
&= \frac{a_{110}}{a_{220}}\Bigg\{\frac{1}{2n_{11}+2}\Big[(x_2-a_{12u_1})^{2n_{11}+2}-(x_{j_2 0}-a_{12u_1})^{2n_{11}+2}\Big]\\
&+ \sum_{\alpha=1}^{2n_{11}-1}\frac{1}{2n_{11}-\alpha+2}b_{12\alpha}\Big[(x_2-a_{12u_1})^{2n_{11}-\alpha+2}-(x_{20}-a_{12u_1})^{2n_{11}-\alpha+2}\Big]\Bigg\},
\end{aligned}
\tag{3.70}
$$

where

$$b_{12\alpha} = \sum_{l_{2n_{11}+1}, t_{2n_{11}+1}=0}^{l_{2n_{11}+1}+t_{2n_{11}+1}=1} \cdots \sum_{l_2, t_2=0}^{l_2+t_2=1} [\prod_{k=2}^{2n_{11}+1} (a_{12u_1} - a_{12u_k})^{l_k}] \delta_{t_2+\cdots+t_{2n_{11}+1}}^{2n_{11}+1-\alpha};$$

$$b_{120} = 1.$$

$$b_{21v_0} = \sum_{t_1, l_1=0}^{t_1+l_1=1} \cdots \sum_{t_{2m_{21}}, l_{2m_{21}}=0}^{t_{2m_{21}}+t_{2m_{21}}=1} \prod_{k=1}^{2m_{21}} (a_{11r_1} - a_{21v_k})^{t_k} \delta_{v_0}^{\Sigma_{r=1}^{2m_{21}} l_r};$$

$$\lim_{v_{2m_{11}-1}\to 0} \frac{1}{v_{2m_{11}-1}} \left[(x_1 - a_{11r_{2m_{11}}})^{v_{2m_{11}-1}} - (x_{j_1 0} - a_{11r_{2m_{11}}})^{v_{2m_{11}-1}}\right] = \ln \frac{|x_1 - a_{11r_{2m_{11}}}|}{|x_{10} - a_{11r_{2m_{11}}}|}; \tag{3.71}$$

$$\begin{aligned} A_{\rho 1}^{(r_k)} &= \frac{1}{\prod_{\sigma=k, \sigma\neq\rho}^{2m_{11}} (a_{11r_\rho} - a_{11r_\sigma})}, \text{ for } \rho \in \{k, k+1, \ldots, 2m_{11}\}, \\ B_{\rho 1}^{(r_k)} &= \frac{(a_{11r_\rho} - a_{11r_k})^{\beta_{k-1}-k}}{\prod_{\sigma=k+1, \sigma\neq\rho}^{2m_{11}} (a_{11r_\rho} - a_{11r_\sigma})}, \text{ for } \rho \in \{k+1, k+2, \ldots, 2m_{11}\}; \\ & k \in \{1, 2, \ldots, 2m_{11}-1\}. \end{aligned} \tag{3.72}$$

For all cases, the $(2n_{11}+1) \times (2m_{21})$-simple equilibriums plus $(2m_{11}) \times (2n_{11}+1)$-hyperbolic flows are based on the following inflection parabola-saddle bifurcations for $2m_{21} \geq 2m_{11}$:

- $([(2m_{11}), (2n_{11}+1)]^{\text{th}}\text{UP}_+\text{:UP}_+, [2(m_{21}-m_{11})]^{\text{th}}{}_{\text{DI:II}}\text{US})$-inflection up-parabola upper-saddle bifurcation for $(a_{110} > 0, a_{220} > 0)$,
- $([(2m_{11}), (2n_{11}+1)]^{\text{th}}\text{DP}_+\text{:DP}_+, [2(m_{21}-m_{11})]^{\text{th}}{}_{\text{II:DI}}\text{US})$-inflection down-parabola upper-saddle bifurcation for $(a_{110} < 0, a_{220} > 0)$,
- $([(2m_{11}), (2n_{11}+1)]^{\text{th}}\text{DP}_-\text{:DP}_-, [2(m_{21}-m_{11})]^{\text{th}}{}_{\text{II:DI}}\text{LS})$-inflection down-parabola lower-saddle bifurcation for $(a_{110} > 0, a_{220} < 0)$,
- $([(2m_{11}), (2n_{11}+1)]^{\text{th}}\text{UP}_-\text{:UP}_-, [2(m_{21}-m_{11})]^{\text{th}}{}_{\text{DI:II}}\text{LS})$-inflection up-parabola lower-saddle bifurcation for $(a_{110} < 0, a_{220} < 0)$;

or based on the following concave-parabola-saddle bifurcations for $2m_{21} < 2m_{11}$:

- $([(2m_{11}), (2n_{11}+1)]^{\text{th}}\text{UP}_+\text{:UP}_+, [2(m_{11}-m_{21})]^{\text{th}}{}_{\text{DC:IC}}\text{US})$-concave up-parabola upper-saddle bifurcation for $(a_{110} > 0, a_{220} > 0)$,
- $([(2m_{11}), (2n_{11}+1)]^{\text{th}}\text{DP}_+\text{:DP}_+, [2(m_{11}-m_{21})]^{\text{th}}{}_{\text{IC:DC}}\text{US})$-concave down-parabola upper-saddle bifurcation for $(a_{110} > 0, a_{220} < 0)$,
- $([(2m_{11}), (2n_{11}+1)]^{\text{th}}\text{DP}_-\text{:DP}_-, [2(m_{11}-m_{21})]^{\text{th}}{}_{\text{IC:DC}}\text{LS})$-concave down-parabola lower-saddle bifurcation for $(a_{110} > 0, a_{220} < 0)$,
- $([(2m_{11}), (2n_{11}+1)]^{\text{th}}\text{UP}_-\text{:UP}_-, [2(m_{11}-m_{21})]^{\text{th}}{}_{\text{DC:IC}}\text{LS})$-concave up-parabola lower-saddle bifurcation for $(a_{110} < 0, a_{220} < 0)$.

(A) Consider $2m_{11}=2m_{21}$ for illustrations, phase portraits for the 2-dimensional systems for the hybrid networks of hyperbolic and hyperbolic-secant flows with the simple equilibriums of the saddles, sinks and sources with $x_1^*=a_{111}, a_{211}, a_{112}, \cdots,$

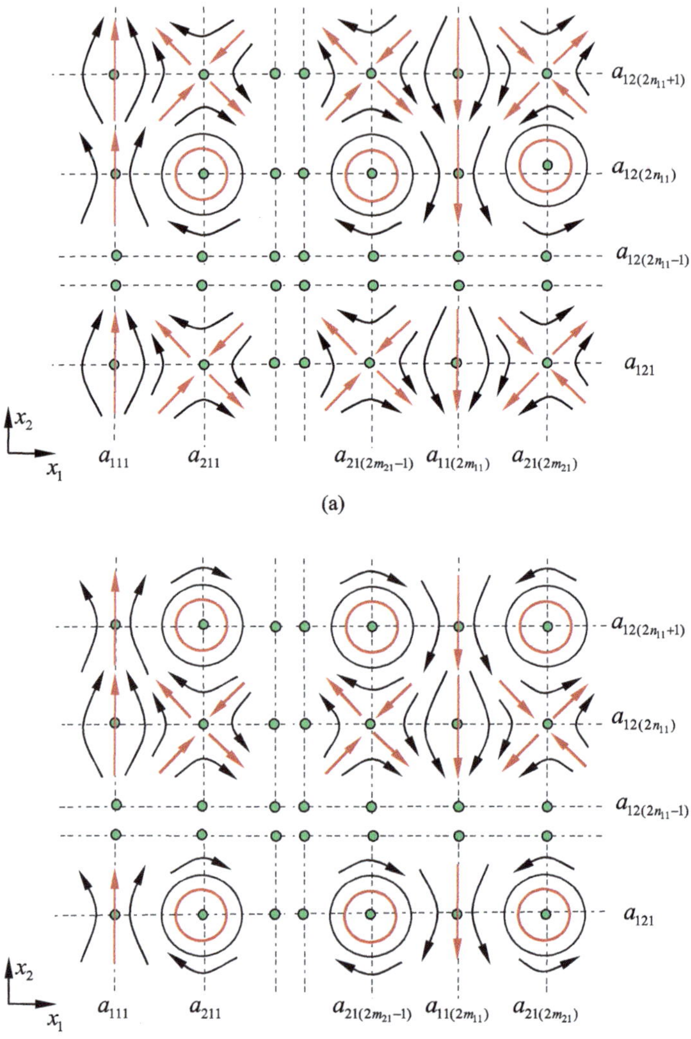

Fig. 3.23 Phase portraits for 2-dimensional systems with $x_1^* = a_{111}, a_{211}, \cdots, a_{11(2m_{11})}, a_{21(2m_{21})}$ and $x_2^* = a_{121}, a_{122}, \cdots, a_{12(2n_{11}+1)}$. The networks of flows with saddles and centers: **a** ($a_{110} > 0, a_{220} > 0$), **b** ($a_{110} < 0, a_{220} > 0$), **c** ($a_{110} > 0, a_{220} < 0$), **d** ($a_{110} < 0, a_{220} < 0$)

$a_{11(2m_{11}+1)}, a_{21(2m_{21}+1)}$ and $x_2^*{=}a_{121}, a_{122}, \cdots, a_{12(2n_{11}+1)}$ are presented in Fig. 3.23a–d for ($a_{110} > 0, a_{220} > 0$)), ($a_{110} < 0, a_{220} > 0$)), ($a_{110} > 0, a_{220} < 0$) and ($a_{110} < 0, a_{220} < 0$)).

(**B**) Consider $2m_{11}{=}2m_{21}$ for illustrations. Phase portraits for the 2-dimensional systems for the hybrid networks of hyperbolic and hyperbolic-secant flows with the simple equilibriums of saddles and centers with $x_1^*{=}a_{211}, a_{111}, \cdots, a_{11(2m_{11})}$,

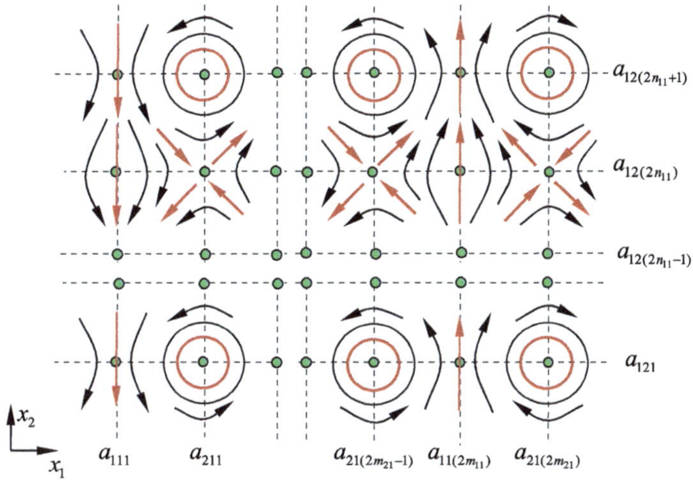

(c)

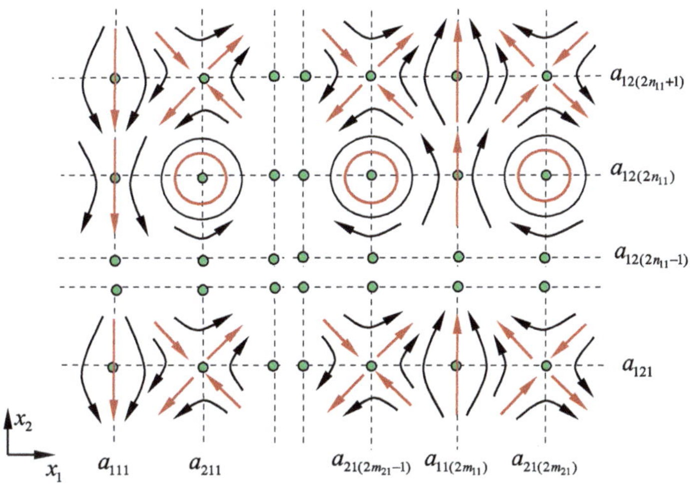

(d)

Fig. 3.23 (continued)

$a_{21(2m_{21}+1)}$, $a_{11(2m_{11}+1)}$ and $x_2^*=a_{121}, a_{122}, \cdots, a_{12(2n_{11}+1)}$ are presented in Fig. 3.24a–d for $(a_{110} > 0, a_{220} > 0)$, $(a_{110} < 0, a_{220} > 0)$, $(a_{110} > 0, a_{220} < 0)$ and $(a_{110} < 0, a_{220} < 0)$.

$$
\begin{Bmatrix}
(a_{111}, a_{12(2n_{11}+1)}) & (a_{12(2n_{11}+1)}, a_{211}) & \cdots & (a_{12(2n_{11}+1)}, a_{21(2m_{21}-1)}) & (a_{11(2m_{11})}, a_{12(2n_{11}+1)}) & (a_{12(2n_{11}+1)}, a_{21(2m_{21})}) \\
(a_{111}, a_{12(2n_{11})}) & (a_{12(2n_{11})}, a_{211}) & \cdots & (a_{12(2n_{11})}, a_{21(2m_{21}-1)}) & (a_{11(2m_{11}+1)}, a_{12(2n_{11})}) & (a_{12(2n_{11})}, a_{21(2m_{21})}) \\
\vdots & \vdots & \vdots & \vdots & \vdots & \vdots \\
(a_{111}, a_{121}) & (a_{121}, a_{211}) & \cdots & (a_{121}, a_{21(2m_{21}-1)}) & (a_{11(2m_{11}1)}, a_{121}) & (a_{121}, a_{21(2m_{21})})
\end{Bmatrix}
$$

$$
= \begin{Bmatrix}
\underbrace{(\text{UP:DP,pF})}_{\text{hyperbolic-secant flow }(+)} & \underbrace{(\text{UP}_-,\text{UP}_-)}_{\text{negative saddle}} & \cdots & \underbrace{(\text{UP}_-,\text{UP}_-)}_{\text{negative saddle}} & \underbrace{(\text{UP:DP,nF})}_{\text{hyperbolic-secant flow }(-)} & \underbrace{(\text{UP}_+,\text{UP}_+)}_{\text{positive saddle}} \\
\underbrace{(\text{DP:UP,pF})}_{\text{hyperbolic flow }(+)} & \underbrace{(\text{DP}_-,\text{DP}_+)}_{\text{CW center}} & \cdots & \underbrace{(\text{DP}_-,\text{DP}_+)}_{\text{CW center}} & \underbrace{(\text{DP:UP,nF})}_{\text{hyperbolic flow }(-)} & \underbrace{(\text{DP}_+,\text{DP}_-)}_{\text{CCW center}} \\
\vdots & \vdots & \vdots & \vdots & \vdots & \vdots \\
\underbrace{(\text{UP:DP,pF})}_{\text{hyperbolic-secant flow }(+)} & \underbrace{(\text{UP}_-,\text{UP}_-)}_{\text{negative saddle}} & \cdots & \underbrace{(\text{UP}_-,\text{UP}_-)}_{\text{negative saddle}} & \underbrace{(\text{UP:DP,nF})}_{\text{hyperbolic-secant flow }(-)} & \underbrace{(\text{UP}_+,\text{UP}_+)}_{\text{positive saddle}}
\end{Bmatrix}_{(2n_{11}+1)\times(2m_{11}+2m_{21})}
$$

for $a_{110} > 0, a_{220} > 0;$ (3.73)

$$
\left\{\begin{array}{cccccc}
(a_{111},a_{12(2n_{11}+1)}) & (a_{12(2n_{11}+1)},a_{211}) & \cdots & (a_{12(2n_{11}+1)},a_{21(2m_{21}-1)}) & (a_{11(2m_{11})},a_{12(2n_{11}+1)}) & (a_{12(2n_{11}+1)},a_{21(2m_{21})}) \\
(a_{111},a_{12(2n_{11})}) & (a_{12(2n_{11})},a_{211}) & \cdots & (a_{12(2n_{11})},a_{21(2m_{21}-1)}) & (a_{11(2m_{11}+1)},a_{12(2n_{11})}) & (a_{12(2n_{11})},a_{21(2m_{21})}) \\
\vdots & \vdots & \vdots & \vdots & \vdots & \vdots \\
(a_{111},a_{121}) & (a_{121},a_{211}) & \cdots & (a_{121},a_{21(2m_{21}-1)}) & (a_{11(2m_{11}1)},a_{121}) & (a_{121},a_{21(2m_{21})})
\end{array}\right\}
$$

$$
=\left\{\begin{array}{cccccc}
\underbrace{(\mathrm{DP{:}UP,pF})}_{\text{hyperbolic flow }(+)} & \underbrace{(\mathrm{DP}_{-},\mathrm{DP}_{+})}_{\text{CW center}} & \cdots & \underbrace{(\mathrm{DP}_{-},\mathrm{DP}_{+})}_{\text{CW center}} & \underbrace{(\mathrm{DP{:}UP,nF})}_{\text{hyperbolic flow }(-)} & \underbrace{(\mathrm{DP}_{+},\mathrm{DP}_{-})}_{\text{CCW center}} \\
\underbrace{(\mathrm{UP{:}DP,pF})}_{\text{hyperbolic-secant flow }(+)} & \underbrace{(\mathrm{UP}_{-},\mathrm{UP}_{-})}_{\text{negative saddle}} & \cdots & \underbrace{(\mathrm{UP}_{-},\mathrm{UP}_{-})}_{\text{negative saddle}} & \underbrace{(\mathrm{UP{:}DP,nF})}_{\text{hyperbolic-secant flow }(-)} & \underbrace{(\mathrm{UP}_{+},\mathrm{UP}_{+})}_{\text{positive saddle}} \\
\vdots & \vdots & \vdots & \vdots & \vdots & \vdots \\
\underbrace{(\mathrm{DP{:}UP,pF})}_{\text{hyperbolic flow }(+)} & \underbrace{(\mathrm{DP}_{-},\mathrm{DP}_{+})}_{\text{CW center}} & \cdots & \underbrace{(\mathrm{DP}_{-},\mathrm{DP}_{+})}_{\text{CW center}} & \underbrace{(\mathrm{DP{:}UP,nF})}_{\text{hyperbolic flow }(-)} & \underbrace{(\mathrm{DP}_{+},\mathrm{DP}_{-})}_{\text{CCW center}}
\end{array}\right\}_{(2n_{11}+1)\times(2m_{11}+2m_{21})}
$$

for $a_{110} < 0, a_{220} > 0$; (3.74)

$$
\left\{\begin{array}{cccccc}
(a_{111},a_{12(2n_{11}+1)}) & (a_{12(2n_{11}+1)},a_{211}) & \cdots & (a_{12(2n_{11}+1)},a_{21(2m_{21}-1)}) & (a_{11(2m_{11})},a_{12(2n_{11}+1)}) & (a_{12(2n_{11}+1)},a_{21(2m_{21})}) \\
(a_{111},a_{12(2n_{11})}) & (a_{12(2n_{11})},a_{211}) & \cdots & (a_{12(2n_{11})},a_{21(2m_{21}-1)}) & (a_{11(2m_{11}+1)},a_{12(2n_{11})}) & (a_{12(2n_{11})},a_{21(2m_{21})}) \\
\vdots & \vdots & \vdots & \vdots & \vdots & \vdots \\
(a_{111},a_{121}) & (a_{121},a_{211}) & \cdots & (a_{121},a_{21(2m_{21}-1)}) & (a_{11(2m_{11}1)},a_{121}) & (a_{121},a_{21(2m_{21})})
\end{array}\right\}
$$

$$
=\left\{\begin{array}{cccccc}
\underbrace{(\text{DP:UP,nF})}_{\text{hyperbolic flow }(-)} & \underbrace{(\text{DP}_{+},\text{DP}_{-})}_{\text{CCW center}} & \cdots & \underbrace{(\text{DP}_{+},\text{DP}_{-})}_{\text{CCW center}} & \underbrace{(\text{DP:UP,pF})}_{\text{hyperbolic flow }(+)} & \underbrace{(\text{DP}_{-},\text{DP}_{+})}_{\text{CW center}} \\
\underbrace{(\text{UP:DP,nF})}_{\text{hyperbolic-secant flow }(-)} & \underbrace{(\text{UP}_{+},\text{UP}_{+})}_{\text{positive saddle}} & \cdots & \underbrace{(\text{UP}_{+},\text{UP}_{+})}_{\text{positive saddle}} & \underbrace{(\text{UP:DP,pF})}_{\text{hyperbolic-secant flow }(+)} & \underbrace{(\text{UP}_{-},\text{UP}_{-})}_{\text{negative saddle}} \\
\vdots & \vdots & \vdots & \vdots & \vdots & \vdots \\
\underbrace{(\text{DP:UP,nF})}_{\text{hyperbolic flow }(-)} & \underbrace{(\text{DP}_{+},\text{DP}_{-})}_{\text{CCW center}} & \cdots & \underbrace{(\text{DP}_{+},\text{DP}_{-})}_{\text{CCW center}} & \underbrace{(\text{DP:UP,pF})}_{\text{hyperbolic flow }(+)} & \underbrace{(\text{DP}_{-},\text{DP}_{+})}_{\text{CW center}}
\end{array}\right\}_{(2n_{11}+1)\times(2m_{11}+2m_{21})}
$$

for $a_{110} > 0, a_{220} < 0$; (3.75)

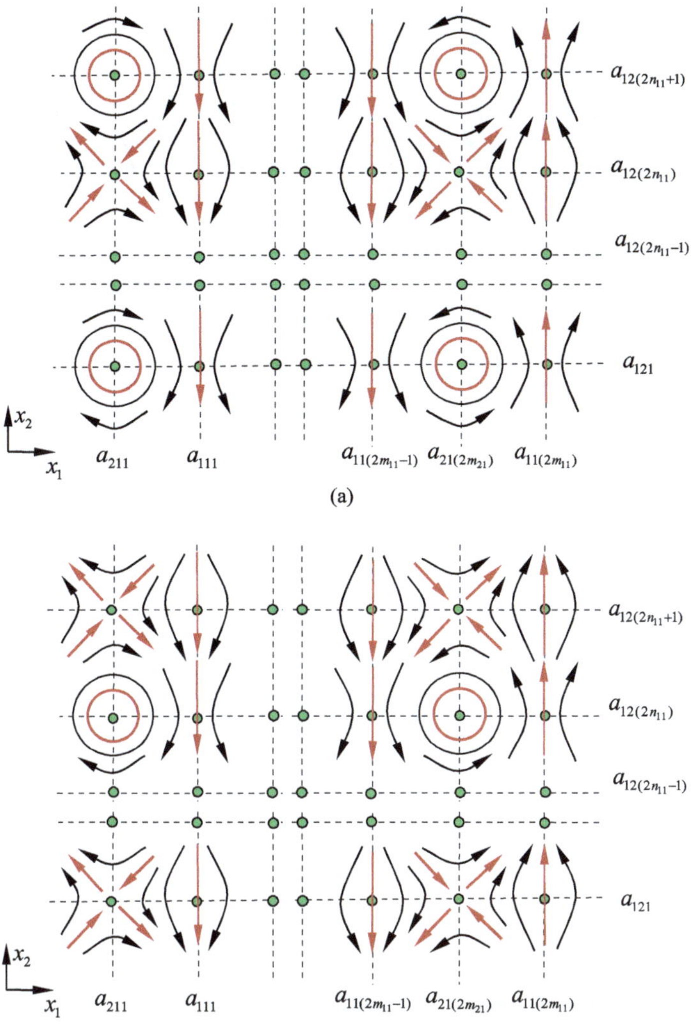

Fig. 3.24 Phase portraits for 2-dimensional systems with $x_1^* = a_{211}, a_{111}, \cdots, a_{21(2m_{21})}, a_{11(2m_{11})}$ and $x_2^* = a_{121}, a_{122}, \cdots, a_{12(2n_{11}+1)}$. The networks of flows with saddles and centers: **a** $(a_{110} > 0, a_{220} > 0)$, **b** $(a_{110} < 0, a_{220} > 0)$, **c** $(a_{110} > 0, a_{220} < 0)$, **d** $(a_{110} < 0, a_{220} < 0)$

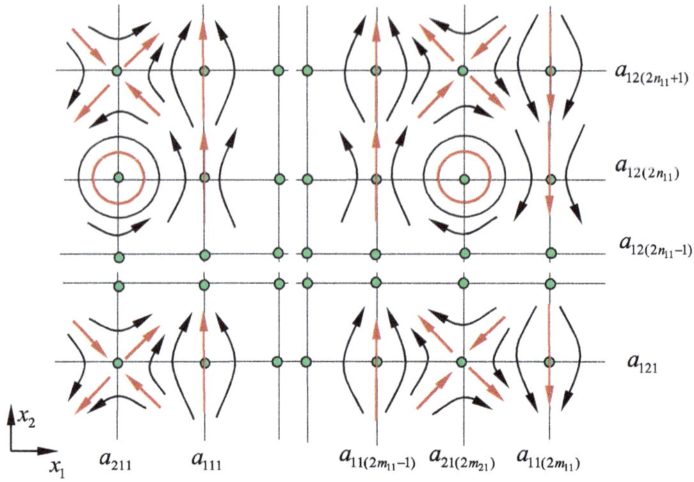

(c)

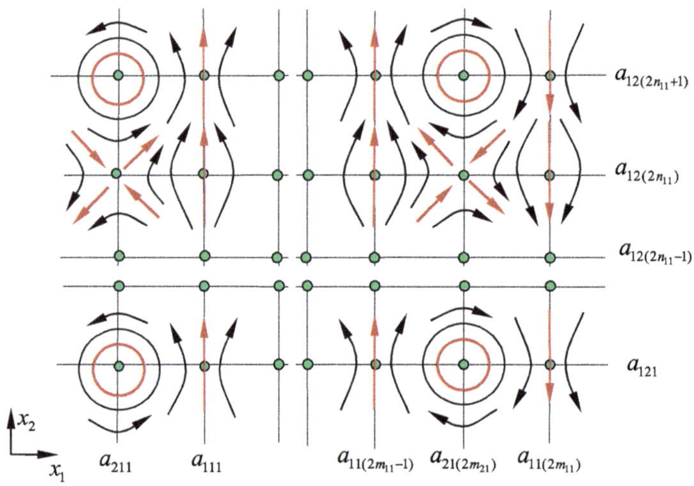

(d)

Fig. 3.24 (continued)

$$
\left\{\begin{array}{cccccc}
(a_{111},a_{12(2n_{11}+1)}) & (a_{12(2n_{11}+1)},a_{211}) & \cdots & (a_{12(2n_{11}+1)},a_{21(2m_{21}-1)}) & (a_{11(2m_{11})},a_{12(2n_{11}+1)}) & (a_{12(2n_{11}+1)},a_{21(2m_{21})}) \\
(a_{111},a_{12(2n_{11})}) & (a_{12(2n_{11})},a_{211}) & \cdots & (a_{12(2n_{11})},a_{21(2m_{21}-1)}) & (a_{11(2m_{11}+1)},a_{12(2n_{11})}) & (a_{12(2n_{11})},a_{21(2m_{21})}) \\
\vdots & \vdots & \vdots & \vdots & \vdots & \vdots \\
(a_{111},a_{121}) & (a_{121},a_{211}) & \cdots & (a_{121},a_{21(2m_{21}-1)}) & (a_{11(2m_{11}1)},a_{121}) & (a_{121},a_{21(2m_{21})})
\end{array}\right\}
$$

$$
= \left\{\begin{array}{cccccc}
\underbrace{(\mathrm{UP{:}DP,nF})}_{\text{hyperbolic-secant flow }(-)} & \underbrace{(\mathrm{UP}_+,\mathrm{UP}_+)}_{\text{positive saddle}} & \cdots & \underbrace{(\mathrm{UP}_+,\mathrm{UP}_+)}_{\text{positive saddle}} & \underbrace{(\mathrm{UP{:}DP,pF})}_{\text{hyperbolic-secant flow }(+)} & \underbrace{(\mathrm{UP}_-,\mathrm{UP}_-)}_{\text{negative saddle}} \\
\underbrace{(\mathrm{DP{:}UP,nF})}_{\text{hyperbolic flow }(-)} & \underbrace{(\mathrm{DP}_+,\mathrm{DP}_-)}_{\text{CCW center}} & \cdots & \underbrace{(\mathrm{DP}_+,\mathrm{DP}_-)}_{\text{CCW center}} & \underbrace{(\mathrm{DP{:}UP,pF})}_{\text{hyperbolic flow }(+)} & \underbrace{(\mathrm{DP}_-,\mathrm{DP}_+)}_{\text{CW center}} \\
\vdots & \vdots & \vdots & \vdots & \vdots & \vdots \\
\underbrace{(\mathrm{UP{:}DP,nF})}_{\text{hyperbolic-secant flow }(-)} & \underbrace{(\mathrm{UP}_+,\mathrm{UP}_+)}_{\text{positive saddle}} & \cdots & \underbrace{(\mathrm{UP}_+,\mathrm{UP}_+)}_{\text{positive saddle}} & \underbrace{(\mathrm{UP{:}DP,pF})}_{\text{hyperbolic-secant flow }(+)} & \underbrace{(\mathrm{UP}_-,\mathrm{UP}_-)}_{\text{negative saddle}}
\end{array}\right\}_{(2n_{11}+1)\times(2m_{11}+2m_{21})}
$$

for $a_{110} < 0, a_{220} < 0$; (3.76)

$$
\left\{\begin{array}{cccccc}
(a_{12(2n_{11}+1)},a_{211}) & (a_{111},a_{12(2n_{11}+1)}) & \cdots & (a_{11(2m_{11}-1)},a_{12(2n_{11}+1)},) & (a_{12(2n_{11}+1)},a_{21(2m_{21})}) & (a_{11(2m_{11})},a_{12(2n_{11}+1)}) \\
(a_{12(2n_{11})},a_{211}) & (a_{111},a_{12(2n_{11})}) & \cdots & (a_{11(2m_{11}-1)},a_{12(2n_{11})}) & (a_{12(2n_{11})},a_{21(2m_{21})}) & (a_{11(2m_{11})},a_{12(2n_{11})}) \\
\vdots & \vdots & \vdots & \vdots & \vdots & \vdots \\
(a_{121},a_{211}) & (a_{111},a_{121}) & \cdots & (a_{11(2m_{11}-1)},a_{121}) & (a_{121},a_{21(2m_{21})}) & (a_{11(2m_{11})},a_{121})
\end{array}\right\}
$$

$$
=\left\{\begin{array}{cccccc}
\underbrace{(\mathrm{DP}_{-},\mathrm{DP}_{+})}_{\text{CW center}} & \underbrace{(\mathrm{DP{:}UP,nF})}_{\text{hyperbolic flow }(-)} & \cdots & \underbrace{(\mathrm{DP{:}UP,nF})}_{\text{hyperbolic flow }(-)} & \underbrace{(\mathrm{DP}_{+},\mathrm{DP}_{-})}_{\text{CCW center}} & \underbrace{(\mathrm{DP{:}UP,pF})}_{\text{hyperbolic flow }(+)} \\
\underbrace{(\mathrm{UP}_{-},\mathrm{UP}_{-})}_{\text{negative saddle}} & \underbrace{(\mathrm{UP{:}DP,nF})}_{\text{hyperbolic-secant flow }(-)} & \cdots & \underbrace{(\mathrm{UP{:}DP,nF})}_{\text{hyperbolic-secant flow }(-)} & \underbrace{(\mathrm{UP}_{+},\mathrm{UP}_{+})}_{\text{positive saddle}} & \underbrace{(\mathrm{UP{:}DP,pF})}_{\text{hyperbolic-secant flow }(+)} \\
\vdots & \vdots & \vdots & \vdots & \vdots & \vdots \\
\underbrace{(\mathrm{DP}_{-},\mathrm{DP}_{+})}_{\text{CW center}} & \underbrace{(\mathrm{DP{:}UP,nF})}_{\text{hyperbolic flow }(-)} & \cdots & \underbrace{(\mathrm{DP{:}UP,nF})}_{\text{hyperbolic flow }(-)} & \underbrace{(\mathrm{DP}_{+},\mathrm{DP}_{-})}_{\text{CCW center}} & \underbrace{(\mathrm{DP{:}UP,pF})}_{\text{hyperbolic flow }(+)}
\end{array}\right\}_{(2n_{11}+1)\times(2m_{11}+2m_{21})}
$$

for $a_{110} > 0, a_{220} > 0$; (3.77)

$$\begin{Bmatrix} (a_{12(2n_{11}+1)},a_{211}) & (a_{111},a_{12(2n_{11}+1)}) & \cdots & (a_{11(2m_{11}-1)},a_{12(2n_{11}+1)},) & (a_{12(2n_{11}+1)},a_{21(2m_{21})}) & (a_{11(2m_{11})},a_{12(2n_{11}+1)}) \\ (a_{12(2n_{11})},a_{211}) & (a_{111},a_{12(2n_{11})}) & \cdots & (a_{11(2m_{11}-1)},a_{12(2n_{11})}) & (a_{12(2n_{11})},a_{21(2m_{21})}) & (a_{11(2m_{11})},a_{12(2n_{11})}) \\ \vdots & \vdots & \vdots & \vdots & \vdots & \vdots \\ (a_{121},a_{211}) & (a_{111},a_{121}) & \cdots & (a_{11(2m_{11}-1)},a_{121}) & (a_{121},a_{21(2m_{21})}) & (a_{11(2m_{11})},a_{121}) \end{Bmatrix}$$

$$= \begin{Bmatrix} \underbrace{(\mathrm{UP}_-,\mathrm{UP}_-)}_{\text{negative saddle}} & \underbrace{(\mathrm{UP{:}DP,nF})}_{\text{hyperbolic-secant flow }(-)} & \cdots & \underbrace{(\mathrm{UP{:}DP,nF})}_{\text{hyperbolic-secant flow }(-)} & \underbrace{(\mathrm{UP}_+,\mathrm{UP}_+)}_{\text{positive saddle}} & \underbrace{(\mathrm{UP{:}DP,pF})}_{\text{hyperbolic-secant flow }(+)} \\ \underbrace{(\mathrm{DP}_-,\mathrm{DP}_+)}_{\text{CW center}} & \underbrace{(\mathrm{DP{:}UP,nF})}_{\text{hyperbolic flow }(-)} & \cdots & \underbrace{(\mathrm{DP{:}UP,nF})}_{\text{hyperbolic flow }(-)} & \underbrace{(\mathrm{DP}_+,\mathrm{DP}_-)}_{\text{CCW center}} & \underbrace{(\mathrm{DP{:}UP,pF})}_{\text{hyperbolic flow }(+)} \\ \vdots & \vdots & \vdots & \vdots & \vdots & \vdots \\ \underbrace{(\mathrm{UP}_-,\mathrm{UP}_-)}_{\text{negative saddle}} & \underbrace{(\mathrm{UP{:}DP,nF})}_{\text{hyperbolic-secant flow }(-)} & \cdots & \underbrace{(\mathrm{UP{:}DP,nF})}_{\text{hyperbolic-secant flow }(-)} & \underbrace{(\mathrm{UP}_+,\mathrm{UP}_+)}_{\text{positive saddle}} & \underbrace{(\mathrm{DP{:}UP,pF})}_{\text{hyperbolic flow }(+)} \end{Bmatrix}_{(2n_{11}+1)\times(2m_{11}+2m_{21})}$$

for $a_{110} < 0, a_{220} > 0$; (3.78)

$$
\begin{Bmatrix}
(a_{12(2n_{11}+1)},a_{211}) & (a_{111},a_{12(2n_{11}+1)}) & \cdots & (a_{11(2m_{11}-1)},a_{12(2n_{11}+1)},) & (a_{12(2n_{11}+1)},a_{21(2m_{21})}) & (a_{11(2m_{11})},a_{12(2n_{11}+1)}) \\
(a_{12(2n_{11})},a_{211}) & (a_{111},a_{12(2n_{11})}) & \cdots & (a_{11(2m_{11}-1)},a_{12(2n_{11})}) & (a_{12(2n_{11})},a_{21(2m_{21})}) & (a_{11(2m_{11})},a_{12(2n_{11})}) \\
\vdots & \vdots & \vdots & \vdots & \vdots & \vdots \\
(a_{121},a_{211}) & (a_{111},a_{121}) & \cdots & (a_{11(2m_{11}-1)},a_{121}) & (a_{121},a_{21(2m_{21})}) & (a_{11(2m_{11})},a_{121})
\end{Bmatrix}
$$

$$
= \begin{Bmatrix}
\underbrace{(\mathrm{UP}_+,\mathrm{UP}_+)}_{\text{positive saddle}} & \underbrace{(\mathrm{UP{:}DP,pF})}_{\text{hyperbolic-secant flow }(+)} & \cdots & \underbrace{(\mathrm{UP{:}DP,pF})}_{\text{hyperbolic-secant flow }(+)} & \underbrace{(\mathrm{UP}_-,\mathrm{UP}_-)}_{\text{negative saddle}} & \underbrace{(\mathrm{UP{:}DP,nF})}_{\text{hyperbolic-secant flow }(-)} \\
\underbrace{(\mathrm{DP}_+,\mathrm{DP}_-)}_{\text{CCW center}} & \underbrace{(\mathrm{DP{:}UP,pF})}_{\text{hyperbolic flow }(+)} & \cdots & \underbrace{(\mathrm{DP{:}UP,pF})}_{\text{hyperbolic flow }(+)} & \underbrace{(\mathrm{DP}_-,\mathrm{DP}_+)}_{\text{CW center}} & \underbrace{(\mathrm{DP{:}UP,nF})}_{\text{hyperbolic flow }(-)} \\
\vdots & \vdots & \vdots & \vdots & \vdots & \vdots \\
\underbrace{(\mathrm{UP}_+,\mathrm{UP}_+)}_{\text{positive saddle}} & \underbrace{(\mathrm{UP{:}DP,pF})}_{\text{hyperbolic-secant flow }(+)} & \cdots & \underbrace{(\mathrm{UP{:}DP,pF})}_{\text{hyperbolic-secant flow }(+)} & \underbrace{(\mathrm{UP}_-,\mathrm{UP}_-)}_{\text{negative saddle}} & \underbrace{(\mathrm{UP{:}DP,nF})}_{\text{hyperbolic-secant flow }(-)}
\end{Bmatrix}_{(2n_{11}+1)\times(2m_{11}+2m_{21})}
\tag{3.79}
$$

for $a_{110} > 0, a_{220} < 0$;

$$
\left\{\begin{array}{cccccc}
(a_{12(2n_{11}+1)},a_{211}) & (a_{111},a_{12(2n_{11}+1)}) & \cdots & (a_{11(2m_{11}-1)},a_{12(2n_{11}+1)},) & (a_{12(2n_{11}+1)},a_{21(2m_{21})}) & (a_{11(2m_{11})},a_{12(2n_{11}+1)}) \\
(a_{12(2n_{11})},a_{211}) & (a_{111},a_{12(2n_{11})}) & \cdots & (a_{11(2m_{11}-1)},a_{12(2n_{11})}) & (a_{12(2n_{11})},a_{21(2m_{21})}) & (a_{11(2m_{11})},a_{12(2n_{11})}) \\
\vdots & \vdots & \vdots & \vdots & \vdots & \vdots \\
(a_{121},a_{211}) & (a_{111},a_{121}) & \cdots & (a_{11(2m_{11}-1)},a_{121}) & (a_{121},a_{21(2m_{21})}) & (a_{11(2m_{11})},a_{121})
\end{array}\right\}
$$

$$
= \left\{\begin{array}{cccccc}
\underbrace{(\mathrm{DP}_+,\mathrm{DP}_-)}_{\text{CCW center}} & \underbrace{(\mathrm{DP{:}UP,pF})}_{\text{hyperbolic flow }(+)} & \cdots & \underbrace{(\mathrm{DP{:}UP,pF})}_{\text{hyperbolic flow }(+)} & \underbrace{(\mathrm{DP}_-,\mathrm{DP}_+)}_{\text{CW center}} & \underbrace{(\mathrm{DP{:}UP,nF})}_{\text{hyperbolic flow }(-)} \\
\underbrace{(\mathrm{UP}_+,\mathrm{UP}_+)}_{\text{positive saddle}} & \underbrace{(\mathrm{UP{:}DP,pF})}_{\text{hyperbolic-secant flow }(+)} & \cdots & \underbrace{(\mathrm{UP{:}DP,pF})}_{\text{hyperbolic-secant flow }(+)} & \underbrace{(\mathrm{UP}_-,\mathrm{UP}_-)}_{\text{negative saddle}} & \underbrace{(\mathrm{UP{:}DP,nF})}_{\text{hyperbolic-secant flow }(-)} \\
\vdots & \vdots & \vdots & \vdots & \vdots & \vdots \\
\underbrace{(\mathrm{DP}_+,\mathrm{DP}_-)}_{\text{CCW center}} & \underbrace{(\mathrm{DP{:}UP,pF})}_{\text{hyperbolic flow }(+)} & \cdots & \underbrace{(\mathrm{DP{:}UP,pF})}_{\text{hyperbolic flow }(+)} & \underbrace{(\mathrm{DP}_-,\mathrm{DP}_+)}_{\text{CW center}} & \underbrace{(\mathrm{DP{:}UP,nF})}_{\text{hyperbolic flow }(-)}
\end{array}\right\}_{(2n_{11}+1)\times(2m_{11}+2m_{21})}
$$

for $a_{110} < 0, a_{220} < 0$; (3.80)

3.5 $[(2m_{11}+1), (2n_{11})]$ and $((2n_{11}), (2m_{21}+1))$-Singular Systems

In this section, there are four global bifurcations of the 1-dimensional flows through singular inflection flows, and four global bifurcations of equilibriums through singular parabola-saddle bifurcations. The corresponding global bifurcations generate networks of hyperbolic and hyperbolic-secant flows with saddles and centers. The infinite-equilibriums are the appearing and switching bifurcations for the singular flows and singular equilibriums and for hybrid networks of hyperbolic flows and saddles and centers.

3.5.1 $[(2m_{11}+1), (2n_{11})]$*-Flows and* $((2n_{11}), (2m_{21}+1))$*-Equilibriums*

Consider a polynomial dynamical system for $a_{111} \neq a_{211}$ as

$$
\begin{aligned}
\dot{x}_2 &= a_{220}(x_1 - a_{211})^{2m_{21}+1},\\
\dot{x}_1 &= a_{110}(x_1 - a_{111})^{2m_{11}+1}(x_2 - a_{121})^{2n_{11}}.
\end{aligned} \tag{3.81}
$$

The first integral manifold is

$$
\begin{aligned}
&\sum_{\alpha,\beta=0}^{\alpha+\beta=2m_{21}+1} \frac{(2m_{21}+1)!}{\alpha!\beta!}\frac{(a_{111}-a_{211})^{\alpha}}{\beta-2m_{11}}\left[(x_1-a_{111})^{\beta-2m_{11}}-(x_{10}-a_{111})^{\beta-2m_{11}}\right]\\
&= \frac{a_{j_1j_10}}{a_{j_2j_20}}\frac{1}{2n_{11}+1}\left[(x_2-a_{121})^{2n_{11}+1}-(x_{20}-a_{121})^{2n_{11}+1}\right],\\
&\text{for } \beta-2m_{11}\neq 0;\\
&\frac{1}{\beta-2m_{11}}\left[(x_1-a_{111})^{\beta-2m_{11}}-(x_{10}-a_{111})^{\beta-2m_{11}}\right] = \ln\frac{|x_1-a_{111}|}{|x_{10}-a_{111}|},\\
&\text{for } \beta-2m_{11}=0.
\end{aligned} \tag{3.82}
$$

The singular hyperbolic and hyperbolic-secant flows with singular positive and negative saddles and CCW and CW centers are illustrated in Figs. 3.25 and 3.26. Through such singular flows, the corresponding global bifurcations generate from a hyperbolic (hyperbolic-secant) flow to a network of singular hyperbolic and hyperbolic-secant flows. Through such singular equilibriums, the corresponding global bifurcations generate from one of saddles and centers to a network of singular saddles and centers equilibriums.

For $(x_1^*, x_2^*) = (a_{111}, a_{121})$ with $\dot{x}_1 = 0$, the $[(2m_{11}+1), (2n_{11})]$-singular flows exist with $\dot{x}_2 = a_{220}(a_{111}-a_{211})^{2m_{21}+1}$. For $(x_1^*, x_2^*) = (a_{111}, a_{121})$, the 1-dimensional

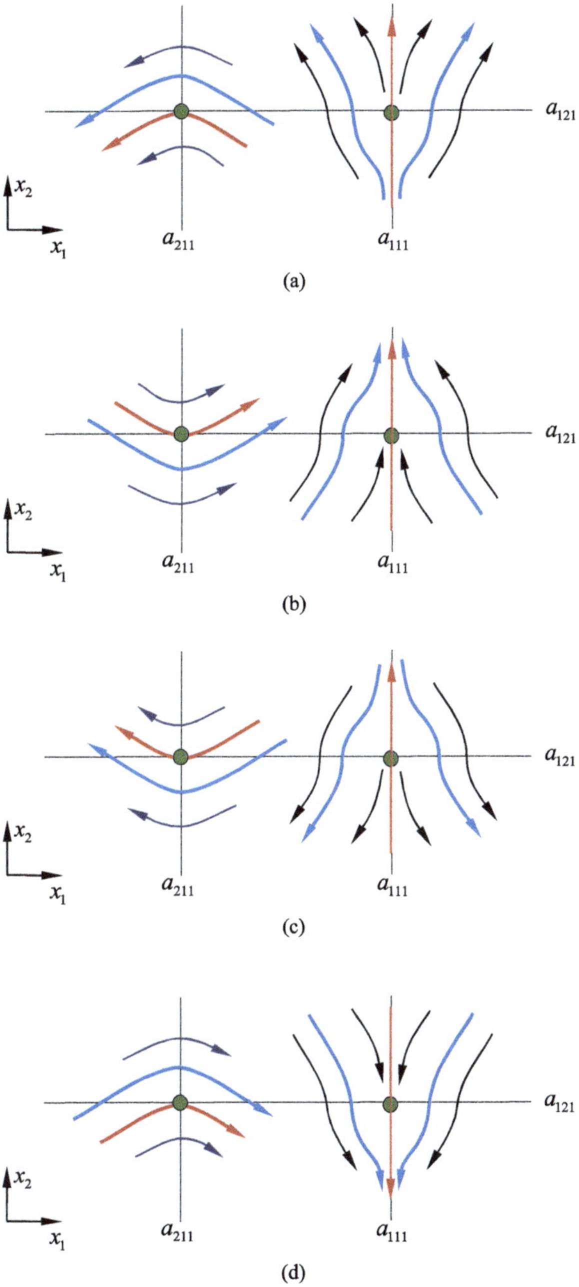

Fig. 3.25 Phase portraits $(a_{111} > a_{211})$ of the $([(2m_{11}+1),(2n_{11})], \text{xF})$-singular inflection-source and sink flows at $(x_1^*, x_2^*) = (a_{111}, a_{121})$ and $((2n_{11}),(2m_{21}+1))$-parabola-saddle at $(x_2^*, x_1^*) = (a_{121}, a_{211})$ for 2-dimensional polynomial systems: **a** $(a_{110} > 0, a_{220} > 0)$, **b** $(a_{110} < 0, a_{220} > 0)$, **c** $(a_{110} > 0, a_{220} < 0)$, **d** $(a_{110} < 0, a_{220} < 0)$

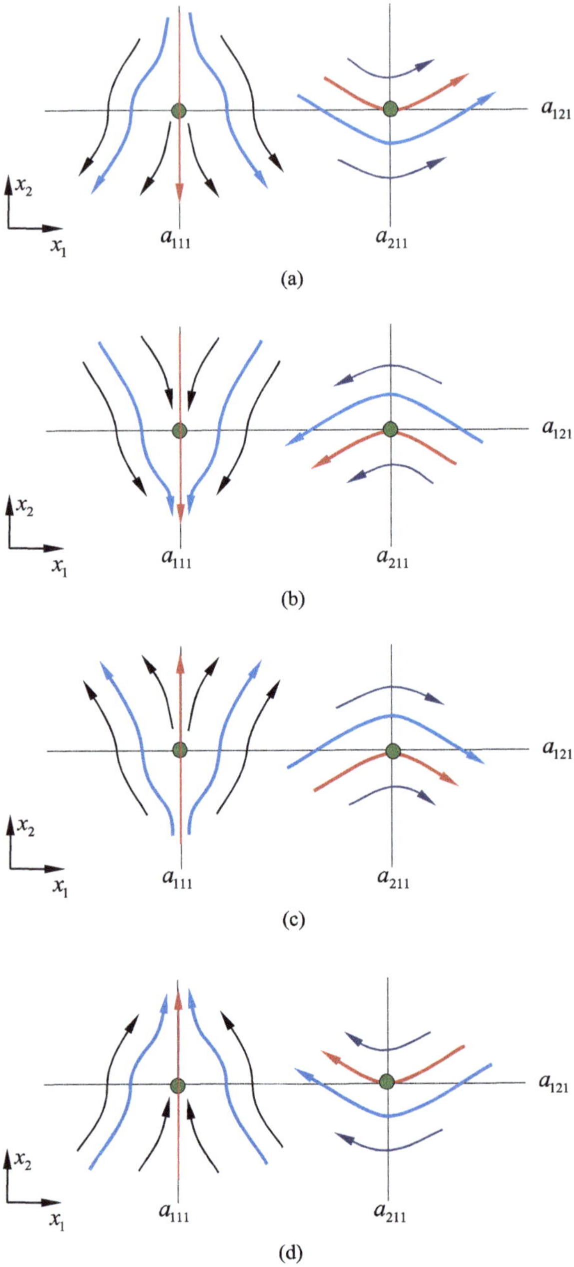

Fig. 3.26 Phase portraits ($a_{111} > a_{211}$) of the ([$(2m_{11}+1)$, $(2n_{11})$], xF)-singular inflection-source and sink flows at $(x_1^*, x_2^*) = (a_{111}, a_{121})$ and ($(2n_{11})$, $(2m_{21}+1)$)-parabola-saddle at $(x_2^*, x_1^*) = (a_{121}, a_{211})$ for 2-dimensional polynomial systems: **a** ($a_{110} > 0, a_{220} > 0$), **b** ($a_{110} < 0, a_{220} > 0$), **c** ($a_{110} > 0, a_{220} < 0$), **d** ($a_{110} < 0, a_{220} < 0$)

flow direction is determined by $\dot{x}_2 \neq 0$. For $(x_2^*, x_1^*) = (a_{121}, a_{211})$ with $\dot{x}_1 = \dot{x}_2 = 0$, the $((2n_{11}), (2m_{21}+1))$-singular equilibriums exist.
(**A**) For $a_{111} > a_{211}$,

$$\begin{Bmatrix} (a_{111}, a_{121}) \\ (a_{121}, a_{211}) \end{Bmatrix} = \begin{Bmatrix} \underbrace{([(2m_{11}+1),(2n_{11})]_{\text{DI:II}}\text{SO,pF})}_{[(2m_{11}+1),(2n_{11})]\text{-inflection-source flow }(+)} \\ \underbrace{((2n_{11})\text{LS},(2m_{21}+1)\text{DP})}_{[(2n_{11}),(2m_{21}+1)]\text{-down-parabola lower-saddle}} \end{Bmatrix},$$
for $a_{110} > 0, a_{220} > 0$;

$$\begin{Bmatrix} (a_{111}, a_{121}) \\ (a_{121}, a_{211}) \end{Bmatrix} = \begin{Bmatrix} \underbrace{([(2m_{11}+1),(2n_{11})]_{\text{II:DI}}\text{SI,pF})}_{[(2m_{11}+1),(2n_{11})]\text{-inflection-sink flow }(+)} \\ \underbrace{((2n_{11})\text{US},(2m_{21}+1)\text{UP})}_{[(2n_{11}),(2m_{21}+1)]\text{-up-parabola upper-saddle}} \end{Bmatrix},$$
for $a_{110} < 0, a_{220} > 0$;

$$\begin{Bmatrix} (a_{111}, a_{121}) \\ (a_{121}, a_{211}) \end{Bmatrix} = \begin{Bmatrix} \underbrace{([(2m_{11}+1),(2n_{11})]_{\text{II:DI}}\text{SO,nF})}_{[(2m_{11}+1),(2n_{11})]\text{-inflection-source flow }(-)} \\ \underbrace{((2n_{11})\text{LS},(2m_{21}+1)\text{UP})}_{[(2n_{11}),(2m_{21}+1)]\text{-up-parabola lower-saddle}} \end{Bmatrix},$$
for $a_{110} > 0, a_{220} < 0$;

$$\begin{Bmatrix} (a_{111}, a_{121}) \\ (a_{121}, a_{211}) \end{Bmatrix} = \begin{Bmatrix} \underbrace{([(2m_{11}+1),(2n_{11})]_{\text{DI:II}}\text{SI,nF})}_{[(2m_{11}+1),(2n_{11})]\text{-inflection-sink flow }(-)} \\ \underbrace{((2n_{11})\text{LS},(2m_{21}+1)\text{DP})}_{[(2n_{11}),(2m_{21}+1)]\text{-down-parabola lower-saddle}} \end{Bmatrix},$$
for $a_{110} < 0, a_{220} < 0$. (3.83)

(**B**) For $a_{111} < a_{211}$,

$$\begin{Bmatrix} (a_{121}, a_{211}) \\ (a_{111}, a_{121}) \end{Bmatrix} = \begin{Bmatrix} \underbrace{((2n_{11})\text{US},(2m_{21}+1)\text{UP})}_{[(2n_{11}),(2m_{21}+1)]\text{-up-parabola upper-saddle}} \\ \underbrace{([(2m_{11}+1),(2n_{11})]_{\text{II:DI}}\text{SO,nF})}_{[(2m_{11}+1),(2n_{11})]\text{-inflection-source flow }(-)} \end{Bmatrix},$$
for $a_{110} > 0, a_{220} > 0$;

$$\begin{Bmatrix} (a_{121}, a_{211}) \\ (a_{111}, a_{121}) \end{Bmatrix} = \begin{Bmatrix} \underbrace{((2n_{11})\text{LS},(2m_{21}+1)\text{DP})}_{[(2n_{11}),(2m_{21}+1)]\text{-down-parabola lower-saddle}} \\ \underbrace{([(2m_{11}+1),(2n_{11})]_{\text{DI:II}}\text{SI,nF})}_{[(2m_{11}+1),(2n_{11})]\text{-infelction-source flow }(-)} \end{Bmatrix},$$
for $a_{110} < 0, a_{220} > 0$;

$$\left\{\begin{array}{l}(a_{121},a_{211})\\(a_{111},a_{121})\end{array}\right\}=\left\{\begin{array}{c}\underbrace{((2n_{11})\text{US},(2m_{21}+1)\text{DP})}_{[(2n_{11}),(2m_{21}+1)]\text{-doen-parabola upper-saddle}}\\\underbrace{([(2m_{11}+1),(2n_{11})]_{\text{DI:II}}\text{SO,pF})}_{[(2m_{11}+1),(2n_{11})]\text{-inflection-source flow }(+)}\end{array}\right\},$$
for $a_{110}>0,a_{220}<0$;
$$\left\{\begin{array}{l}(a_{121},a_{211})\\(a_{111},a_{121})\end{array}\right\}=\left\{\begin{array}{c}\underbrace{((2n_{11})\text{LS},(2m_{21}+1)\text{UP})}_{[(2n_{11}),(2m_{21}+1)]\text{-up-perabola lower-saddle}}\\\underbrace{([(2m_{11}+1),(2n_{11})]_{\text{II:DI}}\text{SI,pF})}_{[(2m_{11}+1),(2n_{11})]\text{-inflection-sink flow }(+)}\end{array}\right\},$$
$$\text{for } a_{110}<0,a_{220}<0. \tag{3.84}$$

- The ($[(2m_{11}+1),(2n_{11})]_{\text{DI:II}}$SO,pF) is for a positive singular inflection-source flow.
- The ($[(2m_{11}+1),(2n_{11})]_{\text{II:DI}}$SI,pF) is for a positive singular inflection-sink flow.
- The ($[(2m_{11}+1),(2n_{11})]_{\text{II:DI}}$SI,nF) is for a negative singular inflection-source flow.
- The ($[(2m_{11}+1),(2n_{11})]_{\text{DI:II}}$SI,nF) is for a negative singular inflection-sink flow.

The acronyms “UP” and “DP” are for up-parabola and down-parabola flows, respectively. “pF” and “nF” are for positive and negative flows, respectively.

- The ($(2n_{11})$US,$(2\text{m}_{21}+1)$UP] is for a ($(2n_{11}),(2m_{21}+1)$]-singular up-parabola upper-saddle.
- The ($(2n_{11})$LS,$(2m_{21}+1)$DP] is for a ($(2n_{11}),(2m_{21}+1)$]-singular down-parabola lower-saddle.
- The ($(2n_{11})$US,$(2\text{m}_{21}+1)$DP] is for a ($(2n_{11}),(2m_{21}+1)$]-singular down-parabola upper-saddle.
- The ($(2n_{11}+1)$LS,$(2\text{m}_{21}+1)$UP] is for a ($(2n_{11}),(2m_{21}+1)$]-singular up-parabola lower-saddle.

3.5.2 $[(2m_{11}+1),(2n_{11})]$ *and* $((2n_{11}),(2m_{21}+1))$*-Infinite-Equilibriums*

Consider a polynomial dynamical system for $a_{111}=a_{211}$ as

$$\begin{aligned}\dot{x}_2&=a_{220}(x_1-a_{211})^{2m_{21}+1},\\\dot{x}_1&=a_{110}(x_1-a_{111})^{2m_{11}+1}(x_2-a_{121})^{2n_{11}}.\end{aligned} \tag{3.85}$$

The first integral manifold is

$$\frac{1}{2(m_{21}-m_{11})+1}\left[(x_1-a_{211})^{2(m_{21}-m_{11})+1}-(x_{10}-a_{211})^{2(m_{21}-m_{11})+1}\right]$$
$$=\frac{a_{110}}{a_{220}}\frac{1}{2n_{11}+1}\left[(x_2-a_{121})^{2n_{11}+1}-(x_{20}-a_{121})^{2n_{11}+1}\right]. \tag{3.86}$$

(**A**) For $2m_{21}+1 \geq 2m_{11}+1$, the corresponding infinite-equilibriums are given.

$$\begin{aligned}
&(a_{121},a_{211})=\underbrace{([(2m_{11}{+}1),(2n_{11})]^{\text{th}}\text{II}_{-}{:}\text{II}_{+},[2(m_{21}-m_{11})]^{\text{th}}{}_{\text{II:II}}\text{SO})}_{([(2m_{11}+1),(2n_{11})],[2(m_{21}-m_{11})])\text{-inflection-source}},\\
&\text{for } a_{110}>0, a_{220}>0;\\
&(a_{121},a_{211})=\underbrace{([(2m_{11}{+}1),(2n_{11})]^{\text{th}}\text{DI}_{-}{:}\text{DI}_{+},[2(m_{21}-m_{11})]^{\text{th}}{}_{\text{DI:DI}}\text{SI})}_{([(2m_{11}+1),(2n_{11})],[2(m_{21}-m_{11})])\text{-inflection-source}},\\
&\text{for } a_{110}<0, a_{220}>0;\\
&(a_{121},a_{211})=\underbrace{([(2m_{11}{+}1),(2n_{11})]^{\text{th}}\text{DI}_{+}{:}\text{DI}_{-},[2(m_{21}-m_{11})]^{\text{th}}{}_{\text{DI:DI}}\text{SO})}_{([(2m_{11}+1),(2n_{11})],[2(m_{21}-m_{11})])\text{-inflection-source}},\\
&\text{for } a_{110}>0, a_{220}<0;\\
&(a_{121},a_{211})=\underbrace{([(2m_{11}{+}1),(2n_{11})]^{\text{th}}\text{II}_{+}{:}\text{II}_{-},[2(m_{21}-m_{11})]^{\text{th}}{}_{\text{II:II}}\text{SI})}_{([(2m_{11}+1),(2n_{11}+1)],[2(m_{21}-m_{11})])\text{-inflection-sink}},\\
&\text{for } a_{110}<0, a_{220}<0.
\end{aligned} \tag{3.87}$$

(**B**) For $2m_{21}+1 < 2m_{11}+1$, the infinite-equilibriums are given as follows.

$$\begin{aligned}
&(a_{121},a_{211})=\underbrace{([(2m_{11}{+}1),(2n_{11})]^{\text{th}}\text{II}_{-}{:}\text{II}_{+},[2(m_{21}-m_{11})]^{\text{th}}{}_{\text{IC:IC}}\text{SO})}_{([(2m_{11}+1),(2n_{11})],[2(m_{21}-m_{11})])\text{-increaing-concave inflection-souce}},\\
&\text{for } a_{110}>0, a_{220}>0;\\
&(a_{121},a_{211})=\underbrace{([(2m_{11}{+}1),(2n_{11})]^{\text{th}}\text{DI}_{-}{:}\text{DI}_{+},[2(m_{21}-m_{11})]^{\text{th}}{}_{\text{DC:DC}}\text{SI})}_{([(2m_{11}+1),(2n_{11})],[2(m_{21}-m_{11})])\text{-decreasing-concave inflection-sink}},\\
&\text{for } a_{110}<0, a_{220}>0;\\
&(a_{121},a_{211})=\underbrace{([(2m_{11}{+}1),(2n_{11})]^{\text{th}}\text{DI}_{+}{:}\text{DI}_{-},[2(m_{21}-m_{11})]^{\text{th}}{}_{\text{DC:DC}}\text{SO})}_{([(2m_{11}+1),(2n_{11})],[2(m_{21}-m_{11})])\text{-decreasing-concave inflection-source}},\\
&\text{for } a_{110}>0, a_{220}<0;\\
&(a_{121},a_{211})=\underbrace{([(2m_{11}{+}1),(2n_{11})]^{\text{th}}\text{II}_{+}{:}\text{II}_{-},[2(m_{21}-m_{11})]^{\text{th}}{}_{\text{IC:IC}}\text{SI})}_{([(2m_{11}+1),(2n_{11})],[2(m_{21}-m_{11})])\text{-increasing-concave inflection-sink}},\\
&\text{for } a_{110}<0, a_{220}<0.
\end{aligned} \tag{3.88}$$

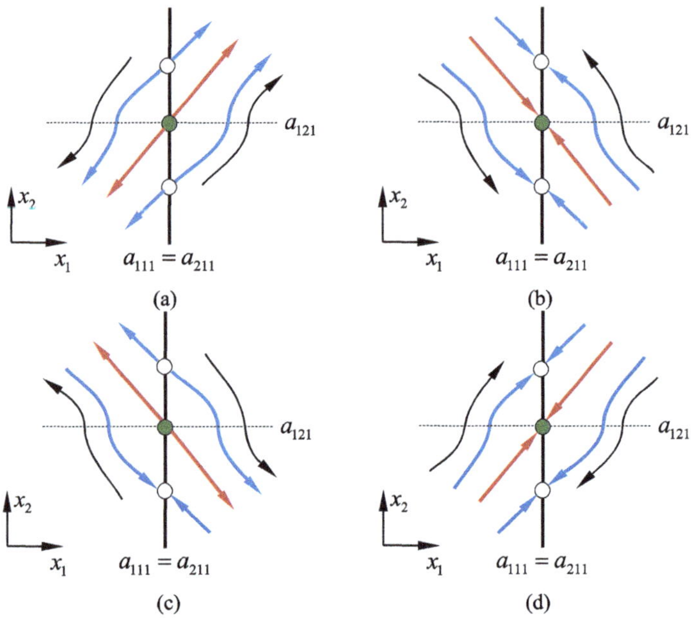

Fig. 3.27 The infinite-equilibriums ($a_{211} = a_{111}, 2m_{21} + 1 \geq 2m_{11} + 1, n_1 = 2n_{11}$) for 2-dimensional polynomial systems: **a** ($a_{110} > 0, a_{220} > 0$), **b** ($a_{110} < 0, a_{220} > 0$), **c** ($a_{110} > 0, a_{220} < 0$), **d** ($a_{110} < 0, a_{220} < 0$)

For $x_1^* = a_{111} = a_{211}$, the infinite-equilibrium exists. The infinite-equilibriums are illustrated in Figs. 3.27 and 3.28 for $m_{21} \geq m_{11}$ and $m_{21} < m_{11}$, respectively. For $m_{21} \geq m_{11}$, the singular inflection-sources and sinks are presented. For $m_{11} > m_{21}$, the concave inflection-sources and sinks are presented.

3.5.3 [$(2m_{11}+1), (2n_{11})$] and (($2n_{11}$), ($2m_{21}+1$))-Hybrid Networks

Consider

$$\dot{x}_2 = a_{220} \prod_{v=1}^{2m_{21}+1} (x_1 - a_{21v}),$$
$$\dot{x}_1 = a_{110} \prod_{r=1}^{2m_{11}+1} (x_1 - a_{11r}) \prod_{u=1}^{2n_{11}} (x_2 - a_{12u}). \tag{3.89}$$

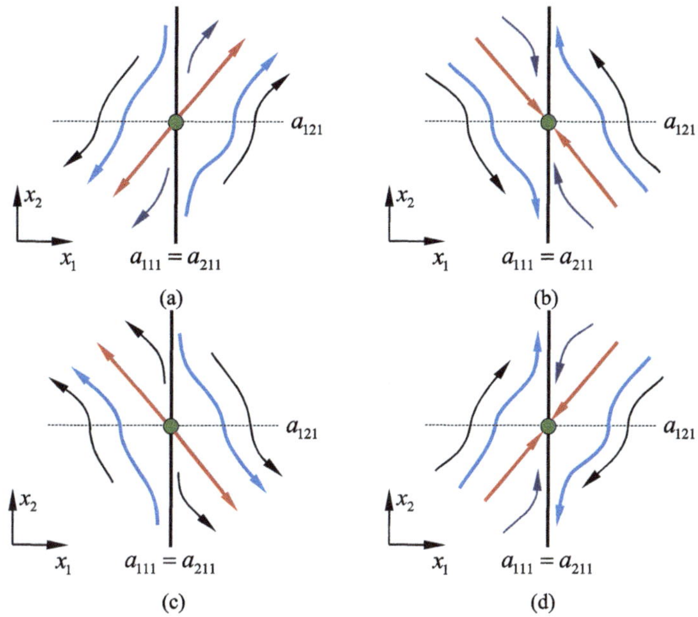

Fig. 3.28 The infinite-equilibriums ($a_{211}=a_{111}, 2m_{21}+1<2m_{11}+1, n_1=2n_{11}$) for 2-dimensional polynomial systems: **a** ($a_{110}>0, a_{220}>0$), **b** ($a_{110}<0, a_{220}>0$), **c** ($a_{110}>0, a_{220}<0$), **d** ($a_{110}<0, a_{220}<0$)

The first integral manifold is

$$\begin{aligned}
&\sum_{\nu_0=2m_{11}+1}^{2m_{21}+1} b_{21\nu_0} \sum_{\substack{\mu_1+\nu_1=\nu_0-1\\ \mu_1\in[0,\beta_0-2m_{11}-1],\\ \nu_1\in[2m_{11},\nu_0-1]}} \sum_{\substack{\mu_2+\nu_2=\nu_1-2\\ \mu_2\in[0,\nu_1-2m_{11}-1]\\ \nu_2\in[2m_{11}-1,\nu_1-2]}} \cdots \sum_{\substack{\mu_{2m_{11}}+\nu_{2m_{11}}=\nu_{2m_{11}-1}-2m_{11}\\ \mu_{2m_{11}}\in[0,\nu_{2m_{11}-1}-2m_{11}-1]\\ \nu_{2m_{11}}\in[1,\nu_{2m_{11}-1}-2m_{11}]}} \prod_{j=1}^{2m_{11}} \frac{(\nu_{j-1}-j)!}{\mu_j!\nu_j!}\\
&\quad\times (a_{11r_{j+1}}-a_{11r_j})^{\mu_j}\frac{1}{\nu_{2m_{11}}}\left[(x_1-a_{11r_{2m_{11}+1}})^{\nu_{2m_{11}}}-(x_{10}-a_{11r_{2m_{11}+1}})^{\nu_{2m_{11}}}\right]\\
&\quad-\sum_{\nu_0=2m_{11}+1}^{2m_{21}+1} b_{21\nu_0}\sum_{k=2}^{2m_{11}}\Bigg[\sum_{\substack{\mu_1+\nu_1=\nu_0-1\\ \mu_1\in[0,\nu_0-2m_{11}-1),\\ \nu_1\in(2m_{11},\nu_0-1]}} \sum_{\substack{\mu_2+\nu_2=\nu_1-2\\ \mu_2\in[0,\nu_1-2m_{11}-1)\\ \nu_2\in(2m_{11}-1,\nu_1-2)}} \cdots\\
&\quad\times\Bigg(\sum_{\substack{\mu_{k-1}+\nu_{k-1}=\nu_{k-2}-k+1\\ \mu_{k-1}\in(\nu_{k-2}-2m_{11}+2,\nu_{k-2}-k+1]\\ \nu_{k-1}\in[0,2m_{11}+1-k)}} \prod_{j=1}^{k-1}\frac{(\nu_{j-1}-j)!}{\mu_j!\nu_j!}(a_{11r_{j+1}}-a_{11r_j})^{\mu_j}\\
&\quad\times\sum_{\rho=k}^{2m_{11}+1} A_{\rho 1}^{(s_k)}\ln\frac{|x_1-a_{11s_\rho}|}{|x_{10}-a_{11s_\rho}|}
\end{aligned}$$

$$+\sum_{\substack{\mu_{k-1}\in(\nu_{k-2}-2m_{11}-1,\nu_{k-2}-2m_{11}]\\ \nu_{k-1}\in[2m_{11}+1-k,2m_{11}-k+2)}}^{\mu_{k-1}+\nu_{k-1}=\nu_{k-2}-k+1}\prod_{j=1}^{k-1}\frac{(\nu_{j-1}-j)!}{\mu_j!\nu_j!}(a_{11r_{j+1}}-a_{11r_j})^{\mu_j}$$

$$\times\sum_{\rho=k+1}^{2m_{11}+1}B_{\rho 1}^{(r_k)}\ln\frac{|x_1-a_{11r_\rho}|}{|x_{10}-a_{11r_\rho}|}\Bigg)\Bigg]+b_{210}\sum_{\rho=1}^{2m_{11}+1}A_{\rho 1}^{(r_1)}\ln\frac{|x_1-a_{11r_\rho}|}{|x_{10}-a_{11r_\rho}|}$$

$$+\sum_{\nu_0=1}^{2m_{11}}b_{21\nu_0}\sum_{\rho=2}^{2m_{11}+1}B_{\rho 1}^{(r_1)}\ln\frac{|x_1-a_{11r_\rho}|}{|x_{10}-a_{11r_\rho}|}$$

$$=\frac{a_{110}}{a_{220}}\Bigg\{\frac{1}{2n_{11}+1}\big[(x_2-a_{12u_1})^{2n_{11}+1}-(x_{j_2 0}-a_{12u_1})^{2n_{11}+1}\big]$$

$$+\sum_{\alpha=1}^{2n_{11}-2}\frac{1}{2n_{11}-\alpha+1}b_{12\alpha}\big[(x_2-a_{12u_1})^{2n_{11}-\alpha+1}-(x_{20}-a_{12u_1})^{2n_{11}-\alpha+1}\big]\Bigg\},\tag{3.90}$$

where

$$b_{12\alpha}=\sum_{l_{2n_{11}},t_{2n_{11}}=0}^{l_{2n_{11}}+t_{2n_{11}}=1}\cdots\sum_{l_2,t_2=0}^{l_2+t_2=1}\prod_{k=2}^{2n_{11}}(a_{12u_1}-a_{12u_k})^{l_k}\delta_{t_2+\cdots+t_{2n_{11}}}^{2n_{11}-\alpha};$$

$$b_{120}=1.$$

$$b_{21\nu_0}=\sum_{t_1,l_1=0}^{t_1+l_1=1}\cdots\sum_{t_{2m_{21}+1},l_{2m_{21}+1}=0}^{t_{2m_{21}+1}+l_{2m_{21}+1}=1}\prod_{k=1}^{2m_{21}+1}(a_{11r_1}-a_{21\nu_k})^{t_k}\delta_{\nu_0}^{\Sigma_{r=1}^{2m_{21}+1}l_r};$$

$$\lim_{\nu_{2m_{11}}\to 0}\frac{1}{\nu_{2m_{11}}}\big[(x_1-a_{11r_{2m_{11}+1}})^{\nu_{2m_{11}}}-(x_{j_1 0}-a_{11r_{2m_{11}+1}})^{\nu_{2m_{11}}}\big]=\ln\frac{|x_1-a_{11r_{2m_{11}+1}}|}{|x_{10}-a_{11r_{2m_{11}+1}}|};\tag{3.91}$$

$$A_{\rho 1}^{(r_k)}=\frac{1}{\prod_{\sigma=k,\sigma\neq\rho}^{2m_{11}+1}(a_{11r_\rho}-a_{11r_\sigma})},\ \text{for }\rho\in\{k,k+1,\ldots,2m_{11}+1\},$$

$$B_{\rho 1}^{(r_k)}=\frac{(a_{11r_\rho}-a_{11r_k})^{\beta_{k-1}-k}}{\prod_{\sigma=k+1,\sigma\neq\rho}^{2m_{11}+1}(a_{11r_\rho}-a_{11r_\sigma})},\ \text{for }\rho\in\{k+1,k+2,\ldots,2m_{11}+1\};$$

$$k\in\{1,2,\ldots,2m_{11}\}.\tag{3.92}$$

For all cases, the $(2n_{11})\times(2m_{21}+1)$-simple equilibriums plus $(2m_{11}+1)\times(2n_{11})$-hyperbolic flows are based on the following inflection sink and source saddle bifurcations for $2m_{21}+1\geq 2m_{11}+1$:

- $([(2m_{11}+1),(2n_{11})]^{\text{th}}\text{II}_{-}{:}\text{II}_{+},[2(m_{21}-m_{11})]^{\text{th}}{}_{\text{DI:II}}\text{SO})$-inflection-source bifurcation for $(a_{110}>0, a_{220}>0)$,
- $([(2m_{11}+1),(2n_{11})]^{\text{th}}\text{DI}_{-}{:}\text{DI}_{+},[2(m_{21}-m_{11})]^{\text{th}}{}_{\text{II:DI}}\text{SI})$-inflection-sink bifurcation for $(a_{110}<0, a_{220}>0)$,
- $([(2m_{11}+1),(2n_{11})]^{\text{th}}\text{DI}_{+}{:}\text{DI}_{-},[2(m_{21}-m_{11})]^{\text{th}}{}_{\text{II:DI}}\text{SO})$-inflection-source bifurcation for $(a_{110}>0, a_{220}<0)$,
- $([(2m_{11}+1),(2n_{11})]^{\text{th}}\text{II}_{+}{:}\text{II}_{-},[2(m_{21}-m_{11})]^{\text{th}}{}_{\text{DI:II}}\text{SI})$-inflection-sink bifurcation for $(a_{110}<0, a_{220}<0)$;

or based on the following concave-source and sink bifurcations for $2m_{21}+1<2m_{11}+1$:

- $([(2m_{11}+1),(2n_{11})]^{\text{th}}\text{II}_{-}{:}\text{II}_{+},[2(m_{11}-m_{21})]^{\text{th}}{}_{\text{IC:IC}}\text{SO})$-increasing-concave source bifurcation for $(a_{110}>0, a_{220}>0)$,
- $([(2m_{11}+1),(2n_{11})]^{\text{th}}\text{DI}_{-}{:}\text{DI}_{+},[2(m_{11}-m_{21})]^{\text{th}}{}_{\text{DC:DC}}\text{SI})$-decreasing-concave sink bifurcation for $(a_{110}>0, a_{220}<0)$,
- $([(2m_{11}+1),(2n_{11})]^{\text{th}}\text{DI}_{+}{:}\text{DI}_{-},[2(m_{11}-m_{21})]^{\text{th}}{}_{\text{DC:DC}}\text{SO})$-decreasing-concave source bifurcation for $(a_{110}>0, a_{220}<0)$,
- $([(2m_{11}+1),(2n_{11})]^{\text{th}}\text{II}_{+}{:}\text{II}_{-},[2(m_{11}-m_{21})]^{\text{th}}{}_{\text{IC:IC}}\text{SI})$-increasing-concave sink bifurcation for $(a_{110}<0, a_{220}<0)$.

(**A**) Consider $2m_{11}+1=2m_{21}+1$ for illustrations, phase portraits for the 2-dimensional systems for the hybrid networks of hyperbolic and hyperbolic-secant flows with the simple equilibriums of the saddles, sinks and sources with $x_1^*=a_{111},a_{211},a_{112},\cdots,a_{11(2m_{11}+1)},a_{21(2m_{21}+1)}$ and $x_2^*=a_{121},a_{122},\cdots,a_{12(2n_{11})}$ are presented in Fig. 3.29a–d for $(a_{110}>0,a_{220}>0)$, $(a_{110}<0,a_{220}>0)$, $(a_{110}>0,a_{220}<0)$ and $(a_{110}<0,a_{220}<0)$.

(**B**) Consider $2m_{11}+1=2m_{21}+1$ for illustrations. Phase portraits for the 2-dimensional systems for the hybrid networks of hyperbolic and hyperbolic-secant flows with the simple equilibriums of saddles and centers with $x_1^*=a_{211},a_{111},\cdots,a_{11(2m_{11})},a_{21(2m_{21}+1)},a_{11(2m_{11}+1)}$ and $x_2^*=a_{121},a_{122},\cdots,a_{12(2n_{11})}$ are presented in Fig. 3.30a–d for $(a_{110}>0,a_{220}>0)$, $(a_{110}<0,a_{220}>0)$, $(a_{110}>0,a_{220}<0)$ and $(a_{110}<0,a_{220}<0)$.

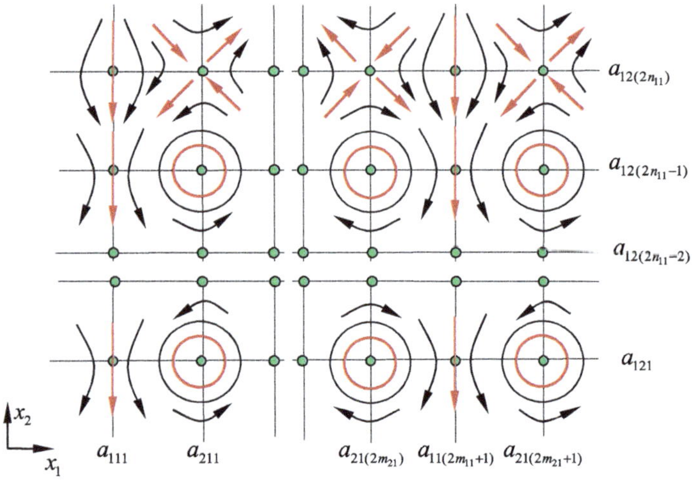

(a)

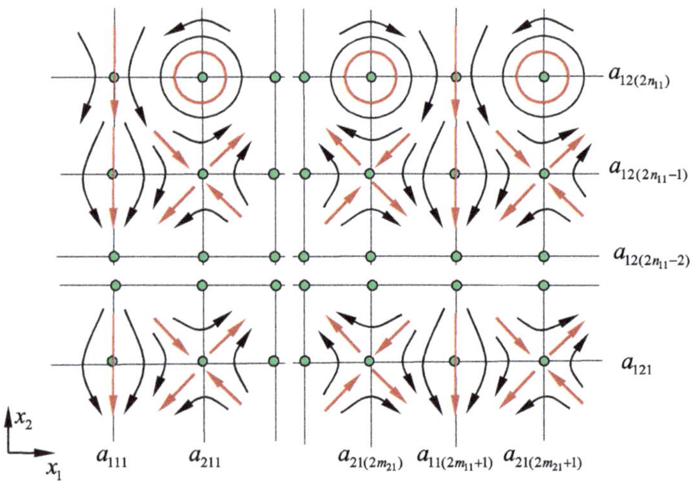

(b)

Fig. 3.29 Phase portraits for 2-dimensional systems with $x_1^* = a_{111}, a_{211}, \cdots, a_{11(2m_{11}+1)}, a_{121(2m_{21}+1)}$ and $x_2^* = a_{121}, a_{122}, \cdots, a_{12(2n_{11})}$. The networks of flows with saddles and centers: **a** ($a_{110} > 0, a_{220} > 0$), **b** ($a_{110} < 0, a_{220} > 0$), **c** ($a_{110} > 0, a_{220} < 0$), **d** ($a_{110} < 0, a_{220} < 0$)

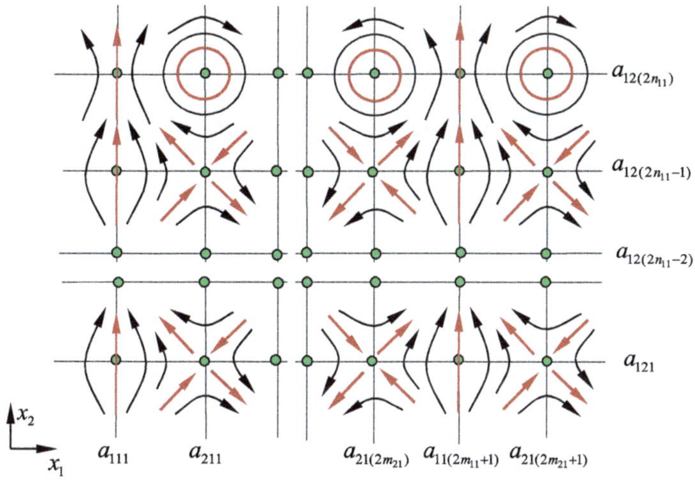

(c)

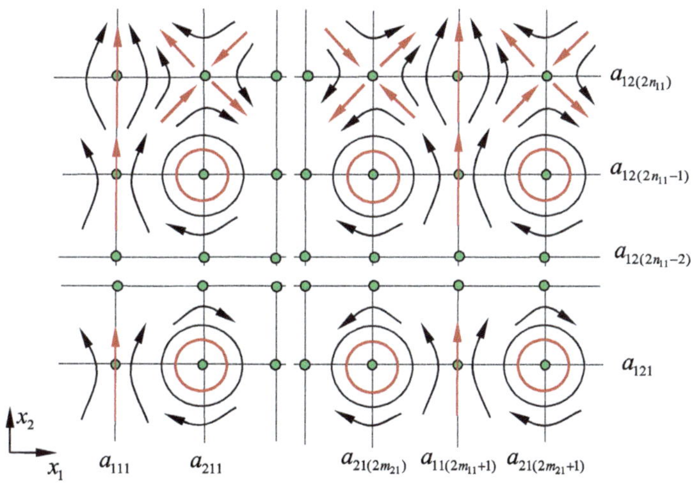

(d)

Fig. 3.29 (continued)

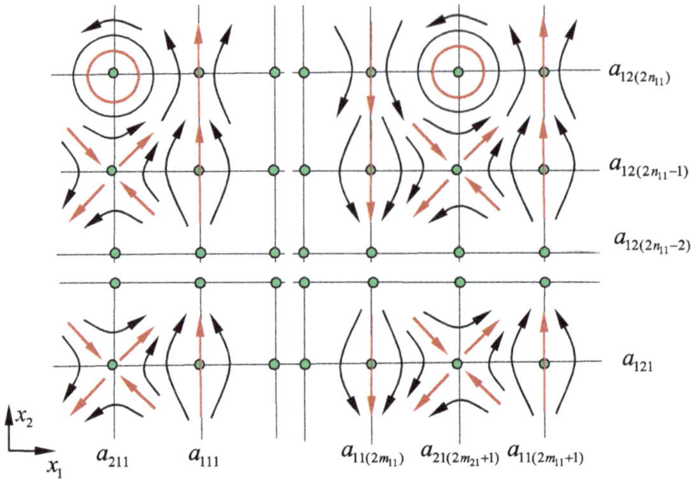

(a)

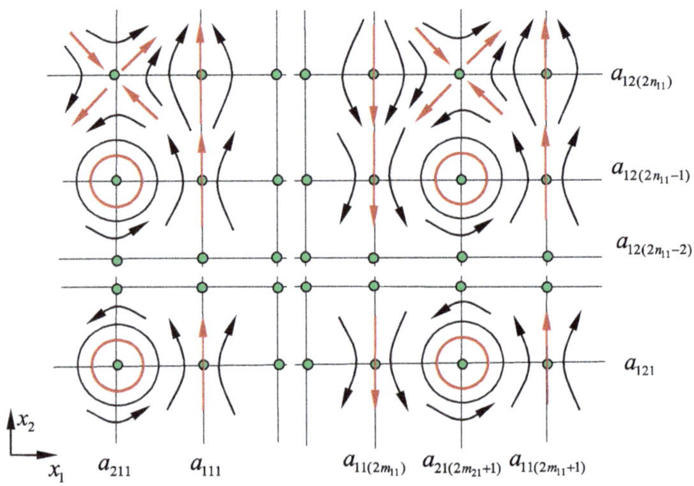

(b)

Fig. 3.30 Phase portraits for 2-dimensional systems with $x_1^* = a_{211}, a_{111}, \cdots, a_{21(2m_{21}+1)}, a_{11(2m_{11}+1)}$ and $x_2^* = a_{121}, a_{122}, \cdots, a_{12(2n_{11})}$. The networks of flows with saddles and centers: **a** $(a_{110} > 0, a_{220} > 0)$, **b** $(a_{110} < 0, a_{220} > 0)$, **c** $(a_{110} > 0, a_{220} < 0)$, **d** $(a_{110} < 0, a_{220} < 0)$

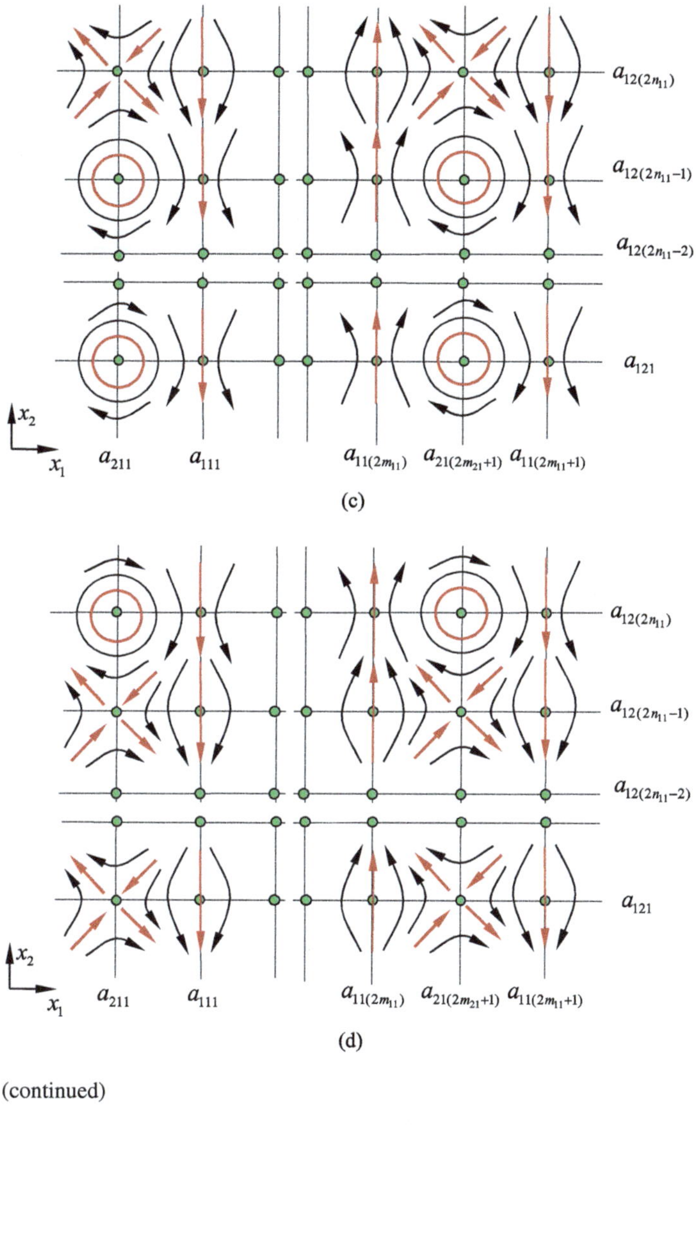

Fig. 3.30 (continued)

$$
\left\{\begin{array}{cccccc}
(a_{111},a_{12(2n_{11})}) & (a_{12(2n_{11})},a_{211}) & \cdots & (a_{12(2n_{11})},a_{21(2m_{21})}) & (a_{11(2m_{11}+1)},a_{12(2n_{11})}) & (a_{12(2n_{11})},a_{21(2m_{21}+1)}) \\
(a_{111},a_{12(2n_{11}-1)}) & (a_{12(2n_{11}-1)},a_{211}) & \cdots & (a_{12(2n_{11}-1)},a_{21(2m_{21})}) & (a_{11(2m_{11}+1)},a_{12(2n_{11}-1)}) & (a_{12(2n_{11}-1)},a_{21(2m_{21}+1)}) \\
\vdots & \vdots & \vdots & \vdots & \vdots & \vdots \\
(a_{111},a_{121}) & (a_{121},a_{211}) & \cdots & (a_{121},a_{21(2m_{21})}) & (a_{11(2m_{11}+1)},a_{121}) & (a_{121},a_{21(2m_{21}+1)})
\end{array}\right\}
$$

$$
=\left\{\begin{array}{cccccc}
\underbrace{(\mathrm{UP{:}DP,nF})}_{\text{hyperbolic-secant flow }(-)} & \underbrace{(\mathrm{UP}_+,\mathrm{UP}_+)}_{\text{positive saddle}} & \cdots & \underbrace{(\mathrm{UP}_-,\mathrm{UP}_-)}_{\text{negative saddle}} & \underbrace{(\mathrm{UP{:}DP,nF})}_{\text{hyperbolic-secant flow }(-)} & \underbrace{(\mathrm{UP}_+,\mathrm{UP}_+)}_{\text{positive saddle}} \\
\underbrace{(\mathrm{DP{:}UP,nF})}_{\text{hyperbolic flow }(-)} & \underbrace{(\mathrm{DP}_+,\mathrm{DP}_-)}_{\text{CCW center}} & \cdots & \underbrace{(\mathrm{DP}_-,\mathrm{DP}_+)}_{\text{CW center}} & \underbrace{(\mathrm{DP{:}UP,nF})}_{\text{hyperbolic flow }(-)} & \underbrace{(\mathrm{DP}_+,\mathrm{DP}_-)}_{\text{CCW center}} \\
\vdots & \vdots & \vdots & \vdots & \vdots & \vdots \\
\underbrace{(\mathrm{DP{:}UP,nF})}_{\text{hyperbolic flow }(-)} & \underbrace{(\mathrm{DP}_+,\mathrm{DP}_-)}_{\text{CCW center}} & \cdots & \underbrace{(\mathrm{DP}_-,\mathrm{DP}_+)}_{\text{CW center}} & \underbrace{(\mathrm{DP{:}UP,nF})}_{\text{hyperbolic flow }(-)} & \underbrace{(\mathrm{DP}_+,\mathrm{DP}_-)}_{\text{CCW center}}
\end{array}\right\}_{(2n_{11})\times(2m_{11}+2m_{21}+2)}
$$

for $a_{110} > 0, a_{220} > 0$; (3.93)

$$
\begin{Bmatrix}
(a_{111},a_{12(2n_{11})}) & (a_{12(2n_{11})},a_{211}) & \cdots & (a_{12(2n_{11})},a_{21(2m_{21})}) & (a_{11(2m_{11}+1)},a_{12(2n_{11})}) & (a_{12(2n_{11})},a_{21(2m_{21}+1)}) \\
(a_{111},a_{12(2n_{11}-1)}) & (a_{12(2n_{11}-1)},a_{211}) & \cdots & (a_{12(2n_{11}-1)},a_{21(2m_{21})}) & (a_{11(2m_{11}+1)},a_{12(2n_{11}-1)}) & (a_{12(2n_{11}-1)},a_{21(2m_{21}+1)}) \\
\vdots & \vdots & \vdots & \vdots & \vdots & \vdots \\
(a_{111},a_{121}) & (a_{121},a_{211}) & \cdots & (a_{121},a_{21(2m_{21})}) & (a_{11(2m_{11}+1)},a_{121}) & (a_{121},a_{21(2m_{21}+1)})
\end{Bmatrix}
$$

$$
= \begin{Bmatrix}
\underbrace{(\text{DP:UP,nF})}_{\text{hyperbolic flow }(-)} & \underbrace{(\text{DP}_+,\text{DP}_-)}_{\text{CCW center}} & \cdots & \underbrace{(\text{DP}_-,\text{DP}_+)}_{\text{CW center}} & \underbrace{(\text{DP:UP,nF})}_{\text{hyperbolic flow }(-)} & \underbrace{(\text{DP}_+,\text{DP}_-)}_{\text{CCW center}} \\
\underbrace{(\text{UP:DP,nF})}_{\text{hyperbolic-secant flow }(-)} & \underbrace{(\text{UP}_+,\text{UP}_+)}_{\text{positive saddle}} & \cdots & \underbrace{(\text{UP}_-,\text{UP}_-)}_{\text{negative saddle}} & \underbrace{(\text{UP:DP,nF})}_{\text{hyperbolic-secant flow }(-)} & \underbrace{(\text{UP}_+,\text{UP}_+)}_{\text{positive saddle}} \\
\vdots & \vdots & \vdots & \vdots & \vdots & \vdots \\
\underbrace{(\text{UP:DP,nF})}_{\text{hyperbolic-secant flow }(-)} & \underbrace{(\text{UP}_+,\text{UP}_+)}_{\text{positive saddle}} & \cdots & \underbrace{(\text{UP}_-,\text{UP}_-)}_{\text{negative saddle}} & \underbrace{(\text{UP:DP,nF})}_{\text{hyperbolic-secant flow }(-)} & \underbrace{(\text{UP}_+,\text{UP}_+)}_{\text{positive saddle}}
\end{Bmatrix}_{(2n_{11})\times(2m_{11}+2m_{21}+2)}
$$

for $a_{110} < 0, a_{220} > 0$; (3.94)

$$
\begin{aligned}
&\left\{\begin{array}{cccccc}
(a_{111},a_{12(2n_{11})}) & (a_{12(2n_{11})},a_{211}) & \cdots & (a_{12(2n_{11})},a_{21(2m_{21})}) & (a_{11(2m_{11}+1)},a_{12(2n_{11})}) & (a_{12(2n_{11})},a_{21(2m_{21}+1)}) \\
(a_{111},a_{12(2n_{11}-1)}) & (a_{12(2n_{11}-1)},a_{211}) & \cdots & (a_{12(2n_{11}-1)},a_{21(2m_{21})}) & (a_{11(2m_{11}+1)},a_{12(2n_{11}-1)}) & (a_{12(2n_{11}-1)},a_{21(2m_{21}+1)}) \\
\vdots & \vdots & \vdots & \vdots & \vdots & \vdots \\
(a_{111},a_{121}) & (a_{121},a_{211}) & \cdots & (a_{121},a_{21(2m_{21})}) & (a_{11(2m_{11}+1)},a_{121}) & (a_{121},a_{21(2m_{21}+1)})
\end{array}\right\} \\
&= \left\{\begin{array}{cccccc}
\underbrace{(\text{DP:UP,pF})}_{\text{hyperbolic flow }(+)} & \underbrace{(\text{DP}_-,\text{DP}_+)}_{\text{CW center}} & \cdots & \underbrace{(\text{DP}_+,\text{DP}_-)}_{\text{CCW center}} & \underbrace{(\text{DP:UP,pF})}_{\text{hyperbolic flow }(+)} & \underbrace{(\text{DP}_-,\text{DP}_+)}_{\text{CW center}} \\
\underbrace{(\text{UP:DP,pF})}_{\text{hyperbolic-secant flow }(+)} & \underbrace{(\text{UP}_-,\text{UP}_-)}_{\text{negative saddle}} & \cdots & \underbrace{(\text{UP}_+,\text{UP}_+)}_{\text{positive saddle}} & \underbrace{(\text{UP:DP,pF})}_{\text{hyperbolic-secant flow }(+)} & \underbrace{(\text{UP}_-,\text{UP}_-)}_{\text{negative saddle}} \\
\vdots & \vdots & \vdots & \vdots & \vdots & \vdots \\
\underbrace{(\text{UP:DP,pF})}_{\text{hyperbolic-secant flow }(+)} & \underbrace{(\text{UP}_-,\text{UP}_-)}_{\text{negative saddle}} & \cdots & \underbrace{(\text{UP}_+,\text{UP}_+)}_{\text{positive saddle}} & \underbrace{(\text{UP:DP,pF})}_{\text{hyperbolic-secant flow }(+)} & \underbrace{(\text{UP}_-,\text{UP}_-)}_{\text{negative saddle}}
\end{array}\right\}_{(2n_{11})\times(2m_{11}+2m_{21}+2)}
\end{aligned}
\tag{3.95}
$$

for $a_{110} > 0, a_{220} < 0$;

$$
\left\{\begin{array}{cccccc}
(a_{111},a_{12(2n_{11})}) & (a_{12(2n_{11})},a_{211}) & \cdots & (a_{12(2n_{11})},a_{21(2m_{21})}) & (a_{11(2m_{11}+1)},a_{12(2n_{11})}) & (a_{12(2n_{11})},a_{21(2m_{21}+1)}) \\
(a_{111},a_{12(2n_{11}-1)}) & (a_{12(2n_{11}-1)},a_{211}) & \cdots & (a_{12(2n_{11}-1)},a_{21(2m_{21})}) & (a_{11(2m_{11}+1)},a_{12(2n_{11}-1)}) & (a_{12(2n_{11}-1)},a_{21(2m_{21}+1)}) \\
\vdots & \vdots & \vdots & \vdots & \vdots & \vdots \\
(a_{111},a_{121}) & (a_{121},a_{211}) & \cdots & (a_{121},a_{21(2m_{21})}) & (a_{11(2m_{11}+1)},a_{121}) & (a_{121},a_{21(2m_{21}+1)})
\end{array}\right\}
$$

$$
=\left\{\begin{array}{cccccc}
\underbrace{(\mathrm{UP{:}DP,pF})}_{\text{hyperbolic-secant flow }(+)} & \underbrace{(\mathrm{UP}_-,\mathrm{UP}_-)}_{\text{negative saddle}} & \cdots & \underbrace{(\mathrm{UP}_+,\mathrm{UP}_+)}_{\text{positive saddle}} & \underbrace{(\mathrm{UP{:}DP,pF})}_{\text{hyperbolic-secant flow }(+)} & \underbrace{(\mathrm{UP}_-,\mathrm{UP}_-)}_{\text{negative saddle}} \\
\underbrace{(\mathrm{DP{:}UP,pF})}_{\text{hyperbolic flow }(+)} & \underbrace{(\mathrm{DP}_-,\mathrm{DP}_+)}_{\text{CW center}} & \cdots & \underbrace{(\mathrm{DP}_+,\mathrm{DP}_-)}_{\text{CCW center}} & \underbrace{(\mathrm{DP{:}UP,pF})}_{\text{hyperbolic flow }(+)} & \underbrace{(\mathrm{DP}_-,\mathrm{DP}_+)}_{\text{CW center}} \\
\vdots & \vdots & \vdots & \vdots & \vdots & \vdots \\
\underbrace{(\mathrm{DP{:}UP,pF})}_{\text{hyperbolic flow }(+)} & \underbrace{(\mathrm{DP}_-,\mathrm{DP}_+)}_{\text{CW center}} & \cdots & \underbrace{(\mathrm{DP}_+,\mathrm{DP}_-)}_{\text{CCW center}} & \underbrace{(\mathrm{DP{:}UP,pF})}_{\text{hyperbolic flow }(+)} & \underbrace{(\mathrm{DP}_-,\mathrm{DP}_+)}_{\text{CW center}}
\end{array}\right\}_{(2n_{11})\times(2m_{11}+2m_{21}+2)}
$$

for $a_{110} < 0, a_{220} < 0$; (3.96)

$$
\begin{Bmatrix}
(a_{12(2n_{11})},a_{211}) & (a_{111},a_{12(2n_{11})}) & \cdots & (a_{11(2m_{11})},a_{12(2n_{11})},) & (a_{12(2n_{11})},a_{21(2m_{21}+1)}) & (a_{11(2m_{11}+1)},a_{12(2n_{11})}) \\
(a_{12(2n_{11}-1)},a_{211}) & (a_{111},a_{12(2n_{11}-1)}) & \cdots & (a_{11(2m_{11})},a_{12(2n_{11}-1)}) & (a_{12(2n_{11}-1)},a_{21(2m_{21}+1)}) & (a_{11(2m_{11}+1)},a_{12(2n_{11}-1)}) \\
\vdots & \vdots & \vdots & \vdots & \vdots & \vdots \\
(a_{121},a_{211}) & (a_{111},a_{121}) & \cdots & (a_{11(2m_{11})},a_{121}) & (a_{121},a_{21(2m_{21}+1)}) & (a_{11(2m_{11}+1)},a_{121})
\end{Bmatrix}
$$

$$
= \begin{Bmatrix}
\underbrace{(\mathrm{DP}_{+},\mathrm{DP}_{-})}_{\text{CCW center}} & \underbrace{(\mathrm{DP{:}UP,pF})}_{\text{hyperbolic flow }(+)} & \cdots & \underbrace{(\mathrm{DP{:}UP,nF})}_{\text{hyperbolic flow }(-)} & \underbrace{(\mathrm{DP}_{+},\mathrm{DP}_{-})}_{\text{CCW center}} & \underbrace{(\mathrm{DP{:}UP,pF})}_{\text{hyperbolic flow }(+)} \\
\underbrace{(\mathrm{UP}_{+},\mathrm{UP}_{+})}_{\text{positive saddle}} & \underbrace{(\mathrm{UP{:}DP,pF})}_{\text{hyperbolic-secant flow }(+)} & \cdots & \underbrace{(\mathrm{UP{:}DP,nF})}_{\text{hyperbolic-secant flow }(-)} & \underbrace{(\mathrm{UP}_{+},\mathrm{UP}_{+})}_{\text{positive saddle}} & \underbrace{(\mathrm{UP{:}DP,pF})}_{\text{hyperbolic-secant flow }(+)} \\
\vdots & \vdots & \vdots & \vdots & \vdots & \vdots \\
\underbrace{(\mathrm{UP}_{+},\mathrm{UP}_{+})}_{\text{positive saddle}} & \underbrace{(\mathrm{UP{:}DP,pF})}_{\text{hyperbolic-secant flow }(+)} & \cdots & \underbrace{(\mathrm{UP{:}DP,nF})}_{\text{hyperbolic-secant flow }(-)} & \underbrace{(\mathrm{UP}_{+},\mathrm{UP}_{+})}_{\text{positive saddle}} & \underbrace{(\mathrm{UP{:}DP,pF})}_{\text{hyperbolic-secant flow }(+)}
\end{Bmatrix}_{(2n_{11})\times(2m_{11}+2m_{21}+2)}
\tag{3.97}
$$

for $a_{110} > 0$, $a_{220} > 0$;

$$
\begin{Bmatrix}
(a_{12(2n_{11})},a_{211}) & (a_{111},a_{12(2n_{11})}) & \cdots & (a_{11(2m_{11})},a_{12(2n_{11})},) & (a_{12(2n_{11})},a_{21(2m_{21}+1)}) & (a_{11(2m_{11}+1)},a_{12(2n_{11})}) \\
(a_{12(2n_{11}-1)},a_{211}) & (a_{111},a_{12(2n_{11}-1)}) & \cdots & (a_{11(2m_{11})},a_{12(2n_{11}-1)}) & (a_{12(2n_{11}-1)},a_{21(2m_{21}+1)}) & (a_{11(2m_{11}+1)},a_{12(2n_{11}-1)}) \\
\vdots & \vdots & \vdots & \vdots & \vdots & \vdots \\
(a_{121},a_{211}) & (a_{111},a_{121}) & \cdots & (a_{11(2m_{11})},a_{121}) & (a_{121},a_{21(2m_{21}+1)}) & (a_{11(2m_{11}+1)},a_{121})
\end{Bmatrix}
$$

$$
= \begin{Bmatrix}
\underbrace{(\mathrm{UP}_+,\mathrm{UP}_+)}_{\text{positive saddle}} & \underbrace{(\mathrm{UP{:}DP,pF})}_{\text{hyperbolic-secant flow }(+)} & \cdots & \underbrace{(\mathrm{UP{:}DP,nF})}_{\text{hyperbolic-secant flow }(-)} & \underbrace{(\mathrm{UP}_+,\mathrm{UP}_+)}_{\text{positive saddle}} & \underbrace{(\mathrm{UP{:}DP,pF})}_{\text{hyperbolic-secant flow }(+)} \\
\underbrace{(\mathrm{DP}_+,\mathrm{DP}_-)}_{\text{CCW center}} & \underbrace{(\mathrm{DP{:}UP,pF})}_{\text{hyperbolic flow }(+)} & \cdots & \underbrace{(\mathrm{DP{:}UP,nF})}_{\text{hyperbolic flow }(-)} & \underbrace{(\mathrm{DP}_+,\mathrm{DP}_-)}_{\text{CCW center}} & \underbrace{(\mathrm{DP{:}UP,pF})}_{\text{hyperbolic flow }(+)} \\
\vdots & \vdots & \vdots & \vdots & \vdots & \vdots \\
\underbrace{(\mathrm{DP}_+,\mathrm{DP}_-)}_{\text{CCW center}} & \underbrace{(\mathrm{DP{:}UP,pF})}_{\text{hyperbolic flow }(+)} & \cdots & \underbrace{(\mathrm{DP{:}UP,nF})}_{\text{hyperbolic flow }(-)} & \underbrace{(\mathrm{DP}_+,\mathrm{DP}_-)}_{\text{CCW center}} & \underbrace{(\mathrm{DP{:}UP,pF})}_{\text{hyperbolic flow }(+)}
\end{Bmatrix}_{(2n_{11})\times(2m_{11}+2m_{21}+2)}
$$

for $a_{110} < 0, a_{220} > 0$; (3.98)

$$\left\{\begin{array}{cccccc}(a_{12(2n_{11})},a_{211}) & (a_{111},a_{12(2n_{11})}) & \cdots & (a_{11(2m_{11})},a_{12(2n_{11})},) & (a_{12(2n_{11})},a_{21(2m_{21}+1)}) & (a_{11(2m_{11}+1)},a_{12(2n_{11})})\\ (a_{12(2n_{11}-1)},a_{211}) & (a_{111},a_{12(2n_{11}-1)}) & \cdots & (a_{11(2m_{11})},a_{12(2n_{11}-1)}) & (a_{12(2n_{11}-1)},a_{21(2m_{21}+1)}) & (a_{11(2m_{11}+1)},a_{12(2n_{11}-1)})\\ \vdots & \vdots & \vdots & \vdots & \vdots & \vdots\\ (a_{121},a_{211}) & (a_{111},a_{121}) & \cdots & (a_{11(2m_{11})},a_{121}) & (a_{121},a_{21(2m_{21}+1)}) & (a_{11(2m_{11}+1)},a_{121})\end{array}\right\}$$
$$=\left\{\begin{array}{cccccc}\underbrace{(\text{UP}_-,\text{UP}_-)}_{\text{negative saddle}} & \underbrace{(\text{UP:DP,nF})}_{\text{hyperbolic-secant flow }(-)} & \cdots & \underbrace{(\text{UP:DP,pF})}_{\text{hyperbolic-secant flow }(+)} & \underbrace{(\text{UP}_-,\text{UP}_-)}_{\text{negative saddle}} & \underbrace{(\text{UP:DP,nF})}_{\text{hyperbolic-secant flow }(-)}\\ \underbrace{(\text{DP}_-,\text{DP}_+)}_{\text{CW center}} & \underbrace{(\text{DP:UP,nF})}_{\text{hyperbolic flow }(-)} & \cdots & \underbrace{(\text{DP:UP,pF})}_{\text{hyperbolic flow }(+)} & \underbrace{(\text{DP}_-,\text{DP}_+)}_{\text{CW center}} & \underbrace{(\text{DP:UP,nF})}_{\text{hyperbolic flow }(-)}\\ \vdots & \vdots & \vdots & \vdots & \vdots & \vdots\\ \underbrace{(\text{DP}_-,\text{DP}_+)}_{\text{CW center}} & \underbrace{(\text{DP:UP,nF})}_{\text{hyperbolic flow }(-)} & \cdots & \underbrace{(\text{DP:UP,pF})}_{\text{hyperbolic flow }(+)} & \underbrace{(\text{DP}_-,\text{DP}_+)}_{\text{CW center}} & \underbrace{(\text{DP:UP,nF})}_{\text{hyperbolic flow }(-)}\end{array}\right\}_{(2n_{11})\times(2m_{11}+2m_{21}+2)}$$

for $a_{110} > 0, a_{220} < 0$; (3.99)

$$
\left\{\begin{matrix}
(a_{12(2n_{11})},a_{211}) & (a_{111},a_{12(2n_{11})}) & \cdots & (a_{11(2m_{11})},a_{12(2n_{11})},) & (a_{12(2n_{11})},a_{21(2m_{21}+1)}) & (a_{11(2m_{11}+1)},a_{12(2n_{11})})\\
(a_{12(2n_{11}-1)},a_{211}) & (a_{111},a_{12(2n_{11}-1)}) & \cdots & (a_{11(2m_{11})},a_{12(2n_{11}-1)}) & (a_{12(2n_{11}-1)},a_{21(2m_{21}+1)}) & (a_{11(2m_{11}+1)},a_{12(2n_{11}-1)})\\
\vdots & \vdots & \vdots & \vdots & \vdots & \vdots\\
(a_{121},a_{211}) & (a_{111},a_{121}) & \cdots & (a_{11(2m_{11})},a_{121}) & (a_{121},a_{21(2m_{21}+1)}) & (a_{11(2m_{11}+1)},a_{121})
\end{matrix}\right\}
$$
$$
=\left\{\begin{matrix}
\underbrace{(\mathrm{DP}_-,\mathrm{DP}_+)}_{\text{CW center}} & \underbrace{(\mathrm{DP{:}UP,nF})}_{\text{hyperbolic flow }(-)} & \cdots & \underbrace{(\mathrm{DP{:}UP,pF})}_{\text{hyperbolic flow }(+)} & \underbrace{(\mathrm{DP}_-,\mathrm{DP}_+)}_{\text{CW center}} & \underbrace{(\mathrm{DP{:}UP,nF})}_{\text{hyperbolic flow }(-)}\\
\underbrace{(\mathrm{UP}_-,\mathrm{UP}_-)}_{\text{negative saddle}} & \underbrace{(\mathrm{UP{:}DP,nF})}_{\text{hyperbolic-secant flow }(-)} & \cdots & \underbrace{(\mathrm{UP{:}DP,pF})}_{\text{hyperbolic-secant flow }(+)} & \underbrace{(\mathrm{UP}_-,\mathrm{UP}_-)}_{\text{negative saddle}} & \underbrace{(\mathrm{UP{:}DP,nF})}_{\text{hyperbolic-secant flow }(-)}\\
\vdots & \vdots & \vdots & \vdots & \vdots & \vdots\\
\underbrace{(\mathrm{UP}_-,\mathrm{UP}_-)}_{\text{negative saddle}} & \underbrace{(\mathrm{UP{:}DP,nF})}_{\text{hyperbolic-secant flow }(-)} & \cdots & \underbrace{(\mathrm{UP{:}DP,pF})}_{\text{hyperbolic-secant flow }(+)} & \underbrace{(\mathrm{UP}_-,\mathrm{UP}_-)}_{\text{negative saddle}} & \underbrace{(\mathrm{UP{:}DP,nF})}_{\text{hyperbolic-secant flow }(-)}
\end{matrix}\right\}_{(2n_{11})\times(2m_{11}+2m_{21}+2)}
$$

for $a_{110}<0,\ a_{220}<0$. (3.100)

3.6 $[(2m_{11}), (2n_{11})]$ and $((2n_{11}), (2m_{21}+1))$-Singular Systems

In this section, there are four global bifurcations of the 1-dimensional flows through singular hyperbolic-to-hyperbolic-secant flows, and four global bifurcations of equilibriums, through singular saddles and centers bifurcations. The corresponding global bifurcations generate networks of hyperbolic and hyperbolic-secant flows with saddles and centers. The infinite-equilibriums are the appearing and switching bifurcations for the singular flows and singular equilibriums and for hybrid networks of hyperbolic flows, saddles and centers.

3.6.1 $[(2m_{11}), (2n_{11})]$-Flows and $((2n_{11}), (2m_{21}+1))$-Equilibriums

Consider a polynomial dynamical system for $a_{111} \neq a_{211}$ as

$$\begin{aligned} \dot{x}_2 &= a_{220}(x_1 - a_{211})^{2m_{21}+1}, \\ \dot{x}_1 &= a_{110}(x_1 - a_{111})^{2m_{11}}(x_2 - a_{121})^{2n_{11}}. \end{aligned} \tag{3.101}$$

The first integral manifold is

$$\begin{aligned} &\sum_{\alpha,\beta=0}^{\alpha+\beta=2m_{21}+1} \frac{(2m_{21}+1)!}{\alpha!\beta!}(a_{111}-a_{211})^{\alpha} \\ &\times \frac{1}{\beta - 2m_{11}+1}\left[(x_1 - a_{111})^{\beta-2m_{11}+1} - (x_{10} - a_{111})^{\beta-2m_{11}+1}\right] \\ &= \frac{a_{j_1 j_1 0}}{a_{j_2 j_2 0}} \frac{1}{2n_{11}+1}\left[(x_2 - a_{121})^{2n_{11}+1} - (x_{20} - a_{121})^{2n_{11}+1}\right], \\ &\text{for } \beta - 2m_{11} + 1 \neq 0; \\ &\frac{1}{\beta - 2m_{11}+1}\left[(x_1 - a_{111})^{\beta-2m_{11}+1} - (x_{10} - a_{111})^{\beta-2m_{11}+1}\right] = \ln\frac{|x_1 - a_{111}|}{|x_{10} - a_{111}|}, \\ &\text{for } \beta - 2m_{11} + 1 = 0. \end{aligned} \tag{3.102}$$

For $(x_1^*, x_2^*) = (a_{111}, a_{121})$ with $\dot{x}_1 = 0$, the $[(2m_{11}), (2n_{11})]$-singular flows exist with $\dot{x}_2 = a_{220}(a_{111} - a_{211})^{2m_{21}+1}$. For $(x_1^*, x_2^*) = (a_{111}, a_{121})$, the 1-dimensional flow direction is determined by $\dot{x}_2 \neq 0$. For $(x_2^*, x_1^*) = (a_{121}, a_{211})$ with $\dot{x}_1 = \dot{x}_2 = 0$, the $((2n_{11}), (2m_{21}+1))$-singular equilibriums exist.

(A) For $a_{111} > a_{211}$,

$$\left\{\begin{matrix}(a_{111}, a_{121})\\(a_{121}, a_{211})\end{matrix}\right\} = \left\{\begin{matrix}\underbrace{([(2m_{11}),(2n_{11})]_{\text{II:II}}\text{US,pF})}_{[(2m_{11}),(2n_{11})]\text{-inflection upper-saddle flow }(+)}\\\underbrace{((2n_{11})\text{US},(2m_{21}+1)\text{UP})}_{[(2n_{11}),(2m_{21}+1)]\text{-up-parabola upper-saddle}}\end{matrix}\right\},$$

for $a_{110} > 0, a_{220} > 0$;

$$\left\{\begin{matrix}(a_{111}, a_{121})\\(a_{121}, a_{211})\end{matrix}\right\} = \left\{\begin{matrix}\underbrace{([(2m_{11}),(2n_{11})]_{\text{DI:DI}}\text{LS,pF})}_{[(2m_{11}),(2n_{11})]\text{-inflection lower-saddle flow }(+)}\\\underbrace{((2n_{11})\text{LS},(2m_{21}+1)\text{DP})}_{[(2n_{11}),(2m_{21}+1)]\text{-down-parabola lower-saddle}}\end{matrix}\right\},$$

for $a_{110} < 0, a_{220} > 0$;

$$\left\{\begin{matrix}(a_{111}, a_{121})\\(a_{121}, a_{211})\end{matrix}\right\} = \left\{\begin{matrix}\underbrace{([(2m_{11}),(2n_{11})]_{\text{DI:DI}}\text{US,nF})}_{[(2m_{11}),(2n_{11})]\text{-inflection upper-saddle flow }(-)}\\\underbrace{((2n_{11})\text{US},(2m_{21}+1)\text{DP})}_{[(2n_{11}),(2m_{21}+1)]\text{-down-parabola upper-saddle}}\end{matrix}\right\},$$

for $a_{110} > 0, a_{220} < 0$;

$$\left\{\begin{matrix}(a_{111}, a_{121})\\(a_{121}, a_{211})\end{matrix}\right\} = \left\{\begin{matrix}\underbrace{([(2m_{11}),(2n_{11})]_{\text{II:II}}\text{LS,nF})}_{[(2m_{11}),(2n_{11})]\text{-inflection lower-saddle flow }(-)}\\\underbrace{((2n_{11})\text{LS},(2m_{21}+1)\text{UP})}_{[(2n_{11}),(2m_{21}+1)]\text{-up-parabola lower-saddle}}\end{matrix}\right\},$$

for $a_{110} < 0, a_{220} < 0$. (3.103)

(**B**) For $a_{111} < a_{211}$,

$$\left\{\begin{matrix}(a_{121}, a_{211})\\(a_{111}, a_{121})\end{matrix}\right\} = \left\{\begin{matrix}\underbrace{((2n_{11})\text{US},(2m_{21}+1)\text{UP})}_{[(2n_{11}),(2m_{21}+1)]\text{-up-parabola upper-saddle}}\\\underbrace{([(2m_{11}),(2n_{11})]_{\text{DI:DI}}\text{US,nF})}_{[(2m_{11}),(2n_{11})]\text{-inflection upper-saddle flow }(-)}\end{matrix}\right\},$$

for $a_{110} > 0, a_{220} > 0$;

$$\left\{\begin{matrix}(a_{121}, a_{211})\\(a_{111}, a_{121})\end{matrix}\right\} = \left\{\begin{matrix}\underbrace{((2n_{11})\text{LS},(2m_{21}+1)\text{DP})}_{[(2n_{11}),(2m_{21}+1)]\text{-down-parabola lower-saddle}}\\\underbrace{([(2m_{11}),(2n_{11})]_{\text{II:II}}\text{LS,nF})}_{[(2m_{11}),(2n_{11})]\text{-inflection lower-saddle flow }(-)}\end{matrix}\right\},$$

for $a_{110} < 0, a_{220} > 0$;

$$\left\{\begin{matrix}(a_{121}, a_{211})\\(a_{111}, a_{121})\end{matrix}\right\} = \left\{\begin{matrix}\underbrace{((2n_{11})\text{US},(2m_{21}+1)\text{DP})}_{[(2n_{11}),(2m_{21}+1)]\text{-down-parabola upper-saddle}}\\\underbrace{([(2m_{11}),(2n_{11})]_{\text{II:II}}\text{US,pF})}_{[(2m_{11}),(2n_{11})]\text{-inflection upper-saddle flow }(+)}\end{matrix}\right\},$$

for $a_{110} > 0, a_{220} < 0$;

$$\left\{ \begin{array}{l} (a_{121}, a_{211}) \\ (a_{111}, a_{121}) \end{array} \right\} = \left\{ \begin{array}{c} \underbrace{((2n_{11})\text{LS},(2m_{21}+1)\text{UP})}_{[(2n_{11}),(2m_{21}+1)]\text{-up-parabola lower-saddle}} \\ \underbrace{([(2m_{11}),(2n_{11})]_{\text{DI:DI}}\text{LS,pF})}_{[(2m_{11}),(2n_{11})]\text{-inflection lower-saddle flow }(+)} \end{array} \right\},$$

for $a_{110} < 0, a_{220} < 0$. (3.104)

The singular inflection upper-saddle flows with singular parabola-saddle are illustrated in Figs. 3.31 and 3.32. Through such singular flows, the corresponding global bifurcations generate from a two-dimensional flow to a network of singular hyperbolic and hyperbolic-secant flows. Through such singular equilibriums, the corresponding global bifurcations generate from a parabola flow to a network of singular saddles and centers.

- The $([(2m_{11}), (2n_{11})]_{\text{II:II}}\text{US,pF})$ is for a positive singular increasing-inflection upper-saddle flow.
- The $([(2m_{11}), (2n_{11})]_{\text{DI:DI}}\text{LS,pF})$ is for a positive singular decreasing-inflection lower-saddle flow.
- The $([(2m_{11}), (2n_{11})]_{\text{DI:DI}}\text{US,nF})$ is for a negative singular decreasing-inflection upper-saddle flow.
- The $([(2m_{11}), (2n_{11})]_{\text{II:II}}\text{LS,nF})$ is for a negative singular increasing-inflection upper-saddle flow.

${}_{\text{II:II}}\text{US}$ and ${}_{\text{II:II}}\text{LS}$ are for an increasing-inflection upper-saddle and lower-saddle flows, and ${}_{\text{DI:DI}}\text{US}$ and ${}_{\text{DI:DI}}\text{LS}$ are for a decreasing-inflection upper-saddle and lower-saddle flows.

3.6.2 $((2m_{11}), (2n_{11}))$ and $((2n_{11}), (2m_{21}+1))$-Infinite-Equilibriums

Consider a polynomial dynamical system for $a_{111} = a_{211}$ as

$$\begin{aligned} \dot{x}_2 &= a_{220}(x_1 - a_{211})^{2m_{21}+1}, \\ \dot{x}_1 &= a_{110}(x_1 - a_{111})^{2m_{11}}(x_2 - a_{121})^{2n_{11}}. \end{aligned} \tag{3.105}$$

The first integral manifold is

$$\frac{1}{2(m_{21}-m_{11})+2}\left[(x_1 - a_{211})^{2(m_{21}-m_{11})+2} - (x_{10} - a_{211})^{2(m_{21}-m_{11})+2}\right]$$

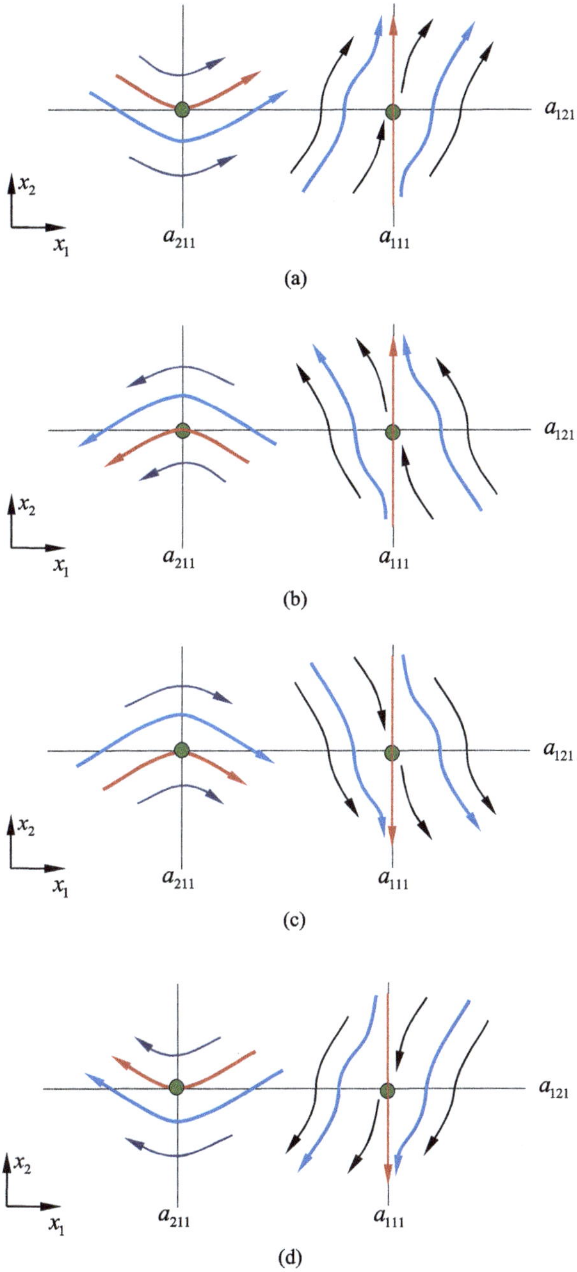

Fig. 3.31 Phase portraits ($a_{111} > a_{211}$) of the ([(2m_{11}), (2n_{11})], xF)-singular inflection-source and sink flows at $(x_1^*, x_2^*) = (a_{111}, a_{121})$ and ((2n_{11}), (2m_{21} + 1))-parabola-saddle at $(x_2^*, x_1^*) = (a_{121}, a_{211})$ for 2-dimensional polynomial systems: **a** ($a_{110} > 0, a_{220} > 0$), **b** ($a_{110} < 0, a_{220} > 0$), **c** ($a_{110} > 0, a_{220} < 0$), **d** ($a_{110} < 0, a_{220} < 0$)

Fig. 3.32 Phase portraits ($a_{111} > a_{211}$) of the ([($2m_{11}$), ($2n_{11}$)], xF)-singular inflection-source and sink flows at $(x_1^*, x_2^*) = (a_{111}, a_{121})$ and (($2n_{11}$), ($2m_{21} + 1$))-parabola-saddle at $(x_2^*, x_1^*) = (a_{121}, a_{211})$ for 2-dimensional polynomial systems:
a ($a_{110} > 0, a_{220} > 0$),
b ($a_{110} < 0, a_{220} > 0$),
c ($a_{110} > 0, a_{220} < 0$),
d ($a_{110} < 0, a_{220} < 0$)

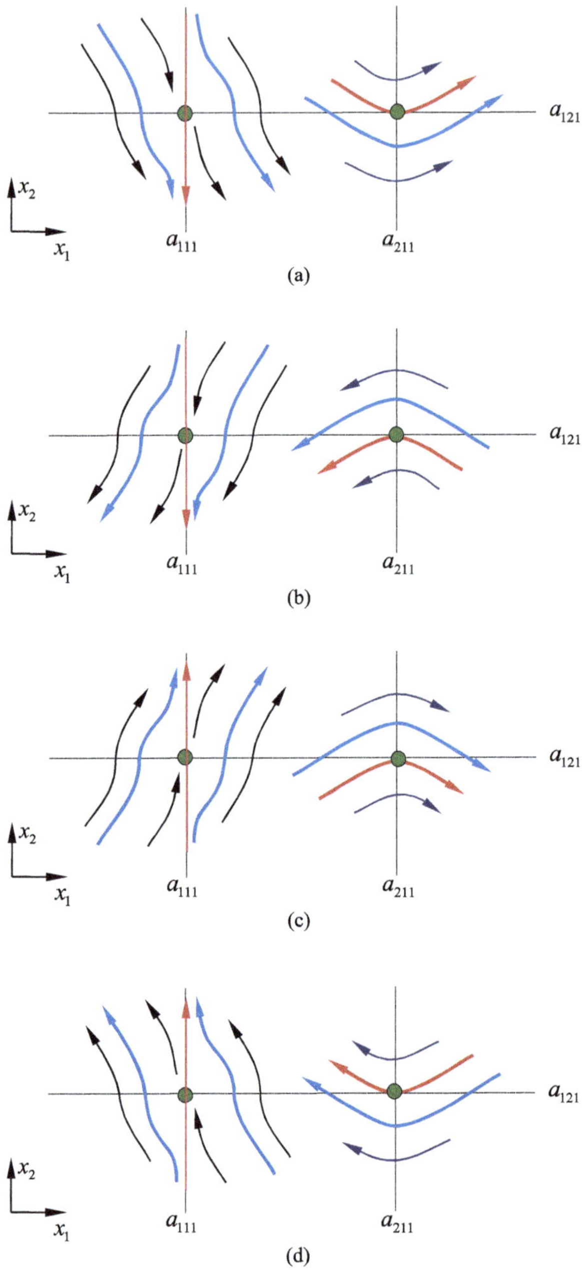

$$
\begin{aligned}
&= \frac{a_{110}}{a_{220}}\frac{1}{2n_{11}+1}\left[(x_2-a_{121})^{2n_{11}+1}-(x_{20}-a_{121})^{2n_{11}+1}\right]\\
&\text{for } 2(m_{21}-m_{11})+2\neq 0;\\
&\frac{1}{2(m_{21}-m_{11})+2}\left[(x_1-a_{211})^{2(m_{21}-m_{11})+2}-(x_{10}-a_{211})^{2(m_{21}-m_{11})+2}\right]\\
&= \ln\frac{|x_1-a_{111}|}{|x_{10}-a_{111}|},\ \text{for } 2(m_{21}-m_{11})+2=0.
\end{aligned}
\tag{3.106}
$$

(**A**) For $2m_{21}+1 \geq 2m_{11}$, the corresponding infinite-equilibriums are given.

$$
\begin{aligned}
&(a_{121},a_{211})\\
&=\underbrace{([(2m_{11}),(2n_{11})]^{\text{th}}\text{DI}_{-}\text{:II}_{+},[2(m_{21}-m_{11})+1]^{\text{th}}{}_{\text{UP:UP}}\text{US})}_{([(2m_{11}),(2n_{11})],[2(m_{21}-m_{11})+1])\text{-double up-parabola upper-saddle}},\\
&\text{for } a_{110}>0, a_{220}>0;\\
&(a_{121},a_{211})\\
&=\underbrace{([(2m_{11}),(2n_{11})]^{\text{th}}\text{II}_{-}\text{:DI}_{+},[2(m_{21}-m_{11})+1]^{\text{th}}{}_{\text{DP:DP}}\text{LS})}_{([(2m_{11}),(2n_{11})],[2(m_{21}-m_{11})+1])\text{-double down-parabola lower-saddle}},\\
&\text{for } a_{110}<0, a_{220}>0;\\
&(a_{121},a_{211})\\
&=\underbrace{([(2m_{11}),(2n_{11})]^{\text{th}}\text{II}_{+}\text{:DI}_{-},[2(m_{21}-m_{11})+1]^{\text{th}}{}_{\text{DP:DP}}\text{US})}_{([(2m_{11}),(2n_{11})],[2(m_{21}-m_{11})+1])\text{-double down-parabola upper-saddle}},\\
&\text{for } a_{110}>0, a_{220}<0;\\
&(a_{121},a_{211})\\
&=\underbrace{([(2m_{11}),(2n_{11})]^{\text{th}}\text{DI}_{-}\text{:II}_{+},[2(m_{21}-m_{11})+1]^{\text{th}}{}_{\text{UP:UP}}\text{LS})}_{([(2m_{11}),(2n_{11})],[2(m_{21}-m_{11})+1])\text{-double up-parabola lower-saddle}},\\
&\text{for } a_{110}<0, a_{220}<0.
\end{aligned}
\tag{3.107}
$$

(**B**) For $2m_{21}+1 < 2m_{11}$, the infinite-equilibriums are given as follows.

$$
\begin{aligned}
&(a_{121},a_{211})=\underbrace{([(2m_{11}),(2n_{11})]^{\text{th}}\text{DI}_{-}\text{:II}_{+},[2(m_{11}-m_{21})-1]^{\text{th}}{}_{\text{DA:DA}}\text{US})}_{([(2m_{11}),(2n_{11})],[2(m_{11}-m_{21})-1])\text{-inflection down-asymptote upper-saddle}},\\
&\text{for } a_{110}>0, a_{220}>0;\\
&(a_{121},a_{211})=\underbrace{([(2m_{11}),(2n_{11})]^{\text{th}}\text{II}_{-}\text{:DI}_{+},[2(m_{11}-m_{21})-1]^{\text{th}}{}_{\text{UA:UA}}\text{LS})}_{([(2m_{11}),(2n_{11})],[2(m_{11}-m_{21})-1])\text{-inflection up-asymptote lower-saddle}},\\
&\text{for } a_{110}<0, a_{220}>0;\\
&(a_{121},a_{211})=\underbrace{([(2m_{11}),(2n_{11})]^{\text{th}}\text{II}_{+}\text{:DI}_{-},[2(m_{11}-m_{21})-1]^{\text{th}}{}_{\text{UA:UA}}\text{US})}_{([(2m_{11}),(2n_{11})],[2(m_{11}-m_{21})-1])\text{-inflection up-asymptote upper-saddle}},
\end{aligned}
$$

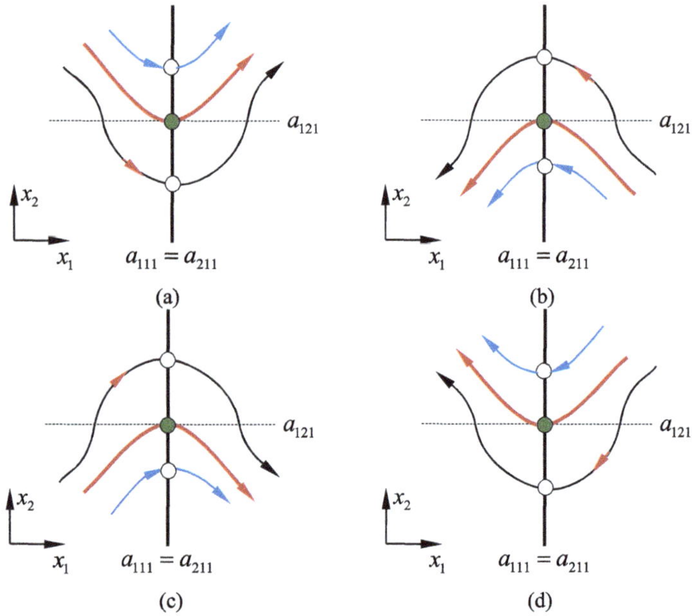

Fig. 3.33 The infinite-equilibriums ($a_{211} = a_{111}, 2m_{21} + 1 \geq 2m_{11}, n_1 = 2n_{11} + 1$) for 2-dimensional polynomial systems: **a** ($a_{110} > 0, a_{220} > 0$), **b** ($a_{110} < 0, a_{220} > 0$), **c** ($a_{110} > 0, a_{220} < 0$), **d** ($a_{110} < 0, a_{220} < 0$)

$$\text{for } a_{110} > 0, a_{220} < 0;$$

$$(a_{121}, a_{211}) = \underbrace{([(2m_{11}), (2n_{11})]^{\text{th}}\text{DI}_{+}\text{:II}_{-},[2(m_{11} - m_{21}) - 1]^{\text{th}}{}_{\text{DA:DA}}\text{LS})}_{([(2m_{11}),(2n_{11})],[2(m_{11}-m_{21})-1])\text{-inflection down-asymptote lower-saddle}},$$

$$\text{for } a_{110} < 0, a_{220} < 0. \tag{3.108}$$

For $x_1^* = a_{111} = a_{211}$, the infinite-equilibrium exists. The infinite-equilibriums are illustrated in Figs. 3.33 and 3.34 for $2m_{21} + 1 \geq 2m_{11}$ and $2m_{21} + 1 < 2m_{11}$, respectively. For $2m_{21} + 1 \geq 2m_{11}$, the singular double parabola-saddles are presented. For $2m_{11} > 2m_{21} + 1$, the inflection asymptote-saddles are presented.

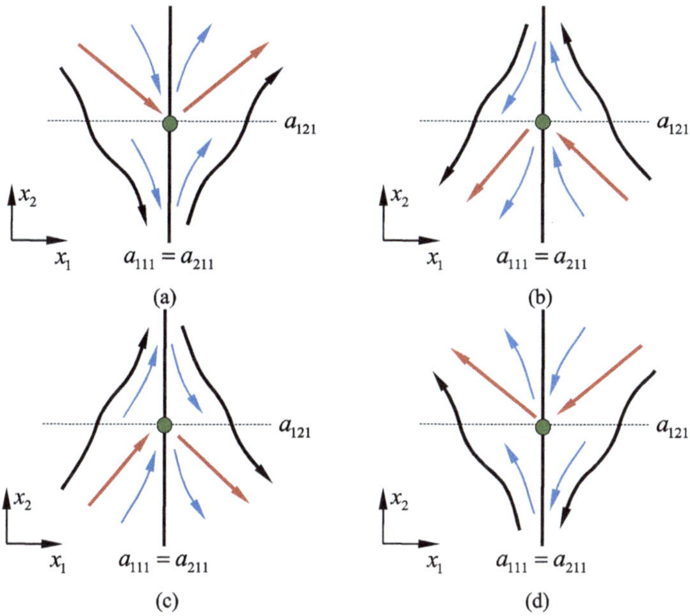

Fig. 3.34 The infinite-equilibriums ($a_{211} = a_{111}, 2m_{21}+1 < 2m_{11}, n_1 = 2n_{11}$) for 2-dimensional polynomial systems: **a** ($a_{110} > 0, a_{220} > 0$), **b** ($a_{110} < 0, a_{220} > 0$), **c** ($a_{110} > 0, a_{220} < 0$), **d** ($a_{110} < 0, a_{220} < 0$)

3.6.3 $[(2m_{11}), (2n_{11})]$ *and* $((2n_{11}), (2m_{21}+1))$*-Hybrid Networks*

Consider

$$\dot{x}_2 = a_{220} \prod_{v=1}^{2m_{21}+1} (x_1 - a_{21v}),$$
$$\dot{x}_1 = a_{110} \prod_{r=1}^{2m_{11}} (x_1 - a_{11r}) \prod_{u=1}^{2n_{11}} (x_2 - a_{12u}). \tag{3.109}$$

The first integral manifold is

$$\sum_{v_0=2m_{11}}^{2m_{21}+1} b_{21v_0} \sum_{\substack{\mu_1\in[0,\beta_0-2m_{11}-2],\\ \nu_1\in[2m_{11}-1,\nu_0-1]}}^{\mu_1+\nu_1=\nu_0-1} \sum_{\substack{\mu_2\in[0,\nu_1-2m_{11}]\\ \nu_2\in[2m_{11}-2,\nu_1-2]}}^{\mu_2+\nu_2=\nu_1-2} \cdots \sum_{\substack{\mu_{2m_{11}}\in[0,\nu_{2m_{11}-1}-2m_{11}]\\ \nu_{2m_{11}}\in[1,\nu_{2m_{11}-1}-2m_{11}+1]}}^{\mu_{2m_{11}}+\nu_{2m_{11}}=\nu_{2m_{11}-1}-2m_{11}+1}$$
$$\prod_{j=1}^{2m_{11}-1} \frac{(\nu_{j-1}-j)!}{\mu_j!\nu_j!} \times (a_{11r_{j+1}} - a_{11r_j})^{\mu_j} \frac{1}{\nu_{2m_{11}-1}} \left[(x_1 - a_{11r_{2m_{11}}})^{\nu_{2m_{11}-1}} - (x_{10} - a_{11r_{2m_{11}}})^{\nu_{2m_{11}-1}}\right]$$

$$
\begin{aligned}
&-\sum_{\nu_0=2m_{11}}^{2m_{21}+1} b_{21\nu_0} \sum_{k=2}^{2m_{11}-1} \Bigg[\sum_{\substack{\mu_1\in[0,\nu_0-2m_{11}),\\ \nu_1\in(2m_{11}-1,\nu_0-1]}}^{\mu_1+\nu_1=\nu_0-1} \sum_{\substack{\mu_2\in[0,\nu_1-2m_{11})\\ \nu_2\in(2m_{11}-2,\nu_1-2]}}^{\mu_2+\nu_2=\nu_1-2} \cdots \\
&\times \Bigg(\sum_{\substack{\mu_{k-1}\in(\nu_{k-2}-2m_{11},\nu_{k-2}-k+1]\\ \nu_{k-1}\in[0,2m_{11}+1-k)}}^{\mu_{k-1}+\nu_{k-1}=\nu_{k-2}-k+1} \prod_{j=1}^{k-1} \frac{(\nu_{j-1}-j)!}{\mu_j!\nu_j!}(a_{11r_{j+1}}-a_{11r_j})^{\mu_j} \\
&\times \sum_{\rho=k}^{2m_{11}} A_{\rho 1}^{(s_k)} \ln \frac{|x_1-a_{11s_\rho}|}{|x_{10}-a_{11s_\rho}|} \\
&+ \sum_{\substack{\mu_{k-1}\in(\nu_{k-2}-2m_{11}-1,\nu_{k-2}-2m_{11}+1]\\ \nu_{k-1}\in[2m_{11}+1-k,2m_{11}-k+2)}}^{\mu_{k-1}+\nu_{k-1}=\nu_{k-2}-k+1} \prod_{j=1}^{k-1} \frac{(\nu_{j-1}-j)!}{\mu_j!\nu_j!}(a_{11r_{j+1}}-a_{11r_j})^{\mu_j} \\
&\times \sum_{\rho=k+1}^{2m_{11}} B_{\rho 1}^{(r_k)} \ln \frac{|x_1-a_{11r_\rho}|}{|x_{10}-a_{11r_\rho}|} \Bigg) \Bigg] + b_{210} \sum_{\rho=1}^{2m_{11}} A_{\rho 1}^{(r_1)} \ln \frac{|x_1-a_{11r_\rho}|}{|x_{10}-a_{11r_\rho}|} \\
&+ \sum_{\nu_0=1}^{2m_{11}-1} b_{21\nu_0} \sum_{\rho=2}^{2m_{11}} B_{\rho 1}^{(r_1)} \ln \frac{|x_1-a_{11r_\rho}|}{|x_{10}-a_{11r_\rho}|} \\
&= \frac{a_{110}}{a_{220}} \Bigg\{ \frac{1}{2n_{11}+1}\Big[(x_2-a_{12u_1})^{2n_{11}+1}-(x_{j_2 0}-a_{12u_1})^{2n_{11}+1}\Big] \\
&+ \sum_{\alpha=1}^{2n_{11}-2} \frac{1}{2n_{11}-\alpha+1} b_{12\alpha}\Big[(x_2-a_{12u_1})^{2n_{11}-\alpha+1}-(x_{20}-a_{12u_1})^{2n_{11}-\alpha+1}\Big] \Bigg\},
\end{aligned}
\tag{3.110}
$$

where

$$
\begin{aligned}
b_{12\alpha} &= \sum_{l_{n_1},t_{n_1}=0}^{l_{n_1}+t_{n_1}=1} \cdots \sum_{l_2,t_2=0}^{l_2+t_2=1} \prod_{k=2}^{2n_{11}} (a_{12u_1}-a_{12u_k})^{l_k} \delta_{t_2+\cdots+t_{2n_{11}}}^{2n_{11}-\alpha}; \\
b_{120} &= 1. \\
b_{21\nu_0} &= \sum_{t_1,l_1=0}^{t_1+l_1=1} \cdots \sum_{t_{2m_{21}+1},l_{2m_{21}+1}=0}^{t_{2m_{21}+1}+l_{2m_{21}+1}=1} \prod_{k=1}^{2m_{21}+1} (a_{11r_1}-a_{21\nu_k})^{t_k} \delta_{\nu_0}^{\Sigma_{r=1}^{2m_{21}+1} l_r}; \\
&\lim_{\nu_{2m_{11}-1}\to 0} \frac{1}{\nu_{2m_{11}-1}}\Big[(x_1-a_{11r_{2m_{11}}})^{\nu_{2m_{11}-1}}-(x_{j_1 0}-a_{11r_{2m_{11}}})^{\nu_{2m_{11}-1}}\Big] = \ln \frac{|x_1-a_{11r_{2m_{11}}}|}{|x_{10}-a_{11r_{2m_{11}}}|};
\end{aligned}
\tag{3.111}
$$

$$
A_{\rho 1}^{(r_k)} = \frac{1}{\prod_{\sigma=k,\sigma\neq\rho}^{2m_{11}} (a_{11r_\rho}-a_{11r_\sigma})}, \text{ for } \rho \in \{k, k+1, \ldots, 2m_{11}\},
$$

$$B_{\rho 1}^{(r_k)} = \frac{(a_{11s_\rho} - a_{11s_k})^{\beta_{k-1}-k}}{\prod_{\sigma=k+1,\sigma\neq\rho}^{2m_{11}} (a_{11r_\rho} - a_{11r_\sigma})}, \text{ for } \rho \in \{k+1, k+2, \dots, 2m_{11}\};$$
$$k \in \{1, 2, \dots, 2m_{11}-1\}. \tag{3.112}$$

For all cases, the $(2n_{11}) \times (2m_{21}+1)$-simple equilibriums plus $(2m_{11}) \times (2n_{11})$-hyperbolic flows are based on the following double-parabola-saddle bifurcations for $2m_{21}+1 \geq 2m_{11}$:

- $([(2m_{11}),(2n_{11})]^{\text{th}}\text{DI}_{-}\text{:II}_{+},[2(m_{21}-m_{11})+1]^{\text{th}}{}_{\text{UP:UP}}\text{US})$-double up-parabola upper-saddle bifurcation for $(a_{110}>0, a_{220}>0)$,
- $([(2m_{11}),(2n_{11})]^{\text{th}}\text{II}_{-}\text{:DI}_{+},[2(m_{21}-m_{11})+1]^{\text{th}}{}_{\text{DP:DP}}\text{LS})$-double down-parabola lower-saddle bifurcation for $(a_{110}<0, a_{220}>0)$,
- $([(2m_{11}),(2n_{11})]^{\text{th}}\text{II}_{+}\text{:DI}_{-},[2(m_{21}-m_{11})+1]^{\text{th}}{}_{\text{DP:DP}}\text{US})$-double down-parabola upper-saddle bifurcation for $(a_{110}>0, a_{220}<0)$,
- $([(2m_{11}),(2n_{11})]^{\text{th}}\text{DI}_{-}\text{:II}_{+},[2(m_{21}-m_{11})+1]^{\text{th}}{}_{\text{UP:UP}}\text{US})$-double up-parabola lower-saddle bifurcation for $(a_{110}<0, a_{220}<0)$;

or based on the following inflection asymptote lower and upper-saddles bifurcations for $2m_{21}+1 < 2m_{11}$:

- $([(2m_{11}),(2n_{11})]^{\text{th}}\text{DI}_{-}\text{:II}_{+},[2(m_{11}-m_{21})-1]^{\text{th}}{}_{\text{DA:DA}}\text{US})$-inflection down-asymptote upper-saddle bifurcation for $(a_{110}>0, a_{220}>0)$,
- $([(2m_{11}),(2n_{11})]^{\text{th}}\text{II}_{-}\text{:DI}_{+},[2(m_{11}-m_{21})-1]^{\text{th}}{}_{\text{UA:UA}}\text{LS})$-inflection up-asymptote lower-saddle bifurcation for $(a_{110}<0, a_{220}>0)$,
- $([(2m_{11}),(2n_{11})]^{\text{th}}\text{II}_{+}\text{:DI}_{-},[2(m_{11}-m_{21})-1]^{\text{th}}{}_{\text{UA:UA}}\text{US})$-inflection up-asymptote upper-saddle bifurcation for $(a_{110}>0, a_{220}<0)$,
- $([(2m_{11}),(2n_{11})]^{\text{th}}\text{DI}_{+}\text{:II}_{-},[2(m_{11}-m_{21})-1]^{\text{th}}{}_{\text{DA:DA}}\text{LS})$-inflection down-asymptote lower-saddle bifurcation for $(a_{110}<0, a_{220}<0)$.

(A) Consider $2m_{11}=2m_{21}$ for illustrations, phase portraits for the 2-dimensional systems for the hybrid networks of hyperbolic and hyperbolic-secant flows with the simple equilibriums of the saddles and centers with $x_1^*=a_{111}, a_{211}, \cdots, a_{11(2m_{11})}, a_{21(2m_{21})}, a_{21(2m_{21}+1)}$ and $x_2^* = a_{121}, a_{122}, \cdots, a_{12(2n_{11})}$ are presented in Fig. 3.35a–d for $(a_{110}>0, a_{220}>0)$, $(a_{110}<0, a_{220}>0)$, $(a_{110}>0, a_{220}<0)$ and $(a_{110}<0, a_{220}<0)$.

(B) Consider $m_{11}=m_{21}$ for illustrations. Phase portraits for the 2-dimensional systems for the hybrid networks of hyperbolic and hyperbolic-secant flows with the simple equilibriums of saddles and centers with $x_1^* = a_{211}, a_{111}, \cdots, a_{11(2m_{11})}, a_{21(2m_{21}+1)}$ and $x_2^* = a_{121}, a_{122}, \cdots, a_{12(2n_{11})}$ are presented in Fig. 3.36a–d for $(a_{110}>0, a_{220}>0)$, $(a_{110}<0, a_{220}>0)$, $(a_{110}>0, a_{220}<0)$ and $(a_{110}<0, a_{220}<0)$.

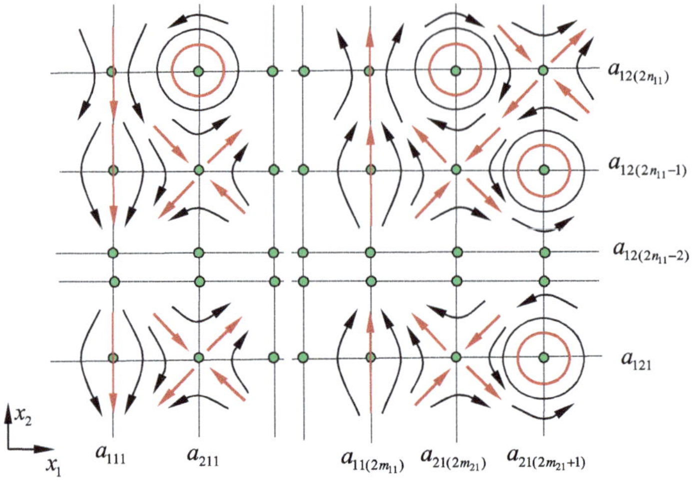

(a)

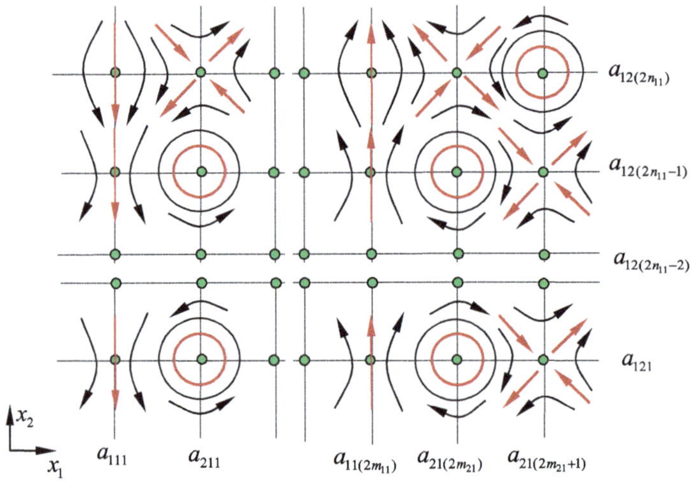

(b)

Fig. 3.35 Phase portraits for 2-dimensional systems with $x_1^* = a_{111}, a_{211}, \cdots, a_{11(2m_{11})}, a_{21(2m_{21})}, a_{21(2m_{21}+1)}$ and $x_2^* = a_{121}, \cdots, a_{12(2n_{11}-1)}, a_{12(2n_{11})}$. The networks of flows with saddles and centers: **a** ($a_{110} > 0, a_{220} > 0$), **b** ($a_{110} < 0, a_{220} > 0$), **c** ($a_{110} > 0, a_{220} < 0$), **d** ($a_{110} < 0, a_{220} < 0$)

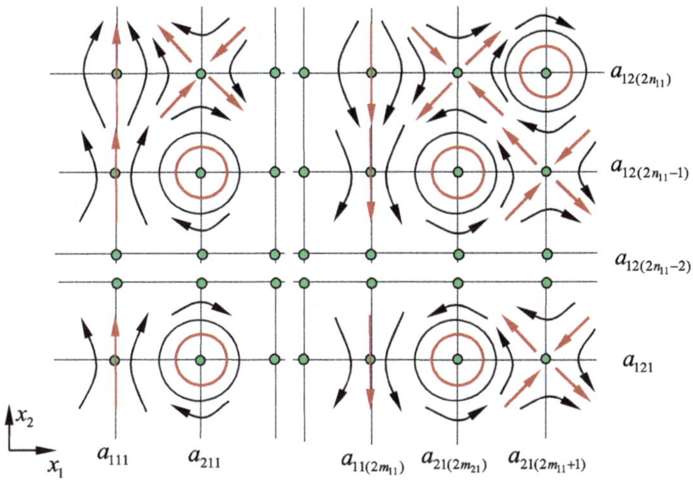

(c)

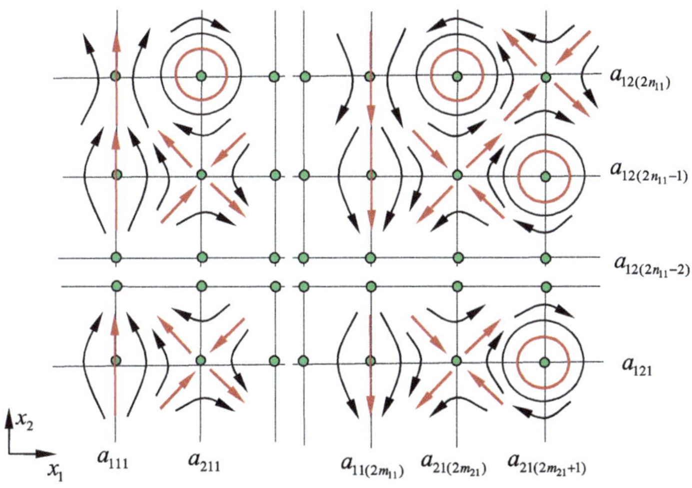

(d)

Fig. 3.35 (continued)

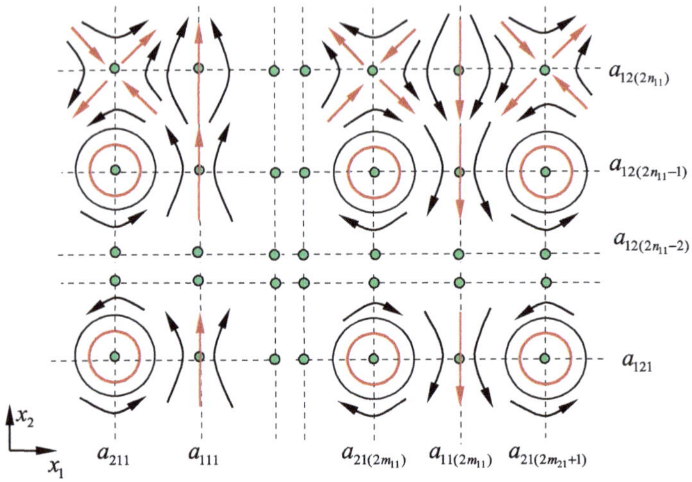

(a)

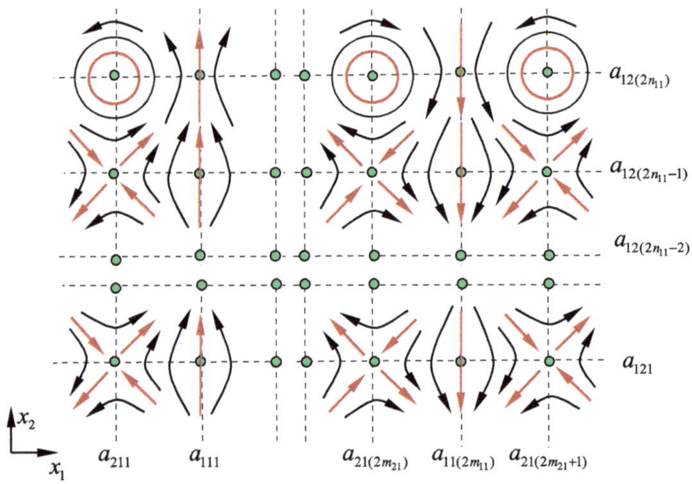

(b)

Fig. 3.36 Phase portraits for 2-dimensional systems with $x_1^*=a_{211}, a_{111}, \cdots, a_{11(2m_{21})}, a_{21(2m_{11}+1)}$ and $x_2^*=a_{121}, a_{122}, \cdots, a_{12(2n_{11})}$. The networks of flows with saddles and centers: **a** ($a_{110} > 0, a_{220} > 0$), **b** ($a_{110} < 0, a_{220} > 0$), **c** ($a_{110} > 0, a_{220} < 0$), **d** ($a_{110} < 0, a_{220} < 0$)

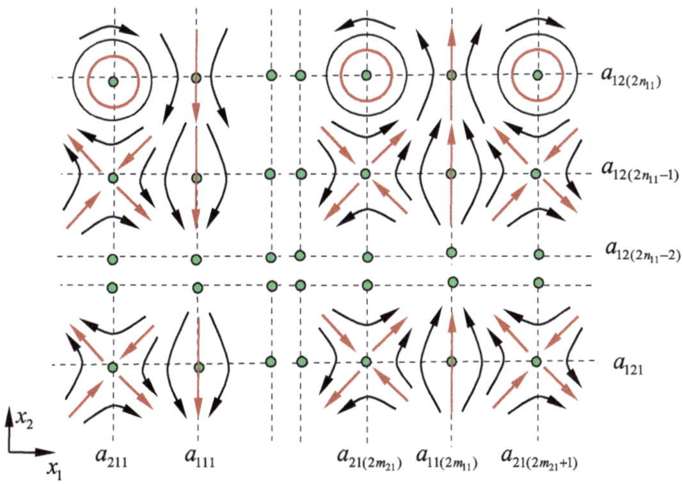

(c)

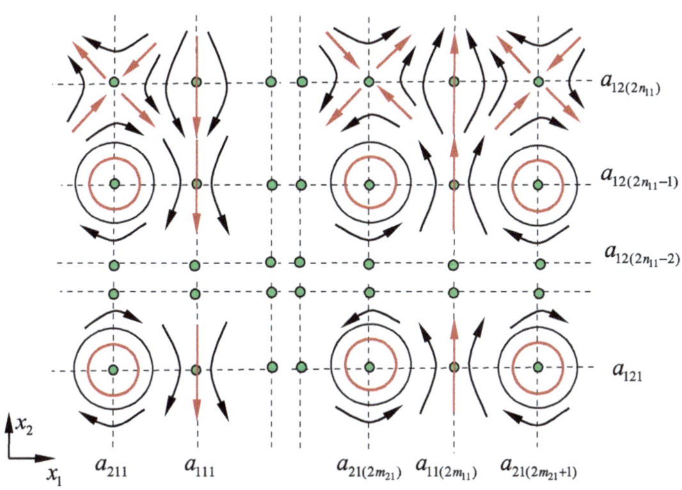

(d)

Fig. 3.36 (continued)

$$
\left\{\begin{array}{cccccc}
(a_{111},a_{12(2n_{11})}) & (a_{12(2n_{11})},a_{211}) & \cdots & (a_{11(2m_{11})},a_{12(2n_{11})}) & (a_{12(2n_{11})},a_{21(2m_{21})}) & (a_{12(2n_{11})},a_{21(2m_{21}+1)}) \\
(a_{111},a_{12(2n_{11}-1)}) & (a_{12(2n_{11}-1)},a_{211}) & \cdots & (a_{11(2m_{11})},a_{12(2n_{11}-1)}) & (a_{12(2n_{11}-1)},a_{21(2m_{21})}) & (a_{12(2n_{11}-1)},a_{21(2m_{21}+1)}) \\
\vdots & \vdots & \vdots & \vdots & \vdots & \vdots \\
(a_{111},a_{121}) & (a_{121},a_{211}) & \cdots & (a_{11(2m_{11})},a_{121}) & (a_{121},a_{21(2m_{21})}) & (a_{121},a_{21(2m_{21}+1)})
\end{array}\right\}
$$

$$
=\left\{\begin{array}{cccccc}
\underbrace{(\mathrm{DP{:}UP,nF})}_{\text{hyperbolic flow }(-)} & \underbrace{(\mathrm{DP}_{+},\mathrm{DP}_{-})}_{\text{CCW center}} & \cdots & \underbrace{(\mathrm{DP{:}UP,pF})}_{\text{hyperbolic flow }(+)} & \underbrace{(\mathrm{DP}_{-},\mathrm{DP}_{+})}_{\text{CW center}} & \underbrace{(\mathrm{UP}_{+},\mathrm{UP}_{+})}_{\text{positive saddle}} \\
\underbrace{(\mathrm{UP{:}DP,nF})}_{\text{hyperbolic-secant flow }(-)} & \underbrace{(\mathrm{UP}_{+},\mathrm{UP}_{+})}_{\text{positive saddle}} & \cdots & \underbrace{(\mathrm{UP{:}DP,pF})}_{\text{hyperbolic-secant flow }(+)} & \underbrace{(\mathrm{UP}_{-},\mathrm{UP}_{-})}_{\text{negative saddle}} & \underbrace{(\mathrm{DP}_{+},\mathrm{DP}_{-})}_{\text{CCW center}} \\
\vdots & \vdots & \vdots & \vdots & \vdots & \vdots \\
\underbrace{(\mathrm{UP{:}DP,nF})}_{\text{hyperbolic-secant flow }(-)} & \underbrace{(\mathrm{UP}_{+},\mathrm{UP}_{+})}_{\text{positive saddle}} & \cdots & \underbrace{(\mathrm{UP{:}DP,pF})}_{\text{hyperbolic-secant flow }(+)} & \underbrace{(\mathrm{UP}_{-},\mathrm{UP}_{-})}_{\text{negative saddle}} & \underbrace{(\mathrm{DP}_{+},\mathrm{DP}_{-})}_{\text{CCW center}}
\end{array}\right\}_{(2n_{11})\times(2m_{11}+2m_{21}+1)}
$$

for $a_{110} > 0, a_{220} > 0;$ (3.113)

$$
\begin{Bmatrix}
(a_{111},a_{12(2n_{11})}) & (a_{12(2n_{11})},a_{211}) & \cdots & (a_{11(2m_{11})},a_{12(2n_{11})}) & (a_{12(2n_{11})},a_{21(2m_{21})}) & (a_{12(2n_{11})},a_{21(2m_{21}+1)}) \\
(a_{111},a_{12(2n_{11}-1)}) & (a_{12(2n_{11}-1)},a_{211}) & \cdots & (a_{11(2m_{11})},a_{12(2n_{11}-1)}) & (a_{12(2n_{11}-1)},a_{21(2m_{21})}) & (a_{12(2n_{11}-1)},a_{21(2m_{21}+1)}) \\
\vdots & \vdots & \vdots & \vdots & \vdots & \vdots \\
(a_{111},a_{121}) & (a_{121},a_{211}) & \cdots & (a_{11(2m_{11})},a_{121}) & (a_{121},a_{21(2m_{21})}) & (a_{121},a_{21(2m_{21}+1)})
\end{Bmatrix}
$$

$$
= \begin{Bmatrix}
\underbrace{(\text{UP:DP,nF})}_{\text{hyperbolic-secant flow }(-)} & \underbrace{(\text{UP}_{+},\text{UP}_{+})}_{\text{positive saddle}} & \cdots & \underbrace{(\text{UP:DP,pF})}_{\text{hyperbolic-secant flow }(+)} & \underbrace{(\text{UP}_{-},\text{UP}_{-})}_{\text{negative saddle}} & \underbrace{(\text{DP}_{+},\text{DP}_{-})}_{\text{CCW center}} \\
\underbrace{(\text{DP:UP,nF})}_{\text{hyperbolic flow }(-)} & \underbrace{(\text{DP}_{+},\text{DP}_{-})}_{\text{CCW center}} & \cdots & \underbrace{(\text{DP:UP,pF})}_{\text{hyperbolic flow }(+)} & \underbrace{(\text{DP}_{-},\text{DP}_{+})}_{\text{CW center}} & \underbrace{(\text{UP}_{+},\text{UP}_{+})}_{\text{positive saddle}} \\
\vdots & \vdots & \vdots & \vdots & \vdots & \vdots \\
\underbrace{(\text{DP:UP,nF})}_{\text{hyperbolic flow }(-)} & \underbrace{(\text{DP}_{+},\text{DP}_{-})}_{\text{CCW center}} & \cdots & \underbrace{(\text{DP:UP,pF})}_{\text{hyperbolic flow }(+)} & \underbrace{(\text{DP}_{-},\text{DP}_{+})}_{\text{CW center}} & \underbrace{(\text{UP}_{+},\text{UP}_{+})}_{\text{positive saddle}}
\end{Bmatrix}_{(2n_{11})\times(2m_{11}+2m_{21}+1)}
$$

for $a_{110} < 0, a_{220} > 0;$ (3.114)

$$
\left\{\begin{array}{cccccc}
(a_{111},a_{12(2n_{11})}) & (a_{12(2n_{11})},a_{211}) & \cdots & (a_{11(2m_{11})},a_{12(2n_{11})}) & (a_{12(2n_{11})},a_{21(2m_{21})}) & (a_{12(2n_{11})},a_{21(2m_{21}+1)}) \\
(a_{111},a_{12(2n_{11}-1)}) & (a_{12(2n_{11}-1)},a_{211}) & \cdots & (a_{11(2m_{11})},a_{12(2n_{11}-1)}) & (a_{12(2n_{11}-1)},a_{21(2m_{21})}) & (a_{12(2n_{11}-1)},a_{21(2m_{21}+1)}) \\
\vdots & \vdots & \vdots & \vdots & \vdots & \vdots \\
(a_{111},a_{121}) & (a_{121},a_{211}) & \cdots & (a_{11(2m_{11})},a_{121}) & (a_{121},a_{21(2m_{21})}) & (a_{121},a_{21(2m_{21}+1)})
\end{array}\right\}
$$

$$
=\left\{\begin{array}{cccccc}
\underbrace{(\text{UP:DP,pF})}_{\text{hyperbolic-secant flow }(+)} & \underbrace{(\text{UP}_-,\text{UP}_-)}_{\text{negative saddle}} & \cdots & \underbrace{(\text{UP:DP,nF})}_{\text{hyperbolic-secant flow }(-)} & \underbrace{(\text{UP}_+,\text{UP}_+)}_{\text{negative saddle}} & \underbrace{(\text{DP}_-,\text{DP}_+)}_{\text{CW center}} \\
\underbrace{(\text{DP:UP,pF})}_{\text{hyperbolic flow }(+)} & \underbrace{(\text{DP}_-,\text{DP}_+)}_{\text{CW center}} & \cdots & \underbrace{(\text{DP:UP,nF})}_{\text{hyperbolic flow }(-)} & \underbrace{(\text{DP}_+,\text{DP}_-)}_{\text{CCW center}} & \underbrace{(\text{UP}_-,\text{UP}_-)}_{\text{negative saddle}} \\
\vdots & \vdots & \vdots & \vdots & \vdots & \vdots \\
\underbrace{(\text{DP:UP,pF})}_{\text{hyperbolic flow }(+)} & \underbrace{(\text{DP}_-,\text{DP}_+)}_{\text{CW center}} & \cdots & \underbrace{(\text{DP:UP,nF})}_{\text{hyperbolic flow }(-)} & \underbrace{(\text{DP}_+,\text{DP}_-)}_{\text{CCW center}} & \underbrace{(\text{UP}_-,\text{UP}_-)}_{\text{negative saddle}}
\end{array}\right\}_{(2n_{11})\times(2m_{11}+2m_{21}+1)}
$$

for $a_{110} > 0, a_{220} < 0$; (3.115)

$$
\left\{\begin{array}{cccccc}
(a_{111},a_{12(2n_{11})}) & (a_{12(2n_{11})},a_{211}) & \cdots & (a_{11(2m_{11})},a_{12(2n_{11})}) & (a_{12(2n_{11})},a_{21(2m_{21})}) & (a_{12(2n_{11})},a_{21(2m_{21}+1)}) \\
(a_{111},a_{12(2n_{11}-1)}) & (a_{12(2n_{11}-1)},a_{211}) & \cdots & (a_{11(2m_{11})},a_{12(2n_{11}-1)}) & (a_{12(2n_{11}-1)},a_{21(2m_{21})}) & (a_{12(2n_{11}-1)},a_{21(2m_{21}+1)}) \\
\vdots & \vdots & \vdots & \vdots & \vdots & \vdots \\
(a_{111},a_{121}) & (a_{121},a_{211}) & \cdots & (a_{11(2m_{11})},a_{121}) & (a_{121},a_{21(2m_{21})}) & (a_{121},a_{21(2m_{21}+1)})
\end{array}\right\}
$$

$$
= \left\{\begin{array}{cccccc}
\underbrace{(\mathrm{DP{:}UP,pF})}_{\text{hyperbolic flow }(+)} & \underbrace{(\mathrm{DP_-,DP_+})}_{\text{CW center}} & \cdots & \underbrace{(\mathrm{DP{:}UP,nF})}_{\text{hyperbolic flow }(-)} & \underbrace{(\mathrm{DP_+,DP_-})}_{\text{CCW center}} & \underbrace{(\mathrm{UP_-,UP_-})}_{\text{negative saddle}} \\
\underbrace{(\mathrm{UP{:}DP,pF})}_{\text{hyperbolic-secant flow }(+)} & \underbrace{(\mathrm{UP_-,UP_-})}_{\text{negative saddle}} & \cdots & \underbrace{(\mathrm{UP{:}DP,nF})}_{\text{hyperbolic-secant flow }(-)} & \underbrace{(\mathrm{UP_+,UP_+})}_{\text{positive saddle}} & \underbrace{(\mathrm{DP_-,DP_+})}_{\text{CW center}} \\
\vdots & \vdots & \vdots & \vdots & \vdots & \vdots \\
\underbrace{(\mathrm{UP{:}DP,pF})}_{\text{hyperbolic-secant flow }(+)} & \underbrace{(\mathrm{UP_-,UP_-})}_{\text{negative saddle}} & \cdots & \underbrace{(\mathrm{UP{:}DP,nF})}_{\text{hyperbolic-secant flow }(-)} & \underbrace{(\mathrm{UP_+,UP_+})}_{\text{positive saddle}} & \underbrace{(\mathrm{DP_-,DP_+})}_{\text{CW center}}
\end{array}\right\}_{(2n_{11})\times(2m_{11}+2m_{21}+1)}
$$

for $a_{110} < 0, a_{220} < 0$. (3.116)

$$
\left\{\begin{array}{cccccc}
(a_{12(2n_{11})},a_{211}) & (a_{111},a_{12(2n_{11})}) & \cdots & (a_{12(2n_{11})},a_{21(2m_{21})}) & (a_{11(2m_{11})},a_{12(2n_{11})}) & (a_{12(2n_{11})},a_{21(2m_{21}+1)}) \\
(a_{12(2n_{11}-1)},a_{211}) & (a_{111},a_{12(2n_{11}-1)}) & \cdots & (a_{12(2n_{11}-1)},a_{21(2m_{21})},) & (a_{11(2m_{11})},a_{12(2n_{11}-1)}) & (a_{12(2n_{11}-1)},a_{21(2m_{21}+1)}) \\
\vdots & \vdots & \vdots & \vdots & \vdots & \vdots \\
(a_{121},a_{211}) & (a_{111},a_{121}) & \cdots & (a_{121},a_{21(2m_{21})},) & (a_{11(2m_{11})},a_{121}) & (a_{121},a_{21(2m_{21}+1)})
\end{array}\right\}
$$
$$
= \left\{\begin{array}{cccccc}
\underbrace{(\mathrm{UP}_{+},\mathrm{UP}_{+})}_{\text{positive saddle}} & \underbrace{(\mathrm{UP{:}DP,pF})}_{\text{hyperbolic-secant flow }(+)} & \cdots & \underbrace{(\mathrm{UP}_{-},\mathrm{UP}_{-})}_{\text{negative saddle}} & \underbrace{(\mathrm{UP{:}DP,nF})}_{\text{hyperbolic-secant flow }(-)} & \underbrace{(\mathrm{UP}_{+},\mathrm{UP}_{+})}_{\text{positive saddle}} \\
\underbrace{(\mathrm{DP}_{+},\mathrm{DP}_{-})}_{\text{CCW center}} & \underbrace{(\mathrm{DP{:}UP,pF})}_{\text{hyperbolic flow }(+)} & \cdots & \underbrace{(\mathrm{DP}_{-},\mathrm{DP}_{+})}_{\text{CW center}} & \underbrace{(\mathrm{DP{:}UP,nF})}_{\text{hyperbolic flow }(-)} & \underbrace{(\mathrm{DP}_{+},\mathrm{DP}_{-})}_{\text{CCW center}} \\
\vdots & \vdots & \vdots & \vdots & \vdots & \vdots \\
\underbrace{(\mathrm{DP}_{+},\mathrm{DP}_{-})}_{\text{CCW center}} & \underbrace{(\mathrm{DP{:}UP,pF})}_{\text{hyperbolic flow }(+)} & \cdots & \underbrace{(\mathrm{DP}_{-},\mathrm{DP}_{+})}_{\text{CW center}} & \underbrace{(\mathrm{DP{:}UP,nF})}_{\text{hyperbolic flow }(-)} & \underbrace{(\mathrm{DP}_{+},\mathrm{DP}_{-})}_{\text{CCW center}}
\end{array}\right\}_{(2n_{21})\times(2m_{11}+2m_{21}+1)}
$$

for $a_{110} > 0, a_{220} > 0;$ (3.117)

$$
\begin{Bmatrix}
(a_{12(2n_{11})},a_{211}) & (a_{111},a_{12(2n_{11})}) & \cdots & (a_{12(2n_{11})},a_{21(2m_{21})}) & (a_{11(2m_{11})},a_{12(2n_{11})}) & (a_{12(2n_{11})},a_{21(2m_{21}+1)}) \\
(a_{12(2n_{11}-1)},a_{211}) & (a_{111},a_{12(2n_{11}-1)}) & \cdots & (a_{12(2n_{11}-1)},a_{21(2m_{21})},) & (a_{11(2m_{11})},a_{12(2n_{11}-1)}) & (a_{12(2n_{11}-1)},a_{21(2m_{21}+1)}) \\
\vdots & \vdots & \vdots & \vdots & \vdots & \vdots \\
(a_{121},a_{211}) & (a_{111},a_{121}) & \cdots & (a_{121},a_{21(2m_{21})},) & (a_{11(2m_{11})},a_{121}) & (a_{121},a_{21(2m_{21}+1)})
\end{Bmatrix}
$$

$$
= \begin{Bmatrix}
\underbrace{(\text{DP}_+,\text{DP}_-)}_{\text{CCW center}} & \underbrace{(\text{DP:UP,pF})}_{\text{hyperbolic flow }(+)} & \cdots & \underbrace{(\text{DP}_-,\text{DP}_+)}_{\text{CW center}} & \underbrace{(\text{DP:UP,nF})}_{\text{hyperbolic flow }(-)} & \underbrace{(\text{DP}_+,\text{DP}_-)}_{\text{CCW center}} \\
\underbrace{(\text{UP}_+,\text{UP}_+)}_{\text{positive saddle}} & \underbrace{(\text{UP:DP,pF})}_{\text{hyperbolic-secant flow }(+)} & \cdots & \underbrace{(\text{UP}_-,\text{UP}_-)}_{\text{negative saddle}} & \underbrace{(\text{UP:DP,nF})}_{\text{hyperbolic-secant flow }(-)} & \underbrace{(\text{UP}_+,\text{UP}_+)}_{\text{positive saddle}} \\
\vdots & \vdots & \vdots & \vdots & \vdots & \vdots \\
\underbrace{(\text{UP}_+,\text{UP}_+)}_{\text{positive saddle}} & \underbrace{(\text{UP:DP,pF})}_{\text{hyperbolic-secant flow }(+)} & \cdots & \underbrace{(\text{UP}_-,\text{UP}_-)}_{\text{negative saddle}} & \underbrace{(\text{UP:DP,nF})}_{\text{hyperbolic-secant flow }(-)} & \underbrace{(\text{UP}_+,\text{UP}_+)}_{\text{positive saddle}}
\end{Bmatrix}_{(2n_{11})\times(2m_{11}+2m_{21}+1)}
$$

for $a_{110} < 0, a_{220} > 0$; (3.118)

$$\left\{\begin{array}{cccccc}
(a_{12(2n_{11})},a_{211}) & (a_{111},a_{12(2n_{11})}) & \cdots & (a_{12(2n_{11})},a_{21(2m_{21})}) & (a_{11(2m_{11})},a_{12(2n_{11})}) & (a_{12(2n_{11})},a_{21(2m_{21}+1)})\\
(a_{12(2n_{11}-1)},a_{211}) & (a_{111},a_{12(2n_{11}-1)}) & \cdots & (a_{12(2n_{11}-1)},a_{21(2m_{21})},) & (a_{11(2m_{11})},a_{12(2n_{11}-1)}) & (a_{12(2n_{11}-1)},a_{21(2m_{21}+1)})\\
\vdots & \vdots & \vdots & \vdots & \vdots & \vdots\\
(a_{121},a_{211}) & (a_{111},a_{121}) & \cdots & (a_{121},a_{21(2m_{21})},) & (a_{11(2m_{11})},a_{121}) & (a_{121},a_{21(2m_{21}+1)})
\end{array}\right\}$$

$$=\left\{\begin{array}{cccccc}
\underbrace{(\mathrm{DP}_-,\mathrm{DP}_+)}_{\text{CW center}} & \underbrace{(\mathrm{DP{:}UP,nF})}_{\text{hyperbolic flow }(-)} & \cdots & \underbrace{(\mathrm{DP}_+,\mathrm{DP}_-)}_{\text{CCW center}} & \underbrace{(\mathrm{DP{:}UP,pF})}_{\text{hyperbolic flow }(+)} & \underbrace{(\mathrm{DP}_-,\mathrm{DP}_+)}_{\text{CW center}}\\
\underbrace{(\mathrm{UP}_-,\mathrm{UP}_-)}_{\text{negative saddle}} & \underbrace{(\mathrm{UP{:}DP,nF})}_{\text{hyperbolic-secant flow }(-)} & \cdots & \underbrace{(\mathrm{UP}_+,\mathrm{UP}_+)}_{\text{positive saddle}} & \underbrace{(\mathrm{UP{:}DP,pF})}_{\text{hyperbolic-secant flow }(+)} & \underbrace{(\mathrm{UP}_-,\mathrm{UP}_-)}_{\text{negative saddle}}\\
\vdots & \vdots & \vdots & \vdots & \vdots & \vdots\\
\underbrace{(\mathrm{UP}_-,\mathrm{UP}_-)}_{\text{negative saddle}} & \underbrace{(\mathrm{UP{:}DP,nF})}_{\text{hyperbolic-secant flow }(-)} & \cdots & \underbrace{(\mathrm{UP}_+,\mathrm{UP}_+)}_{\text{positive saddle}} & \underbrace{(\mathrm{UP{:}DP,pF})}_{\text{hyperbolic-secant flow }(+)} & \underbrace{(\mathrm{UP}_-,\mathrm{UP}_-)}_{\text{negative saddle}}
\end{array}\right\}_{(2n_{11}+1)\times(2m_{11}+2m_{21}+1)}$$

for $a_{110} > 0$, $a_{220} < 0$; (3.119)

$$\left\{\begin{array}{cccccc}(a_{12(2n_{11})},a_{211}) & (a_{111},a_{12(2n_{11})}) & \cdots & (a_{12(2n_{11})},a_{21(2m_{21})}) & (a_{11(2m_{11})},a_{12(2n_{11})}) & (a_{12(2n_{11})},a_{21(2m_{21}+1)})\\ (a_{12(2n_{11}-1)},a_{211}) & (a_{111},a_{12(2n_{11}-1)}) & \cdots & (a_{12(2n_{11}-1)},a_{21(2m_{21})},) & (a_{11(2m_{11})},a_{12(2n_{11}-1)}) & (a_{12(2n_{11}-1)},a_{21(2m_{21}+1)})\\ \vdots & \vdots & \vdots & \vdots & \vdots & \vdots\\ (a_{121},a_{211}) & (a_{111},a_{121}) & \cdots & (a_{121},a_{21(2m_{21})},) & (a_{11(2m_{11})},a_{121}) & (a_{121},a_{21(2m_{21}+1)})\end{array}\right\}$$
$$=\left\{\begin{array}{cccccc}\underbrace{(\mathrm{UP}_-,\mathrm{UP}_-)}_{\text{negative saddle}} & \underbrace{(\mathrm{UP{:}DP,nF})}_{\text{hyperbolic-secant flow }(-)} & \cdots & \underbrace{(\mathrm{UP}_+,\mathrm{UP}_+)}_{\text{positive saddle}} & \underbrace{(\mathrm{UP{:}DP,pF})}_{\text{hyperbolic-secant flow }(+)} & \underbrace{(\mathrm{UP}_-,\mathrm{UP}_-)}_{\text{negative saddle}}\\ \underbrace{(\mathrm{DP}_-,\mathrm{DP}_+)}_{\text{CCW center}} & \underbrace{(\mathrm{DP{:}UP,nF})}_{\text{hyperbolic flow }(-)} & \cdots & \underbrace{(\mathrm{DP}_+,\mathrm{DP}_-)}_{\text{CW center}} & \underbrace{(\mathrm{DP{:}UP,pF})}_{\text{hyperbolic flow }(+)} & \underbrace{(\mathrm{DP}_-,\mathrm{DP}_+)}_{\text{CCW center}}\\ \vdots & \vdots & \vdots & \vdots & \vdots & \vdots\\ \underbrace{(\mathrm{DP}_-,\mathrm{DP}_+)}_{\text{CCW center}} & \underbrace{(\mathrm{DP{:}UP,nF})}_{\text{hyperbolic flow }(-)} & \cdots & \underbrace{(\mathrm{DP}_+,\mathrm{DP}_-)}_{\text{CW center}} & \underbrace{(\mathrm{DP{:}UP,pF})}_{\text{hyperbolic flow }(+)} & \underbrace{(\mathrm{DP}_-,\mathrm{DP}_+)}_{\text{CCW center}}\end{array}\right\}_{(2n_{11})\times(2m_{11}+2m_{21}+1)}$$

for $a_{110} < 0, a_{220} < 0$; (3.120)

3.7 $[(2m_{11}+1),(2n_{11})]$ and $((2n_{11}),(2m_{21}))$-Singular Systems

In this section, there are four global bifurcations of the 1-dimensional flows through singular inflection-source and sink flows, and four global bifurcations of equilibriums, through singular double inflection-saddles bifurcations. The corresponding global bifurcations generate networks of hyperbolic and hyperbolic-secant flows with saddles and centers. The infinite-equilibriums are the appearing and switching bifurcations for the singular flows and singular equilibriums and for hybrid networks of hyperbolic flows and saddles and centers.

3.7.1 $[(2m_{11}+1),(2n_{11})]$-Flows and $((2n_{11}),(2m_{21}))$-Equilibriums

Consider a polynomial dynamical system for $a_{111}\neq a_{211}$ as

$$\begin{aligned}\dot{x}_2 &= a_{220}(x_1-a_{211})^{2m_{21}},\\ \dot{x}_1 &= a_{110}(x_1-a_{111})^{2m_{11}+1}(x_2-a_{121})^{2n_{11}}.\end{aligned}\tag{3.121}$$

The first integral manifold is

$$\begin{aligned}&\sum_{\alpha,\beta=0}^{\alpha+\beta=2m_{21}}\frac{(2m_{21})!}{\alpha!\beta!}\frac{(a_{111}-a_{211})^{\alpha}}{\beta-2m_{11}}\left[(x_1-a_{111})^{\beta-2m_{11}}-(x_{10}-a_{111})^{\beta-2m_{11}}\right]\\ &=\frac{a_{j_1j_10}}{a_{j_2j_20}}\frac{1}{2n_{11}+1}\left[(x_2-a_{121})^{2n_{11}+1}-(x_{20}-a_{121})^{2n_{11}+1}\right],\\ &\text{for } \beta-2m_{11}\neq 0;\\ &\frac{1}{\beta-2m_{11}}\left[(x_1-a_{111})^{\beta-2m_{11}}-(x_{10}-a_{111})^{\beta-2m_{11}}\right]=\ln\frac{|x_1-a_{111}|}{|x_{10}-a_{111}|},\\ &\text{for } \beta-2m_{11}=0.\end{aligned}\tag{3.122}$$

The singular hyperbolic and hyperbolic-secant flows with singular parabola upper and lower-saddle are illustrated in Figs. 3.37 and 3.38. Through such singular flows, the corresponding global bifurcations generate from a hyperbolic (hyperbolic-secant) flow to a network of singular hyperbolic and hyperbolic-secant flows. Through such singular equilibriums, the corresponding global bifurcations generate from one of 1-dimensional flows to a network of singular saddles and centers equilibriums.

For $(x_1^*,x_2^*)=(a_{111},a_{121})$ with $\dot{x}_1=0$, the $[(2m_{11}+1),(2n_{11})]$-singular flows exist with $\dot{x}_2=a_{220}(a_{111}-a_{211})^{2m_{21}}$. For $(x_1^*,x_2^*)=(a_{111},a_{121})$, the 1-dimensional flow direction is determined by $\dot{x}_2\neq 0$. For $(x_2^*,x_1^*)=(a_{121},a_{211})$ with $\dot{x}_1=\dot{x}_2=0$, the $((2n_{11}),(2m_{21}))$-singular equilibriums exist.

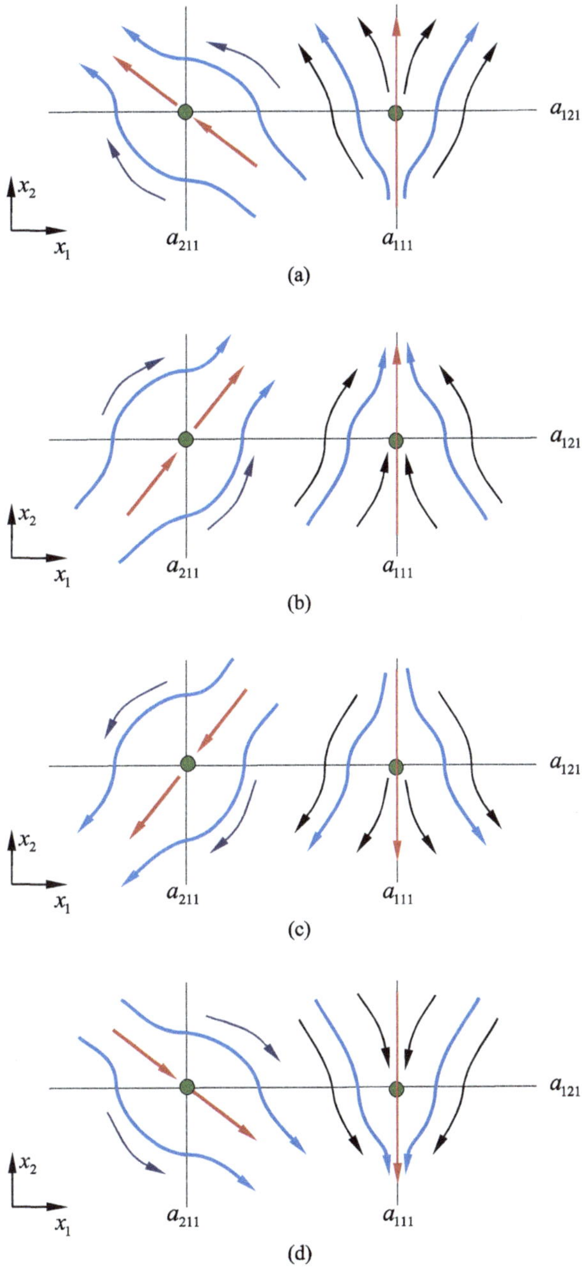

Fig. 3.37 Phase portraits $(a_{111} > a_{211})$ of the $([(2m_{11}+1),(2n_{11})], \text{xF})$-singular inflection-source and sink flows at $(x_1^*, x_2^*) = (a_{111}, a_{121})$ and $((2n_{11}),(2m_{21}))$-parabola-saddle at $(x_2^*, x_1^*) = (a_{121}, a_{211})$ for 2-dimensional polynomial systems: **a** $(a_{110} > 0, a_{220} > 0)$, **b** $(a_{110} < 0, a_{220} > 0)$, **c** $(a_{110} > 0, a_{220} < 0)$, **d** $(a_{110} < 0, a_{220} < 0)$

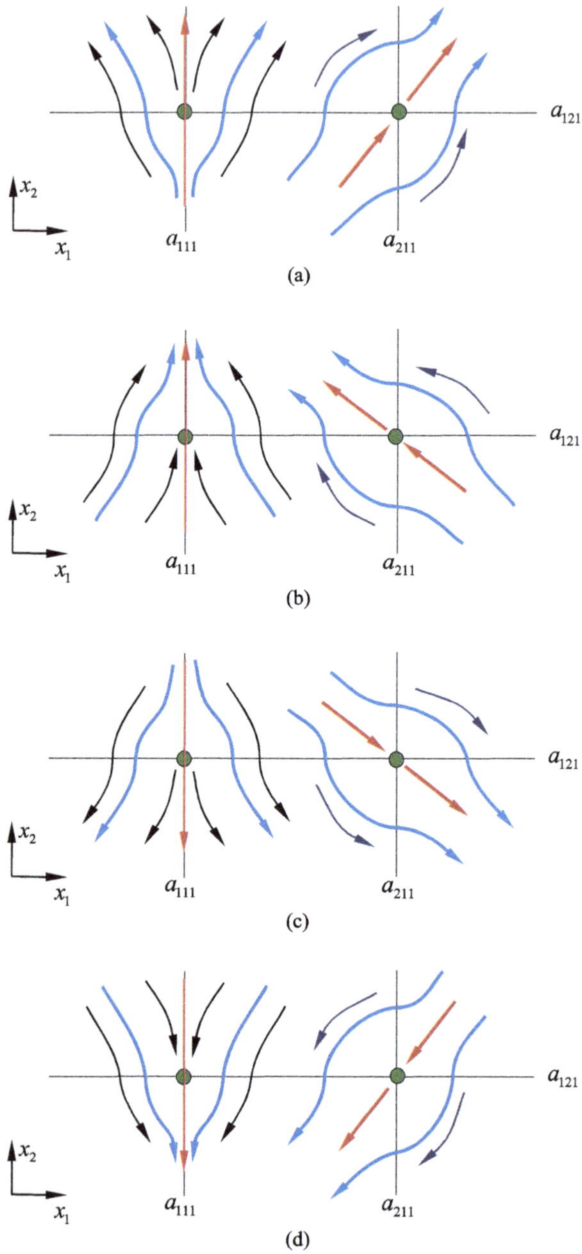

Fig. 3.38 Phase portraits ($a_{111} > a_{211}$) of the $([(2m_{11}+1),(2n_{11})], \text{xF})$-singular inflection-source and sink flows at $(x_1^*, x_2^*) = (a_{111}, a_{121})$ and $((2n_{11}), (2m_{21}))$-parabola-saddle at $(x_2^*, x_1^*) = (a_{121}, a_{211})$ for 2-dimensional polynomial systems: **a** ($a_{110} > 0, a_{220} > 0$), **b** ($a_{110} < 0, a_{220} > 0$), **c** ($a_{110} > 0, a_{220} < 0$), **d** ($a_{110} < 0, a_{220} < 0$)

For $a_{111} > a_{211}$,

$$\begin{aligned}
&\left\{\begin{matrix}(a_{111},a_{121})\\(a_{121},a_{211})\end{matrix}\right\}=\left\{\begin{matrix}\underbrace{([(2m_{11}+1),(2n_{11})]_{\mathrm{DI:II}}\mathrm{SO,pF})}_{[(2m_{11}+1),(2n_{11})]\text{-inflection-source flow }(+)}\\\underbrace{((2n_{11})\mathrm{DI}_{+},(2m_{21})\mathrm{DI}_{-})}_{[(2n_{11}),(2m_{21})]\text{-double inflection-saddle}}\end{matrix}\right\},\ \text{for } a_{110}>0, a_{220}>0;\\
&\left\{\begin{matrix}(a_{111},a_{121})\\(a_{121},a_{211})\end{matrix}\right\}=\left\{\begin{matrix}\underbrace{([(2m_{11}+1),(2n_{11})]_{\mathrm{II:DI}}\mathrm{SI,pF})}_{[(2m_{11}+1),(2n_{11})]\text{-inflection-sink flow }(+)}\\\underbrace{((2n_{11})\mathrm{II}_{+},(2m_{21})\mathrm{II}_{+})}_{[(2n_{11}),(2m_{21})]\text{-double inflection-saddle}}\end{matrix}\right\},\ \text{for } a_{110}<0, a_{220}>0;\\
&\left\{\begin{matrix}(a_{111},a_{121})\\(a_{121},a_{211})\end{matrix}\right\}=\left\{\begin{matrix}\underbrace{([(2m_{11}+1),(2n_{11})]_{\mathrm{II:DI}}\mathrm{SO,nF})}_{[(2m_{11}+1),(2n_{11})]\text{-inflection-source flow }(-)}\\\underbrace{((2n_{11})\mathrm{II}_{-},(2m_{21})\mathrm{II}_{-})}_{[(2n_{11}),(2m_{21})]\text{-double infection-saddle}}\end{matrix}\right\},\ \text{for } a_{110}>0, a_{220}<0;\\
&\left\{\begin{matrix}(a_{111},a_{121})\\(a_{121},a_{211})\end{matrix}\right\}=\left\{\begin{matrix}\underbrace{([(2m_{11}+1),(2n_{11})]_{\mathrm{DI:II}}\mathrm{SI,nF})}_{[(2m_{11}+1),(2n_{11})]\text{-inflection-sink flow }(-)}\\\underbrace{((2n_{11})\mathrm{DI}_{-},(2m_{21})\mathrm{DI}_{+})}_{[(2n_{11}),(2m_{21})]\text{-double inflection-saddle}}\end{matrix}\right\},\ \text{for } a_{110}<0, a_{220}<0.
\end{aligned}\tag{3.123}$$

For $a_{111} < a_{211}$,

$$\begin{aligned}
&\left\{\begin{matrix}(a_{121},a_{211})\\(a_{111},a_{121})\end{matrix}\right\}=\left\{\begin{matrix}\underbrace{((2n_{11})\mathrm{II}_{+},(2m_{21})\mathrm{II}_{+})}_{[(2n_{11}),(2m_{21})]\text{-double inflection-saddle}}\\\underbrace{([(2m_{11}+1),(2n_{11})]_{\mathrm{DI:II}}\mathrm{SO,pF})}_{[(2m_{11}+1),(2n_{11})]\text{-inflection-source flow }(+)}\end{matrix}\right\},\ \text{for } a_{110}>0, a_{220}>0;\\
&\left\{\begin{matrix}(a_{121},a_{211})\\(a_{111},a_{121})\end{matrix}\right\}=\left\{\begin{matrix}\underbrace{((2n_{11})\mathrm{DI}_{+},(2m_{21})\mathrm{DI}_{-})}_{[(2n_{11}),(2m_{21})]\text{-double inflection-saddle}}\\\underbrace{([(2m_{11}+1),(2n_{11})]_{\mathrm{II:DI}}\mathrm{SI,pF})}_{[(2m_{11}+1),(2n_{11})]\text{-inflection-sink flow }(+)}\end{matrix}\right\},\ \text{for } a_{110}<0, a_{220}>0;\\
&\left\{\begin{matrix}(a_{121},a_{211})\\(a_{111},a_{121})\end{matrix}\right\}=\left\{\begin{matrix}\underbrace{((2n_{11})\mathrm{DI}_{-},(2m_{21})\mathrm{DI}_{+})}_{[(2n_{11}),(2m_{21})]\text{-double inflection-saddle}}\\\underbrace{([(2m_{11}+1),(2n_{11})]_{\mathrm{II:DI}}\mathrm{SO,nF})}_{[(2m_{11}+1),(2n_{11})]\text{-inflection-source flow }(-)}\end{matrix}\right\},\ \text{for } a_{110}>0, a_{220}<0;
\end{aligned}$$

$$\left\{\begin{matrix}(a_{121}, a_{211})\\(a_{111}, a_{121})\end{matrix}\right\} = \left\{\begin{matrix}\underbrace{((2n_{11})\text{II}_{-}, (2m_{21})\text{II}_{-})}_{[(2n_{11}),(2m_{21})]\text{-double inflection-saddle}}\\\underbrace{([(2m_{11}+1), (2n_{11})]_{\text{DI:II}}\text{SI,nF})}_{[(2m_{11}+1),(2n_{11})]\text{-infelction-source flow }(-)}\end{matrix}\right\}, \text{ for } a_{110} < 0, a_{220} < 0. \tag{3.124}$$

- The $((2n_{11})\text{II}_{+}, (2m_{21})\text{II}_{+})$ is for a $((2n_{11}), (2m_{21}))$-singular double increasing-inflection-saddle.
- The $((2n_{11})\text{DI}_{+}, (2m_{21})\text{DI}_{-})$ is for a $((2n_{11}), (2m_{21}))$-singular double decreasing-inflection-saddle.
- The $((2n_{11})\text{DI}_{-}, (2m_{21})\text{DI}_{+})$ is for a $((2n_{11}), (2m_{21}))$-singular double - decreasing inflection-saddle.
- The $((2n_{11})\text{II}_{-}, (2m_{21})\text{II}_{-})$ is for a $((2n_{11}), (2m_{21}))$-singular double increasing-inflection-saddle.

3.7.2 $[(2m_{11}+1), (2n_{11})]$ *and* $((2n_{11}), (2m_{21}))$*-Infinite-Equilibriums*

Consider a polynomial dynamical system for $a_{111} = a_{211}$ as

$$\begin{aligned}\dot{x}_2 &= a_{220}(x_1 - a_{211})^{2m_{21}},\\ \dot{x}_1 &= a_{110}(x_1 - a_{111})^{2m_{11}+1}(x_2 - a_{121})^{2n_{11}}.\end{aligned} \tag{3.125}$$

The first integral manifold is

$$\begin{aligned}&\frac{1}{2(m_{21}-m_{11})}\left[(x_1 - a_{111})^{2(m_{21}-m_{11})} - (x_{10} - a_{111})^{2(m_{21}-m_{11})}\right]\\ &= \frac{a_{110}}{a_{220}}\frac{1}{2n_{11}+1}\left[(x_2 - a_{121})^{2n_{11}+1} - (x_{20} - a_{121})^{2n_{11}+1}\right],\\ &\text{for } 2(m_{21}-m_{11}) \neq 0;\\ &\frac{1}{2(m_{21}-m_{11})}\left[(x_1 - a_{111})^{2(m_{21}-m_{11})} - (x_{10} - a_{111})^{2(m_{21}-m_{11})}\right]\\ &= \ln\frac{|x_1 - a_{111}|}{|x_{10} - a_{111}|}, \text{ for } 2(m_{21}-m_{11}) = 0.\end{aligned} \tag{3.126}$$

(**A**) For $2m_{21} + 1 \geq 2m_{11}$, the corresponding infinite-equilibriums are given.

$$\begin{aligned}&(a_{121}, a_{211})\\ &= \underbrace{([(2m_{11}+1), (2n_{11})]^{\text{th}}\text{DI}_{+}\text{:II}_{+},[2(m_{21}-m_{11})-1]^{\text{th}}{}_{\text{UP:UP}}\text{SO})}_{([(2m_{11}+1),(2n_{11})],[2(m_{21}-m_{11})\text{-}1])\text{-double up-parabola source}},\end{aligned}$$

$$\text{for } a_{110} > 0, a_{220} > 0;$$

$$(a_{121}, a_{211}) = \underbrace{([(2m_{11}+1), (2n_{11})]^{\text{th}}\text{II}_{+}\text{:DI}_{+}, [2(m_{21}-m_{11})-1]^{\text{th}}{}_{\text{DP:DP}}\text{SI})}_{([(2m_{11}),(2n_{11})],[2(m_{21}-m_{11})-1])\text{-double down-parabola sink}},$$

$$\text{for } a_{110} < 0, a_{220} > 0;$$

$$(a_{121}, a_{211}) = \underbrace{([(2m_{11}+1), (2n_{11})]^{\text{th}}\text{II}_{-}\text{:DI}_{-}, [2(m_{21}-m_{11})-1]^{\text{th}}{}_{\text{DP:DP}}\text{SO})}_{([(2m_{11}+1),(2n_{11})],[2(m_{21}-m_{11})-1])\text{-double down-parabola source}},$$

$$\text{for } a_{110} > 0, a_{220} < 0;$$

$$(a_{121}, a_{211}) = \underbrace{([(2m_{11}+1), (2n_{11})]^{\text{th}}\text{DI}_{-}\text{:II}_{-}, [2(m_{21}-m_{11})-1]^{\text{th}}{}_{\text{UP:UP}}\text{SI})}_{([(2m_{11}+1),(2n_{11})],[2(m_{21}-m_{11})-1])\text{-double up-parabola sink}},$$

$$\text{for } a_{110} < 0, a_{220} < 0. \tag{3.127}$$

(**B**) For $2m_{21} + 1 < 2m_{11}$, the infinite-equilibriums are given as follows.

$$(a_{121}, a_{211}) = \underbrace{([(2m_{11}+1), (2n_{11})]^{\text{th}}\text{DI}_{+}\text{:II}_{+}, [2(m_{11}-m_{21})+1]^{\text{th}}{}_{\text{DA:DA}}\text{SO})}_{([(2m_{11}+1),(2n_{11})],[2(m_{11}-m_{21})+1])\text{-inflection down-asymptote source}},$$

$$\text{for } a_{110} > 0, a_{220} > 0;$$

$$(a_{121}, a_{211}) = \underbrace{([(2m_{11}+1), (2n_{11})]^{\text{th}}\text{II}_{+}\text{:DI}_{+}, [2(m_{11}-m_{21})+1]^{\text{th}}{}_{\text{UA:UA}}\text{SI})}_{([(2m_{11}+1),(2n_{11})],[2(m_{11}-m_{21})+1])\text{-inflection up-asymptote sink}},$$

$$\text{for } a_{110} < 0, a_{220} > 0;$$

$$(a_{121}, a_{211}) = \underbrace{([(2m_{11}+1), (2n_{11})]^{\text{th}}\text{II}_{+}\text{:DI}_{-}, [2(m_{11}-m_{21})+1]^{\text{th}}{}_{\text{UA:UA}}\text{SO})}_{([(2m_{11}+1),(2n_{11})],[2(m_{11}-m_{21})+1])\text{-inflection up-asymptote source}},$$

$$\text{for } a_{110} > 0, a_{220} < 0;$$

$$(a_{121}, a_{211}) = \underbrace{([(2m_{11}+1), (2n_{11})]^{\text{th}}\text{DI}_{+}\text{:II}_{-}, [2(m_{11}-m_{21})+1]^{\text{th}}{}_{\text{DA:DA}}\text{SI})}_{([(2m_{11}+1),(2n_{11})],[2(m_{11}-m_{21})+1])\text{-inflection down-asymptote sink}},$$

$$\text{for } a_{110} < 0, a_{220} < 0. \tag{3.128}$$

For $x_1^* = a_{111} = a_{211}$, the infinite-equilibrium exists. The infinite-equilibriums are illustrated in Figs. 3.39 and 3.40 for $2m_{21} \geq 2m_{11} + 1$ and $2m_{21} < 2m_{11} + 1$, respectively. For $2m_{21} \geq 2m_{11}+1$, the singular inflection parabola-sources and sinks are presented. For $2m_{11} + 1 > 2m_{21}$, the inflection asymptote-sources and sinks are presented.

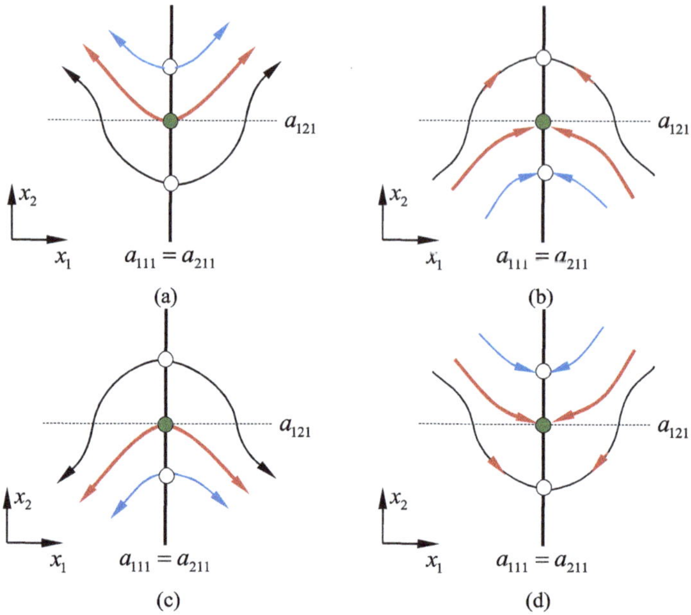

Fig. 3.39 The infinite-equilibriums ($a_{211} = a_{111}, 2m_{21} + 1 \geq 2m_{11} + 1, n_1 = 2n_{11}$) for 2-dimensional polynomial systems: **a** ($a_{110} > 0, a_{220} > 0$), **b** ($a_{110} < 0, a_{220} > 0$), **c** ($a_{110} > 0, a_{220} < 0$), **d** ($a_{110} < 0, a_{220} < 0$)

3.7.3 $[(2m_{11} + 1), (2n_{11})]$ *and* $((2n_{11}), (2m_{21}))$*-Hybrid Networks*

Consider

$$\begin{aligned}\dot{x}_2 &= a_{220} \prod_{v=1}^{2m_{21}} (x_1 - a_{21v}),\\ \dot{x}_1 &= a_{110} \prod_{r=1}^{2m_{11}+1} (x_1 - a_{11r}) \prod_{u=1}^{2n_{11}} (x_2 - a_{12u}).\end{aligned} \tag{3.129}$$

The first integral manifold is

$$\sum_{v_0=2m_{11}+1}^{2m_{21}} b_{21v_0} \sum_{\substack{\mu_1 \in [0, \beta_0 - 2m_{11} - 1],\\ v_1 \in [2m_{11}, v_0 - 1]}}^{\mu_1 + v_1 = v_0 - 1} \sum_{\substack{\mu_2 \in [0, v_1 - 2m_{11} - 1]\\ v_2 \in [2m_{11} - 1, v_1 - 2]}}^{\mu_2 + v_2 = v_1 - 2} \cdots \sum_{\substack{\mu_{2m_{11}} \in [0, v_{2m_{11}-1} - 2m_{11} - 1]\\ v_{2m_{11}} \in [1, v_{2m_{11}-1} - 2m_{11}]}}^{\mu_{2m_{11}} + v_{2m_{11}} = v_{2m_{11}-1} - 2m_{11}} \prod_{j=1}^{2m_{11}} \frac{(v_{j-1} - j)!}{\mu_j! v_j!}$$
$$\times (a_{11r_{j+1}} - a_{11r_j})^{\mu_j} \frac{1}{v_{2m_{11}}} \left[(x_1 - a_{11r_{2m_{11}+1}})^{v_{2m_{11}}} - (x_{10} - a_{11r_{2m_{11}+1}})^{v_{2m_{11}}} \right]$$

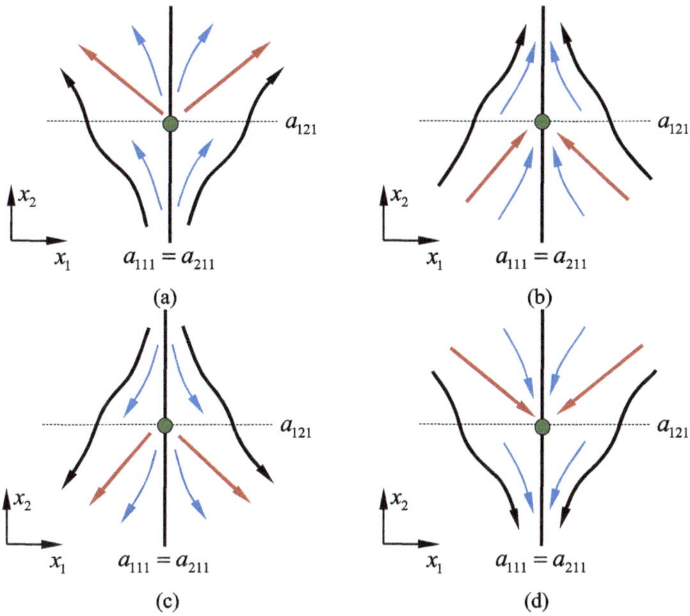

Fig. 3.40 The infinite-equilibriums ($a_{211} = a_{111}, 2m_{21}+1 < 2m_{11}+1, n_1 = 2n_{11}$) for 2-dimensional polynomial systems: **a** ($a_{110} > 0, a_{220} > 0$), **b** ($a_{110} < 0, a_{220} > 0$), **c** ($a_{110} > 0, a_{220} < 0$), **d** ($a_{110} < 0, a_{220} < 0$)

$$
\begin{aligned}
&-\sum_{\nu_0=2m_{11}+1}^{2m_{21}} b_{21\nu_0} \sum_{k=2}^{2m_{11}} \Bigg[\sum_{\substack{\mu_1+\nu_1=\nu_0-1 \\ \mu_1\in[0,\nu_0-2m_{11}-1), \\ \nu_1\in(2m_{11},\nu_0-1]}} \sum_{\substack{\mu_2+\nu_2=\nu_1-2 \\ \mu_2\in[0,\nu_1-2m_{11}-1] \\ \nu_2\in(2m_{11}-1,\nu_1-2)}} \cdots \\
&\times \Bigg(\sum_{\substack{\mu_{k-1}+\nu_{k-1}=\nu_{k-2}-k+1 \\ \mu_{k-1}\in(\nu_{k-2}-2m_{11},\nu_{k-2}-k+1] \\ \nu_{k-1}\in[0,2m_{11}+1-k)}} \prod_{j=1}^{k-1} \frac{(\nu_{j-1}-j)!}{\mu_j!\nu_j!} (a_{11r_{j+1}} - a_{11r_j})^{\mu_j} \\
&\times \sum_{\rho=k}^{2m_{11}+1} A_{\rho 1}^{(r_k)} \ln \frac{|x_1 - a_{11r_\rho}|}{|x_{10} - a_{11r_\rho}|} \\
&+ \sum_{\substack{\mu_{k-1}+\nu_{k-1}=\nu_{k-2}-k+1 \\ \mu_{k-1}\in(\nu_{k-2}-2m_{11}-1,\nu_{k-2}-2m_{11}] \\ \nu_{k-1}\in[2m_{11}+1-k,2m_{11}-k+2)}} \prod_{j=1}^{k-1} \frac{(\nu_{j-1}-j)!}{\mu_j!\nu_j!} (a_{11r_{j+1}} - a_{11r_j})^{\mu_j} \\
&\times \sum_{\rho=k+1}^{2m_{11}+1} B_{\rho 1}^{(r_k)} \ln \frac{|x_1 - a_{11r_\rho}|}{|x_{10} - a_{11r_\rho}|} \Bigg) \Bigg] + b_{210} \sum_{\rho=1}^{2m_{11}+1} A_{\rho 1}^{(r_1)} \ln \frac{|x_1 - a_{11r_\rho}|}{|x_{10} - a_{11r_\rho}|} \\
&+ \sum_{\nu_0=1}^{2m_{11}} b_{21\nu_0} \sum_{\rho=2}^{2m_{11}+1} B_{\rho 1}^{(r_1)} \ln \frac{|x_1 - a_{11s_\rho}|}{|x_{10} - a_{11s_\rho}|}
\end{aligned}
$$

$$= \frac{a_{110}}{a_{220}} \left\{ \frac{1}{2n_{11}+1} \left[(x_2 - a_{12u_1})^{2n_{11}+1} - (x_{j_20} - a_{12u_1})^{2n_{11}+1} \right] \right.$$
$$\left. + \sum_{\alpha=1}^{2n_{11}-2} \frac{1}{2n_{11}-\alpha+1} b_{12\alpha} \left[(x_2 - a_{12u_1})^{2n_{11}-\alpha+1} - (x_{20} - a_{12u_1})^{2n_{11}-\alpha+1} \right] \right\}, \tag{3.130}$$

where

$$b_{12\alpha} = \sum_{l_{n_1},t_{n_1}=0}^{l_{n_1}+t_{n_1}=1} \cdots \sum_{l_2,t_2=0}^{l_2+t_2=1} \prod_{k=2}^{2n_{11}} (a_{12u_1} - a_{12u_k})^{l_k} \delta_{t_2+\cdots+t_{2n_{11}}}^{2n_{11}-\alpha};$$
$$b_{120} = 1.$$
$$b_{21\nu_0} = \sum_{t_1,l_1=0}^{t_1+l_1=1} \cdots \sum_{t_{2m_{21}},l_{2m_{21}}=0}^{t_{2m_{21}}+l_{2m_{21}}=1} \prod_{k=1}^{2m_{21}} (a_{11r_1} - a_{21\nu_k})^{t_k} \delta_{\nu_0}^{\Sigma_{r=1}^{2m_{21}} l_r};$$
$$\lim_{\nu_{2m_{11}}\to 0} \frac{1}{\nu_{2m_{11}}} \left[(x_1 - a_{11r_{2m_{11}+1}})^{\nu_{2m_{11}}} - (x_{j_10} - a_{11r_{2m_{11}+1}})^{\nu_{2m_{11}}} \right] = \ln \frac{|x_1 - a_{11r_{2m_{11}+1}}|}{|x_{10} - a_{11r_{2m_{11}+1}}|}; \tag{3.131}$$

$$A_{\rho 1}^{(r_k)} = \frac{1}{\prod_{\sigma=k,\sigma\neq\rho}^{2m_{11}+1} (a_{11r_\rho} - a_{11r_\sigma})}, \text{ for } \rho \in \{k, k+1, \ldots, 2m_{11}+1\},$$
$$B_{\rho 1}^{(r_k)} = \frac{(a_{11s_\rho} - a_{11s_k})^{\beta_{k-1}-k}}{\prod_{\sigma=k+1,\sigma\neq\rho}^{2m_{11}+1} (a_{11r_\rho} - a_{11r_\sigma})}, \text{ for } \rho \in \{k+1, k+2, \ldots, 2m_{11}+1\};$$
$$k \in \{1, 2, \ldots, 2m_{11}\}. \tag{3.132}$$

For all cases, the $(2n_{11}+1)\times(2m_{21})$-simple equilibriums plus $(2m_{11}+1)\times(2n_{11}+1)$-hyperbolic flows are based on the following inflection parabola-source and sink bifurcations for $2m_{21} \geq 2m_{11}+1$:

- $([(2m_{11}+1), (2n_{11})]^{\text{th}}\text{DI}_+\text{:II}_+, [2(m_{21}-m_{11})-1]^{\text{th}}{}_{\text{UP:UP}}\text{SO})$-inflection up-parabola-source bifurcation for $(a_{110} > 0, a_{220} > 0)$,
- $([(2m_{11}+1), (2n_{11})]^{\text{th}}\text{II}_+\text{:DI}_+, [2(m_{21}-m_{11})-1]^{\text{th}}{}_{\text{DP:DP}}\text{SI})$-inflection down-parabola-sink bifurcation for $(a_{110} > 0, a_{220} > 0)$,
- $([(2m_{11}+1), (2n_{11})]^{\text{th}}\text{II}_-\text{:DI}_-, [2(m_{21}-m_{11})-1]^{\text{th}}{}_{\text{DP:DP}}\text{SO})$-inflection down-parabola-source bifurcation for $(a_{110} > 0, a_{220} < 0)$,
- $([(2m_{11}+1), (2n_{11})]^{\text{th}}\text{DI}_-\text{:II}_-, [2(m_{21}-m_{11})-1]^{\text{th}}{}_{\text{UP:UP}}\text{SI})$-inflection up-parabola-sink bifurcation for $(a_{110} < 0, a_{220} < 0)$;

or based on the following inflections asymptote-source and sink bifurcations for $2m_{21} < 2m_{11}+1$:

- $([(2m_{11}+1),(2n_{11})]^{\text{th}}\text{II}_{+}{:}\text{II}_{+},[2(m_{11}-m_{21})+1]^{\text{th}}{}_{\text{DA:DA}}\text{SO})$-inflection down-asymptote-source bifurcation for $(a_{110}>0, a_{220}>0)$,
- $([(2m_{11}+1),(2n_{11})]^{\text{th}}\text{DI}_{+}{:}\text{DI}_{+},[2(m_{11}-m_{21})+1]^{\text{th}}{}_{\text{UA:UA}}\text{SI})$-inflection up-asymptote-sink bifurcation for $(a_{110}>0, a_{220}>0)$,
- $([(2m_{11}+1),(2n_{11})]^{\text{th}}\text{DI}_{-}{:}\text{DI}_{-},[2(m_{11}-m_{21})+1]^{\text{th}}{}_{\text{UA:UA}}\text{SO})$-inflection up-asymptote-source bifurcation for $(a_{110}>0, a_{220}<0)$,
- $([(2m_{11}+1),(2n_{11})]^{\text{th}}\text{II}_{-}{:}\text{II}_{-},[2(m_{11}-m_{21})+1]^{\text{th}}{}_{\text{DA:DA}}\text{SI})$-inflection down-asymptote-source bifurcation for $(a_{110}<0, a_{220}<0)$.

(**A**) Consider $2m_{11}{=}2m_{21}$ for illustrations, phase portraits for the 2-dimensional systems for the hybrid networks of hyperbolic and hyperbolic-secant flows with the simple equilibriums of the saddles, and centers with $x_1^*=a_{111},a_{211},a_{112},\cdots,a_{21(2m_{21})},a_{11(2m_{11}+1)}$ and $x_2^*{=}a_{121},a_{122},\cdots,a_{12(2n_{11})}$ are presented in Fig. 3.41a–d for $(a_{110}{>}0, a_{220}{>}0)$, $(a_{110}{<}0, a_{220}{>}0)$, $(a_{110}>0, a_{220}<0)$ and $(a_{110}<0, a_{220}<0)$.
(**B**) Consider $2m_{11}{=}2m_{21}$ for illustrations, phase portraits for the 2-dimensional systems for the hybrid networks of hyperbolic and hyperbolic-secant flows with the simple equilibriums of saddles and centers with $x_1^*=a_{211},a_{111},\cdots,a_{21(2m_{21})},a_{11(2m_{11})},a_{11(2m_{11}+1)}$ and $x_2^*{=}a_{121},a_{122},\cdots,a_{12(2n_{11}+1)}$ are presented in Fig. 3.42a–d for $(a_{110}{>}0, a_{220}{>}0)$, $(a_{110}{<}0, a_{220}{>}0)$, $(a_{110}{>}0, a_{220}{<}0)$ and $(a_{110}{<}0, a_{220}{<}0)$.

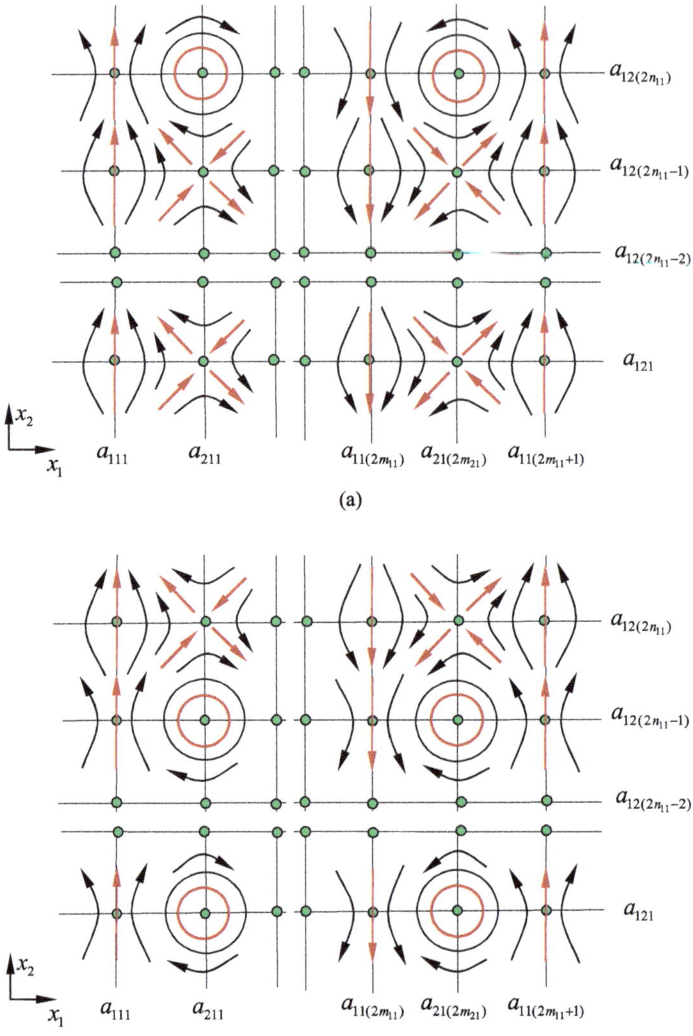

Fig. 3.41 Phase portraits for 2-dimensional systems with $x_1^* = a_{111}, a_{211}, \cdots, a_{11(2m_{11})}, a_{21(2m_{21})}, a_{11(2m_{11}+1)}$ and $x_2^* = a_{121}, a_{122}, \cdots, a_{12(2n_{11}+1)}$. The networks of flows with saddles and centers: **a** ($a_{110} > 0, a_{220} > 0$), **b** ($a_{110} < 0, a_{220} > 0$), **c** ($a_{110} > 0, a_{220} < 0$), **d** ($a_{110} < 0, a_{220} < 0$)

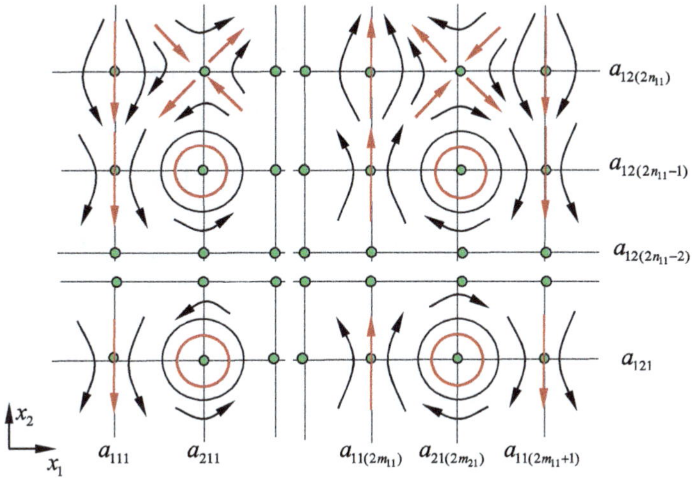

(c)

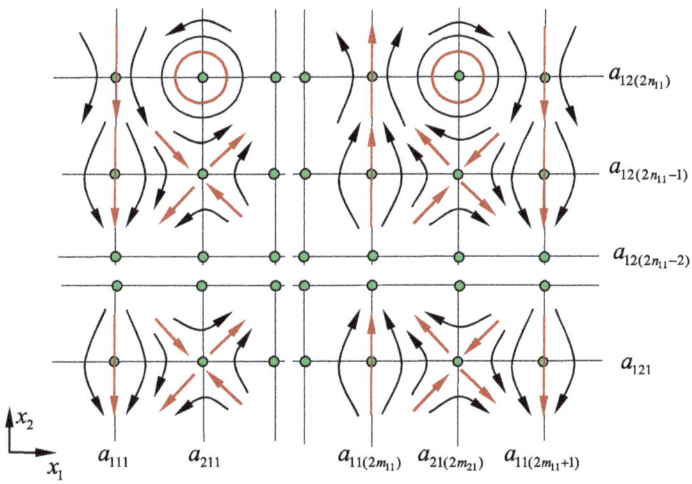

(d)

Fig. 3.41 (continued)

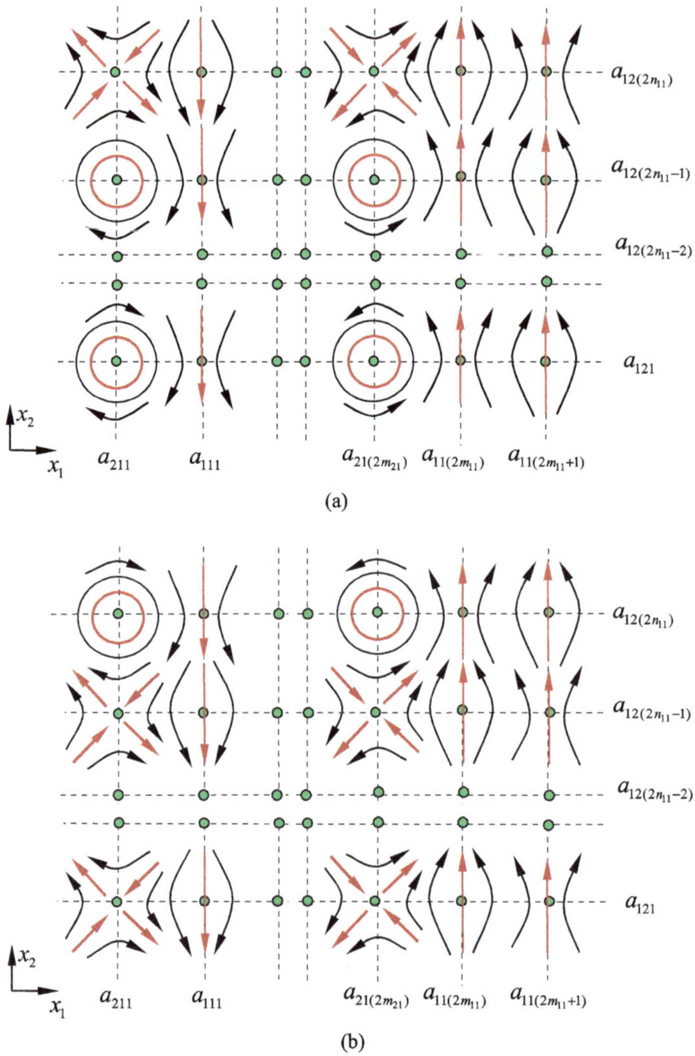

Fig. 3.42 Phase portraits for 2-dimensional systems with $x_1^* = a_{211}, a_{111}, \cdots, a_{21(2m_{21})}, a_{11(2m_{11})}, a_{11(2m_{11}+1)}$ and $x_2^* = a_{121}, a_{122}, \cdots, a_{12(2n_{11})}$. The networks of flows with saddles and centers: **a** ($a_{110} > 0, a_{220} > 0$), **b** ($a_{110} < 0, a_{220} > 0$), **c** ($a_{110} > 0, a_{220} < 0$), **d** ($a_{110} < 0, a_{220} < 0$)

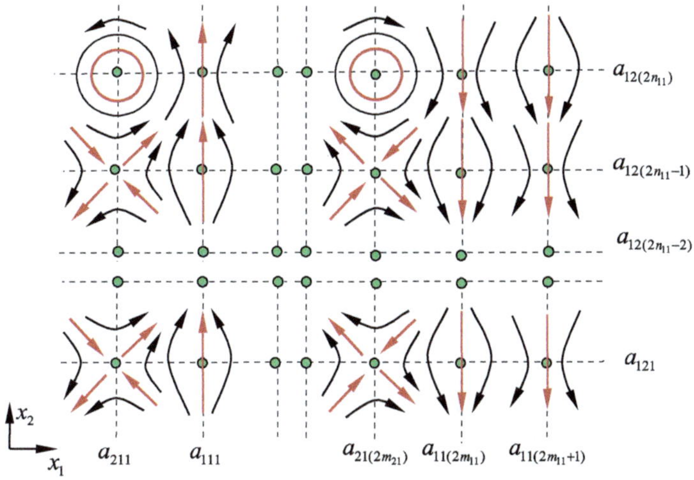

(c)

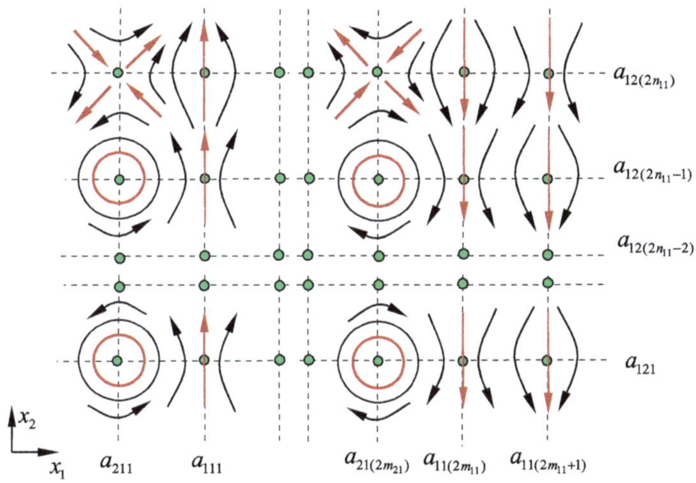

(d)

Fig. 3.42 (continued)

$$
\begin{Bmatrix}
(a_{111},a_{12(2n_{11})}) & (a_{12(2n_{11})},a_{211}) & \cdots & (a_{11(2m_{11})},a_{12(2n_{11})}) & (a_{12(2n_{11})},a_{21(2m_{21})}) & (a_{11(2m_{11}+1)},a_{12(2n_{11})}) \\
(a_{111},a_{12(2n_{11}-1)}) & (a_{12(2n_{11}-1)},a_{211}) & \cdots & (a_{11(2m_{11})},a_{12(2n_{11}-1)}) & (a_{12(2n_{11}-1)},a_{21(2m_{21})}) & (a_{11(2m_{11}+1)},a_{12(2n_{11}-1)}) \\
\vdots & \vdots & \vdots & \vdots & \vdots & \vdots \\
(a_{111},a_{121}) & (a_{121},a_{211}) & \cdots & (a_{11(2m_{11})},a_{121}) & (a_{121},a_{21(2m_{21})}) & (a_{11(2m_{11}+1)},a_{121})
\end{Bmatrix}
$$

$$
= \begin{Bmatrix}
\underbrace{(\mathrm{DP:UP,pF})}_{\text{hyperbolic flow }(+)} & \underbrace{(\mathrm{DP}_-,\mathrm{DP}_+)}_{\text{CW center}} & \cdots & \underbrace{(\mathrm{UP:DP,nF})}_{\text{hyperbolic-secant flow }(-)} & \underbrace{(\mathrm{DP}_+,\mathrm{DP}_-)}_{\text{CCW center}} & \underbrace{(\mathrm{DP:UP,pF})}_{\text{hyperbolic flow }(+)} \\
\underbrace{(\mathrm{UP:DP,pF})}_{\text{hyperbolic-secant flow }(+)} & \underbrace{(\mathrm{UP}_-,\mathrm{UP}_-)}_{\text{negative saddle}} & \cdots & \underbrace{(\mathrm{DP:UP,nF})}_{\text{hyperbolic flow }(-)} & \underbrace{(\mathrm{UP}_+,\mathrm{UP}_+)}_{\text{positive saddle}} & \underbrace{(\mathrm{UP:DP,pF})}_{\text{hyperbolic-secant flow }(+)} \\
\vdots & \vdots & \vdots & \vdots & \vdots & \vdots \\
\underbrace{(\mathrm{UP:DP,pF})}_{\text{hyperbolic-secant flow }(+)} & \underbrace{(\mathrm{UP}_-,\mathrm{UP}_-)}_{\text{negative saddle}} & \cdots & \underbrace{(\mathrm{DP:UP,nF})}_{\text{hyperbolic flow }(-)} & \underbrace{(\mathrm{UP}_+,\mathrm{UP}_+)}_{\text{positive saddle}} & \underbrace{(\mathrm{UP:DP,pF})}_{\text{hyperbolic-secant flow }(+)}
\end{Bmatrix}_{(2n_{11})\times(2m_{11}+2m_{21}+1)}
$$

for $a_{110} > 0, a_{220} > 0$; (3.133)

$$
\begin{Bmatrix}
(a_{111},a_{12(2n_{11})}) & (a_{12(2n_{11})},a_{211}) & \cdots & (a_{11(2m_{11})},a_{12(2n_{11})}) & (a_{12(2n_{11})},a_{21(2m_{21})}) & (a_{11(2m_{11}+1)},a_{12(2n_{11})}) \\
(a_{111},a_{12(2n_{11}-1)}) & (a_{12(2n_{11}-1)},a_{211}) & \cdots & (a_{11(2m_{11})},a_{12(2n_{11}-1)}) & (a_{12(2n_{11}-1)},a_{21(2m_{21})}) & (a_{11(2m_{11}+1)},a_{12(2n_{11}-1)}) \\
\vdots & \vdots & \vdots & \vdots & \vdots & \vdots \\
(a_{111},a_{121}) & (a_{121},a_{211}) & \cdots & (a_{11(2m_{11})},a_{121}) & (a_{121},a_{21(2m_{21})}) & (a_{11(2m_{11}+1)},a_{121})
\end{Bmatrix}
$$

$$
= \begin{Bmatrix}
\underbrace{(\text{UP:DP,pF})}_{\text{hyperbolic-secant flow }(+)} & \underbrace{(\text{UP}_-,\text{UP}_-)}_{\text{negative saddle}} & \cdots & \underbrace{(\text{UP:DP,nF})}_{\text{hyperbolic-secant flow }(-)} & \underbrace{(\text{UP}_+,\text{UP}_+)}_{\text{positive saddle}} & \underbrace{(\text{UP:DP,pF})}_{\text{hyperbolic-secant flow }(+)} \\
\underbrace{(\text{DP:UP,pF})}_{\text{hyperbolic flow }(+)} & \underbrace{(\text{DP}_-,\text{DP}_+)}_{\text{CW center}} & \cdots & \underbrace{(\text{DP:UP,nF})}_{\text{hyperbolic flow }(-)} & \underbrace{(\text{DP}_+,\text{DP}_-)}_{\text{CCW center}} & \underbrace{(\text{DP:UP,pF})}_{\text{hyperbolic flow }(+)} \\
\vdots & \vdots & \vdots & \vdots & \vdots & \vdots \\
\underbrace{(\text{DP:UP,pF})}_{\text{hyperbolic flow }(+)} & \underbrace{(\text{DP}_-,\text{DP}_+)}_{\text{CW center}} & \cdots & \underbrace{(\text{DP:UP,nF})}_{\text{hyperbolic flow }(-)} & \underbrace{(\text{DP}_+,\text{DP}_-)}_{\text{CCW center}} & \underbrace{(\text{DP:UP,pF})}_{\text{hyperbolic flow }(+)}
\end{Bmatrix}_{(2n_{11})\times(2m_{11}+2m_{21}+1)}
$$

for $a_{110} < 0, a_{220} > 0$; (3.134)

$$
\left\{\begin{array}{cccccc}
(a_{111},a_{12(2n_{11})}) & (a_{12(2n_{11})},a_{211}) & \cdots & (a_{11(2m_{11})},a_{12(2n_{11})}) & (a_{12(2n_{11})},a_{21(2m_{21})}) & (a_{11(2m_{11}+1)},a_{12(2n_{11})}) \\
(a_{111},a_{12(2n_{11}-1)}) & (a_{12(2n_{11}-1)},a_{211}) & \cdots & (a_{11(2m_{11})},a_{12(2n_{11}-1)}) & (a_{12(2n_{11}-1)},a_{21(2m_{21})}) & (a_{11(2m_{11}+1)},a_{12(2n_{11}-1)}) \\
\vdots & \vdots & \vdots & \vdots & \vdots & \vdots \\
(a_{111},a_{121}) & (a_{121},a_{211}) & \cdots & (a_{11(2m_{11})},a_{121}) & (a_{121},a_{21(2m_{21})}) & (a_{11(2m_{11}+1)},a_{121})
\end{array}\right\}
$$

$$
=\left\{\begin{array}{cccccc}
\underbrace{(\text{UP:DP,nF})}_{\text{hyperbolic-secant flow }(-)} & \underbrace{(\text{UP}_+,\text{UP}_+)}_{\text{positive saddle}} & \cdots & \underbrace{(\text{UP:DP,pF})}_{\text{hyperbolic-secant flow }(+)} & \underbrace{(\text{UP}_-,\text{UP}_-)}_{\text{negative saddle}} & \underbrace{(\text{UP:DP,nF})}_{\text{hyperbolic-secant flow }(-)} \\
\underbrace{(\text{DP:UP,nF})}_{\text{hyperbolic flow }(-)} & \underbrace{(\text{DP}_+,\text{DP}_-)}_{\text{CCW center}} & \cdots & \underbrace{(\text{DP:UP,pF})}_{\text{hyperbolic flow }(+)} & \underbrace{(\text{DP}_-,\text{DP}_+)}_{\text{CW center}} & \underbrace{(\text{DP:UP,nF})}_{\text{hyperbolic flow }(-)} \\
\vdots & \vdots & \vdots & \vdots & \vdots & \vdots \\
\underbrace{(\text{DP:UP,nF})}_{\text{hyperbolic flow }(-)} & \underbrace{(\text{DP}_+,\text{DP}_-)}_{\text{CCW center}} & \cdots & \underbrace{(\text{DP:UP,pF})}_{\text{hyperbolic flow }(+)} & \underbrace{(\text{DP}_-,\text{DP}_+)}_{\text{CW center}} & \underbrace{(\text{DP:UP,nF})}_{\text{hyperbolic flow }(-)}
\end{array}\right\}_{(2n_{11})\times(2m_{11}+2m_{21}+1)}
$$

for $a_{110} < 0, a_{220} > 0;$ (3.135)

$$\left\{\begin{array}{cccccc}
(a_{111},a_{12(2n_{11})}) & (a_{12(2n_{11})},a_{211}) & \cdots & (a_{11(2m_{11})},a_{12(2n_{11})}) & (a_{12(2n_{11})},a_{21(2m_{21})}) & (a_{11(2m_{11}+1)},a_{12(2n_{11})})\\
(a_{111},a_{12(2n_{11}-1)}) & (a_{12(2n_{11}-1)},a_{211}) & \cdots & (a_{11(2m_{11})},a_{12(2n_{11}-1)}) & (a_{12(2n_{11}-1)},a_{21(2m_{21})}) & (a_{11(2m_{11}+1)},a_{12(2n_{11}-1)})\\
\vdots & \vdots & \vdots & \vdots & \vdots & \vdots\\
(a_{111},a_{121}) & (a_{121},a_{211}) & \cdots & (a_{11(2m_{11})},a_{121}) & (a_{121},a_{21(2m_{21})}) & (a_{11(2m_{11}+1)},a_{121})
\end{array}\right\}$$
$$=\left\{\begin{array}{cccccc}
\underbrace{(\text{DP:UP,nF})}_{\text{hyperbolic flow }(-)} & \underbrace{(\text{DP}_+,\text{DP}_-)}_{\text{CCW center}} & \cdots & \underbrace{(\text{DP:UP,pF})}_{\text{hyperbolic flow }(+)} & \underbrace{(\text{DP}_-,\text{DP}_+)}_{\text{CW center}} & \underbrace{(\text{DP:UP,nF})}_{\text{hyperbolic flow }(-)}\\
\underbrace{(\text{UP:DP,nF})}_{\text{hyperbolic-secant flow }(-)} & \underbrace{(\text{UP}_+,\text{UP}_+)}_{\text{positive saddle}} & \cdots & \underbrace{(\text{UP:DP,pF})}_{\text{hyperbolic-secant flow }(+)} & \underbrace{(\text{UP}_-,\text{UP}_-)}_{\text{negative saddle}} & \underbrace{(\text{UP:DP,nF})}_{\text{hyperbolic-secant flow }(-)}\\
\vdots & \vdots & \vdots & \vdots & \vdots & \vdots\\
\underbrace{(\text{UP:DP,nF})}_{\text{hyperbolic-secant flow }(-)} & \underbrace{(\text{UP}_+,\text{UP}_+)}_{\text{positive saddle}} & \cdots & \underbrace{(\text{UP:DP,pF})}_{\text{hyperbolic-secant flow }(+)} & \underbrace{(\text{UP}_-,\text{UP}_-)}_{\text{negative saddle}} & \underbrace{(\text{UP:DP,nF})}_{\text{hyperbolic-secant flow }(-)}
\end{array}\right\}_{(2n_{11})\times(2m_{11}+2m_{21}+1)}$$

for $a_{110} < 0, a_{220} < 0$; (3.136)

$$
\left\{\begin{array}{cccccc}
(a_{12(2n_{11})},a_{211}) & (a_{111},a_{12(2n_{11})}) & \cdots & (a_{12(2n_{11})},a_{21(2m_{21})}) & (a_{11(2m_{11})},a_{12(2n_{11})}) & (a_{11(2m_{11}+1)},a_{12(2n_{11})}) \\
(a_{12(2n_{11}-1)},a_{211}) & (a_{111},a_{12(2n_{11}-1)}) & \cdots & (a_{12(2n_{11}-1)},a_{21(2m_{21})}) & (a_{11(2m_{11})},a_{12(2n_{11}-1)}) & (a_{11(2m_{11}+1)},a_{12(2n_{11}-1)}) \\
\vdots & \vdots & \vdots & \vdots & \vdots & \vdots \\
(a_{121},a_{211}) & (a_{111},a_{121}) & \cdots & (a_{121},a_{21(2m_{21})}) & (a_{11(2m_{11})},a_{121}) & (a_{11(2m_{11}+1)},a_{121})
\end{array}\right\}
$$

$$
=\left\{\begin{array}{cccccc}
\underbrace{(\mathrm{UP}_-,\mathrm{UP}_-)}_{\text{positive saddle}} & \underbrace{(\mathrm{UP{:}DP,nF})}_{\text{hyperbolic-secant flow }(-)} & \cdots & \underbrace{(\mathrm{UP}_+,\mathrm{UP}_+)}_{\text{positive saddle}} & \underbrace{(\mathrm{UP{:}DP,pF})}_{\text{hyperbolic-secant flow }(+)} & \underbrace{(\mathrm{DP{:}UP,pF})}_{\text{hyperbolic flow }(+)} \\
\underbrace{(\mathrm{DP}_-,\mathrm{DP}_+)}_{\text{CW center}} & \underbrace{(\mathrm{DP{:}UP,nF})}_{\text{hyperbolic flow }(-)} & \cdots & \underbrace{(\mathrm{DP}_+,\mathrm{DP}_-)}_{\text{CCW center}} & \underbrace{(\mathrm{DP{:}UP,pF})}_{\text{hyperbolic flow }(+)} & \underbrace{(\mathrm{UP{:}DP,pF})}_{\text{hyperbolic-secant flow }(+)} \\
\vdots & \vdots & \vdots & \vdots & \vdots & \vdots \\
\underbrace{(\mathrm{DP}_-,\mathrm{DP}_+)}_{\text{CW center}} & \underbrace{(\mathrm{DP{:}UP,nF})}_{\text{hyperbolic flow }(-)} & \cdots & \underbrace{(\mathrm{DP}_+,\mathrm{DP}_-)}_{\text{CCW center}} & \underbrace{(\mathrm{DP{:}UP,pF})}_{\text{hyperbolic flow }(+)} & \underbrace{(\mathrm{UP{:}DP,pF})}_{\text{hyperbolic-secant flow }(+)}
\end{array}\right\}_{(2n_{11})\times(2m_{11}+2m_{21}+1)}
$$

for $a_{110} > 0, a_{220} > 0$; (3.137)

$$\left\{\begin{array}{cccccc}
(a_{12(2n_{11})},a_{211}) & (a_{111},a_{12(2n_{11})}) & \cdots & (a_{12(2n_{11})},a_{21(2m_{21})}) & (a_{11(2m_{11})},a_{12(2n_{11})}) & (a_{11(2m_{11}+1)},a_{12(2n_{11})}) \\
(a_{12(2n_{11}-1)},a_{211}) & (a_{111},a_{12(2n_{11}-1)}) & \cdots & (a_{12(2n_{11}-1)},a_{21(2m_{21})}) & (a_{11(2m_{11})},a_{12(2n_{11}-1)}) & (a_{11(2m_{11}+1)},a_{12(2n_{11}-1)}) \\
\vdots & \vdots & \vdots & \vdots & \vdots & \vdots \\
(a_{121},a_{211}) & (a_{111},a_{121}) & \cdots & (a_{121},a_{21(2m_{21})}) & (a_{11(2m_{11})},a_{121}) & (a_{11(2m_{11}+1)},a_{121})
\end{array}\right\}$$
$$= \left\{\begin{array}{cccccc}
\underbrace{(\mathrm{DP}_-,\mathrm{DP}_+)}_{\text{CW center}} & \underbrace{(\mathrm{DP{:}UP,nF})}_{\text{hyperbolic flow }(-)} & \cdots & \underbrace{(\mathrm{DP}_+,\mathrm{DP}_-)}_{\text{CCW center}} & \underbrace{(\mathrm{DP{:}UP,pF})}_{\text{hyperbolic flow }(+)} & \underbrace{(\mathrm{UP{:}DP,pF})}_{\text{hyperbolic-secant flow }(+)} \\
\underbrace{(\mathrm{UP}_-,\mathrm{UP}_-)}_{\text{negative saddle}} & \underbrace{(\mathrm{UP{:}DP,nF})}_{\text{hyperbolic-secant flow }(-)} & \cdots & \underbrace{(\mathrm{UP}_+,\mathrm{UP}_+)}_{\text{positive saddle}} & \underbrace{(\mathrm{UP{:}DP,pF})}_{\text{hyperbolic-secant flow }(+)} & \underbrace{(\mathrm{DP{:}UP,pF})}_{\text{hyperbolic flow }(+)} \\
\vdots & \vdots & \vdots & \vdots & \vdots & \vdots \\
\underbrace{(\mathrm{UP}_-,\mathrm{UP}_-)}_{\text{negative saddle}} & \underbrace{(\mathrm{UP{:}DP,nF})}_{\text{hyperbolic-secant flow }(-)} & \cdots & \underbrace{(\mathrm{UP}_+,\mathrm{UP}_+)}_{\text{positive saddle}} & \underbrace{(\mathrm{UP{:}DP,pF})}_{\text{hyperbolic-secant flow }(+)} & \underbrace{(\mathrm{DP{:}UP,pF})}_{\text{hyperbolic flow }(+)}
\end{array}\right\}_{(2n_{11})\times(2m_{11}+2m_{21}+1)}$$

for $a_{110}<0,\ a_{220}>0$; (3.138)

$$\begin{Bmatrix} (a_{12(2n_{11})},a_{211}) & (a_{111},a_{12(2n_{11})}) & \cdots & (a_{12(2n_{11})},a_{21(2m_{21})}) & (a_{11(2m_{11})},a_{12(2n_{11})}) & (a_{11(2m_{11}+1)},a_{12(2n_{11})}) \\ (a_{12(2n_{11}-1)},a_{211}) & (a_{111},a_{12(2n_{11}-1)}) & \cdots & (a_{12(2n_{11}-1)},a_{21(2m_{21})}) & (a_{11(2m_{11})},a_{12(2n_{11}-1)}) & (a_{11(2m_{11}+1)},a_{12(2n_{11}-1)}) \\ \vdots & \vdots & \vdots & \vdots & \vdots & \vdots \\ (a_{121},a_{211}) & (a_{111},a_{121}) & \cdots & (a_{121},a_{21(2m_{21})}) & (a_{11(2m_{11})},a_{121}) & (a_{11(2m_{11}+1)},a_{121}) \end{Bmatrix}$$

$$= \begin{Bmatrix} \underbrace{(\mathrm{DP}_+,\mathrm{DP}_-)}_{\text{CCW center}} & \underbrace{(\mathrm{DP{:}UP,pF})}_{\text{hyperbolic flow }(+)} & \cdots & \underbrace{(\mathrm{DP}_-,\mathrm{DP}_+)}_{\text{CW center}} & \underbrace{(\mathrm{DP{:}UP,nF})}_{\text{hyperbolic flow }(-)} & \underbrace{(\mathrm{UP{:}DP,nF})}_{\text{hyperbolic-secant flow }(-)} \\ \underbrace{(\mathrm{UP}_+,\mathrm{UP}_+)}_{\text{positive saddle}} & \underbrace{(\mathrm{UP{:}DP,pF})}_{\text{hyperbolic-secant flow }(+)} & \cdots & \underbrace{(\mathrm{UP}_-,\mathrm{UP}_-)}_{\text{negative saddle}} & \underbrace{(\mathrm{UP{:}DP,nF})}_{\text{hyperbolic-secant flow }(-)} & \underbrace{(\mathrm{DP{:}UP,nF})}_{\text{hyperbolic flow }(-)} \\ \vdots & \vdots & \vdots & \vdots & \vdots & \vdots \\ \underbrace{(\mathrm{UP}_+,\mathrm{UP}_+)}_{\text{positive saddle}} & \underbrace{(\mathrm{UP{:}DP,pF})}_{\text{hyperbolic-secant flow }(+)} & \cdots & \underbrace{(\mathrm{UP}_-,\mathrm{UP}_-)}_{\text{negative saddle}} & \underbrace{(\mathrm{UP{:}DP,nF})}_{\text{hyperbolic-secant flow }(-)} & \underbrace{(\mathrm{DP{:}UP,nF})}_{\text{hyperbolic flow }(-)} \end{Bmatrix}_{(2n_{11})\times(2m_{11}+2m_{21}+1)} \tag{3.139}$$

for $a_{110} > 0$, $a_{220} < 0$;

$$
\left\{\begin{matrix}
(a_{12(2n_{11})},a_{211}) & (a_{111},a_{12(2n_{11})}) & \cdots & (a_{12(2n_{11})},a_{21(2m_{21})}) & (a_{11(2m_{11})},a_{12(2n_{11})}) & (a_{11(2m_{11}+1)},a_{12(2n_{11})}) \\
(a_{12(2n_{11}-1)},a_{211}) & (a_{111},a_{12(2n_{11}-1)}) & \cdots & (a_{12(2n_{11}-1)},a_{21(2m_{21})}) & (a_{11(2m_{11})},a_{12(2n_{11}-1)}) & (a_{11(2m_{11}+1)},a_{12(2n_{11}-1)}) \\
\vdots & \vdots & \vdots & \vdots & \vdots & \vdots \\
(a_{121},a_{211}) & (a_{111},a_{121}) & \cdots & (a_{121},a_{21(2m_{21})}) & (a_{11(2m_{11})},a_{121}) & (a_{11(2m_{11}+1)},a_{121})
\end{matrix}\right\}
$$
$$
=\left\{\begin{matrix}
\underbrace{(\mathrm{UP}_+,\mathrm{UP}_+)}_{\text{positive saddle}} & \underbrace{(\mathrm{UP{:}DP,pF})}_{\text{hyperbolic-secant flow }(+)} & \cdots & \underbrace{(\mathrm{UP}_-,\mathrm{UP}_-)}_{\text{negative saddle}} & \underbrace{(\mathrm{UP{:}DP,nF})}_{\text{hyperbolic-secant flow }(-)} & \underbrace{(\mathrm{DP{:}UP,nF})}_{\text{hyperbolic flow }(-)} \\
\underbrace{(\mathrm{DP}_-,\mathrm{DP}_+)}_{\text{CW center}} & \underbrace{(\mathrm{DP{:}UP,pF})}_{\text{hyperbolic flow }(+)} & \cdots & \underbrace{(\mathrm{DP}_-,\mathrm{DP}_+)}_{\text{CCW center}} & \underbrace{(\mathrm{DP{:}UP,nF})}_{\text{hyperbolic flow }(-)} & \underbrace{(\mathrm{UP{:}DP,nF})}_{\text{hyperbolic-secant flow }(-)} \\
\vdots & \vdots & \vdots & \vdots & \vdots & \vdots \\
\underbrace{(\mathrm{DP}_-,\mathrm{DP}_+)}_{\text{CW center}} & \underbrace{(\mathrm{DP{:}UP,pF})}_{\text{hyperbolic flow }(+)} & \cdots & \underbrace{(\mathrm{DP}_-,\mathrm{DP}_+)}_{\text{CCW center}} & \underbrace{(\mathrm{DP{:}UP,nF})}_{\text{hyperbolic flow }(-)} & \underbrace{(\mathrm{UP{:}DP,nF})}_{\text{hyperbolic-secant flow }(-)}
\end{matrix}\right\}_{(2n_{11})\times(2m_{11}+2m_{21}+1)}
$$

for $a_{110} < 0, a_{220} < 0$. (3.140)

3.8 $[(2m_{11}), (2n_{11})]$ and $((2n_{11}), (2m_{21}))$-Singular Systems

In this section, there are four global bifurcations of the 1-dimensional flows through singular inflection upper-saddle and lower-saddle flows, and four global bifurcations of equilibriums, through singular double inflection-saddle bifurcations. The corresponding global bifurcations generate networks of hyperbolic and hyperbolic-secant flows with saddles and centers. The infinite-equilibriums are the appearing and switching bifurcations for the singular flows and singular equilibriums and for hybrid networks of hyperbolic flows and saddles and centers.

3.8.1 $[(2m_{11}), (2n_{11})]$-*Flows and* $((2n_{11}), (2m_{21}))$-*Equilibriums*

Consider a polynomial dynamical system for $a_{111} \neq a_{211}$ as

$$\begin{aligned}\dot{x}_2 &= a_{220}(x_1 - a_{211})^{2m_{21}},\\ \dot{x}_1 &= a_{110}(x_1 - a_{111})^{2m_{11}}(x_2 - a_{121})^{2n_{11}}.\end{aligned} \tag{3.141}$$

The first integral manifold is

$$\begin{aligned}&\sum_{\alpha,\beta=0}^{\alpha+\beta=2m_{21}} \frac{(2m_{21})!}{\alpha!\beta!}\frac{(a_{111}-a_{211})^{\alpha}}{\beta-2m_{11}+1}\left[(x_1-a_{111})^{\beta-2m_{11}+1}-(x_{10}-a_{111})^{\beta-2m_{11}+1}\right]\\ &=\frac{a_{j_1j_10}}{a_{j_2j_20}}\frac{1}{2n_{11}+1}\left[(x_2-a_{121})^{2n_{11}+1}-(x_{20}-a_{121})^{2n_{11}+1}\right],\\ &\text{for } \beta-2m_{11}+1\neq 0;\\ &\frac{1}{\beta-2m_{11}+1}\left[(x_1-a_{111})^{\beta-2m_{11}+1}-(x_{10}-a_{111})^{\beta-2m_{11}+1}\right]=\ln\frac{|x_1-a_{111}|}{|x_{10}-a_{111}|},\\ &\text{for } \beta-2m_{11}+1=0.\end{aligned} \tag{3.142}$$

The singular hyperbolic-to-hyperbolic-secant flows with singular parabola-saddle are illustrated in Figs. 3.43 and 3.44. Through such singular flows, the corresponding global bifurcations generate from a hyperbolic (hyperbolic-secant) flow to a network of singular hyperbolic and hyperbolic-secant flows. Through such singular equilibriums, the corresponding global bifurcations generate from one of a flow to a network of singular saddles and centers equilibriums.

For $(x_1^*, x_2^*) = (a_{111}, a_{121})$ with $\dot{x}_1 = 0$, the $[(2m_{11}), (2n_{11})]$-singular flows exist with $\dot{x}_2 = a_{220}(a_{111} - a_{211})^{2m_{21}}$. For $(x_1^*, x_2^*) = (a_{111}, a_{121})$, the 1-dimensional flow direction is determined by $\dot{x}_2 \neq 0$. For $(x_2^*, x_1^*) = (a_{121}, a_{211})$ with $\dot{x}_1 = \dot{x}_2 = 0$, the $((2n_{11}), (2m_{21}))$-singular equilibriums exist.

(A) For $a_{111} > a_{211}$,

Fig. 3.43 Phase portraits ($a_{111} > a_{211}$) of the ($[(2m_{11}), (2n_{11})]$, xF)-singular inflection upper-saddle and lower-saddle flows at $(x_1^*, x_2^*) = (a_{111}, a_{121})$ and $((2n_{11}), (2m_{21}))$-parabola-saddle at $(x_2^*, x_1^*) = (a_{121}, a_{211})$ for 2-dimensional polynomial systems:
a ($a_{110} > 0, a_{220} > 0$),
b ($a_{110} < 0, a_{220} > 0$),
c ($a_{110} > 0, a_{220} < 0$),
d ($a_{110} < 0, a_{220} < 0$)

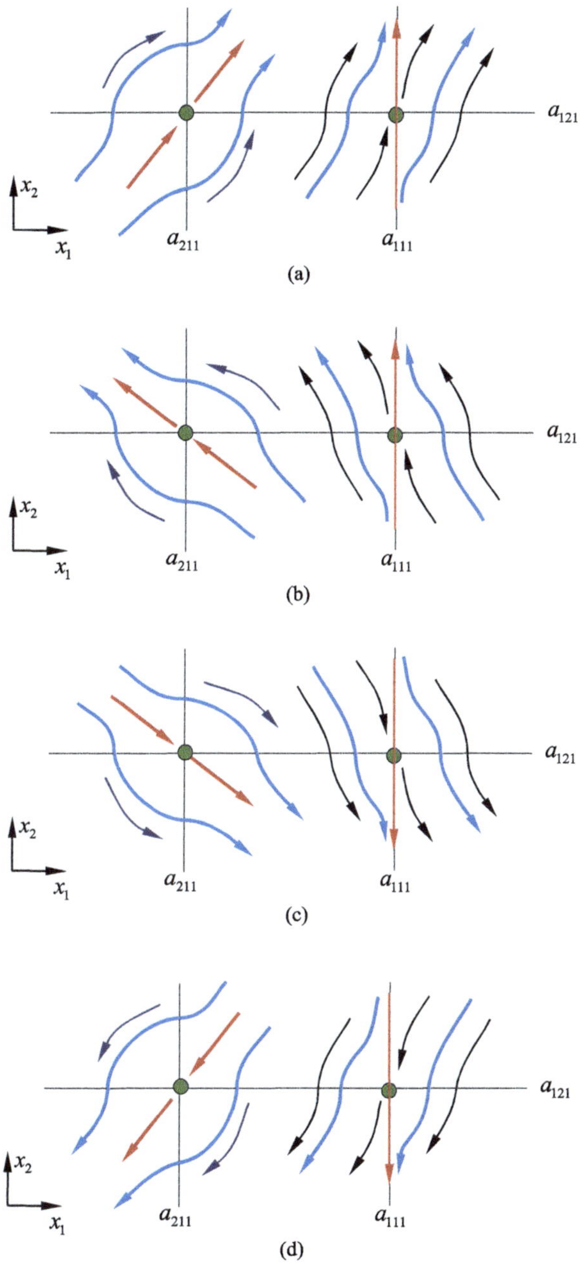

Fig. 3.44 Phase portraits ($a_{111} > a_{211}$) of the $([(2m_{11}), (2n_{11})], \text{xF})$-singular inflection upper-saddle and lower-saddle flows at $(x_1^*, x_2^*) = (a_{111}, a_{121})$ and $((2n_{11}), (2m_{21}))$-parabola-saddle at $(x_2^*, x_1^*) = (a_{121}, a_{211})$ for 2-dimensional polynomial systems: **a** ($a_{110} > 0, a_{220} > 0$), **b** ($a_{110} < 0, a_{220} > 0$), **c** ($a_{110} > 0, a_{220} < 0$), **d** ($a_{110} < 0, a_{220} < 0$)

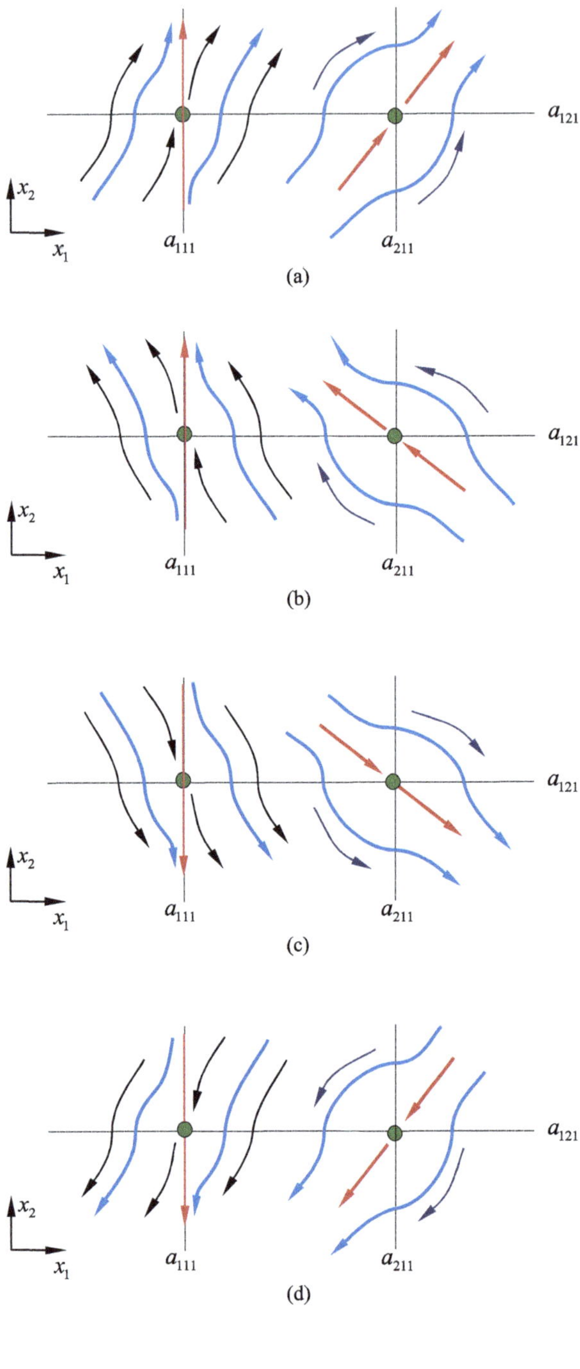

$$\left\{\begin{matrix}(a_{111},a_{121})\\(a_{121},a_{211})\end{matrix}\right\}=\left\{\begin{matrix}\underbrace{([(2m_{11}),(2n_{11})]_{\mathrm{II:II}}\mathrm{US,pF})}_{[(2m_{11}),(2n_{11})]\text{-increasing-inflection upper-saddle flow }(+)}\\\underbrace{((2n_{11})\mathrm{II}_{+},(2m_{21})\mathrm{II}_{+})}_{[(2n_{11}),(2m_{21})]\text{-double inflection-saddle}}\end{matrix}\right\},$$
for $a_{110}>0, a_{220}>0$;

$$\left\{\begin{matrix}(a_{111},a_{121})\\(a_{121},a_{211})\end{matrix}\right\}=\left\{\begin{matrix}\underbrace{([(2m_{11}),(2n_{11})]_{\mathrm{DI:DI}}\mathrm{LS,pF})}_{[(2m_{11}),(2n_{11})]\text{-decreasing-inflection lower-saddle flow }(+)}\\\underbrace{((2n_{11})\mathrm{DI}_{+},(2m_{21})\mathrm{DI}_{-})}_{[(2n_{11}),(2m_{21})]\text{-double inflection-saddle}}\end{matrix}\right\},$$
for $a_{110}<0, a_{220}>0$;

$$\left\{\begin{matrix}(a_{111},a_{121})\\(a_{121},a_{211})\end{matrix}\right\}=\left\{\begin{matrix}\underbrace{([(2m_{11}),(2n_{11})]_{\mathrm{DI:DI}}\mathrm{US,nF})}_{[(2m_{11}),(2n_{11})]\text{-decreasing-inflection upper-saddle flow }(-)}\\\underbrace{((2n_{11})\mathrm{DI}_{-},(2m_{21})\mathrm{DI}_{+})}_{[(2n_{11}),(2m_{21})]\text{-double inflection-saddle}}\end{matrix}\right\},$$
for $a_{110}>0, a_{220}<0$;

$$\left\{\begin{matrix}(a_{111},a_{121})\\(a_{121},a_{211})\end{matrix}\right\}=\left\{\begin{matrix}\underbrace{([(2m_{11}),(2n_{11})]_{\mathrm{II:II}}\mathrm{LS,nF})}_{[(2m_{11}),(2n_{11})]\text{-increasing-inflection lower-saddle flow }(-)}\\\underbrace{((2n_{11})\mathrm{II}_{-},(2m_{21})\mathrm{II}_{-})}_{[(2n_{11}),(2m_{21})]\text{-double infection-saddle}}\end{matrix}\right\},$$
for $a_{110}<0, a_{220}<0$. (3.143)

(**B**) For $a_{211}>a_{111}$,

$$\left\{\begin{matrix}(a_{121},a_{211})\\(a_{111},a_{121})\end{matrix}\right\}=\left\{\begin{matrix}\underbrace{((2n_{11})\mathrm{II}_{+},(2m_{21})\mathrm{II}_{+})}_{[(2n_{11}),(2m_{21})]\text{-double inflection-saddle}}\\\underbrace{([(2m_{11}),(2n_{11})]_{\mathrm{II:II}}\mathrm{US,pF})}_{[(2m_{11}),(2n_{11})]\text{-increasing-inflection upper-saddle flow }(+)}\end{matrix}\right\},$$
for $a_{110}>0, a_{220}>0$;

$$\left\{\begin{matrix}(a_{121},a_{211})\\(a_{111},a_{121})\end{matrix}\right\}=\left\{\begin{matrix}\underbrace{((2n_{11})\mathrm{DI}_{+},(2m_{21})\mathrm{DI}_{-})}_{[(2n_{11}),(2m_{21})]\text{-double inflection-saddle}}\\\underbrace{([(2m_{11}),(2n_{11})]_{\mathrm{DI:DI}}\mathrm{LS,pF})}_{[(2m_{11}),(2n_{11})]\text{-decreasing-inflection lower-saddle flow }(+)}\end{matrix}\right\},$$
for $a_{110}<0, a_{220}>0$;

$$\left\{\begin{matrix}(a_{121},a_{211})\\(a_{111},a_{121})\end{matrix}\right\}=\left\{\begin{matrix}\underbrace{((2n_{11})\mathrm{DI}_{-},(2m_{21})\mathrm{DI}_{+})}_{[(2n_{11}),(2m_{21})]\text{-double inflection-saddle}}\\\underbrace{([(2m_{11}),(2n_{11})]_{\mathrm{DI:DI}}\mathrm{US,nF})}_{[(2m_{11}),(2n_{11})]\text{-decreasing-inflection upper-saddle flow }(-)}\end{matrix}\right\},$$

for $a_{110} > 0, a_{220} < 0$;

$$\left\{\begin{matrix}(a_{121}, a_{211})\\(a_{111}, a_{121})\end{matrix}\right\} = \left\{\begin{matrix}\underbrace{((2n_{11})\text{II}_{-},(2m_{21})\text{II}_{-})}_{[(2n_{11}),(2m_{21})]\text{-double infection-saddle}}\\ \underbrace{([(2m_{11}),(2n_{11})]_{\text{II:II}}\text{LS,nF})}_{[(2m_{11}),(2n_{11})]\text{-increasing-inflection lower-saddle flow }(-)}\end{matrix}\right\},$$

for $a_{110} < 0, a_{220} < 0$. (3.144)

3.8.2 $[(2m_{11}), (2n_{11})]$ and $((2n_{11}), (2m_{21}))$-Infinite-Equilibriums

Consider a polynomial dynamical system for $a_{111} = a_{211}$ as

$$\begin{aligned}\dot{x}_2 &= a_{220}(x_1 - a_{211})^{2m_{21}},\\ \dot{x}_1 &= a_{110}(x_1 - a_{111})^{2m_{11}}(x_2 - a_{121})^{2n_{11}}.\end{aligned} \tag{3.145}$$

The first integral manifold is.

$$\begin{aligned}&\frac{1}{2(m_{21} - m_{11}) + 1}\left[(x_1 - a_{211})^{2(m_{21}-m_{11})+1} - (x_{10} - a_{211})^{2(m_{21}-m_{11})+1}\right]\\ &= \frac{a_{110}}{a_{220}}\frac{1}{2n_{11} + 1}\left[(x_2 - a_{121})^{2n_{11}+1} - (x_{20} - a_{121})^{2n_{11}+1}\right].\end{aligned} \tag{3.146}$$

(**A**) For $2m_{21} \geq 2m_{11}$, the corresponding infinite-equilibriums are given.

$$(a_{121}, a_{211}) = \underbrace{([(2m_{11}), (2n_{11})]^{\text{th}}\text{II}_{+}\text{:II}_{+},[2(m_{21} - m_{11})]^{\text{th}}{}_{\text{II:II}}\text{US})}_{([(2m_{11}),(2n_{11})],[2(m_{21}-m_{11})])\text{-increasing-inflection upper-saddle}},$$

for $a_{110} > 0, a_{220} > 0$;

$$(a_{121}, a_{211}) = \underbrace{([(2m_{11}), (2n_{11})]^{\text{th}}\text{DI}_{+}\text{:DI}_{+},[2(m_{21} - m_{11})]^{\text{th}}{}_{\text{DI:DI}}\text{LS})}_{([(2m_{11}),(2n_{11})],[2(m_{21}-m_{11})])\text{-decreasing-inflection lower-saddle}},$$

for $a_{110} < 0, a_{220} > 0$;

$$(a_{121}, a_{211}) = \underbrace{([(2m_{11}), (2n_{11})]^{\text{th}}\text{DI}_{-}\text{:DI}_{-},[2(m_{21} - m_{11})]^{\text{th}}{}_{\text{DI:DI}}\text{US})}_{([(2m_{11}+1),(2n_{11})],[2(m_{21}-m_{11})])\text{-decreasing-inflection upper-saddle}},$$

for $a_{110} > 0, a_{220} < 0$;

$$(a_{121}, a_{211}) = \underbrace{([(2m_{11}), (2n_{11})]^{\text{th}}\text{II}_{-}\text{:II}_{-},[2(m_{21} - m_{11})]^{\text{th}}{}_{\text{II:II}}\text{LS})}_{([(2m_{11}+1),(2n_{11})],[2(m_{21}-m_{11})])\text{-increasing-inflection lower-saddle}},$$

for $a_{110} < 0, a_{220} < 0$. (3.147)

(**B**) For $2m_{21} < 2m_{11}$, the infinite-equilibriums are given as follows.

$$\begin{aligned}
(a_{121}, a_{211}) &= \underbrace{([(2m_{11}), (2n_{11})]^{\text{th}}\text{II}_{+}\text{:II}_{+}, [2(m_{11} - m_{21})]^{\text{th}}{}_{\text{IC:IC}}\text{US})}_{([(2m_{11}),(2n_{11})],[2(m_{11}-m_{21})])\text{-increasing-concave inflection-upper-saddle}},\\
&\text{for } a_{110} > 0, a_{220} > 0;\\
(a_{121}, a_{211}) &= \underbrace{([(2m_{11}), (2n_{11})]^{\text{th}}\text{DI}_{+}\text{:DI}_{+}, [2(m_{11} - m_{21})]^{\text{th}}{}_{\text{DC:DC}}\text{LS})}_{([(2m_{11}),(2n_{11})],[2(m_{11}-m_{21})])\text{-decreasing-concacve inflection-lower-saddle}},\\
&\text{for } a_{110} < 0, a_{220} > 0;\\
(a_{121}, a_{211}) &= \underbrace{([(2m_{11}), (2n_{11})]^{\text{th}}\text{DI}_{-}\text{:DI}_{-}, [2(m_{11} - m_{21})]^{\text{th}}{}_{\text{DC:DC}}\text{US})}_{([(2m_{11}+1),(2n_{11})],[2(m_{11}-m_{21})])\text{-decreasing-concave inflection-upper-saddle}},\\
&\text{for } a_{110} > 0, a_{220} < 0;\\
(a_{121}, a_{211}) &= \underbrace{([(2m_{11}), (2n_{11})]^{\text{th}}\text{II}_{-}\text{:II}_{-}, [2(m_{11} - m_{21})]^{\text{th}}{}_{\text{IC:IC}}\text{LS})}_{([(2m_{11}+1),(2n_{11})],[2(m_{11}-m_{21})])\text{-increasing-concave inflection-lower-saddle}},\\
&\text{for } a_{110} < 0, a_{220} < 0.
\end{aligned} \tag{3.148}$$

For $x_1^* = a_{111} = a_{211}$, the infinite-equilibrium exists. The infinite-equilibriums are illustrated in Figs. 3.45 and 3.46 for $m_{21} \geq m_{11}$ and $m_{21} < m_{11}$, respectively. For $m_{21} \geq m_{11}$, the singular inflection saddles are presented. For $m_{11} > m_{21}$, the concave inflection-saddles are presented.

3.8.3 [(2m_{11}), (2n_{11})] and ((2n_{11}), (2m_{21}))-Hybrid Networks

Consider

$$\begin{aligned}
\dot{x}_2 &= a_{220} \prod_{v=1}^{2m_{21}} (x_1 - a_{21v}),\\
\dot{x}_1 &= a_{110} \prod_{r=1}^{2m_{11}} (x_1 - a_{11r}) \prod_{u=1}^{2n_{11}} (x_2 - a_{12u}).
\end{aligned} \tag{3.149}$$

The first integral manifold is

$$\sum_{v_0=2m_{11}+1}^{2m_{21}} b_{21v_0} \sum_{\substack{\mu_1\in[0,\beta_0-2m_{11}],\\ v_1\in[2m_{11}-1,v_0-1]}}^{\mu_1+v_1=v_0-1} \sum_{\substack{\mu_2\in[0,v_1-2m_{11}]\\ v_2\in[2m_{11}-2,v_1-2]}}^{\mu_2+v_2=v_1-2} \cdots \sum_{\substack{\mu_{2m_{11}}\in[0,v_{2m_{11}-1}-2m_{11}]\\ v_{2m_{11}}\in[1,v_{2m_{11}-1}-2m_{11}+1]}}^{\mu_{2m_{11}}+v_{2m_{11}}=v_{2m_{11}-1}-2m_{11}+1}$$

$$\prod_{j=1}^{2m_{11}-1} \frac{(v_{j-1}-j)!}{\mu_j! v_j!} \times (a_{11r_{j+1}}[-a_{11r_j})^{\mu_j} \frac{1}{v_{2m_{11}-1}} \left[(x_1-a_{11r_{2m_{11}}})^{v_{2m_{11}-1}} - (x_{10}-a_{11r_{2m_{11}}})^{v_{2m_{11}-1}}\right]$$

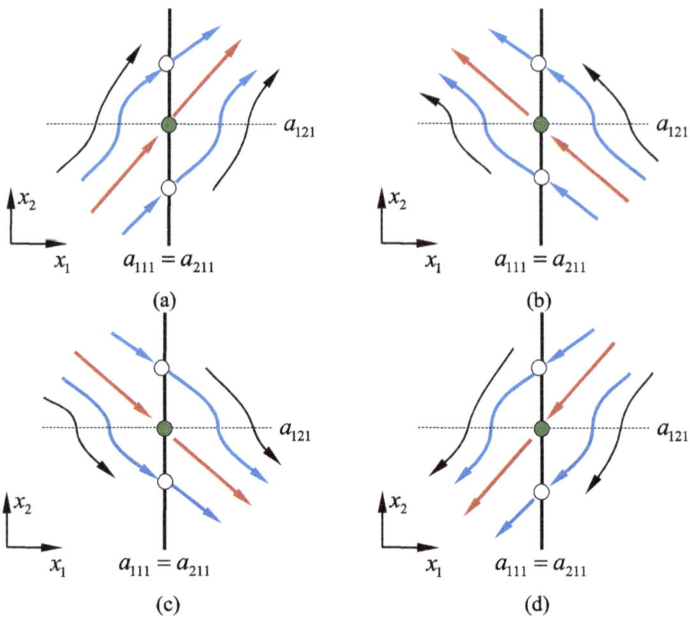

Fig. 3.45 The infinite-equilibriums ($a_{211} = a_{111}, 2m_{21} \geq 2m_{11}, n_1 = 2n_{11}$) for 2-dimensional polynomial systems: **a** ($a_{110} > 0, a_{220} > 0$), **b** ($a_{110} < 0, a_{220} > 0$), **c** ($a_{110} > 0, a_{220} < 0$), **d** ($a_{110} < 0, a_{220} < 0$)

$$
\begin{aligned}
&- \sum_{\nu_0=2m_{11}+1}^{2m_{21}} b_{21\nu_0} \sum_{k=2}^{2m_{11}-1} \Bigg[\sum_{\substack{\mu_1+\nu_1=\nu_0-1 \\ \mu_1\in[0,\nu_0-2m_{11}-1),\\ \nu_1\in(2m_{11},\nu_0-1]}} \sum_{\substack{\mu_2+\nu_2=\nu_1-2 \\ \mu_2\in[0,\nu_1-2m_{11}-1)\\ \nu_2\in(2m_{11}-1,\nu_1-2]}} \cdots \\
&\times \Bigg(\sum_{\substack{\mu_{k-1}+\nu_{k-1}=\nu_{k-2}-k+1 \\ \mu_{k-1}\in(\nu_{k-2}-2m_{11}+1,\nu_{k-2}-k+1] \\ \nu_{k-1}\in[0,2m_{11}-k)}} \prod_{j=1}^{k-1} \frac{(\nu_{j-1}-j)!}{\mu_j!\nu_j!}(a_{11r_{j+1}} - a_{11r_j})^{\mu_j} \\
&\times \sum_{\rho=k}^{2m_{11}} A_{\rho1}^{(s_k)} \ln \frac{|x_1 - a_{11s_\rho}|}{|x_{10} - a_{11s_\rho}|} \\
&+ \sum_{\substack{\mu_{k-1}+\nu_{k-1}=\nu_{k-2}-k+1 \\ \mu_{k-1}\in(\nu_{k-2}-2m_{11},\nu_{k-2}-2m_{11}+1] \\ \nu_{k-1}\in[2m_{11}-k,2m_{11}-k+1)}} \prod_{j=1}^{k-1} \frac{(\nu_{j-1}-j)!}{\mu_j!\nu_j!}(a_{11r_{j+1}} - a_{11r_j})^{\mu_j} \\
&\times \sum_{\rho=k+1}^{2m_{11}} B_{\rho1}^{(r_k)} \ln \frac{|x_1 - a_{11r_\rho}|}{|x_{10} - a_{11r_\rho}|} \Bigg) \Bigg] + b_{210} \sum_{\rho=1}^{2m_{11}} A_{\rho1}^{(r_1)} \ln \frac{|x_1 - a_{11r_\rho}|}{|x_{10} - a_{11r_\rho}|} \\
&+ \sum_{\nu_0=1}^{2m_{11}-1} b_{21\nu_0} \sum_{\rho=2}^{2m_{11}} B_{\rho1}^{(r_1)} \ln \frac{|x_1 - a_{11r_\rho}|}{|x_{10} - a_{11r_\rho}|}
\end{aligned}
$$

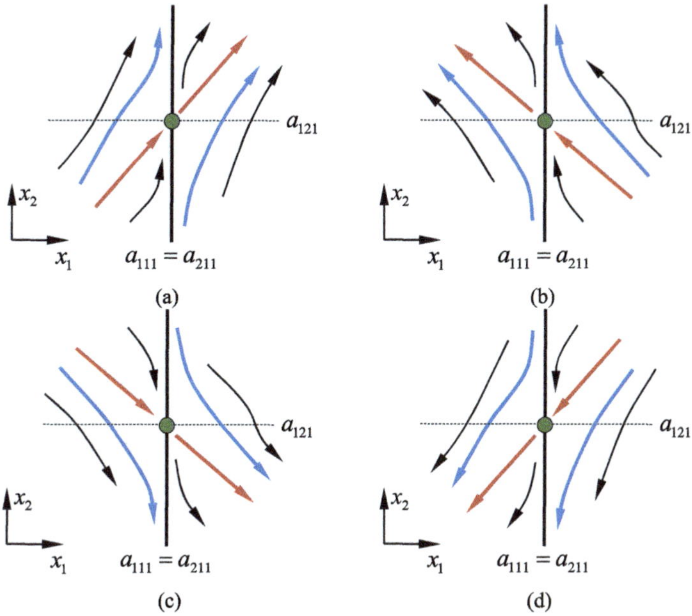

Fig. 3.46 The infinite-equilibriums ($a_{211} = a_{111}, 2m_{21} < 2m_{11}, n_1 = 2n_{11}+1$) for 2-dimensional polynomial systems: **a** ($a_{110} > 0, a_{220} > 0$), **b** ($a_{110} < 0, a_{220} > 0$), **c** ($a_{110} > 0, a_{220} < 0$), **d** ($a_{110} < 0, a_{220} < 0$)

$$= \frac{a_{110}}{a_{220}}\left\{ \frac{1}{2n_{11}+1}\left[(x_2 - a_{12u_1})^{2n_{11}+1} - (x_{j_2 0} - a_{12u_1})^{2n_{11}+1}\right] \right.$$
$$\left. + \sum_{\alpha=1}^{2n_{11}-2} \frac{1}{2n_{11}-\alpha+1} b_{12\alpha}\left[(x_2 - a_{12u_1})^{2n_{11}-\alpha+1} - (x_{20} - a_{12u_1})^{n_{11}-\alpha+1}\right] \right\}, \tag{3.150}$$

where

$$b_{12\alpha} = \sum_{l_{n_1},t_{n_1}=0}^{l_{n_1}+t_{n_1}=1} \cdots \sum_{l_2,t_2=0}^{l_2+t_2=1} \prod_{k=2}^{2n_{11}} (a_{12u_1} - a_{12u_k})^{l_k} \delta_{t_2+\cdots+t_{2n_{11}}}^{2n_{11}-\alpha};$$
$$b_{120} = 1.$$
$$b_{21\nu_0} = \sum_{t_1,l_1=0}^{t_1+l_1=1} \cdots \sum_{t_{2m_{21}},l_{2m_{21}}=0}^{t_{2m_{21}}+l_{2m_{21}}=1} \prod_{k=1}^{2m_{21}} (a_{11r_1} - a_{21\nu_k})^{t_k} \delta_{\nu_0}^{\Sigma_{r=1}^{2m_{21}} l_r};$$
$$\lim_{\nu_{2m_{11}-1}\to 0} \frac{1}{\nu_{2m_{11}-1}}\left[(x_1 - a_{11r_{2m_{11}}})^{\nu_{2m_{11}-1}} - (x_{j_1 0} - a_{11r_{2m_{11}}})^{\nu_{2m_{11}-1}}\right] = \ln \frac{|x_1 - a_{11r_{2m_{11}}}|}{|x_{10} - a_{11r_{2m_{11}}}|}; \tag{3.151}$$

$$
\begin{aligned}
A_{\rho 1}^{(r_k)} &= \frac{1}{\prod_{\sigma=k,\sigma\neq\rho}^{2m_{11}}(a_{11r_\rho}-a_{11r_\sigma})}, \text{ for } \rho\in\{k,k+1,\ldots,2m_{11}\},\\
B_{\rho 1}^{(r_k)} &= \frac{(a_{11s_\rho}-a_{11s_k})^{\beta_{k-1}-k}}{\prod_{\sigma=k+1,\sigma\neq\rho}^{2m_{11}}(a_{11r_\rho}-a_{11r_\sigma})}, \text{ for } \rho\in\{k+1,k+2,\ldots,2m_{11}\};\\
& k\in\{1,2,\ldots,2m_{11}-1\}.
\end{aligned}
\tag{3.152}
$$

For all cases, the $(2n_{11}+1)\times(2m_{21})$-simple equilibriums plus $(2m_{11})\times(2n_{11}+1)$-hyperbolic flows are based on the following inflections-saddle bifurcations for $2m_{21}\geq 2m_{11}$:

- $([(2m_{11}),(2n_{11})]^{\text{th}}\text{II}_{+}\text{:II}_{+},[2(m_{21}-m_{11})]^{\text{th}}{}_{\text{II:II}}\text{US})$-increasing-inflection upper-saddle bifurcation for $(a_{110}>0, a_{220}>0)$,
- $([(2m_{11}),(2n_{11})]^{\text{th}}\text{DI}_{+}\text{:DI}_{+},[2(m_{21}-m_{11})]^{\text{th}}{}_{\text{DI:DI}}\text{US})$-decreasing-inflection upper-saddle bifurcation for $(a_{110}<0, a_{220}>0)$,
- $([(2m_{11}),(2n_{11})]^{\text{th}}\text{DI}_{-}\text{:DI}_{-},[2(m_{21}-m_{11})]^{\text{th}}{}_{\text{DI:DI}}\text{LS})$-decreasing-inflection lower-saddle bifurcation for $(a_{110}>0, a_{220}<0)$,
- $([(2m_{11}),(2n_{11})]^{\text{th}}\text{II}_{-}\text{:II}_{-},[2(m_{21}-m_{11})]^{\text{th}}{}_{\text{II:II}}\text{LS})$-increasing-inflection lower-saddle bifurcation for $(a_{110}<0, a_{220}<0)$;

or based on the following concave-inflection-saddle bifurcations for $2m_{21}<2m_{11}$:

- $([(2m_{11}),(2n_{11})]^{\text{th}}\text{II}_{+}\text{:II}_{+},[2(m_{21}-m_{11})]^{\text{th}}{}_{\text{IC:IC}}\text{US})$-increasing-concave inflection upper-saddle bifurcation for $(a_{110}>0, a_{220}>0)$,
- $([(2m_{11}),(2n_{11})]^{\text{th}}\text{DI}_{+}\text{:DI}_{+},[2(m_{21}-m_{11})]^{\text{th}}{}_{\text{DC:DC}}\text{US})$-decreasing-concave inflection upper-saddle bifurcation for $(a_{110}<0, a_{220}>0)$,
- $([(2m_{11}),(2n_{11})]^{\text{th}}\text{DI}_{-}\text{:DI}_{-},[2(m_{21}-m_{11})]^{\text{th}}{}_{\text{DC:DC}}\text{LS})$-decreasing-concave inflection lower-saddle bifurcation for $(a_{110}>0, a_{220}<0)$,
- $([(2m_{11}),(2n_{11})]^{\text{th}}\text{II}_{-}\text{:II}_{-},[2(m_{21}-m_{11})]^{\text{th}}{}_{\text{IC:IC}}\text{LS})$-increasing-concave inflection lower-saddle bifurcation for $(a_{110}<0, a_{220}<0)$.

(**A**) Consider $2m_{11}=2m_{21}$ for illustrations, phase portraits for the 2-dimensional systems for the hybrid networks of hyperbolic and hyperbolic-secant flows with the simple equilibriums of the saddles, sinks and sources with $x_1^*=a_{111},a_{211},a_{112},\cdots,a_{11(2m_{11})},a_{21(2m_{21})}$ and $x_2^*=a_{121},a_{122},\cdots,a_{12(2n_{11})}$ are presented in Fig. 3.47a–d for $(a_{110}>0, a_{220}>0)$, $(a_{110}<0, a_{220}>0)$, $(a_{110}>0, a_{220}<0)$ and $(a_{110}<0, a_{220}<0)$.

(**B**) Consider $2m_{11}=2m_{21}$ for illustrations, phase portraits for the 2-dimensional systems for the hybrid networks of hyperbolic and hyperbolic-secant flows with the simple equilibriums of saddles and centers with $x_1^*=a_{211},a_{111},\cdots,a_{11(2m_{11})},a_{21(2m_{21})},a_{11(2m_{11})}$ and $x_2^*=a_{121},a_{122},\cdots,a_{12(2n_{11})}$ are presented in Fig. 3.48a-d for $(a_{110}>0, a_{220}>0)$, $(a_{110}<0, a_{220}>0)$, $(a_{110}>0, a_{220}<0)$ and $(a_{110}<0, a_{220}<0)$.

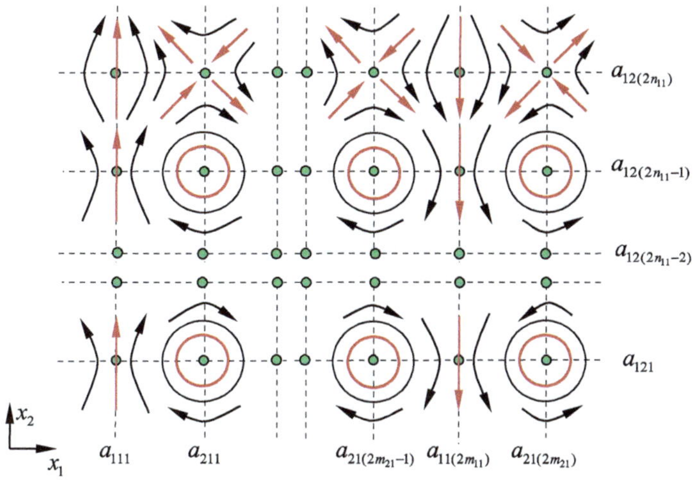

(a)

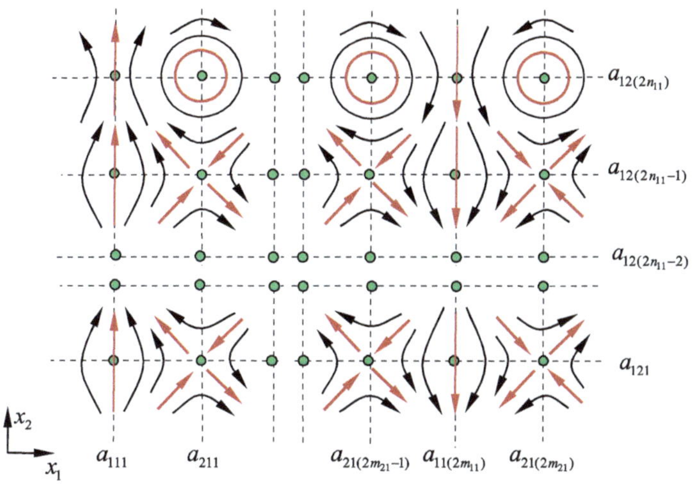

(b)

Fig. 3.47 Phase portraits for 2-dimensional systems with $x_1^* = a_{111}, a_{211}, \cdots, a_{11(2m_{11})}, a_{21(2m_{21})}$ and $x_2^* = a_{121}, a_{122}, \cdots, a_{12(2n_{11}+1)}$. The networks of flows with saddles and centers: **a** ($a_{110} > 0, a_{220} > 0$), **b** ($a_{110} < 0, a_{220} > 0$), **c** ($a_{110} > 0, a_{220} < 0$), **d** ($a_{110} < 0, a_{220} < 0$)

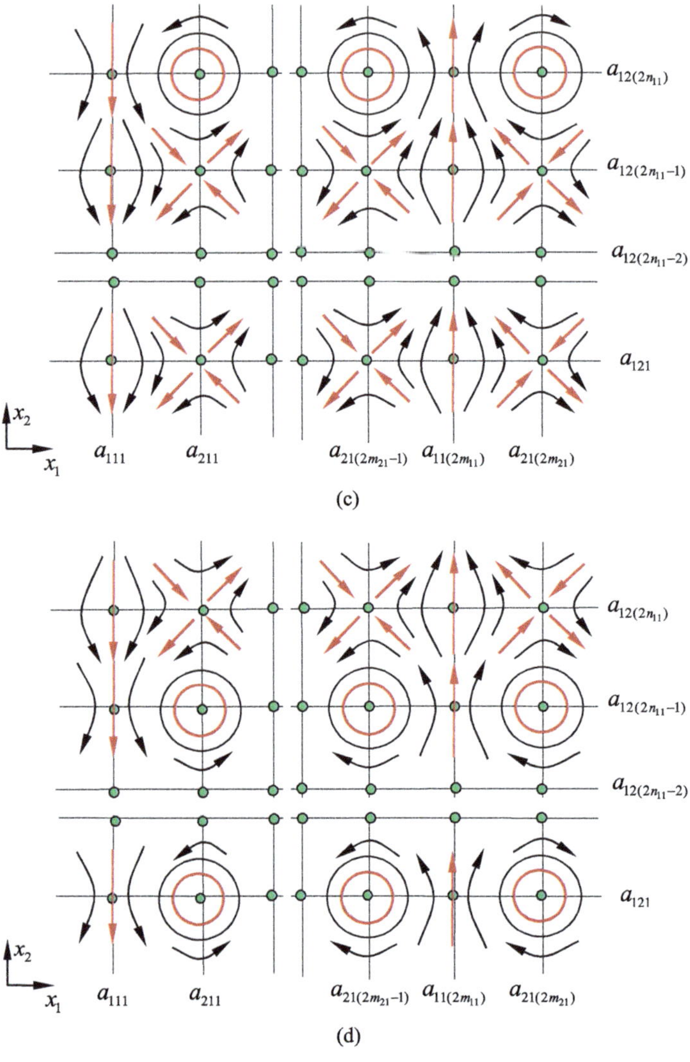

Fig. 3.47 (continued)

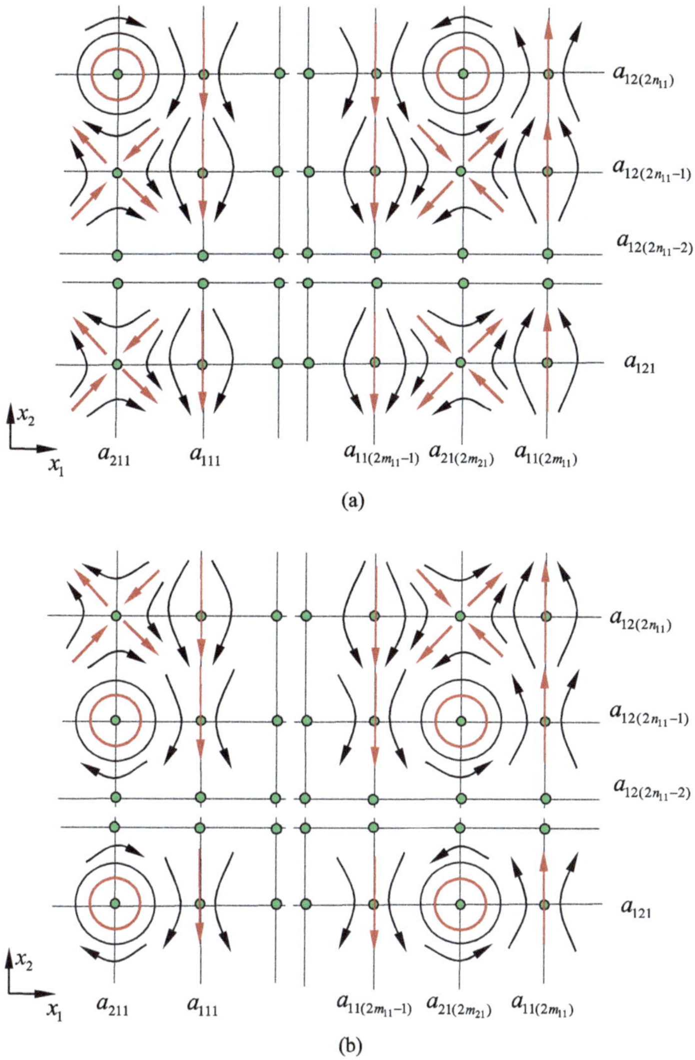

Fig. 3.48 Phase portraits for 2-dimensional systems with $x_1^* = a_{211}, a_{111}, \cdots, a_{21(2m_{21})}, a_{11(2m_{11})}$ and $x_2^* = a_{121}, a_{122}, \cdots, a_{12(2n_{11})}$. The networks of flows with saddles and centers: **a** $(a_{110} > 0, a_{220} > 0)$, **b** $(a_{110} < 0, a_{220} > 0)$, **c** $(a_{110} > 0, a_{220} < 0)$, **d** $(a_{110} < 0, a_{220} < 0)$

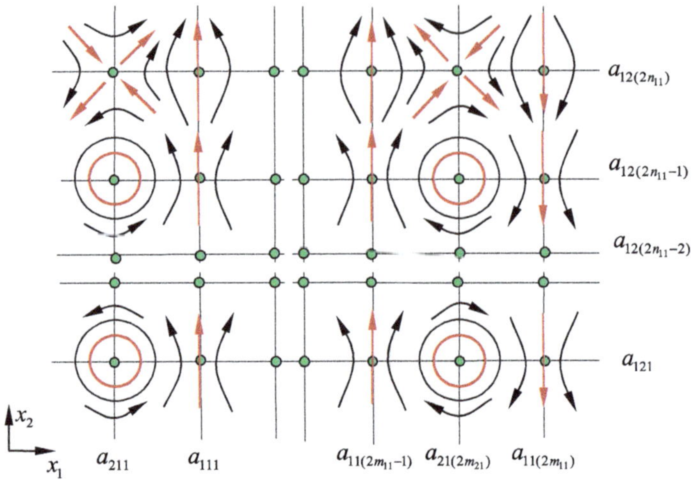

(c)

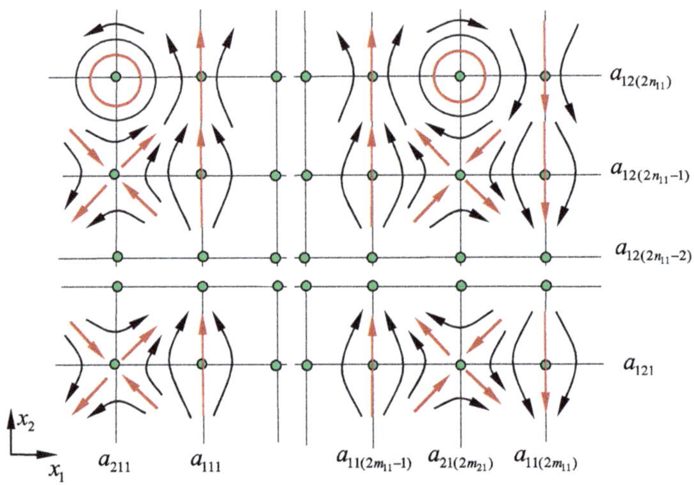

(d)

Fig. 3.48 (continued)

$$
\begin{aligned}
&\left\{\begin{array}{cccccc}
(a_{111},a_{12(2n_{11})}) & (a_{12(2n_{11})},a_{211}) & \cdots & (a_{12(2n_{11})},a_{21(2m_{21}-1)}) & (a_{11(2m_{11})},a_{12(2n_{11})}) & (a_{12(2n_{11})},a_{21(2m_{21})}) \\
(a_{111},a_{12(2n_{11}-1)}) & (a_{12(2n_{11}-1)},a_{211}) & \cdots & (a_{12(2n_{11}-1)},a_{21(2m_{21}-1)}) & (a_{11(2m_{11}+1)},a_{12(2n_{11}-1)}) & (a_{12(2n_{11}-1)},a_{21(2m_{21})}) \\
\vdots & \vdots & \vdots & \vdots & \vdots & \vdots \\
(a_{111},a_{121}) & (a_{121},a_{211}) & \cdots & (a_{121},a_{21(2m_{21}-1)}) & (a_{11(2m_{11}1)},a_{121}) & (a_{121},a_{21(2m_{21})})
\end{array}\right\} \\
&= \left\{\begin{array}{cccccc}
\underbrace{(\text{UP:DP,pF})}_{\text{hyperbolic-secant flow }(+)} & \underbrace{(\text{UP}_-,\text{UP}_-)}_{\text{negative saddle}} & \cdots & \underbrace{(\text{UP}_-,\text{UP}_-)}_{\text{negative saddle}} & \underbrace{(\text{UP:DP,nF})}_{\text{hyperbolic-secant flow }(-)} & \underbrace{(\text{UP}_+,\text{UP}_+)}_{\text{positive saddle}} \\
\underbrace{(\text{DP:UP,pF})}_{\text{hyperbolic flow }(+)} & \underbrace{(\text{DP}_-,\text{DP}_+)}_{\text{CW center}} & \cdots & \underbrace{(\text{DP}_-,\text{DP}_+)}_{\text{CW center}} & \underbrace{(\text{DP:UP,nF})}_{\text{hyperbolic flow }(-)} & \underbrace{(\text{DP}_+,\text{DP}_-)}_{\text{CCW center}} \\
\vdots & \vdots & \vdots & \vdots & \vdots & \vdots \\
\underbrace{(\text{DP:UP,pF})}_{\text{hyperbolic flow }(+)} & \underbrace{(\text{DP}_-,\text{DP}_+)}_{\text{CW center}} & \cdots & \underbrace{(\text{DP}_-,\text{DP}_+)}_{\text{CW center}} & \underbrace{(\text{DP:UP,nF})}_{\text{hyperbolic flow }(-)} & \underbrace{(\text{DP}_+,\text{DP}_-)}_{\text{CCW center}}
\end{array}\right\}_{(2n_{11})\times(2m_{11}+2m_{21})}
\end{aligned}
\tag{3.153}
$$

for $a_{110} > 0$, $a_{220} > 0$;

$$
\left\{\begin{array}{cccccc}
(a_{111},a_{12(2n_{11})}) & (a_{12(2n_{11})},a_{211}) & \cdots & (a_{12(2n_{11})},a_{21(2m_{21}-1)}) & (a_{11(2m_{11})},a_{12(2n_{11})}) & (a_{12(2n_{11})},a_{21(2m_{21})}) \\
(a_{111},a_{12(2n_{11}-1)}) & (a_{12(2n_{11}-1)},a_{211}) & \cdots & (a_{12(2n_{11}-1)},a_{21(2m_{21}-1)}) & (a_{11(2m_{11}+1)},a_{12(2n_{11}-1)}) & (a_{12(2n_{11}-1)},a_{21(2m_{21})}) \\
\vdots & \vdots & \vdots & \vdots & \vdots & \vdots \\
(a_{111},a_{121}) & (a_{121},a_{211}) & \cdots & (a_{121},a_{21(2m_{21}-1)}) & (a_{11(2m_{11}1)},a_{121}) & (a_{121},a_{21(2m_{21})})
\end{array}\right\}
$$

$$
= \left\{\begin{array}{cccccc}
\underbrace{(\mathrm{DP{:}UP,pF})}_{\text{hyperbolic flow }(+)} & \underbrace{(\mathrm{DP}_{-},\mathrm{DP}_{+})}_{\text{CW center}} & \cdots & \underbrace{(\mathrm{DP}_{-},\mathrm{DP}_{+})}_{\text{CW center}} & \underbrace{(\mathrm{DP{:}UP,nF})}_{\text{hyperbolic flow }(-)} & \underbrace{(\mathrm{DP}_{+},\mathrm{DP}_{-})}_{\text{CCW center}} \\
\underbrace{(\mathrm{UP{:}DP,pF})}_{\text{hyperbolic-secant flow }(+)} & \underbrace{(\mathrm{UP}_{-},\mathrm{UP}_{-})}_{\text{negative saddle}} & \cdots & \underbrace{(\mathrm{UP}_{-},\mathrm{UP}_{-})}_{\text{negative saddle}} & \underbrace{(\mathrm{UP{:}DP,nF})}_{\text{hyperbolic-secant flow }(-)} & \underbrace{(\mathrm{UP}_{+},\mathrm{UP}_{+})}_{\text{positive saddle}} \\
\vdots & \vdots & \vdots & \vdots & \vdots & \vdots \\
\underbrace{(\mathrm{UP{:}DP,pF})}_{\text{hyperbolic-secant flow }(+)} & \underbrace{(\mathrm{UP}_{-},\mathrm{UP}_{-})}_{\text{negative saddle}} & \cdots & \underbrace{(\mathrm{UP}_{-},\mathrm{UP}_{-})}_{\text{negative saddle}} & \underbrace{(\mathrm{UP{:}DP,nF})}_{\text{hyperbolic-secant flow }(-)} & \underbrace{(\mathrm{UP}_{+},\mathrm{UP}_{+})}_{\text{positive saddle}}
\end{array}\right\}_{(2n_{11})\times(2m_{11}+2m_{21})}
$$

for $a_{110} < 0, a_{220} > 0;$ (3.154)

$$
\left\{\begin{array}{cccccc}
(a_{111},a_{12(2n_{11}+1)}) & (a_{12(2n_{11}+1)},a_{211}) & \cdots & (a_{12(2n_{11}+1)},a_{21(2m_{21}-1)}) & (a_{11(2m_{11})},\, a_{12(2n_{11}+1)}) & (a_{12(2n_{11}+1)},a_{21(2m_{21})}) \\
(a_{111},a_{12(2n_{11})}) & (a_{12(2n_{11})},a_{211}) & \cdots & (a_{12(2n_{11})},a_{21(2m_{21}-1)}) & (a_{11(2m_{11}+1)},\, a_{12(2n_{11})}) & (a_{12(2n_{11})},a_{21(2m_{21})}) \\
\vdots & \vdots & \vdots & \vdots & \vdots & \vdots \\
(a_{111},a_{121}) & (a_{121},a_{211}) & \cdots & (a_{121},a_{21(2m_{21}-1)}) & (a_{11(2m_{11}1)},\, a_{121}) & (a_{121},a_{21(2m_{21})})
\end{array}\right\}
$$

$$
=\left\{\begin{array}{cccccc}
\underbrace{\text{(DP:UP,nF)}}_{\text{hyperbolic flow }(-)} & \underbrace{(\text{DP}_+,\text{DP}_-)}_{\text{CCW center}} & \cdots & \underbrace{(\text{DP}_+,\text{DP}_-)}_{\text{CCW center}} & \underbrace{\text{(DP:UP,pF)}}_{\text{hyperbolic flow }(+)} & \underbrace{(\text{DP}_-,\text{DP}_+)}_{\text{CW center}} \\
\underbrace{\text{(UP:DP,nF)}}_{\text{hyperbolic-secant flow }(-)} & \underbrace{(\text{UP}_+,\text{UP}_+)}_{\text{positive saddle}} & \cdots & \underbrace{(\text{UP}_+,\text{UP}_+)}_{\text{positive saddle}} & \underbrace{\text{(UP:DP,pF)}}_{\text{hyperbolic-secant flow }(+)} & \underbrace{(\text{UP}_-,\text{UP}_-)}_{\text{negative saddle}} \\
\vdots & \vdots & \vdots & \vdots & \vdots & \vdots \\
\underbrace{\text{(UP:DP,nF)}}_{\text{hyperbolic-secant flow }(-)} & \underbrace{(\text{UP}_+,\text{UP}_+)}_{\text{positive saddle}} & \cdots & \underbrace{(\text{UP}_+,\text{UP}_+)}_{\text{positive center}} & \underbrace{\text{(UP:DP,pF)}}_{\text{hyperbolic-secant flow }(+)} & \underbrace{(\text{UP}_-,\text{UP}_-)}_{\text{negative center}}
\end{array}\right\}_{(2n_{11})\times(2m_{11}+2m_{21})}
$$

for $a_{110} > 0$, $a_{220} < 0$; (3.155)

$$
\left\{\begin{matrix}
(a_{111},a_{12(2n_{11}+1)}) & (a_{12(2n_{11}+1)},a_{211}) & \cdots & (a_{12(2n_{11}+1)},a_{21(2m_{21}-1)}) & (a_{11(2m_{11})}, a_{12(2n_{11}+1)}) & (a_{12(2n_{11}+1)},a_{21(2m_{21})}) \\
(a_{111},a_{12(2n_{11})}) & (a_{12(2n_{11})},a_{211}) & \cdots & (a_{12(2n_{11})},a_{21(2m_{21}-1)}) & (a_{11(2m_{11}+1)}, a_{12(2n_{11})}) & (a_{12(2n_{11})},a_{21(2m_{21})}) \\
\vdots & \vdots & \vdots & \vdots & \vdots & \vdots \\
(a_{111},a_{121}) & (a_{121},a_{211}) & \cdots & (a_{121},a_{21(2m_{21}-1)}) & (a_{11(2m_{11}1)}, a_{121}) & (a_{121},a_{21(2m_{21})})
\end{matrix}\right\}
$$

$$
= \left\{\begin{matrix}
\underbrace{(\mathrm{UP{:}DP,nF})}_{\text{hyperbolic-secant flow }(-)} & \underbrace{(\mathrm{UP}_+,\mathrm{UP}_+)}_{\text{positive saddle}} & \cdots & \underbrace{(\mathrm{UP}_+,\mathrm{UP}_+)}_{\text{positive saddle}} & \underbrace{(\mathrm{UP{:}DP,pF})}_{\text{hyperbolic-secant flow }(+)} & \underbrace{(\mathrm{UP}_-,\mathrm{UP}_-)}_{\text{negative saddle}} \\
\underbrace{(\mathrm{DP{:}UP,nF})}_{\text{hyperbolic flow }(-)} & \underbrace{(\mathrm{DP}_+,\mathrm{DP}_-)}_{\text{CCW center}} & \cdots & \underbrace{(\mathrm{DP}_+,\mathrm{DP}_-)}_{\text{CCW center}} & \underbrace{(\mathrm{DP{:}UP,pF})}_{\text{hyperbolic flow }(+)} & \underbrace{(\mathrm{DP}_-,\mathrm{DP}_+)}_{\text{CW center}} \\
\vdots & \vdots & \vdots & \vdots & \vdots & \vdots \\
\underbrace{(\mathrm{DP{:}UP,nF})}_{\text{hyperbolic flow }(-)} & \underbrace{(\mathrm{DP}_+,\mathrm{DP}_-)}_{\text{CCW center}} & \cdots & \underbrace{(\mathrm{DP}_+,\mathrm{DP}_-)}_{\text{CCW center}} & \underbrace{(\mathrm{DP{:}UP,pF})}_{\text{hyperbolic flow }(+)} & \underbrace{(\mathrm{DP}_-,\mathrm{DP}_+)}_{\text{CW center}}
\end{matrix}\right\}_{(2n_{11})\times(2m_{11}+2m_{21})}
$$

for $a_{110} < 0, a_{220} < 0;$ (3.156)

$$
\left\{\begin{array}{cccccc}
(a_{12(2n_{11})},a_{211}) & (a_{111},a_{12(2n_{11})}) & \cdots & (a_{11(2m_{11}-1)},a_{12(2n_{11})},) & (a_{12(2n_{11})},a_{21(2m_{21})}) & (a_{11(2m_{11})},a_{12(2n_{11})}) \\
(a_{12(2n_{11}-1)},a_{211}) & (a_{111},a_{12(2n_{11}-1)}) & \cdots & (a_{11(2m_{11}-1)},a_{12(2n_{11}-1)}) & (a_{12(2n_{11}-1)},a_{21(2m_{21})}) & (a_{11(2m_{11})},a_{12(2n_{11}-1)}) \\
\vdots & \vdots & \vdots & \vdots & \vdots & \vdots \\
(a_{121},a_{211}) & (a_{111},a_{121}) & \cdots & (a_{11(2m_{11}-1)},a_{121}) & (a_{121},a_{21(2m_{21})}) & (a_{11(2m_{11})},a_{121})
\end{array}\right\}
$$

$$
= \left\{\begin{array}{cccccc}
\underbrace{(\mathrm{DP}_-,\mathrm{DP}_+)}_{\text{CW center}} & \underbrace{(\mathrm{DP{:}UP,nF})}_{\text{hyperbolic flow }(-)} & \cdots & \underbrace{(\mathrm{DP{:}UP,nF})}_{\text{hyperbolic flow }(-)} & \underbrace{(\mathrm{DP}_+,\mathrm{DP}_-)}_{\text{CCW center}} & \underbrace{(\mathrm{DP{:}UP,pF})}_{\text{hyperbolic flow }(+)} \\
\underbrace{(\mathrm{UP}_-,\mathrm{UP}_-)}_{\text{negative saddle}} & \underbrace{(\mathrm{UP{:}DP,nF})}_{\text{hyperbolic-secant flow }(-)} & \cdots & \underbrace{(\mathrm{UP{:}DP,nF})}_{\text{hyperbolic-secant flow }(-)} & \underbrace{(\mathrm{UP}_+,\mathrm{UP}_+)}_{\text{positive saddle}} & \underbrace{(\mathrm{UP{:}DP,pF})}_{\text{hyperbolic-secant flow }(+)} \\
\vdots & \vdots & \vdots & \vdots & \vdots & \vdots \\
\underbrace{(\mathrm{UP}_-,\mathrm{UP}_-)}_{\text{negative saddle}} & \underbrace{(\mathrm{DP{:}UP,nF})}_{\text{hyperbolic flow }(-)} & \cdots & \underbrace{(\mathrm{DP{:}UP,pF})}_{\text{hyperbolic flow }(-)} & \underbrace{(\mathrm{UP}_+,\mathrm{UP}_+)}_{\text{positive saddle}} & \underbrace{(\mathrm{UP{:}DP,pF})}_{\text{hyperbolic-secant flow }(+)}
\end{array}\right\}_{(2n_{11})\times(2m_{11}+2m_{21})} \tag{3.157}
$$

for $a_{110} > 0, a_{220} > 0$;

$$
\left\{\begin{array}{cccccc}
(a_{12(2n_{11})},a_{211}) & (a_{111},a_{12(2n_{11})}) & \cdots & (a_{11(2m_{11}-1)},a_{12(2n_{11})},) & (a_{12(2n_{11})},a_{21(2m_{21})}) & (a_{11(2m_{11})},a_{12(2n_{11})}) \\
(a_{12(2n_{11}-1)},a_{211}) & (a_{111},a_{12(2n_{11}-1)}) & \cdots & (a_{11(2m_{11}-1)},a_{12(2n_{11}-1)}) & (a_{12(2n_{11}-1)},a_{21(2m_{21})}) & (a_{11(2m_{11})},a_{12(2n_{11}-1)}) \\
\vdots & \vdots & \vdots & \vdots & \vdots & \vdots \\
(a_{121},a_{211}) & (a_{111},a_{121}) & \cdots & (a_{11(2m_{11}-1)},a_{121}) & (a_{121},a_{21(2m_{21})}) & (a_{11(2m_{11})},a_{121})
\end{array}\right\}
$$

$$
=\left\{\begin{array}{cccccc}
\underbrace{(\mathrm{UP}_-,\mathrm{UP}_-)}_{\text{negative saddle}} & \underbrace{(\mathrm{UP{:}DP,nF})}_{\text{hyperbolic-secant flow }(-)} & \cdots & \underbrace{(\mathrm{UP{:}DP,nF})}_{\text{hyperbolic-secant flow }(-)} & \underbrace{(\mathrm{UP}_+,\mathrm{UP}_+)}_{\text{positive saddle}} & \underbrace{(\mathrm{UP{:}DP,pF})}_{\text{hyperbolic-secant flow }(+)} \\
\underbrace{(\mathrm{DP}_-,\mathrm{DP}_+)}_{\text{CW center}} & \underbrace{(\mathrm{DP{:}UP,nF})}_{\text{hyperbolic flow }(-)} & \cdots & \underbrace{(\mathrm{DP{:}UP,nF})}_{\text{hyperbolic flow }(-)} & \underbrace{(\mathrm{DP}_+,\mathrm{DP}_-)}_{\text{CCW center}} & \underbrace{(\mathrm{DP{:}UP,pF})}_{\text{hyperbolic flow }(+)} \\
\vdots & \vdots & \vdots & \vdots & \vdots & \vdots \\
\underbrace{(\mathrm{UP}_-,\mathrm{UP}_-)}_{\text{negative saddle}} & \underbrace{(\mathrm{UP{:}DP,nF})}_{\text{hyperbolic-secant flow }(-)} & \cdots & \underbrace{(\mathrm{UP{:}DP,nF})}_{\text{hyperbolic-secant flow }(-)} & \underbrace{(\mathrm{DP}_+,\mathrm{DP}_-)}_{\text{CCW center}} & \underbrace{(\mathrm{DP{:}UP,pF})}_{\text{hyperbolic flow }(+)}
\end{array}\right\}_{(2n_{11})\times(2m_{11}+2m_{21})}
\tag{3.158}
$$

for $a_{110} < 0, a_{220} > 0$;

$$
\left\{\begin{array}{cccccc}
(a_{12(2n_{11})},a_{211}) & (a_{111},a_{12(2n_{11})}) & \cdots & (a_{11(2m_{11}-1)},a_{12(2n_{11})},) & (a_{12(2n_{11})},a_{21(2m_{21})}) & (a_{11(2m_{11})},a_{12(2n_{11})})\\
(a_{12(2n_{11}-1)},a_{211}) & (a_{111},a_{12(2n_{11}-1)}) & \cdots & (a_{11(2m_{11}-1)},a_{12(2n_{11}-1)}) & (a_{12(2n_{11}-1)},a_{21(2m_{21})}) & (a_{11(2m_{11})},a_{12(2n_{11}-1)})\\
\vdots & \vdots & \vdots & \vdots & \vdots & \vdots\\
(a_{121},a_{211}) & (a_{111},a_{121}) & \cdots & (a_{11(2m_{11}-1)},a_{121}) & (a_{121},a_{21(2m_{21})}) & (a_{11(2m_{11})},a_{121})
\end{array}\right\}
$$

$$
=\left\{\begin{array}{cccccc}
\underbrace{(\mathrm{UP}_+,\mathrm{UP}_+)}_{\text{positive saddle}} & \underbrace{(\mathrm{UP{:}DP,pF})}_{\text{hyperbolic-secant flow }(+)} & \cdots & \underbrace{(\mathrm{UP{:}DP,pF})}_{\text{hyperbolic-secant flow }(+)} & \underbrace{(\mathrm{UP}_-,\mathrm{UP}_-)}_{\text{negative saddle}} & \underbrace{(\mathrm{UP{:}DP,nF})}_{\text{hyperbolic-secant flow }(-)}\\
\underbrace{(\mathrm{DP}_+,\mathrm{DP}_-)}_{\text{CCW center}} & \underbrace{(\mathrm{DP{:}UP,pF})}_{\text{hyperbolic flow }(+)} & \cdots & \underbrace{(\mathrm{DP{:}UP,pF})}_{\text{hyperbolic flow }(+)} & \underbrace{(\mathrm{DP}_-,\mathrm{DP}_+)}_{\text{CW center}} & \underbrace{(\mathrm{DP{:}UP,nF})}_{\text{hyperbolic flow }(-)}\\
\vdots & \vdots & \vdots & \vdots & \vdots & \vdots\\
\underbrace{(\mathrm{DP}_+,\mathrm{DP}_-)}_{\text{CCW center}} & \underbrace{(\mathrm{DP{:}UP,pF})}_{\text{hyperbolic flow }(+)} & \cdots & \underbrace{(\mathrm{DP{:}UP,pF})}_{\text{hyperbolic flow }(+)} & \underbrace{(\mathrm{DP}_-,\mathrm{DP}_+)}_{\text{CW center}} & \underbrace{(\mathrm{DP{:}UP,nF})}_{\text{hyperbolic flow }(-)}
\end{array}\right\}_{(2n_{11})\times(2m_{11}+2m_{21})}
\tag{3.159}
$$

for $a_{110} > 0, a_{220} < 0$;

$$\left\{\begin{array}{cccccc}
(a_{12(2n_{11})},a_{211}) & (a_{111},a_{12(2n_{11})}) & \cdots & (a_{11(2m_{11}-1)},a_{12(2n_{11})},) & (a_{12(2n_{11})},a_{21(2m_{21})}) & (a_{11(2m_{11})},a_{12(2n_{11})}) \\
(a_{12(2n_{11}-1)},a_{211}) & (a_{111},a_{12(2n_{11}-1)}) & \cdots & (a_{11(2m_{11}-1)},a_{12(2n_{11}-1)}) & (a_{12(2n_{11}-1)},a_{21(2m_{21})}) & (a_{11(2m_{11})},a_{12(2n_{11}-1)}) \\
\vdots & \vdots & \vdots & \vdots & \vdots & \vdots \\
(a_{121},a_{211}) & (a_{111},a_{121}) & \cdots & (a_{11(2m_{11}-1)},a_{121}) & (a_{121},a_{21(2m_{21})}) & (a_{11(2m_{11})},a_{121})
\end{array}\right\}$$

$$= \left\{\begin{array}{cccccc}
\underbrace{(\text{DP}_+,\text{DP}_-)}_{\text{CCW center}} & \underbrace{(\text{UP:DP,pF})}_{\text{hyperbolic-secant flow }(+)} & \cdots & \underbrace{(\text{DP:UP,pF})}_{\text{hyperbolic flow }(+)} & \underbrace{(\text{DP}_-,\text{DP}_+)}_{\text{CW center}} & \underbrace{(\text{DP:UP,pF})}_{\text{hyperbolic flow }(+)} \\
\underbrace{(\text{UP}_+,\text{UP}_+)}_{\text{positive saddle}} & \underbrace{(\text{DP:UP,pF})}_{\text{hyperbolic flow }(+)} & \cdots & \underbrace{(\text{UP:DP,pF})}_{\text{hyperbolic-secant flow }(+)} & \underbrace{(\text{UP}_-,\text{UP}_-)}_{\text{negative saddle}} & \underbrace{(\text{UP:DP,nF})}_{\text{hyperbolic-secant flow }(-)} \\
\vdots & \vdots & \vdots & \vdots & \vdots & \vdots \\
\underbrace{(\text{UP}_+,\text{UP}_+)}_{\text{positive saddle}} & \underbrace{(\text{DP:UP,pF})}_{\text{hyperbolic flow }(+)} & \cdots & \underbrace{(\text{UP:DP,pF})}_{\text{hyperbolic-secant flow }(+)} & \underbrace{(\text{UP}_-,\text{UP}_-)}_{\text{negative saddle}} & \underbrace{(\text{UP:DP,nF})}_{\text{hyperbolic-secant flow }(-)}
\end{array}\right\}_{(2n_{11})\times(2m_{11}+2m_{21})}$$

for $a_{110} < 0, a_{220} < 0;$ (3.160)

Index

A

A. C. J. Luo, *Two-dimensional Crossing and Product Polynomial Systems*,
https://doi.org/10.1007/978-981-96-5715-5

C

D

H

I

N

P

U

Y

GPSR Compliance
The European Union's (EU) General Product Safety Regulation (GPSR) is a set of rules that requires consumer products to be safe and our obligations to ensure this.

If you have any concerns about our products, you can contact us on

ProductSafety@springernature.com

In case Publisher is established outside the EU, the EU authorized representative is:

Springer Nature Customer Service Center GmbH
Europaplatz 3
69115 Heidelberg, Germany

www.ingramcontent.com/pod-product-compliance
Ingram Content Group UK Ltd.
Pitfield, Milton Keynes, MK11 3LW, UK
UKHW021614260926
475829UK00010B/14

9789819657148